$48.00

Craftsman

# 1998 NATIONAL PAINTING COST ESTIMATOR

by Dennis D. Gleason, CPE

**Includes inside the back cover:**
An estimating CD with all the costs in this book plus an interactive multimedia video that shows how to use this disk to compile construction cost estimates for your jobs.

Craftsman

**Craftsman Book Company**
6058 Corte del Cedro / P.O. Box 6500 / Carlsbad, CA 92018

# Acknowledgments

The author thanks the following individuals and organizations for furnishing materials and information used in the preparation of various portions of this book.

**Howard Shahan,** *American Design Painting & Drywall,* Poway, CA
**American Society of Professional Estimators (ASPE),** Wheaton, MD
**Ameritone Paint,** San Diego, CA
**Benjamin Moore Paints,** San Diego, CA
**Gordon H. Brevcort,** *Brevcort Consulting Associates,* Ridgewood, NJ
**Paul Ross,** *Deen Pierce Paint Company,* Pleasant Hill, CA
**John San Marcos,** *Devoe Coatings,* San Diego Marine Hardware, San Diego, CA
**Gene Christopher,** *Dunn-Edwards Paints & Wallcovering,* Daly City, CA
**Dick Stevenson,** *Dunn-Edwards Paints & Wallcovering,* Concord, CA
**Randy Martin,** *Dunn-Edwards Paints & Wallcovering,* San Diego, CA
**Bob Young,** *E. Turman & Co. Painting,* Livermore, CA
**Expo Builders Supply,** San Diego, CA
**Ken Hall,** *Frazee Paint & Wallcovering,* San Diego, CA
**Mike Crowley,** *Fuller O'Brien Paints,* Cupertino, CA
**John Christiansen,** *Fuller O'Brien Paints,* South San Francisco, CA
**The Glidden Company,** San Diego, CA
**Bill Killeen,** *ICI Paint Stores,* San Jose, CA
**Hugh Champeny,** *Kelly Moore Paint Company,* San Carlos, CA
**Tim Boland,** *Kelly Moore Paint Company,* Walnut Creek, CA
**Dennis Cripe,** *R.W. Little Co, Inc., Sandblasting,* San Diego, CA
**Bruce McMullan,** *McMullan & Son Painting,* San Diego, CA
**Joe Garrigan,** *Mr. Paints,* San Diego, CA
**Bernard Aborita,** *Old Quaker Paint Company*, Escondido, CA
**John Horne,** *Old Quaker Paint Company,* Escondido, CA
**PPG Industries, Inc.,** *Pittsburgh Paints,* Torrance, CA
**Chris Sheehan,** *Rich Paint Company,* Oakland, CA
**Richardson Engineering Services, Inc.,** Mesa, AZ
**Rust-Oleum Protective Coatings,** Los Angeles, CA
**Keith Braswell,** *Scotty Rents,* Pleasant Hill, CA
**Tom Hartrick,** *The Sherwin-Williams Company,* Dublin, CA
**The Sherwin-Williams Company,** San Diego, CA
**Al Chavez,** *Sinclair Paints,* San Diego, CA
**Squires-Belt Material Co.,** San Diego, CA
**Steel Structures Painting Council,** Pittsburgh, PA
**John Meyer,** *U.S. Government, Department of the Navy, Public Works,* San Diego, CA
**Vista Paint Centers,** Fullerton, CA
**John Johnston,** *Vista Paint Centers,* San Diego, CA
**Jerry Rittgarn,** *Waco-Arise Scaffolding & Equipment,* San Diego, CA

Cover photos by: Tom Page, www.redrom.com/tom

Painting contractor: Keith Monroe Inc., San Diego, CA

Software resource: Robert W. Kramer III

Sheila M. Scott, Calligraphy


***ISBN 1-57218-049-8 ISSN 1092-6852***

# Contents

# List of Figures

# How to Use This Book

Paint estimating is more of an art than a science. There's no price that's exactly right for every job and for every bidder. That's because every painting job is unique. No single material cost, no labor estimate, no pricing system fits all types of work. And just as every job varies, so do painting companies. No two painting contractors have the same productivity rates, the same labor burden, the same overhead expense and the same profit requirements.

The best paint estimates are always custom-made for a particular job. They're based on the contractor's actual productivity rate, material cost, labor cost, overhead percentage and profit expectations. No estimating book, no computerized estimating system, no estimating service can possibly account for all the variables that make every job and every painting company different. Only a skilled estimator using professional judgment and a proven estimating system can produce consistently reliable estimates on a wide variety of painting jobs.

## So, Why Buy This Book?

That's easy. This is the most complete, authoritative and reliable unit cost guide ever made available to paint estimators. No matter what types of work you estimate, no matter what your costs are, this book will help produce consistently accurate painting cost estimates in dollars and cents. But it isn't a substitute for expertise. It's not a simple way to do in minutes what an experienced paint estimator might not be able to do in hours. Instead, this unit cost guide will aid you in developing a good estimate of costs for any painting operation on any project. Think of this manual as *one* good estimating tool. But it's not (or at least shouldn't be) the only estimating tool you'll use.

For most jobs, I expect that the figures you see here will prove to be good estimates. But anyone who understands paint estimating will understand why judgment is needed when applying figures from this manual – or any other paint estimating guide. It's your responsibility to decide which conditions on the job you're bidding are like conditions assumed in this manual, and which conditions are different. Where conditions are different, you'll need good professional judgment to arrive at a realistic estimated cost.

Inside the back cover of this book you'll find an envelope with a compact disk that you can install on your computer's hard drive. The disk has *National Estimator*, an easy-to-use estimating program, and all the cost data in this book. An interactive, multimedia video shows how to use the software to create painting

| | Manhour productivity | Labor cost per hour | Labor burden percent | Labor burden dollars | Labor cost plus burden | Material price discount | Overhead percent | Profit |
|---|---|---|---|---|---|---|---|---|
| Slow (1P) | Low | $10.25 | 27.25% | $2.79 | $13.04 | 20% | 31.0% | 15% |
| Medium (2P) | Average | 16.50 | 33.65% | 5.55 | 22.05 | 30% | 24.5% | 11% |
| Fast (3P) | High | 22.00 | 40.55% | 8.92 | 30.92 | 40% | 18.5% | 6% |

**Note:** These rates are for painters. Hourly rates for wallcovering are different. See page 29. Slow, Medium and Fast jobs are defined on page 13. National Estimator computer software program uses hourly rates in the Labor cost plus burden column. National Estimator shows productivity rates (Slow, Medium and Fast) and copies the words Slow, Medium or Fast to your estimate. It also copies the crew productivity code, either 1P (Slow), 2P (Medium), or 3P (Fast) to your estimating form. National Estimator allows you to enter any percentage you select for overhead and profit.

**Figure 1**
The basis for cost estimates in this book

estimates for your jobs. Instructions for using National Estimator begin on page 431.

## How to Use the Tables

The estimating tables in this book show typical costs and bid prices for every painting operation you're likely to encounter, whether paint is applied by brush, roller, mitt or spray. Selecting the right cost table and the correct application method is easy. Tables are divided into four parts:

***Part I:*** General Painting Costs

***Part II:*** Preparation Costs

***Part III:*** Industrial, Institutional and Heavy Commercial Painting Costs

***Part IV:*** Wallcovering Costs

Each section is arranged alphabetically by operation. If you have trouble finding the tables you need, use the Table of Contents at the front of the book or the Index at the back of the book.

Once you've found the right table and the appropriate application method, you have to select the correct application rate. For each of the application methods (brush, roll, mitt or spray), the tables show three application rates: "Slow," "Medium," or "Fast." That's a very important decision when using this book, because each application rate assumes different manhour productivity, material coverage, material cost per gallon, hourly labor cost, labor burden, overhead and profit. Your decision on the application rate to use (or which combination of rates to use) has to be based on your evaluation of the job, your painters and your company. That's where good common sense is needed.

Figure 1 shows crew codes, labor costs, labor burdens, material discounts, and profit for each of the three production rates for painting.

The "Slow" application rate in Figure 1 assumes lower productivity (less area covered per manhour), a lower labor cost (due to a less skilled crew), a lower labor burden (due to lower fringe benefits), a lower discount on materials (because of low volume), higher overhead (due to lower volume) and a higher profit margin (typical on small repaint or custom jobs). Figures in this "Slow" application row will apply where painters with lower skill levels are working on smaller or more difficult repaint jobs.

Look at the "Fast" row in Figure 1. These estimates will apply where a skilled crew (higher hourly rate and larger fringe benefits) is working under good supervision and good conditions (more area covered per manhour) on larger (volume discount on materials) and more competitive jobs (lower profit margin). Figures in the "Fast" application row assume high productivity and lower material coverage (unpainted surfaces absorb more paint), like that of a residential tract job.

Each of the three application rates is described more completely later in this section.

| | Pricing variables | | | Unit cost estimate | | | | | |
|---|---|---|---|---|---|---|---|---|---|
| | 1 | 2 | 3 | 4 | 5 | 6 | 7 | 8 | 9 |
| | Labor SF per manhour | Material coverage SF/gallon | Material cost per gallon | Labor cost per 100 SF | Labor burden 100 SF | Material cost per 100 SF | Overhead per 100 SF | Profit per 100 SF | Total cost per 100 SF |
| **Walls, gypsum drywall, orange peel or knock-down, roll, per 100 sq ft of wall** | | | | | | | | | |
| Flat latex, water base (material #5) | | | | | | | | | |
| Roll 1st coat | | | | | | | | | |
| Slow | 400 | 300 | 16.20 | 2.56 | .81 | 5.40 | 2.68 | 1.70 | 13.15 |
| Medium | 538 | 275 | 14.20 | 3.07 | 1.13 | 5.16 | 2.27 | 1.27 | 12.90 |
| Fast | 675 | 250 | 12.10 | 3.26 | 1.46 | 4.84 | 1.74 | .67 | 11.97 |
| | | | | 2.81 | .77 | 5.16 | 2.14 | 1.20 | 12.08 |

**Figure 2**
Customize the tables

## The Easy Case: No Adjustments

Let's suppose the "Slow" application rate fits the job you're estimating almost perfectly. Your crew's productivity is expected to be low. The labor cost will be $10.25 per hour. Labor burden (fringes, taxes and insurance) will be 27.25 percent. Discount on materials will be 20 percent. Overhead will be 31 percent and profit will be 15 percent. Then your task is easy. All of your costs match the costs in the "Slow" row. No modifications are needed. The same is true if your costs fit the "Medium" or "Fast" rows.

But that's not always going to happen. More often, the job, your crew and your company won't fit exactly into any of the three rows. What then? More evaluation is required. You'll combine costs from several application rate rows to reach an accurate bid price. I call that *customizing your costs* and it's nearly always required for an accurate estimate.

## Customizing Your Costs

Every company has a different combination of worker speed and experience, taxes, benefits, spread rates, equipment needs, percentage for overhead, and profit margin. These are the cost variables in paint estimating. This book is designed so you can quickly and easily adjust estimates to reflect actual costs on the job you're estimating. It's important that you *read the rest of this section before using the cost tables in this book.* That's the only way to get from this manual all the accuracy and flexibility that's built into it.

In the remainder of this section I'll describe the assumptions I've made and the methods I used to compile the cost tables in this manual. Once you understand them, you'll be able to combine and modify costs in the estimating tables so your bids fit the job, your crew and your company as closely as possible.

When you start using the cost tables in this book, I suggest you circle numbers in the "Slow," "Medium," or "Fast" application rate rows that best fit your company and your jobs. To improve accuracy even more, write your own figures in the blank row below the "Fast" row in each table, like I've done in Figure 2.

### *A Practical Example*

Figure 2 is part of an estimating table from Part I of this book, General Painting Costs. I'm going to use it to show how to customize estimates to match your actual costs. In Figure 2 I've circled some of the costs I plan to use in a sample estimate and calculated others.

In column 1, *Labor SF per manhour*, I've circled 675 because I feel the journeyman painter assigned to this job can paint walls at the "Fast" rate of 675 square feet per hour. That's the number I plan to use for my estimate.

For column 2, *Material coverage SF/gallon*, I've reviewed my past performance and I expect coverage will be about 275 square feet per gallon of paint. So I've circled that figure.

In column 3, *Material cost per gallon*, I've circled 14.20 for my cost per gallon for flat water base latex (including tax and an allowance for consumable supplies), based on a 30 percent discount from the retail price.

So far, so good. That completes the first three columns, what I call the *pricing variables*. Now we can begin on the *unit cost estimate*, columns 4 through 9. Each of these columns show a price per 100 square feet of wall.

We'll start with column 4, *Labor cost per 100 SF*. Notice that I've entered 2.81 for this column. Here's why. Look back at Figure 1. Throughout this book the painting labor rate for "Fast" work is assumed to be $22.00 per hour. See page 29 for the wall covering application rate. I can't use the labor cost per 100 SF for "Fast" work because the journeymen on my job earn $19.00 per hour. That pay rate is between the "Medium" and "Fast" pay rates as shown in Figure 1. To calculate the labor cost per 100 SF, divide $19.00 by 675 and multiply by 100: 19/675 = .0281 x 100 = 2.81.

In column 5, *Labor burden 100 SF*, I've entered .77. This figure is a result of my labor cost at $2.81 x 27.25 percent, my labor burden (taxes, insurance and benefits) from the "Slow" row of Figure 1. Even though the labor rate is "Fast" and the labor cost is between "Fast" and "Medium," for this example labor burden will be most like work done at the "Slow" rate because this company doesn't offer many benefits.

In column 6, *Material cost per 100 SF*, I've circled 5.16, the number in the "Medium" row. Since I've used numbers in the "Medium" row in both columns 2 and 3, I can take the figure in column 6 for material costs directly from the table, without any calculations.

In column 7, *Overhead per 100 SF*, I've calculated the overhead dollar value by adding the labor cost, labor burden and material cost then multiplying that sum by the "Medium" overhead at 24.5 percent: $2.81 + $.77 + $5.16 = $8.74 x .245 = $2.14.

In column 8, *Profit per 100 SF*, I've calculated the profit dollar value by adding the labor cost, labor burden, material cost and overhead then multiplying that sum by the "Medium" profit at 11 percent from Figure 1. The result is $2.81 + $.77+ $5.16 + $2.14 = $10.88 x .11 = $1.20.

Column 9, *Total cost per 100 SF*, is the bid price – it's the sum of columns 4 through 8 for each row. Because I've circled costs that fall in more than one row, I can't use any figure in column 9. Instead, I simply add the circled or calculated figures in columns 4 through 8: $2.81 + $.77 + $5.16 + $2.14 + $1.20 = $12.08. That's my bid price per 100 square feet on this job. It's the combination of costs that fit my company, my painters and the job.

### Using Your Good Judgment

Of course, judgment is required when using these tables, as it is when making any estimate. For example, if your journeymen painters earn the top rate of $22.00 but work at the "Medium" production rate or slower, your labor cost per unit will be higher than the highest cost listed in column 4. An adjustment will be required.

Because figures in columns 7 and 8 are percentages of figures in columns 4, 5 and 6, you have to be careful when you blend costs from different rows. Let's look at an extreme (and unlikely) example.

Suppose you use costs from the "Slow" application row for columns 4 (2.56), 5 (.81) and 6 (5.40) of Figure 2. The total of those three costs is 8.77. Then you decide to use overhead from the "Fast" row because your overhead is about 18.5 percent of cost, not 31 percent of cost as in the "Slow" row. "Fast" overhead is listed as 1.74 in Figure 2. The correct overhead figure is $1.62, 18.5 percent of the sum of "Slow" costs in columns 4, 5 and 6. Be aware of this small discrepancy and calculate figures for all the categories yourself if extreme accuracy is essential.

## Converting Unit Prices

The last column in Figure 2 shows the total cost per 100 square feet of wall. Some estimating tables in this book show a total cost per 100 linear feet (such as for baseboard) or total costs per unit (such as for doors). To convert a cost per 100 square feet to a cost per square foot, move the decimal point two places to the left. Thus the cost per 100 square feet for the "Fast" rate in Figure 2 is $11.97 or about 12 cents per square foot.

## General Qualifications

It's important that you understand the conditions the tables are based upon. I call these conditions the job *qualifications*. A qualifications statement follows each estimating table to help you understand what's included and what's excluded. Please read those qualifications before using costs from this manual in your estimates. The following points apply to *all* tables in this book:

### *Included Costs*

- Minor preparation, both time and material. Normal preparation for new residential construction is included in the "Fast" row and for new commercial jobs in the "Medium" row. Minimal preparation is included for repaint jobs in the "Slow" row.
- Minimum setup and cleanup
- Equipment such as ladders, spray rigs and brushes are included in overhead for the "Fast" rate (residential tracts) or "Medium" (commercial) work. Add equipment costs at their rental rate for "Slow" (repaint) jobs.

### *Excluded Costs*

- Equipment costs such as ladders, spray rigs, etc. for "Slow" (repaint) jobs. Add these at their rental rate whether or not you own the equipment.
- Extensive surface preparation. Add the cost of time and materials needed for more than "normal" preparation work. Also add time to remove and replace hardware and accessories, protect adjacent surfaces, and do any extensive setup, cleanup, or touchup. (See the discussion of SURRPTUCU on the next page.)
- Mobilization or demobilization
- Supervision
- Material handling, delivery, or storage
- Sample preparation
- Mixing coatings
- Excessive material waste or spillage
- Equipment rental or placement costs
- Scaffolding rental and erection costs
- Subcontract costs
- Contingency allowance
- Owner allowances
- Commissions, bonuses, overtime, premium pay for shift adjustments (evening work), travel time or per diem.
- Bonds, fees, or permits
- Additional insurance to meet owner requirements
- Work at heights above 8 feet or beyond the reach of a wand or extension pole. (See the table for High Time Difficulty Factors on page 137.)

## Surface Preparation

The Preparation estimating tables that follow Part I: General Painting Costs, apply to both interior and exterior surfaces.

Surface preparation is one of the hardest parts of the job to estimate accurately. Any experienced painter can make a reasonably good estimate of the quantity of paint and time needed for application. But the amount of prep work needed will vary widely – especially for repaint jobs. Some will need very little work. Others will take more time for prep than for painting.

Preparation work for new construction jobs is relatively standard and consistent. You'll have to mask cabinets before spraying sealer on wet area walls, caulk at the baseboards, putty the nail holes in wood trim, and occasionally use a wire brush to smooth and clean a surface. The time required for this work is fairly predictable.

Labor cost for normal preparation of unpainted surfaces in new residential construction is included in the "Fast" *labor* costs and for new commercial construction in the "Medium" *labor* cost. The cost of *materials* for normal surface preparation on unpainted surfaces is included in the sundries allowance that's part of the "Fast" or "Medium" material cost.

But if more than normal surface prep work is needed, estimate the extra manhours and materials required and add these costs to your estimate.

### *Add for Repaint Preparation*

The "Slow" unit costs include no surface preparation other than a quick wipedown. Preparation on a repaint job may take longer than the painting itself. That's why you have to estimate surface prep as a separate item and add that cost to your estimate.

A misjudgment in estimating preparation work can be very expensive. That's why I recommend that you bid surface preparation by the hour, using your shop rate for "time and material" jobs, or some other specified hourly rate. That protects you against cost overruns if the preparation takes longer than anticipated. But there's a danger here. Owners may be angry about the cost because they don't understand what's involved in preparation and why it takes so long. You can avoid this with a "not to exceed" bid that contains a maximum price for the prep work. Your bid should define the scope of preparation work in detail and list exactly what's included and excluded. Be sure to consider all the labor, material, and equipment costs involved.

If you have to bid repaint work, be sure to include all the miscellaneous costs. The acronym I use to identify these miscellaneous costs is SURRPTUCU: Setup (SU), Remove and Replace (RR), Protection (P), Touchup (TU) and Cleanup (CU). Add these costs to your repaint estimate if they require anything beyond minimum attention.

1) *Setup* includes unloading the vehicle, spreading the tarp and setting up the tools – everything that has to be done before prep or painting can begin.

2) *Remove and replace* everything that will interfere with painting, including door and cabinet hardware, the contents of cabinets, light fixtures, bathroom accessories, switch covers and outlet plates, among others.

3) *Protection* for furniture and adjacent surfaces such as floors, cabinets, plumbing or electrical fixtures, windows, and doors. Protection methods include masking, applying visqueen, laying drop cloths and applying a protective coating on windows.

4) *Touchup* time varies with the speed and quality of the painting job and how fussy the owner is. The more careful your painters are, the less touchup time needed. You can estimate touchup time accurately only if you know how well your crews perform. The Touchup table in this book is based on a percentage of total job cost.

5) *Cleanup* time is usually about the same as setup time, about 20 to 30 minutes each day for repaint jobs. Cleanup time begins when work stops for the day and ends when the crew is back in the truck and ready to go home. It includes cleaning tools, dismantling the paint shop and loading the vehicle.

## Subcontractors

Painting contractors don't hire many subcontractors. But once in a while you'll need a specialist for sandblasting, waterblasting, wallcovering, scaffolding or pavement marking. Subcontract costs are not included in the estimating tables. Add the cost of any subcontract work that will be required.

Figure 3 shows some typical rates quoted by sandblasting subcontractors. Of course, prices in your area will probably be different. You could also figure sandblasting unit costs from the sandblasting estimating tables included in Part II, Preparation Costs, in this book.

Figure 4 shows typical subcontract bids for pavement marking. Again, prices in your area may be different.

| | |
|---|---|
| **Minimum charges:** $500.00, scaffolding not included | |
| **Additional insurance:** May be required to cover adjacent personal and real property which may not be protected. | |
| **Sandblasting water soluble paints** | $.80 to .90/SF |
| **Sandblasting oil paints** | .85 to .95/SF |
| **Sandblasting heavy mastic** | |
| (depends on coating thickness) | 1.10 to 1.20/SF |
| **Sandblasting brick** - light blast | .80 to .90/SF |
| **Sandblasting masonry block walls** | |
| Clean up & remove grime - light | .75 to .80/SF |
| - heavy | 1.15 to 1.25/SF |
| **Sandblasting structural steel** | |
| Pricing rules of thumb: | |
| Pipe up to 12" O.D. | 1.15 to 1.70/SF |
| Structural steel up to 2 SF/LF | 1.05 to 1.15/SF |
| Structural steel from 2 to 5 SF/LF | 1.25 to 1.40/SF |
| Structural steel over 5 SF/LF | (depends on shape) |
| Tanks and vessels up to 12'0" O.D. | 1.65 to 1.90/SF |
| Tanks and vessels over 12'0" O.D. | 1.65 to 1.90/SF |
| **Brush off blast** - light blast (loose mill scale) | |
| Field welded, new, uncoated | |
| ground runs | .50 to .65/SF |
| above ground | .75 to 1.40/SF |
| Previously painted surfaces - add | .50 to .90/SF |
| Epoxy coated - add | .95 to 1.05/SF |
| With portable equipment - add | .55 to .80/SF |
| **Commercial blast** - 67% white stage | |
| Flat welded, new, uncoated | |
| ground runs | .85 to 1.00/SF |
| above ground | 1.05 to 1.65/SF |
| Previously painted surfaces - add | .50 to .90/SF |
| Epoxy coated - add | .90 to 1.05/SF |
| With portable equipment - add | .65 to .80/SF |
| **Near white blast** - 95% white stage | |
| Field welded, new, uncoated | |
| ground runs | 1.00 to 1.15/SF |
| above ground | 1.15 to 1.75/SF |
| Previously painted surfaces - add | .50 to .90/SF |
| Epoxy coated - add | .90 to 1.05/SF |
| With portable equipment - add | .65 to .80/SF |
| **White blast** - 100% uniform white stage | |
| Field welded, new, uncoated | |
| ground runs | 1.50 to 1.75/SF |
| above ground | 1.65 to 1.95/SF |
| Previously painted surfaces - add | .50 to .85/SF |
| Epoxy coated - add | .90 to 1.05/SF |
| With portable equipment - add | .50 to .75/SF |

**Figure 3**
Sandblasting pricing table

If you do much repainting, you'll probably want to buy a waterblasting rig. Even if you own the blaster, include a charge in each estimate for the equipment as though you rented it from a rental yard just for that job. Figure the unit costs for waterblasting from Part II of this book, Preparation Costs.

Consider using a waterblasting subcontractor if you don't need the service often. Figure 5 shows some typical rates for waterblasting. Make up a table like this based on quotes from subcontractors in your area. For a more detailed table, see Sandblasting in the Preparation section, page 301.

When you hire a subcontractor, make sure the quoted price includes everything that contractor has to do – all labor, material (with tax, if applicable), equipment, overhead and profit. Add your overhead and profit percentage to the subcontractor's bid price when you enter that item on the estimate.

## Contingencies

Occasionally you'll add a contingency allowance on bids for repaint projects where there are unknowns that can't be forecast before work actually begins. Contingency allowances are rarely needed when estimating new construction. When necessary, the contingency amount is usually from 3 to 5 percent. It can go higher, however, if there are unusual conditions or unknowns that make it hard to produce an accurate estimate. Include a contingency allowance in your estimates only if you have reason to expect:

- An uncertain scope of work (unknown job conditions)
- An inexperienced owner or general contractor
- Incomplete drawings
- Delays in beginning the project
- Owner involvement in supervision
- Below-standard working conditions

Don't use contingency allowances as a substitute for complete estimates. Include contingency only to cover what can't be estimated, not what you don't have time to estimate accurately.

| | |
|---|---|
| **Pricing rule of thumb:** | |
| Number of parking spaces: Figure on one space per 300 SF of pavement | |
| Single line striping with light graphics apply | $ 7.50 per space |
| Single line striping with heavy graphics apply | 13.00 per space |
| Single striping, light graphics and 3' wheel stop | 18.50 per space |
| Single striping, heavy graphics and 3' wheel stop | 24.00 per space |
| **Equipment pricing:** | |
| Simple "inverted spray can" approximate cost | $185.00 |
| Professional striping machine cost range | $3,800 to 4,200 |
| Professional road/highway striper | $209,000 |
| **Subcontractor pricing:** | |
| Move on | $125.00 to 150.00 |
| **Striping prices:** | |
| Single line striping | $.40 to .50 per lineal foot |
| Bike lane striping | .50 to .60 per lineal foot |
| Fire lane, red curb | .40 to .50 per lineal foot |
| **Symbol pricing:** | |
| Templates - 8'0" template | $145.00 to 170.00 each |
| Arrows | 32.00 to 37.00 each |
| Handicap symbol, one color | 13.00 to 18.00 each |
| two color | 24.00 to 29.00 each |
| No parking fire lane stencil | 2.55 to 3.10 each |
| **Wheel stops:** | |
| 3'0" stops | $18.00 to 23.00 each if pinned on asphalt |
| | 24.00 to 29.00 each if glued and pinned |
| 6'0" stops | 29.00 to 34.00 each if pinned on asphalt |
| | 34.00 to 39.00 each if glued and pinned |
| | (add for wheel stops pinned to concrete) |
| **Signs and posts:** | |
| Sign only 12" x 18" | $40.00 to 56.00 |
| Post mounted 12" x 18" | 105.00 to 145.00 |
| **Pavement markers:** | |
| One way pavement markers | $8.50 each |
| Two way pavement markers | 11.00 each |

**Figure 4**
Pavement marking pricing table

| | |
|---|---|
| **Minimum charges:** $500.00, scaffolding not included | |
| **Additional insurance:** May be required to cover adjacent personal and real property | |
| **Pricing rules of thumb:** | |
| Up to 5,000 PSI blast | 4 hour minimum $100.00/hour |
| 5,000 to 10,000 PSI blast | 8 hour minimum $145.00/hour |
| Over 10,000 PSI blast | 8 hour minimum $180.00/hour |
| Wet sandblasting | 4 hour minimum $115.00/hour |

**Figure 5**
Waterblasting pricing table

## Column Headings Defined

Take another look at Figure 2. The heading describes the surface to be coated: the type, texture, and often, condition. Sections within each surface heading are divided according to coating material, then by application method, and further into the "Slow," "Medium," and "Fast" application rates.

### Column 1: Labor Productivity

This column shows units of work completed per manhour. My estimates assume that painters are experienced and motivated professionals. The labor productivity categories are shown in Figure 6.

My experience is that a painting company that can handle larger projects will have more skilled, better qualified and more productive painters. The estimating tables also assume that repainting a surface usually takes about 35 percent more time than painting newly constructed surfaces. Much of this extra time is spent protecting adjacent areas.

To establish your company's production levels, ask your field superintendent to monitor the time needed to complete each task and to keep records of crew productivity. Your best guide to productivity on future jobs is productivity on jobs already completed. The more you know about your painters' performance, the more accurate your estimates will be. But don't expect your estimates and actual production to match exactly. Painters are human beings, not robots. You can't expect them to work at the same rate at all times.

| **Slow** | **Medium** | **Fast** |
|---|---|---|
| Repaint jobs | New commercial projects | New residential production |
| Custom painting | Industrial painting | Repetitious painting |
| Tenant improvements | — | — |
| Small jobs | Medium-sized jobs | Large projects |
| Single units | Two to four units | Five or more units |
| Low production | Average production | High production |
| High difficulty | Average difficulty | Low difficulty |
| Poor conditions | Average conditions | Good conditions |
| High quality | Average quality | Minimum quality |
| Unskilled crew | Semi-skilled crew | Skilled crew |
| No supervision | Some supervision | Good supervision |

**Figure 6**
Labor productivity categories

### Reduced Productivity

The tables assume no overtime work. Excessive overtime puts a strain on your craftsmen and reduces productivity. A few consecutive days of overtime can drag productivity down to well below average. It's good practice not to assign overtime work on more than two consecutive days.

Work efficiency is also lower when men, materials and equipment are confined in a small area or required to work in cluttered, poorly lit or dirty rooms. Painters need elbow room to work efficiently and get maximum productivity. They're also more productive in a clean environment where they can see what they're doing. It's easier – and safer – to work in a well-lighted area that's relatively clear of debris. If the work area is confined or dirty, reduce estimated productivity accordingly.

### Supervision

Supervision expense is *not* included in the cost tables. Add the cost of supervision to your estimates.

Most supervision is done by foremen. Every crew should have a project foreman designated, usually the most experienced and reliable painter on the job. When not supervising, project foremen should be painting. Thus the project foreman is a working supervisor. Part of the foreman's time will be productive (applying coatings) and part will be nonproductive (directing the work).

If you have more than three or four jobs going at one time, you need a field superintendent. The field superintendent is the foreman's supervisor. His or her primary responsibility is to be sure that each foreman has the manpower, materials and equipment needed to get the job done. The field superintendent should monitor job progress to be sure manhour productivity and materials used are in line with estimates. Field superintendents usually are not working supervisors; all their time is nonproductive. Figure the field superintendent's salary as overhead expense, because you can't charge his salary to a specific job.

Your project foremen and field superintendent can make or break a job. The better they are, the more work will be done. You want a field superintendent who assigns the right painters to the right foreman, and a foremen who puts the right painters on the right tasks. The most experienced tradesmen should work on tasks that require more skill. Other painters should be used where less skill is needed. The project foreman is also responsible for job safety and quality control.

Your estimates will be more competitive if you can assume high productivity. That's only possible when you have good supervision, from both foremen and superintendent and motivated crews.

### Allowances for Supervision

Supervision isn't considered productive labor. A foreman isn't painting when he's scheduling, organizing a job and instructing his workers. Here are my rule-of-thumb allowances for nonproductive labor on painting jobs.

*Custom homes.* Allow 2.5 hours of nonproductive supervision for a home up to 1,500 square feet, 3 hours on a home between 1,500 and 2,000 square feet, 4 hours on a custom home between 2,000 and 2,500 SF, and 5 hours on a larger home.

*Model homes* in a tract. One hour of nonproductive supervision for each day your crew will be on the job.

*Most tract homes.* One hour per house.

*Higher-quality tract homes.* Two hours per house.

*Apartments and condos.* Allow 1 hour per unit if there are 10 units or less. For 11 to 30 units, allow 0.75 hour of nonproductive time per unit. If there are more than 30 units, allow 0.5 hour per unit.

Nonproductive labor on commercial, industrial, institutional and government projects varies considerably. More complex jobs will require proportionately more nonproductive labor. Use your knowledge based on past experience to estimate supervision either as a percentage of job cost or by the square foot of floor.

| Slow application and light coverage (Repaint jobs) | Medium application and medium coverage (Commercial projects) | Fast application and heavy coverage (Residential tracts) |
|---|---|---|
| Repaint jobs | Commercial projects | Residential production |
| Light usage | Moderate usage | Heavy usage |
| Low absorption | Moderate absorption | High absorption |
| Light application | Medium application | Heavy application |
| Low waste | Moderate waste | High waste |
| Quality paint | Standard paint | Production paint |
| Unskilled painters | Semi-skilled crew | Skilled crew |

**Figure 7**
Material coverage rates

## Column 2: Material Coverage

The second column in the cost tables shows the estimated material coverage in units (usually square feet or linear feet) per gallon. Figure 7 shows the conditions likely to apply for each of the three material coverage rates. Every condition listed in each of these categories won't necessarily occur on every painting operation. For example, it's possible to have high waste and use low quality paint on a repaint job. But it's more likely that waste will be low and paint quality high on jobs like that.

The "Slow" (repaint) application rate assumes light coverage, "Medium" (commercial project) application rate assumes medium coverage and "Fast" (residential tract) application rate assumes heavy coverage. Light coverage is typical on "Slow" (repaint) jobs because previously painted surfaces usually absorb 10 to 15 percent less paint than an unpainted surface. All coverage rates are based on paint that's been thinned according to the manufacturer's recommendations.

Of course, coverage varies with the paint you're using and the surface you're painting. Paint manufacturers usually list the recommended coverage rate on the container label. I've listed estimated coverage rates in the tables throughout this book.

### *Calculating Film Thickness*

Many project specifications for commercial, industrial and government jobs identify the coating (film) thickness you have to apply to each surface. The thickness is given in *mils*, or thousandths of an inch. One mil is 0.001 inch.

The thickness of the dry paint film depends on the percentage of solids in the paint. If you apply a gallon of paint containing 100 percent solids over 1,600 square feet, the dry film will be 1 mil thick – that is, if 100 percent of the paint adheres to the wall. But if there's 10 percent waste (because of paint that's left in the can, on brushes, or spilled), only 90 percent of the material ends up on the surface.

Here's a formula for coverage rates that makes it easy to calculate mil thickness, including the waste factor. Coverage rate equals:

$$\frac{\text{\% of solids x 1600}}{\text{mil thickness}} \text{ x (1.00 - waste factor)}$$

Here's an example. Assume you're applying paint with 40 percent solids (by volume), using a roller. The waste factor is 10 percent. You need a thickness of 5 mils.

| Slow application | Medium application | Fast application |
|---|---|---|
| Repaint jobs | Commercial projects | Residential tracts |
| Low volume | Medium volume | High volume |
| 20% discount | 30% discount | 40% discount |

**Figure 8**
Material price discounts

Here's the calculation for the coverage rate:

$$\frac{.40 \times 1600}{5} \times (1.00 - .10) = 115.2 \text{ SF per gallon}$$

You may have to apply several coats to get a thickness of 5 mils. In any case, you'll have to use one gallon of paint for each 115.2 square feet of surface.

### *Waste Factors*

Be sure to consider waste and spillage when you figure coverage rates. Professional painters waste very little paint. They rarely kick over a five-gallon paint bucket. But there's always some waste. My material coverage formulas include a typical waste allowance for each application method, whether it's brush, roller or spray. Of course, actual waste depends on the skill of your painters no matter what application method they use.

These are the waste factors I've built into the tables:

Brush . . . . . . . . . . . . . . . . . . . . . . . . . .3 to 5%

Roll . . . . . . . . . . . . . . . . . . . . . . . . . . .5 to 10%

Airless spray . . . . . . . . . . . . . . . . . . .20 to 25%

Conventional spray . . . . . . . . . . . . . .25 to 35%

### *Changes in Paint Formulation*

In the late 1970s, the California State Air Resources Board established a "model rule" for lowering the solvent in oil-based paints. They mandated replacing solvent-based paint with water-based formulas. The objective was to lower the amount of solvents escaping into the air. This change in the formulation of oil-based paints is being adopted nationwide.

Changes in paint formulation will affect coverage rates and the cost for non-flat paints. Review actual coverage rates and paint prices and make adjustments where necessary before using the estimates in this book.

## Column 3: Material Pricing

The third column in the cost tables shows the cost of materials. The "Slow," "Medium," and "Fast" prices in each table are based on the discounts usually offered by suppliers for volume purchases by contractor customers. The material discounts used in this book are defined in Figure 8.

The more paint a contractor buys over a given period, the greater the discount that contractor can expect. Most paint contractors get a discount of at least 20 percent off retail. Contractors buying in heavy volume usually get discounts that approach 40 percent off retail.

### *Material Pricing Tables*

Figures 9, 10 and 11 show the material prices I've used for each of three application rates throughout this book. In the cost estimating tables each coating is identified by a material number. To find out more about the cost of any of these coatings, refer to the material number listed in Figure 9, 10 or 11.

Figure 9 shows prices at a 20 percent discount off retail. It applies to "Slow" work and assumes light coverage on a previously painted surface. These costs would be typical for a lower-volume company handling mostly repaint or custom work.

## Material prices at 20% discount

All pricing is based on production grade material purchased in 5 gallon quantities.

| | | Retail price guide | Contractor price at a 20% discount | Add 15% sundries & escalation | Price with sales tax at 7% | Estimating prices with tax |
|---|---|---|---|---|---|---|
| **Interior:** | | | | | | |
| | Sealer, off white (wet area walls & ceilings) | | | | | |
| #1 - | Water base | 15.70 | 12.56 | 14.44 | 15.45 | 15.50 |
| #2 - | Oil base | 21.00 | 16.80 | 19.32 | 20.67 | 20.70 |
| | Undercoat (doors, casings and other paint grade wood) | | | | | |
| #3 - | Water base | 18.85 | 15.08 | 17.34 | 18.55 | 18.60 |
| #4 - | Oil base | 20.45 | 16.36 | 18.81 | 20.13 | 20.10 |
| | Flat latex (walls, ceilings & paint grade baseboard) | | | | | |
| #5 - | Water base latex paint | 16.45 | 13.16 | 15.13 | 16.19 | 16.20 |
| | Acoustic spray-on texture | | | | | |
| #6 - | Primer | 17.65 | 14.12 | 16.24 | 17.38 | 17.40 |
| #7 - | Finish | 15.40 | 12.32 | 14.17 | 15.16 | 15.20 |
| #8 - | Dripowder mixed (pound) | .35 | .28 | .32 | .34 | .34 |
| | Enamel (wet area walls & ceilings and openings) | | | | | |
| #9 - | Water base | 23.15 | 18.52 | 21.30 | 22.79 | 22.80 |
| #10 - | Oil base | 26.30 | 21.04 | 24.20 | 25.89 | 25.90 |
| | System estimate (cabinets, bookshelves, molding, interior windows) | | | | | |
| #11a - | Wiping stain, oil base | 27.85 | 22.28 | 25.62 | 27.41 | 27.40 |
| #11b - | Sanding sealer, lacquer | 17.55 | 14.04 | 16.15 | 17.28 | 17.30 |
| #11c - | Lacquer, semi gloss | 18.25 | 14.60 | 16.79 | 17.97 | 18.00 |
| #11 - | Stain, seal & 2 coat lacquer system | | | | | |
| | Average cost [11a + b + (2 x c)] | | 16.38 | 18.84 | 20.16 | 20.20 |
| #12 - | Shellac, clear | 22.95 | 18.36 | 21.11 | 22.59 | 22.60 |
| #13 - | Penetrating oil stain | 25.35 | 20.28 | 23.32 | 24.95 | 25.00 |
| #14 - | Penetrating stain wax (molding) | 29.00 | 23.20 | 26.68 | 28.55 | 28.60 |
| #15 - | Wax, per pound (floors) | 7.90 | 6.32 | 7.27 | 7.78 | 7.80 |
| #16 - | Glazing (mottling over enamel) | 29.50 | 23.60 | 27.14 | 29.04 | 29.00 |
| #17 - | Spray can, each (HVAC registers) | 3.70 | 2.96 | 3.40 | 3.64 | 3.60 |

**Figure 9**
Material prices at 20% discount

## Material prices at 20% discount (cont.)

| | | Retail price guide | Contractor price at a 20% discount | Add 15% sundries & escalation | Price with sales tax at 7% | Estimating prices with tax |
|---|---|---|---|---|---|---|
| **Exterior:** | | | | | | |
| | Solid body/color stain (beams, light valance, fascia, overhang, siding, plant-on trim, wood shelves) | | | | | |
| #18 - | Water base stain | 19.70 | 15.76 | 18.12 | 19.39 | 19.40 |
| #19 - | Oil base | 23.45 | 18.76 | 21.57 | 23.08 | 23.10 |
| | Semi-transparent stain (beams, siding, T & G ceiling) | | | | | |
| #20 - | Water base stain | 19.90 | 15.92 | 18.31 | 19.59 | 19.60 |
| #21 - | Oil base stain | 23.80 | 19.04 | 21.90 | 23.43 | 23.40 |
| #22 - | Polyurethane (exterior doors) | 31.05 | 24.84 | 28.57 | 30.57 | 30.60 |
| #23 - | Marine spar varnish, flat or gloss (exterior doors) | | | | | |
| | Interior or exterior | 36.50 | 29.20 | 33.58 | 35.93 | 35.90 |
| | Exterior enamel (exterior doors & trim) | | | | | |
| #24 - | Water base | 25.35 | 20.28 | 23.32 | 24.95 | 25.00 |
| #25 - | Oil base | 27.65 | 22.12 | 25.44 | 27.22 | 27.20 |
| | Porch & deck enamel - interior or exterior | | | | | |
| #26 - | Water base | 27.20 | 21.76 | 25.02 | 26.77 | 26.80 |
| #27 - | Oil base | 27.55 | 22.04 | 25.35 | 27.12 | 27.10 |
| #28 - | Epoxy, 1 part, water base | 29.87 | 23.90 | 27.49 | 29.41 | 29.40 |
| #29 - | Epoxy, 2 part system | 47.15 | 37.72 | 43.38 | 46.42 | 46.40 |
| | System estimate (exterior windows) | | | | | |
| #30a - | Wiping stain, oil base | 26.55 | 21.24 | 24.43 | 26.14 | 26.10 |
| #30b - | Sanding sealer, varnish | 27.80 | 22.24 | 25.58 | 27.37 | 27.40 |
| #30c - | Varnish, flat or gloss | 34.45 | 27.56 | 31.69 | 33.91 | 33.90 |
| #30 - | Stain, seal & 1 coat varnish system | | | | | |
| | Average cost (30a + b + c) | | 23.68 | 27.23 | 29.14 | 29.10 |
| | Masonry paint (masonry, concrete, plaster) | | | | | |
| #31 - | Water base, flat or gloss | 19.80 | 15.84 | 18.22 | 19.50 | 19.50 |
| #32 - | Oil base paint | 25.30 | 20.24 | 23.28 | 24.91 | 24.90 |
| #33 - | Masonry block filler | 14.30 | 11.44 | 13.16 | 14.08 | 14.10 |
| #34 - | Waterproofing, clear hydro seal | 18.30 | 14.64 | 16.84 | 18.02 | 18.00 |
| | Metal primer, rust inhibitor | | | | | |
| #35 - | Clean metal | 28.60 | 22.88 | 26.31 | 28.15 | 28.20 |
| #36 - | Rusty metal | 33.85 | 27.08 | 31.14 | 33.32 | 33.30 |
| | Metal finish, synthetic enamel, gloss, interior or exterior | | | | | |
| #37 - | Off white | 26.35 | 21.08 | 24.24 | 25.94 | 25.90 |
| #38 - | Colors (except orange/red) | 35.60 | 28.48 | 32.75 | 35.04 | 35.00 |
| | Anti-graffiti stain eliminator | | | | | |
| #39 - | Water base primer & sealer | 30.90 | 24.72 | 28.43 | 30.42 | 30.40 |
| #40 - | Oil base primer & sealer | 28.65 | 22.92 | 26.36 | 28.21 | 28.20 |
| #41 - | Polyurethane 2 part system | 64.26 | 51.41 | 59.12 | 63.26 | 63.30 |

**Figure 9 (continued)**
Material prices at 20% discount

## Material pricing at 20% discount (cont.)

| | Retail price guide | Contractor price at a 20% discount | Add 15% sundries & escalation | Price with sales tax at 7% | Estimating prices with tax |
|---|---|---|---|---|---|
| **Preparation:** | | | | | |
| #42 - Caulking, per fluid ounce | .17 | .14 | .16 | .17 | .17 |
| Paint remover, per gallon | | | | | |
| #43 - Light duty | 16.65 | 13.32 | 15.32 | 16.39 | 16.40 |
| #44 - Heavy duty | 22.70 | 18.16 | 20.88 | 22.34 | 22.30 |
| #45 - Putty, per pound | 6.95 | 5.56 | 6.39 | 6.84 | 6.80 |
| #46 - Silica sand, per pound | .04 | .03 | .03 | .03 | .03 |
| #47 - Window protective coating (wax) | 12.35 | 9.88 | 11.36 | 12.16 | 12.20 |
| #48 - Wood filler, per gallon | 29.95 | 23.96 | 27.55 | 29.48 | 29.50 |
| **Industrial:** | | | | | |
| #49 - Acid wash (muriatic acid) | 5.25 | 4.20 | 4.83 | 5.17 | 5.20 |
| #50 - Aluminum base paint | 34.45 | 27.56 | 31.69 | 33.91 | 33.90 |
| Epoxy coating, 2 part system | | | | | |
| #51 - Clear | 57.90 | 46.32 | 53.27 | 57.00 | 57.00 |
| #52 - White | 58.30 | 46.64 | 53.64 | 57.39 | 57.40 |
| Heat resistant enamel | | | | | |
| #53 - 800 to 1200 degree range | 70.30 | 56.24 | 64.68 | 69.21 | 69.20 |
| #54 - 300 to 800 degree range | 52.95 | 42.36 | 48.71 | 52.12 | 52.10 |
| #55 - Industrial bonding & penetrating oil paint | 27.30 | 21.84 | 25.12 | 26.88 | 26.90 |
| Industrial enamel, oil base, high gloss | | | | | |
| #56 - Light colors | 29.30 | 23.44 | 26.96 | 28.85 | 28.90 |
| #57 - Dark (OSHA) colors | 35.40 | 28.32 | 32.57 | 34.85 | 34.90 |
| #58 - Industrial waterproofing | 21.95 | 17.56 | 20.19 | 21.60 | 21.60 |
| #59 - Vinyl coating (tanks) | 60.15 | 48.12 | 55.34 | 59.21 | 59.20 |
| **Wallcovering:** | | | | | |
| Ready-mix | | | | | |
| #60 - Light-weight vinyl (gal) | 6.45 | 5.16 | 5.93 | 6.35 | 6.40 |
| #61 - Heavy weight vinyl (gal) | 6.75 | 5.40 | 6.21 | 6.64 | 6.60 |
| #62 - Cellulose, clear (gal) | 6.55 | 5.24 | 6.03 | 6.45 | 6.50 |
| #63 - Vinyl to vinyl (gal) | 15.50 | 12.40 | 14.26 | 15.26 | 15.30 |
| #64 - Powdered cellulose, 2 - 4 ounces | 2.15 | 1.72 | 1.98 | 2.12 | 2.10 |
| #65 - Powdered vinyl, 2 - 4 ounces | 2.50 | 2.00 | 2.30 | 2.46 | 2.50 |
| #66 - Powdered wheat paste, 2 - 4 ounces | 2.20 | 1.76 | 2.02 | 2.16 | 2.20 |

**Note:** Typically, powdered paste is in 2 to 4 ounce packages which will adhere 6 to 12 rolls of wallcovering.

**Figure 9 (continued)**
Material prices at 20% discount

## Material pricing at 30% discount

All pricing is based on production grade material purchased in 5 gallon quantities.

| | | Retail price guide | Contractor price at a 30% discount | Add 15% sundries & escalation | Price with sales tax at 7% | Estimating prices with tax |
|---|---|---|---|---|---|---|
| **Interior:** | | | | | | |
| | Sealer, off white (wet area walls & ceilings) | | | | | |
| #1 - | Water base | 15.70 | 10.99 | 12.64 | 13.52 | 13.50 |
| #2 - | Oil base | 21.00 | 14.70 | 16.91 | 18.09 | 18.10 |
| | Undercoat (doors, casings and other paint grade wood) | | | | | |
| #3 - | Water base | 18.85 | 13.20 | 15.18 | 16.24 | 16.20 |
| #4 - | Oil base | 20.45 | 14.32 | 16.47 | 17.62 | 17.60 |
| | Flat latex (walls, ceilings & paint grade baseboard) | | | | | |
| #5 - | Water base latex paint | 16.45 | 11.52 | 13.25 | 14.18 | 14.20 |
| | Acoustic spray-on texture | | | | | |
| #6 - | Primer | 17.65 | 12.36 | 14.21 | 15.20 | 15.20 |
| #7 - | Finish | 15.40 | 10.78 | 12.40 | 13.27 | 13.30 |
| #8 - | Dripowder mixed (pound) | .35 | .25 | .29 | .31 | .30 |
| | Enamel (wet area walls & ceilings and openings) | | | | | |
| #9 - | Water base enamel | 23.15 | 16.21 | 18.64 | 19.94 | 19.90 |
| #10 - | Oil base enamel | 26.30 | 18.41 | 21.17 | 22.65 | 22.70 |
| | System estimate (cabinets, bookshelves, molding, interior windows) | | | | | |
| #11a - | Wiping stain, oil base | 27.85 | 19.50 | 22.43 | 24.00 | 24.00 |
| #11b - | Sanding sealer, lacquer | 17.55 | 12.29 | 14.13 | 15.12 | 15.10 |
| #11c - | Lacquer, semi gloss | 18.25 | 12.78 | 14.70 | 15.73 | 15.70 |
| #11 - | Stain, seal & 2 coat lacquer system | | | | | |
| | Average cost [11a + b + (2 x c)] | | 14.86 | 17.09 | 18.29 | 18.30 |
| #12 - | Shellac, clear | 22.95 | 16.07 | 18.48 | 19.77 | 19.80 |
| #13 - | Penetrating oil stain | 25.35 | 17.75 | 20.41 | 21.84 | 21.80 |
| #14 - | Penetrating stain wax (molding) | 29.00 | 20.30 | 23.35 | 24.98 | 25.00 |
| #15 - | Wax, per pound (floors) | 7.90 | 5.53 | 6.36 | 6.81 | 6.80 |
| #16 - | Glazing (mottling over enamel) | 29.50 | 20.65 | 23.75 | 25.41 | 25.40 |
| #17 - | Spray can, each (HVAC registers) | 3.70 | 2.59 | 2.98 | 3.19 | 3.20 |

**Figure 10**
Material prices at 30% discount

## Material pricing at 30% discount (cont.)

| | | Retail price guide | Contractor price at a 30% discount | Add 15% sundries & escalation | Price with sales tax at 7% | Estimating prices with tax |
|---|---|---|---|---|---|---|
| **Exterior:** | | | | | | |
| | Solid body/color stain (beams, light valance, fascia, overhang, siding, plant-on trim, wood shelves) | | | | | |
| #18 - | Water base stain | 19.70 | 13.79 | 15.86 | 16.97 | 17.00 |
| #19 - | Oil base stain | 23.45 | 16.42 | 18.88 | 20.20 | 20.20 |
| | Semi-transparent stain (beams, siding, T & G ceiling) | | | | | |
| #20 - | Water base stain | 19.90 | 13.93 | 16.02 | 17.14 | 17.10 |
| #21 - | Oil base stain | 23.80 | 16.66 | 19.16 | 20.50 | 20.50 |
| #22 - | Polyurethane (exterior doors) | 31.05 | 21.74 | 25.00 | 26.75 | 26.80 |
| #23 - | Marine spar varnish, flat or gloss (exterior doors) | | | | | |
| | Interior or exterior | 36.50 | 25.55 | 29.38 | 31.44 | 31.40 |
| | Exterior enamel (exterior doors & trim) | | | | | |
| #24 - | Water base | 25.35 | 17.75 | 20.41 | 21.84 | 21.80 |
| #25 - | Oil base | 27.65 | 19.36 | 22.26 | 23.82 | 23.80 |
| | Porch & deck enamel - interior or exterior | | | | | |
| #26 - | Water base | 27.20 | 19.04 | 21.90 | 23.43 | 23.40 |
| #27 - | Oil base | 27.55 | 19.29 | 22.18 | 23.73 | 23.70 |
| #28 - | Epoxy, 1 part, water base | 29.87 | 20.91 | 24.05 | 25.73 | 25.70 |
| #29 - | Epoxy, 2 part system | 47.15 | 33.01 | 37.96 | 40.62 | 40.60 |
| | System estimate (exterior windows) | | | | | |
| #30a - | Wiping stain, oil base | 26.55 | 18.59 | 21.38 | 22.88 | 22.90 |
| #30b - | Sanding sealer, varnish | 27.80 | 19.46 | 22.38 | 23.95 | 24.00 |
| #30c - | Varnish, flat or gloss | 34.45 | 24.12 | 27.74 | 29.68 | 29.70 |
| #30 - | Stain, seal & 1 coat varnish system | | | | | |
| | Average cost (30a + b + c) | | 20.72 | 23.83 | 25.50 | 25.50 |
| | Masonry paint (masonry, concrete, plaster) | | | | | |
| #31 - | Water base, flat or gloss | 19.80 | 13.86 | 15.94 | 17.06 | 17.10 |
| #32 - | Oil base paint | 25.30 | 17.71 | 20.37 | 21.80 | 21.80 |
| #33 - | Block filler | 14.30 | 10.01 | 11.51 | 12.32 | 12.30 |
| #34 - | Waterproofing, clear hydro seal | 18.30 | 12.81 | 14.73 | 15.76 | 15.80 |
| | Metal primer, rust inhibitor | | | | | |
| #35 - | Clean metal | 28.60 | 20.02 | 23.02 | 24.63 | 24.60 |
| #36 - | Rusty metal | 33.85 | 23.70 | 27.26 | 29.17 | 29.20 |
| | Metal finish, synthetic enamel, gloss, interior or exterior | | | | | |
| #37 - | Off white | 26.35 | 18.45 | 21.22 | 22.71 | 22.70 |
| #38 - | Colors (except orange/red) | 35.60 | 24.92 | [illegible]8.66 | 30.67 | 30.70 |
| | Anti-graffiti stain eliminator | | | | | |
| #39 - | Water base primer & sealer | 30.90 | 21.63 | 24.87 | 26.61 | 26.60 |
| #40 - | Oil base primer & sealer | 28.65 | 20.06 | 23.07 | 24.68 | 24.70 |
| #41 - | Polyurethane 2 part system | 64.26 | 44.98 | 51.73 | 55.35 | 55.40 |

**Figure 10 (continued)**
Material prices at 30% discount

## Material pricing at 30% discount (cont.)

| | Retail price guide | Contractor price at a 30% discount | Add 15% sundries & escalation | Price with sales tax at 7% | Estimating prices with tax |
|---|---|---|---|---|---|
| **Preparation:** | | | | | |
| #42 - Caulking, per fluid ounce | .17 | .12 | .14 | .15 | .20 |
| Paint remover, per gallon | | | | | |
| #43 - Light duty | 16.65 | 11.66 | 13.41 | 14.35 | 14.40 |
| #44 - Heavy duty | 22.70 | 15.89 | 18.27 | 19.55 | 19.60 |
| #45 - Putty, per pound | 6.95 | 4.87 | 5.60 | 5.99 | 6.00 |
| #46 - Silica sand, per pound | .04 | .03 | .03 | .03 | .03 |
| #47 - Window protective coating (wax) | 12.35 | 8.65 | 9.95 | 10.65 | 10.70 |
| #48 - Wood filler, per gallon | 29.95 | 20.97 | 24.12 | 25.81 | 25.80 |
| **Industrial:** | | | | | |
| #49 - Acid wash (muriatic acid) | 5.25 | 3.68 | 4.23 | 4.53 | 4.50 |
| #50 - Aluminum base paint | 34.45 | 24.12 | 27.74 | 29.68 | 29.70 |
| Epoxy coating, 2 part system | | | | | |
| #51 - Clear | 57.90 | 40.53 | 46.61 | 49.87 | 49.90 |
| #52 - White | 58.30 | 40.81 | 46.93 | 50.22 | 50.20 |
| Heat resistant enamel | | | | | |
| #53 - 800 to 1200 degree range | 70.30 | 49.21 | 56.59 | 60.55 | 60.60 |
| #54 - 300 to 800 degree range | 52.95 | 37.07 | 42.63 | 45.61 | 45.60 |
| #55 - Industrial bonding & penetrating oil paint | 27.30 | 19.11 | 21.98 | 23.52 | 23.50 |
| Industrial enamel, oil base, high gloss | | | | | |
| #56 - Light colors | 29.30 | 20.51 | 23.59 | 25.24 | 25.20 |
| #57 - Dark (OSHA) colors | 35.40 | 24.78 | 28.50 | 30.50 | 30.50 |
| #58 - Industrial waterproofing | 21.95 | 15.37 | 17.68 | 18.92 | 18.90 |
| #59 - Vinyl coating (tanks) | 60.15 | 42.11 | 48.43 | 51.82 | 51.80 |
| **Wallcovering:** | | | | | |
| Ready-mix: | | | | | |
| #60 - Light-weight vinyl (gal) | 6.45 | 4.52 | 5.20 | 5.56 | 5.60 |
| #61 - Heavy weight vinyl (gal) | 6.75 | 4.73 | 5.44 | 5.82 | 5.80 |
| #62 - Cellulose, clear (gal) | 6.55 | 4.59 | 5.28 | 5.65 | 5.70 |
| #63 - Vinyl to vinyl (gal) | 15.50 | 10.85 | 12.48 | 13.35 | 13.40 |
| #64 - Powdered cellulose, 2 - 4 ounces | 2.15 | 1.51 | 1.74 | 1.86 | 1.90 |
| #65 - Powdered vinyl, 2 - 4 ounces | 2.50 | 1.75 | 2.01 | 2.15 | 2.20 |
| #66 - Powdered wheat paste, 2 - 4 ounces | 2.20 | 1.54 | 1.77 | 1.89 | 1.90 |

**Note:** Typically, powdered paste is in 2 to 4 ounce packages which will adhere 6 to 12 rolls of wallcovering.

**Figure 10 (continued)**
Material prices at 30% discount

## Material pricing at 40% discount

All pricing is based on production grade material purchased in 5 gallon quantities.

| | | Retail price guide | Contractor price at a 40% discount | Add 15% sundries & escalation | Price with sales tax at 7% | Estimating prices with tax |
|---|---|---|---|---|---|---|
| **Interior:** | | | | | | |
| | Sealer, off white (wet area walls & ceilings) | | | | | |
| #1 - | Water base | 15.70 | 9.42 | 10.83 | 11.59 | 11.60 |
| #2 - | Oil base | 21.00 | 12.60 | 14.49 | 15.50 | 15.50 |
| | Undercoat (doors, casings and other paint grade wood) | | | | | |
| #3 - | Water base | 18.85 | 11.31 | 13.01 | 13.92 | 13.90 |
| #4 - | Oil base | 20.45 | 12.27 | 14.11 | 15.10 | 15.10 |
| | Flat latex (walls, ceilings & paint grade baseboard) | | | | | |
| #5 - | Water base latex paint | 16.45 | 9.87 | 11.35 | 12.14 | 12.10 |
| | Acoustic spray-on texture | | | | | |
| #6 - | Primer | 17.65 | 10.59 | 12.18 | 13.03 | 13.00 |
| #7 - | Finish | 15.40 | 9.24 | 10.63 | 11.37 | 11.40 |
| #8 - | Dripowder mixed (pound) | .35 | .21 | .24 | .26 | .30 |
| | Enamel (wet area walls & ceilings and openings) | | | | | |
| #9 - | Water base enamel | 23.15 | 13.89 | 15.97 | 17.09 | 17.10 |
| #10 - | Oil base enamel | 26.30 | 15.78 | 18.15 | 19.42 | 19.40 |
| | System estimate (cabinets, bookshelves, molding, interior windows) | | | | | |
| #11a - | Wiping stain, oil base | 27.85 | 16.71 | 19.22 | 20.57 | 20.60 |
| #11b - | Sanding sealer, lacquer | 17.55 | 10.53 | 12.11 | 12.96 | 13.00 |
| #11c - | Lacquer, semi gloss | 18.25 | 10.95 | 12.59 | 13.47 | 13.50 |
| #11 - | Stain, seal & 2 coat lacquer system | | | | | |
| | Average cost [11a + b + (2 x c)] | | 12.29 | 14.13 | 15.12 | 15.10 |
| #12 - | Shellac, clear | 22.95 | 13.77 | 15.84 | 16.95 | 17.00 |
| #13 - | Penetrating oil stain | 25.35 | 15.21 | 17.49 | 18.71 | 18.70 |
| #14 - | Penetrating stain wax (molding) | 29.00 | 17.40 | 20.01 | 21.41 | 21.40 |
| #15 - | Wax, per pound (floors) | 7.90 | 4.74 | 5.45 | 5.83 | 5.80 |
| #16 - | Glazing (mottling over enamel) | 29.50 | 17.70 | 20.36 | 21.79 | 21.80 |
| #17 - | Spray can, each (HVAC registers) | 3.70 | 2.22 | 2.55 | 2.73 | 2.70 |

**Figure 11**
Material prices at 40% discount

## Material pricing at 40% discount (cont.)

| | | Retail price guide | Contractor price at a 40% discount | Add 15% sundries & escalation | Price with sales tax at 7% | Estimating prices with tax |
|---|---|---|---|---|---|---|
| **Exterior:** | | | | | | |
| | Solid body/color stain (beams, light valance, fascia, overhang, siding, plant-on trim, wood shelves) | | | | | |
| #18 - | Water base stain | 19.70 | 11.82 | 13.59 | 14.54 | 14.50 |
| #19 - | Oil base stain | 23.45 | 14.07 | 16.18 | 17.31 | 17.30 |
| | Semi-transparent stain (beams, siding, T & G ceiling) | | | | | |
| #20 - | Water base stain | 19.90 | 11.94 | 13.73 | 14.69 | 14.70 |
| #21 - | Oil base stain | 23.80 | 14.28 | 16.42 | 17.57 | 17.60 |
| #22 - | Polyurethane (exterior doors) | 31.05 | 18.63 | 21.42 | 22.92 | 22.90 |
| #23 - | Marine spar varnish, flat or gloss (exterior doors) | | | | | |
| | Interior or exterior | 36.50 | 21.90 | 25.19 | 26.95 | 27.00 |
| | Exterior enamel (exterior doors & trim) | | | | | |
| #24 - | Water base | 25.35 | 15.21 | 17.49 | 18.71 | 18.70 |
| #25 - | Oil base | 27.65 | 16.59 | 19.08 | 20.42 | 20.40 |
| | Porch & deck enamel - interior or exterior | | | | | |
| #26 - | Water base enamel | 27.20 | 16.32 | 18.77 | 20.08 | 20.10 |
| #27 - | Oil base enamel | 27.55 | 16.53 | 19.01 | 20.34 | 20.30 |
| #28 - | Epoxy, 1 part, water base | 29.87 | 17.92 | 20.61 | 22.05 | 22.10 |
| #29 - | Epoxy, 2 part system | 47.15 | 28.29 | 32.53 | 34.81 | 34.80 |
| | System estimate (exterior windows) | | | | | |
| #30a - | Wiping stain, oil base | 26.55 | 15.93 | 18.32 | 19.60 | 19.60 |
| #30b - | Sanding sealer, varnish | 27.80 | 16.68 | 19.18 | 20.52 | 20.50 |
| #30c - | Varnish, flat or gloss | 34.45 | 20.67 | 23.77 | 25.43 | 25.40 |
| #30 - | Stain, seal & 1 coat varnish system | | | | | |
| | Average cost (30a + b + c) | | 17.76 | 20.42 | 21.85 | 21.90 |
| | Masonry paint (masonry, concrete, plaster) | | | | | |
| #31 - | Water base, flat or gloss | 19.80 | 11.88 | 13.66 | 14.62 | 14.60 |
| #32 - | Oil base paint | 25.30 | 15.18 | 17.46 | 18.68 | 18.70 |
| #33 - | Block filler | 14.30 | 8.58 | 9.87 | 10.56 | 10.60 |
| #34 - | Waterproofing, clear hydro seal | 18.30 | 10.98 | 12.63 | 13.51 | 13.50 |
| | Metal primer, rust inhibitor | | | | | |
| #35 - | Clean metal | 28.60 | 17.16 | 19.73 | 21.11 | 21.10 |
| #36 - | Rusty metal | 33.85 | 20.31 | 23.36 | 25.00 | 25.00 |
| | Metal finish, synthetic enamel, gloss, interior or exterior | | | | | |
| #37 - | Off white | 26.35 | 15.81 | 18.18 | 19.45 | 19.50 |
| #38 - | Colors (except orange/red) | 35.60 | 21.36 | 24.56 | 26.28 | 26.30 |
| | Anti-graffiti stain eliminator | | | | | |
| #39 - | Water base primer & sealer | 30.90 | 18.54 | 21.32 | 22.81 | 22.80 |
| #40 - | Oil base primer & sealer | 28.65 | 17.19 | 19.77 | 21.15 | 21.20 |
| #41 - | Polyurethane 2 part system | 64.26 | 38.56 | 44.34 | 47.44 | 47.40 |

**Figure 11 (continued)**
Material prices at 40% discount

## Material pricing at 40% discount (cont.)

| | Retail price guide | Contractor price at a 40% discount | Add 15% sundries & escalation | Price with sales tax at 7% | Estimating prices with tax |
|---|---|---|---|---|---|
| **Preparation:** | | | | | |
| #42 - Caulking, per fluid ounce | .17 | .10 | .12 | .13 | .10 |
| Paint remover, per gallon | | | | | |
| #43 - Light duty | 16.65 | 9.99 | 11.49 | 12.29 | 12.30 |
| #44 - Heavy duty | 22.70 | 13.62 | 15.66 | 16.76 | 16.80 |
| #45 - Putty, per pound | 6.95 | 4.17 | 4.80 | 5.14 | 5.10 |
| #46 - Silica sand, per pound | .04 | .02 | .02 | .02 | .02 |
| #47 - Window protective coating (wax) | 12.35 | 7.41 | 8.52 | 9.12 | 9.10 |
| #48 - Wood filler, per gallon | 29.95 | 17.97 | 20.67 | 22.12 | 22.10 |
| **Industrial:** | | | | | |
| #49 - Acid wash (muriatic acid) | 5.25 | 3.15 | 3.62 | 3.87 | 3.90 |
| #50 - Aluminum base paint | 34.45 | 20.67 | 23.77 | 25.43 | 25.40 |
| Epoxy coating, 2 part system | | | | | |
| #51 - Clear | 57.90 | 34.74 | 39.95 | 42.75 | 42.80 |
| #52 - White | 58.30 | 34.98 | 40.23 | 43.05 | 43.10 |
| Heat resistant enamel | | | | | |
| #53 - 800 to 1200 degree range | 70.30 | 42.18 | 48.51 | 51.91 | 51.90 |
| #54 - 300 to 800 degree range | 52.95 | 31.77 | 36.54 | 39.10 | 39.10 |
| #55 - Industrial bonding & penetrating oil paint | 27.30 | 16.38 | 18.84 | 20.16 | 20.20 |
| Industrial enamel, oil base, high gloss | | | | | |
| #56 - Light colors | 29.30 | 17.58 | 20.22 | 21.64 | 21.60 |
| #57 - Dark (OSHA) colors | 35.40 | 21.24 | 24.43 | 26.14 | 26.10 |
| #58 - Industrial waterproofing | 21.95 | 13.17 | 15.15 | 16.21 | 16.20 |
| #59 - Vinyl coating (tanks) | 60.15 | 36.09 | 41.50 | 44.41 | 44.40 |
| **Wallcovering:** | | | | | |
| Ready-mix: | | | | | |
| #60 - Light-weight vinyl (gal) | 6.45 | 3.87 | 4.45 | 4.76 | 4.80 |
| #61 - Heavy weight vinyl (gal) | 6.75 | 4.05 | 4.66 | 4.99 | 5.00 |
| #62 - Cellulose, clear (gal) | 6.55 | 3.93 | 4.52 | 4.84 | 4.80 |
| #63 - Vinyl to vinyl (gal) | 15.50 | 9.30 | 10.70 | 11.45 | 11.50 |
| #64 - Powdered cellulose, 2 - 4 ounces | 2.15 | 1.29 | 1.48 | 1.58 | 1.60 |
| #65 - Powdered vinyl, 2 - 4 ounces | 2.50 | 1.50 | 1.73 | 1.85 | 1.90 |
| #66 - Powdered wheat paste, 2 - 4 ounces | 2.20 | 1.32 | 1.52 | 1.63 | 1.60 |

**Note:** Typically, powdered paste is in 2 to 4 ounce packages which will adhere 6 to 12 rolls of wallcovering

**Figure 11 (continued)**
Material prices at 40% discount

Figure 10 reflects a 30 percent discount. It applies to "Medium" work and assumes medium coverage, as in commercial work.

Figure 11 is the 40 percent discount table. It applies to "Fast" work and assumes heavier coverage typically required on unpainted surfaces in new construction. This discount is usually available only to large, high-volume painting companies that purchase materials in large quantities.

Here's an explanation of the columns in Figures 9, 10 and 11:

**Retail price guide:** This is an average based on a survey of eleven paint manufacturers or distributors, for standard grade, construction-quality paint, purchased in five gallon quantities.

Material pricing and discount percentages will vary from supplier to supplier and from area to area. Always keep your supplier's current price list handy. It should show your current cost for all the coatings and supplies you use. Also post a list of all suppliers, their phone numbers, and the salesperson's name beside your phone.

Prices change frequently. Paint quality, your supplier's discount programs, their marketing strategy and competition from other paint manufacturers will influence the price you pay. Never guess about paint prices – especially about less commonly used coatings. Don't assume that a product you haven't used before costs about the same as similar products. It might not. A heavy-duty urethane finish, for example, will cost about twice as much as a heavy-duty vinyl coating. If you don't know that, your profit for the job can disappear very quickly.

**Prices at discount:** The retail price, less the appropriate discount.

**Allowance for sundries:** It's not practical to figure the cost of every sheet of sandpaper and every rag you'll use on a job. And there's no way to accurately predict how many jobs you'll get out of each brush or roller pole, roller handle, ladder, or drop cloth. But don't let that keep you from including an allowance for these important costs in your estimates. If you leave them out, it's the same as estimating the cost of those items as *zero*. That's a 100 percent miss. Too many of those, and you're out of the painting business. It's better to estimate any amount than to omit some costs entirely.

Figure 12 is a sundries inventory checklist. Use it to keep track of the actual cost of expendable tools and equipment.

I've added 15 percent to the paint cost to cover expendable tools and supplies. This is enough for sundries on most jobs. There is one exception, however. On repaint jobs where there's extensive prep work, the cost of sundries may be more than 15 percent of the paint cost. When preparation work is extensive, figure the actual cost of supplies. Then add to the estimate that portion of the sundries cost that exceeds 15 percent of the paint cost. You might have to double the normal sundries allowance. When it comes to prep work, make sure your estimate covers *all* your supplies.

**Price with sales tax at 7 percent:** This column increases the material cost, including sundries, by 7 percent to cover sales tax. If sales tax in your area is more or less than 7 percent, you can adjust the material cost, or use the price that's closest to your actual cost.

In most cases contractors have to pay sales tax. If you don't pay the tax yourself, you may have to collect it from the building owner or general contractor and remit it to the state taxing authority. In either case, include sales tax in your estimate.

**Estimating prices with tax:** The figures in the last column of Figures 9 through 11 are rounded to the nearest dime unless the total is under a dollar. Those prices are rounded to the nearest penny.

This system for pricing materials isn't exact. But it's quick, easy and flexible. Compare your current material costs with costs in Figures 9, 10 and 11. If your costs are more than a few percent higher or lower than my costs, make a note on the blank line below "Fast" in the estimating tables.

### *Price Escalation*

Escalation is the change in prices between the time you bid a job and the time you pay for labor and materials. Painting contractors seldom include escalation clauses in their bids because they don't expect lengthy delays. That's why escalation isn't included as a separate item in the estimating forms, Figures 18 and 19.

## Sundry Inventory Checklist

**Suppliers:** D-Dumphy Paints
F-Fisher Paints
S-Superior Paints
P-Pioneer Paints

| Supplier | Product number | Product | Inventory quantity | Unit | Cost | 7/22 | 7/29 | 8/5 | 8/12 |
|---|---|---|---|---|---|---|---|---|---|
| D | # — | Bender paint pads | 3 | Each | $3.50 | | | | |
| D | #792 | Brush - 3" nylon Peacock | 2 | Each | $18.85 | | 1 | | |
| D | #783 | Brush - 4" nylon Scooter | 2 | Each | $27.85 | | | 1 | |
| D | #115 | Brush - 5" nylon Pacer | 2 | Each | $44.60 | | | 1 | |
| D | #784 | Brush - 3" bristle | 2 | Each | $17.55 | | | 1 | |
| D | #2170 | Caulking bags | 2 | Each | $3.75 | | | | |
| D | Latex | Caulking-DAP Acrylic latex | 12 | Each | $2.05 | | 12 | | |
| D | #2172 | Caulking gun (Newborn) | 2 | Each | $ 7.15 | | 1 | | |
| P | # — | Hydraulic fluid | 2 | Qt | $8.05 | | | | |
| P | # — | Lemon oil | 2 | Pint | $4.05 | | 1 | | |
| F | # — | Masking paper 18" wide | 3 | Roll | $20.60 | | | | |
| F | Anchor | Masking tape 1½" | 24 | Roll | $3.05 | | 12 | | 12 |
| P | #2176 | Lacquer | 2 | 5's | $24.90 | | | 1 | |
| P | #2173 | Sanding sealer | 2 | 5's | $20.35 | | 1 | | |
| P | #9850 | Resin sealer | 2 | 5's | $17.75 | | | | |
| P | #131 | PVA sealer (clear) | 2 | 5's | $18.80 | | 1 | | |
| F | #8500 | Particle masks 100/box | 1 | Box | $13.45 | | | 1 | |
| P | # — | Putty (Crawfords) | 3 | Qt | $9.20 | | 2 | | |
| F | #R-10 | Respirators | 1 | Each | $37.65 | | | | 1 |
| F | #R-49 | Respirator cartridges 20/box | 2 | Box | $43.90 | | | | |
| F | #R-51 | Respirator filters 20/box | 2 | Box | $31.35 | | | 1 | |
| P | # — | Rags | 10 | Lb | $2.25 | | | | |
| F | #AR 691 | Roller covers 9" x 3/4" | 6 | Each | $4.20 | | 2 | | |
| F | #AR 692 | Roller covers 9" x 3/8" | 6 | Each | $3.50 | 3 | | | 2 |
| F | #AR 671 | Roller covers 7" x 3/4" | 3 | Each | $3.45 | | | 1 | |
| F | #AR 672 | Roller covers 7" x 3/8" | 3 | Each | $3.75 | | 1 | | |

**Figure 12**
Sundry inventory checklist

| Supplier | Product number | Product | Inventory quantity | Unit | Cost | 7/22 | 7/29 | 8/5 | 8/12 |
|---|---|---|---|---|---|---|---|---|---|
| F | #AR 611 | Roller covers mini | 3 | Each | $2.90 | | | 1 | |
| F | #95 | Roller frames 9" | 6 | Each | $5.40 | 1 | 2 | | |
| F | #75 | Roller frames 7" | 5 | Each | $5.20 | 3 | | 3 | |
| F | #TSR | Roller frames mini | 2 | Each | $2.95 | | | | |
| D | #40 | Roller poles 4' wood tip | 3 | Each | $2.60 | | 1 | | |
| D | #10 | Roller poles 6' wood tip | 10 | Each | $4.05 | | | | |
| P | # 1 | Roller pole tips metal | 2 | Each | $3.30 | | | 2 | |
| P | # — | Sandpaper (120C production) | 2 | Slve | $48.65 | | | 2 | |
| P | # — | Sandpaper (220A trimite) | 2 | Slve | $37.85 | | | | 1 |
| P | # — | Sandpaper (220A garnet) | 1 | Slve | $34.45 | | 1 | | |
| D | # — | Spackle (Synkloid) | 3 | Qt | $5.30 | 1 | | 1 | |
| D | #42/61 | Spray bombs (black$^{B}$/white$^{W}$) | 12 | Each | $3.10 | $^{B}$12 | | | $^{W}$12 |
| F | # — | Spray gun tips #3 or #4 | 10 | Each | $7.65 | | | 3 | |
| F | #2762 | Spray gun couplers | 10 | Each | $2.10 | | | 5 | |
| F | #S-27 | Spray socks 48/Box | 1 | Box | $16.70 | | | | |
| D | #5271 | Stip fill | 1 | Gal | $9.00 | | | 1 | |
| D | #5927 | Strainer bags | 2 | Each | $1.45 | 1 | | | |
| D | #JT-21 | Staples - 5/16" | 2 | Box | $2.35 | | | | |
| P | 50 Gal | Thinner, lacquer | 1 | Drum | $420.00 | | | | |
| P | 50 Gal | Thinner, paint | 1 | Drum | $210.00 | | | | 1 |
| P | # — | Thinner, shellac (alcohol) | 1 | Gal | $9.80 | | | | |
| D | # — | Visqueen 1.5 mil 12' x 200' | 3 | Roll | $24.00 | | | | |
| D | #5775 | Work pots (2 gal. plastic) | 3 | Each | $2.80 | | 1 | | 2 |
| | # | | | | $ | | | | |
| | # | | | | $ | | | | |
| | # | | | | $ | | | | |
| | | Order date: | | | | 7/21 | 7/27 | 8/2 | 8/10 |
| | | Ordered by: (initials) | | | | jj | jj | jj | jj |
| | | Purchase order no. | | | | 0352 | 0356 | 0361 | 0371 |

**Figure 12 (continued)**
Sundry inventory checklist

| | Residential Wallcovering | | | | Commercial Wallcovering | | | | Flexible Wood Wallcovering | | | |
|---|---|---|---|---|---|---|---|---|---|---|---|---|
| Production Rate | Computer Program Crew Code | Labor Cost per Hour | Labor Burden per Hour | Labor Cost + Burden | Computer Program Crew Code | Labor Cost per Hour | Labor Burden per Hour | Labor Cost + Burden | Computer Program Crew Code | Labor Cost per Hour | Labor Burden per Hour | Labor Cost + Burden |
| Slow | 1W | $9.75 | $2.66 | $12.41 | 4W | $8.75 | $2.38 | $11.13 | 7W | $9.25 | $2.52 | $11.77 |
| Medium | 2W | 15.50 | 5.22 | 20.72 | 5W | 14.00 | 4.71 | 18.71 | 8W | 14.75 | 4.96 | 19.71 |
| Fast | 3W | 20.50 | 8.31 | 28.81 | 6W | 18.50 | 7.50 | 26.00 | 9W | 19.50 | 7.91 | 27.41 |

**Figure 13**
Hourly wage rates for wallcovering application

Any minor price escalation will be covered by the 15 percent added to material prices for sundries. But don't rely on that small cushion to absorb major inflationary cost increases. Plan ahead if prices are rising. In that case, add escalation as a separate item in the estimate.

Many formal construction contracts include an escalator clause that allows the contractor to recover for cost increases during the time of construction – especially if there was an unreasonable delay through no fault of the subcontractor. This clause may give you the right to collect for increases in both labor and material costs.

If work is delayed after you've been awarded the contract, you may be able to recover for cost increases under the escalator clause. This is more likely on public projects than on private jobs. Also, if there's a significant delay due to weather, you may have a good argument for adjusting the contract amount.

You can protect yourself against escalation if you include an expiration date on your bids. If the contract award is delayed beyond your expiration date, you can review your costs and make necessary adjustments. But be careful here. Increase the bid too much and you'll probably lose the contract. So raise your bid only if necessary, and then only by the amount of the actual cost increases. Don't try to make a killing on the job just because the bid prices have expired.

## Column 4: Labor Cost

Column 4 in Figure 2 on page 7 shows the labor cost per unit. This figure is based on the productivity rate in column 1 and the wage rate in Figure 1. The wage rate for "Slow" (repaint) work is assumed to be $10.25 per hour. The wage rate for "Medium" (commercial) work is $16.50 per hour. The wage rate for "Fast" (residential tract) work is $22.00 per hour. Wage rates for wallcovering are different.

### *Wage Rates Vary*

Wages vary from city to city. Recently I saw a survey of hourly union rates for painters in U.S. cities. The lowest rate shown was $16.28 an hour for painters in Raleigh, North Carolina. The highest rate was $41.95 for painters in Nome, Alaska. You might ask, "Why don't all the painters in Raleigh move to Nome?"

I don't know the answer, except to suggest that painters aren't starving in Raleigh. Nor are they getting rich in Nome. Working conditions and the cost of living are very different in those two cities. However, on private jobs using non-union tradesmen, wage rates usually don't vary as much from city to city. The wage you pay depends on the demand for painting and how many painters are available for work.

Wages also change over time. For example, wage rates increased between 1986 and 1996. The national average union wage (including fringes) for painters in large cities went from $19.50 in 1986 to $27.60 per hour in 1996. Since then, wage rates haven't changed as quickly. Always base your estimates on the actual wages you'll pay your experienced painters.

| | Fixed burden | | | | | Fringe benefits | | | | | |
|---|---|---|---|---|---|---|---|---|---|---|---|
| | FICA | FUTA | SUI | WCI | Liab. Ins. | Vac | Med | Life | Pension | Training | Total |
| Slow | 7.65% | .8% | 3.0% | 9.5% | 6.3% | 0 | 0 | 0 | 0 | 0 | 27.25% |
| Medium | 7.65% | .8% | 4.5% | 11.5% | 6.7% | 1.0% | 1.0% | .25% | .25% | 0 | 33.65% |
| Fast | 7.65% | .8% | 6.0% | 13.5% | 7.1% | 2.5% | 2.0% | .25% | .5% | .25% | 40.55% |

**Figure 14**
Labor burden percentages

### *Wages for Higher Skilled Specialists*

Wages also vary with a workers' skill, dependability and with job difficulty. Generally higher paid painters are more productive than lower paid painters. Here's a chart I use to determine how much more per hour I should estimate for painting and surface preparation specialists. These figures are in addition to the basic journeyman rate.

Foremen . . . . . . . . . . . . . . . . . . . . .$.50 to 2.00

Field superintendents . . . . . . . . . .$3.00 to 4.50

Swing stage brush painters,

spray painters, or paperhangers . . . . . . .$.25

Iron, steel and bridge painters

(ground work) . . . . . . . . . . . . . . . . . . . . .$.50

Sandblasters, iron, steel, or

bridge painters (swing stage) . . . . . . . . .$1.00

Steeplejacks . . . . . . . . . . . . . . . . . . . . . . .$1.50

Most government and defense painting contracts require compliance with the Davis Bacon Act, which specifies that contractors pay at least the prevailing wage for each trade in the area where the job is located.

### *Calculate Your Labor Rate*

Use the wage rate in Figure 1 ($10.25, $16.50 or $22.00 for "Slow," "Medium," or "Fast") that's appropriate for your company. Or, use a rate somewhere in between the rates listed. If you use your own wage rate, divide the hourly wage by the labor productivity (such as square feet per manhour in column 1). That's your labor cost per unit. Multiply by 100 if the units used are 100 linear feet or 100 square feet. ($10 ÷ 400 x 100 = $2.50.)

## Column 5: Labor Burden

For each dollar of wages your company pays, at least another 28 cents has to be paid in payroll tax and for insurance. That's part of your labor burden. The rest is fringe benefits such as vacation pay, health benefits and pension plans.

Federal taxes are the same for all employers. State taxes vary from state to state. Fringe benefits vary the most. Generally, larger companies with more skilled painters offer considerably more fringe benefits than smaller companies.

In the estimating tables, the labor burden percentage varies with the application rate. For "Slow" (repaint) work, it's assumed to be 27.25 percent of $10.25 or $2.80 per hour. For "Medium" (commercial) work, the estimating tables use 33.65 percent of $16.50 or $5.55 per hour. For "Fast" (residential tract) work, the labor burden is 40.55 percent of $22.00 or $8.92 per hour.

Figure 14 shows how the labor burden percentages were compiled for each application rate.

**FICA - Social Security tax:** This is the portion paid by employers and is set by federal law. A similar amount is withheld from each employee's wage and deposited with a Federal Reserve bank by the employer.

**FUTA - Federal Unemployment Insurance tax:** Paid entirely by the employer and set by federal law. No portion is deducted from employee wages.

**SUI - State Unemployment Insurance:** Varies from state to state.

**WCI - Workers' Compensation Insurance:** Provides benefits for employees in case of injury on the job. Workers' comp is required by state law. Rates vary by state, job description and the loss experience of the employer.

**Liab. Ins. - Liability Insurance:** Covers injury or damage done to the public by employees. Comprehensive contractor's liability insurance includes current operations, completed operations, bodily injury, property damage, protective and contractual coverages with a $1,000,000 policy limit.

**Fringe benefits:** *Vac* is vacation pay. *Med* is medical insurance. *Life* is life insurance contribution. *Pension* is a pension plan contribution. *Training* is an apprentice training fund.

Vacation, life, pension and training payments depend on the agreement between employers and employees. These are voluntary contributions if not required by a collective bargaining agreement. Smaller companies are less likely to provide these benefits. The cost of fringe benefits in a painting company can range from zero to more than 10 percent of wages.

## Column 6: Material Cost per Unit

This column is the result of dividing column 3 (material cost) by column 2 (material coverage) for each application rate. For example, in Figure 2 in the "Medium" row, a material cost of 14.20 is divided by material coverage of 275, then multiplied by 100 to arrive at $5.16 per 100 square feet. That's the figure listed for "Medium" work in column 6.

## Column 7: Overhead

The overhead rate for "Slow" (repaint) jobs is assumed to be 31 percent. For "Medium" (commercial projects), overhead is 24.5 percent. For "Fast" (residential tracts), overhead is 18.5 percent. The overhead cost per unit in each row is calculated by adding the labor cost per unit, labor burden per unit, and material cost per unit and then multiplying by the appropriate overhead percentage.

There are two types of overhead, direct overhead and indirect overhead. Only indirect overhead is included in the "Overhead" column of the estimating cost tables. Enter your direct overhead costs on a separate line on your take-off sheet.

*Direct overhead* is job site overhead, expenses you charge to a specific job. Examples include performance bonds, special insurance premiums, or rental of a job site storage trailer. These expenses are not included in the estimating tables and have to be added to your estimates. On many jobs, there may be little or no direct overhead.

*Indirect overhead* is office overhead, expenses that aren't related to any particular job and that tend to continue whether the volume of work increases or decreases. Examples are non-trade salaries, office rent, vehicles, sales and financial expenses, insurance, taxes and licenses.

The percentage of income spent on overhead is assumed to be lower for high volume companies and higher for low volume companies. A large company working many projects at the same time can spread overhead costs over many projects – charging a smaller percentage of overhead to each job. The more jobs, the lower the overhead per job – assuming overhead doesn't increase faster than business volume.

On the other hand, a small business may have to absorb all overhead on a single job. Even painting contractors who work out of their homes have overhead expenses.

Here's one overhead expense every paint contractor has and that you might overlook: the cost of estimating jobs. That's part of the salary cost of the employee who does the estimating.

### Figure Overhead Carefully

Estimating indirect (office) overhead isn't as easy as estimating labor and material. There aren't as many clear-cut answers. That's why indirect overhead is often underestimated. Don't make that mistake in your estimates. Underestimating overhead is the same as giving up part of your profit. After all, indirect overhead expenses are real costs, just like paint, labor and taxes.

In large painting companies, management accumulates indirect overhead costs and translates them into a percentage the estimator should add to the costs of each job. In smaller companies, the estimator should keep a record of indirect overhead expenses. With a good record of overhead expense, you can calculate your overhead percentage for future periods very accurately. Then it's easy to add a percentage for indirect overhead costs into your estimate.

### Computing Your Overhead Percentage

Here's how to decide which overhead rate to use in the cost estimating tables:

1) List all your overhead expenses for at least the last six months; a year would be better. You need overhead cost information that goes back far enough to eliminate the effect of seasonal changes in business volume.

   If your company is new, estimate your annual overhead by projecting overhead costs for the first full year. For example, if you've been in business for five months and overhead has been $5,500 so far, you can expect annual overhead to be about $13,200 ($5,500 divided by 5 and multiplied by 12).

2) Here's how to calculate your indirect overhead percentage:

$$\frac{\text{Annual indirect overhead}}{\text{Annual direct overhead (job expenses)}} = \text{Overhead percentage}$$

Calculate your annual indirect overhead by adding together your real (or anticipated) annual expenses for the following:

*Salaries.* Include what you pay for all employees except trade workers, plus payroll-related expenses for all employees.

*Office and shop expense.* Rent or mortgage, utilities, furniture and equipment, maintenance, office supplies and postage, storage sheds, warehouses, fences or yard maintenance.

*Vehicles.* Lease or purchase payments, maintenance, repairs and fuel.

*Sales promotion.* Advertising, entertainment and sales-related travel.

*Taxes.* Property tax and income tax, and sales tax (if not included in your material prices).

*Licenses.* Contractor's and business licenses.

*Insurance.* General liability, property and vehicle policies.

*Interest expense.* Loan interest and bank charges. Also consider loss of interest on payments retained by the general contractor until the job is finished.

*Miscellaneous expenses.* This could include depreciation and amortization on building and vehicles, bad debts, legal and accounting fees, and educational expenses.

Direct overhead is easier to figure. It's all job expenses except tradesman labor, payroll taxes and insurance, materials, equipment, subcontracts, and contingency expenses. Permits, bonds, fees and special insurance policies for property owners are also examples of direct overhead. Add the direct overhead expense on the appropriate lines in your estimate. Direct overhead is not included in the estimating tables in this manual.

### Field Equipment May Be Part of Overhead

As you may have noticed, there's no equipment cost column in the estimating tables. Instead, field equipment expense is included in the overhead percentage for "Fast" and "Medium" work but not "Slow" work.

**New Construction and Commercial Work:** The overhead percentage for "Fast" (residential tract) work and "Medium" (commercial) projects *includes* equipment costs such as ladders, spray

## Equipment Rental Rates

Use the following rates only as a guide. They may not be accurate for your area.
Verify equipment rental rates at your local yard.

| | Rental | | |
|---|---|---|---|
| | $/day | $/week | $/month |
| **Acoustical sprayer** | 45.00 | 135.00 | 337.50 |
| **Air compressors** | | | |
| Electric or gasoline, wheel mounted | | | |
| 5 CFM, 1.5 HP, electric | 28.00 | 84.00 | 210.00 |
| 8 CFM, 1.5 HP, electric | 32.00 | 96.00 | 240.00 |
| 10 CFM, 5.5 HP, gasoline | 37.00 | 111.00 | 277.50 |
| 15 CFM, shop type, electric | 42.00 | 126.00 | 315.00 |
| 50 CFM, shop type, electric | 55.00 | 165.00 | 412.50 |
| 100 CFM, gasoline | 75.00 | 225.00 | 562.50 |
| 125 CFM, gasoline | 85.00 | 255.00 | 637.50 |
| 150 CFM, gasoline | 95.00 | 285.00 | 712.50 |
| 175 CFM, gasoline | 105.00 | 315.00 | 787.50 |
| 190 CFM, gasoline | 115.00 | 345.00 | 862.50 |
| Diesel, wheel mounted | | | |
| to 159 CFM | 85.00 | 255.00 | 765.00 |
| 160 to 249 CFM | 105.00 | 315.00 | 945.00 |
| 250 to 449 CFM | 155.00 | 465.00 | 1,395.00 |
| 450 to 749 CFM | 230.00 | 690.00 | 2,070.00 |
| 750 to 1199 CFM | 315.00 | 945.00 | 2,835.00 |
| 1200 CFM & over | 460.00 | 1,380.00 | 4,140.00 |
| **Air hose** - with coupling, 50' lengths | | | |
| 1/4" I.D. | 6.00 | 18.00 | 45.00 |
| 3/8" I.D. | 7.00 | 21.00 | 52.50 |
| 1/2" I.D. | 8.00 | 24.00 | 60.00 |
| 5/8" I.D. | 9.00 | 27.00 | 67.50 |
| 3/4" I.D. | 10.00 | 30.00 | 75.00 |
| 1" I.D. | 11.00 | 33.00 | 82.50 |
| 1 1/2" I.D. | 16.00 | 48.00 | 120.00 |
| **Boomlifts** | | | |
| 3' x 4' to 3' x 8' basket | | | |
| 20' two wheel drive | 140.00 | 420.00 | 1,260.00 |
| 30' two wheel drive | 170.00 | 510.00 | 1,530.00 |
| 40' four wheel drive | 195.00 | 585.00 | 1,755.00 |
| 50' - 1000 lb. | 320.00 | 960.00 | 2,880.00 |
| Telescoping and articulating booms, self propelled, gas or diesel powered, 2-wheel drive | | | |
| 21' to 30' high | 200.00 | 600.00 | 1,800.00 |
| 31' to 40' high | 250.00 | 750.00 | 2,250.00 |
| 41' to 50' high | 325.00 | 975.00 | 2,925.00 |
| 51' to 60' high | 400.00 | 1,200.00 | 3,600.00 |
| **Burner, paint** | 12.00 | 36.00 | 90.00 |
| **Dehumidifier** - 5000 Btu, 89 lb, 8.7 amp | 55.00 | 165.00 | 412.50 |
| **Ladders** | | | |
| Aluminum extension | | | |
| 16' to 36' | 30.00 | 90.00 | 225.00 |
| 40' to 60' | 45.00 | 135.00 | 337.50 |

| | Rental | | |
|---|---|---|---|
| | $/day | $/week | $/month |
| **Ladders (Continued)** | | | |
| Step - fiberglass or wood | | | |
| 6' | 8.00 | 24.00 | 60.00 |
| 8' | 10.00 | 30.00 | 75.00 |
| 10' | 12.00 | 36.00 | 90.00 |
| 12' | 14.00 | 42.00 | 105.00 |
| 14' | 16.00 | 48.00 | 120.00 |
| 16' | 20.00 | 60.00 | 150.00 |
| 20' | 26.00 | 78.00 | 195.00 |
| **Ladder jacks** - No guardrail. | 8.00 | 20.00 | 50.00 |
| **Masking paper dispenser** | 20.00 | 50.00 | 125.00 |
| **Painter's pic** (walkboards); No guardrail. (Also known as airplane planks, toothpicks and banana boards) | | | |
| 16' long | 8.00 | 24.00 | 60.00 |
| 20' long | 16.00 | 48.00 | 120.00 |
| 24' long | 20.00 | 60.00 | 150.00 |
| 28' long | 24.00 | 72.00 | 180.00 |
| 32' long | 28.00 | 84.00 | 210.00 |
| **Planks** - plain end microlam scaffold plank | | | |
| 9" wide | 10.00 | 30.00 | 75.00 |
| 10" wide | 12.00 | 36.00 | 90.00 |
| 12" wide | 14.00 | 42.00 | 105.00 |
| **Pressure washers** (See Water pressure washers) | | | |
| **Sandblast compressor and hopper** | | | |
| To 250 PSI | 60.00 | 180.00 | 450.00 |
| Over 250 to 300 PSI | 85.00 | 255.00 | 637.50 |
| Over 600 to 1000 PSI | 110.00 | 330.00 | 825.00 |
| **Sandblast machines** | | | |
| 150 lb pot with hood, 175 CFM compressor | 230.00 | 690.00 | 1,725.00 |
| 300 lb pot with hood, 325 CFM compressor | 410.00 | 1,230.00 | 3,075.00 |
| 600 lb pot with hood, 600 CFM compressor | 745.00 | 2,235.00 | 5,587.50 |
| **Sandblast hoses** - 50' lengths, coupled | | | |
| 3/8" I.D. | 10.00 | 30.00 | 75.00 |
| 3/4" I.D. | 14.00 | 42.00 | 105.00 |
| 1" I.D. | 18.00 | 54.00 | 135.00 |
| 1 1/4" I.D. | 20.00 | 60.00 | 150.00 |
| 1 1/2" I.D. | 22.00 | 66.00 | 165.00 |
| **Sandblast accessories** | | | |
| Nozzles, all types | 18.00 | 54.00 | 135.00 |
| Hood, air-fed | 28.00 | 84.00 | 210.00 |
| Valves, remote control (deadman, all sizes) | 30.00 | 90.00 | 225.00 |

**Figure 15**
Typical equipment purchase and rental prices

| | Rental | | |
|---|---|---|---|
| | $/day | $/week | $/month |
| **Sanders** | | | |
| Belt - 3" | 14.00 | 42.00 | 105.00 |
| Belt - 4" x 24" | 17.00 | 51.00 | 127.50 |
| Disc - 7" | 22.00 | 66.00 | 165.00 |
| Finish sander, 6" | 12.00 | 36.00 | 90.00 |
| Floor edger, 7" disk, 29#, 15 amp. | 20.00 | 60.00 | 150.00 |
| Floor sander, 8" drum, 118#, 14 amp | 45.00 | 135.00 | 337.50 |
| Palm sander, 4" x 4" | 10.00 | 30.00 | 75.00 |
| Palm sander, 4½" x 9¼" | 12.00 | 36.00 | 90.00 |
| **Scaffolding:** rolling stage, caster mounted, 30" wide by 7' or 10' long | | | |
| 4' to 6' reach | 40.00 | 80.00 | 160.00 |
| 7' to 11' reach | 50.00 | 100.00 | 200.00 |
| 12' to 16' reach | 70.00 | 140.00 | 280.00 |
| 17' to 21' reach | 95.00 | 190.00 | 380.00 |
| 22' to 26' reach | 105.00 | 210.00 | 420.00 |
| 27' to 30' reach | 115.00 | 230.00 | 460.00 |
| Casters - each | 10.00 | 20.00 | 30.00 |
| **Scissor lifts** | | | |
| Electric powered, rolling with 2' x 3' platform, 650 lb capacity | | | |
| 30' high | 75.00 | 225.00 | 675.00 |
| 40' high | 130.00 | 390.00 | 1,170.00 |
| 50' high | 150.00 | 450.00 | 1,350.00 |
| Rolling, self-propelled, hydraulic, electric powered | | | |
| to 20' high | 110.00 | 330.00 | 990.00 |
| 21' to 30' high | 135.00 | 405.00 | 1,215.00 |
| 31' to 40' high | 170.00 | 510.00 | 1,530.00 |
| Rolling, self-propelled, hydraulic, diesel powered | | | |
| to 20' high | 125.00 | 375.00 | 1,125.00 |
| 21' to 30' high | 155.00 | 465.00 | 1,395.00 |
| 31' to 40' high | 200.00 | 600.00 | 1,800.00 |
| **Spray rigs** | | | |
| Airless pumps, complete with gun and 50' of line | | | |
| Titan 447, ⅞ HP, electric | 70.00 | 210.00 | 630.00 |
| Titan 660, 1 HP, electric | 80.00 | 240.00 | 720.00 |
| Gasoline, .75 gpm | 85.00 | 255.00 | 765.00 |
| Emulsion pumps | | | |
| 65 gal, 5 HP engine | 70.00 | 210.00 | 630.00 |
| 200 gal, 5 HP engine | 80.00 | 240.00 | 720.00 |
| Emulsion airless, 1.25 gpm, gasoline | 85.00 | 255.00 | 765.00 |

| | Rental | | |
|---|---|---|---|
| | $/day | $/week | $/month |
| **Spray rigs (Continued)** | | | |
| Conventional pumps, gas, portable | | | |
| High pressure, low volume (HVLP) | 45.00 | 135.00 | 405.00 |
| 8 CFM complete | 60.00 | 180.00 | 540.00 |
| 17 CFM complete | 65.00 | 195.00 | 585.00 |
| 85 CFM complete | 75.00 | 225.00 | 675.00 |
| 150 CFM complete | 110.00 | 330.00 | 990.00 |
| Spray rig accessories: 6' wand | 7.00 | 21.00 | 52.50 |
| **Striper, paint** (parking lot striping) | | | |
| Aerosol | 20.00 | 60.00 | 150.00 |
| Pressure regulated | 28.00 | 84.00 | 210.00 |
| **Swing stage** | | | |
| Any length drop, motor operated, excluding safety gear and installation or dismantling. Note: Must be set up by a professional to ensure safety. | | | |
| Swing stage | 100.00 | 300.00 | 900.00 |
| Basket | 50.00 | 150.00 | 450.00 |
| Bosun's chair | 50.00 | 150.00 | 450.00 |
| Swing stage safety gear, purchase only | | | |
| Safety harness | (90.00) | | |
| 4' lanyard with locking snap at each end | (65.00) | | |
| DBI rope grab for ⅝" safety line | (70.00) | | |
| Komet rope grab for ¾" safety line | (100.00) | | |
| **Texturing equipment** | | | |
| Texturing gun - w/ hopper, no compressor | 5.00 | 15.00 | 45.00 |
| Texturing mud paddle mixer | 7.00 | 21.00 | 63.00 |
| Texturing outfit - 1 HP w/ gun, 50' hose, 75 PSI | 11.00 | 33.00 | 99.00 |
| **Wallpaper hanging kit** | 17.00 | 51.00 | 153.00 |
| **Wallpaper steamer** | | | |
| Electric, small, 10 amp | 20.00 | 60.00 | 180.00 |
| Electric, 15 amp | 30.00 | 90.00 | 270.00 |
| Pressurized, electric | 38.00 | 114.00 | 342.00 |
| **Water pressure washer** (pressure washer, water blaster, power washer) | | | |
| 1000 PSI, electric, 15 amp | 45.00 | 135.00 | 405.00 |
| 2000 PSI, gas | 75.00 | 225.00 | 675.00 |
| 2500 PSI, gas | 80.00 | 240.00 | 720.00 |
| 3500 PSI, gas | 88.00 | 264.00 | 792.00 |

**Figure 15 (continued)**
Typical equipment purchase and rental prices

equipment, and masking paper holders. Those items are used on many jobs, not just one specific job. The overhead allowance covers equipment purchase payments, along with maintenance, repairs and fuel. If you have to rent equipment for a specific new construction project, *add* that rental expense as a separate cost item in your estimate.

**Repaint Jobs:** Overhead rates for "Slow" (repaint) work do *not* include equipment costs. When you estimate a repaint job, any small or short-term job, or a job that uses only a small quantity of materials, *add* the cost of equipment at the rental rate – even if the equipment is owned by your company.

Rental yards quote daily, weekly and monthly equipment rental rates. Figure 15 shows typical rental costs for painting equipment. Your actual equipment costs may be different. Here's a suggestion that can save you more than a few minutes on the telephone collecting rental rates. Make up a blank form like Figure 15 and give it to your favorite rental equipment suppliers. Ask each supplier to fill in current rental costs. Use the completed forms until you notice that rates have changed. Then ask for a new set of rental rates.

### *Commissions and Bonuses*

Any commissions or bonuses you have to pay on a job aren't included in the estimating tables. You must add these expenses to your bid.

Painting contractors rarely have a sales staff, so there won't be sales commissions to pay on most jobs. There's one exception, however. Most room addition and remodeling contractors have salespeople. And many of their remodeling projects exclude painting. In fact, their contract may specify that the owner is responsible for the painting. These jobs may be a good source of leads for a painting contractor. Develop a relationship with the remodeling contractor's sales staff (with the remodeling contractor's approval, of course). If you have to pay a sales commission for the referral, this is direct overhead and has to be added to the estimate.

Some painting contractors pay their estimators a bonus of 1 to 3 percent per job in addition to their salary. If you offer an incentive like this, add the cost to your estimate, again as a direct overhead item.

### *An Example of Overhead*

Here's an example of how overhead is added into an estimate. A painting company completed 20 new housing projects in the last year. Average revenue per project was $50,000. Gross receipts were $1,000.000 and the company made a 5 percent profit.

| | |
|---|---|
| Gross income | $1,000,000 |
| Less the profit earned (5%) | 50,000 |
| Gross expenses | 950,000 |
| Less total direct job cost | 825,000 |
| Indirect overhead expense | 125,000 |

$$\frac{125{,}000 \text{ (overhead cost)}}{825{,}000 \text{ (direct job cost)}} = 0.1515 \text{ or } 15.15\%$$

When you've calculated indirect overhead as a percentage of direct job cost, add that percentage to your estimates. If you leave indirect overhead out of your estimates, you've left out some very significant costs.

## Column 8: Profit

The estimating tables assume that profit on "Slow" (repaint) jobs is 15 percent, profit on "Medium" (commercial) projects is 11 percent and profit on "Fast" (residential tract) jobs is 6 percent. Calculate the profit per unit by first adding together the costs in columns 4 (labor cost per unit), column 5 (labor burden per unit), column 6 (material costs per unit), and column 7 (overhead per unit). Then multiply the total by the appropriate profit percentage to find the profit per unit.

It's my experience that larger companies with larger projects can survive with a smaller profit percentage. Stiff competition for high volume tract work forces bidders to trim their profit margin. Many smaller companies doing custom work earn a higher profit margin because they produce better quality work, have fewer jobs, and face less competition.

| Risk factor | Normal profit (assume 10%) | | Difficulty factor | | Proposed profit range |
|---|---|---|---|---|---|
| High risk | 10% | x | 1.5 to 3.5 | = | 15% to 35% |
| Average risk | 10% | x | 1.3 to 1.4 | = | 13% to 14% |
| Moderate risk | 10% | x | 1.0 to 1.2 | = | 10% to 12% |
| Low risk | 10% | x | 0.5 to 0.9 | = | 5% to 9% |

**Figure 16**
Risk factors and profit margin

### *Profit and Risk*

Profit is usually proportionate to risk. The more risk, the greater the potential profit has to be to attract bidders. Smaller companies handling custom or repaint work have more risk of a major cost overrun because there are many more variables in that type of work. It's usually safe to estimate a smaller profit on new work because new work tends to be more predictable. The risk of loss smaller.

How do you define risk? Here's my definition: Risk is the *headache factor*, the number and size of potential problems you could face in completing the project. Repaint jobs have more unknowns, so they're a greater risk. And dealing with an indecisive or picky homeowner can be the greatest headache of all. You may need to use a profit margin even higher than the 15 to 35 percent range indicated for high-risk work in Figure 16.

### *Tailoring Your Profit Margin*

Of course, your profit margin has to be based on the job, your company and the competition. But don't cut your profit to the bone just to get more work. Instead, review your bid to see if there are reasons why the standard costs wouldn't apply.

I use the term *standard base* bid to refer to my usual charge for all the estimated costs, including my standard profit. Before submitting any bid, spend a minute or two deciding whether your standard base bid will apply.

### *Risk Factors*

Your assessment of the difficulty of the job may favor assigning a risk factor that could be used to modify your profit percentage. The higher the risk, the higher potential profit should be. My suggestions are in Figure 16.

As you might expect, opinions on difficulty factors can vary greatly. There's a lot of knowledge involved. You need experience and good judgment to apply these factors effectively.

### *Bidding Variables*

Of course, your profit may be affected by an error in evaluating the job risk factor. You can greatly reduce the risk by accurately evaluating the bidding variables in Figure 17. Make adjustments to your standard base bid for example, if you expect your crews to be more or less efficient on this project, or if you expect competition to be intense. If there are logical reasons to modify your standard base bid, make those changes.

But remember, if you adjust your standard base bid, you're not changing your profit margin. You're only allowing for cost variables in the job. Adjust your standard base costs for unusual labor productivity, material or equipment cost changes, or because of unusual overhead conditions. Review the following bidding variables when deciding how to adjust your standard base bid.

### *The Bottom Line*

The profit margin you include in estimates depends on the way you do business, the kind of work you do, and your competition. Only you can decide what percentage is right for your bids. Don't take another paint estimator's advice on the "correct" profit margin. There's no single correct answer. Use your own judgment. But here are some typical profit margins for the kinds of work most painting contractors do.

| | | |
|---|---|---|
| Repaints: | Custom | 20 to 35% |
| | Average | 15 to 20% |
| Commercial or industrial | | 10 to 15% |
| New residential: | 1 to 4 units | 10 to 12% |
| | 5 or more | 5 to 7% |
| Government work | | 5 to 7% |

## Column 9: Total Cost

The costs in Column 9 of Figure 2, and all the estimating tables in this book, are the totals per unit for each application rate in columns 4, 5, 6, 7, and 8. That includes labor, labor burden, material cost, overhead and profit.

## Sample Estimate

Figure 18 is a sample repaint estimate, using the slow production rate, for a small house with many amenities. The final bid total is the bid price. Figure 19 is a blank estimating form for your use.

**REPUTATIONS AND ATTITUDES**

- Owner
- Architect
- General Contractor
- Lender
- Inspector

**THE PROJECT**

- Building type
- Project size
- Your financial limits
- Start date
- Weather conditions
- Manpower availability and capability

**THE SITE**

- Location (distance from shop and suppliers)
- Accessibility
- Working conditions
- Security requirements
- Safety considerations

**COMPETITION**

- Number bidding
- Their strength, size and competence

**DESIRE FOR THE WORK**

**Figure 17**
Bidding variables

Date 6/7/97
Customer Dan Gleason
Address 3333 A Street
City/State/Zip Yourtown, USA 77777
Phone (619)555-1212
Estimated by CHS

Due date 6/15/97
Job name Gleason Repaint
Job location 3333 A Street
Estimate # 97-051
Total square feet 1,020 SF (5 rooms)
Checked by Jack

## Interior Costs

| | Operation | Material | Application method | Dimensions | Quantity SF/LF/Each | | Unit cost | | Total cost |
|---|---|---|---|---|---|---|---|---|---|
| 1 | Ceiling-T&G | Semi-Trans | R + B | 17.5 x 15.3 x 1.3 | 348 SF | x | .20 | = $ | 70.00 |
| 2 | Beams to 13'H | Solid | R + B | 17.5x7 | 122.5 LF | x | 1.16 | = $ | 142.00 |
| 3 | Ceiling-Drywall | Flat | R | 127 + 127 | 254 SF | x | .14 | = $ | 36.00 |
| 4 | Ceiling-Drywall | Sealer-water | R | 75 + 15 + 40 | 130 SF | x | .12 | = $ | 16.00 |
| 5 | Ceiling-Drywall | Enamel-water | R | 75 + 15 + 40 | 130 SF | x | .15 | = $ | 20.00 |
| 6 | Walls-Drywall | Flat | R | 675 + 392 + 392 | 1,459 SF | x | .13 | = $ | 190.00 |
| 7 | Walls-Above 8' (clip) | Flat | R + B | 70 + 85 = 155x1.3 | 201.5 SF | x | .13 | = $ | 26.00 |
| 8 | Walls-Drywall | Sealer-water | R | 280 + 128 + 208 | 616 SF | x | .15 | = $ | 92.00 |
| 9 | Walls-Drywall | Enamel-water | R | 280 + 128 + 208 | 616 SF | x | .19 | = $ | 117.00 |
| 10 | Doors-Finish | Enamel-water | R + B | Opening Count | 10 Ea | x | 90.77 | = $ | 91.00 |
| 11 | Baseboard-Prime | Flat w/ walls | R + B | 64 + 49 + 49 | 162 LF | x | .05 | = $ | 8.00 |
| 12 | Baseboard-Finish | Enamel (cut-in) | B | 11 + 16 + 35 | 62 LF | x | .25 | = $ | 16.00 |
| 13 | Railing-W.I. | Enamel | B | 42" High | 15 LF | x | 1.36 | = $ | 20.00 |
| 14 | Light Valance | Solid | B | 2 x 8 | 10 LF | x | .99 | = $ | 10.00 |
| 15 | Registers | Can | Spray | 1,020 SF Home | 1,020 SF | x | .03 | = $ | 31.00 |
| 16 | | | | | | x | | = $ | |
| 17 | | | | | | x | | = $ | |
| 18 | | | | | | x | | = $ | |
| | | | | Total Interior Costs (includes overhead and profit) | | | | = $ | 885.00 |

## Exterior Costs

| | Operation | Material | Application method | Dimensions | Quantity SF/LF/Each | | Unit cost | | Total cost |
|---|---|---|---|---|---|---|---|---|---|
| 1 | S.M. Diverter-3'W | Finish | B | 14 | 14 LF | x | .13 | = $ | 2.00 |
| 2 | Roof Jacks | Finish | B | 1 Story | 1,020 SF | x | .01 | = $ | 10.00 |
| 3 | SM Vents & Flashings | Finish | B | 1 Story | 1 Ea | x | 29.68 | = $ | 30.00 |
| 4 | Fascia-2 x 4 | Solid | Roll | 66 + 59 | 125 LF | x | .34 | = $ | 43.00 |
| 5 | Overhang-24" | Solid | R + B | (132 + 76) x 1.5 | 312 SF | x | .37 | = $ | 115.00 |
| 6 | Siding-R.S. Wood | Solid | Roll | (½x24x4.5)x2 | 108 SF | x | .26 | = $ | 28.00 |
| 7 | Plaster/Stucco | Masonry | Spray | 255+255+204+204 | 918 SF | x | .18 | = $ | 165.00 |
| 8 | Door-Panel (Entry) | Enamel 2 coats | R + B | Entry | 1 Ea | x | 37.59 | = $ | 38.00 |
| 9 | Door - Flush | Enamel 2 coats | R + B | Exterior | 1 Ea | x | 15.91 | = $ | 16.00 |
| 10 | Plant-On Trim-2x4 | Solid | R + B | 66 + 62 + 52 | 180 LF | x | .35 | = $ | 63.00 |
| 11 | Pot Shelf | Solid | R + B | 27 | 27 LF | x | 1.06 | = $ | 29.00 |
| 12 | Pass Through | Finish | B | 10 | 10 LF | x | 1.01 | = $ | 10.00 |
| 13 | | | | | | x | | = $ | |
| 14 | | | | | | x | | = $ | |
| 15 | | | | | | x | | = $ | |
| 16 | | | | | | x | | = $ | |
| 17 | | | | | | x | | = $ | |
| 18 | | | | | | x | | = $ | |
| | | | | Total Exterior Costs (includes overhead and profit) | | | | = $ | 549.00 |

**Figure 18**
Sample painting estimate

## Preparation Costs

| | Operation | Dimensions | Quantity SF/LF/Each | | Unit cost | | Total cost |
|---|---|---|---|---|---|---|---|
| 1 | Sand Wood Ceiling | 17.5 x 15.3 x 1.3 | 348 SF | x | .11 | = | $ 38.00 |
| 2 | Sand and Putty Int. Wall | 675 + 392 + 392 | 1,459 SF | x | .11 | = | $ 160.00 |
| 3 | Sand Openings | 14 Ea x 42 SF | 588 SF | x | .08 | = | $ 47.00 |
| 4 | Wash Int. Walls / Ceil-Enam | 280 + 128 + 208 | 616 SF | x | .11 | = | $ 68.00 |
| 5 | Waterblast | 125 + 210 + 108 + 918 | 1,361 SF | x | .045 | = | $ 61.00 |
| 6 | Sand and Putty Ext. Trim | 125 + 210 + 108 | 443 SF | x | .20 | = | $ 89.00 |
| 7 | Caulk Ext. Windows-1/8" gap | 20 + 15 + 10 + 20 + 12 | 77 SF | x | .35 | = | $ 27.00 |
| 8 | Window Protective Coating | 34+25+14+5+48+18 | 144 SF | x | .28 | = | $ 40.00 |
| 9 | | | | x | | = | $ |
| 10 | | | | x | | = | $ |
| | Total Preparation Costs (includes overhead and profit) | | | | | = | $ 530.00 |

## SURRPTUCU Costs

| Operation | Description | Labor hours | Labor cost (at $ 16.50) | Material cost | Totals |
|---|---|---|---|---|---|
| **S**et**U**p | 2 Days @.5 hr/day | 1.0 | 16.50 | — | = $ 17.00 |
| **R**emove/**R**eplace | Hardware & SW Plates | 1.25 | 20.63 | — | = $ 21.00 |
| **P**rotection | Furniture & Floors | 1.0 | 16.50 | 15.00 | = $ 32.00 |

**T**ouch**U**p is applied as a percentage of the total costs. See *Extensions*

| Operation | Description | Labor hours | Labor cost (at $ 16.50) | Material cost | Totals |
|---|---|---|---|---|---|
| **C**lean**U**p | 2 Days @.5 hr/day | 1.0 | 16.50 | — | = $ 17.00 |

## Equipment Costs

| Equipment description | Rental days | Daily cost | Total cost |
|---|---|---|---|
| Waterblast | 1 | 50.00 | $ 50.00 |
| Ladders, 2 Ea. | 1 | 5.00 | $ 10.00 |
| Sander Belt | 1 | 14.00 | $ 14.00 |
| | | | $ |
| | | | $ |
| | | | $ |
| | | Total Equipment Costs | $ 74.00 |

## Subcontractor Costs

| Trade | Bid amount |
|---|---|
| Pavement marking | $ 0 |
| Sandblasting | $ 0 |
| Scaffolding | $ 0 |
| Wallcovering | $ 0 |
| Waterblasting | $ 0 |
| Other ___ | $ 0 |
| Other ___ | $ 0 |
| Other ___ | $ 0 |
| Total Subcontractor Costs | $ — |

## Extensions

| | |
|---|---|
| Supervision 1Hr. | $ 18.00 |
| Setup | 17.00 |
| Remove/replace | 21.00 |
| Protection | 32.00 |
| Cleanup | 17.00 |
| Equipment | 74.00 |
| Subcontracts | 0 |
| Commissions | 0 |
| Other costs | 0 |
| Subtotal | 179.00 |
| Overhead (24.5%) | 44.00 |
| Profit (11%) | 20.00 |
| Subtotal | 243.00 |
| Preparation | 530.00 |
| Interior total | 885.00 |
| Exterior total | 549.00 |
| Subtotal | 1,964.00 |
| Touchup (10%) | 196.00 |
| Contingency (0%) | 0 |
| Total base bid | 2,160.00 |
| Adjustment (-2%) | 43.00 |
| Final bid total | 2,117.00 |
| Price per SF | 2.08 |
| Price per room | 423.00 |

**Figure 18 (continued)**
Sample painting estimate

Date ______________________ Due date ______________________

Customer ______________________ Job name ______________________

Address ______________________ Job location ______________________

City/State/Zip ______________________ Estimate # ______________________

Phone ______________________ Total square feet ______________________

Estimated by ______________________ Checked by ______________________

## Interior Costs

| Operation | Material | Application method | Dimensions | Quantity SF/LF/Each | | Unit cost | Total cost |
|---|---|---|---|---|---|---|---|
| 1 | | | | | x | | = $ |
| 2 | | | | | x | | = $ |
| 3 | | | | | x | | = $ |
| 4 | | | | | x | | = $ |
| 5 | | | | | x | | = $ |
| 6 | | | | | x | | = $ |
| 7 | | | | | x | | = $ |
| 8 | | | | | x | | = $ |
| 9 | | | | | x | | = $ |
| 10 | | | | | x | | = $ |
| 11 | | | | | x | | = $ |
| 12 | | | | | x | | = $ |
| 13 | | | | | x | | = $ |
| 14 | | | | | x | | = $ |
| 15 | | | | | x | | = $ |
| 16 | | | | | x | | = $ |
| 17 | | | | | x | | = $ |
| 18 | | | | | x | | = $ |
| | | | Total Interior Costs (includes overhead and profit) | | | | = $ |

## Exterior Costs

| Operation | Material | Application method | Dimensions | Quantity SF/LF/Each | | Unit cost | Total cost |
|---|---|---|---|---|---|---|---|
| 1 | | | | | x | | = $ |
| 2 | | | | | x | | = $ |
| 3 | | | | | x | | = $ |
| 4 | | | | | x | | = $ |
| 5 | | | | | x | | = $ |
| 6 | | | | | x | | = $ |
| 7 | | | | | x | | = $ |
| 8 | | | | | x | | = $ |
| 9 | | | | | x | | = $ |
| 10 | | | | | x | | = $ |
| 11 | | | | | x | | = $ |
| 12 | | | | | x | | = $ |
| 13 | | | | | x | | = $ |
| 14 | | | | | x | | = $ |
| 15 | | | | | x | | = $ |
| 16 | | | | | x | | = $ |
| 17 | | | | | x | | = $ |
| 18 | | | | | x | | = $ |
| | | | Total Exterior Costs (includes overhead and profit) | | | | = $ |

**Figure 19**
Blank painting estimate

## Preparation Costs

| | Operation | Dimensions | Quantity SF/LF/Each | | Unit cost | | Total cost |
|---|---|---|---|---|---|---|---|
| 1 | | | | x | | = | $ |
| 2 | | | | x | | = | $ |
| 3 | | | | x | | = | $ |
| 4 | | | | x | | = | $ |
| 5 | | | | x | | = | $ |
| 6 | | | | x | | = | $ |
| 7 | | | | x | | = | $ |
| 8 | | | | x | | = | $ |
| 9 | | | | x | | = | $ |
| 10 | | | | x | | = | $ |

Total Preparation Costs (includes overhead and profit) = $ ______

## SURRPTUCU Costs

| Operation | Description | Labor hours | Labor cost (at $ ____) | Material cost | Totals |
|---|---|---|---|---|---|
| SetUp | | | | | = $ |
| Remove/Replace | | | | | = $ |
| Protection | | | | | = $ |

TouchUp is applied as a percentage of the total costs. See *Extensions*

| Operation | Description | Labor hours | Labor cost | Material cost | Totals |
|---|---|---|---|---|---|
| CleanUp | | | | | = $ |

## Equipment Costs

| Equipment description | Rental days | Daily cost | Total cost |
|---|---|---|---|
| | | | $ |
| | | | $ |
| | | | $ |
| | | | $ |
| | | | $ |
| | | | $ |
| | | Total Equipment Costs | $ |

## Subcontractor Costs

| Trade | Bid amount |
|---|---|
| Pavement marking | $ |
| Sandblasting | $ |
| Scaffolding | $ |
| Wallcovering | $ |
| Waterblasting | $ |
| Other ___ | $ |
| Other ___ | $ |
| Other ___ | $ |
| Total Subcontractor Costs | $ |

## Extensions

| | |
|---|---|
| Supervision ____ | $ |
| Setup | |
| Remove/replace | |
| Protection | |
| Cleanup | |
| Equipment | |
| Subcontracts | |
| Commissions | |
| Other costs | |
| Subtotal | |
| Overhead (__%) | |
| Profit (__%) | |
| Subtotal | |
| Preparation | |
| Interior total | |
| Exterior total | |
| Subtotal | |
| Touchup (___%) | |
| Contingency (__%) | |
| Total base bid | |
| Adjustment (__%) | |
| Final bid total | |
| Price per SF | |
| Price per room | |

**Figure 19 (continued)**
Blank painting estimate

# Part I

# GENERAL Painting COSTS

| | Labor LF per manhour | Material coverage LF/gallon | Material cost per gallon | Labor cost per 100 LF | Labor burden 100 LF | Material cost per 100 LF | Overhead per 100 LF | Profit per 100 LF | Total price per 100 LF |
|---|---|---|---|---|---|---|---|---|---|
| **Baseboard, per linear foot** | | | | | | | | | |
| Roll 1 coat with walls, brush touchup, paint grade base | | | | | | | | | |
| Flat latex, water base, (material #5) | | | | | | | | | |
| Slow | 900 | 800 | 16.20 | 1.14 | .31 | 2.03 | 1.08 | .68 | 5.24 |
| Medium | 1200 | 750 | 14.20 | 1.38 | .45 | 1.89 | .91 | .51 | 5.14 |
| Fast | 1500 | 700 | 12.10 | 1.47 | .60 | 1.73 | .70 | .27 | 4.77 |
| Enamel, water base (material #9) | | | | | | | | | |
| Slow | 600 | 750 | 22.80 | 1.71 | .47 | 3.04 | 1.62 | 1.03 | 7.87 |
| Medium | 800 | 725 | 19.90 | 2.06 | .70 | 2.74 | 1.35 | .75 | 7.60 |
| Fast | 1000 | 700 | 17.10 | 2.20 | .89 | 2.44 | 1.02 | .39 | 6.94 |
| Enamel, oil base (material #10) | | | | | | | | | |
| Slow | 600 | 750 | 25.90 | 1.71 | .47 | 3.45 | 1.75 | 1.11 | 8.49 |
| Medium | 800 | 725 | 22.70 | 2.06 | .70 | 3.13 | 1.44 | .81 | 8.14 |
| Fast | 1000 | 700 | 19.40 | 2.20 | .89 | 2.77 | 1.08 | .42 | 7.36 |
| Brush 1 coat, cut-in, paint grade base | | | | | | | | | |
| Enamel, water base (material #9) | | | | | | | | | |
| Slow | 100 | 700 | 22.80 | 10.25 | 2.79 | 3.26 | 5.05 | 3.20 | 24.55 |
| Medium | 120 | 675 | 19.90 | 13.75 | 4.62 | 2.95 | 5.23 | 2.92 | 29.47 |
| Fast | 140 | 650 | 17.10 | 15.71 | 6.37 | 2.63 | 4.57 | 1.76 | 31.04 |
| Enamel, oil base (material #10) | | | | | | | | | |
| Slow | 100 | 700 | 25.90 | 10.25 | 2.79 | 3.70 | 5.19 | 3.29 | 25.22 |
| Medium | 120 | 675 | 22.70 | 13.75 | 4.62 | 3.36 | 5.33 | 2.98 | 30.04 |
| Fast | 140 | 650 | 19.40 | 15.71 | 6.37 | 2.98 | 4.64 | 1.78 | 31.48 |
| Spray 1 coat, stain in boneyard, stain grade base | | | | | | | | | |
| Wiping stain (material #11a) | | | | | | | | | |
| Slow | -- | -- | -- | -- | -- | -- | -- | -- | -- |
| Medium | 1500 | 1750 | 24.00 | 1.10 | .38 | 1.37 | .70 | .39 | 3.94 |
| Fast | 2000 | 1500 | 20.60 | 1.10 | .45 | 1.37 | .54 | .21 | 3.67 |

Use these figures for 1-1/2 inch to 3 inch baseboard stock, painted or stained on one side. Measurements are based on linear feet of baseboard. Paint grade base is painted after it is installed but stain grade base is usually stained in a boneyard. Typically, finger joint stock is paint grade and butt joint stock is stain grade. These figures include minimal preparation time and material. Add for extensive preparation. "Slow" work is based on a $10.25 hourly wage, "Medium" work on a $16.50 hourly wage, and "Fast" work on a $22.00 hourly wage. Other qualifications that apply to this table are on page 9.

| | Labor SF per manhour | Material coverage SF/gallon | Material cost per gallon | Labor cost per 100 SF | Labor burden 100 SF | Material cost per 100 SF | Overhead per 100 SF | Profit per 100 SF | Total price per 100 SF |
|---|---|---|---|---|---|---|---|---|---|
| **Baseboard, per square foot of floor area** | | | | | | | | | |
| Roll 1 coat with walls, brush touchup, paint grade base | | | | | | | | | |
| Flat latex, water base, (material #5) | | | | | | | | | |
| Slow | 2500 | 1500 | 16.20 | .41 | .11 | 1.08 | .50 | .32 | 2.42 |
| Medium | 2750 | 1250 | 14.20 | .60 | .19 | 1.14 | .48 | .27 | 2.68 |
| Fast | 3000 | 1000 | 12.10 | .73 | .29 | 1.21 | .41 | .16 | 2.80 |
| Enamel, water base (material #9) | | | | | | | | | |
| Slow | 2000 | 1000 | 22.80 | .51 | .14 | 2.28 | .91 | .58 | 4.42 |
| Medium | 2200 | 900 | 19.90 | .75 | .24 | 2.21 | .79 | .44 | 4.43 |
| Fast | 2400 | 800 | 17.10 | .92 | .38 | 2.14 | .63 | .24 | 4.31 |
| Enamel, oil base (material #10) | | | | | | | | | |
| Slow | 2000 | 1000 | 25.90 | .51 | .14 | 2.59 | 1.00 | .64 | 4.88 |
| Medium | 2200 | 900 | 22.70 | .75 | .24 | 2.52 | .86 | .48 | 4.85 |
| Fast | 2400 | 800 | 19.40 | .92 | .38 | 2.43 | .69 | .26 | 4.68 |
| Brush 1 coat, cut-in, paint grade base | | | | | | | | | |
| Enamel, water base (material #9) | | | | | | | | | |
| Slow | 500 | 1500 | 22.80 | 2.05 | .56 | 1.52 | 1.28 | .81 | 6.22 |
| Medium | 550 | 1350 | 19.90 | 3.00 | 1.01 | 1.47 | 1.34 | .75 | 7.57 |
| Fast | 600 | 1200 | 17.10 | 3.67 | 1.49 | 1.43 | 1.22 | .47 | 8.28 |
| Enamel, oil base (material #10) | | | | | | | | | |
| Slow | 500 | 1500 | 25.90 | 2.05 | .56 | 1.73 | 1.35 | .85 | 6.54 |
| Medium | 550 | 1350 | 22.70 | 3.00 | 1.01 | 1.68 | 1.39 | .78 | 7.86 |
| Fast | 600 | 1200 | 19.40 | 3.67 | 1.49 | 1.62 | 1.25 | .48 | 8.51 |
| Spray 1 coat, stain in boneyard, stain grade base | | | | | | | | | |
| Wiping stain (material #11a) | | | | | | | | | |
| Slow | -- | -- | -- | -- | -- | -- | -- | -- | -- |
| Medium | 4000 | 1350 | 24.00 | .41 | .14 | 1.78 | .57 | .32 | 3.22 |
| Fast | 5000 | 1200 | 20.60 | .44 | .18 | 1.72 | .43 | .17 | 2.94 |

Baseboard measurements are based on square feet of floor area. Use these figures for 1-1/2 inch to 3 inch stock, painted or stained on one side. Stain grade base is to be stained in a boneyard. Typically, finger joint stock is paint grade and butt joint stock is stain grade. These figures include minimal preparation time and material. Add for extensive preparation. "Slow" work is based on a $10.25 hourly wage, "Medium" work on a $16.50 hourly wage, and "Fast" work on a $22.00 hourly wage. Other qualifications that apply to this table are on page 9.

| | Labor LF per manhour | Material coverage LF/gallon | Material cost per gallon | Labor cost per 100 LF | Labor burden 100 LF | Material cost per 100 LF | Overhead per 100 LF | Profit per 100 LF | Total price per 100 LF |
|---|---|---|---|---|---|---|---|---|---|
| **Beams, per linear foot, heights to 13 feet** | | | | | | | | | |
| Solid body stain, water base (material #18) | | | | | | | | | |
| Roll & brush each coat | | | | | | | | | |
| Slow | 35 | 50 | 19.40 | 29.29 | 7.97 | 38.80 | 23.58 | 14.95 | 114.59 |
| Medium | 40 | 45 | 17.00 | 41.25 | 13.88 | 37.78 | 22.76 | 12.72 | 128.39 |
| Fast | 45 | 40 | 14.50 | 48.89 | 19.81 | 36.25 | 19.42 | 7.46 | 131.83 |
| Solid body stain, oil base (material #19) | | | | | | | | | |
| Roll & brush each coat | | | | | | | | | |
| Slow | 35 | 50 | 23.10 | 29.29 | 7.97 | 46.20 | 25.88 | 16.40 | 125.74 |
| Medium | 40 | 45 | 20.20 | 41.25 | 13.88 | 44.89 | 24.50 | 13.70 | 138.22 |
| Fast | 45 | 40 | 17.30 | 48.89 | 19.81 | 43.25 | 20.71 | 7.96 | 140.62 |
| Semi-transparent stain, water base (material #20) | | | | | | | | | |
| Roll & brush each coat | | | | | | | | | |
| Slow | 40 | 55 | 19.60 | 25.63 | 6.97 | 35.64 | 21.16 | 13.41 | 102.81 |
| Medium | 45 | 50 | 17.10 | 36.67 | 12.33 | 34.20 | 20.39 | 11.40 | 114.99 |
| Fast | 50 | 45 | 14.70 | 44.00 | 17.84 | 32.67 | 17.48 | 6.72 | 118.71 |
| Semi-transparent stain, oil base (material #21) | | | | | | | | | |
| Roll & brush each coat | | | | | | | | | |
| Slow | 40 | 55 | 23.40 | 25.63 | 6.97 | 42.55 | 23.30 | 14.77 | 113.22 |
| Medium | 45 | 50 | 20.50 | 36.67 | 12.33 | 41.00 | 22.05 | 12.33 | 124.38 |
| Fast | 50 | 45 | 17.60 | 44.00 | 17.84 | 39.11 | 18.68 | 7.18 | 126.81 |

Beam measurements are based on linear feet of installed 4" x 6" to 8" x 14" beams. High time difficulty factors are already figured into the formula. "Slow" work is based on a $10.25 hourly wage, "Medium" work on a $16.50 hourly wage, and "Fast" work on a $22.00 hourly wage. Other qualifications that apply to this table are on page 9.

| | Labor LF per manhour | Material coverage LF/gallon | Material cost per gallon | Labor cost per 100 LF | Labor burden 100 LF | Material cost per 100 LF | Overhead per 100 LF | Profit per 100 LF | Total price per 100 LF |
|---|---|---|---|---|---|---|---|---|---|
| **Beams, per linear foot, heights from 13 to 17 feet** | | | | | | | | | |
| Solid body stain, water base (material #18) | | | | | | | | | |
| Roll & brush each coat | | | | | | | | | |
| Slow | 24 | 50 | 19.40 | 42.71 | 11.63 | 38.80 | 28.88 | 18.30 | 140.32 |
| Medium | 27 | 45 | 17.00 | 61.11 | 20.56 | 37.78 | 29.27 | 16.36 | 165.08 |
| Fast | 30 | 40 | 14.50 | 73.33 | 29.73 | 36.25 | 25.77 | 9.91 | 174.99 |
| Solid body stain, oil base (material #19) | | | | | | | | | |
| Roll & brush each coat | | | | | | | | | |
| Slow | 24 | 50 | 23.10 | 42.71 | 11.63 | 46.20 | 31.17 | 19.76 | 151.47 |
| Medium | 27 | 45 | 20.20 | 61.11 | 20.56 | 44.89 | 31.01 | 17.33 | 174.90 |
| Fast | 30 | 40 | 17.30 | 73.33 | 29.73 | 43.25 | 27.07 | 10.40 | 183.78 |
| Semi-transparent stain, water base (material #20) | | | | | | | | | |
| Roll & brush each coat | | | | | | | | | |
| Slow | 28 | 55 | 19.60 | 36.61 | 9.96 | 35.64 | 25.49 | 16.16 | 123.86 |
| Medium | 31 | 50 | 17.10 | 53.23 | 17.90 | 34.20 | 25.81 | 14.43 | 145.57 |
| Fast | 34 | 45 | 14.70 | 64.71 | 26.23 | 32.67 | 22.87 | 8.79 | 155.27 |
| Semi-transparent stain, oil base (material #21) | | | | | | | | | |
| Roll & brush each coat | | | | | | | | | |
| Slow | 28 | 55 | 23.40 | 36.61 | 9.96 | 42.55 | 27.63 | 17.52 | 134.27 |
| Medium | 31 | 50 | 20.50 | 53.23 | 17.90 | 41.00 | 27.47 | 15.36 | 154.96 |
| Fast | 34 | 45 | 17.60 | 64.71 | 26.23 | 39.11 | 24.06 | 9.25 | 163.36 |

Beam measurements are based on linear feet of installed 4" x 6" to 8" x 14" beams. High time difficulty factors are already figured into the formula. "Slow" work is based on a $10.25 hourly wage, "Medium" work on a $16.50 hourly wage, and "Fast" work on a $22.00 hourly wage. Other qualifications that apply to this table are on page 9.

| | Labor LF per manhour | Material coverage LF/gallon | Material cost per gallon | Labor cost per 100 LF | Labor burden 100 LF | Material cost per 100 LF | Overhead per 100 LF | Profit per 100 LF | Total price per 100 LF |
|---|---|---|---|---|---|---|---|---|---|
| **Beams, per linear foot, heights above 18 feet** | | | | | | | | | |
| Solid body stain, water base (material #18) | | | | | | | | | |
| Roll & brush each coat | | | | | | | | | |
| Slow | 16 | 50 | 19.40 | 64.06 | 17.44 | 38.80 | 37.30 | 23.64 | 181.24 |
| Medium | 18 | 45 | 17.00 | 91.67 | 30.84 | 37.78 | 39.27 | 21.95 | 221.51 |
| Fast | 20 | 40 | 14.50 | 110.00 | 44.60 | 36.25 | 35.31 | 13.57 | 239.73 |
| Solid body stain, oil base (material #19) | | | | | | | | | |
| Roll & brush each coat | | | | | | | | | |
| Slow | 16 | 50 | 23.10 | 64.06 | 17.44 | 46.20 | 39.59 | 25.10 | 192.39 |
| Medium | 18 | 45 | 20.20 | 91.67 | 30.84 | 44.89 | 41.02 | 22.93 | 231.35 |
| Fast | 20 | 40 | 17.30 | 110.00 | 44.60 | 43.25 | 36.60 | 14.07 | 248.52 |
| Semi-transparent stain, water base (material #20) | | | | | | | | | |
| Roll & brush each coat | | | | | | | | | |
| Slow | 19 | 55 | 19.60 | 53.95 | 14.68 | 35.64 | 32.33 | 20.49 | 157.09 |
| Medium | 21 | 50 | 17.10 | 78.57 | 26.43 | 34.20 | 34.11 | 19.07 | 192.38 |
| Fast | 23 | 45 | 14.70 | 95.65 | 38.79 | 32.67 | 30.92 | 11.88 | 209.91 |
| Semi-transparent stain, oil base (material #21) | | | | | | | | | |
| Roll & brush each coat | | | | | | | | | |
| Slow | 19 | 55 | 23.40 | 53.95 | 14.68 | 42.55 | 34.47 | 21.85 | 167.50 |
| Medium | 21 | 50 | 20.50 | 78.57 | 26.43 | 41.00 | 35.77 | 20.00 | 201.77 |
| Fast | 23 | 45 | 17.60 | 95.65 | 38.79 | 39.11 | 32.11 | 12.34 | 218.00 |

Beam measurements are based on linear feet of installed 4" x 6" to 8" x 14" beams. High time difficulty factors are already figured into the formula. "Slow" work is based on a $10.25 hourly wage, "Medium" work on a $16.50 hourly wage, and "Fast" work on a $22.00 hourly wage. Other qualifications that apply to this table are on page 9.

| | Labor SF per manhour | Material coverage SF/gallon | Material cost per gallon | Labor cost per 100 SF | Labor burden 100 SF | Material cost per 100 SF | Overhead per 100 SF | Profit per 100 SF | Total price per 100 SF |
|---|---|---|---|---|---|---|---|---|---|
| **Bookcases and shelves, paint grade, brush application** | | | | | | | | | |
| Undercoat, water base (material #3) | | | | | | | | | |
| Brush 1 coat | | | | | | | | | |
| Slow | 25 | 300 | 18.60 | 41.00 | 11.16 | 6.20 | 18.09 | 11.47 | 87.92 |
| Medium | 30 | 280 | 16.20 | 55.00 | 18.49 | 5.79 | 19.43 | 10.86 | 109.57 |
| Fast | 35 | 260 | 13.90 | 62.86 | 25.48 | 5.35 | 17.33 | 6.66 | 117.68 |
| Undercoat, oil base (material #4) | | | | | | | | | |
| Brush 1 coat | | | | | | | | | |
| Slow | 25 | 340 | 20.10 | 41.00 | 11.16 | 5.91 | 18.00 | 11.41 | 87.48 |
| Medium | 30 | 318 | 17.60 | 55.00 | 18.49 | 5.53 | 19.36 | 10.82 | 109.20 |
| Fast | 35 | 295 | 15.10 | 62.86 | 25.48 | 5.12 | 17.29 | 6.65 | 117.40 |
| Split coat (1/2 undercoat + 1/2 enamel), water base (material #3 + #9) | | | | | | | | | |
| Brush each coat | | | | | | | | | |
| Slow | 40 | 350 | 20.70 | 25.63 | 6.97 | 5.91 | 11.94 | 7.57 | 58.02 |
| Medium | 45 | 328 | 18.05 | 36.67 | 12.33 | 5.50 | 13.35 | 7.46 | 75.31 |
| Fast | 50 | 305 | 15.50 | 44.00 | 17.84 | 5.08 | 12.38 | 4.76 | 84.06 |
| Split coat (1/2 undercoat + 1/2 enamel), oil base (material #4 + #10) | | | | | | | | | |
| Brush each coat | | | | | | | | | |
| Slow | 40 | 350 | 23.00 | 25.63 | 6.97 | 6.57 | 12.15 | 7.70 | 59.02 |
| Medium | 45 | 328 | 18.05 | 36.67 | 12.33 | 5.50 | 13.35 | 7.46 | 75.31 |
| Fast | 50 | 305 | 15.50 | 44.00 | 17.84 | 5.08 | 12.38 | 4.76 | 84.06 |
| Enamel, water base (material #9) | | | | | | | | | |
| Brush 1st finish coat | | | | | | | | | |
| Slow | 35 | 340 | 22.80 | 29.29 | 7.97 | 6.71 | 13.63 | 8.64 | 66.24 |
| Medium | 40 | 318 | 19.90 | 41.25 | 13.88 | 6.26 | 15.04 | 8.41 | 84.84 |
| Fast | 45 | 295 | 17.10 | 48.89 | 19.81 | 5.80 | 13.78 | 5.30 | 93.58 |
| Brush 2nd or additional finish coats | | | | | | | | | |
| Slow | 40 | 350 | 22.80 | 25.63 | 6.97 | 6.51 | 12.13 | 7.69 | 58.93 |
| Medium | 45 | 328 | 19.90 | 36.67 | 12.33 | 6.07 | 13.49 | 7.54 | 76.10 |
| Fast | 50 | 305 | 17.10 | 44.00 | 17.84 | 5.61 | 12.48 | 4.80 | 84.73 |
| Enamel, oil base (material #10) | | | | | | | | | |
| Brush 1st finish coat | | | | | | | | | |
| Slow | 35 | 340 | 25.90 | 29.29 | 7.97 | 7.62 | 13.92 | 8.82 | 67.62 |
| Medium | 40 | 318 | 22.70 | 41.25 | 13.88 | 7.14 | 15.26 | 8.53 | 86.06 |
| Fast | 45 | 295 | 19.40 | 48.89 | 19.81 | 6.58 | 13.93 | 5.35 | 94.56 |
| Brush 2nd or additional finish coats | | | | | | | | | |
| Slow | 40 | 350 | 25.90 | 25.63 | 6.97 | 7.40 | 12.40 | 7.86 | 60.26 |
| Medium | 45 | 318 | 22.70 | 36.67 | 12.33 | 7.14 | 13.76 | 7.69 | 77.59 |
| Fast | 50 | 305 | 19.40 | 44.00 | 17.84 | 6.36 | 12.62 | 4.85 | 85.67 |

Bookcase and shelf estimates are based on overall dimensions (length times width) to 8 feet high and include painting all exposed surfaces (including stiles, interior shelves and backs). For heights above 8 feet, use the High Time Difficulty Factors on page 137. "Slow" work is based on a $10.25 hourly wage, "Medium" work on a $16.50 hourly wage, and "Fast" work on a $22.00 hourly wage. Other qualifications that apply to this table are on page 9.

| | Labor SF per manhour | Material coverage SF/gallon | Material cost per gallon | Labor cost per 100 SF | Labor burden 100 SF | Material cost per 100 SF | Overhead per 100 SF | Profit per 100 SF | Total price per 100 SF |
|---|---|---|---|---|---|---|---|---|---|
| **Bookcases and shelves, paint grade, spray application** | | | | | | | | | |
| Undercoat, water base (material #3) | | | | | | | | | |
| Spray 1 coat | | | | | | | | | |
| Slow | 150 | 145 | 18.60 | 6.83 | 1.87 | 12.83 | 6.67 | 4.23 | 32.43 |
| Medium | 165 | 133 | 16.20 | 10.00 | 3.36 | 12.18 | 6.26 | 3.50 | 35.30 |
| Fast | 175 | 120 | 13.90 | 12.57 | 5.09 | 11.58 | 5.41 | 2.08 | 36.73 |
| Undercoat, oil base (material #4) | | | | | | | | | |
| Spray 1 coat | | | | | | | | | |
| Slow | 150 | 145 | 20.10 | 6.83 | 1.87 | 13.86 | 6.99 | 4.43 | 33.98 |
| Medium | 165 | 133 | 17.60 | 10.00 | 3.36 | 13.23 | 6.52 | 3.64 | 36.75 |
| Fast | 175 | 120 | 15.10 | 12.57 | 5.09 | 12.58 | 5.60 | 2.15 | 37.99 |
| Split coat (1/2 undercoat + 1/2 enamel), water base (material #3 + #9) | | | | | | | | | |
| Spray each coat | | | | | | | | | |
| Slow | 245 | 195 | 20.70 | 4.18 | 1.14 | 10.62 | 4.94 | 3.13 | 24.01 |
| Medium | 270 | 183 | 18.05 | 6.11 | 2.05 | 9.86 | 4.42 | 2.47 | 24.91 |
| Fast | 295 | 170 | 15.50 | 7.46 | 3.02 | 9.12 | 3.63 | 1.39 | 24.62 |
| Split coat (1/2 undercoat + 1/2 enamel), oil base (material #4 + #10) | | | | | | | | | |
| Spray each coat | | | | | | | | | |
| Slow | 245 | 195 | 23.00 | 4.18 | 1.14 | 11.79 | 5.30 | 3.36 | 25.77 |
| Medium | 270 | 183 | 20.15 | 6.11 | 2.05 | 11.01 | 4.70 | 2.63 | 26.50 |
| Fast | 295 | 170 | 17.25 | 7.46 | 3.02 | 10.15 | 3.82 | 1.47 | 25.92 |
| Enamel, water base (material #9) | | | | | | | | | |
| Spray 1st finish coat | | | | | | | | | |
| Slow | 225 | 170 | 22.80 | 4.56 | 1.23 | 13.41 | 5.96 | 3.78 | 28.94 |
| Medium | 250 | 158 | 19.90 | 6.60 | 2.22 | 12.59 | 5.25 | 2.93 | 29.59 |
| Fast | 275 | 145 | 17.10 | 8.00 | 3.25 | 11.79 | 4.26 | 1.64 | 28.94 |
| Spray 2nd or additional finish coats | | | | | | | | | |
| Slow | 245 | 195 | 22.80 | 4.18 | 1.14 | 11.69 | 5.27 | 3.34 | 25.62 |
| Medium | 270 | 183 | 19.90 | 6.11 | 2.05 | 10.87 | 4.66 | 2.61 | 26.30 |
| Fast | 295 | 170 | 17.10 | 7.46 | 3.02 | 10.06 | 3.80 | 1.46 | 25.80 |
| Enamel, oil base (material #10) | | | | | | | | | |
| Spray 1st finish coat | | | | | | | | | |
| Slow | 225 | 170 | 25.90 | 4.56 | 1.23 | 15.24 | 6.52 | 4.13 | 31.68 |
| Medium | 250 | 158 | 22.70 | 6.60 | 2.22 | 14.37 | 5.68 | 3.18 | 32.05 |
| Fast | 275 | 145 | 19.40 | 8.00 | 3.25 | 13.38 | 4.55 | 1.75 | 30.93 |
| Spray 2nd or additional finish coats | | | | | | | | | |
| Slow | 245 | 195 | 25.90 | 4.18 | 1.14 | 13.28 | 5.77 | 3.66 | 28.03 |
| Medium | 270 | 183 | 22.70 | 6.11 | 2.05 | 12.40 | 5.04 | 2.82 | 28.42 |
| Fast | 295 | 170 | 19.40 | 7.46 | 3.02 | 11.41 | 4.05 | 1.56 | 27.50 |

Bookcase and shelf estimates are based on overall dimensions (length times width) to 8 feet high and include painting all exposed surfaces (including stiles, interior shelves and backs). For heights above 8 feet, use the High Time Difficulty Factors on page 137. "Slow" work is based on a $10.25 hourly wage, "Medium" work on a $16.50 hourly wage, and "Fast" work on a $22.00 hourly wage. Other qualifications that apply to this table are on page 9.

| | Labor SF per manhour | Material coverage SF/gallon | Material cost per gallon | Labor cost per 100 SF | Labor burden 100 SF | Material cost per 100 SF | Overhead per 100 SF | Profit per 100 SF | Total price per 100 SF |
|---|---|---|---|---|---|---|---|---|---|
| **Bookcases and shelves, stain grade** | | | | | | | | | |
| Stain, seal & lacquer (7 step process) | | | | | | | | | |
| STEP 1: Sand & putty; | | | | | | | | | |
| Slow | 100 | -- | -- | 10.25 | 2.79 | -- | 4.04 | 2.56 | 19.64 |
| Medium | 125 | -- | -- | 13.20 | 4.44 | -- | 4.32 | 2.42 | 24.38 |
| Fast | 150 | -- | -- | 14.67 | 5.95 | -- | 3.81 | 1.47 | 25.90 |
| STEP 2 & 3: Stain (material #11a) & wipe | | | | | | | | | |
| Brush 1 coat & wipe | | | | | | | | | |
| Slow | 75 | 500 | 27.40 | 13.67 | 3.71 | 5.48 | 7.09 | 4.50 | 34.45 |
| Medium | 85 | 475 | 24.00 | 19.41 | 6.52 | 5.05 | 7.59 | 4.24 | 42.81 |
| Fast | 95 | 450 | 20.60 | 23.16 | 9.40 | 4.58 | 6.87 | 2.64 | 46.65 |
| Spray 1 coat & wipe | | | | | | | | | |
| Slow | 300 | 175 | 27.40 | 3.42 | .92 | 15.66 | 6.20 | 3.93 | 30.13 |
| Medium | 400 | 138 | 24.00 | 4.13 | 1.38 | 17.39 | 5.61 | 3.14 | 31.65 |
| Fast | 500 | 100 | 20.60 | 4.40 | 1.78 | 20.60 | 4.95 | 1.90 | 33.63 |
| STEP 4: Sanding sealer (material #11b) | | | | | | | | | |
| Brush 1 coat | | | | | | | | | |
| Slow | 130 | 550 | 17.30 | 7.88 | 2.15 | 3.15 | 4.09 | 2.59 | 19.86 |
| Medium | 140 | 525 | 15.10 | 11.79 | 3.95 | 2.88 | 4.57 | 2.55 | 25.74 |
| Fast | 150 | 500 | 13.00 | 14.67 | 5.95 | 2.60 | 4.30 | 1.65 | 29.17 |
| Spray 1 coat | | | | | | | | | |
| Slow | 375 | 175 | 17.30 | 2.73 | .75 | 9.89 | 4.14 | 2.63 | 20.14 |
| Medium | 475 | 138 | 15.10 | 3.47 | 1.18 | 10.94 | 3.82 | 2.13 | 21.54 |
| Fast | 575 | 100 | 13.00 | 3.83 | 1.55 | 13.00 | 3.40 | 1.31 | 23.09 |
| STEP 5: Sand lightly | | | | | | | | | |
| Slow | 175 | -- | -- | 5.86 | 1.59 | -- | 2.31 | 1.47 | 11.23 |
| Medium | 225 | -- | -- | 7.33 | 2.46 | -- | 2.40 | 1.34 | 13.53 |
| Fast | 275 | -- | -- | 8.00 | 3.25 | -- | 2.08 | .80 | 14.13 |
| STEP 6 & 7: Lacquer (material #11c), 2 coats | | | | | | | | | |
| Brush 1st coat | | | | | | | | | |
| Slow | 140 | 400 | 18.00 | 7.32 | 1.99 | 4.50 | 4.28 | 2.71 | 20.80 |
| Medium | 185 | 375 | 15.70 | 8.92 | 3.01 | 4.19 | 3.95 | 2.21 | 22.28 |
| Fast | 245 | 350 | 13.50 | 8.98 | 3.64 | 3.86 | 3.05 | 1.17 | 20.70 |
| Brush 2nd coat | | | | | | | | | |
| Slow | 155 | 425 | 18.00 | 6.61 | 1.80 | 4.24 | 3.92 | 2.49 | 19.06 |
| Medium | 208 | 413 | 15.70 | 7.93 | 2.68 | 3.80 | 3.53 | 1.97 | 19.91 |
| Fast | 260 | 400 | 13.50 | 8.46 | 3.44 | 3.38 | 2.82 | 1.09 | 19.19 |
| Spray 1st coat | | | | | | | | | |
| Slow | 340 | 175 | 18.00 | 3.01 | .82 | 10.29 | 4.38 | 2.78 | 21.28 |
| Medium | 458 | 138 | 15.70 | 3.60 | 1.21 | 11.38 | 3.97 | 2.22 | 22.38 |
| Fast | 575 | 100 | 13.50 | 3.83 | 1.55 | 13.50 | 3.49 | 1.34 | 23.71 |

| | Labor SF per manhour | Material coverage SF/gallon | Material cost per gallon | Labor cost per 100 SF | Labor burden 100 SF | Material cost per 100 SF | Overhead per 100 SF | Profit per 100 SF | Total price per 100 SF |
|---|---|---|---|---|---|---|---|---|---|
| Spray 2nd coat | | | | | | | | | |
| Slow | 430 | 200 | 18.00 | 2.38 | .66 | 9.00 | 3.73 | 2.36 | 18.13 |
| Medium | 530 | 163 | 15.70 | 3.11 | 1.06 | 9.63 | 3.38 | 1.89 | 19.07 |
| Fast | 630 | 125 | 13.50 | 3.49 | 1.43 | 10.80 | 4.87 | 3.09 | 23.68 |
| Complete 7 step stain, seal & lacquer process (material #11) | | | | | | | | | |
| Brush all coats | | | | | | | | | |
| Slow | 30 | 160 | 20.20 | 34.17 | 9.29 | 12.63 | 17.39 | 11.03 | 84.51 |
| Medium | 35 | 150 | 18.30 | 47.14 | 15.86 | 12.20 | 23.31 | 14.78 | 113.29 |
| Fast | 40 | 140 | 15.10 | 55.00 | 22.30 | 10.79 | 21.58 | 12.06 | 121.73 |
| Spray all coats | | | | | | | | | |
| Slow | 65 | 60 | 20.20 | 15.77 | 4.29 | 33.67 | 13.17 | 7.36 | 74.26 |
| Medium | 83 | 48 | 18.30 | 19.88 | 6.69 | 38.13 | 20.06 | 12.71 | 97.47 |
| Fast | 100 | 35 | 15.10 | 22.00 | 8.92 | 43.14 | 18.14 | 10.14 | 102.34 |
| Shellac, clear (material #12) | | | | | | | | | |
| Brush each coat | | | | | | | | | |
| Slow | 205 | 570 | 22.60 | 5.00 | 1.36 | 3.96 | 3.20 | 2.03 | 15.55 |
| Medium | 230 | 545 | 19.80 | 7.17 | 2.42 | 3.63 | 3.24 | 1.81 | 18.27 |
| Fast | 255 | 520 | 17.00 | 8.63 | 3.49 | 3.27 | 2.85 | 1.10 | 19.34 |
| Varnish, flat or gloss (material #30c) | | | | | | | | | |
| Brush each coat | | | | | | | | | |
| Slow | 175 | 450 | 33.90 | 5.86 | 1.59 | 7.53 | 4.65 | 2.95 | 22.58 |
| Medium | 200 | 438 | 29.70 | 8.25 | 2.78 | 6.78 | 4.36 | 2.44 | 24.61 |
| Fast | 225 | 425 | 25.40 | 9.78 | 3.95 | 5.98 | 3.65 | 1.40 | 24.76 |
| Penetrating stain wax (material #14) & polish | | | | | | | | | |
| Brush 1st coat | | | | | | | | | |
| Slow | 150 | 595 | 28.60 | 6.83 | 1.87 | 4.81 | 4.19 | 2.65 | 20.35 |
| Medium | 175 | 558 | 25.00 | 9.43 | 3.16 | 4.48 | 4.18 | 2.34 | 23.59 |
| Fast | 200 | 520 | 21.40 | 11.00 | 4.46 | 4.12 | 3.62 | 1.39 | 24.59 |
| Brush 2nd or additional coats | | | | | | | | | |
| Slow | 175 | 600 | 28.60 | 5.86 | 1.59 | 4.77 | 3.00 | 1.68 | 16.90 |
| Medium | 200 | 575 | 25.00 | 8.25 | 2.78 | 4.35 | 4.77 | 3.02 | 23.17 |
| Fast | 225 | 550 | 21.40 | 9.78 | 3.95 | 3.89 | 4.32 | 2.42 | 24.36 |

Bookcase and shelf estimates are based on overall dimensions (length times width) to 8 feet high and include painting all exposed surfaces (including stiles, interior shelves and backs). For heights above 8 feet, use the High Time Difficulty Factors on page 137. "Slow" work is based on a $10.25 hourly wage, "Medium" work on a $16.50 hourly wage, and "Fast" work on a $22.00 hourly wage. Other qualifications that apply to this table are on page 9.

| | Labor SF per manhour | Material coverage SF/gallon | Material cost per gallon | Labor cost per 100 SF | Labor burden 100 SF | Material cost per 100 SF | Overhead per 100 SF | Profit per 100 SF | Total price per 100 SF |
|---|---|---|---|---|---|---|---|---|---|
| **Cabinet backs, paint grade, brush** | | | | | | | | | |
| Flat latex, water base (material #5) | | | | | | | | | |
| Brush each coat | | | | | | | | | |
| Slow | 100 | 300 | 16.20 | 10.25 | 2.79 | 5.40 | 5.72 | 3.62 | 27.78 |
| Medium | 150 | 275 | 14.20 | 11.00 | 3.71 | 5.16 | 4.87 | 2.72 | 27.46 |
| Fast | 200 | 250 | 12.10 | 11.00 | 4.46 | 4.84 | 3.76 | 1.44 | 25.50 |
| Enamel, water base (material #9) | | | | | | | | | |
| Brush each coat | | | | | | | | | |
| Slow | 80 | 275 | 22.80 | 12.81 | 3.49 | 8.29 | 7.62 | 4.83 | 37.04 |
| Medium | 130 | 250 | 19.90 | 12.69 | 4.27 | 7.96 | 6.11 | 3.41 | 34.44 |
| Fast | 175 | 225 | 17.10 | 12.57 | 5.09 | 7.60 | 4.67 | 1.80 | 31.73 |
| Enamel, oil base (material #10) | | | | | | | | | |
| Brush each coat | | | | | | | | | |
| Slow | 80 | 275 | 25.90 | 12.81 | 3.49 | 9.42 | 7.97 | 5.05 | 38.74 |
| Medium | 130 | 250 | 22.70 | 12.69 | 4.27 | 9.08 | 6.38 | 3.57 | 35.99 |
| Fast | 175 | 225 | 19.40 | 12.57 | 5.09 | 8.62 | 4.86 | 1.87 | 33.01 |

Cabinet back estimates are based on overall dimensions (length times width) to 8 feet high and include painting the inside back wall of paint grade or stain grade cabinets. ADD for preparation time. For heights above 8 feet, use the High Time Difficulty Factors on page 137. Measurements are based on total area of cabinet faces. "Slow" work is based on a $10.25 hourly wage, "Medium" work on a $16.50 hourly wage, and "Fast" work on a $22.00 hourly wage. Other qualifications that apply to this table are on page 9.

| | Labor SF per manhour | Material coverage SF/gallon | Material cost per gallon | Labor cost per 100 SF | Labor burden 100 SF | Material cost per 100 SF | Overhead per 100 SF | Profit per 100 SF | Total price per 100 SF |
|---|---|---|---|---|---|---|---|---|---|
| **Cabinet faces, stain grade** | | | | | | | | | |
| Complete 7 step stain, seal & 2 coat lacquer system (material #11) | | | | | | | | | |
| Brush all coats | | | | | | | | | |
| Slow | 20 | 190 | 20.20 | 51.25 | 13.95 | 10.63 | 23.51 | 14.90 | 114.24 |
| Medium | 35 | 178 | 18.30 | 47.14 | 15.86 | 10.28 | 17.95 | 10.04 | 101.27 |
| Fast | 50 | 165 | 15.10 | 44.00 | 17.84 | 9.15 | 13.13 | 5.05 | 89.17 |
| Spray all coats | | | | | | | | | |
| Slow | 85 | 67 | 20.20 | 12.06 | 3.28 | 30.15 | 14.11 | 8.94 | 68.54 |
| Medium | 110 | 51 | 18.30 | 15.00 | 5.04 | 35.88 | 13.70 | 7.66 | 77.28 |
| Fast | 135 | 35 | 15.10 | 16.30 | 6.61 | 43.14 | 12.22 | 4.70 | 82.97 |

Cabinet face estimates are based on overall dimensions (length times width) to 8 feet high. Use these figures to estimate finishing the faces of stain grade kitchen, bar, linen, pullman or vanity cabinets. For the stain, seal and lacquer process, the figures include finishing the cabinet faces only. ADD for preparation time. For heights above 8 feet, use the High Time Difficulty Factors on page 137. Measurements are based on total area of cabinet faces. "Slow" work is based on a $10.25 hourly wage, "Medium" work on a $16.50 hourly wage, and "Fast" work on a $22.00 hourly wage. Other qualifications that apply to this table are on page 9.

| | Labor SF per manhour | Material coverage SF/gallon | Material cost per gallon | Labor cost per 100 SF | Labor burden 100 SF | Material cost per 100 SF | Overhead per 100 SF | Profit per 100 SF | Total price per 100 SF |
|---|---|---|---|---|---|---|---|---|---|
| **Cabinets, paint grade, roll and brush** | | | | | | | | | |
| Undercoat, water base (material #3) | | | | | | | | | |
| Roll & brush, 1 coat | | | | | | | | | |
| Slow | 75 | 260 | 18.60 | 13.67 | 3.71 | 7.15 | 7.61 | 4.82 | 36.96 |
| Medium | 93 | 250 | 16.20 | 17.74 | 5.96 | 6.48 | 7.40 | 4.13 | 41.71 |
| Fast | 110 | 240 | 13.90 | 20.00 | 8.11 | 5.79 | 6.27 | 2.41 | 42.58 |
| Undercoat, oil base (material #4) | | | | | | | | | |
| Roll & brush, 1 coat | | | | | | | | | |
| Slow | 75 | 275 | 20.10 | 13.67 | 3.71 | 7.31 | 7.66 | 4.86 | 37.21 |
| Medium | 93 | 268 | 17.60 | 17.74 | 5.96 | 6.57 | 7.42 | 4.15 | 41.84 |
| Fast | 110 | 250 | 15.10 | 20.00 | 8.11 | 6.04 | 6.32 | 2.43 | 42.90 |
| Split coat (1/2 undercoat + 1/2 enamel), water base (material #3 + #9) | | | | | | | | | |
| Roll & brush each coat | | | | | | | | | |
| Slow | 95 | 310 | 20.70 | 10.79 | 2.94 | 6.68 | 6.33 | 4.01 | 30.75 |
| Medium | 113 | 298 | 18.05 | 14.60 | 4.91 | 6.06 | 6.26 | 3.50 | 35.33 |
| Fast | 130 | 285 | 15.50 | 16.92 | 6.86 | 5.44 | 5.41 | 2.08 | 36.71 |
| Split coat (1/2 undercoat + 1/2 enamel), oil base (material #4 + #10) | | | | | | | | | |
| Roll & brush each coat | | | | | | | | | |
| Slow | 95 | 310 | 23.00 | 10.79 | 2.94 | 7.42 | 6.56 | 4.16 | 31.87 |
| Medium | 113 | 298 | 20.15 | 14.60 | 4.91 | 6.76 | 6.44 | 3.60 | 36.31 |
| Fast | 130 | 285 | 17.25 | 16.92 | 6.86 | 6.05 | 5.52 | 2.12 | 37.47 |
| Enamel, water base (material #9) | | | | | | | | | |
| Roll & brush 1st finish coat | | | | | | | | | |
| Slow | 85 | 300 | 22.80 | 12.06 | 3.28 | 7.60 | 7.11 | 4.51 | 34.56 |
| Medium | 103 | 288 | 19.90 | 16.02 | 5.39 | 6.91 | 6.94 | 3.88 | 39.14 |
| Fast | 120 | 275 | 17.10 | 18.33 | 7.43 | 6.22 | 5.92 | 2.27 | 40.17 |
| Roll & brush 2nd or additional finish coats | | | | | | | | | |
| Slow | 95 | 310 | 22.80 | 10.79 | 2.94 | 7.35 | 6.53 | 4.14 | 31.75 |
| Medium | 113 | 298 | 19.90 | 14.60 | 4.91 | 6.68 | 6.42 | 3.59 | 36.20 |
| Fast | 130 | 285 | 17.10 | 16.92 | 6.86 | 6.00 | 5.51 | 2.12 | 37.41 |
| Enamel, oil base (material #10) | | | | | | | | | |
| Roll & brush 1st finish coat | | | | | | | | | |
| Slow | 85 | 300 | 25.90 | 12.06 | 3.28 | 8.63 | 7.43 | 4.71 | 36.11 |
| Medium | 103 | 288 | 22.70 | 16.02 | 5.39 | 7.88 | 7.18 | 4.01 | 40.48 |
| Fast | 120 | 275 | 19.40 | 18.33 | 7.43 | 7.05 | 6.07 | 2.33 | 41.21 |
| Roll & brush 2nd or additional finish coats | | | | | | | | | |
| Slow | 95 | 310 | 25.90 | 10.79 | 2.94 | 8.35 | 6.84 | 4.34 | 33.26 |
| Medium | 113 | 298 | 22.70 | 14.60 | 4.91 | 7.62 | 6.65 | 3.72 | 37.50 |
| Fast | 130 | 285 | 19.40 | 16.92 | 6.86 | 6.81 | 5.66 | 2.18 | 38.43 |

Cabinet estimates are based on overall dimensions (length times width) to 8 feet high and include painting the cabinet face, back of doors, stiles and rails. See Cabinet backs for painting the inside back wall of the cabinets. Use these figures to estimate paint grade kitchen cabinets. Use the Opening Count Method to estimate paint grade pullmans, vanities, bars or linen cabinets. For heights above 8 feet, apply the High time difficulty factor to labor costs and the labor burden cost categories and add these figures to the total cost. Measurements are based on total area of cabinet faces. "Slow" work is based on a $10.25 hourly wage, "Medium" work on a $16.50 hourly wage, and "Fast" work on a $22.00 hourly wage. Other qualifications that apply to this table are on page 9.

| | Labor SF per manhour | Material coverage SF/gallon | Material cost per gallon | Labor cost per 100 SF | Labor burden 100 SF | Material cost per 100 SF | Overhead per 100 SF | Profit per 100 SF | Total price per 100 SF |
|---|---|---|---|---|---|---|---|---|---|
| **Cabinets, paint grade, spray application** | | | | | | | | | |
| Undercoat, water base (material #3) | | | | | | | | | |
| Spray 1 coat | | | | | | | | | |
| Slow | 125 | 125 | 18.60 | 8.20 | 2.23 | 14.88 | 7.85 | 4.97 | 38.13 |
| Medium | 140 | 113 | 16.20 | 11.79 | 3.95 | 14.34 | 7.37 | 4.12 | 41.57 |
| Fast | 155 | 100 | 13.90 | 14.19 | 5.75 | 13.90 | 6.26 | 2.41 | 42.51 |
| Undercoat, oil base (material #4) | | | | | | | | | |
| Spray 1 coat | | | | | | | | | |
| Slow | 125 | 135 | 20.10 | 8.20 | 2.23 | 14.89 | 7.85 | 4.98 | 38.15 |
| Medium | 140 | 123 | 17.60 | 11.79 | 3.95 | 14.31 | 7.37 | 4.12 | 41.54 |
| Fast | 155 | 110 | 15.10 | 14.19 | 5.75 | 13.73 | 6.23 | 2.39 | 42.29 |
| Split coat (1/2 undercoat + 1/2 enamel), water base (material #3 + #9) | | | | | | | | | |
| Spray each coat | | | | | | | | | |
| Slow | 200 | 175 | 20.70 | 5.13 | 1.39 | 11.83 | 5.69 | 3.61 | 27.65 |
| Medium | 225 | 163 | 18.05 | 7.33 | 2.46 | 11.07 | 5.11 | 2.86 | 28.83 |
| Fast | 250 | 150 | 15.50 | 8.80 | 3.57 | 10.33 | 4.20 | 1.61 | 28.51 |
| Split coat (1/2 undercoat + 1/2 enamel), oil base (material #4 + #10) | | | | | | | | | |
| Spray each coat | | | | | | | | | |
| Slow | 200 | 175 | 23.00 | 5.13 | 1.39 | 13.14 | 6.10 | 3.87 | 29.63 |
| Medium | 225 | 163 | 20.15 | 7.33 | 2.46 | 12.36 | 5.43 | 3.03 | 30.61 |
| Fast | 250 | 150 | 17.25 | 8.80 | 3.57 | 11.50 | 4.42 | 1.70 | 29.99 |
| Enamel, water base (material #9) | | | | | | | | | |
| Spray 1st finish coat | | | | | | | | | |
| Slow | 185 | 150 | 22.80 | 5.54 | 1.51 | 15.20 | 6.90 | 4.37 | 33.52 |
| Medium | 210 | 138 | 19.90 | 7.86 | 2.64 | 14.42 | 6.11 | 3.41 | 34.44 |
| Fast | 235 | 125 | 17.10 | 9.36 | 3.81 | 13.68 | 4.97 | 1.91 | 33.73 |
| Spray 2nd or additional finish coats | | | | | | | | | |
| Slow | 200 | 175 | 22.80 | 5.13 | 1.39 | 13.03 | 6.06 | 3.84 | 29.45 |
| Medium | 225 | 163 | 19.90 | 7.33 | 2.46 | 12.21 | 5.39 | 3.01 | 30.40 |
| Fast | 250 | 150 | 17.10 | 8.80 | 3.57 | 11.40 | 4.40 | 1.69 | 29.86 |
| Enamel, oil base (material #10) | | | | | | | | | |
| Spray 1st finish coat | | | | | | | | | |
| Slow | 185 | 160 | 25.90 | 5.54 | 1.51 | 16.19 | 7.20 | 4.57 | 35.01 |
| Medium | 210 | 148 | 22.70 | 7.86 | 2.64 | 15.34 | 6.33 | 3.54 | 35.71 |
| Fast | 235 | 135 | 19.40 | 9.36 | 3.81 | 14.37 | 5.09 | 1.96 | 34.59 |
| Spray 2nd or additional finish coats | | | | | | | | | |
| Slow | 200 | 185 | 25.90 | 5.13 | 1.39 | 14.00 | 6.36 | 4.03 | 30.91 |
| Medium | 225 | 173 | 22.70 | 7.33 | 2.46 | 13.12 | 5.62 | 3.14 | 31.67 |
| Fast | 250 | 160 | 19.40 | 8.80 | 3.57 | 12.13 | 4.53 | 1.74 | 30.77 |

Cabinet estimates are based on overall dimensions (length times width) to 8 feet high and include painting the cabinet face, back of doors, stiles and rails. See Cabinet backs for painting the inside back wall of the cabinets. Use these figures to estimate paint grade kitchen cabinets. Use the Opening Count Method to estimate paint grade pullmans, vanities, bars or linen cabinets. For heights above 8 feet, apply the High time difficulty factor to labor costs and the labor burden cost categories and add these figures to the total cost. Measurements are based on total area of cabinet faces. "Slow" work is based on a $10.25 hourly wage, "Medium" work on a $16.50 hourly wage, and "Fast" work on a $22.00 hourly wage. Other qualifications that apply to this table are on page 9.

| | Labor SF per manhour | Material coverage SF/gallon | Material cost per gallon | Labor cost per 100 SF | Labor burden 100 SF | Material cost per 100 SF | Overhead per 100 SF | Profit per 100 SF | Total price per 100 SF |
|---|---|---|---|---|---|---|---|---|---|
| **Cabinets, stain grade** | | | | | | | | | |
| Stain, seal & 2 coats lacquer system (7 step process) | | | | | | | | | |
| STEP 1: Sand & putty; | | | | | | | | | |
| Slow | 125 | -- | -- | 8.20 | 2.23 | -- | 3.23 | 2.05 | 15.71 |
| Medium | 150 | -- | -- | 11.00 | 3.71 | -- | 3.60 | 2.01 | 20.32 |
| Fast | 175 | -- | -- | 12.57 | 5.09 | -- | 3.27 | 1.26 | 22.19 |
| STEP 2 & 3: Stain (material #11a) & wipe | | | | | | | | | |
| Brush 1 coat & wipe | | | | | | | | | |
| Slow | 65 | 450 | 27.40 | 15.77 | 4.29 | 6.09 | 8.11 | 5.14 | 39.40 |
| Medium | 75 | 400 | 24.00 | 22.00 | 7.39 | 6.00 | 8.67 | 4.85 | 48.91 |
| Fast | 85 | 350 | 20.60 | 25.88 | 10.48 | 5.89 | 7.82 | 3.00 | 53.07 |
| Spray 1 coat & wipe | | | | | | | | | |
| Slow | 250 | 175 | 27.40 | 4.10 | 1.12 | 15.66 | 6.47 | 4.10 | 31.45 |
| Medium | 350 | 138 | 24.00 | 4.71 | 1.60 | 17.39 | 5.80 | 3.24 | 32.74 |
| Fast | 450 | 100 | 20.60 | 4.89 | 1.97 | 20.60 | 5.08 | 1.95 | 34.49 |
| STEP 4: Sanding sealer (material #11b) | | | | | | | | | |
| Brush 1 coat | | | | | | | | | |
| Slow | 110 | 450 | 17.30 | 9.32 | 2.53 | 3.84 | 4.87 | 3.09 | 23.65 |
| Medium | 120 | 425 | 15.10 | 13.75 | 4.62 | 3.55 | 5.37 | 3.00 | 30.29 |
| Fast | 130 | 400 | 13.00 | 16.92 | 6.86 | 3.25 | 5.00 | 1.92 | 33.95 |
| Spray 1 coat | | | | | | | | | |
| Slow | 330 | 175 | 17.30 | 3.11 | .84 | 9.89 | 4.29 | 2.72 | 20.85 |
| Medium | 430 | 138 | 15.10 | 3.84 | 1.30 | 10.94 | 3.94 | 2.20 | 22.22 |
| Fast | 530 | 100 | 13.00 | 4.15 | 1.69 | 13.00 | 3.48 | 1.34 | 23.66 |
| STEP 5: Sand lightly | | | | | | | | | |
| Slow | 200 | -- | -- | 5.13 | 1.39 | -- | 2.02 | 1.28 | 9.82 |
| Medium | 250 | -- | -- | 6.60 | 2.22 | -- | 2.16 | 1.21 | 12.19 |
| Fast | 300 | -- | -- | 7.33 | 2.97 | -- | 1.91 | .73 | 12.94 |
| STEP 6 & 7: Lacquer (material #11c), 2 coats | | | | | | | | | |
| Brush 1st coat | | | | | | | | | |
| Slow | 120 | 375 | 18.00 | 8.54 | 2.32 | 4.80 | 4.86 | 3.08 | 23.60 |
| Medium | 165 | 350 | 15.70 | 10.00 | 3.36 | 4.49 | 4.38 | 2.45 | 24.68 |
| Fast | 215 | 325 | 13.50 | 10.23 | 4.15 | 4.15 | 3.43 | 1.32 | 23.28 |
| Brush 2nd coat | | | | | | | | | |
| Slow | 130 | 400 | 18.00 | 7.88 | 2.15 | 4.50 | 4.50 | 2.85 | 21.88 |
| Medium | 173 | 388 | 15.70 | 9.54 | 3.20 | 4.05 | 4.12 | 2.30 | 23.21 |
| Fast | 225 | 375 | 13.50 | 9.78 | 3.95 | 3.60 | 3.21 | 1.23 | 21.77 |
| Spray 1st coat | | | | | | | | | |
| Slow | 275 | 150 | 18.00 | 3.73 | 1.02 | 12.00 | 5.19 | 3.29 | 25.23 |
| Medium | 388 | 100 | 15.70 | 4.25 | 1.44 | 15.70 | 5.24 | 2.93 | 29.56 |
| Fast | 500 | 75 | 13.50 | 4.40 | 1.78 | 18.00 | 4.47 | 1.72 | 30.37 |

| | Labor SF per manhour | Material coverage SF/gallon | Material cost per gallon | Labor cost per 100 SF | Labor burden 100 SF | Material cost per 100 SF | Overhead per 100 SF | Profit per 100 SF | Total price per 100 SF |
|---|---|---|---|---|---|---|---|---|---|
| Spray 2nd coat | | | | | | | | | |
| Slow | 350 | 200 | 18.00 | 2.93 | .80 | 9.00 | 3.95 | 2.50 | 19.18 |
| Medium | 475 | 163 | 15.70 | 3.47 | 1.18 | 9.63 | 3.50 | 1.95 | 19.73 |
| Fast | 600 | 125 | 13.50 | 3.67 | 1.49 | 10.80 | 2.95 | 1.13 | 20.04 |
| Complete 7 step stain, seal & 2 coat lacquer system (material #11) | | | | | | | | | |
| Brush all coats | | | | | | | | | |
| Slow | 20 | 125 | 20.20 | 51.25 | 13.95 | 16.16 | 25.23 | 15.99 | 122.58 |
| Medium | 25 | 113 | 18.30 | 66.00 | 22.20 | 16.19 | 25.58 | 14.30 | 144.27 |
| Fast | 30 | 100 | 15.10 | 73.33 | 29.73 | 15.10 | 21.86 | 8.40 | 148.42 |
| Spray all coats | | | | | | | | | |
| Slow | 40 | 40 | 20.20 | 25.63 | 6.97 | 50.50 | 25.76 | 16.33 | 125.19 |
| Medium | 50 | 30 | 18.30 | 33.00 | 11.10 | 61.00 | 25.75 | 14.39 | 145.24 |
| Fast | 60 | 21 | 15.10 | 36.67 | 14.87 | 71.90 | 22.84 | 8.78 | 155.06 |
| Shellac, clear (material #12) | | | | | | | | | |
| Brush each coat | | | | | | | | | |
| Slow | 175 | 525 | 22.60 | 5.86 | 1.59 | 4.30 | 2.88 | 1.61 | 16.24 |
| Medium | 200 | 513 | 19.80 | 8.25 | 2.78 | 3.86 | 2.75 | 1.06 | 18.70 |
| Fast | 225 | 500 | 17.00 | 9.78 | 3.95 | 3.40 | 3.17 | 1.22 | 21.52 |
| Varnish, flat or gloss (material #30c) | | | | | | | | | |
| Brush each coat | | | | | | | | | |
| Slow | 155 | 475 | 33.90 | 6.61 | 1.80 | 7.14 | 2.88 | 1.11 | 19.54 |
| Medium | 180 | 463 | 29.70 | 9.17 | 3.09 | 6.41 | 5.79 | 3.67 | 28.13 |
| Fast | 205 | 450 | 25.40 | 10.73 | 4.36 | 5.64 | 5.08 | 2.84 | 28.65 |
| Penetrating stain wax (material #14) & polish | | | | | | | | | |
| Brush 1st coat | | | | | | | | | |
| Slow | 125 | 575 | 28.60 | 8.20 | 2.23 | 4.97 | 2.85 | 1.10 | 19.35 |
| Medium | 150 | 538 | 25.00 | 11.00 | 3.71 | 4.65 | 6.00 | 3.80 | 29.16 |
| Fast | 175 | 500 | 21.40 | 12.57 | 5.09 | 4.28 | 5.38 | 3.01 | 30.33 |
| Brush 2nd or additional coats | | | | | | | | | |
| Slow | 150 | 600 | 28.60 | 6.83 | 1.87 | 4.77 | 2.49 | .96 | 16.92 |
| Medium | 175 | 575 | 25.00 | 9.43 | 3.16 | 4.35 | 3.14 | 1.21 | 21.29 |
| Fast | 200 | 550 | 21.40 | 11.00 | 4.46 | 3.89 | 6.00 | 3.80 | 29.15 |

Cabinet estimates are based on overall dimensions (length times width) to 8 feet high. Use these figures to estimate stain grade kitchen, bar, linen, pullman or vanity cabinets. For the stain, seal and lacquer process, the figures include finishing both sides of cabinet doors, stiles and rails with a fog coat of stain on shelves and the wall behind the cabinet (cabinet back). See Cabinet backs for painting the inside back wall of the cabinets. For heights above 8 feet, use the High Time Difficulty Factors on page 137. Measurements are based on total area of cabinet faces. "Slow" work is based on a $10.25 hourly wage, "Medium" work on a $16.50 hourly wage, and "Fast" work on a $22.00 hourly wage. Other qualifications that apply to this table are on page 9.

| | Labor SF per manhour | Material coverage SF/gallon | Material cost per gallon | Labor cost per 100 SF | Labor burden 100 SF | Material cost per 100 SF | Overhead per 100 SF | Profit per 100 SF | Total price per 100 SF |
|---|---|---|---|---|---|---|---|---|---|
| **Ceiling panels, suspended, fiber panels in T-bar frames, brush application** | | | | | | | | | |
| Flat latex, water base (material #5) | | | | | | | | | |
| Brush 1st coat | | | | | | | | | |
| Slow | 80 | 260 | 16.20 | 12.81 | 3.49 | 6.23 | 6.98 | 4.43 | 33.94 |
| Medium | 110 | 230 | 14.20 | 15.00 | 5.04 | 6.17 | 6.42 | 3.59 | 36.22 |
| Fast | 140 | 200 | 12.10 | 15.71 | 6.37 | 6.05 | 5.20 | 2.00 | 35.33 |
| Brush 2nd or additional coats | | | | | | | | | |
| Slow | 130 | 300 | 16.20 | 7.88 | 2.15 | 5.40 | 4.78 | 3.03 | 23.24 |
| Medium | 150 | 275 | 14.20 | 11.00 | 3.71 | 5.16 | 4.87 | 2.72 | 27.46 |
| Fast | 170 | 250 | 12.10 | 12.94 | 5.24 | 4.84 | 4.26 | 1.64 | 28.92 |
| Enamel, water base (material #9) | | | | | | | | | |
| Brush 1st coat | | | | | | | | | |
| Slow | 65 | 260 | 22.80 | 15.77 | 4.29 | 8.77 | 8.94 | 5.67 | 43.44 |
| Medium | 100 | 230 | 19.90 | 16.50 | 5.55 | 8.65 | 7.52 | 4.20 | 42.42 |
| Fast | 125 | 200 | 17.10 | 17.60 | 7.14 | 8.55 | 6.16 | 2.37 | 41.82 |
| Brush 2nd or additional coats | | | | | | | | | |
| Slow | 115 | 300 | 22.80 | 8.91 | 2.43 | 7.60 | 5.87 | 3.72 | 28.53 |
| Medium | 135 | 275 | 19.90 | 12.22 | 4.12 | 7.24 | 5.77 | 3.23 | 32.58 |
| Fast | 155 | 250 | 17.10 | 14.19 | 5.75 | 6.84 | 4.95 | 1.90 | 33.63 |
| Enamel, oil base (material #10) | | | | | | | | | |
| Brush 1st coat | | | | | | | | | |
| Slow | 65 | 250 | 25.90 | 15.77 | 4.29 | 10.36 | 9.43 | 5.98 | 45.83 |
| Medium | 95 | 213 | 22.70 | 17.37 | 5.85 | 10.66 | 8.30 | 4.64 | 46.82 |
| Fast | 125 | 175 | 19.40 | 17.60 | 7.14 | 11.09 | 6.63 | 2.55 | 45.01 |
| Brush 2nd or additional coats | | | | | | | | | |
| Slow | 115 | 275 | 25.90 | 8.91 | 2.43 | 9.42 | 6.44 | 4.08 | 31.28 |
| Medium | 135 | 260 | 22.70 | 12.22 | 4.12 | 8.73 | 6.14 | 3.43 | 34.64 |
| Fast | 155 | 240 | 19.40 | 14.19 | 5.75 | 8.08 | 5.18 | 1.99 | 35.19 |

Ceiling panel estimates are based on overall dimensions (length times width) to 8 feet high. For heights above 8 feet, use the High Time Difficulty Factors on page 137. "Slow" work is based on a $10.25 hourly wage, "Medium" work on a $16.50 hourly wage, and "Fast" work on a $22.00 hourly wage. Other qualifications that apply to this table are on page 9.

| | Labor SF per manhour | Material coverage SF/gallon | Material cost per gallon | Labor cost per 100 SF | Labor burden 100 SF | Material cost per 100 SF | Overhead per 100 SF | Profit per 100 SF | Total price per 100 SF |
|---|---|---|---|---|---|---|---|---|---|

## Ceiling panels, suspended, fiber panels in T-bar frame, roll application

| | Labor SF per manhour | Material coverage SF/gallon | Material cost per gallon | Labor cost per 100 SF | Labor burden 100 SF | Material cost per 100 SF | Overhead per 100 SF | Profit per 100 SF | Total price per 100 SF |
|---|---|---|---|---|---|---|---|---|---|
| Flat latex, water base (material #5) | | | | | | | | | |
| Roll 1st coat | | | | | | | | | |
| Slow | 150 | 270 | 16.20 | 6.83 | 1.87 | 6.00 | 4.55 | 2.89 | 22.14 |
| Medium | 215 | 235 | 14.20 | 7.67 | 2.58 | 6.04 | 3.99 | 2.23 | 22.51 |
| Fast | 280 | 200 | 12.10 | 7.86 | 3.18 | 6.05 | 3.16 | 1.22 | 21.47 |
| Roll 2nd or additional coats | | | | | | | | | |
| Slow | 225 | 280 | 16.20 | 4.56 | 1.23 | 5.79 | 3.59 | 2.28 | 17.45 |
| Medium | 288 | 260 | 14.20 | 5.73 | 1.92 | 5.46 | 3.21 | 1.80 | 18.12 |
| Fast | 350 | 240 | 12.10 | 6.29 | 2.55 | 5.04 | 2.57 | .99 | 17.44 |
| Enamel, water base (material #9) | | | | | | | | | |
| Roll 1st coat | | | | | | | | | |
| Slow | 135 | 250 | 22.80 | 7.59 | 2.07 | 9.12 | 5.82 | 3.69 | 28.29 |
| Medium | 200 | 220 | 19.90 | 8.25 | 2.78 | 9.05 | 4.92 | 2.75 | 27.75 |
| Fast | 265 | 190 | 17.10 | 8.30 | 3.36 | 9.00 | 3.82 | 1.47 | 25.95 |
| Roll 2nd or additional finish coats | | | | | | | | | |
| Slow | 210 | 280 | 22.80 | 4.88 | 1.33 | 8.14 | 4.45 | 2.82 | 21.62 |
| Medium | 273 | 260 | 19.90 | 6.04 | 2.03 | 7.65 | 3.85 | 2.15 | 21.72 |
| Fast | 335 | 240 | 17.10 | 6.57 | 2.68 | 7.13 | 3.03 | 1.16 | 20.57 |
| Enamel, oil base (material #10) | | | | | | | | | |
| Roll 1st coat | | | | | | | | | |
| Slow | 135 | 240 | 25.90 | 7.59 | 2.07 | 10.79 | 6.34 | 4.02 | 30.81 |
| Medium | 200 | 230 | 22.70 | 8.25 | 2.78 | 9.87 | 5.12 | 2.86 | 28.88 |
| Fast | 265 | 210 | 19.40 | 8.30 | 3.36 | 9.24 | 3.87 | 1.49 | 26.26 |
| Roll 2nd or additional finish coats | | | | | | | | | |
| Slow | 210 | 275 | 25.90 | 4.88 | 1.33 | 9.42 | 4.85 | 3.07 | 23.55 |
| Medium | 273 | 250 | 22.70 | 6.04 | 2.03 | 9.08 | 4.20 | 2.35 | 23.70 |
| Fast | 335 | 230 | 19.40 | 6.57 | 2.68 | 8.43 | 3.27 | 1.26 | 22.21 |

Ceiling panel estimates are based on overall dimensions (length times width) to 8 feet high. For heights above 8 feet, use the High Time Difficulty Factors on page 137. "Slow" work is based on a $10.25 hourly wage, "Medium" work on a $16.50 hourly wage, and "Fast" work on a $22.00 hourly wage. Other qualifications that apply to this table are on page 9.

| | Labor SF per manhour | Material coverage SF/gallon | Material cost per gallon | Labor cost per 100 SF | Labor burden 100 SF | Material cost per 100 SF | Overhead per 100 SF | Profit per 100 SF | Total price per 100 SF |
|---|---|---|---|---|---|---|---|---|---|
| **Ceiling panels, suspended, fiber panels in T-bar frame, spray application** | | | | | | | | | |
| Flat latex, water base (material #5) | | | | | | | | | |
| Spray 1st coat | | | | | | | | | |
| Slow | 300 | 250 | 16.20 | 3.42 | .92 | 6.48 | 3.36 | 2.13 | 16.31 |
| Medium | 345 | 238 | 14.20 | 4.78 | 1.61 | 5.97 | 3.03 | 1.69 | 17.08 |
| Fast | 390 | 225 | 12.10 | 5.64 | 2.28 | 5.38 | 2.46 | .95 | 16.71 |
| Spray 2nd or additional coats | | | | | | | | | |
| Slow | 500 | 300 | 16.20 | 2.05 | .56 | 5.40 | 2.48 | 1.57 | 12.06 |
| Medium | 545 | 288 | 14.20 | 3.03 | 1.01 | 4.93 | 2.20 | 1.23 | 12.40 |
| Fast | 590 | 275 | 12.10 | 3.73 | 1.50 | 4.40 | 1.78 | .69 | 12.10 |
| Enamel, water base (material #9) | | | | | | | | | |
| Spray 1st coat | | | | | | | | | |
| Slow | 275 | 250 | 22.80 | 3.73 | 1.02 | 9.12 | 4.30 | 2.73 | 20.90 |
| Medium | 325 | 238 | 19.90 | 5.08 | 1.71 | 8.36 | 3.71 | 2.07 | 20.93 |
| Fast | 375 | 225 | 17.10 | 5.87 | 2.39 | 7.60 | 2.93 | 1.13 | 19.92 |
| Spray 2nd or additional coats | | | | | | | | | |
| Slow | 450 | 325 | 22.80 | 2.28 | .61 | 7.02 | 3.08 | 1.95 | 14.94 |
| Medium | 500 | 313 | 19.90 | 3.30 | 1.11 | 6.36 | 2.64 | 1.48 | 14.89 |
| Fast | 550 | 300 | 17.10 | 4.00 | 1.63 | 5.70 | 2.09 | .80 | 14.22 |
| Enamel, oil base (material #10) | | | | | | | | | |
| Spray 1st coat | | | | | | | | | |
| Slow | 275 | 240 | 25.90 | 3.73 | 1.02 | 10.79 | 4.82 | 3.05 | 23.41 |
| Medium | 325 | 220 | 22.70 | 5.08 | 1.71 | 10.32 | 4.19 | 2.34 | 23.64 |
| Fast | 375 | 200 | 19.40 | 5.87 | 2.39 | 9.70 | 3.32 | 1.28 | 22.56 |
| Spray 2nd or additional coats | | | | | | | | | |
| Slow | 450 | 300 | 25.90 | 2.28 | .61 | 8.63 | 3.57 | 2.27 | 17.36 |
| Medium | 500 | 288 | 22.70 | 3.30 | 1.11 | 7.88 | 3.01 | 1.68 | 16.98 |
| Fast | 550 | 275 | 19.40 | 4.00 | 1.63 | 7.05 | 2.34 | .90 | 15.92 |

Ceiling panel estimates are based on overall dimensions (length times width) to 8 feet high. For heights above 8 feet, use the High Time Difficulty Factors on page 137. "Slow" work is based on a $10.25 hourly wage, "Medium" work on a $16.50 hourly wage, and "Fast" work on a $22.00 hourly wage. Other qualifications that apply to this table are on page 9.

| | Labor SF per manhour | Material coverage SF/gallon | Material cost per gallon | Labor cost per 100 SF | Labor burden 100 SF | Material cost per 100 SF | Overhead per 100 SF | Profit per 100 SF | Total price per 100 SF |
|---|---|---|---|---|---|---|---|---|---|
| **Ceiling pans, metal, exterior enamel finish** | | | | | | | | | |
| Enamel, water base (material #24) | | | | | | | | | |
| Brush each coat | | | | | | | | | |
| Slow | 80 | 450 | 25.00 | 12.81 | 3.49 | 5.56 | 6.78 | 4.30 | 32.94 |
| Medium | 100 | 388 | 21.80 | 16.50 | 5.55 | 5.62 | 6.78 | 3.79 | 38.24 |
| Fast | 125 | 325 | 18.70 | 17.60 | 7.14 | 5.75 | 5.64 | 2.17 | 38.30 |
| Enamel, oil base (material #25) | | | | | | | | | |
| Brush each coat | | | | | | | | | |
| Slow | 80 | 400 | 27.20 | 12.81 | 3.49 | 6.80 | 7.16 | 4.54 | 34.80 |
| Medium | 103 | 338 | 23.80 | 16.02 | 5.39 | 7.04 | 6.97 | 3.90 | 39.32 |
| Fast | 125 | 275 | 20.40 | 17.60 | 7.14 | 7.42 | 5.95 | 2.29 | 40.40 |
| Enamel, water base (material #24) | | | | | | | | | |
| Roll each coat | | | | | | | | | |
| Slow | 175 | 425 | 25.00 | 5.86 | 1.59 | 5.88 | 4.14 | 2.62 | 20.09 |
| Medium | 200 | 368 | 21.80 | 8.25 | 2.78 | 5.92 | 4.15 | 2.32 | 23.42 |
| Fast | 225 | 300 | 18.70 | 9.78 | 3.95 | 6.23 | 3.70 | 1.42 | 25.08 |
| Enamel, oil base (material #25) | | | | | | | | | |
| Roll each coat | | | | | | | | | |
| Slow | 175 | 375 | 27.20 | 5.86 | 1.59 | 7.25 | 4.56 | 2.89 | 22.15 |
| Medium | 200 | 313 | 23.80 | 8.25 | 2.78 | 7.60 | 4.56 | 2.55 | 25.74 |
| Fast | 225 | 250 | 20.40 | 9.78 | 3.95 | 8.16 | 4.05 | 1.56 | 27.50 |
| Enamel, water base (material #24) | | | | | | | | | |
| Spray each coat | | | | | | | | | |
| Slow | 550 | 380 | 25.00 | 1.86 | .51 | 6.58 | 2.77 | 1.76 | 13.48 |
| Medium | 600 | 370 | 21.80 | 2.75 | .93 | 5.89 | 2.34 | 1.31 | 13.22 |
| Fast | 650 | 260 | 18.70 | 3.38 | 1.38 | 7.19 | 2.21 | .85 | 15.01 |
| Enamel, oil base (material #25) | | | | | | | | | |
| Spray each coat | | | | | | | | | |
| Slow | 550 | 330 | 27.20 | 1.86 | .51 | 8.24 | 3.29 | 2.09 | 15.99 |
| Medium | 600 | 270 | 23.80 | 2.75 | .93 | 8.81 | 3.06 | 1.71 | 17.26 |
| Fast | 650 | 210 | 20.40 | 3.38 | 1.38 | 9.71 | 2.68 | 1.03 | 18.18 |

Ceiling panel estimates are based on overall dimensions (length times width) to 8 feet high. For heights above 8 feet, use the High Time Difficulty Factors on page 137. "Slow" work is based on a $10.25 hourly wage, "Medium" work on a $16.50 hourly wage, and "Fast" work on a $22.00 hourly wage. Other qualifications that apply to this table are on page 9.

| | Labor SF per manhour | Material coverage SF/gallon | Material cost per gallon | Labor cost per 100 SF | Labor burden 100 SF | Material cost per 100 SF | Overhead per 100 SF | Profit per 100 SF | Total price per 100 SF |
|---|---|---|---|---|---|---|---|---|---|
| **Ceilings, acoustic spray-on texture on drywall** | | | | | | | | | |
| Acoustic spray-on texture, primer (material #6) | | | | | | | | | |
| Spray prime coat | | | | | | | | | |
| Slow | 250 | 100 | 17.40 | 4.10 | 1.12 | 17.40 | 7.01 | 4.44 | 34.07 |
| Medium | 300 | 90 | 15.20 | 5.50 | 1.84 | 16.89 | 5.94 | 3.32 | 33.49 |
| Fast | 350 | 80 | 13.00 | 6.29 | 2.55 | 16.25 | 4.64 | 1.78 | 31.51 |
| Acoustic spray-on texture, finish (material #7) | | | | | | | | | |
| Spray 1st finish coat | | | | | | | | | |
| Slow | 400 | 180 | 15.20 | 2.56 | .70 | 8.44 | 3.63 | 2.30 | 17.63 |
| Medium | 450 | 170 | 13.30 | 3.67 | 1.23 | 7.82 | 3.12 | 1.74 | 17.58 |
| Fast | 500 | 160 | 11.40 | 4.40 | 1.78 | 7.13 | 2.46 | .95 | 16.72 |
| Spray 2nd or additional finish coats | | | | | | | | | |
| Slow | 500 | 200 | 15.20 | 2.05 | .56 | 7.60 | 3.17 | 2.01 | 15.39 |
| Medium | 550 | 188 | 13.30 | 3.00 | 1.01 | 7.07 | 2.71 | 1.52 | 15.31 |
| Fast | 600 | 175 | 11.40 | 3.67 | 1.49 | 6.51 | 2.16 | .83 | 14.66 |

Ceiling texture estimates are based on overall dimensions (length times width) to 8 feet high. For heights above 8 feet, use the High Time Difficulty Factors on page 137. "Slow" work is based on a $10.25 hourly wage, "Medium" work on a $16.50 hourly wage, and "Fast" work on a $22.00 hourly wage. Other qualifications that apply to this table are on page 9.

| | Labor SF per manhour | Material coverage SF/pound | Material cost per pound | Labor cost per 100 SF | Labor burden 100 SF | Material cost per 100 SF | Overhead per 100 SF | Profit per 100 SF | Total price per 100 SF |
|---|---|---|---|---|---|---|---|---|---|
| **Ceilings, drywall, stipple finish, spray** | | | | | | | | | |
| Stipple finish texture paint, Dripowder mix (material #8) | | | | | | | | | |
| Spray each coat | | | | | | | | | |
| Slow | 225 | 10.0 | .34 | 4.56 | 1.23 | 3.40 | 2.85 | 1.81 | 13.85 |
| Medium | 250 | 7.5 | .30 | 6.60 | 2.22 | 4.00 | 3.14 | 1.76 | 17.72 |
| Fast | 275 | 5.0 | .30 | 8.00 | 3.25 | 6.00 | 3.19 | 1.23 | 21.67 |

Ceiling texture estimates are based on overall dimensions (length times width) to 8 feet high. For heights above 8 feet, use the High Time Difficulty Factors on page 137. "Slow" work is based on a $10.25 hourly wage, "Medium" work on a $16.50 hourly wage, and "Fast" work on a $22.00 hourly wage. Other qualifications that apply to this table are on page 9.

| | Labor SF per manhour | Material coverage SF/gallon | Material cost per gallon | Labor cost per 100 SF | Labor burden 100 SF | Material cost per 100 SF | Overhead per 100 SF | Profit per 100 SF | Total price per 100 SF |
|---|---|---|---|---|---|---|---|---|---|
| **Ceilings, drywall, smooth finish, brush** | | | | | | | | | |
| Flat latex, water base (material #5) | | | | | | | | | |
| Brush 1st coat | | | | | | | | | |
| Slow | 175 | 325 | 16.20 | 5.86 | 1.59 | 4.98 | 3.86 | 2.45 | 18.74 |
| Medium | 200 | 313 | 14.20 | 8.25 | 2.78 | 4.54 | 3.81 | 2.13 | 21.51 |
| Fast | 225 | 300 | 12.10 | 9.78 | 3.95 | 4.03 | 3.29 | 1.26 | 22.31 |
| Brush 2nd coat | | | | | | | | | |
| Slow | 225 | 400 | 16.20 | 4.56 | 1.23 | 4.05 | 3.05 | 1.94 | 14.83 |
| Medium | 250 | 375 | 14.20 | 6.60 | 2.22 | 3.79 | 3.09 | 1.73 | 17.43 |
| Fast | 275 | 350 | 12.10 | 8.00 | 3.25 | 3.46 | 2.72 | 1.05 | 18.48 |
| Brush 3rd or additional coats | | | | | | | | | |
| Slow | 250 | 425 | 16.20 | 4.10 | 1.12 | 3.81 | 2.80 | 1.77 | 13.60 |
| Medium | 275 | 400 | 14.20 | 6.00 | 2.03 | 3.55 | 2.83 | 1.58 | 15.99 |
| Fast | 300 | 375 | 12.10 | 7.33 | 2.97 | 3.23 | 2.50 | .96 | 16.99 |
| Sealer, water base (material #1) | | | | | | | | | |
| Brush prime coat | | | | | | | | | |
| Slow | 200 | 325 | 15.50 | 5.13 | 1.39 | 4.77 | 3.50 | 2.22 | 17.01 |
| Medium | 225 | 313 | 13.50 | 7.33 | 2.46 | 4.31 | 3.46 | 1.93 | 19.49 |
| Fast | 250 | 300 | 11.60 | 8.80 | 3.57 | 3.87 | 3.00 | 1.15 | 20.39 |
| Sealer, oil base (material #2) | | | | | | | | | |
| Brush prime coat | | | | | | | | | |
| Slow | 200 | 350 | 20.70 | 5.13 | 1.39 | 5.91 | 3.86 | 2.45 | 18.74 |
| Medium | 225 | 338 | 18.10 | 7.33 | 2.46 | 5.36 | 3.71 | 2.08 | 20.94 |
| Fast | 250 | 325 | 15.50 | 8.80 | 3.57 | 4.77 | 3.17 | 1.22 | 21.53 |
| Enamel, water base (material #9) | | | | | | | | | |
| Brush 1st finish coat | | | | | | | | | |
| Slow | 200 | 400 | 22.80 | 5.13 | 1.39 | 5.70 | 3.79 | 2.40 | 18.41 |
| Medium | 225 | 375 | 19.90 | 7.33 | 2.46 | 5.31 | 3.70 | 2.07 | 20.87 |
| Fast | 250 | 350 | 17.10 | 8.80 | 3.57 | 4.89 | 3.19 | 1.23 | 21.68 |
| Brush 2nd and additional finish coats | | | | | | | | | |
| Slow | 225 | 425 | 22.80 | 4.56 | 1.23 | 5.36 | 3.46 | 2.19 | 16.80 |
| Medium | 250 | 400 | 19.90 | 6.60 | 2.22 | 4.98 | 3.38 | 1.89 | 19.07 |
| Fast | 275 | 375 | 17.10 | 8.00 | 3.25 | 4.56 | 2.92 | 1.12 | 19.85 |
| Enamel, oil base (material #10) | | | | | | | | | |
| Brush 1st finish coat | | | | | | | | | |
| Slow | 200 | 400 | 25.90 | 5.13 | 1.39 | 6.48 | 4.03 | 2.56 | 19.59 |
| Medium | 225 | 388 | 22.70 | 7.33 | 2.46 | 5.85 | 3.83 | 2.14 | 21.61 |
| Fast | 250 | 375 | 19.40 | 8.80 | 3.57 | 5.17 | 3.24 | 1.25 | 22.03 |

| | Labor SF per manhour | Material coverage SF/gallon | Material cost per gallon | Labor cost per 100 SF | Labor burden 100 SF | Material cost per 100 SF | Overhead per 100 SF | Profit per 100 SF | Total price per 100 SF |
|---|---|---|---|---|---|---|---|---|---|
| Brush 2nd or additional finish coats | | | | | | | | | |
| Slow | 225 | 425 | 25.90 | 4.56 | 1.23 | 6.09 | 3.69 | 2.34 | 17.91 |
| Medium | 250 | 413 | 22.70 | 6.60 | 2.22 | 5.50 | 3.51 | 1.96 | 19.79 |
| Fast | 275 | 400 | 19.40 | 8.00 | 3.25 | 4.85 | 2.98 | 1.14 | 20.22 |
| Epoxy coating, white (material #52) | | | | | | | | | |
| Brush 1st coat | | | | | | | | | |
| Slow | 175 | 425 | 57.40 | 5.86 | 1.59 | 13.51 | 6.50 | 4.12 | 31.58 |
| Medium | 200 | 400 | 50.20 | 8.25 | 2.78 | 12.55 | 5.78 | 3.23 | 32.59 |
| Fast | 225 | 375 | 43.10 | 9.78 | 3.95 | 11.49 | 4.67 | 1.79 | 31.68 |
| Brush 2nd or additional coats | | | | | | | | | |
| Slow | 200 | 450 | 57.40 | 5.13 | 1.39 | 12.76 | 5.98 | 3.79 | 29.05 |
| Medium | 225 | 425 | 50.20 | 7.33 | 2.46 | 11.81 | 5.29 | 2.96 | 29.85 |
| Fast | 250 | 400 | 43.10 | 8.80 | 3.57 | 10.78 | 4.28 | 1.65 | 29.08 |
| Stipple finish | | | | | | | | | |
| Slow | 200 | -- | -- | 5.13 | 1.39 | -- | 2.02 | 1.28 | 9.82 |
| Medium | 225 | -- | -- | 7.33 | 2.46 | -- | 2.40 | 1.34 | 13.53 |
| Fast | 250 | -- | -- | 8.80 | 3.57 | -- | 2.29 | .88 | 15.54 |

Ceiling estimates are based on overall dimensions (length times width) to 8 feet high. For heights above 8 feet, use the High Time Difficulty Factors on page 137. "Slow" work is based on a $10.25 hourly wage, "Medium" work on a $16.50 hourly wage, and "Fast" work on a $22.00 hourly wage. Other qualifications that apply to this table are on page 9.

| | Labor SF per manhour | Material coverage SF/gallon | Material cost per gallon | Labor cost per 100 SF | Labor burden 100 SF | Material cost per 100 SF | Overhead per 100 SF | Profit per 100 SF | Total price per 100 SF |
|---|---|---|---|---|---|---|---|---|---|
| **Ceilings, drywall, smooth finish, roll** | | | | | | | | | |
| Flat latex, water base (material #5) | | | | | | | | | |
| Roll 1st coat | | | | | | | | | |
| Slow | 325 | 350 | 16.20 | 3.15 | .87 | 4.63 | 2.68 | 1.70 | 13.03 |
| Medium | 375 | 325 | 14.20 | 4.40 | 1.49 | 4.37 | 2.51 | 1.40 | 14.17 |
| Fast | 425 | 300 | 12.10 | 5.18 | 2.09 | 4.03 | 2.09 | .80 | 14.19 |
| Roll 2nd coat | | | | | | | | | |
| Slow | 375 | 375 | 16.20 | 2.73 | .75 | 4.32 | 2.41 | 1.53 | 11.74 |
| Medium | 413 | 363 | 14.20 | 4.00 | 1.34 | 3.91 | 2.27 | 1.27 | 12.79 |
| Fast | 450 | 350 | 12.10 | 4.89 | 1.97 | 3.46 | 1.91 | .73 | 12.96 |
| Roll 3rd or additional coats | | | | | | | | | |
| Slow | 425 | 400 | 16.20 | 2.41 | .65 | 4.05 | 2.21 | 1.40 | 10.72 |
| Medium | 450 | 388 | 14.20 | 3.67 | 1.23 | 3.66 | 2.10 | 1.17 | 11.83 |
| Fast | 475 | 375 | 12.10 | 4.63 | 1.89 | 3.23 | 1.80 | .69 | 12.24 |
| Sealer, water base (material #1) | | | | | | | | | |
| Roll prime coat | | | | | | | | | |
| Slow | 350 | 350 | 15.50 | 2.93 | .80 | 4.43 | 2.53 | 1.60 | 12.29 |
| Medium | 400 | 325 | 13.50 | 4.13 | 1.38 | 4.15 | 2.37 | 1.32 | 13.35 |
| Fast | 450 | 300 | 11.60 | 4.89 | 1.97 | 3.87 | 1.99 | .76 | 13.48 |
| Sealer, oil base (material #2) | | | | | | | | | |
| Roll prime coat | | | | | | | | | |
| Slow | 350 | 300 | 20.70 | 2.93 | .80 | 6.90 | 3.30 | 2.09 | 16.02 |
| Medium | 400 | 288 | 18.10 | 4.13 | 1.38 | 6.28 | 2.89 | 1.62 | 16.30 |
| Fast | 450 | 275 | 15.50 | 4.89 | 1.97 | 5.64 | 2.31 | .89 | 15.70 |
| Enamel, water base, (material #9) | | | | | | | | | |
| Roll 1st finish coat | | | | | | | | | |
| Slow | 350 | 375 | 22.80 | 2.93 | .80 | 6.08 | 3.04 | 1.93 | 14.78 |
| Medium | 400 | 363 | 19.90 | 4.13 | 1.38 | 5.48 | 2.70 | 1.51 | 15.20 |
| Fast | 450 | 350 | 17.10 | 4.89 | 1.97 | 4.89 | 2.18 | .84 | 14.77 |
| Roll 2nd or additional finish coats | | | | | | | | | |
| Slow | 425 | 400 | 22.80 | 2.41 | .65 | 5.70 | 2.72 | 1.72 | 13.20 |
| Medium | 450 | 388 | 19.90 | 3.67 | 1.23 | 5.13 | 2.46 | 1.37 | 13.86 |
| Fast | 475 | 375 | 17.10 | 4.63 | 1.89 | 4.56 | 2.05 | .79 | 13.92 |
| Enamel, oil base (material #10) | | | | | | | | | |
| Roll 1st finish coat | | | | | | | | | |
| Slow | 350 | 350 | 25.90 | 2.93 | .80 | 7.40 | 3.45 | 2.19 | 16.77 |
| Medium | 400 | 338 | 22.70 | 4.13 | 1.38 | 6.72 | 3.00 | 1.68 | 16.91 |
| Fast | 450 | 325 | 19.40 | 4.89 | 1.97 | 5.97 | 2.38 | .91 | 16.12 |

| | Labor SF per manhour | Material coverage SF/gallon | Material cost per gallon | Labor cost per 100 SF | Labor burden 100 SF | Material cost per 100 SF | Overhead per 100 SF | Profit per 100 SF | Total price per 100 SF |
|---|---|---|---|---|---|---|---|---|---|
| Roll 2nd or additional finish coats | | | | | | | | | |
| Slow | 425 | 375 | 25.90 | 2.41 | .65 | 6.91 | 3.09 | 1.96 | 15.02 |
| Medium | 450 | 363 | 22.70 | 3.67 | 1.23 | 6.25 | 2.73 | 1.53 | 15.41 |
| Fast | 475 | 350 | 19.40 | 4.63 | 1.89 | 5.54 | 2.23 | .86 | 15.15 |
| Epoxy coating, white (material #52) | | | | | | | | | |
| Roll 1st finish coat | | | | | | | | | |
| Slow | 325 | 400 | 57.40 | 3.15 | .87 | 14.35 | 5.69 | 3.61 | 27.67 |
| Medium | 363 | 375 | 50.20 | 4.55 | 1.51 | 13.39 | 4.77 | 2.67 | 26.89 |
| Fast | 400 | 350 | 43.10 | 5.50 | 2.23 | 12.31 | 3.71 | 1.43 | 25.18 |
| Roll 2nd or additional finish coats | | | | | | | | | |
| Slow | 400 | 425 | 57.40 | 2.56 | .70 | 13.51 | 5.20 | 3.30 | 25.27 |
| Medium | 425 | 400 | 50.20 | 3.88 | 1.30 | 12.55 | 4.35 | 2.43 | 24.51 |
| Fast | 450 | 375 | 43.10 | 4.89 | 1.97 | 11.49 | 3.40 | 1.31 | 23.06 |

Ceiling estimates are based on overall dimensions (length times width) to 8 feet high. For heights above 8 feet, use the High Time Difficulty Factors on page 137. "Slow" work is based on a $10.25 hourly wage, "Medium" work on a $16.50 hourly wage, and "Fast" work on a $22.00 hourly wage. Other qualifications that apply to this table are on page 9.

| | Labor SF per manhour | Material coverage SF/gallon | Material cost per gallon | Labor cost per 100 SF | Labor burden 100 SF | Material cost per 100 SF | Overhead per 100 SF | Profit per 100 SF | Total price per 100 SF |
|---|---|---|---|---|---|---|---|---|---|
| **Ceilings, drywall, smooth finish, spray** | | | | | | | | | |
| Flat latex, water base (material #5) | | | | | | | | | |
| Spray 1st coat | | | | | | | | | |
| Slow | 750 | 300 | 16.20 | 1.37 | .36 | 5.40 | 2.21 | 1.40 | 10.74 |
| Medium | 850 | 275 | 14.20 | 1.94 | .66 | 5.16 | 1.90 | 1.06 | 10.72 |
| Fast | 950 | 250 | 12.10 | 2.32 | .93 | 4.84 | 1.50 | .58 | 10.17 |
| Spray 2nd coat | | | | | | | | | |
| Slow | 850 | 350 | 16.20 | 1.21 | .33 | 4.63 | 1.91 | 1.21 | 9.29 |
| Medium | 950 | 325 | 14.20 | 1.74 | .58 | 4.37 | 1.64 | .92 | 9.25 |
| Fast | 1050 | 300 | 12.10 | 2.10 | .84 | 4.03 | 1.29 | .50 | 8.76 |
| Spray 3rd or additional coats | | | | | | | | | |
| Slow | 900 | 350 | 16.20 | 1.14 | .31 | 4.63 | 1.88 | 1.19 | 9.15 |
| Medium | 1000 | 338 | 14.20 | 1.65 | .56 | 4.20 | 1.57 | .88 | 8.86 |
| Fast | 1100 | 325 | 12.10 | 2.00 | .81 | 3.72 | 1.21 | .46 | 8.20 |
| Sealer, water base (material #1) | | | | | | | | | |
| Spray prime coat | | | | | | | | | |
| Slow | 800 | 300 | 15.50 | 1.28 | .35 | 5.17 | 2.11 | 1.34 | 10.25 |
| Medium | 900 | 275 | 13.50 | 1.83 | .62 | 4.91 | 1.80 | 1.01 | 10.17 |
| Fast | 1000 | 250 | 11.60 | 2.20 | .89 | 4.64 | 1.43 | .55 | 9.71 |
| Sealer, oil base (material #2) | | | | | | | | | |
| Spray prime coat | | | | | | | | | |
| Slow | 800 | 250 | 20.70 | 1.28 | .35 | 8.28 | 3.07 | 1.95 | 14.93 |
| Medium | 900 | 238 | 18.10 | 1.83 | .62 | 7.61 | 2.46 | 1.38 | 13.90 |
| Fast | 1000 | 225 | 15.50 | 2.20 | .89 | 6.89 | 1.85 | .71 | 12.54 |
| Enamel, water base (material #9) | | | | | | | | | |
| Spray 1st finish coat | | | | | | | | | |
| Slow | 800 | 350 | 22.80 | 1.28 | .35 | 6.51 | 2.52 | 1.60 | 12.26 |
| Medium | 900 | 325 | 19.90 | 1.83 | .62 | 6.12 | 2.10 | 1.17 | 11.84 |
| Fast | 1000 | 300 | 17.10 | 2.20 | .89 | 5.70 | 1.63 | .63 | 11.05 |
| Spray 2nd or additional finish coats | | | | | | | | | |
| Slow | 850 | 350 | 22.80 | 1.21 | .33 | 6.51 | 2.50 | 1.58 | 12.13 |
| Medium | 950 | 338 | 19.90 | 1.74 | .58 | 5.89 | 2.01 | 1.13 | 11.35 |
| Fast | 1050 | 325 | 17.10 | 2.10 | .84 | 5.26 | 1.52 | .58 | 10.30 |
| Enamel, oil base (material #10) | | | | | | | | | |
| Spray 1st finish coat | | | | | | | | | |
| Slow | 800 | 300 | 25.90 | 1.28 | .35 | 8.63 | 3.18 | 2.02 | 15.46 |
| Medium | 900 | 280 | 22.70 | 1.83 | .62 | 8.11 | 2.59 | 1.45 | 14.60 |
| Fast | 1000 | 260 | 19.40 | 2.20 | .89 | 7.46 | 1.95 | .75 | 13.25 |
| Spray 2nd or additional finish coats | | | | | | | | | |
| Slow | 850 | 325 | 25.90 | 1.21 | .33 | 7.97 | 2.95 | 1.87 | 14.33 |
| Medium | 950 | 313 | 22.70 | 1.74 | .58 | 7.25 | 2.35 | 1.31 | 13.23 |
| Fast | 1050 | 300 | 19.40 | 2.10 | .84 | 6.47 | 1.74 | .67 | 11.82 |

Ceiling estimates are based on overall dimensions (length times width) to 8 feet high. For heights above 8 feet, use the High Time Difficulty Factors on page 137. "Slow" work is based on a $10.25 hourly wage, "Medium" work on a $16.50 hourly wage, and "Fast" work on a $22.00 hourly wage. Other qualifications that apply to this table are on page 9.

| | Labor SF per manhour | Material coverage SF/gallon | Material cost per gallon | Labor cost per 100 SF | Labor burden 100 SF | Material cost per 100 SF | Overhead per 100 SF | Profit per 100 SF | Total price per 100 SF |
|---|---|---|---|---|---|---|---|---|---|
| **Ceilings, gypsum drywall, anti-graffiti stain eliminator** | | | | | | | | | |
| Water base primer and pigmented sealer (material #39) | | | | | | | | | |
| Roll & brush each coat | | | | | | | | | |
| Slow | 350 | 450 | 30.40 | 2.93 | .80 | 6.76 | 3.25 | 2.06 | 15.80 |
| Medium | 375 | 425 | 26.60 | 4.40 | 1.49 | 6.26 | 2.97 | 1.66 | 16.78 |
| Fast | 400 | 400 | 22.80 | 5.50 | 2.23 | 5.70 | 2.48 | .95 | 16.86 |
| Oil base primer and pigmented sealer (material #40) | | | | | | | | | |
| Roll & brush each coat | | | | | | | | | |
| Slow | 350 | 400 | 28.20 | 2.93 | .80 | 7.05 | 3.34 | 2.12 | 16.24 |
| Medium | 375 | 388 | 24.70 | 4.40 | 1.49 | 6.37 | 3.00 | 1.68 | 16.94 |
| Fast | 400 | 375 | 21.20 | 5.50 | 2.23 | 5.65 | 2.48 | .95 | 16.81 |
| Polyurethane 2 part system (material #41) | | | | | | | | | |
| Roll & brush each coat | | | | | | | | | |
| Slow | 300 | 400 | 63.30 | 3.42 | .92 | 15.83 | 6.26 | 3.97 | 30.40 |
| Medium | 325 | 375 | 55.40 | 5.08 | 1.71 | 14.77 | 5.28 | 2.95 | 29.79 |
| Fast | 350 | 350 | 47.40 | 6.29 | 2.55 | 13.54 | 4.14 | 1.59 | 28.11 |

Ceiling estimates are based on overall dimensions (length times width) to 8 feet high. For heights above 8 feet, use the High Time Difficulty Factors on page 137. "Slow" work is based on a $10.25 hourly wage, "Medium" work on a $16.50 hourly wage, and "Fast" work on a $22.00 hourly wage. Other qualifications that apply to this table are on page 9.

| | Labor SF per manhour | Material coverage SF/gallon | Material cost per gallon | Labor cost per 100 SF | Labor burden 100 SF | Material cost per 100 SF | Overhead per 100 SF | Profit per 100 SF | Total price per 100 SF |
|---|---|---|---|---|---|---|---|---|---|
| **Ceilings, gypsum drywall, orange peel or knock-down texture, brush** | | | | | | | | | |
| Flat latex, water base (material #5) | | | | | | | | | |
| Brush 1st coat | | | | | | | | | |
| Slow | 150 | 300 | 16.20 | 6.83 | 1.87 | 5.40 | 4.37 | 2.77 | 21.24 |
| Medium | 175 | 288 | 14.20 | 9.43 | 3.16 | 4.93 | 4.29 | 2.40 | 24.21 |
| Fast | 200 | 275 | 12.10 | 11.00 | 4.46 | 4.40 | 3.67 | 1.41 | 24.94 |
| Brush 2nd coat | | | | | | | | | |
| Slow | 175 | 350 | 16.20 | 5.86 | 1.59 | 4.63 | 3.75 | 2.38 | 18.21 |
| Medium | 200 | 338 | 14.20 | 8.25 | 2.78 | 4.20 | 3.73 | 2.09 | 21.05 |
| Fast | 225 | 325 | 12.10 | 9.78 | 3.95 | 3.72 | 3.23 | 1.24 | 21.92 |
| Brush 3rd or additional coats | | | | | | | | | |
| Slow | 200 | 400 | 16.20 | 5.13 | 1.39 | 4.05 | 3.28 | 2.08 | 15.93 |
| Medium | 225 | 375 | 14.20 | 7.33 | 2.46 | 3.79 | 3.33 | 1.86 | 18.77 |
| Fast | 250 | 350 | 12.10 | 8.80 | 3.57 | 3.46 | 2.93 | 1.13 | 19.89 |
| Sealer, water base (material #1) | | | | | | | | | |
| Brush prime coat | | | | | | | | | |
| Slow | 175 | 300 | 15.50 | 5.86 | 1.59 | 5.17 | 3.92 | 2.48 | 19.02 |
| Medium | 200 | 288 | 13.50 | 8.25 | 2.78 | 4.69 | 3.85 | 2.15 | 21.72 |
| Fast | 225 | 275 | 11.60 | 9.78 | 3.95 | 4.22 | 3.32 | 1.28 | 22.55 |
| Sealer, oil base (material #2) | | | | | | | | | |
| Brush prime coat | | | | | | | | | |
| Slow | 175 | 250 | 20.70 | 5.86 | 1.59 | 8.28 | 4.88 | 3.09 | 23.70 |
| Medium | 200 | 238 | 18.10 | 8.25 | 2.78 | 7.61 | 4.57 | 2.55 | 25.76 |
| Fast | 225 | 225 | 15.50 | 9.78 | 3.95 | 6.89 | 3.82 | 1.47 | 25.91 |
| Enamel, water base (material #9) | | | | | | | | | |
| Brush 1st finish coat | | | | | | | | | |
| Slow | 150 | 350 | 22.80 | 6.83 | 1.87 | 6.51 | 4.71 | 2.99 | 22.91 |
| Medium | 175 | 338 | 19.90 | 9.43 | 3.16 | 5.89 | 4.53 | 2.53 | 25.54 |
| Fast | 200 | 325 | 17.10 | 11.00 | 4.46 | 5.26 | 3.83 | 1.47 | 26.02 |
| Brush 2nd or additional finish coats | | | | | | | | | |
| Slow | 175 | 400 | 22.80 | 5.86 | 1.59 | 5.70 | 4.08 | 2.59 | 19.82 |
| Medium | 200 | 375 | 19.90 | 8.25 | 2.78 | 5.31 | 4.00 | 2.24 | 22.58 |
| Fast | 225 | 350 | 17.10 | 9.78 | 3.95 | 4.89 | 3.45 | 1.33 | 23.40 |

| | Labor SF per manhour | Material coverage SF/gallon | Material cost per gallon | Labor cost per 100 SF | Labor burden 100 SF | Material cost per 100 SF | Overhead per 100 SF | Profit per 100 SF | Total price per 100 SF |
|---|---|---|---|---|---|---|---|---|---|
| Enamel, oil base (material #10) | | | | | | | | | |
| Brush 1st finish coat | | | | | | | | | |
| Slow | 150 | 325 | 25.90 | 6.83 | 1.87 | 7.97 | 5.16 | 3.27 | 25.10 |
| Medium | 175 | 313 | 22.70 | 9.43 | 3.16 | 7.25 | 4.86 | 2.72 | 27.42 |
| Fast | 200 | 300 | 19.40 | 11.00 | 4.46 | 6.47 | 4.06 | 1.56 | 27.55 |
| Brush 2nd or additional finish coats | | | | | | | | | |
| Slow | 150 | 400 | 25.90 | 6.83 | 1.87 | 6.48 | 4.70 | 2.98 | 22.86 |
| Medium | 175 | 375 | 22.70 | 9.43 | 3.16 | 6.05 | 4.57 | 2.55 | 25.76 |
| Fast | 200 | 350 | 19.40 | 11.00 | 4.46 | 5.54 | 3.89 | 1.49 | 26.38 |
| Epoxy coating, white (material #52) | | | | | | | | | |
| Brush 1st coat | | | | | | | | | |
| Slow | 125 | 350 | 57.40 | 8.20 | 2.23 | 16.40 | 8.32 | 5.27 | 40.42 |
| Medium | 150 | 325 | 50.20 | 11.00 | 3.71 | 15.45 | 7.39 | 4.13 | 41.68 |
| Fast | 175 | 300 | 43.10 | 12.57 | 5.09 | 14.37 | 5.93 | 2.28 | 40.24 |
| Brush 2nd or additional coats | | | | | | | | | |
| Slow | 175 | 375 | 57.40 | 5.86 | 1.59 | 15.31 | 7.06 | 4.47 | 34.29 |
| Medium | 200 | 350 | 50.20 | 8.25 | 2.78 | 14.34 | 6.22 | 3.47 | 35.06 |
| Fast | 225 | 325 | 43.10 | 9.78 | 3.95 | 13.26 | 5.00 | 1.92 | 33.91 |

Ceiling estimates are based on overall dimensions (length times width) to 8 feet high. For heights above 8 feet, use the High Time Difficulty Factors on page 137. "Slow" work is based on a $10.25 hourly wage, "Medium" work on a $16.50 hourly wage, and "Fast" work on a $22.00 hourly wage. Other qualifications that apply to this table are on page 9.

| | Labor SF per manhour | Material coverage SF/gallon | Material cost per gallon | Labor cost per 100 SF | Labor burden 100 SF | Material cost per 100 SF | Overhead per 100 SF | Profit per 100 SF | Total price per 100 SF |
|---|---|---|---|---|---|---|---|---|---|
| **Ceilings, gypsum drywall, orange peel or knock-down texture, roll** | | | | | | | | | |
| Flat latex, water base (material #5) | | | | | | | | | |
| Roll 1st coat | | | | | | | | | |
| Slow | 325 | 300 | 16.20 | 3.15 | .87 | 5.40 | 2.92 | 1.85 | 14.19 |
| Medium | 350 | 275 | 14.20 | 4.71 | 1.60 | 5.16 | 2.81 | 1.57 | 15.85 |
| Fast | 375 | 250 | 12.10 | 5.87 | 2.39 | 4.84 | 2.42 | .93 | 16.45 |
| Roll 2nd coat | | | | | | | | | |
| Slow | 350 | 325 | 16.20 | 2.93 | .80 | 4.98 | 2.70 | 1.71 | 13.12 |
| Medium | 375 | 313 | 14.20 | 4.40 | 1.49 | 4.54 | 2.55 | 1.43 | 14.41 |
| Fast | 400 | 300 | 12.10 | 5.50 | 2.23 | 4.03 | 2.18 | .84 | 14.78 |
| Roll 3rd or additional coats | | | | | | | | | |
| Slow | 400 | 350 | 16.20 | 2.56 | .70 | 4.63 | 2.45 | 1.55 | 11.89 |
| Medium | 425 | 338 | 14.20 | 3.88 | 1.30 | 4.20 | 2.30 | 1.29 | 12.97 |
| Fast | 450 | 325 | 12.10 | 4.89 | 1.97 | 3.72 | 1.96 | .75 | 13.29 |
| Sealer, water base (material #1) | | | | | | | | | |
| Roll prime coat | | | | | | | | | |
| Slow | 350 | 300 | 15.50 | 2.93 | .80 | 5.17 | 2.76 | 1.75 | 13.41 |
| Medium | 375 | 275 | 13.50 | 4.40 | 1.49 | 4.91 | 2.64 | 1.48 | 14.92 |
| Fast | 400 | 250 | 11.60 | 5.50 | 2.23 | 4.64 | 2.29 | .88 | 15.54 |
| Sealer, oil base (material #2) | | | | | | | | | |
| Roll prime coat | | | | | | | | | |
| Slow | 350 | 275 | 20.70 | 2.93 | .80 | 7.53 | 3.49 | 2.21 | 16.96 |
| Medium | 375 | 250 | 18.10 | 4.40 | 1.49 | 7.24 | 3.21 | 1.80 | 18.14 |
| Fast | 400 | 225 | 15.50 | 5.50 | 2.23 | 6.89 | 2.70 | 1.04 | 18.36 |
| Enamel, water base (material #9) | | | | | | | | | |
| Roll 1st finish coat | | | | | | | | | |
| Slow | 325 | 325 | 22.80 | 3.15 | .87 | 7.02 | 3.42 | 2.17 | 16.63 |
| Medium | 350 | 313 | 19.90 | 4.71 | 1.60 | 6.36 | 3.10 | 1.73 | 17.50 |
| Fast | 375 | 300 | 17.10 | 5.87 | 2.39 | 5.70 | 2.58 | .99 | 17.53 |
| Roll 2nd or additional finish coats | | | | | | | | | |
| Slow | 375 | 350 | 22.80 | 2.73 | .75 | 6.51 | 3.09 | 1.96 | 15.04 |
| Medium | 400 | 338 | 19.90 | 4.13 | 1.38 | 5.89 | 2.80 | 1.56 | 15.76 |
| Fast | 425 | 325 | 17.10 | 5.18 | 2.09 | 5.26 | 2.32 | .89 | 15.74 |

| | Labor SF per manhour | Material coverage SF/gallon | Material cost per gallon | Labor cost per 100 SF | Labor burden 100 SF | Material cost per 100 SF | Overhead per 100 SF | Profit per 100 SF | Total price per 100 SF |
|---|---|---|---|---|---|---|---|---|---|
| Enamel, oil base (material #10) | | | | | | | | | |
| Roll 1st finish coat | | | | | | | | | |
| Slow | 325 | 300 | 25.90 | 3.15 | .87 | 8.63 | 3.92 | 2.48 | 19.05 |
| Medium | 350 | 275 | 22.70 | 4.71 | 1.60 | 8.25 | 3.56 | 1.99 | 20.11 |
| Fast | 375 | 250 | 19.40 | 5.87 | 2.39 | 7.76 | 2.96 | 1.14 | 20.12 |
| Roll 2nd or additional finish coats | | | | | | | | | |
| Slow | 375 | 300 | 25.90 | 2.73 | .75 | 8.63 | 3.75 | 2.38 | 18.24 |
| Medium | 400 | 288 | 22.70 | 4.13 | 1.38 | 7.88 | 3.28 | 1.83 | 18.50 |
| Fast | 425 | 275 | 19.40 | 5.18 | 2.09 | 7.05 | 2.65 | 1.02 | 17.99 |
| Epoxy coating, white (material #52) | | | | | | | | | |
| Roll 1st coat | | | | | | | | | |
| Slow | 300 | 300 | 57.40 | 3.42 | .92 | 19.13 | 7.28 | 4.61 | 35.36 |
| Medium | 325 | 288 | 50.20 | 5.08 | 1.71 | 17.43 | 5.93 | 3.32 | 33.47 |
| Fast | 350 | 275 | 43.10 | 6.29 | 2.55 | 15.67 | 4.53 | 1.74 | 30.78 |
| Roll 2nd or additional coats | | | | | | | | | |
| Slow | 350 | 300 | 57.40 | 2.93 | .80 | 19.13 | 7.09 | 4.49 | 34.44 |
| Medium | 375 | 288 | 50.20 | 4.40 | 1.49 | 17.43 | 5.71 | 3.19 | 32.22 |
| Fast | 400 | 275 | 43.10 | 5.50 | 2.23 | 15.67 | 4.33 | 1.66 | 29.39 |

Ceiling estimates are based on overall dimensions (length times width) to 8 feet high. For heights above 8 feet, use the High Time Difficulty Factors on page 137. "Slow" work is based on a $10.25 hourly wage, "Medium" work on a $16.50 hourly wage, and "Fast" work on a $22.00 hourly wage. Other qualifications that apply to this table are on page 9.

| | Labor SF per manhour | Material coverage SF/gallon | Material cost per gallon | Labor cost per 100 SF | Labor burden 100 SF | Material cost per 100 SF | Overhead per 100 SF | Profit per 100 SF | Total price per 100 SF |
|---|---|---|---|---|---|---|---|---|---|
| **Ceilings, gypsum drywall, orange peel or knock-down texture, spray** | | | | | | | | | |
| Flat latex, water base (material #5) | | | | | | | | | |
| Spray 1st coat | | | | | | | | | |
| Slow | 650 | 225 | 16.20 | 1.58 | .43 | 7.20 | 2.86 | 1.81 | 13.88 |
| Medium | 750 | 200 | 14.20 | 2.20 | .73 | 7.10 | 2.46 | 1.38 | 13.87 |
| Fast | 850 | 175 | 12.10 | 2.59 | 1.06 | 6.91 | 1.95 | .75 | 13.26 |
| Spray 2nd coat | | | | | | | | | |
| Slow | 775 | 250 | 16.20 | 1.32 | .36 | 6.48 | 2.53 | 1.60 | 12.29 |
| Medium | 875 | 225 | 14.20 | 1.89 | .62 | 6.31 | 2.17 | 1.21 | 12.20 |
| Fast | 975 | 200 | 12.10 | 2.26 | .92 | 6.05 | 1.71 | .66 | 11.60 |
| Spray 3rd or additional coats | | | | | | | | | |
| Slow | 825 | 275 | 16.20 | 1.24 | .34 | 5.89 | 2.32 | 1.47 | 11.26 |
| Medium | 925 | 250 | 14.20 | 1.78 | .60 | 5.68 | 1.97 | 1.10 | 11.13 |
| Fast | 1025 | 225 | 12.10 | 2.15 | .88 | 5.38 | 1.55 | .60 | 10.56 |
| Sealer, water base (material #1) | | | | | | | | | |
| Spray prime coat | | | | | | | | | |
| Slow | 700 | 225 | 15.50 | 1.46 | .40 | 6.89 | 2.71 | 1.72 | 13.18 |
| Medium | 800 | 200 | 13.50 | 2.06 | .70 | 6.75 | 2.33 | 1.30 | 13.14 |
| Fast | 900 | 175 | 11.60 | 2.44 | .99 | 6.63 | 1.86 | .72 | 12.64 |
| Sealer, oil base (material #2) | | | | | | | | | |
| Spray prime coat | | | | | | | | | |
| Slow | 700 | 200 | 20.70 | 1.46 | .40 | 10.35 | 3.79 | 2.40 | 18.40 |
| Medium | 800 | 188 | 18.10 | 2.06 | .70 | 9.63 | 3.03 | 1.70 | 17.12 |
| Fast | 900 | 175 | 15.50 | 2.44 | .99 | 8.86 | 2.27 | .87 | 15.43 |
| Enamel, water base (material #9) | | | | | | | | | |
| Spray 1st finish coat | | | | | | | | | |
| Slow | 725 | 250 | 22.80 | 1.41 | .39 | 9.12 | 3.38 | 2.14 | 16.44 |
| Medium | 825 | 225 | 19.90 | 2.00 | .67 | 8.84 | 2.82 | 1.58 | 15.91 |
| Fast | 925 | 200 | 17.10 | 2.38 | .96 | 8.55 | 2.20 | .85 | 14.94 |
| Spray 2nd or additional finish coat | | | | | | | | | |
| Slow | 775 | 275 | 22.80 | 1.32 | .36 | 8.29 | 3.09 | 1.96 | 15.02 |
| Medium | 875 | 250 | 19.90 | 1.89 | .62 | 7.96 | 2.57 | 1.44 | 14.48 |
| Fast | 975 | 225 | 17.10 | 2.26 | .92 | 7.60 | 1.99 | .77 | 13.54 |

| | Labor SF per manhour | Material coverage SF/gallon | Material cost per gallon | Labor cost per 100 SF | Labor burden 100 SF | Material cost per 100 SF | Overhead per 100 SF | Profit per 100 SF | Total price per 100 SF |
|---|---|---|---|---|---|---|---|---|---|
| Enamel, oil base (material #10) | | | | | | | | | |
| Spray 1st finish coat | | | | | | | | | |
| Slow | 725 | 225 | 25.90 | 1.41 | .39 | 11.51 | 4.12 | 2.61 | 20.04 |
| Medium | 825 | 213 | 22.70 | 2.00 | .67 | 10.66 | 3.27 | 1.83 | 18.43 |
| Fast | 925 | 200 | 19.40 | 2.38 | .96 | 9.70 | 2.41 | .93 | 16.38 |
| Spray 2nd or additional finish coat | | | | | | | | | |
| Slow | 775 | 250 | 25.90 | 1.32 | .36 | 10.36 | 3.73 | 2.37 | 18.14 |
| Medium | 875 | 238 | 22.70 | 1.89 | .62 | 9.54 | 2.96 | 1.65 | 16.66 |
| Fast | 975 | 225 | 19.40 | 2.26 | .92 | 8.62 | 2.18 | .84 | 14.82 |

Ceiling estimates are based on overall dimensions (length times width) to 8 feet high. For heights above 8 feet, use the High Time Difficulty Factors on page 137. "Slow" work is based on a $10.25 hourly wage, "Medium" work on a $16.50 hourly wage, and "Fast" work on a $22.00 hourly wage. Other qualifications that apply to this table are on page 9.

| | Labor SF per manhour | Material coverage SF/gallon | Material cost per gallon | Labor cost per 100 SF | Labor burden 100 SF | Material cost per 100 SF | Overhead per 100 SF | Profit per 100 SF | Total price per 100 SF |
|---|---|---|---|---|---|---|---|---|---|
| **Ceilings, gypsum drywall, sand finish or skip trowel texture, brush** | | | | | | | | | |
| Flat latex, water base (material #5) | | | | | | | | | |
| Brush 1st coat | | | | | | | | | |
| Slow | 175 | 325 | 16.20 | 5.86 | 1.59 | 4.98 | 3.86 | 2.45 | 18.74 |
| Medium | 200 | 313 | 14.20 | 8.25 | 2.78 | 4.54 | 3.81 | 2.13 | 21.51 |
| Fast | 225 | 300 | 12.10 | 9.78 | 3.95 | 4.03 | 3.29 | 1.26 | 22.31 |
| Brush 2nd coat | | | | | | | | | |
| Slow | 200 | 400 | 16.20 | 5.13 | 1.39 | 4.05 | 3.28 | 2.08 | 15.93 |
| Medium | 238 | 375 | 14.20 | 6.93 | 2.33 | 3.79 | 3.20 | 1.79 | 18.04 |
| Fast | 275 | 350 | 12.10 | 8.00 | 3.25 | 3.46 | 2.72 | 1.05 | 18.48 |
| Brush 3rd or additional coats | | | | | | | | | |
| Slow | 225 | 425 | 16.20 | 4.56 | 1.23 | 3.81 | 2.98 | 1.89 | 14.47 |
| Medium | 263 | 400 | 14.20 | 6.27 | 2.11 | 3.55 | 2.92 | 1.63 | 16.48 |
| Fast | 300 | 375 | 12.10 | 7.33 | 2.97 | 3.23 | 2.50 | .96 | 16.99 |
| Sealer, water base (material #1) | | | | | | | | | |
| Brush prime coat | | | | | | | | | |
| Slow | 200 | 325 | 15.50 | 5.13 | 1.39 | 4.77 | 3.50 | 2.22 | 17.01 |
| Medium | 225 | 313 | 13.50 | 7.33 | 2.46 | 4.31 | 3.46 | 1.93 | 19.49 |
| Fast | 250 | 300 | 11.60 | 8.80 | 3.57 | 3.87 | 3.00 | 1.15 | 20.39 |
| Sealer, oil base (material #2) | | | | | | | | | |
| Brush prime coat | | | | | | | | | |
| Slow | 200 | 325 | 20.70 | 5.13 | 1.39 | 6.37 | 4.00 | 2.54 | 19.43 |
| Medium | 225 | 313 | 18.10 | 7.33 | 2.46 | 5.78 | 3.82 | 2.13 | 21.52 |
| Fast | 250 | 300 | 15.50 | 8.80 | 3.57 | 5.17 | 3.24 | 1.25 | 22.03 |
| Enamel, water base (material #9) | | | | | | | | | |
| Brush 1st finish coat | | | | | | | | | |
| Slow | 200 | 400 | 22.80 | 5.13 | 1.39 | 5.70 | 3.79 | 2.40 | 18.41 |
| Medium | 225 | 375 | 19.90 | 7.33 | 2.46 | 5.31 | 3.70 | 2.07 | 20.87 |
| Fast | 250 | 350 | 17.10 | 8.80 | 3.57 | 4.89 | 3.19 | 1.23 | 21.68 |
| Brush 2nd or additional finish coats | | | | | | | | | |
| Slow | 225 | 425 | 22.80 | 4.56 | 1.23 | 5.36 | 3.46 | 2.19 | 16.80 |
| Medium | 263 | 400 | 19.90 | 6.27 | 2.11 | 4.98 | 3.27 | 1.83 | 18.46 |
| Fast | 300 | 375 | 17.10 | 7.33 | 2.97 | 4.56 | 2.75 | 1.06 | 18.67 |

| | Labor SF per manhour | Material coverage SF/gallon | Material cost per gallon | Labor cost per 100 SF | Labor burden 100 SF | Material cost per 100 SF | Overhead per 100 SF | Profit per 100 SF | Total price per 100 SF |
|---|---|---|---|---|---|---|---|---|---|
| Enamel, oil base (material #10) | | | | | | | | | |
| Brush 1st finish coat | | | | | | | | | |
| Slow | 200 | 375 | 25.90 | 5.13 | 1.39 | 6.91 | 4.17 | 2.64 | 20.24 |
| Medium | 225 | 350 | 22.70 | 7.33 | 2.46 | 6.49 | 3.99 | 2.23 | 22.50 |
| Fast | 250 | 325 | 19.40 | 8.80 | 3.57 | 5.97 | 3.39 | 1.30 | 23.03 |
| Brush 2nd or additional finish coats | | | | | | | | | |
| Slow | 225 | 400 | 25.90 | 4.56 | 1.23 | 6.48 | 3.81 | 2.41 | 18.49 |
| Medium | 263 | 375 | 22.70 | 6.27 | 2.11 | 6.05 | 3.54 | 1.98 | 19.95 |
| Fast | 300 | 350 | 19.40 | 7.33 | 2.97 | 5.54 | 2.93 | 1.13 | 19.90 |
| Epoxy coating, white (material #52) | | | | | | | | | |
| Brush 1st coat | | | | | | | | | |
| Slow | 150 | 375 | 57.40 | 6.83 | 1.87 | 15.31 | 7.44 | 4.72 | 36.17 |
| Medium | 175 | 350 | 50.20 | 9.43 | 3.16 | 14.34 | 6.60 | 3.69 | 37.22 |
| Fast | 225 | 325 | 43.10 | 9.78 | 3.95 | 13.26 | 5.00 | 1.92 | 33.91 |
| Brush 2nd or additional coats | | | | | | | | | |
| Slow | 175 | 400 | 57.40 | 5.86 | 1.59 | 14.35 | 6.76 | 4.29 | 32.85 |
| Medium | 200 | 375 | 50.20 | 8.25 | 2.78 | 13.39 | 5.98 | 3.34 | 33.74 |
| Fast | 225 | 350 | 43.10 | 9.78 | 3.95 | 12.31 | 4.82 | 1.85 | 32.71 |

Ceiling estimates are based on overall dimensions (length times width) to 8 feet high. For heights above 8 feet, use the High Time Difficulty Factors on page 137. "Slow" work is based on a $10.25 hourly wage, "Medium" work on a $16.50 hourly wage, and "Fast" work on a $22.00 hourly wage. Other qualifications that apply to this table are on page 9.

| | Labor SF per manhour | Material coverage SF/gallon | Material cost per gallon | Labor cost per 100 SF | Labor burden 100 SF | Material cost per 100 SF | Overhead per 100 SF | Profit per 100 SF | Total price per 100 SF |
|---|---|---|---|---|---|---|---|---|---|

## Ceilings, gypsum drywall, sand finish or skip trowel texture, roll

| | Labor SF per manhour | Material coverage SF/gallon | Material cost per gallon | Labor cost per 100 SF | Labor burden 100 SF | Material cost per 100 SF | Overhead per 100 SF | Profit per 100 SF | Total price per 100 SF |
|---|---|---|---|---|---|---|---|---|---|
| **Flat latex, water base (material #5)** | | | | | | | | | |
| Roll 1st coat | | | | | | | | | |
| Slow | 300 | 325 | 16.20 | 3.42 | .92 | 4.98 | 2.89 | 1.83 | 14.04 |
| Medium | 350 | 300 | 14.20 | 4.71 | 1.60 | 4.73 | 2.70 | 1.51 | 15.25 |
| Fast | 400 | 275 | 12.10 | 5.50 | 2.23 | 4.40 | 2.24 | .86 | 15.23 |
| Roll 2nd coat | | | | | | | | | |
| Slow | 350 | 350 | 16.20 | 2.93 | .80 | 4.63 | 2.59 | 1.64 | 12.59 |
| Medium | 388 | 338 | 14.20 | 4.25 | 1.44 | 4.20 | 2.42 | 1.35 | 13.66 |
| Fast | 425 | 325 | 12.10 | 5.18 | 2.09 | 3.72 | 2.04 | .78 | 13.81 |
| Roll 3rd or additional coats | | | | | | | | | |
| Slow | 425 | 350 | 16.20 | 2.41 | .65 | 4.63 | 2.39 | 1.51 | 11.59 |
| Medium | 450 | 338 | 14.20 | 3.67 | 1.23 | 4.20 | 2.23 | 1.25 | 12.58 |
| Fast | 475 | 325 | 12.10 | 4.63 | 1.89 | 3.72 | 1.89 | .73 | 12.86 |
| **Sealer, water base (material #1)** | | | | | | | | | |
| Roll prime coat | | | | | | | | | |
| Slow | 325 | 325 | 15.50 | 3.15 | .87 | 4.77 | 2.72 | 1.73 | 13.24 |
| Medium | 375 | 300 | 13.50 | 4.40 | 1.49 | 4.50 | 2.54 | 1.42 | 14.35 |
| Fast | 425 | 275 | 11.60 | 5.18 | 2.09 | 4.22 | 2.13 | .82 | 14.44 |
| **Sealer, oil base (material #2)** | | | | | | | | | |
| Roll prime coat | | | | | | | | | |
| Slow | 325 | 300 | 20.70 | 3.15 | .87 | 6.90 | 3.38 | 2.14 | 16.44 |
| Medium | 375 | 275 | 18.10 | 4.40 | 1.49 | 6.58 | 3.05 | 1.71 | 17.23 |
| Fast | 425 | 250 | 15.50 | 5.18 | 2.09 | 6.20 | 2.49 | .96 | 16.92 |
| **Enamel, water base (material #9)** | | | | | | | | | |
| Roll 1st finish coat | | | | | | | | | |
| Slow | 325 | 350 | 22.80 | 3.15 | .87 | 6.51 | 3.26 | 2.07 | 15.86 |
| Medium | 363 | 338 | 19.90 | 4.55 | 1.51 | 5.89 | 2.93 | 1.64 | 16.52 |
| Fast | 400 | 325 | 17.10 | 5.50 | 2.23 | 5.26 | 2.40 | .92 | 16.31 |
| Roll 2nd or additional finish coats | | | | | | | | | |
| Slow | 400 | 350 | 22.80 | 2.56 | .70 | 6.51 | 3.03 | 1.92 | 14.72 |
| Medium | 425 | 338 | 19.90 | 3.88 | 1.30 | 5.89 | 2.71 | 1.52 | 15.30 |
| Fast | 450 | 325 | 17.10 | 4.89 | 1.97 | 5.26 | 2.24 | .86 | 15.22 |
| **Enamel, oil base (material #10)** | | | | | | | | | |
| Roll 1st finish coat | | | | | | | | | |
| Slow | 325 | 325 | 25.90 | 3.15 | .87 | 7.97 | 3.71 | 2.35 | 18.05 |
| Medium | 363 | 313 | 22.70 | 4.55 | 1.51 | 7.25 | 3.27 | 1.83 | 18.41 |
| Fast | 400 | 300 | 19.40 | 5.50 | 2.23 | 6.47 | 2.63 | 1.01 | 17.84 |

| | Labor SF per manhour | Material coverage SF/gallon | Material cost per gallon | Labor cost per 100 SF | Labor burden 100 SF | Material cost per 100 SF | Overhead per 100 SF | Profit per 100 SF | Total price per 100 SF |
|---|---|---|---|---|---|---|---|---|---|
| Roll 2nd or additional finish coats | | | | | | | | | |
| Slow | 400 | 350 | 25.90 | 2.56 | .70 | 7.40 | 3.30 | 2.09 | 16.05 |
| Medium | 425 | 338 | 22.70 | 3.88 | 1.30 | 6.72 | 2.92 | 1.63 | 16.45 |
| Fast | 450 | 325 | 19.40 | 4.89 | 1.97 | 5.97 | 2.38 | .91 | 16.12 |
| Epoxy coating, white (material #52) | | | | | | | | | |
| Roll 1st coat | | | | | | | | | |
| Slow | 300 | 350 | 57.40 | 3.42 | .92 | 16.40 | 6.43 | 4.08 | 31.25 |
| Medium | 350 | 325 | 50.20 | 4.71 | 1.60 | 15.45 | 5.33 | 2.98 | 30.07 |
| Fast | 375 | 300 | 43.10 | 5.87 | 2.39 | 14.37 | 4.18 | 1.61 | 28.42 |
| Roll 2nd or additional coats | | | | | | | | | |
| Slow | 375 | 375 | 57.40 | 2.73 | .75 | 15.31 | 5.82 | 3.69 | 28.30 |
| Medium | 400 | 350 | 50.20 | 4.13 | 1.38 | 14.34 | 4.87 | 2.72 | 27.44 |
| Fast | 425 | 325 | 43.10 | 5.18 | 2.09 | 13.26 | 3.80 | 1.46 | 25.79 |

Ceiling estimates are based on overall dimensions (length times width) to 8 feet high. For heights above 8 feet, use the High Time Difficulty Factors on page 137. "Slow" work is based on a $10.25 hourly wage, "Medium" work on a $16.50 hourly wage, and "Fast" work on a $22.00 hourly wage. Other qualifications that apply to this table are on page 9.

| | Labor SF per manhour | Material coverage SF/gallon | Material cost per gallon | Labor cost per 100 SF | Labor burden 100 SF | Material cost per 100 SF | Overhead per 100 SF | Profit per 100 SF | Total price per 100 SF |
|---|---|---|---|---|---|---|---|---|---|

## Ceilings, gypsum drywall, sand finish or skip trowel texture, spray

| | | | | | | | | | |
|---|---|---|---|---|---|---|---|---|---|
| Flat latex, water base (material #5) | | | | | | | | | |
| Spray 1st coat | | | | | | | | | |
| Slow | 700 | 275 | 16.20 | 1.46 | .40 | 5.89 | 2.40 | 1.52 | 11.67 |
| Medium | 800 | 250 | 14.20 | 2.06 | .70 | 5.68 | 2.07 | 1.16 | 11.67 |
| Fast | 900 | 225 | 12.10 | 2.44 | .99 | 5.38 | 1.63 | .63 | 11.07 |
| Spray 2nd coat | | | | | | | | | |
| Slow | 800 | 325 | 16.20 | 1.28 | .35 | 4.98 | 2.05 | 1.30 | 9.96 |
| Medium | 900 | 300 | 14.20 | 1.83 | .62 | 4.73 | 1.76 | .98 | 9.92 |
| Fast | 1000 | 275 | 12.10 | 2.20 | .89 | 4.40 | 1.39 | .53 | 9.41 |
| Spray 3rd or additional coats | | | | | | | | | |
| Slow | 850 | 325 | 16.20 | 1.21 | .33 | 4.98 | 2.02 | 1.28 | 9.82 |
| Medium | 950 | 313 | 14.20 | 1.74 | .58 | 4.54 | 1.68 | .94 | 9.48 |
| Fast | 1050 | 300 | 12.10 | 2.10 | .84 | 4.03 | 1.29 | .50 | 8.76 |
| Sealer, water base (material #1) | | | | | | | | | |
| Spray prime coat | | | | | | | | | |
| Slow | 750 | 275 | 15.50 | 1.37 | .36 | 5.64 | 2.29 | 1.45 | 11.11 |
| Medium | 850 | 250 | 13.50 | 1.94 | .66 | 5.40 | 1.96 | 1.09 | 11.05 |
| Fast | 950 | 225 | 11.60 | 2.32 | .93 | 5.16 | 1.56 | .60 | 10.57 |
| Sealer, oil base (material #2) | | | | | | | | | |
| Spray prime coat | | | | | | | | | |
| Slow | 750 | 225 | 20.70 | 1.37 | .36 | 9.20 | 3.39 | 2.15 | 16.47 |
| Medium | 850 | 213 | 18.10 | 1.94 | .66 | 8.50 | 2.72 | 1.52 | 15.34 |
| Fast | 950 | 200 | 15.50 | 2.32 | .93 | 7.75 | 2.04 | .78 | 13.82 |
| Enamel, water base (material #9) | | | | | | | | | |
| Spray 1st finish coat | | | | | | | | | |
| Slow | 750 | 325 | 22.80 | 1.37 | .36 | 7.02 | 2.72 | 1.72 | 13.19 |
| Medium | 850 | 300 | 19.90 | 1.94 | .66 | 6.63 | 2.26 | 1.26 | 12.75 |
| Fast | 950 | 275 | 17.10 | 2.32 | .93 | 6.22 | 1.75 | .67 | 11.89 |
| Spray 2nd or additional finish coat | | | | | | | | | |
| Slow | 800 | 325 | 22.80 | 1.28 | .35 | 7.02 | 2.68 | 1.70 | 13.03 |
| Medium | 900 | 313 | 19.90 | 1.83 | .62 | 6.36 | 2.16 | 1.21 | 12.18 |
| Fast | 1000 | 300 | 17.10 | 2.20 | .89 | 5.70 | 1.63 | .63 | 11.05 |

| | Labor SF per manhour | Material coverage SF/gallon | Material cost per gallon | Labor cost per 100 SF | Labor burden 100 SF | Material cost per 100 SF | Overhead per 100 SF | Profit per 100 SF | Total price per 100 SF |
|---|---|---|---|---|---|---|---|---|---|
| Enamel, oil base (material #10) | | | | | | | | | |
| Spray 1st finish coat | | | | | | | | | |
| Slow | 750 | 300 | 25.90 | 1.37 | .36 | 8.63 | 3.21 | 2.04 | 15.61 |
| Medium | 850 | 288 | 22.70 | 1.94 | .66 | 7.88 | 2.57 | 1.43 | 14.48 |
| Fast | 950 | 275 | 19.40 | 2.32 | .93 | 7.05 | 1.91 | .73 | 12.94 |
| Spray 2nd or additional finish coat | | | | | | | | | |
| Slow | 800 | 325 | 25.90 | 1.28 | .35 | 7.97 | 2.98 | 1.89 | 14.47 |
| Medium | 900 | 313 | 22.70 | 1.83 | .62 | 7.25 | 2.38 | 1.33 | 13.41 |
| Fast | 1000 | 300 | 19.40 | 2.20 | .89 | 6.47 | 1.77 | .68 | 12.01 |

Ceiling estimates are based on overall dimensions (length times width) to 8 feet high. For heights above 8 feet, use the High Time Difficulty Factors on page 137. "Slow" work is based on a $10.25 hourly wage, "Medium" work on a $16.50 hourly wage, and "Fast" work on a $22.00 hourly wage. Other qualifications that apply to this table are on page 9.

| | Labor SF per manhour | Material coverage SF/gallon | Material cost per gallon | Labor cost per 100 SF | Labor burden 100 SF | Material cost per 100 SF | Overhead per 100 SF | Profit per 100 SF | Total price per 100 SF |
|---|---|---|---|---|---|---|---|---|---|
| **Ceilings, tongue & groove, paint grade, brush** | | | | | | | | | |
| Flat latex, water base (material #5) | | | | | | | | | |
| Brush 1st coat | | | | | | | | | |
| Slow | 55 | 300 | 16.20 | 18.64 | 5.07 | 5.40 | 9.03 | 5.72 | 43.86 |
| Medium | 65 | 288 | 14.20 | 25.38 | 8.53 | 4.93 | 9.52 | 5.32 | 53.68 |
| Fast | 75 | 275 | 12.10 | 29.33 | 11.89 | 4.40 | 8.44 | 3.24 | 57.30 |
| Brush 2nd coat | | | | | | | | | |
| Slow | 65 | 350 | 16.20 | 15.77 | 4.29 | 4.63 | 7.66 | 4.85 | 37.20 |
| Medium | 75 | 338 | 14.20 | 22.00 | 7.39 | 4.20 | 8.23 | 4.60 | 46.42 |
| Fast | 85 | 325 | 12.10 | 25.88 | 10.48 | 3.72 | 7.42 | 2.85 | 50.35 |
| Brush 3rd or additional coats | | | | | | | | | |
| Slow | 80 | 375 | 16.20 | 12.81 | 3.49 | 4.32 | 6.39 | 4.05 | 31.06 |
| Medium | 90 | 363 | 14.20 | 18.33 | 6.17 | 3.91 | 6.96 | 3.89 | 39.26 |
| Fast | 100 | 350 | 12.10 | 22.00 | 8.92 | 3.46 | 6.36 | 2.44 | 43.18 |
| Sealer, water base (material #1) | | | | | | | | | |
| Brush prime coat | | | | | | | | | |
| Slow | 60 | 300 | 15.50 | 17.08 | 4.66 | 5.17 | 8.34 | 5.29 | 40.54 |
| Medium | 70 | 288 | 13.50 | 23.57 | 7.94 | 4.69 | 8.87 | 4.96 | 50.03 |
| Fast | 80 | 275 | 11.60 | 27.50 | 11.15 | 4.22 | 7.93 | 3.05 | 53.85 |
| Sealer, oil base (material #2) | | | | | | | | | |
| Brush prime coat | | | | | | | | | |
| Slow | 60 | 375 | 20.70 | 17.08 | 4.66 | 5.52 | 8.45 | 5.36 | 41.07 |
| Medium | 70 | 363 | 18.10 | 23.57 | 7.94 | 4.99 | 8.94 | 5.00 | 50.44 |
| Fast | 80 | 350 | 15.50 | 27.50 | 11.15 | 4.43 | 7.97 | 3.06 | 54.11 |
| Enamel, water base (material #9) | | | | | | | | | |
| Brush 1st finish coat | | | | | | | | | |
| Slow | 55 | 300 | 22.80 | 18.64 | 5.07 | 7.60 | 9.71 | 6.15 | 47.17 |
| Medium | 65 | 288 | 19.90 | 25.38 | 8.53 | 6.91 | 10.00 | 5.59 | 56.41 |
| Fast | 75 | 275 | 17.10 | 29.33 | 11.89 | 6.22 | 8.78 | 3.37 | 59.59 |
| Brush 2nd or additional finish coats | | | | | | | | | |
| Slow | 70 | 375 | 22.80 | 14.64 | 3.99 | 6.08 | 7.66 | 4.86 | 37.23 |
| Medium | 80 | 363 | 19.90 | 20.63 | 6.93 | 5.48 | 8.10 | 4.53 | 45.67 |
| Fast | 90 | 350 | 17.10 | 24.44 | 9.91 | 4.89 | 7.26 | 2.79 | 49.29 |

| | Labor SF per manhour | Material coverage SF/gallon | Material cost per gallon | Labor cost per 100 SF | Labor burden 100 SF | Material cost per 100 SF | Overhead per 100 SF | Profit per 100 SF | Total price per 100 SF |
|---|---|---|---|---|---|---|---|---|---|
| Enamel, oil base (material #10) | | | | | | | | | |
| Brush 1st finish coat | | | | | | | | | |
| Slow | 55 | 375 | 25.90 | 18.64 | 5.07 | 6.91 | 9.50 | 6.02 | 46.14 |
| Medium | 65 | 363 | 22.70 | 25.38 | 8.53 | 6.25 | 9.84 | 5.50 | 55.50 |
| Fast | 75 | 350 | 19.40 | 29.33 | 11.89 | 5.54 | 8.65 | 3.32 | 58.73 |
| Brush 2nd or additional finish coats | | | | | | | | | |
| Slow | 70 | 425 | 25.90 | 14.64 | 3.99 | 6.09 | 7.66 | 4.86 | 37.24 |
| Medium | 80 | 413 | 22.70 | 20.63 | 6.93 | 5.50 | 8.10 | 4.53 | 45.69 |
| Fast | 90 | 400 | 19.40 | 24.44 | 9.91 | 4.85 | 7.25 | 2.79 | 49.24 |

Ceiling estimates are based on overall dimensions (length times width) to 8 feet high. For heights above 8 feet, use the High Time Difficulty Factors on page 137. "Slow" work is based on a $10.25 hourly wage, "Medium" work on a $16.50 hourly wage, and "Fast" work on a $22.00 hourly wage. Other qualifications that apply to this table are on page 9.

| | Labor SF per manhour | Material coverage SF/gallon | Material cost per gallon | Labor cost per 100 SF | Labor burden 100 SF | Material cost per 100 SF | Overhead per 100 SF | Profit per 100 SF | Total price per 100 SF |
|---|---|---|---|---|---|---|---|---|---|
| **Ceilings, tongue & groove, paint grade, roll** | | | | | | | | | |
| Flat latex, water base (material #5) | | | | | | | | | |
| Roll 1st coat | | | | | | | | | |
| Slow | 90 | 275 | 16.20 | 11.39 | 3.10 | 5.89 | 6.32 | 4.01 | 30.71 |
| Medium | 110 | 263 | 14.20 | 15.00 | 5.04 | 5.40 | 6.24 | 3.49 | 35.17 |
| Fast | 130 | 250 | 12.10 | 16.92 | 6.86 | 4.84 | 5.29 | 2.03 | 35.94 |
| Roll 2nd coat | | | | | | | | | |
| Slow | 140 | 325 | 16.20 | 7.32 | 1.99 | 4.98 | 4.43 | 2.81 | 21.53 |
| Medium | 155 | 313 | 14.20 | 10.65 | 3.57 | 4.54 | 4.60 | 2.57 | 25.93 |
| Fast | 170 | 300 | 12.10 | 12.94 | 5.24 | 4.03 | 4.11 | 1.58 | 27.90 |
| Roll 3rd or additional coats | | | | | | | | | |
| Slow | 190 | 350 | 16.20 | 5.39 | 1.47 | 4.63 | 3.56 | 2.26 | 17.31 |
| Medium | 200 | 338 | 14.20 | 8.25 | 2.78 | 4.20 | 3.73 | 2.09 | 21.05 |
| Fast | 210 | 325 | 12.10 | 10.48 | 4.24 | 3.72 | 3.41 | 1.31 | 23.16 |
| Sealer, water base (material #1) | | | | | | | | | |
| Roll prime coat | | | | | | | | | |
| Slow | 100 | 275 | 15.50 | 10.25 | 2.79 | 5.64 | 5.79 | 3.67 | 28.14 |
| Medium | 120 | 263 | 13.50 | 13.75 | 4.62 | 5.13 | 5.76 | 3.22 | 32.48 |
| Fast | 150 | 250 | 11.60 | 14.67 | 5.95 | 4.64 | 4.67 | 1.80 | 31.73 |
| Sealer, oil base (material #2) | | | | | | | | | |
| Roll prime coat | | | | | | | | | |
| Slow | 100 | 350 | 20.70 | 10.25 | 2.79 | 5.91 | 5.87 | 3.72 | 28.54 |
| Medium | 120 | 325 | 18.10 | 13.75 | 4.62 | 5.57 | 5.87 | 3.28 | 33.09 |
| Fast | 150 | 300 | 15.50 | 14.67 | 5.95 | 5.17 | 4.77 | 1.83 | 32.39 |
| Enamel, water base (material #9) | | | | | | | | | |
| Roll 1st finish coat | | | | | | | | | |
| Slow | 130 | 325 | 22.80 | 7.88 | 2.15 | 7.02 | 5.29 | 3.35 | 25.69 |
| Medium | 145 | 313 | 19.90 | 11.38 | 3.83 | 6.36 | 5.28 | 2.95 | 29.80 |
| Fast | 160 | 300 | 17.10 | 13.75 | 5.58 | 5.70 | 4.63 | 1.78 | 31.44 |
| Roll 2nd or additional finish coats | | | | | | | | | |
| Slow | 180 | 350 | 22.80 | 5.69 | 1.56 | 6.51 | 4.26 | 2.70 | 20.72 |
| Medium | 190 | 338 | 19.90 | 8.68 | 2.92 | 5.89 | 4.29 | 2.40 | 24.18 |
| Fast | 200 | 325 | 17.10 | 11.00 | 4.46 | 5.26 | 3.83 | 1.47 | 26.02 |

| | Labor SF per manhour | Material coverage SF/gallon | Material cost per gallon | Labor cost per 100 SF | Labor burden 100 SF | Material cost per 100 SF | Overhead per 100 SF | Profit per 100 SF | Total price per 100 SF |
|---|---|---|---|---|---|---|---|---|---|
| Enamel, oil base (material #10) | | | | | | | | | |
| Roll 1st finish coat | | | | | | | | | |
| Slow | 130 | 375 | 25.90 | 7.88 | 2.15 | 6.91 | 5.25 | 3.33 | 25.52 |
| Medium | 145 | 363 | 22.70 | 11.38 | 3.83 | 6.25 | 5.26 | 2.94 | 29.66 |
| Fast | 160 | 350 | 19.40 | 13.75 | 5.58 | 5.54 | 4.60 | 1.77 | 31.24 |
| Roll 2nd or additional finish coats | | | | | | | | | |
| Slow | 180 | 400 | 25.90 | 5.69 | 1.56 | 6.48 | 4.25 | 2.70 | 20.68 |
| Medium | 190 | 388 | 22.70 | 8.68 | 2.92 | 5.85 | 4.28 | 2.39 | 24.12 |
| Fast | 200 | 375 | 19.40 | 11.00 | 4.46 | 5.17 | 3.82 | 1.47 | 25.92 |

Ceiling estimates are based on overall dimensions (length times width) to 8 feet high. For heights above 8 feet, use the High Time Difficulty Factors on page 137. "Slow" work is based on a $10.25 hourly wage, "Medium" work on a $16.50 hourly wage, and "Fast" work on a $22.00 hourly wage. Other qualifications that apply to this table are on page 9.

| | Labor SF per manhour | Material coverage SF/gallon | Material cost per gallon | Labor cost per 100 SF | Labor burden 100 SF | Material cost per 100 SF | Overhead per 100 SF | Profit per 100 SF | Total price per 100 SF |
|---|---|---|---|---|---|---|---|---|---|

## Ceilings, tongue & groove, paint grade, spray

| | Labor SF per manhour | Material coverage SF/gallon | Material cost per gallon | Labor cost per 100 SF | Labor burden 100 SF | Material cost per 100 SF | Overhead per 100 SF | Profit per 100 SF | Total price per 100 SF |
|---|---|---|---|---|---|---|---|---|---|
| Flat latex, water base (material #5) | | | | | | | | | |
| Spray 1st coat | | | | | | | | | |
| Slow | 300 | 180 | 16.20 | 3.42 | .92 | 9.00 | 4.14 | 2.62 | 20.10 |
| Medium | 360 | 155 | 14.20 | 4.58 | 1.55 | 9.16 | 3.74 | 2.09 | 21.12 |
| Fast | 420 | 125 | 12.10 | 5.24 | 2.12 | 9.68 | 3.15 | 1.21 | 21.40 |
| Spray 2nd coat | | | | | | | | | |
| Slow | 420 | 250 | 16.20 | 2.44 | .66 | 6.48 | 2.97 | 1.88 | 14.43 |
| Medium | 470 | 225 | 14.20 | 3.51 | 1.19 | 6.31 | 2.70 | 1.51 | 15.22 |
| Fast | 520 | 200 | 12.10 | 4.23 | 1.71 | 6.05 | 2.22 | .85 | 15.06 |
| Spray 3rd or additional coats | | | | | | | | | |
| Slow | 520 | 350 | 16.20 | 1.97 | .53 | 4.63 | 2.21 | 1.40 | 10.74 |
| Medium | 570 | 338 | 14.20 | 2.89 | .97 | 4.20 | 1.97 | 1.10 | 11.13 |
| Fast | 620 | 325 | 12.10 | 3.55 | 1.43 | 3.72 | 1.61 | .62 | 10.93 |
| Sealer, water base (material #1) | | | | | | | | | |
| Spray prime coat | | | | | | | | | |
| Slow | 320 | 180 | 15.50 | 3.20 | .88 | 8.61 | 3.93 | 2.49 | 19.11 |
| Medium | 380 | 155 | 13.50 | 4.34 | 1.46 | 8.71 | 3.55 | 1.99 | 20.05 |
| Fast | 440 | 125 | 11.60 | 5.00 | 2.02 | 9.28 | 3.02 | 1.16 | 20.48 |
| Sealer, oil base (material #2) | | | | | | | | | |
| Spray prime coat | | | | | | | | | |
| Slow | 320 | 200 | 20.70 | 3.20 | .88 | 10.35 | 4.47 | 2.83 | 21.73 |
| Medium | 380 | 190 | 18.10 | 4.34 | 1.46 | 9.53 | 3.76 | 2.10 | 21.19 |
| Fast | 440 | 180 | 15.50 | 5.00 | 2.02 | 8.61 | 2.89 | 1.11 | 19.63 |
| Enamel, water base (material #9) | | | | | | | | | |
| Spray 1st finish coat | | | | | | | | | |
| Slow | 400 | 250 | 22.80 | 2.56 | .70 | 9.12 | 3.84 | 2.43 | 18.65 |
| Medium | 450 | 225 | 19.90 | 3.67 | 1.23 | 8.84 | 3.37 | 1.88 | 18.99 |
| Fast | 500 | 200 | 17.10 | 4.40 | 1.78 | 8.55 | 2.73 | 1.05 | 18.51 |
| Spray 2nd or additional finish coat | | | | | | | | | |
| Slow | 500 | 350 | 22.80 | 2.05 | .56 | 6.51 | 2.83 | 1.79 | 13.74 |
| Medium | 550 | 325 | 19.90 | 3.00 | 1.01 | 6.12 | 2.48 | 1.39 | 14.00 |
| Fast | 600 | 300 | 17.10 | 3.67 | 1.49 | 5.70 | 2.01 | .77 | 13.64 |

| | Labor SF per manhour | Material coverage SF/gallon | Material cost per gallon | Labor cost per 100 SF | Labor burden 100 SF | Material cost per 100 SF | Overhead per 100 SF | Profit per 100 SF | Total price per 100 SF |
|---|---|---|---|---|---|---|---|---|---|
| Enamel, oil base (material #10) | | | | | | | | | |
| Spray 1st finish coat | | | | | | | | | |
| Slow | 400 | 270 | 25.90 | 2.56 | .70 | 9.59 | 3.98 | 2.52 | 19.35 |
| Medium | 450 | 250 | 22.70 | 3.67 | 1.23 | 9.08 | 3.43 | 1.92 | 19.33 |
| Fast | 500 | 230 | 19.40 | 4.40 | 1.78 | 8.43 | 2.70 | 1.04 | 18.35 |
| Spray 2nd or additional finish coat | | | | | | | | | |
| Slow | 500 | 325 | 25.90 | 2.05 | .56 | 7.97 | 3.28 | 2.08 | 15.94 |
| Medium | 550 | 313 | 22.70 | 3.00 | 1.01 | 7.25 | 2.76 | 1.54 | 15.56 |
| Fast | 600 | 300 | 19.40 | 3.67 | 1.49 | 6.47 | 2.15 | .83 | 14.61 |

Ceiling estimates are based on overall dimensions (length times width) to 8 feet high. For heights above 8 feet, use the High Time Difficulty Factors on page 137. "Slow" work is based on a $10.25 hourly wage, "Medium" work on a $16.50 hourly wage, and "Fast" work on a $22.00 hourly wage. Other qualifications that apply to this table are on page 9.

| | Labor SF per manhour | Material coverage SF/gallon | Material cost per gallon | Labor cost per 100 SF | Labor burden 100 SF | Material cost per 100 SF | Overhead per 100 SF | Profit per 100 SF | Total price per 100 SF |
|---|---|---|---|---|---|---|---|---|---|
| **Ceilings, tongue & groove, stain grade** | | | | | | | | | |
| Semi-transparent stain, water base (material #20) | | | | | | | | | |
| Roll & brush each coat | | | | | | | | | |
| Slow | 200 | 300 | 19.60 | 5.13 | 1.39 | 6.53 | 4.05 | 2.57 | 19.67 |
| Medium | 240 | 275 | 17.10 | 6.88 | 2.31 | 6.22 | 3.78 | 2.11 | 21.30 |
| Fast | 280 | 250 | 14.70 | 7.86 | 3.18 | 5.88 | 3.13 | 1.20 | 21.25 |
| Semi-transparent stain, oil base (material #21) | | | | | | | | | |
| Roll & brush each coat | | | | | | | | | |
| Slow | 200 | 280 | 23.40 | 5.13 | 1.39 | 8.36 | 4.62 | 2.93 | 22.43 |
| Medium | 240 | 260 | 20.50 | 6.88 | 2.31 | 7.88 | 4.18 | 2.34 | 23.59 |
| Fast | 280 | 240 | 17.60 | 7.86 | 3.18 | 7.33 | 3.40 | 1.31 | 23.08 |
| Semi-transparent stain, water base (material #20) | | | | | | | | | |
| Spray each coat | | | | | | | | | |
| Slow | 300 | 220 | 19.60 | 3.42 | .92 | 8.91 | 4.11 | 2.61 | 19.97 |
| Medium | 350 | 200 | 17.10 | 4.71 | 1.60 | 8.55 | 3.64 | 2.03 | 20.53 |
| Fast | 400 | 180 | 14.70 | 5.50 | 2.23 | 8.17 | 2.94 | 1.13 | 19.97 |
| Semi-transparent stain, oil base (material #21) | | | | | | | | | |
| Spray each coat | | | | | | | | | |
| Slow | 300 | 200 | 23.40 | 3.42 | .92 | 11.70 | 4.98 | 3.15 | 24.17 |
| Medium | 350 | 188 | 20.50 | 4.71 | 1.60 | 10.90 | 4.21 | 2.35 | 23.77 |
| Fast | 400 | 175 | 17.60 | 5.50 | 2.23 | 10.06 | 3.29 | 1.26 | 22.34 |
| Stain, seal and 2 coat lacquer system (7 step process) | | | | | | | | | |
| STEP 1: Sand & putty | | | | | | | | | |
| Slow | 100 | -- | -- | 10.25 | 2.79 | -- | 4.04 | 2.56 | 19.64 |
| Medium | 125 | -- | -- | 13.20 | 4.44 | -- | 4.32 | 2.42 | 24.38 |
| Fast | 150 | -- | -- | 14.67 | 5.95 | -- | 3.81 | 1.47 | 25.90 |
| STEP 2 & 3: Wiping stain, oil base (material #11a) & wipe | | | | | | | | | |
| Roll & brush, 1 coat & wipe | | | | | | | | | |
| Slow | 75 | 300 | 27.40 | 13.67 | 3.71 | 9.13 | 8.22 | 5.21 | 39.94 |
| Medium | 100 | 275 | 24.00 | 16.50 | 5.55 | 8.73 | 7.54 | 4.22 | 42.54 |
| Fast | 125 | 250 | 20.60 | 17.60 | 7.14 | 8.24 | 6.10 | 2.34 | 41.42 |
| Spray, 1 coat & wipe | | | | | | | | | |
| Slow | 275 | 150 | 27.40 | 3.73 | 1.02 | 18.27 | 7.14 | 4.52 | 34.68 |
| Medium | 200 | 125 | 24.00 | 8.25 | 2.78 | 19.20 | 7.41 | 4.14 | 41.78 |
| Fast | 325 | 100 | 20.60 | 6.77 | 2.75 | 20.60 | 5.57 | 2.14 | 37.83 |

| | Labor SF per manhour | Material coverage SF/gallon | Material cost per gallon | Labor cost per 100 SF | Labor burden 100 SF | Material cost per 100 SF | Overhead per 100 SF | Profit per 100 SF | Total price per 100 SF |
|---|---|---|---|---|---|---|---|---|---|
| **STEP 4: Sanding sealer (material #11b)** | | | | | | | | | |
| Brush, 1 coat | | | | | | | | | |
| Slow | 125 | 325 | 17.30 | 8.20 | 2.23 | 5.32 | 4.88 | 3.09 | 23.72 |
| Medium | 150 | 300 | 15.10 | 11.00 | 3.71 | 5.03 | 4.83 | 2.70 | 27.27 |
| Fast | 175 | 275 | 13.00 | 12.57 | 5.09 | 4.73 | 4.14 | 1.59 | 28.12 |
| Spray, 1 coat | | | | | | | | | |
| Slow | 350 | 150 | 17.30 | 2.93 | .80 | 11.53 | 4.73 | 3.00 | 22.99 |
| Medium | 400 | 125 | 15.10 | 4.13 | 1.38 | 12.08 | 4.31 | 2.41 | 24.31 |
| Fast | 450 | 100 | 13.00 | 4.89 | 1.97 | 13.00 | 3.68 | 1.41 | 24.95 |
| **STEP 5: Sand lightly** | | | | | | | | | |
| Slow | 175 | -- | -- | 5.86 | 1.59 | -- | 2.31 | 1.47 | 11.23 |
| Medium | 225 | -- | -- | 7.33 | 2.46 | -- | 2.40 | 1.34 | 13.53 |
| Fast | 275 | -- | -- | 8.00 | 3.25 | -- | 2.08 | .80 | 14.13 |
| **STEP 6 & 7: Lacquer, 2 coats (material #11c)** | | | | | | | | | |
| Brush, 1st coat | | | | | | | | | |
| Slow | 150 | 350 | 18.00 | 6.83 | 1.87 | 5.14 | 4.29 | 2.72 | 20.85 |
| Medium | 200 | 338 | 15.70 | 8.25 | 2.78 | 4.64 | 3.84 | 2.15 | 21.66 |
| Fast | 275 | 325 | 13.50 | 8.00 | 3.25 | 4.15 | 2.85 | 1.09 | 19.34 |
| Brush, 2nd coat | | | | | | | | | |
| Slow | 200 | 400 | 18.00 | 5.13 | 1.39 | 4.50 | 3.42 | 2.17 | 16.61 |
| Medium | 250 | 375 | 15.70 | 6.60 | 2.22 | 4.19 | 3.19 | 1.78 | 17.98 |
| Fast | 325 | 350 | 13.50 | 6.77 | 2.75 | 3.86 | 2.48 | .95 | 16.81 |
| Spray, 1st coat | | | | | | | | | |
| Slow | 425 | 350 | 18.00 | 2.41 | .65 | 5.14 | 2.55 | 1.61 | 12.36 |
| Medium | 525 | 300 | 15.70 | 3.14 | 1.05 | 5.23 | 2.31 | 1.29 | 13.02 |
| Fast | 625 | 250 | 13.50 | 3.52 | 1.43 | 5.40 | 1.91 | .74 | 13.00 |
| Spray, 2nd coat | | | | | | | | | |
| Slow | 475 | 425 | 18.00 | 2.16 | .59 | 4.24 | 2.17 | 1.37 | 10.53 |
| Medium | 588 | 375 | 15.70 | 2.81 | .94 | 4.19 | 1.95 | 1.09 | 10.98 |
| Fast | 650 | 325 | 13.50 | 3.38 | 1.38 | 4.15 | 1.65 | .63 | 11.19 |

| | Labor SF per manhour | Material coverage SF/gallon | Material cost per gallon | Labor cost per 100 SF | Labor burden 100 SF | Material cost per 100 SF | Overhead per 100 SF | Profit per 100 SF | Total price per 100 SF |
|---|---|---|---|---|---|---|---|---|---|
| Complete 7 step stain, seal & 2 coat lacquer system (material #11) | | | | | | | | | |
| Brush all coats | | | | | | | | | |
| Slow | 30 | 165 | 20.20 | 34.17 | 9.29 | 12.24 | 17.27 | 10.95 | 83.92 |
| Medium | 35 | 155 | 18.30 | 47.14 | 15.86 | 11.81 | 18.33 | 10.25 | 103.39 |
| Fast | 40 | 145 | 15.10 | 55.00 | 22.30 | 10.41 | 16.23 | 6.24 | 110.18 |
| Spray all coats | | | | | | | | | |
| Slow | 60 | 60 | 20.20 | 17.08 | 4.66 | 33.67 | 17.17 | 10.89 | 83.47 |
| Medium | 70 | 50 | 18.30 | 23.57 | 7.94 | 36.60 | 16.68 | 9.33 | 94.12 |
| Fast | 80 | 40 | 15.10 | 27.50 | 11.15 | 37.75 | 14.13 | 5.43 | 95.96 |

Ceiling estimates are based on overall dimensions (length times width) to 8 feet high. For heights above 8 feet, use the High Time Difficulty Factors on page 137. "Slow" work is based on a $10.25 hourly wage, "Medium" work on a $16.50 hourly wage, and "Fast" work on a $22.00 hourly wage. Other qualifications that apply to this table are on page 9.

## Closet pole, stain grade

| | Labor SF per manhour | Material coverage SF/gallon | Material cost per gallon | Labor cost per 100 SF | Labor burden 100 SF | Material cost per 100 SF | Overhead per 100 SF | Profit per 100 SF | Total price per 100 SF |
|---|---|---|---|---|---|---|---|---|---|
| Penetrating oil stain, (material #13) | | | | | | | | | |
| Brush & wipe, 1 coat | | | | | | | | | |
| Slow | 40 | 225 | 25.00 | 25.63 | 6.97 | 11.11 | 13.55 | 8.59 | 65.85 |
| Medium | 50 | 213 | 21.80 | 33.00 | 11.10 | 10.23 | 13.31 | 7.44 | 75.08 |
| Fast | 60 | 200 | 18.70 | 36.67 | 14.87 | 9.35 | 11.26 | 4.33 | 76.48 |

To stain poles in new construction, apply stain before installation. On repaints, remove the pole before staining. When estimating by the Opening Count Method, count one opening for each 10 linear feet of pole. "Slow" work is based on a $10.25 hourly wage, "Medium" work on a $16.50 hourly wage, and "Fast" work on a $22.00 hourly wage. Other qualifications that apply to this table are on page 9.

| | Labor LF per manhour | Material coverage LF/gallon | Material cost per gallon | Labor cost per 100 LF | Labor burden 100 LF | Material cost per 100 LF | Overhead per 100 LF | Profit per 100 LF | Total price per 100 LF |
|---|---|---|---|---|---|---|---|---|---|
| **Closet shelf & pole, paint grade** | | | | | | | | | |
| Undercoat, water base (material #3) | | | | | | | | | |
| Brush 1 coat | | | | | | | | | |
| Slow | 17 | 80 | 18.60 | 60.29 | 16.41 | 23.25 | 30.99 | 19.64 | 150.58 |
| Medium | 22 | 70 | 16.20 | 75.00 | 25.22 | 23.14 | 30.23 | 16.90 | 170.49 |
| Fast | 33 | 60 | 13.90 | 66.67 | 27.02 | 23.17 | 21.62 | 8.31 | 146.79 |
| Undercoat, oil base (material #4) | | | | | | | | | |
| Brush 1 coat | | | | | | | | | |
| Slow | 17 | 90 | 20.10 | 60.29 | 16.41 | 22.33 | 30.71 | 19.46 | 149.20 |
| Medium | 22 | 80 | 17.60 | 75.00 | 25.22 | 22.00 | 29.95 | 16.74 | 168.91 |
| Fast | 33 | 65 | 15.10 | 66.67 | 27.02 | 23.23 | 21.63 | 8.31 | 146.86 |
| Split coat (1/2 undercoat + 1/2 enamel), water base (material #3 & #9) | | | | | | | | | |
| Brush 1 coat | | | | | | | | | |
| Slow | 16 | 80 | 20.70 | 64.06 | 17.44 | 25.88 | 33.29 | 21.10 | 161.77 |
| Medium | 21 | 70 | 18.05 | 78.57 | 26.43 | 25.79 | 32.05 | 17.91 | 180.75 |
| Fast | 32 | 60 | 15.50 | 68.75 | 27.88 | 25.83 | 22.66 | 8.71 | 153.83 |
| Split coat (1/2 undercoat + 1/2 enamel), oil base (material #4 & #10) | | | | | | | | | |
| Brush 1 coat | | | | | | | | | |
| Slow | 16 | 90 | 23.00 | 64.06 | 17.44 | 25.56 | 33.19 | 21.04 | 161.29 |
| Medium | 21 | 80 | 20.15 | 78.57 | 26.43 | 25.19 | 31.90 | 17.83 | 179.92 |
| Fast | 32 | 65 | 17.25 | 68.75 | 27.88 | 26.54 | 22.79 | 8.76 | 154.72 |
| Enamel, water base (material #9) | | | | | | | | | |
| Brush each coat | | | | | | | | | |
| Slow | 15 | 80 | 22.80 | 68.33 | 18.61 | 28.50 | 35.79 | 22.69 | 173.92 |
| Medium | 20 | 70 | 19.90 | 82.50 | 27.75 | 28.43 | 33.98 | 18.99 | 191.65 |
| Fast | 30 | 60 | 17.10 | 73.33 | 29.73 | 28.50 | 24.34 | 9.35 | 165.25 |
| Enamel, oil base (material #10) | | | | | | | | | |
| Brush each coat | | | | | | | | | |
| Slow | 15 | 90 | 25.90 | 68.33 | 18.61 | 28.78 | 35.88 | 22.74 | 174.34 |
| Medium | 20 | 80 | 22.70 | 82.50 | 27.75 | 28.38 | 33.97 | 18.99 | 191.59 |
| Fast | 30 | 65 | 19.40 | 73.33 | 29.73 | 29.85 | 24.59 | 9.45 | 166.95 |

Use these costs for painting the wardrobe closet shelves and poles with an undercoat and enamel system. If painting wardrobe closet shelves and poles with flat latex paint along with walls, use the Opening count method discribed under Doors, interior openings. Measurements are based on linear feet (LF) of shelves and poles. "Slow" work is based on a $10.25 hourly wage, "Medium" work on a $16.50 hourly wage, and "Fast" work on a $22.00 hourly wage. Other qualifications that apply to this table are on page 9.

| | Labor LF per manhour | Material coverage LF/gallon | Material cost per gallon | Labor cost per 100 LF | Labor burden 100 LF | Material cost per 100 LF | Overhead per 100 LF | Profit per 100 LF | Total price per 100 LF |
|---|---|---|---|---|---|---|---|---|---|
| **Closet shelves, paint grade** | | | | | | | | | |
| Undercoat, water base (material #3) | | | | | | | | | |
| Brush 1 coat | | | | | | | | | |
| Slow | 30 | 100 | 18.60 | 34.17 | 9.29 | 18.60 | 19.24 | 12.20 | 93.50 |
| Medium | 37 | 90 | 16.20 | 44.59 | 15.01 | 18.00 | 19.01 | 10.63 | 107.24 |
| Fast | 44 | 80 | 13.90 | 50.00 | 20.28 | 17.38 | 16.22 | 6.23 | 110.11 |
| Undercoat, oil base (material #4) | | | | | | | | | |
| Brush 1 coat | | | | | | | | | |
| Slow | 30 | 110 | 20.10 | 34.17 | 9.29 | 18.27 | 19.14 | 12.13 | 93.00 |
| Medium | 37 | 100 | 17.60 | 44.59 | 15.01 | 17.60 | 18.91 | 10.57 | 106.68 |
| Fast | 44 | 90 | 15.10 | 50.00 | 20.28 | 16.78 | 16.11 | 6.19 | 109.36 |
| Split coat (1/2 undercoat + 1/2 enamel), water base (material #3 & #9) | | | | | | | | | |
| Brush 1 coat | | | | | | | | | |
| Slow | 27 | 100 | 20.70 | 37.96 | 10.34 | 20.70 | 21.39 | 13.56 | 103.95 |
| Medium | 35 | 90 | 18.05 | 47.14 | 15.86 | 20.06 | 20.35 | 11.38 | 114.79 |
| Fast | 42 | 80 | 15.50 | 52.38 | 21.24 | 19.38 | 17.21 | 6.61 | 116.82 |
| Split coat (1/2 undercoat + 1/2 enamel), oil base (material #4 & #10) | | | | | | | | | |
| Brush 1 coat | | | | | | | | | |
| Slow | 27 | 110 | 23.00 | 37.96 | 10.34 | 20.91 | 21.46 | 13.60 | 104.27 |
| Medium | 35 | 100 | 20.15 | 47.14 | 15.86 | 20.15 | 20.37 | 11.39 | 114.91 |
| Fast | 42 | 90 | 17.25 | 52.38 | 21.24 | 19.17 | 17.17 | 6.60 | 116.56 |
| Enamel, water base (material #9) | | | | | | | | | |
| Brush each coat | | | | | | | | | |
| Slow | 25 | 100 | 22.80 | 41.00 | 11.16 | 22.80 | 23.24 | 14.73 | 112.93 |
| Medium | 33 | 90 | 19.90 | 50.00 | 16.81 | 22.11 | 21.79 | 12.18 | 122.89 |
| Fast | 40 | 80 | 17.10 | 55.00 | 22.30 | 21.38 | 18.26 | 7.02 | 123.96 |
| Enamel, oil base (material #10) | | | | | | | | | |
| Brush each coat | | | | | | | | | |
| Slow | 25 | 110 | 25.90 | 41.00 | 11.16 | 23.55 | 23.47 | 14.88 | 114.06 |
| Medium | 33 | 100 | 22.70 | 50.00 | 16.81 | 22.70 | 21.93 | 12.26 | 123.70 |
| Fast | 40 | 90 | 19.40 | 55.00 | 22.30 | 21.56 | 18.29 | 7.03 | 124.18 |

Use these costs for painting the wardrobe closet shelves with an undercoat and enamel system. If painting wardrobe closet shelves with flat latex paint along with walls, use the Opening Count Method described under Doors, interior openings. Measurements are based on linear feet (LF) of shelves. "Slow" work is based on a $10.25 hourly wage, "Medium" work on a $16.50 hourly wage, and "Fast" work on a $22.00 hourly wage. Other qualifications that apply to this table are on page 9.

| | Labor LF per manhour | Material coverage LF/gallon | Material cost per gallon | Labor cost per 100 LF | Labor burden 100 LF | Material cost per 100 LF | Overhead per 100 LF | Profit per 100 LF | Total price per 100 LF |
|---|---|---|---|---|---|---|---|---|---|
| **Closets, molding at perimeter, paint grade** | | | | | | | | | |
| Undercoat, water base (material #3) | | | | | | | | | |
| Brush 1 coat | | | | | | | | | |
| Slow | 80 | 225 | 18.60 | 12.81 | 3.49 | 8.27 | 7.62 | 4.83 | 37.02 |
| Medium | 95 | 213 | 16.20 | 17.37 | 5.85 | 7.61 | 7.55 | 4.22 | 42.60 |
| Fast | 110 | 200 | 13.90 | 20.00 | 8.11 | 6.95 | 6.49 | 2.49 | 44.04 |
| Undercoat, oil base (material #4) | | | | | | | | | |
| Brush 1 coat | | | | | | | | | |
| Slow | 80 | 250 | 20.10 | 12.81 | 3.49 | 8.04 | 7.55 | 4.78 | 36.67 |
| Medium | 95 | 238 | 17.60 | 17.37 | 5.85 | 7.39 | 7.50 | 4.19 | 42.30 |
| Fast | 110 | 220 | 15.10 | 20.00 | 8.11 | 6.86 | 6.47 | 2.49 | 43.93 |
| Split coat (1/2 undercoat + 1/2 enamel), water base (material #3 & #9) | | | | | | | | | |
| Brush 1 coat | | | | | | | | | |
| Slow | 75 | 225 | 20.70 | 13.67 | 3.71 | 9.20 | 8.25 | 5.23 | 40.06 |
| Medium | 90 | 213 | 18.05 | 18.33 | 6.17 | 8.47 | 8.08 | 4.52 | 45.57 |
| Fast | 105 | 200 | 15.50 | 20.95 | 8.49 | 7.75 | 6.88 | 2.64 | 46.71 |
| Split coat (1/2 undercoat + 1/2 enamel), oil base (material #4 & #10) | | | | | | | | | |
| Brush 1 coat | | | | | | | | | |
| Slow | 75 | 250 | 23.00 | 13.67 | 3.71 | 9.20 | 8.25 | 5.23 | 40.06 |
| Medium | 90 | 238 | 20.15 | 18.33 | 6.17 | 8.47 | 8.08 | 4.52 | 45.57 |
| Fast | 105 | 220 | 17.25 | 20.95 | 8.49 | 7.84 | 6.90 | 2.65 | 46.83 |
| Enamel, water base (material #9) | | | | | | | | | |
| Brush each coat | | | | | | | | | |
| Slow | 70 | 225 | 22.80 | 14.64 | 3.99 | 10.13 | 8.92 | 5.65 | 43.33 |
| Medium | 85 | 213 | 19.90 | 19.41 | 6.52 | 9.34 | 8.64 | 4.83 | 48.74 |
| Fast | 100 | 200 | 17.10 | 22.00 | 8.92 | 8.55 | 7.30 | 2.81 | 49.58 |
| Enamel, oil base (material #10) | | | | | | | | | |
| Brush each coat | | | | | | | | | |
| Slow | 70 | 250 | 25.90 | 14.64 | 3.99 | 10.36 | 8.99 | 5.70 | 43.68 |
| Medium | 85 | 238 | 22.70 | 19.41 | 6.52 | 9.54 | 8.69 | 4.86 | 49.02 |
| Fast | 100 | 220 | 19.40 | 22.00 | 8.92 | 8.82 | 7.35 | 2.83 | 49.92 |

Use these costs for molding around wardrobe closets. Measurements are based on linear feet (LF) of molding. "Slow" work is based on a $10.25 hourly wage, "Medium" work on a $16.50 hourly wage, and "Fast" work on a $22.00 hourly wage. Other qualifications that apply to this table are on page 9.

| | Labor LF per manhour | Material coverage LF/gallon | Material cost per gallon | Labor cost per 100 LF | Labor burden 100 LF | Material cost per 100 LF | Overhead per 100 LF | Profit per 100 LF | Total price per 100 LF |
|---|---|---|---|---|---|---|---|---|---|
| **Corbels, wood trim, stain grade, average size 4" x 8"** | | | | | | | | | |
| Solid body stain, water base (material #18) | | | | | | | | | |
| Brush each coat | | | | | | | | | |
| Slow | 15 | 50 | 19.40 | 68.33 | 18.61 | 38.80 | 38.98 | 24.71 | 189.43 |
| Medium | 20 | 48 | 17.00 | 82.50 | 27.75 | 35.42 | 35.69 | 19.95 | 201.31 |
| Fast | 25 | 45 | 14.50 | 88.00 | 35.68 | 32.22 | 28.84 | 11.08 | 195.82 |
| Solid body stain, oil base (material #19) | | | | | | | | | |
| Brush each coat | | | | | | | | | |
| Slow | 15 | 55 | 23.10 | 68.33 | 18.61 | 42.00 | 39.97 | 25.34 | 194.25 |
| Medium | 20 | 53 | 20.20 | 82.50 | 27.75 | 38.11 | 36.35 | 20.32 | 205.03 |
| Fast | 25 | 50 | 17.30 | 88.00 | 35.68 | 34.60 | 29.28 | 11.25 | 198.81 |
| Semi-transparent stain, water base (material #20) | | | | | | | | | |
| Brush each coat | | | | | | | | | |
| Slow | 18 | 55 | 19.60 | 56.94 | 15.51 | 35.64 | 33.51 | 21.24 | 162.84 |
| Medium | 22 | 53 | 17.10 | 75.00 | 25.22 | 32.26 | 32.46 | 18.15 | 183.09 |
| Fast | 28 | 50 | 14.70 | 78.57 | 31.85 | 29.40 | 25.87 | 9.94 | 175.63 |
| Semi-transparent stain, oil base (material #21) | | | | | | | | | |
| Brush each coat | | | | | | | | | |
| Slow | 18 | 60 | 23.40 | 56.94 | 15.51 | 39.00 | 34.55 | 21.90 | 167.90 |
| Medium | 22 | 58 | 20.50 | 75.00 | 25.22 | 35.34 | 33.22 | 18.57 | 187.35 |
| Fast | 28 | 55 | 17.60 | 78.57 | 31.85 | 32.00 | 26.35 | 10.13 | 178.90 |

Use these costs for painting corbels averaging 4" x 8" in size. Measurements are based on linear feet (LF) of corbels that are painted or stained with a different material or color than the surface they extend from. "Slow" work is based on a $10.25 hourly wage, "Medium" work on a $16.50 hourly wage, and "Fast" work on a $22.00 hourly wage. Other qualifications that apply to this table are on page 9.

## Deck overhang, wood

Multiply the horizontal surface area by 1.5 to allow for painting floor joists and use the overhang table for areas greater than 2.5 feet to determine pricing.

## Deck surfaces, steps, stair treads & porches, wood

Measure the surface area and apply the prices for smooth siding.

| | Labor LF per manhour | Material coverage LF/gallon | Material cost per gallon | Labor cost per 100 LF | Labor burden 100 LF | Material cost per 100 LF | Overhead per 100 LF | Profit per 100 LF | Total price per 100 LF |
|---|---|---|---|---|---|---|---|---|---|
| **Door frames and trim only, per 100 linear feet** | | | | | | | | | |
| Undercoat, water base (material #3) | | | | | | | | | |
| Brush 1 coat | | | | | | | | | |
| Slow | 220 | 510 | 18.60 | 4.66 | 1.27 | 3.65 | 2.97 | 1.88 | 14.43 |
| Medium | 270 | 465 | 16.20 | 6.11 | 2.05 | 3.48 | 2.85 | 1.60 | 16.09 |
| Fast | 320 | 425 | 13.90 | 6.88 | 2.80 | 3.27 | 2.39 | .92 | 16.26 |
| Undercoat, oil base (material #4) | | | | | | | | | |
| Brush 1 coat | | | | | | | | | |
| Slow | 220 | 560 | 20.10 | 4.66 | 1.27 | 3.59 | 2.95 | 1.87 | 14.34 |
| Medium | 270 | 510 | 17.60 | 6.11 | 2.05 | 3.45 | 2.85 | 1.59 | 16.05 |
| Fast | 320 | 465 | 15.10 | 6.88 | 2.80 | 3.25 | 2.39 | .92 | 16.24 |
| Split coat (1/2 undercoat + 1/2 enamel), water base (material #3 & #9) | | | | | | | | | |
| Brush 1 coat | | | | | | | | | |
| Slow | 210 | 510 | 20.70 | 4.88 | 1.33 | 4.06 | 3.18 | 2.02 | 15.47 |
| Medium | 258 | 465 | 18.05 | 6.40 | 2.16 | 3.88 | 3.05 | 1.70 | 17.19 |
| Fast | 305 | 425 | 15.50 | 7.21 | 2.93 | 3.65 | 2.55 | .98 | 17.32 |
| Split coat (1/2 undercoat + 1/2 enamel), oil base (material #4 & #10) | | | | | | | | | |
| Brush 1 coat | | | | | | | | | |
| Slow | 210 | 560 | 23.00 | 4.88 | 1.33 | 4.11 | 3.20 | 2.03 | 15.55 |
| Medium | 258 | 510 | 20.15 | 6.40 | 2.16 | 3.95 | 3.06 | 1.71 | 17.28 |
| Fast | 305 | 465 | 17.25 | 7.21 | 2.93 | 3.71 | 2.56 | .98 | 17.39 |
| Enamel, water base (material #9) | | | | | | | | | |
| Brush each coat | | | | | | | | | |
| Slow | 200 | 510 | 22.80 | 5.13 | 1.39 | 4.47 | 3.41 | 2.16 | 16.56 |
| Medium | 245 | 465 | 19.90 | 6.73 | 2.27 | 4.28 | 3.25 | 1.82 | 18.35 |
| Fast | 290 | 425 | 17.10 | 7.59 | 3.08 | 4.02 | 2.72 | 1.04 | 18.45 |
| Enamel, oil base (material #10) | | | | | | | | | |
| Brush each coat | | | | | | | | | |
| Slow | 200 | 560 | 25.90 | 5.13 | 1.39 | 4.63 | 3.46 | 2.19 | 16.80 |
| Medium | 245 | 510 | 22.70 | 6.73 | 2.27 | 4.45 | 3.29 | 1.84 | 18.58 |
| Fast | 290 | 465 | 19.40 | 7.59 | 3.08 | 4.17 | 2.75 | 1.06 | 18.65 |

Use these costs for painting door frames and wood trim on all sides when doors are not to be painted. ADD for additional preparation time, as needed, including setup, remove & replace (hardware), protection (mask walls and visqueen floors), putty, caulking, sanding and cleanup. When doors are painted along with the frames and trim use the Opening Count Method under Doors, interior openings, and/or the exterior door costs under Doors, exterior. Measurements for door frames and trim are based on linear feet (LF) of one side of the frame. Prices are for about 17 linear feet of frame at each opening. "Slow" work is based on a \$10.25 hourly wage, "Medium" work on a \$16.50 hourly wage, and "Fast" work on a \$22.00 hourly wage. Other qualifications that apply to this table are on page 9.

| | Openings per manhour | Openings per gallon | Material cost per gallon | Labor cost per opening | Labor burden opening | Material cost per opening | Overhead per opening | Profit per opening | Total price per opening |
|---|---|---|---|---|---|---|---|---|---|
| **Door frames and trim only, per opening** | | | | | | | | | |
| Undercoat, water base (material #3) | | | | | | | | | |
| Brush 1 coat | | | | | | | | | |
| Slow | 13 | 30 | 18.60 | .79 | .21 | .62 | .51 | .32 | 2.45 |
| Medium | 16 | 28 | 16.20 | 1.03 | .36 | .58 | .48 | .27 | 2.72 |
| Fast | 18 | 25 | 13.90 | 1.22 | .51 | .56 | .42 | .16 | 2.87 |
| Undercoat, oil base (material #4) | | | | | | | | | |
| Brush 1 coat | | | | | | | | | |
| Slow | 13 | 33 | 20.10 | .79 | .21 | .61 | .50 | .32 | 2.43 |
| Medium | 16 | 31 | 17.60 | 1.03 | .36 | .57 | .48 | .27 | 2.71 |
| Fast | 18 | 28 | 15.10 | 1.22 | .51 | .54 | .42 | .16 | 2.85 |
| Split coat (1/2 undercoat + 1/2 enamel), water base (material #3 & #9) | | | | | | | | | |
| Brush 1 coat | | | | | | | | | |
| Slow | 12 | 30 | 20.70 | .85 | .23 | .69 | .55 | .35 | 2.67 |
| Medium | 14 | 28 | 18.05 | 1.18 | .39 | .64 | .54 | .30 | 3.05 |
| Fast | 16 | 25 | 15.50 | 1.38 | .57 | .62 | .47 | .18 | 3.22 |
| Split coat (1/2 undercoat + 1/2 enamel), oil base (material #4 & #10) | | | | | | | | | |
| Brush 1 coat | | | | | | | | | |
| Slow | 12 | 33 | 23.00 | .85 | .23 | .70 | .55 | .35 | 2.68 |
| Medium | 14 | 31 | 20.15 | 1.18 | .39 | .65 | .55 | .31 | 3.08 |
| Fast | 16 | 28 | 17.25 | 1.38 | .57 | .62 | .47 | .18 | 3.22 |
| Enamel, water base (material #9) | | | | | | | | | |
| Brush each coat | | | | | | | | | |
| Slow | 11 | 30 | 22.80 | .93 | .26 | .76 | .60 | .38 | 2.93 |
| Medium | 14 | 28 | 19.90 | 1.18 | .39 | .71 | .56 | .31 | 3.15 |
| Fast | 16 | 25 | 17.10 | 1.38 | .57 | .68 | .48 | .19 | 3.30 |
| Enamel, oil base (material #10) | | | | | | | | | |
| Brush each coat | | | | | | | | | |
| Slow | 11 | 33 | 25.90 | .93 | .26 | .78 | .61 | .39 | 2.97 |
| Medium | 14 | 31 | 22.70 | 1.18 | .39 | .73 | .57 | .32 | 3.19 |
| Fast | 16 | 28 | 19.40 | 1.38 | .57 | .69 | .49 | .19 | 3.32 |

Use these costs for painting door frames and wood trim on all sides when doors are not to be painted. ADD for additional preparation time, as needed, including setup, remove & replace (hardware), protection (mask walls and visqueen floors); putty, caulking, sanding and cleanup. When doors are painted along with the frames and trim, use the Opening Count Method under Doors, interior openings, and/or the exterior door costs under Doors, exterior. These costs are based on a count of the openings requiring paint. Prices are for about 17 linear feet of at each opening. "Slow" work is based on a $10.25 hourly wage, "Medium" work on a $16.50 hourly wage, frame and "Fast" work on a $22.00 hourly wage. Other qualifications that apply to this table are on page 9.

## Doors, exterior

The tables that follow include costs for both time and material needed to apply two coats of a high quality finish to all six sides of each exterior door, finish the jamb and trim, and lay-off each door smoothly. These costs are in addition to those shown under Doors, interior openings, for both the *Opening Count Method*, and the *Per Door Method* which include one coat of undercoat and one coat of enamel for each exterior door along with the interior doors. New exterior paint grade doors actually receive two coats of exterior enamel. New exterior stain grade doors actually receive a coat of stain, sealer, and then a coat of either marine spar varnish or polyurethane finish. The following examples give total cost to finish a flush exterior door with polyurethane, and includes the cost for the two coats from the interior take-off.

| OPENING COUNT METHOD (Interior doors) | |
|---|---|
| Included in the interior take-off | |
| Opening Count Method, interior undercoat cost - 1 coat, slow | $ 10.28 |
| Opening Count Method, interior enamel cost - 1 coat, slow | $ 10.77 |
| Included in the exterior take-off | |
| Exterior, flush, two coat system, polyurethane cost - 2 coats, slow | $ 23.29 |
| Total to finish the exterior door | $ 44.34 |

***OR***

| PER DOOR METHOD (Interior doors) | |
|---|---|
| Included in the interior take-off | |
| Per Door Method, interior, flush, undercoat cost - 1 coat, slow | $ 10.19 |
| Per Door Method, interior enamel cost - 1 coat, slow | $ 9.09 |
| Included in the exterior take-off | |
| Exterior, flush, two coat system, polyurethane cost - 2 coats, slow | $ 23.29 |
| Total to finish the exterior door | $ 42.57 |

Under this system, almost half the cost to paint the exterior door is included in the interior take-off. When counting interior doors, be sure you include all the exterior doors whether you use either the *Opening Count Method* or the *Per Door Method* to estimate doors.

| | Manhours per door | Doors per gallon | Material cost per gallon | Labor cost per door | Labor burden door | Material cost per door | Overhead per door | Profit per door | Total price per door |
|---|---|---|---|---|---|---|---|---|---|
| **Doors, exterior, flush, two coat system** | | | | | | | | | |
| Exterior enamel, 2 coat system, water base (material #24) | | | | | | | | | |
| Roll & brush 2 coats | | | | | | | | | |
| Slow | 0.5 | 7 | 25.00 | 5.13 | 1.39 | 3.57 | 3.13 | 1.98 | 15.20 |
| Medium | 0.4 | 6 | 21.80 | 6.60 | 2.22 | 3.63 | 3.05 | 1.71 | 17.21 |
| Fast | 0.3 | 5 | 18.70 | 6.60 | 2.68 | 3.74 | 2.41 | .93 | 16.36 |
| Exterior enamel, 2 coat system, oil base (material #25) | | | | | | | | | |
| Roll & brush 2 coats | | | | | | | | | |
| Slow | 0.5 | 7 | 27.20 | 5.13 | 1.39 | 3.89 | 3.23 | 2.05 | 15.69 |
| Medium | 0.4 | 6 | 23.80 | 6.60 | 2.22 | 3.97 | 3.13 | 1.75 | 17.67 |
| Fast | 0.3 | 5 | 20.40 | 6.60 | 2.68 | 4.08 | 2.47 | .95 | 16.78 |
| Polyurethane (material #22) | | | | | | | | | |
| Brush 2 coats | | | | | | | | | |
| Slow | 0.7 | 5.0 | 30.60 | 7.18 | 1.95 | 6.12 | 4.73 | 3.00 | 22.98 |
| Medium | 0.6 | 4.5 | 26.80 | 9.90 | 3.33 | 5.96 | 4.70 | 2.63 | 26.52 |
| Fast | 0.5 | 4.0 | 22.90 | 11.00 | 4.46 | 5.73 | 3.92 | 1.51 | 26.62 |
| Marine spar varnish, flat or gloss (material #23) | | | | | | | | | |
| Brush 2 coats | | | | | | | | | |
| Slow | 0.5 | 6 | 35.90 | 5.13 | 1.39 | 5.98 | 3.88 | 2.46 | 18.84 |
| Medium | 0.4 | 5 | 31.40 | 6.60 | 2.22 | 6.28 | 3.70 | 2.07 | 20.87 |
| Fast | 0.3 | 4 | 27.00 | 6.60 | 2.68 | 6.75 | 2.97 | 1.14 | 20.14 |
| ADD - Preparation for spar varnish | | | | | | | | | |
| Steel wool buff | 0.2 | -- | -- | 3.30 | 1.11 | -- | 1.08 | .60 | 6.09 |
| Wax application | 0.2 | -- | -- | 3.30 | 1.11 | -- | 1.08 | .60 | 6.09 |

Use these figures for painting flush exterior doors, other than entry doors. These doors require a two coat system, and these costs are included with the exterior door take-off. ADD for additional preparation time, as needed, including setup, remove & replace (hardware), protection (mask walls and visqueen floors); putty, caulking, sanding and cleanup. Add minimum preparation time for varnishing as indicated above. These costs are in addition to the costs that are included in the interior take-off as explained by the example in the previous section. "Slow" work is based on a $10.25 hourly wage, "Medium" work on a $16.50 hourly wage, and "Fast" work on a $22.00 hourly wage. Other qualifications that apply to this table are on page 9.

| | Manhours per door | Doors per gallon | Material cost per gallon | Labor cost per door | Labor burden door | Material cost per door | Overhead per door | Profit per door | Total price per door |
|---|---|---|---|---|---|---|---|---|---|

## Doors, exterior, French, two coat system

| | Manhours per door | Doors per gallon | Material cost per gallon | Labor cost per door | Labor burden door | Material cost per door | Overhead per door | Profit per door | Total price per door |
|---|---|---|---|---|---|---|---|---|---|
| Exterior enamel, 2 coat system, water base (material #24) | | | | | | | | | |
| Roll & brush 2 coats | | | | | | | | | |
| Slow | 1.0 | 12 | 25.00 | 10.25 | 2.79 | 2.08 | 4.69 | 2.97 | 22.78 |
| Medium | 0.8 | 10 | 21.80 | 13.20 | 4.44 | 2.18 | 4.86 | 2.71 | 27.39 |
| Fast | 0.6 | 8 | 18.70 | 13.20 | 5.35 | 2.34 | 3.86 | 1.49 | 26.24 |
| Exterior enamel, 2 coat system, oil base (material #25) | | | | | | | | | |
| Roll & brush 2 coats | | | | | | | | | |
| Slow | 1.0 | 12 | 27.20 | 10.25 | 2.79 | 2.27 | 4.75 | 3.01 | 23.07 |
| Medium | 0.8 | 10 | 23.80 | 13.20 | 4.44 | 2.38 | 4.90 | 2.74 | 27.66 |
| Fast | 0.6 | 8 | 20.40 | 13.20 | 5.35 | 2.55 | 3.90 | 1.50 | 26.50 |
| Polyurethane (material #22) | | | | | | | | | |
| Brush 2 coats | | | | | | | | | |
| Slow | 1.5 | 8.0 | 30.60 | 15.38 | 4.18 | 3.83 | 7.25 | 4.60 | 35.24 |
| Medium | 1.3 | 7.5 | 26.80 | 21.45 | 7.22 | 3.57 | 7.90 | 4.42 | 44.56 |
| Fast | 1.0 | 7.0 | 22.90 | 22.00 | 8.92 | 3.27 | 6.33 | 2.43 | 42.95 |
| Marine spar varnish, flat or gloss (material #23) | | | | | | | | | |
| Brush 2 coats | | | | | | | | | |
| Slow | 1.0 | 12 | 35.90 | 10.25 | 2.79 | 2.99 | 4.97 | 3.15 | 24.15 |
| Medium | 0.8 | 10 | 31.40 | 13.20 | 4.44 | 3.14 | 5.09 | 2.85 | 28.72 |
| Fast | 0.6 | 8 | 27.00 | 13.20 | 5.35 | 3.38 | 4.06 | 1.56 | 27.55 |
| ADD - Preparation for spar varnish | | | | | | | | | |
| Steel wool buff | 0.3 | -- | -- | 4.95 | 1.67 | -- | 1.62 | .91 | 9.15 |
| Wax application | 0.3 | -- | -- | 4.95 | 1.67 | -- | 1.62 | .91 | 9.15 |

Use these figures for painting exterior French doors that have 10 to 15 lites. These doors require a two coat system, and these costs are included with the exterior door take-off. ADD for additional preparation time, as needed, including setup, remove & replace (hardware), protection (mask walls and visqueen floors); putty, caulking, sanding and cleanup. Add minimum preparation time for varnishing as indicated above. These costs are in addition to the costs that are included in the interior take-off as explained by the example in the previous section. "Slow" work is based on a $10.25 hourly wage, "Medium" work on a $16.50 hourly wage, and "Fast" work on a $22.00 hourly wage. Other qualifications that apply to this table are on page 9.

| | Manhours per door | Doors per gallon | Material cost per gallon | Labor cost per door | Labor burden door | Material cost per door | Overhead per door | Profit per door | Total price per door |
|---|---|---|---|---|---|---|---|---|---|

## Doors, exterior, louvered, two coat system

| | Manhours per door | Doors per gallon | Material cost per gallon | Labor cost per door | Labor burden door | Material cost per door | Overhead per door | Profit per door | Total price per door |
|---|---|---|---|---|---|---|---|---|---|
| Exterior enamel, 2 coat system, water base (material #24) | | | | | | | | | |
| Roll & brush 2 coats | | | | | | | | | |
| Slow | 1.4 | 7 | 25.00 | 14.35 | 3.91 | 3.57 | 6.77 | 4.29 | 32.89 |
| Medium | 1.1 | 6 | 21.80 | 18.15 | 6.11 | 3.63 | 6.83 | 3.82 | 38.54 |
| Fast | 0.7 | 5 | 18.70 | 15.40 | 6.24 | 3.74 | 4.70 | 1.80 | 31.88 |
| Exterior enamel, 2 coat system, oil base (material #25) | | | | | | | | | |
| Roll & brush 2 coats | | | | | | | | | |
| Slow | 1.4 | 7 | 27.20 | 14.35 | 3.91 | 3.89 | 6.87 | 4.35 | 33.37 |
| Medium | 1.1 | 6 | 23.80 | 18.15 | 6.11 | 3.97 | 6.92 | 3.87 | 39.02 |
| Fast | 0.7 | 5 | 20.40 | 15.40 | 6.24 | 4.08 | 4.76 | 1.83 | 32.31 |
| Polyurethane (material #22) | | | | | | | | | |
| Brush 2 coats | | | | | | | | | |
| Slow | 1.7 | 5.0 | 30.60 | 17.43 | 4.74 | 6.12 | 8.77 | 5.56 | 42.62 |
| Medium | 1.5 | 4.5 | 26.80 | 24.75 | 8.33 | 5.96 | 9.56 | 5.35 | 53.95 |
| Fast | 1.2 | 4.0 | 22.90 | 26.40 | 10.70 | 5.73 | 7.93 | 3.05 | 53.81 |
| Marine spar varnish, flat or gloss (material #23) | | | | | | | | | |
| Brush 2 coats | | | | | | | | | |
| Slow | 1.4 | 7 | 35.90 | 14.35 | 3.91 | 5.13 | 7.25 | 4.60 | 35.24 |
| Medium | 1.1 | 6 | 31.40 | 18.15 | 6.11 | 5.23 | 7.23 | 4.04 | 40.76 |
| Fast | 0.7 | 5 | 27.00 | 15.40 | 6.24 | 5.40 | 5.00 | 1.92 | 33.96 |
| ADD - Preparation for spar varnish | | | | | | | | | |
| Steel wool buff | 0.4 | -- | -- | 6.60 | 2.22 | -- | 2.16 | 1.21 | 12.19 |
| Wax application | 0.4 | -- | -- | 6.60 | 2.22 | -- | 2.16 | 1.21 | 12.19 |

Use these figures for painting exterior louvered doors. These doors require a two coat system, and these costs are included with the exterior door take-off. ADD for additional preparation time, as needed, including setup, remove & replace (hardware), protection (mask walls and visqueen floors); putty, caulking, sanding and cleanup. Add minimum preparation time for varnishing as indicated above. These costs are in addition to the costs that are included in the interior take-off as explained by the example in the previous section. "Slow" work is based on a \$10.25 hourly wage, "Medium" work on a \$16.50 hourly wage, and "Fast" work on a \$22.00 hourly wage. Other qualifications that apply to this table are on page 9.

| | Manhours per door | Doors per gallon | Material cost per gallon | Labor cost per door | Labor burden door | Material cost per door | Overhead per door | Profit per door | Total price per door |
|---|---|---|---|---|---|---|---|---|---|
| **Doors, exterior, panel (entry), two coat system** | | | | | | | | | |
| Exterior enamel, 2 coat system, water base (material #24) | | | | | | | | | |
| Roll & brush 2 coats | | | | | | | | | |
| Slow | 1.4 | 4 | 25.00 | 14.35 | 3.91 | 6.25 | 7.60 | 4.82 | 36.93 |
| Medium | 1.1 | 3 | 21.80 | 18.15 | 6.11 | 7.27 | 7.72 | 4.32 | 43.57 |
| Fast | 0.8 | 2 | 18.70 | 17.60 | 7.14 | 9.35 | 6.31 | 2.42 | 42.82 |
| Exterior enamel, 2 coat system, oil base (material #25) | | | | | | | | | |
| Roll & brush 2 coats | | | | | | | | | |
| Slow | 1.4 | 4 | 27.20 | 14.35 | 3.91 | 6.80 | 7.77 | 4.92 | 37.75 |
| Medium | 1.1 | 3 | 23.80 | 18.15 | 6.11 | 7.93 | 7.89 | 4.41 | 44.49 |
| Fast | 0.8 | 2 | 20.40 | 17.60 | 7.14 | 10.20 | 6.46 | 2.48 | 43.88 |
| Polyurethane (material #22) | | | | | | | | | |
| Brush 2 coats | | | | | | | | | |
| Slow | 1.7 | 4 | 30.60 | 17.43 | 4.74 | 7.65 | 9.25 | 5.86 | 44.93 |
| Medium | 1.5 | 3 | 26.80 | 24.75 | 8.33 | 8.93 | 10.29 | 5.75 | 58.05 |
| Fast | 1.2 | 2 | 22.90 | 26.40 | 10.70 | 11.45 | 8.98 | 3.45 | 60.98 |
| Marine spar varnish, flat or gloss (material #23) | | | | | | | | | |
| Brush 2 coats | | | | | | | | | |
| Slow | 1.4 | 4 | 35.90 | 14.35 | 3.91 | 8.98 | 8.44 | 5.35 | 41.03 |
| Medium | 1.1 | 3 | 31.40 | 18.15 | 6.11 | 10.47 | 8.51 | 4.76 | 48.00 |
| Fast | 0.8 | 2 | 14.70 | 17.60 | 7.14 | 7.35 | 5.94 | 2.28 | 40.31 |
| ADD - Preparation for spar varnish | | | | | | | | | |
| Steel wool buff | 0.3 | -- | -- | 4.95 | 1.67 | -- | 1.62 | .91 | 9.15 |
| Wax application | 0.3 | -- | -- | 4.95 | 1.67 | -- | 1.62 | .91 | 9.15 |

Use these figures for painting typical exterior paneled doors. These doors require a two coat system, and these costs are included with the exterior door take-off. ADD for additional preparation time, as needed, including setup, remove & replace (hardware), protection (mask walls and visqueen floors); putty, caulking, sanding and cleanup. Add minimum preparation time for varnishing as indicated above. These costs are in addition to the costs that are included in the interior take-off as explained by the example in the previous section. "Slow" work is based on a $10.25 hourly wage, "Medium" work on a $16.50 hourly wage, and "Fast" work on a $22.00 hourly wage. Other qualifications that apply to this table are on page 9.

## Doors, Opening Count Method

Many painting companies estimate paint grade doors, jambs, frames, wood windows, pullmans, linens, bookcases, wine racks and other interior surfaces that take an undercoat and enamel finish by the "opening." Each opening type is considered to take the same time regardless of whether it's a door, window, pullman, etc. These figures are based on the number of openings finished per 8-hour day and the material required per opening. The Opening Count Method of estimating involves counting the quantity of all openings (including all exterior doors) based on the opening allowance table at Figure 20 below. After you determine the number of openings use the following table in accumulated multiples of 10 for applying undercoat and enamel. The undercoat process is based on 11 to 13 openings per gallon and enamel is based on 10 to 12 openings per gallon. As an example, using the medium rate for water based material on 12 openings with 1 coat of undercoat and 1 coat of enamel, add the 10 opening figures for each coat to the 2 opening figures for each coat as follows:

**Interior Take-off**

| | **Undercoat** | **Enamel** |
|---|---|---|
| 10 openings | $83.85 | $90.77 |
| 2 openings | 18.56 | 19.92 |
| 12 openings total | $102.41 | $110.69 |

| | **Item** | **Opening Count** |
|---|---|---|
| **Closets** | Molding at closet perimeter | Count 1 opening per 25'0" length |
| | Poles, stain | Count 1 opening per 10'0" length |
| | Shelf & pole (undercoat or enamel) | Count 1 opening per 6'0" length |
| | Shelves (undercoat or enamel) | Count 1 opening per 10'0" length |
| **Doors** | Bifold doors & frames | Count 1 opening per door |
| | Dutch doors & frames | Count 2 openings per door |
| | Entry doors & frames | Count 1 opening per door |
| | Forced air unit doors & frames | Count 1 opening per door |
| | French doors & frames | Count 1.5 openings per door |
| | Linen doors with face frame | Count 1 opening per 2'0" width |
| | Louvered bifold doors & frames | Count 1 opening per door panel |
| | false | Count 1 opening per door panel |
| | real | Count 1 opening per door or per 1'6" width |
| | Passage doors & frames | |
| | flush | Count 1 opening per door |
| | paneled | Count 1.25 openings per door |
| | Wardrobe doors | Count 1 opening per door |
| | Split coat operation, doors & frames | Count 1 opening per door |
| | Tipoff operation (doors only) | Count .5 opening per door |
| **Pullman cabinets** | | Count 1 opening per lavatory or per 4'0" width |
| **Windows, wood** | | Count 1 opening per 6 SF of window |

**Figure 20**
Interior opening count allowance table

| | Manhours per opening(s) | Gallons per opening(s) | Material cost per gallon | Labor cost per opening(s) | Labor burden opening(s) | Material cost per opening(s) | Overhead per opening(s) | Profit per opening(s) | Total price per opening(s) |
|---|---|---|---|---|---|---|---|---|---|
| **Doors, interior openings, *Opening Count Method*** | | | | | | | | | |
| **1 opening total** | | | | | | | | | |
| Undercoat, water base (material #3) | | | | | | | | | |
| Roll & brush 1 coat | | | | | | | | | |
| Slow | 0.4 | 0.080 | 18.60 | 4.10 | 1.12 | 1.49 | 2.08 | 1.32 | 10.11 |
| Medium | 0.3 | 0.085 | 16.20 | 4.95 | 1.67 | 1.38 | 1.96 | 1.10 | 11.06 |
| Fast | 0.2 | 0.090 | 13.90 | 4.40 | 1.78 | 1.25 | 1.37 | .53 | 9.33 |
| Undercoat, oil base (material #4) | | | | | | | | | |
| Roll & brush 1 coat | | | | | | | | | |
| Slow | 0.4 | 0.080 | 20.10 | 4.10 | 1.12 | 1.61 | 2.12 | 1.34 | 10.29 |
| Medium | 0.3 | 0.085 | 17.60 | 4.95 | 1.67 | 1.50 | 1.99 | 1.11 | 11.22 |
| Fast | 0.2 | 0.090 | 15.10 | 4.40 | 1.78 | 1.36 | 1.39 | .54 | 9.47 |
| Enamel, water base (material #9) | | | | | | | | | |
| Roll & brush 1 coat | | | | | | | | | |
| Slow | 0.4 | 0.08 | 22.80 | 4.10 | 1.12 | 1.82 | 2.18 | 1.38 | 10.60 |
| Medium | 0.3 | 0.09 | 19.90 | 4.95 | 1.67 | 1.79 | 2.06 | 1.15 | 11.62 |
| Fast | 0.2 | 0.10 | 17.10 | 4.40 | 1.78 | 1.71 | 1.46 | .56 | 9.91 |
| Enamel, oil base (material #10) | | | | | | | | | |
| Roll & brush 1 coat | | | | | | | | | |
| Slow | 0.4 | 0.08 | 25.90 | 4.10 | 1.12 | 2.07 | 2.26 | 1.43 | 10.98 |
| Medium | 0.3 | 0.09 | 22.70 | 4.95 | 1.67 | 2.04 | 2.12 | 1.19 | 11.97 |
| Fast | 0.2 | 0.10 | 19.40 | 4.40 | 1.78 | 1.94 | 1.50 | .58 | 10.20 |
| **2 openings total** | | | | | | | | | |
| Undercoat, water base (material #3) | | | | | | | | | |
| Roll & brush 1 coat | | | | | | | | | |
| Slow | 0.7 | 0.16 | 18.60 | 7.18 | 1.95 | 2.98 | 3.76 | 2.38 | 18.25 |
| Medium | 0.5 | 0.17 | 16.20 | 8.25 | 2.78 | 2.75 | 3.38 | 1.89 | 19.05 |
| Fast | 0.3 | 0.18 | 13.90 | 6.60 | 2.68 | 2.50 | 2.18 | .84 | 14.80 |
| Undercoat, oil base (material #4) | | | | | | | | | |
| Roll & brush 1 coat | | | | | | | | | |
| Slow | 0.7 | 0.16 | 20.10 | 7.18 | 1.95 | 3.22 | 3.83 | 2.43 | 18.61 |
| Medium | 0.5 | 0.17 | 17.60 | 8.25 | 2.78 | 2.99 | 3.43 | 1.92 | 19.37 |
| Fast | 0.3 | 0.18 | 15.10 | 6.60 | 2.68 | 2.72 | 2.22 | .85 | 15.07 |
| Enamel, water base (material #9) | | | | | | | | | |
| Roll & brush 1 coat | | | | | | | | | |
| Slow | 0.7 | 0.17 | 22.80 | 7.18 | 1.95 | 3.88 | 4.04 | 2.56 | 19.61 |
| Medium | 0.5 | 0.18 | 19.90 | 8.25 | 2.78 | 3.58 | 3.58 | 2.00 | 20.19 |
| Fast | 0.3 | 0.20 | 17.10 | 6.60 | 2.68 | 3.42 | 2.35 | .90 | 15.95 |

| | Manhours per opening(s) | Gallons per opening(s) | Material cost per gallon | Labor cost per opening(s) | Labor burden opening(s) | Material cost per opening(s) | Overhead per opening(s) | Profit per opening(s) | Total price per opening(s) |
|---|---|---|---|---|---|---|---|---|---|
| Enamel, oil base (material #10) | | | | | | | | | |
| Roll & brush 1 coat | | | | | | | | | |
| Slow | 0.7 | 0.17 | 25.90 | 7.18 | 1.95 | 4.40 | 4.20 | 2.66 | 20.39 |
| Medium | 0.5 | 0.18 | 22.70 | 8.25 | 2.78 | 4.09 | 3.70 | 2.07 | 20.89 |
| Fast | 0.3 | 0.20 | 19.40 | 6.60 | 2.68 | 3.88 | 2.43 | .94 | 16.53 |
| **3 openings total** | | | | | | | | | |
| Undercoat, water base (material #3) | | | | | | | | | |
| Roll & brush 1 coat | | | | | | | | | |
| Slow | 1.00 | 0.23 | 18.60 | 10.25 | 2.79 | 4.28 | 5.37 | 3.40 | 26.09 |
| Medium | 0.75 | 0.25 | 16.20 | 12.38 | 4.16 | 4.05 | 5.05 | 2.82 | 28.46 |
| Fast | 0.50 | 0.27 | 13.90 | 11.00 | 4.46 | 3.75 | 3.55 | 1.37 | 24.13 |
| Undercoat, oil base (material #4) | | | | | | | | | |
| Roll & brush 1 coat | | | | | | | | | |
| Slow | 1.00 | 0.23 | 20.10 | 10.25 | 2.79 | 4.62 | 5.47 | 3.47 | 26.60 |
| Medium | 0.75 | 0.25 | 17.60 | 12.38 | 4.16 | 4.40 | 5.13 | 2.87 | 28.94 |
| Fast | 0.50 | 0.27 | 15.10 | 11.00 | 4.46 | 4.08 | 3.61 | 1.39 | 24.54 |
| Enamel, water base (material #9) | | | | | | | | | |
| Roll & brush 1 coat | | | | | | | | | |
| Slow | 1.00 | 0.25 | 22.80 | 10.25 | 2.79 | 5.70 | 5.81 | 3.68 | 28.23 |
| Medium | 0.75 | 0.27 | 19.90 | 12.38 | 4.16 | 5.37 | 5.37 | 3.00 | 30.28 |
| Fast | 0.50 | 0.30 | 17.10 | 11.00 | 4.46 | 5.13 | 3.81 | 1.46 | 25.86 |
| Enamel, oil base (material #10) | | | | | | | | | |
| Roll & brush 1 coat | | | | | | | | | |
| Slow | 1.00 | 0.25 | 25.90 | 10.25 | 2.79 | 6.48 | 6.05 | 3.84 | 29.41 |
| Medium | 0.75 | 0.27 | 22.70 | 12.38 | 4.16 | 6.13 | 5.56 | 3.11 | 31.34 |
| Fast | 0.50 | 0.30 | 19.40 | 11.00 | 4.46 | 5.82 | 3.94 | 1.51 | 26.73 |
| **4 openings total** | | | | | | | | | |
| Undercoat, water base (material #3) | | | | | | | | | |
| Roll & brush 1 coat | | | | | | | | | |
| Slow | 1.3 | 0.31 | 18.60 | 13.33 | 3.62 | 5.77 | 7.05 | 4.47 | 34.24 |
| Medium | 1.0 | 0.33 | 16.20 | 16.50 | 5.55 | 5.35 | 6.71 | 3.75 | 37.86 |
| Fast | 0.7 | 0.36 | 13.90 | 15.40 | 6.24 | 5.00 | 4.93 | 1.89 | 33.46 |
| Undercoat, oil base (material #4) | | | | | | | | | |
| Roll & brush 1 coat | | | | | | | | | |
| Slow | 1.3 | 0.31 | 20.10 | 13.33 | 3.62 | 6.23 | 7.19 | 4.56 | 34.93 |
| Medium | 1.0 | 0.33 | 17.60 | 16.50 | 5.55 | 5.81 | 6.83 | 3.82 | 38.51 |
| Fast | 0.7 | 0.36 | 15.10 | 15.40 | 6.24 | 5.44 | 5.01 | 1.93 | 34.02 |

| | Manhours per opening(s) | Gallons per opening(s) | Material cost per gallon | Labor cost per opening(s) | Labor burden opening(s) | Material cost per opening(s) | Overhead per opening(s) | Profit per opening(s) | Total price per opening(s) |
|---|---|---|---|---|---|---|---|---|---|
| Enamel, water base (material #9) | | | | | | | | | |
| Roll & brush 1 coat | | | | | | | | | |
| Slow | 1.3 | 0.33 | 22.80 | 13.33 | 3.62 | 7.52 | 7.59 | 4.81 | 36.87 |
| Medium | 1.0 | 0.36 | 19.90 | 16.50 | 5.55 | 7.16 | 7.16 | 4.00 | 40.37 |
| Fast | 0.7 | 0.40 | 17.10 | 15.40 | 6.24 | 6.84 | 5.27 | 2.03 | 35.78 |
| Enamel, oil base (material #10) | | | | | | | | | |
| Roll & brush 1 coat | | | | | | | | | |
| Slow | 1.3 | 0.33 | 25.90 | 13.33 | 3.62 | 8.55 | 7.91 | 5.01 | 38.42 |
| Medium | 1.0 | 0.36 | 22.70 | 16.50 | 5.55 | 8.17 | 7.40 | 4.14 | 41.76 |
| Fast | 0.7 | 0.40 | 19.40 | 15.40 | 6.24 | 7.76 | 5.44 | 2.09 | 36.93 |
| **5 openings total** | | | | | | | | | |
| Undercoat, water base (material #3) | | | | | | | | | |
| Roll & brush 1 coat | | | | | | | | | |
| Slow | 1.6 | 0.38 | 18.60 | 16.40 | 4.46 | 7.07 | 8.66 | 5.49 | 42.08 |
| Medium | 1.3 | 0.41 | 16.20 | 21.45 | 7.22 | 6.64 | 8.65 | 4.84 | 48.80 |
| Fast | 0.9 | 0.45 | 13.90 | 19.80 | 8.03 | 6.26 | 6.31 | 2.42 | 42.82 |
| Undercoat, oil base (material #4) | | | | | | | | | |
| Roll & brush 1 coat | | | | | | | | | |
| Slow | 1.6 | 0.38 | 20.10 | 16.40 | 4.46 | 7.64 | 8.84 | 5.60 | 42.94 |
| Medium | 1.3 | 0.41 | 17.60 | 21.45 | 7.22 | 7.22 | 8.79 | 4.91 | 49.59 |
| Fast | 0.9 | 0.45 | 15.10 | 19.80 | 8.03 | 6.80 | 6.41 | 2.46 | 43.50 |
| Enamel, water base (material #9) | | | | | | | | | |
| Roll & brush 1 coat | | | | | | | | | |
| Slow | 1.6 | 0.42 | 22.80 | 16.40 | 4.46 | 9.58 | 9.44 | 5.98 | 45.86 |
| Medium | 1.3 | 0.46 | 19.90 | 21.45 | 7.22 | 9.15 | 9.27 | 5.18 | 52.27 |
| Fast | 0.9 | 0.50 | 17.10 | 19.80 | 8.03 | 8.55 | 6.73 | 2.59 | 45.70 |
| Enamel, oil base (material #10) | | | | | | | | | |
| Roll & brush 1 coat | | | | | | | | | |
| Slow | 1.6 | 0.42 | 25.90 | 16.40 | 4.46 | 10.88 | 9.84 | 6.24 | 47.82 |
| Medium | 1.3 | 0.46 | 22.70 | 21.45 | 7.22 | 10.44 | 9.58 | 5.36 | 54.05 |
| Fast | 0.9 | 0.50 | 19.40 | 19.80 | 8.03 | 9.70 | 6.94 | 2.67 | 47.14 |
| **6 openings total** | | | | | | | | | |
| Undercoat, water base (material #3) | | | | | | | | | |
| Roll & brush 1 coat | | | | | | | | | |
| Slow | 1.90 | 0.46 | 18.60 | 19.48 | 5.30 | 8.56 | 10.34 | 6.55 | 50.23 |
| Medium | 1.45 | 0.50 | 16.20 | 23.93 | 8.04 | 8.10 | 9.82 | 5.49 | 55.38 |
| Fast | 1.00 | 0.54 | 13.90 | 22.00 | 8.92 | 7.51 | 7.11 | 2.73 | 48.27 |

| | Manhours per opening(s) | Gallons per opening(s) | Material cost per gallon | Labor cost per opening(s) | Labor burden opening(s) | Material cost per opening(s) | Overhead per opening(s) | Profit per opening(s) | Total price per opening(s) |
|---|---|---|---|---|---|---|---|---|---|
| Undercoat, oil base (material #4) | | | | | | | | | |
| Roll & brush 1 coat | | | | | | | | | |
| Slow | 1.90 | 0.46 | 20.10 | 19.48 | 5.30 | 9.25 | 10.55 | 6.69 | 51.27 |
| Medium | 1.45 | 0.50 | 17.60 | 23.93 | 8.04 | 8.80 | 9.99 | 5.58 | 56.34 |
| Fast | 1.00 | 0.54 | 15.10 | 22.00 | 8.92 | 8.15 | 7.23 | 2.78 | 49.08 |
| Enamel, water base (material #9) | | | | | | | | | |
| Roll & brush 1 coat | | | | | | | | | |
| Slow | 1.90 | 0.50 | 22.80 | 19.48 | 5.30 | 11.40 | 11.22 | 7.11 | 54.51 |
| Medium | 1.45 | 0.55 | 19.90 | 23.93 | 8.04 | 10.95 | 10.52 | 5.88 | 59.32 |
| Fast | 1.00 | 0.60 | 17.10 | 22.00 | 8.92 | 10.26 | 7.62 | 2.93 | 51.73 |
| Enamel, oil base (material #10) | | | | | | | | | |
| Roll & brush 1 coat | | | | | | | | | |
| Slow | 1.90 | 0.50 | 25.90 | 19.48 | 5.30 | 12.95 | 11.70 | 7.42 | 56.85 |
| Medium | 1.45 | 0.55 | 22.70 | 23.93 | 8.04 | 12.49 | 10.90 | 6.09 | 61.45 |
| Fast | 1.00 | 0.60 | 19.40 | 22.00 | 8.92 | 11.64 | 7.87 | 3.03 | 53.46 |
| **7 openings total** | | | | | | | | | |
| Undercoat, water base (material #3) | | | | | | | | | |
| Roll & brush 1 coat | | | | | | | | | |
| Slow | 2.2 | 0.54 | 18.60 | 22.55 | 6.14 | 10.04 | 12.01 | 7.61 | 58.35 |
| Medium | 1.7 | 0.59 | 16.20 | 28.05 | 9.44 | 9.56 | 11.53 | 6.44 | 65.02 |
| Fast | 1.2 | 0.64 | 13.90 | 26.40 | 10.70 | 8.90 | 8.51 | 3.27 | 57.78 |
| Undercoat, oil base (material #4) | | | | | | | | | |
| Roll & brush 1 coat | | | | | | | | | |
| Slow | 2.2 | 0.54 | 20.10 | 22.55 | 6.14 | 10.85 | 12.26 | 7.77 | 59.57 |
| Medium | 1.7 | 0.59 | 17.60 | 28.05 | 9.44 | 10.38 | 11.73 | 6.56 | 66.16 |
| Fast | 1.2 | 0.64 | 15.10 | 26.40 | 10.70 | 9.66 | 8.65 | 3.33 | 58.74 |
| Enamel, water base (material #9) | | | | | | | | | |
| Roll & brush 1 coat | | | | | | | | | |
| Slow | 2.2 | 0.58 | 22.80 | 22.55 | 6.14 | 13.22 | 12.99 | 8.24 | 63.14 |
| Medium | 1.7 | 0.64 | 19.90 | 28.05 | 9.44 | 12.74 | 12.31 | 6.88 | 69.42 |
| Fast | 1.2 | 0.70 | 17.10 | 26.40 | 10.70 | 11.97 | 9.08 | 3.49 | 61.64 |
| Enamel, oil base (material #10) | | | | | | | | | |
| Roll & brush 1 coat | | | | | | | | | |
| Slow | 2.2 | 0.58 | 25.90 | 22.55 | 6.14 | 15.02 | 13.55 | 8.59 | 65.85 |
| Medium | 1.7 | 0.64 | 22.70 | 28.05 | 9.44 | 14.53 | 12.74 | 7.12 | 71.88 |
| Fast | 1.2 | 0.70 | 19.40 | 26.40 | 10.70 | 13.58 | 9.38 | 3.60 | 63.66 |

| | Manhours per opening(s) | Gallons per opening(s) | Material cost per gallon | Labor cost per opening(s) | Labor burden opening(s) | Material cost per opening(s) | Overhead per opening(s) | Profit per opening(s) | Total price per opening(s) |
|---|---|---|---|---|---|---|---|---|---|
| **8 openings total** | | | | | | | | | |
| Undercoat, water base (material #3) | | | | | | | | | |
| Roll & brush 1 coat | | | | | | | | | |
| Slow | 2.50 | 0.62 | 18.60 | 25.63 | 6.97 | 11.53 | 13.68 | 8.67 | 66.48 |
| Medium | 1.95 | 0.67 | 16.20 | 32.18 | 10.82 | 10.85 | 13.20 | 7.38 | 74.43 |
| Fast | 1.40 | 0.73 | 13.90 | 30.80 | 12.49 | 10.15 | 9.89 | 3.80 | 67.13 |
| Undercoat, oil base (material #4) | | | | | | | | | |
| Roll & brush 1 coat | | | | | | | | | |
| Slow | 2.50 | 0.62 | 20.10 | 25.63 | 6.97 | 12.46 | 13.97 | 8.86 | 67.89 |
| Medium | 1.95 | 0.67 | 17.60 | 32.18 | 10.82 | 11.79 | 13.43 | 7.51 | 75.73 |
| Fast | 1.40 | 0.73 | 15.10 | 30.80 | 12.49 | 11.02 | 10.05 | 3.86 | 68.22 |
| Enamel, water base (material #9) | | | | | | | | | |
| Roll & brush 1 coat | | | | | | | | | |
| Slow | 2.50 | 0.67 | 22.80 | 25.63 | 6.97 | 15.28 | 14.85 | 9.41 | 72.14 |
| Medium | 1.95 | 0.74 | 19.90 | 32.18 | 10.82 | 14.73 | 14.15 | 7.91 | 79.79 |
| Fast | 1.40 | 0.80 | 17.10 | 30.80 | 12.49 | 13.68 | 10.54 | 4.05 | 71.56 |
| Enamel, oil base (material #10) | | | | | | | | | |
| Roll & brush 1 coat | | | | | | | | | |
| Slow | 2.50 | 0.67 | 25.90 | 25.63 | 6.97 | 17.35 | 15.49 | 9.82 | 75.26 |
| Medium | 1.95 | 0.74 | 22.70 | 32.18 | 10.82 | 16.80 | 14.65 | 8.19 | 82.64 |
| Fast | 1.40 | 0.80 | 19.40 | 30.80 | 12.49 | 15.52 | 10.88 | 4.18 | 73.87 |
| **9 openings total** | | | | | | | | | |
| Undercoat, water base (material #3) | | | | | | | | | |
| Roll & brush 1 coat | | | | | | | | | |
| Slow | 2.80 | 0.69 | 18.60 | 28.70 | 7.81 | 12.83 | 15.30 | 9.70 | 74.34 |
| Medium | 2.15 | 0.75 | 16.20 | 35.48 | 11.93 | 12.15 | 14.59 | 8.16 | 82.31 |
| Fast | 1.50 | 0.81 | 13.90 | 33.00 | 13.38 | 11.26 | 10.66 | 4.10 | 72.40 |
| Undercoat, oil base (material #4) | | | | | | | | | |
| Roll & brush 1 coat | | | | | | | | | |
| Slow | 2.80 | 0.69 | 20.10 | 28.70 | 7.81 | 13.87 | 15.62 | 9.90 | 75.90 |
| Medium | 2.15 | 0.75 | 17.60 | 35.48 | 11.93 | 13.20 | 14.85 | 8.30 | 83.76 |
| Fast | 1.50 | 0.81 | 15.10 | 33.00 | 13.38 | 12.23 | 10.84 | 4.17 | 73.62 |
| Enamel, water base (material #9) | | | | | | | | | |
| Roll & brush 1 coat | | | | | | | | | |
| Slow | 2.80 | 0.75 | 22.80 | 28.70 | 7.81 | 17.10 | 16.62 | 10.54 | 80.77 |
| Medium | 2.15 | 0.82 | 19.90 | 35.48 | 11.93 | 16.32 | 15.62 | 8.73 | 88.08 |
| Fast | 1.50 | 0.90 | 17.10 | 33.00 | 13.38 | 15.39 | 11.43 | 4.39 | 77.59 |

| | Manhours per opening(s) | Gallons per opening(s) | Material cost per gallon | Labor cost per opening(s) | Labor burden opening(s) | Material cost per opening(s) | Overhead per opening(s) | Profit per opening(s) | Total price per opening(s) |
|---|---|---|---|---|---|---|---|---|---|
| Enamel, oil base (material #10) | | | | | | | | | |
| Roll & brush 1 coat | | | | | | | | | |
| Slow | 2.80 | 0.75 | 25.90 | 28.70 | 7.81 | 19.43 | 17.34 | 10.99 | 84.27 |
| Medium | 2.15 | 0.82 | 22.70 | 35.48 | 11.93 | 18.61 | 16.18 | 9.04 | 91.24 |
| Fast | 1.50 | 0.90 | 19.40 | 33.00 | 13.38 | 17.46 | 11.81 | 4.54 | 80.19 |
| **10 openings total** | | | | | | | | | |
| Undercoat, water base (material #3) | | | | | | | | | |
| Roll & brush 1 coat | | | | | | | | | |
| Slow | 3.1 | 0.77 | 18.60 | 31.78 | 8.64 | 14.32 | 16.98 | 10.76 | 82.48 |
| Medium | 2.4 | 0.84 | 16.20 | 39.60 | 13.32 | 13.61 | 16.30 | 9.11 | 91.94 |
| Fast | 1.7 | 0.90 | 13.90 | 37.40 | 15.16 | 12.51 | 12.04 | 4.63 | 81.74 |
| Undercoat, oil base (material #4) | | | | | | | | | |
| Roll & brush 1 coat | | | | | | | | | |
| Slow | 3.1 | 0.77 | 20.10 | 31.78 | 8.64 | 15.48 | 17.34 | 10.99 | 84.23 |
| Medium | 2.4 | 0.84 | 17.60 | 39.60 | 13.32 | 14.78 | 16.59 | 9.27 | 93.56 |
| Fast | 1.7 | 0.90 | 15.10 | 37.40 | 15.16 | 13.59 | 12.24 | 4.70 | 83.09 |
| Enamel, water base (material #9) | | | | | | | | | |
| Slow | 3.1 | 0.83 | 22.80 | 31.78 | 8.64 | 18.92 | 18.40 | 11.66 | 89.40 |
| Medium | 2.4 | 0.92 | 19.90 | 39.60 | 13.32 | 18.31 | 17.45 | 9.76 | 98.44 |
| Fast | 1.7 | 1.00 | 17.10 | 37.40 | 15.16 | 17.10 | 12.89 | 4.95 | 87.50 |
| Enamel, oil base (material #10) | | | | | | | | | |
| Roll & brush 1 coat | | | | | | | | | |
| Slow | 3.1 | 0.83 | 25.90 | 31.78 | 8.64 | 21.50 | 19.20 | 12.17 | 93.29 |
| Medium | 2.4 | 0.92 | 22.70 | 39.60 | 13.32 | 20.88 | 18.08 | 10.11 | 101.99 |
| Fast | 1.7 | 1.00 | 19.40 | 37.40 | 15.16 | 19.40 | 13.31 | 5.12 | 90.39 |

Use these figures for painting interior doors, pullmans, linens and other surfaces described in Figure 20 on page 100. ADD for additional preparation time, as needed, including setup, remove & replace (hardware), protection (mask walls and visqueen floors); putty, caulking, sanding and cleanup.
Slow work is based on a $10.25 hourly wage, "Medium" work on a $16.50 hourly wage, and "Fast" work on a $22.00 hourly wage. Other qualifications that apply to this table are on page 9.

| | Manhours per door | Doors per gallon | Material cost per gallon | Labor cost per door | Labor burden door | Material cost per door | Overhead per door | Profit per door | Total price per door |
|---|---|---|---|---|---|---|---|---|---|
| **Doors, interior, flush, paint grade, roll & brush, based on per door method** | | | | | | | | | |
| Undercoat, water base (material #3) | | | | | | | | | |
| Roll & brush 1 coat | | | | | | | | | |
| Slow | 0.40 | 13.0 | 18.60 | 4.10 | 1.12 | 1.43 | 2.06 | 1.31 | 10.02 |
| Medium | 0.30 | 11.5 | 16.20 | 4.95 | 1.67 | 1.41 | 1.97 | 1.10 | 11.10 |
| Fast | 0.20 | 10.0 | 13.90 | 4.40 | 1.78 | 1.39 | 1.40 | .54 | 9.51 |
| Undercoat, oil base (material #4) | | | | | | | | | |
| Roll & brush 1 coat | | | | | | | | | |
| Slow | 0.40 | 13.0 | 20.10 | 4.10 | 1.12 | 1.55 | 2.10 | 1.33 | 10.20 |
| Medium | 0.30 | 11.5 | 17.60 | 4.95 | 1.67 | 1.53 | 2.00 | 1.12 | 11.27 |
| Fast | 0.20 | 10.0 | 15.10 | 4.40 | 1.78 | 1.51 | 1.42 | .55 | 9.66 |
| Enamel, water base (material #9) | | | | | | | | | |
| Roll & brush 1st finish coat | | | | | | | | | |
| Slow | 0.33 | 14.0 | 22.80 | 3.38 | .92 | 1.63 | 1.84 | 1.17 | 8.94 |
| Medium | 0.25 | 12.5 | 19.90 | 4.13 | 1.38 | 1.59 | 1.74 | .97 | 9.81 |
| Fast | 0.17 | 11.0 | 17.10 | 3.74 | 1.52 | 1.55 | 1.26 | .48 | 8.55 |
| Roll & brush additional finish coats | | | | | | | | | |
| Slow | 0.25 | 15.0 | 22.80 | 2.56 | .70 | 1.52 | 1.48 | .94 | 7.20 |
| Medium | 0.20 | 13.5 | 19.90 | 3.30 | 1.11 | 1.47 | 1.44 | .81 | 8.13 |
| Fast | 0.15 | 12.0 | 17.10 | 3.30 | 1.34 | 1.43 | 1.12 | .43 | 7.62 |
| Enamel, oil base (material #10) | | | | | | | | | |
| Roll & brush 1st finish coat | | | | | | | | | |
| Slow | 0.33 | 14.0 | 25.90 | 3.38 | .92 | 1.85 | 1.91 | 1.21 | 9.27 |
| Medium | 0.25 | 12.5 | 22.70 | 4.13 | 1.38 | 1.82 | 1.80 | 1.01 | 10.14 |
| Fast | 0.17 | 11.0 | 19.40 | 3.74 | 1.52 | 1.76 | 1.30 | .50 | 8.82 |
| Roll & brush additional finish coats | | | | | | | | | |
| Slow | 0.25 | 15.0 | 25.90 | 2.56 | .70 | 1.73 | 1.55 | .98 | 7.52 |
| Medium | 0.20 | 13.5 | 22.70 | 3.30 | 1.11 | 1.68 | 1.49 | .83 | 8.41 |
| Fast | 0.15 | 12.0 | 19.40 | 3.30 | 1.34 | 1.62 | 1.16 | .45 | 7.87 |

This table is not to be confused with the Opening Count Method for estimating doors. Use these figures for painting both interior and exterior doors. These costs are in addition to the costs for applying the finish coats to exterior doors as explained on page 95. These figures include coating all six (6) sides of each door along with the frame and the jamb on both sides. These figures include minimal preparation time. ADD for additional preparation time, as needed, including setup, remove & replace (hardware), protection (mask walls and visqueen floors); putty, caulking, sanding and cleanup. "Slow" work is based on a $10.25 hourly wage, "Medium" work on a $16.50 hourly wage, and "Fast" work on a $22.00 hourly wage. Other qualifications that apply to this table are on page 9.

| | Manhours per door | Doors per gallon | Material cost per gallon | Labor cost per door | Labor burden door | Material cost per door | Overhead per door | Profit per door | Total price per door |
|---|---|---|---|---|---|---|---|---|---|
| **Doors, interior, flush, paint grade, spray application, based on per door method** | | | | | | | | | |
| Undercoat, water base (material #3) | | | | | | | | | |
| Spray 1 coat | | | | | | | | | |
| Slow | 0.10 | 17 | 18.60 | 1.03 | .27 | 1.09 | .74 | .47 | 3.60 |
| Medium | 0.09 | 16 | 16.20 | 1.49 | .49 | 1.01 | .74 | .41 | 4.14 |
| Fast | 0.08 | 15 | 13.90 | 1.76 | .71 | .93 | .63 | .24 | 4.27 |
| Undercoat, oil base (material #4) | | | | | | | | | |
| Spray 1 coat | | | | | | | | | |
| Slow | 0.10 | 17 | 20.10 | 1.03 | .27 | 1.18 | .77 | .49 | 3.74 |
| Medium | 0.09 | 16 | 17.60 | 1.49 | .49 | 1.10 | .76 | .42 | 4.26 |
| Fast | 0.08 | 15 | 15.10 | 1.76 | .71 | 1.01 | .64 | .25 | 4.37 |
| Enamel, water base (material #9) | | | | | | | | | |
| Spray 1st finish coat | | | | | | | | | |
| Slow | 0.09 | 18 | 22.80 | .92 | .25 | 1.27 | .76 | .48 | 3.68 |
| Medium | 0.08 | 17 | 19.90 | 1.32 | .44 | 1.17 | .72 | .40 | 4.05 |
| Fast | 0.07 | 16 | 17.10 | 1.54 | .62 | 1.07 | .60 | .23 | 4.06 |
| Spray 2nd or additional finish coats | | | | | | | | | |
| Slow | 0.08 | 19 | 22.80 | .82 | .22 | 1.20 | .69 | .44 | 3.37 |
| Medium | 0.07 | 18 | 19.90 | 1.16 | .38 | 1.11 | .65 | .36 | 3.66 |
| Fast | 0.06 | 17 | 17.10 | 1.32 | .54 | 1.01 | .53 | .20 | 3.60 |
| Enamel, oil base (material #10) | | | | | | | | | |
| Spray 1st finish coat | | | | | | | | | |
| Slow | 0.09 | 18 | 25.90 | .92 | .25 | 1.44 | .81 | .51 | 3.93 |
| Medium | 0.08 | 17 | 22.70 | 1.32 | .44 | 1.34 | .76 | .42 | 4.28 |
| Fast | 0.07 | 16 | 19.40 | 1.54 | .62 | 1.21 | .62 | .24 | 4.23 |
| Spray 2nd or additional finish coats | | | | | | | | | |
| Slow | 0.08 | 19 | 25.90 | .82 | .22 | 1.36 | .74 | .47 | 3.61 |
| Medium | 0.07 | 18 | 22.70 | 1.16 | .38 | 1.26 | .69 | .39 | 3.88 |
| Fast | 0.06 | 17 | 19.40 | 1.32 | .54 | 1.14 | .56 | .21 | 3.77 |

This table is not to be confused with the Opening Count Method for estimating doors. Use these figures for painting both interior and exterior doors. These costs are in addition to the costs for applying the finish coats to exterior doors as explained on page 95. These figures include coating all six (6) sides of each door along with the frame and the jamb on both sides. These figures include minimal preparation time. ADD for additional preparation time, as needed, including setup, remove & replace (hardware), protection (mask walls and visqueen floors); putty, caulking, sanding and cleanup. "Slow" work is based on a $10.25 hourly wage, "Medium" work on a $16.50 hourly wage, and "Fast" work on a $22.00 hourly wage. Other qualifications that apply to this table are on page 9.

| | Manhours per door | Doors per gallon | Material cost per gallon | Labor cost per door | Labor burden door | Material cost per door | Overhead per door | Profit per door | Total price per door |
|---|---|---|---|---|---|---|---|---|---|

## Doors, interior, flush, stain grade, spray application, based on per door method

| | Manhours per door | Doors per gallon | Material cost per gallon | Labor cost per door | Labor burden door | Material cost per door | Overhead per door | Profit per door | Total price per door |
|---|---|---|---|---|---|---|---|---|---|
| Complete 7 step stain, seal & 2 coat lacquer system (material #11) | | | | | | | | | |
| Spray all coats | | | | | | | | | |
| Slow | 0.90 | 6 | 20.20 | 9.23 | 2.51 | 3.37 | 4.69 | 2.97 | 22.77 |
| Medium | 0.80 | 5 | 18.30 | 13.20 | 4.44 | 3.66 | 5.22 | 2.92 | 29.44 |
| Fast | 0.70 | 4 | 15.10 | 15.40 | 6.24 | 3.78 | 4.70 | 1.81 | 31.93 |

This table is not to be confused with the Opening Count Method for estimating doors. Use these figures for painting both interior and exterior doors. These costs are in addition to the costs for applying the finish coats to exterior doors as explained on page 95. These figures include coating all six (6) sides of each door along with the frame and the jamb on both sides. These figures include minimal preparation time. ADD for preparation time, as needed, including setup, remove & replace (hardware), protection (mask walls and visqueen floors); putty, caulking, sanding and cleanup. "Slow" work is based on a $10.25 hourly wage, "Medium" work on a $16.50 hourly wage, and "Fast" work on a $22.00 hourly wage. Other qualifications that apply to this table are on page 9.

## Doors, interior, French, paint grade, roll & brush, based on per door method

| | Manhours per door | Doors per gallon | Material cost per gallon | Labor cost per door | Labor burden door | Material cost per door | Overhead per door | Profit per door | Total price per door |
|---|---|---|---|---|---|---|---|---|---|
| Undercoat, water base (material #3) | | | | | | | | | |
| Roll & brush 1 coat | | | | | | | | | |
| Slow | 0.45 | 14 | 18.60 | 4.61 | 1.26 | 1.33 | 2.23 | 1.41 | 10.84 |
| Medium | 0.38 | 13 | 16.20 | 6.27 | 2.11 | 1.25 | 2.36 | 1.32 | 13.31 |
| Fast | 0.30 | 12 | 13.90 | 6.60 | 2.68 | 1.16 | 1.93 | .74 | 13.11 |
| Undercoat, oil base (material #4) | | | | | | | | | |
| Roll & brush 1 coat | | | | | | | | | |
| Slow | 0.45 | 14 | 20.10 | 4.61 | 1.26 | 1.44 | 2.27 | 1.44 | 11.02 |
| Medium | 0.38 | 13 | 17.60 | 6.27 | 2.11 | 1.35 | 2.38 | 1.33 | 13.44 |
| Fast | 0.30 | 12 | 15.10 | 6.60 | 2.68 | 1.26 | 1.95 | .75 | 13.24 |
| Enamel, water base (material #9) | | | | | | | | | |
| Roll & brush 1st finish coat | | | | | | | | | |
| Slow | 0.43 | 15 | 22.80 | 4.41 | 1.20 | 1.52 | 2.21 | 1.40 | 10.74 |
| Medium | 0.35 | 14 | 19.90 | 5.78 | 1.94 | 1.42 | 2.24 | 1.25 | 12.63 |
| Fast | 0.28 | 13 | 17.10 | 6.16 | 2.50 | 1.32 | 1.85 | .71 | 12.54 |
| Roll & brush 2nd or additional finish coats | | | | | | | | | |
| Slow | 0.40 | 16 | 22.80 | 4.10 | 1.12 | 1.43 | 2.06 | 1.31 | 10.02 |
| Medium | 0.33 | 15 | 19.90 | 5.45 | 1.83 | 1.33 | 2.11 | 1.18 | 11.90 |
| Fast | 0.25 | 14 | 17.10 | 5.50 | 2.23 | 1.22 | 1.66 | .64 | 11.25 |

| | Manhours per door | Doors per gallon | Material cost per gallon | Labor cost per door | Labor burden door | Material cost per door | Overhead per door | Profit per door | Total price per door |
|---|---|---|---|---|---|---|---|---|---|
| Enamel, oil base (material #10) | | | | | | | | | |
| Roll & brush 1st finish coat | | | | | | | | | |
| Slow | 0.43 | 15 | 25.90 | 4.41 | 1.20 | 1.73 | 2.28 | 1.44 | 11.06 |
| Medium | 0.35 | 14 | 22.70 | 5.78 | 1.94 | 1.62 | 2.29 | 1.28 | 12.91 |
| Fast | 0.28 | 13 | 19.40 | 6.16 | 2.50 | 1.49 | 1.88 | .72 | 12.75 |
| Roll & brush 2nd or additional finish coats | | | | | | | | | |
| Slow | 0.40 | 16 | 25.90 | 4.10 | 1.12 | 1.62 | 2.12 | 1.34 | 10.30 |
| Medium | 0.33 | 15 | 22.70 | 5.45 | 1.83 | 1.51 | 2.15 | 1.20 | 12.14 |
| Fast | 0.25 | 14 | 19.40 | 5.50 | 2.23 | 1.39 | 1.69 | .65 | 11.46 |

This table is not to be confused with the Opening Count Method for estimating doors. Use these figures for painting both interior and exterior doors. These costs are in addition to the costs for applying the finish coats to exterior doors as explained on page 95. These figures include coating all six (6) sides of each door along with the frame and the jamb on both sides. These figures include minimal preparation time. ADD for additional preparation time, as needed, including setup, remove & replace (hardware), protection (mask walls and visqueen floors); putty, caulking, sanding and cleanup. "Slow" work is based on a $10.25 hourly wage, "Medium" work on a $16.50 hourly wage, and "Fast" work on a $22.00 hourly wage. Other qualifications that apply to this table are on page 9.

## Doors, interior, French, stain grade, spray application, based on per door method

| | Manhours per door | Doors per gallon | Material cost per gallon | Labor cost per door | Labor burden door | Material cost per door | Overhead per door | Profit per door | Total price per door |
|---|---|---|---|---|---|---|---|---|---|
| Complete 7 step stain, seal & 2 coat lacquer system (material #11) | | | | | | | | | |
| Spray all coats | | | | | | | | | |
| Slow | 1.50 | 12 | 20.20 | 15.38 | 4.18 | 1.68 | 6.59 | 4.18 | 32.01 |
| Medium | 1.25 | 11 | 18.30 | 20.63 | 6.93 | 1.66 | 7.16 | 4.00 | 40.38 |
| Fast | 1.00 | 10 | 15.10 | 22.00 | 8.92 | 1.51 | 6.00 | 2.31 | 40.74 |

This table is not to be confused with the Opening Count Method for estimating doors. Use these figures for painting both interior and exterior doors. These costs are in addition to the costs for applying the finish coats to exterior doors as explained on page 95. These figures include coating all six (6) sides of each door along with the frame and the jamb on both sides. These figures include minimal preparation time. ADD for additional preparation time, as needed, including setup, remove & replace (hardware), protection (mask walls and visqueen floors); putty, caulking, sanding and cleanup. "Slow" work is based on a $10.25 hourly wage, "Medium" work on a $16.50 hourly wage, and "Fast" work on a $22.00 hourly wage. Other qualifications that apply to this table are on page 9.

| | Manhours per door | Doors per gallon | Material cost per gallon | Labor cost per door | Labor burden door | Material cost per door | Overhead per door | Profit per door | Total price per door |
|---|---|---|---|---|---|---|---|---|---|

## Doors, interior, louvered, paint grade, roll & brush, based on per door method

| | Manhours per door | Doors per gallon | Material cost per gallon | Labor cost per door | Labor burden door | Material cost per door | Overhead per door | Profit per door | Total price per door |
|---|---|---|---|---|---|---|---|---|---|
| Undercoat, water base (material #3) | | | | | | | | | |
| Roll & brush 1 coat | | | | | | | | | |
| Slow | 0.67 | 8 | 18.60 | 6.87 | 1.87 | 2.33 | 3.43 | 2.18 | 16.68 |
| Medium | 0.54 | 7 | 16.20 | 8.91 | 3.00 | 2.31 | 3.48 | 1.95 | 19.65 |
| Fast | 0.40 | 6 | 13.90 | 8.80 | 3.57 | 2.32 | 2.72 | 1.04 | 18.45 |
| Undercoat, oil base (material #4) | | | | | | | | | |
| Roll & brush 1 coat | | | | | | | | | |
| Slow | 0.67 | 8 | 20.10 | 6.87 | 1.87 | 2.51 | 3.49 | 2.21 | 16.95 |
| Medium | 0.54 | 7 | 17.60 | 8.91 | 3.00 | 2.51 | 3.53 | 1.97 | 19.92 |
| Fast | 0.40 | 6 | 15.10 | 8.80 | 3.57 | 2.52 | 2.75 | 1.06 | 18.70 |
| Enamel, water base (material #9) | | | | | | | | | |
| Roll & brush 1st finish coat | | | | | | | | | |
| Slow | 0.50 | 9 | 22.80 | 5.13 | 1.39 | 2.53 | 2.81 | 1.78 | 13.64 |
| Medium | 0.42 | 8 | 19.90 | 6.93 | 2.33 | 2.49 | 2.88 | 1.61 | 16.24 |
| Fast | 0.33 | 7 | 17.10 | 7.26 | 2.94 | 2.44 | 2.34 | .90 | 15.88 |
| Roll & brush 2nd or additional finish coats | | | | | | | | | |
| Slow | 0.40 | 10 | 22.80 | 4.10 | 1.12 | 2.28 | 2.33 | 1.47 | 11.30 |
| Medium | 0.30 | 9 | 19.90 | 4.95 | 1.67 | 2.21 | 2.16 | 1.21 | 12.20 |
| Fast | 0.20 | 8 | 17.10 | 4.40 | 1.78 | 2.14 | 1.54 | .59 | 10.45 |
| Enamel, oil base (material #10) | | | | | | | | | |
| Roll & brush 1st finish coat | | | | | | | | | |
| Slow | 0.50 | 9 | 25.90 | 5.13 | 1.39 | 2.88 | 2.92 | 1.85 | 14.17 |
| Medium | 0.42 | 8 | 22.70 | 6.93 | 2.33 | 2.84 | 2.96 | 1.66 | 16.72 |
| Fast | 0.33 | 7 | 19.40 | 7.26 | 2.94 | 2.77 | 2.40 | .92 | 16.29 |
| Roll & brush 2nd or additional finish coats | | | | | | | | | |
| Slow | 0.40 | 10 | 25.90 | 4.10 | 1.12 | 2.59 | 2.42 | 1.53 | 11.76 |
| Medium | 0.30 | 9 | 22.70 | 4.95 | 1.67 | 2.52 | 2.24 | 1.25 | 12.63 |
| Fast | 0.20 | 8 | 19.40 | 4.40 | 1.78 | 2.43 | 1.59 | .61 | 10.81 |

This table is not to be confused with the Opening Count Method for estimating doors. Use these figures for painting both interior and exterior doors. These costs are in addition to the costs for applying the finish coats to exterior doors as explained on page 95. These figures include coating all six (6) sides of each door along with the frame and the jamb on both sides. These figures include minimal preparation time. ADD for additional preparation time, as needed, including setup, remove & replace (hardware), protection (mask walls and visqueen floors); putty, caulking, sanding and cleanup. "Slow" work is based on a $10.25 hourly wage, "Medium" work on a $16.50 hourly wage, and "Fast" work on a $22.00 hourly wage. Other qualifications that apply to this table are on page 9.

| | Manhours per door | Doors per gallon | Material cost per gallon | Labor cost per door | Labor burden door | Material cost per door | Overhead per door | Profit per door | Total price per door |
|---|---|---|---|---|---|---|---|---|---|
| **Doors, interior, louvered, paint grade, spray application, based on per door method** | | | | | | | | | |
| Undercoat, water base (material #3) | | | | | | | | | |
| Spray 1 coat | | | | | | | | | |
| Slow | 0.17 | 12 | 18.60 | 1.74 | .48 | 1.55 | 1.17 | .74 | 5.68 |
| Medium | 0.14 | 11 | 16.20 | 2.31 | .78 | 1.47 | 1.12 | .62 | 6.30 |
| Fast | 0.12 | 10 | 13.90 | 2.64 | 1.07 | 1.39 | .94 | .36 | 6.40 |
| Undercoat, oil base (material #4) | | | | | | | | | |
| Spray 1 coat | | | | | | | | | |
| Slow | 0.17 | 12 | 20.10 | 1.74 | .48 | 1.68 | 1.21 | .77 | 5.88 |
| Medium | 0.14 | 11 | 17.60 | 2.31 | .78 | 1.60 | 1.15 | .64 | 6.48 |
| Fast | 0.12 | 10 | 15.10 | 2.64 | 1.07 | 1.51 | .97 | .37 | 6.56 |
| Enamel, water base (material #9) | | | | | | | | | |
| Spray 1st finish coat | | | | | | | | | |
| Slow | 0.13 | 13 | 22.80 | 1.33 | .37 | 1.75 | 1.07 | .68 | 5.20 |
| Medium | 0.11 | 12 | 19.90 | 1.82 | .61 | 1.66 | 1.00 | .56 | 5.65 |
| Fast | 0.09 | 11 | 17.10 | 1.98 | .80 | 1.55 | .80 | .31 | 5.44 |
| Spray 2nd or additional finish coats | | | | | | | | | |
| Slow | 0.10 | 14 | 22.80 | 1.03 | .27 | 1.63 | .91 | .58 | 4.42 |
| Medium | 0.09 | 13 | 19.90 | 1.49 | .49 | 1.53 | .86 | .48 | 4.85 |
| Fast | 0.08 | 12 | 17.10 | 1.76 | .71 | 1.43 | .72 | .28 | 4.90 |
| Enamel, oil base (material #10) | | | | | | | | | |
| Spray 1st finish coat | | | | | | | | | |
| Slow | 0.13 | 13 | 25.90 | 1.33 | .37 | 1.99 | 1.14 | .72 | 5.55 |
| Medium | 0.11 | 12 | 22.70 | 1.82 | .61 | 1.89 | 1.06 | .59 | 5.97 |
| Fast | 0.09 | 11 | 19.40 | 1.98 | .80 | 1.76 | .84 | .32 | 5.70 |
| Spray 2nd or additional finish coats | | | | | | | | | |
| Slow | 0.10 | 14 | 25.90 | 1.03 | .27 | 1.85 | .98 | .62 | 4.75 |
| Medium | 0.09 | 13 | 22.70 | 1.49 | .49 | 1.75 | .92 | .51 | 5.16 |
| Fast | 0.08 | 12 | 19.40 | 1.76 | .71 | 1.62 | .76 | .29 | 5.14 |

This table is not to be confused with the Opening Count Method for estimating doors. Use these figures for painting both interior and exterior doors. These costs are in addition to the costs for applying the finish coats to exterior doors as explained on page 95. These figures include coating all six (6) sides of each door along with the frame and the jamb on both sides. These figures include minimal preparation time. ADD for additional preparation time, as needed, including setup, remove & replace (hardware), protection (mask walls and visqueen floors); putty, caulking, sanding and cleanup. "Slow" work is based on a $10.25 hourly wage, "Medium" work on a $16.50 hourly wage, and "Fast" work on a $22.00 hourly wage. Other qualifications that apply to this table are on page 9.

| | Manhours per door | Doors per gallon | Material cost per gallon | Labor cost per door | Labor burden door | Material cost per door | Overhead per door | Profit per door | Total price per door |
|---|---|---|---|---|---|---|---|---|---|

## Doors, interior, louvered, stain grade, spray application, based on per door method

Complete 7 step stain, seal & 2 coat lacquer system (material #11)

| Spray all coats | | | | | | | | | |
|---|---|---|---|---|---|---|---|---|---|
| Slow | 1.70 | 5 | 20.20 | 17.43 | 4.74 | 4.04 | 8.13 | 5.15 | 39.49 |
| Medium | 1.45 | 4 | 18.30 | 23.93 | 8.04 | 4.58 | 8.96 | 5.01 | 50.52 |
| Fast | 1.20 | 3 | 15.10 | 26.40 | 10.70 | 5.03 | 7.80 | 3.00 | 52.93 |

This table is not to be confused with the Opening Count Method for estimating doors. Use these figures for painting both interior and exterior doors. These costs are in addition to the costs for applying the finish coats to exterior doors as explained on page 95. These figures include coating all six (6) sides of each door along with the frame and the jamb on both sides. These figures include minimal preparation time. ADD for additional preparation time, as needed, including setup, remove & replace (hardware), protection (mask walls and visqueen floors); putty, caulking, sanding and cleanup. "Slow" work is based on a $10.25 hourly wage, "Medium" work on a $16.50 hourly wage, and "Fast" work on a $22.00 hourly wage. Other qualifications that apply to this table are on page 9.

| | Manhours per door | Doors per gallon | Material cost per gallon | Labor cost per door | Labor burden door | Material cost per door | Overhead per door | Profit per door | Total price per door |
|---|---|---|---|---|---|---|---|---|---|
| **Doors, interior, panel, paint grade, roll & brush, based on per door method** | | | | | | | | | |
| Undercoat, water base (material #3) | | | | | | | | | |
| Roll & brush 1 coat | | | | | | | | | |
| Slow | 0.50 | 8 | 18.60 | 5.13 | 1.39 | 2.33 | 2.75 | 1.74 | 13.34 |
| Medium | 0.33 | 7 | 16.20 | 5.45 | 1.83 | 2.31 | 2.35 | 1.31 | 13.25 |
| Fast | 0.25 | 6 | 13.90 | 5.50 | 2.23 | 2.32 | 1.86 | .71 | 12.62 |
| Undercoat, oil base (material #4) | | | | | | | | | |
| Roll & brush 1 coat | | | | | | | | | |
| Slow | 0.50 | 8 | 20.10 | 5.13 | 1.39 | 2.51 | 2.80 | 1.78 | 13.61 |
| Medium | 0.33 | 7 | 17.60 | 5.45 | 1.83 | 2.51 | 2.40 | 1.34 | 13.53 |
| Fast | 0.25 | 6 | 15.10 | 5.50 | 2.23 | 2.52 | 1.90 | .73 | 12.88 |
| Enamel, water base (material #9) | | | | | | | | | |
| Roll & brush 1st finish coat | | | | | | | | | |
| Slow | 0.40 | 11 | 22.80 | 4.10 | 1.12 | 2.07 | 2.26 | 1.43 | 10.98 |
| Medium | 0.30 | 10 | 19.90 | 4.95 | 1.67 | 1.99 | 2.11 | 1.18 | 11.90 |
| Fast | 0.20 | 9 | 17.10 | 4.40 | 1.78 | 1.90 | 1.49 | .57 | 10.14 |
| Roll & brush 2nd or additional finish coats | | | | | | | | | |
| Slow | 0.33 | 12 | 22.80 | 3.38 | .92 | 1.90 | 1.92 | 1.22 | 9.34 |
| Medium | 0.25 | 11 | 19.90 | 4.13 | 1.38 | 1.81 | 1.80 | 1.00 | 10.12 |
| Fast | 0.17 | 10 | 17.10 | 3.74 | 1.52 | 1.71 | 1.29 | .50 | 8.76 |
| Enamel, oil base (material #10) | | | | | | | | | |
| Roll & brush 1st finish coat | | | | | | | | | |
| Slow | 0.40 | 11 | 25.90 | 4.10 | 1.12 | 2.35 | 2.35 | 1.49 | 11.41 |
| Medium | 0.30 | 10 | 22.70 | 4.95 | 1.67 | 2.27 | 2.18 | 1.22 | 12.29 |
| Fast | 0.20 | 9 | 19.40 | 4.40 | 1.78 | 2.16 | 1.54 | .59 | 10.47 |
| Roll & brush 2nd or additional finish coats | | | | | | | | | |
| Slow | 0.33 | 12 | 25.90 | 3.38 | .92 | 2.16 | 2.00 | 1.27 | 9.73 |
| Medium | 0.25 | 11 | 22.70 | 4.13 | 1.38 | 2.06 | 1.86 | 1.04 | 10.47 |
| Fast | 0.17 | 10 | 19.40 | 3.74 | 1.52 | 1.94 | 1.33 | .51 | 9.04 |

This table is not to be confused with the Opening Count Method for estimating doors. Use these figures for painting both interior and exterior doors. These costs are in addition to the costs for applying the finish coats to exterior doors as explained on page 95. These figures include coating all six (6) sides of each door along with the frame and the jamb on both sides. These figures include minimal preparation time. ADD for additional preparation time, as needed, including setup, remove & replace (hardware), protection (mask walls and visqueen floors); putty, caulking, sanding and cleanup. "Slow" work is based on a $10.25 hourly wage, "Medium" work on a $16.50 hourly wage, and "Fast" work on a $22.00 hourly wage. Other qualifications that apply to this table are on page 9.

| | Manhours per door | Doors per gallon | Material cost per gallon | Labor cost per door | Labor burden door | Material cost per door | Overhead per door | Profit per door | Total price per door |
|---|---|---|---|---|---|---|---|---|---|
| **Doors, interior, panel, paint grade, spray application, based on per door method** | | | | | | | | | |
| Undercoat, water base (material #3) | | | | | | | | | |
| Spray 1 coat | | | | | | | | | |
| Slow | 0.13 | 15 | 18.60 | 1.33 | .37 | 1.24 | .91 | .58 | 4.43 |
| Medium | 0.11 | 14 | 16.20 | 1.82 | .61 | 1.16 | .88 | .49 | 4.96 |
| Fast | 0.10 | 13 | 13.90 | 2.20 | .89 | 1.07 | .77 | .30 | 5.23 |
| Undercoat, oil base (material #4) | | | | | | | | | |
| Spray 1 coat | | | | | | | | | |
| Slow | 0.13 | 15 | 20.10 | 1.33 | .37 | 1.34 | .94 | .60 | 4.58 |
| Medium | 0.11 | 14 | 17.60 | 1.82 | .61 | 1.26 | .90 | .50 | 5.09 |
| Fast | 0.10 | 13 | 15.10 | 2.20 | .89 | 1.16 | .79 | .30 | 5.34 |
| Enamel, water base (material #9) | | | | | | | | | |
| Spray 1st finish coat | | | | | | | | | |
| Slow | 0.09 | 16 | 22.80 | .92 | .25 | 1.43 | .81 | .51 | 3.92 |
| Medium | 0.08 | 15 | 19.90 | 1.32 | .44 | 1.33 | .76 | .42 | 4.27 |
| Fast | 0.08 | 14 | 17.10 | 1.76 | .71 | 1.22 | .68 | .26 | 4.63 |
| Spray 2nd or additional finish coats | | | | | | | | | |
| Slow | 0.08 | 17 | 22.80 | .82 | .22 | 1.34 | .74 | .47 | 3.59 |
| Medium | 0.08 | 16 | 19.90 | 1.32 | .44 | 1.24 | .74 | .41 | 4.15 |
| Fast | 0.07 | 15 | 17.10 | 1.54 | .62 | 1.14 | .61 | .23 | 4.14 |
| Enamel, oil base (material #10) | | | | | | | | | |
| Spray 1st finish coat | | | | | | | | | |
| Slow | 0.09 | 16 | 25.90 | .92 | .25 | 1.62 | .86 | .55 | 4.20 |
| Medium | 0.08 | 15 | 22.70 | 1.32 | .44 | 1.51 | .80 | .45 | 4.52 |
| Fast | 0.08 | 14 | 19.40 | 1.76 | .71 | 1.39 | .71 | .27 | 4.84 |
| Spray 2nd or additional finish coats | | | | | | | | | |
| Slow | 0.08 | 17 | 25.90 | .82 | .22 | 1.52 | .79 | .50 | 3.85 |
| Medium | 0.08 | 16 | 22.70 | 1.32 | .44 | 1.42 | .78 | .44 | 4.40 |
| Fast | 0.07 | 15 | 19.40 | 1.54 | .62 | 1.29 | .64 | .25 | 4.34 |

This table is not to be confused with the Opening Count Method for estimating doors. Use these figures for painting both interior and exterior doors. These costs are in addition to the costs for applying the finish coats to exterior doors as explained on page 95. These figures include coating all six (6) sides of each door along with the frame and the jamb on both sides. These figures include minimal preparation time. ADD for additional preparation time, as needed, including setup, remove & replace (hardware), protection (mask walls and visqueen floors); putty, caulking, sanding and cleanup. "Slow" work is based on a $10.25 hourly wage, "Medium" work on a $16.50 hourly wage, and "Fast" work on a $22.00 hourly wage. Other qualifications that apply to this table are on page 9.

| | Manhours per door | Doors per gallon | Material cost per gallon | Labor cost per door | Labor burden door | Material cost per door | Overhead per door | Profit per door | Total price per door |
|---|---|---|---|---|---|---|---|---|---|

## Doors, interior, panel, stain grade, spray application, based on per door method

| | | | | | | | | | |
|---|---|---|---|---|---|---|---|---|---|
| Complete 7 step stain, seal & 2 coat lacquer system (material #11) | | | | | | | | | |
| Spray all coats | | | | | | | | | |
| Slow | 1.25 | 6 | 20.20 | 12.81 | 3.49 | 3.37 | 6.10 | 3.87 | 29.64 |
| Medium | 1.08 | 5 | 18.30 | 17.82 | 5.99 | 3.66 | 6.73 | 3.76 | 37.96 |
| Fast | 0.90 | 4 | 15.10 | 19.80 | 8.03 | 3.78 | 5.85 | 2.25 | 39.71 |

This table is not to be confused with the Opening Count Method for estimating doors. Use these figures for painting both interior and exterior doors. These costs are in addition to the costs for applying the finish coats to exterior doors as explained on page 95. These figures include coating all six (6) sides of each door along with the frame and the jamb on both sides. These figures include minimal preparation time. ADD for additional preparation time, as needed, including setup, remove & replace (hardware), protection (mask walls and visqueen floors); putty, caulking, sanding and cleanup. "Slow" work is based on a $10.25 hourly wage, "Medium" work on a $16.50 hourly wage, and "Fast" work on a $22.00 hourly wage. Other qualifications that apply to this table are on page 9.

| | Labor LF per manhour | Material coverage LF/gallon | Material cost per gallon | Labor cost per 100 LF | Labor burden 100 LF | Material cost per 100 LF | Overhead per 100 LF | Profit per 100 LF | Total price per 100 LF |
|---|---|---|---|---|---|---|---|---|---|

## Fascia, 2" x 4", brush one coat stain "to cover"

| | | | | | | | | | |
|---|---|---|---|---|---|---|---|---|---|
| Solid body stain, water or oil base (material #18 or #19) | | | | | | | | | |
| Brush each coat | | | | | | | | | |
| Slow | 80 | 170 | 21.25 | 12.81 | 3.49 | 12.50 | 8.93 | 5.66 | 43.39 |
| Medium | 105 | 160 | 18.60 | 15.71 | 5.28 | 11.63 | 7.99 | 4.47 | 45.08 |
| Fast | 130 | 150 | 15.90 | 16.92 | 6.86 | 10.60 | 6.36 | 2.44 | 43.18 |
| Semi-transparent stain, water or oil base (material #20 or #21) | | | | | | | | | |
| Brush each coat | | | | | | | | | |
| Slow | 95 | 195 | 21.50 | 10.79 | 2.94 | 11.03 | 7.68 | 4.87 | 37.31 |
| Medium | 120 | 185 | 18.80 | 13.75 | 4.62 | 10.16 | 6.99 | 3.91 | 39.43 |
| Fast | 145 | 175 | 16.15 | 15.17 | 6.16 | 9.23 | 5.65 | 2.17 | 38.38 |

Use these figures for brushing stain on 2" x 4" fascia board on two sides - the face and the lower edge. The back side is calculated with the overhang (eaves) operation. These figures include one full coat and any touchup required to meet the "to cover" specification. They also include minimum preparation time. ADD for additional preparation time, as needed, including setup, remove & replace (hardware), protection (masking and visqueen); putty, caulking, sanding and cleanup. Measurements are based on continuous linear feet. "Slow" work is based on a $10.25 hourly wage, "Medium" work on a $16.50 hourly wage, and "Fast" work on a $22.00 hourly wage. Other qualifications that apply to this table are on page 9.

| | Labor LF per manhour | Material coverage LF/gallon | Material cost per gallon | Labor cost per 100 LF | Labor burden 100 LF | Material cost per 100 LF | Overhead per 100 LF | Profit per 100 LF | Total price per 100 LF |
|---|---|---|---|---|---|---|---|---|---|

## Fascia, 2" x 4", roll one coat stain "to cover"

| | | | | | | | | | |
|---|---|---|---|---|---|---|---|---|---|
| Solid body stain, water or oil base (material #18 or #19) | | | | | | | | | |
| Roll each coat | | | | | | | | | |
| Slow | 180 | 140 | 21.25 | 5.69 | 1.56 | 15.18 | 6.95 | 4.41 | 33.79 |
| Medium | 205 | 130 | 18.60 | 8.05 | 2.71 | 14.31 | 6.14 | 3.43 | 34.64 |
| Fast | 230 | 120 | 15.90 | 9.57 | 3.88 | 13.25 | 4.94 | 1.90 | 33.54 |
| Semi-transparent stain, water or oil base (material #20 or #21) | | | | | | | | | |
| Roll each coat | | | | | | | | | |
| Slow | 200 | 160 | 21.50 | 5.13 | 1.39 | 13.44 | 6.19 | 3.92 | 30.07 |
| Medium | 225 | 150 | 18.80 | 7.33 | 2.46 | 12.53 | 5.47 | 3.06 | 30.85 |
| Fast | 250 | 140 | 16.15 | 8.80 | 3.57 | 11.54 | 4.42 | 1.70 | 30.03 |

Use these figures for rolling stain on 2" x 4" fascia board on two sides - the face and the lower edge. The back side is calculated with the overhang (eaves) operation. These figures include one full coat and any touchup required to meet the "to cover" specification. They also include minimum preparation time. Add time for extensive preparation. Measurements are based on continuous linear feet. "Slow" work is based on a $10.25 hourly wage, "Medium" work on a $16.50 hourly wage, and "Fast" work on a $22.00 hourly wage. Other qualifications that apply to this table are on page 9.

| | Labor LF per manhour | Material coverage LF/gallon | Material cost per gallon | Labor cost per 100 LF | Labor burden 100 LF | Material cost per 100 LF | Overhead per 100 LF | Profit per 100 LF | Total price per 100 LF |
|---|---|---|---|---|---|---|---|---|---|

## Fascia, 2" x 4", spray one coat stain "to cover"

| | | | | | | | | | |
|---|---|---|---|---|---|---|---|---|---|
| Solid body stain, water or oil base (material #18 or #19) | | | | | | | | | |
| Spray each coat | | | | | | | | | |
| Slow | 275 | 110 | 21.25 | 3.73 | 1.02 | 19.32 | 7.46 | 4.73 | 36.26 |
| Medium | 325 | 100 | 18.60 | 5.08 | 1.71 | 18.60 | 6.22 | 3.48 | 35.09 |
| Fast | 375 | 90 | 15.90 | 5.87 | 2.39 | 17.67 | 4.80 | 1.84 | 32.57 |
| Semi-transparent stain, water or oil base (material #20 or #21) | | | | | | | | | |
| Spray each coat | | | | | | | | | |
| Slow | 300 | 125 | 21.50 | 3.42 | .92 | 17.20 | 6.68 | 4.23 | 32.45 |
| Medium | 350 | 115 | 18.80 | 4.71 | 1.60 | 16.35 | 5.55 | 3.10 | 31.31 |
| Fast | 400 | 105 | 16.15 | 5.50 | 2.23 | 15.38 | 4.28 | 1.64 | 29.03 |

Use these figures for spraying stain on 2" x 4" fascia board on two sides - the face and the lower edge. The back side is calculated with the overhang (eaves) operation. These figures include one full coat and any touchup required to meet the "to cover" specification. They also include minimum preparation time. Add time for extensive preparation. Measurements are based on continuous linear feet. "Slow" work is based on a $10.25 hourly wage, "Medium" work on a $16.50 hourly wage, and "Fast" work on a $22.00 hourly wage. Other qualifications that apply to this table are on page 9.

| | Labor LF per manhour | Material coverage LF/gallon | Material cost per gallon | Labor cost per 100 LF | Labor burden 100 LF | Material cost per 100 LF | Overhead per 100 LF | Profit per 100 LF | Total price per 100 LF |
|---|---|---|---|---|---|---|---|---|---|
| **Fascia, 2" x 6" to 2" x 10", brush one coat stain "to cover"** | | | | | | | | | |
| Solid body stain, water or oil base (material #18 or #19) | | | | | | | | | |
| Brush each coat | | | | | | | | | |
| Slow | 70 | 140 | 21.25 | 14.64 | 3.99 | 15.18 | 10.48 | 6.64 | 50.93 |
| Medium | 90 | 130 | 18.60 | 18.33 | 6.17 | 14.31 | 9.51 | 5.32 | 53.64 |
| Fast | 110 | 120 | 15.90 | 20.00 | 8.11 | 13.25 | 7.65 | 2.94 | 51.95 |
| Semi-transparent stain, water or oil base (material #20 or #21) | | | | | | | | | |
| Brush each coat | | | | | | | | | |
| Slow | 85 | 165 | 21.50 | 12.06 | 3.28 | 13.03 | 8.80 | 5.58 | 42.75 |
| Medium | 105 | 155 | 18.80 | 15.71 | 5.28 | 12.13 | 8.12 | 4.54 | 45.78 |
| Fast | 125 | 145 | 16.15 | 17.60 | 7.14 | 11.14 | 6.64 | 2.55 | 45.07 |

Use these figures for brushing stain on 2" x 6" to 2" x 10" fascia board on two sides - the face and the lower edge. The back side is calculated with the overhang (eaves) operation. These figures include one full coat and any touchup required to meet the "to cover" specification. They also include minimum preparation time. Add time for extensive preparation. Measurements are based on continuous linear feet. "Slow" work is based on a $10.25 hourly wage, "Medium" work on a $16.50 hourly wage, and "Fast" work on a $22.00 hourly wage. Other qualifications that apply to this table are on page 9.

| | Labor LF per manhour | Material coverage LF/gallon | Material cost per gallon | Labor cost per 100 LF | Labor burden 100 LF | Material cost per 100 LF | Overhead per 100 LF | Profit per 100 LF | Total price per 100 LF |
|---|---|---|---|---|---|---|---|---|---|
| **Fascia, 2" x 6" to 2" x 10", roll one coat stain "to cover"** | | | | | | | | | |
| Solid body stain, water or oil base (material #18 or #19) | | | | | | | | | |
| Roll each coat | | | | | | | | | |
| Slow | 150 | 120 | 21.25 | 6.83 | 1.87 | 17.71 | 8.18 | 5.19 | 39.78 |
| Medium | 175 | 110 | 18.60 | 9.43 | 3.16 | 16.91 | 7.23 | 4.04 | 40.77 |
| Fast | 200 | 100 | 15.90 | 11.00 | 4.46 | 15.90 | 5.80 | 2.23 | 39.39 |
| Semi-transparent stain, water or oil base (material #20 or #21) | | | | | | | | | |
| Roll each coat | | | | | | | | | |
| Slow | 170 | 140 | 21.50 | 6.03 | 1.64 | 15.36 | 7.14 | 4.53 | 34.70 |
| Medium | 195 | 130 | 18.80 | 8.46 | 2.85 | 14.46 | 6.31 | 3.53 | 35.61 |
| Fast | 220 | 120 | 16.15 | 10.00 | 4.07 | 13.46 | 5.09 | 1.96 | 34.58 |

Use these figures for rolling stain on 2" x 6" to 2" x 10" fascia board on two sides - the face and the lower edge. The back side is calculated with the overhang (eaves) operation. These figures include one full coat and any touchup required to meet the "to cover" specification. They also include minimum preparation time. Add time for extensive preparation. Measurements are based on continuous linear feet. "Slow" work is based on a $10.25 hourly wage, "Medium" work on a $16.50 hourly wage, and "Fast" work on a $22.00 hourly wage. Other qualifications that apply to this table are on page 9.

| | Labor LF per manhour | Material coverage LF/gallon | Material cost per gallon | Labor cost per 100 LF | Labor burden 100 LF | Material cost per 100 LF | Overhead per 100 LF | Profit per 100 LF | Total price per 100 LF |
|---|---|---|---|---|---|---|---|---|---|
| **Fascia, 2" x 6" to 2" x 10", spray one coat stain "to cover"** | | | | | | | | | |
| Solid body stain, water or oil base (material #18 or #19) | | | | | | | | | |
| Spray each coat | | | | | | | | | |
| Slow | 225 | 90 | 21.25 | 4.56 | 1.23 | 23.61 | 9.12 | 5.78 | 44.30 |
| Medium | 300 | 80 | 18.60 | 5.50 | 1.84 | 23.25 | 7.50 | 4.19 | 42.28 |
| Fast | 350 | 70 | 15.90 | 6.29 | 2.55 | 22.71 | 5.84 | 2.24 | 39.63 |
| Semi-transparent stain, water or oil base (material #20 or #21) | | | | | | | | | |
| Spray each coat | | | | | | | | | |
| Slow | 250 | 105 | 21.50 | 4.10 | 1.12 | 20.48 | 7.97 | 5.05 | 38.72 |
| Medium | 313 | 95 | 18.80 | 5.27 | 1.76 | 19.79 | 6.57 | 3.67 | 37.06 |
| Fast | 375 | 85 | 16.15 | 5.87 | 2.39 | 19.00 | 5.04 | 1.94 | 34.24 |

Use these figures for spraying stain on 2" x 6" to 2" x 10" fascia board on two sides - the face and the lower edge. The back side is calculated with the overhang (eaves) operation. These figures include one full coat and any touchup required to meet the "to cover" specification. They also include minimum preparation. Add time for extensive preparation. Measurements are based on continuous linear feet. "Slow" work is based on a $10.25 hourly wage, "Medium" work on a $16.50 hourly wage, and "Fast" work on a $22.00 hourly wage. Other qualifications that apply to this table are on page 9.

| | Labor LF per manhour | Material coverage LF/gallon | Material cost per gallon | Labor cost per 100 LF | Labor burden 100 LF | Material cost per 100 LF | Overhead per 100 LF | Profit per 100 LF | Total price per 100 LF |
|---|---|---|---|---|---|---|---|---|---|
| **Fascia, 2" x 12", brush one coat stain "to cover"** | | | | | | | | | |
| Solid body stain, water or oil base (material #18 or #19) | | | | | | | | | |
| Brush each coat | | | | | | | | | |
| Slow | 60 | 100 | 21.25 | 17.08 | 4.66 | 21.25 | 13.32 | 8.45 | 64.76 |
| Medium | 80 | 90 | 18.60 | 20.63 | 6.93 | 20.67 | 11.82 | 6.61 | 66.66 |
| Fast | 100 | 80 | 15.90 | 22.00 | 8.92 | 19.88 | 9.40 | 3.61 | 63.81 |
| Semi-transparent stain, water or oil base (material #20 or #21) | | | | | | | | | |
| Brush each coat | | | | | | | | | |
| Slow | 75 | 125 | 21.50 | 13.67 | 3.71 | 17.20 | 10.73 | 6.80 | 52.11 |
| Medium | 95 | 115 | 18.80 | 17.37 | 5.85 | 16.35 | 9.69 | 5.42 | 54.68 |
| Fast | 105 | 105 | 16.15 | 20.95 | 8.49 | 15.38 | 8.29 | 3.19 | 56.30 |

Use these figures for brushing stain on 2" x 12" fascia board on two sides - the face and the lower edge. The back side is calculated with the overhang (eaves) operation. These figures include one full coat and any touchup required to meet the "to cover" specification. They also include minimum preparation. Add time for extensive preparation. Measurements are based on continuous linear feet. "Slow" work is based on a $10.25 hourly wage, "Medium" work on a $16.50 hourly wage, and "Fast" work on a $22.00 hourly wage. Other qualifications that apply to this table are on page 9.

| | Labor LF per manhour | Material coverage LF/gallon | Material cost per gallon | Labor cost per 100 LF | Labor burden 100 LF | Material cost per 100 LF | Overhead per 100 LF | Profit per 100 LF | Total price per 100 LF |
|---|---|---|---|---|---|---|---|---|---|

## Fascia, 2" x 12", roll one coat stain "to cover"

| | | | | | | | | | |
|---|---|---|---|---|---|---|---|---|---|
| Solid body stain, water or oil base (material #18 or #19) | | | | | | | | | |
| Roll each coat | | | | | | | | | |
| Slow | 110 | 90 | 21.25 | 9.32 | 2.53 | 23.61 | 11.00 | 6.97 | 53.43 |
| Medium | 130 | 75 | 18.60 | 12.69 | 4.27 | 24.80 | 10.23 | 5.72 | 57.71 |
| Fast | 150 | 60 | 15.90 | 14.67 | 5.95 | 26.50 | 8.72 | 3.35 | 59.19 |
| Semi-transparent stain, water or oil base (material #20 or #21) | | | | | | | | | |
| Roll each coat | | | | | | | | | |
| Slow | 130 | 110 | 21.50 | 7.88 | 2.15 | 19.55 | 9.17 | 5.81 | 44.56 |
| Medium | 150 | 95 | 18.80 | 11.00 | 3.71 | 19.79 | 8.45 | 4.72 | 47.67 |
| Fast | 170 | 80 | 16.15 | 12.94 | 5.24 | 20.19 | 7.10 | 2.73 | 48.20 |

Use these figures for rolling stain on 2" x 12" fascia board on two sides - the face and the lower edge. The back side is calculated with the overhang (eaves) operation. These figures include one full coat and any touchup required to meet the "to cover" specification. They also include minimum preparation. Add time for extensive preparation. Measurements are based on continuous linear feet. "Slow" work is based on a $10.25 hourly wage, "Medium" work on a $16.50 hourly wage, and "Fast" work on a $22.00 hourly wage. Other qualifications that apply to this table are on page 9.

| | Labor LF per manhour | Material coverage LF/gallon | Material cost per gallon | Labor cost per 100 LF | Labor burden 100 LF | Material cost per 100 LF | Overhead per 100 LF | Profit per 100 LF | Total price per 100 LF |
|---|---|---|---|---|---|---|---|---|---|

## Fascia, 2" x 12", spray one coat stain "to cover"

| | | | | | | | | | |
|---|---|---|---|---|---|---|---|---|---|
| Solid body stain, water or oil base (material #18 or #19) | | | | | | | | | |
| Spray each coat | | | | | | | | | |
| Slow | 200 | 60 | 21.25 | 5.13 | 1.39 | 35.42 | 13.00 | 8.24 | 63.18 |
| Medium | 263 | 50 | 18.60 | 6.27 | 2.11 | 37.20 | 11.17 | 6.24 | 62.99 |
| Fast | 325 | 40 | 15.90 | 6.77 | 2.75 | 39.75 | 9.11 | 3.50 | 61.88 |
| Semi-transparent stain, water or oil base (material #20 or #21) | | | | | | | | | |
| Spray each coat | | | | | | | | | |
| Slow | 225 | 75 | 21.50 | 4.56 | 1.23 | 28.67 | 10.69 | 6.77 | 51.92 |
| Medium | 288 | 65 | 18.80 | 5.73 | 1.92 | 28.92 | 8.96 | 5.01 | 50.54 |
| Fast | 350 | 55 | 16.15 | 6.29 | 2.55 | 29.36 | 7.07 | 2.72 | 47.99 |

Use these figures for spraying stain on 2" x 12" fascia board on two sides - the face and the lower edge. The back side is calculated with the overhang (eaves) operation. These figures include one full coat and any touchup required to meet the "to cover" specification. They also include minimum preparation. Add time for extensive preparation. Measurements are based on continuous linear feet. "Slow" work is based on a $10.25 hourly wage, "Medium" work on a $16.50 hourly wage, and "Fast" work on a $22.00 hourly wage. Other qualifications that apply to this table are on page 9.

| | Labor SF per manhour | Material coverage SF/gallon | Material cost per gallon | Labor cost per 100 SF | Labor burden 100 SF | Material cost per 100 SF | Overhead per 100 SF | Profit per 100 SF | Total price per 100 SF |
|---|---|---|---|---|---|---|---|---|---|
| **Fence, chain link or wire mesh** | | | | | | | | | |
| Solid body stain, water or oil base (material #18 or #19) | | | | | | | | | |
| Brush 1st coat | | | | | | | | | |
| Slow | 90 | 600 | 21.25 | 11.39 | 3.10 | 3.54 | 5.59 | 3.54 | 27.16 |
| Medium | 110 | 550 | 18.60 | 15.00 | 5.04 | 3.38 | 5.74 | 3.21 | 32.37 |
| Fast | 125 | 500 | 15.90 | 17.60 | 7.14 | 3.18 | 5.17 | 1.99 | 35.08 |
| Brush 2nd coat | | | | | | | | | |
| Slow | 130 | 650 | 21.25 | 7.88 | 2.15 | 3.27 | 4.12 | 2.61 | 20.03 |
| Medium | 145 | 600 | 18.60 | 11.38 | 3.83 | 3.10 | 4.49 | 2.51 | 25.31 |
| Fast | 160 | 550 | 15.90 | 13.75 | 5.58 | 2.89 | 4.11 | 1.58 | 27.91 |
| Roll 1st coat | | | | | | | | | |
| Slow | 260 | 575 | 21.25 | 3.94 | 1.08 | 3.70 | 2.70 | 1.71 | 13.13 |
| Medium | 275 | 525 | 18.60 | 6.00 | 2.03 | 3.54 | 2.83 | 1.58 | 15.98 |
| Fast | 290 | 475 | 15.90 | 7.59 | 3.08 | 3.35 | 2.59 | 1.00 | 17.61 |
| Roll 2nd coat | | | | | | | | | |
| Slow | 280 | 625 | 21.25 | 3.66 | 1.00 | 3.40 | 2.50 | 1.58 | 12.14 |
| Medium | 300 | 575 | 18.60 | 5.50 | 1.84 | 3.23 | 2.59 | 1.45 | 14.61 |
| Fast | 320 | 525 | 15.90 | 6.88 | 2.80 | 3.03 | 2.35 | .90 | 15.96 |

Use these figures for chain link or wire mesh fencing. The figures are based on painting both sides to meet the "to cover" specification. These figures include minimum preparation time. Add time for extensive preparation. To calculate the area, base measurements on the square feet (length times width) of one side of the fence then multiply by a difficulty factor of 3 and use the figures in the above table. For example, if the fence is 100' long x 3' high, the area is 300 SF. Multiply 300 SF x 3 to arrive at 900 SF which is the total to be used with the table. "Slow" work is based on a $10.25 hourly wage, "Medium" work on a $16.50 hourly wage, and "Fast" work on a $22.00 hourly wage. Other qualifications that apply to this table are on page 9.

## Fence, wood

For solid plank fence, measure the surface area of one side and multiply by 2 to find the area for both sides. Then use the cost table for Siding, exterior. For good neighbor fence (planks on alternate sides of the rail), measure the surface area of one side and multiply by 2 to find the area for both sides. Then multiply by a difficulty factor of 1.5 and use the pricing table for Siding, exterior.

| | Labor SF per manhour | Material coverage SF/gallon | Material cost per gallon | Labor cost per 100 SF | Labor burden 100 SF | Material cost per 100 SF | Overhead per 100 SF | Profit per 100 SF | Total price per 100 SF |
|---|---|---|---|---|---|---|---|---|---|
| **Fence, picket, brush application** | | | | | | | | | |
| Solid body or semi-transparent stain, water base (material #18 or #20) | | | | | | | | | |
| Brush 1st coat | | | | | | | | | |
| Slow | 75 | 400 | 19.50 | 13.67 | 3.71 | 4.88 | 6.91 | 4.38 | 33.55 |
| Medium | 113 | 388 | 17.05 | 14.60 | 4.91 | 4.39 | 5.86 | 3.27 | 33.03 |
| Fast | 150 | 375 | 14.60 | 14.67 | 5.95 | 3.89 | 4.53 | 1.74 | 30.78 |
| Brush 2nd or additional coats | | | | | | | | | |
| Slow | 120 | 450 | 19.50 | 8.54 | 2.32 | 4.33 | 4.71 | 2.99 | 22.89 |
| Medium | 145 | 438 | 17.05 | 11.38 | 3.83 | 3.89 | 4.68 | 2.62 | 26.40 |
| Fast | 170 | 425 | 14.60 | 12.94 | 5.24 | 3.44 | 4.00 | 1.54 | 27.16 |
| Solid body or semi-transparent stain, oil base (material #19 or #21) | | | | | | | | | |
| Brush 1st coat | | | | | | | | | |
| Slow | 75 | 450 | 23.25 | 13.67 | 3.71 | 5.17 | 7.00 | 4.44 | 33.99 |
| Medium | 113 | 438 | 20.35 | 14.60 | 4.91 | 4.65 | 5.92 | 3.31 | 33.39 |
| Fast | 150 | 425 | 17.45 | 14.67 | 5.95 | 4.11 | 4.58 | 1.76 | 31.07 |
| Brush 2nd or additional coats | | | | | | | | | |
| Slow | 120 | 500 | 23.25 | 8.54 | 2.32 | 4.65 | 4.81 | 3.05 | 23.37 |
| Medium | 145 | 488 | 20.35 | 11.38 | 3.83 | 4.17 | 4.75 | 2.65 | 26.78 |
| Fast | 170 | 475 | 17.45 | 12.94 | 5.24 | 3.67 | 4.04 | 1.55 | 27.44 |

For picket fence, measure the overall area of one side and multiply by 4 for painting both sides. Then apply these cost figures. "Slow" work is based on a $10.25 hourly wage, "Medium" work on a $16.50 hourly wage, and "Fast" work on a $22.00 hourly wage. Other qualifications that apply to this table are on page 9.

| | Labor SF per manhour | Material coverage SF/gallon | Material cost per gallon | Labor cost per 100 SF | Labor burden 100 SF | Material cost per 100 SF | Overhead per 100 SF | Profit per 100 SF | Total price per 100 SF |
|---|---|---|---|---|---|---|---|---|---|
| **Fence, picket, roll application** | | | | | | | | | |
| Solid body or semi-transparent stain, water base (material #18 or #20) | | | | | | | | | |
| Roll 1st coat | | | | | | | | | |
| Slow | 120 | 360 | 19.50 | 8.54 | 2.32 | 5.42 | 5.05 | 3.20 | 24.53 |
| Medium | 145 | 343 | 17.05 | 11.38 | 3.83 | 4.97 | 4.94 | 2.76 | 27.88 |
| Fast | 170 | 325 | 14.60 | 12.94 | 5.24 | 4.49 | 4.20 | 1.61 | 28.48 |
| Roll 2nd or additional coats | | | | | | | | | |
| Slow | 200 | 400 | 19.50 | 5.13 | 1.39 | 4.88 | 3.54 | 2.24 | 17.18 |
| Medium | 225 | 388 | 17.05 | 7.33 | 2.46 | 4.39 | 3.48 | 1.94 | 19.60 |
| Fast | 250 | 375 | 14.60 | 8.80 | 3.57 | 3.89 | 3.01 | 1.16 | 20.43 |
| Solid body or semi-transparent stain, oil base (material #19 or #21) | | | | | | | | | |
| Roll 1st coat | | | | | | | | | |
| Slow | 120 | 400 | 23.25 | 8.54 | 2.32 | 5.81 | 5.17 | 3.28 | 25.12 |
| Medium | 145 | 388 | 20.35 | 11.38 | 3.83 | 5.24 | 5.01 | 2.80 | 28.26 |
| Fast | 170 | 375 | 17.45 | 12.94 | 5.24 | 4.65 | 4.23 | 1.62 | 28.68 |
| Roll 2nd or additional coats | | | | | | | | | |
| Slow | 200 | 450 | 23.25 | 5.13 | 1.39 | 5.17 | 3.63 | 2.30 | 17.62 |
| Medium | 225 | 438 | 20.35 | 7.33 | 2.46 | 4.65 | 3.54 | 1.98 | 19.96 |
| Fast | 250 | 425 | 17.45 | 8.80 | 3.57 | 4.11 | 3.05 | 1.17 | 20.70 |

For picket fence, measure the overall area of one side and multiply by 4 for painting both sides. Then apply these cost figures. "Slow" work is based on a $10.25 hourly wage, "Medium" work on a $16.50 hourly wage, and "Fast" work on a $22.00 hourly wage. Other qualifications that apply to this table are on page 9.

| | Labor SF per manhour | Material coverage SF/gallon | Material cost per gallon | Labor cost per 100 SF | Labor burden 100 SF | Material cost per 100 SF | Overhead per 100 SF | Profit per 100 SF | Total price per 100 SF |
|---|---|---|---|---|---|---|---|---|---|

## Fence, picket, spray application

| | | | | | | | | | |
|---|---|---|---|---|---|---|---|---|---|
| Solid body or semi-transparent stain, water base (material #18 or #20) | | | | | | | | | |
| Spray 1st coat | | | | | | | | | |
| Slow | 400 | 300 | 19.50 | 2.56 | .70 | 6.50 | 3.03 | 1.92 | 14.71 |
| Medium | 500 | 275 | 17.05 | 3.30 | 1.11 | 6.20 | 2.60 | 1.45 | 14.66 |
| Fast | 600 | 250 | 14.60 | 3.67 | 1.49 | 5.84 | 2.04 | .78 | 13.82 |
| Spray 2nd or additional coats | | | | | | | | | |
| Slow | 500 | 350 | 19.50 | 2.05 | .56 | 5.57 | 2.54 | 1.61 | 12.33 |
| Medium | 600 | 325 | 17.05 | 2.75 | .93 | 5.25 | 2.19 | 1.22 | 12.34 |
| Fast | 700 | 300 | 14.60 | 3.14 | 1.28 | 4.87 | 1.72 | .66 | 11.67 |
| Solid body or semi-transparent stain, oil base (material #19 or #21) | | | | | | | | | |
| Spray 1st coat | | | | | | | | | |
| Slow | 400 | 350 | 23.25 | 2.56 | .70 | 6.64 | 3.07 | 1.95 | 14.92 |
| Medium | 500 | 325 | 20.35 | 3.30 | 1.11 | 6.26 | 2.61 | 1.46 | 14.74 |
| Fast | 600 | 300 | 17.45 | 3.67 | 1.49 | 5.82 | 2.03 | .78 | 13.79 |
| Spray 2nd or additional coats | | | | | | | | | |
| Slow | 500 | 425 | 23.25 | 2.05 | .56 | 5.47 | 2.50 | 1.59 | 12.17 |
| Medium | 600 | 400 | 20.35 | 2.75 | .93 | 5.09 | 2.15 | 1.20 | 12.12 |
| Fast | 700 | 375 | 17.45 | 3.14 | 1.28 | 4.65 | 1.68 | .64 | 11.39 |

For picket fence, measure the overall area of one side and multiply by 4 for painting both sides. Then apply these cost figures. "Slow" work is based on a $10.25 hourly wage, "Medium" work on a $16.50 hourly wage, and "Fast" work on a $22.00 hourly wage. Other qualifications that apply to this table are on page 9.

| | Labor SF per manhour | Material coverage SF/gallon | Material cost per gallon | Labor cost per 100 SF | Labor burden 100 SF | Material cost per 100 SF | Overhead per 100 SF | Profit per 100 SF | Total price per 100 SF |
|---|---|---|---|---|---|---|---|---|---|
| **Fireplace masonry, interior, smooth surface masonry** | | | | | | | | | |
| Masonry paint, water base (material #31) | | | | | | | | | |
| Brush each coat | | | | | | | | | |
| Slow | 70 | 140 | 19.50 | 14.64 | 3.99 | 13.93 | 10.09 | 6.40 | 49.05 |
| Medium | 75 | 130 | 17.10 | 22.00 | 7.39 | 13.15 | 10.42 | 5.83 | 58.79 |
| Fast | 80 | 120 | 14.60 | 27.50 | 11.15 | 12.17 | 9.40 | 3.61 | 63.83 |
| Masonry paint, oil base (material #32) | | | | | | | | | |
| Brush each coat | | | | | | | | | |
| Slow | 70 | 165 | 24.90 | 14.64 | 3.99 | 15.09 | 10.45 | 6.63 | 50.80 |
| Medium | 75 | 155 | 21.80 | 22.00 | 7.39 | 14.06 | 10.65 | 5.95 | 60.05 |
| Fast | 80 | 145 | 18.70 | 27.50 | 11.15 | 12.90 | 9.54 | 3.67 | 64.76 |
| Masonry paint, water base (material #31) | | | | | | | | | |
| Roll each coat | | | | | | | | | |
| Slow | 140 | 120 | 19.50 | 7.32 | 1.99 | 16.25 | 7.92 | 5.02 | 38.50 |
| Medium | 150 | 110 | 17.10 | 11.00 | 3.71 | 15.55 | 7.41 | 4.14 | 41.81 |
| Fast | 160 | 100 | 14.60 | 13.75 | 5.58 | 14.60 | 6.28 | 2.41 | 42.62 |
| Masonry paint, oil base (material #32) | | | | | | | | | |
| Roll each coat | | | | | | | | | |
| Slow | 140 | 140 | 24.90 | 7.32 | 1.99 | 17.79 | 8.40 | 5.33 | 40.83 |
| Medium | 150 | 130 | 21.80 | 11.00 | 3.71 | 16.77 | 7.71 | 4.31 | 43.50 |
| Fast | 160 | 120 | 18.70 | 13.75 | 5.58 | 15.58 | 6.46 | 2.48 | 43.85 |
| Masonry paint, water base (material #31) | | | | | | | | | |
| Spray each coat | | | | | | | | | |
| Slow | 400 | 105 | 19.50 | 2.56 | .70 | 18.57 | 6.77 | 4.29 | 32.89 |
| Medium | 450 | 100 | 17.10 | 3.67 | 1.23 | 17.10 | 5.39 | 3.01 | 30.40 |
| Fast | 500 | 90 | 14.60 | 4.40 | 1.78 | 16.22 | 4.14 | 1.59 | 28.13 |
| Masonry paint, oil base (material #32) | | | | | | | | | |
| Spray each coat | | | | | | | | | |
| Slow | 400 | 125 | 24.90 | 2.56 | .70 | 19.92 | 7.19 | 4.56 | 34.93 |
| Medium | 450 | 120 | 21.80 | 3.67 | 1.23 | 18.17 | 5.65 | 3.16 | 31.88 |
| Fast | 500 | 110 | 18.70 | 4.40 | 1.78 | 17.00 | 4.29 | 1.65 | 29.12 |

Measurements are based on square feet of the surface area (length times width) to be painted. For fireplace exteriors, use the Masonry cost table which applies. "Slow" work is based on a $10.25 hourly wage, "Medium" work on a $16.50 hourly wage, and "Fast" work on a $22.00 hourly wage. Other qualifications that apply to this table are on page 9.

| | Labor LF per manhour | Material coverage LF/gallon | Material cost per gallon | Labor cost per 100 LF | Labor burden 100 LF | Material cost per 100 LF | Overhead per 100 LF | Profit per 100 LF | Total price per 100 LF |
|---|---|---|---|---|---|---|---|---|---|
| **Fireplace trim, wood, roll & brush each coat** | | | | | | | | | |
| Mantle, rough sawn 4' x 12" | | | | | | | | | |
| Solid body or semi-transparent stain, water or oil base (Material #18 or #19 or #20 or #21) | | | | | | | | | |
| Roll & brush each coat | | | | | | | | | |
| Slow | 15 | 50 | 21.38 | 68.33 | 18.61 | 42.76 | 40.21 | 25.49 | 195.40 |
| Medium | 18 | 45 | 18.70 | 91.67 | 30.84 | 41.56 | 40.20 | 22.47 | 226.74 |
| Fast | 20 | 40 | 16.03 | 110.00 | 44.60 | 40.08 | 36.02 | 13.84 | 244.54 |
| Plant-on trim, interior | | | | | | | | | |
| Solid body or semi-transparent stain, water or oil base (Material #18 or #19 or #20 or #21) | | | | | | | | | |
| Roll & brush each coat | | | | | | | | | |
| Slow | 75 | 135 | 21.38 | 13.67 | 3.71 | 15.84 | 10.30 | 6.53 | 50.05 |
| Medium | 80 | 130 | 18.70 | 20.63 | 6.93 | 14.38 | 10.28 | 5.75 | 57.97 |
| Fast | 85 | 125 | 16.03 | 25.88 | 10.48 | 12.82 | 9.10 | 3.50 | 61.78 |
| Siding, interior, tongue & groove | | | | | | | | | |
| Solid body or semi-transparent stain, water or oil base (Material #18 or #19 or #20 or #21) | | | | | | | | | |
| Roll & brush each coat | | | | | | | | | |
| Slow | 50 | 100 | 21.38 | 20.50 | 5.58 | 21.38 | 14.72 | 9.33 | 71.51 |
| Medium | 75 | 95 | 18.70 | 22.00 | 7.39 | 19.68 | 12.02 | 6.72 | 67.81 |
| Fast | 100 | 90 | 16.03 | 22.00 | 8.92 | 17.81 | 9.02 | 3.47 | 61.22 |

"Slow" work is based on a $10.25 hourly wage, "Medium" work on a $16.50 hourly wage, and "Fast" work on a $22.00 hourly wage. Other qualifications that apply to this table are on page 9.

| | Manhours per box | Material coverage gallons/box | Material cost per gallon | Labor cost per box | Labor burden box | Material cost per box | Overhead per box | Profit per box | Total price per box |
|---|---|---|---|---|---|---|---|---|---|
| **Firewood boxes, wood, brush each coat** | | | | | | | | | |
| Boxes of rough sawn wood, 3'0" x 3'0" x 3'0" deep | | | | | | | | | |
| Solid body or semi-transparent stain, water or oil base (Material #18 or #19 or #20 or #21) | | | | | | | | | |
| Roll & brush each coat | | | | | | | | | |
| Slow | 0.40 | 0.20 | 21.38 | 4.10 | 1.12 | 4.28 | 2.95 | 1.87 | 14.32 |
| Medium | 0.35 | 0.23 | 18.70 | 5.78 | 1.94 | 4.30 | 2.94 | 1.65 | 16.61 |
| Fast | 0.30 | 0.25 | 16.03 | 6.60 | 2.68 | 4.01 | 2.46 | .95 | 16.70 |

"Slow" work is based on a $10.25 hourly wage, "Medium" work on a $16.50 hourly wage, and "Fast" work on a $22.00 hourly wage. Other qualifications that apply to this table are on page 9.

| | Labor SF per manhour | Material coverage SF/gallon | Material cost per gallon | Labor cost per 100 SF | Labor burden 100 SF | Material cost per 100 SF | Overhead per 100 SF | Profit per 100 SF | Total price per 100 SF |
|---|---|---|---|---|---|---|---|---|---|
| **Floors, concrete, brush, interior or exterior** | | | | | | | | | |
| Masonry (concrete) paint, water base (material #31) | | | | | | | | | |
| Brush 1st coat | | | | | | | | | |
| Slow | 90 | 250 | 19.50 | 11.39 | 3.10 | 7.80 | 6.91 | 4.38 | 33.58 |
| Medium | 145 | 238 | 17.10 | 11.38 | 3.83 | 7.18 | 5.49 | 3.07 | 30.95 |
| Fast | 200 | 225 | 14.60 | 11.00 | 4.46 | 6.49 | 4.06 | 1.56 | 27.57 |
| Brush 2nd coat | | | | | | | | | |
| Slow | 125 | 375 | 19.50 | 8.20 | 2.23 | 5.20 | 4.85 | 3.07 | 23.55 |
| Medium | 200 | 325 | 17.10 | 8.25 | 2.78 | 5.26 | 3.99 | 2.23 | 22.51 |
| Fast | 275 | 275 | 14.60 | 8.00 | 3.25 | 5.31 | 3.06 | 1.18 | 20.80 |
| Brush 3rd or additional coats | | | | | | | | | |
| Slow | 150 | 335 | 19.50 | 6.83 | 1.87 | 5.82 | 4.50 | 2.85 | 21.87 |
| Medium | 225 | 310 | 17.10 | 7.33 | 2.46 | 5.52 | 3.75 | 2.10 | 21.16 |
| Fast | 300 | 285 | 14.60 | 7.33 | 2.97 | 5.12 | 2.85 | 1.10 | 19.37 |
| Masonry (concrete) paint, oil base (material #32) | | | | | | | | | |
| Brush 1st coat | | | | | | | | | |
| Slow | 90 | 300 | 24.90 | 11.39 | 3.10 | 8.30 | 7.06 | 4.48 | 34.33 |
| Medium | 145 | 288 | 21.80 | 11.38 | 3.83 | 7.57 | 5.58 | 3.12 | 31.48 |
| Fast | 200 | 275 | 18.70 | 11.00 | 4.46 | 6.80 | 4.12 | 1.58 | 27.96 |
| Brush 2nd coat | | | | | | | | | |
| Slow | 125 | 400 | 24.90 | 8.20 | 2.23 | 6.23 | 5.16 | 3.27 | 25.09 |
| Medium | 200 | 388 | 21.80 | 8.25 | 2.78 | 5.62 | 4.08 | 2.28 | 23.01 |
| Fast | 275 | 375 | 18.70 | 8.00 | 3.25 | 4.99 | 3.00 | 1.15 | 20.39 |
| Brush 3rd or additional coats | | | | | | | | | |
| Slow | 150 | 550 | 24.90 | 6.83 | 1.87 | 4.53 | 4.10 | 2.60 | 19.93 |
| Medium | 225 | 525 | 21.80 | 7.33 | 2.46 | 4.15 | 3.42 | 1.91 | 19.27 |
| Fast | 300 | 500 | 18.70 | 7.33 | 2.97 | 3.74 | 2.60 | 1.00 | 17.64 |
| Epoxy, 1 part, water base (material #28) | | | | | | | | | |
| Brush each coat | | | | | | | | | |
| Slow | 125 | 400 | 29.40 | 8.20 | 2.23 | 7.35 | 5.51 | 3.49 | 26.78 |
| Medium | 163 | 388 | 25.70 | 10.12 | 3.40 | 6.62 | 4.94 | 2.76 | 27.84 |
| Fast | 200 | 375 | 22.10 | 11.00 | 4.46 | 5.89 | 3.95 | 1.52 | 26.82 |
| Epoxy, 2 part system (material #29) | | | | | | | | | |
| Brush each coat | | | | | | | | | |
| Slow | 100 | 400 | 46.40 | 10.25 | 2.79 | 11.60 | 7.64 | 4.84 | 37.12 |
| Medium | 138 | 388 | 40.60 | 11.96 | 4.03 | 10.46 | 6.48 | 3.62 | 36.55 |
| Fast | 175 | 375 | 34.80 | 12.57 | 5.09 | 9.28 | 4.99 | 1.92 | 33.85 |

Also use these formulas for concrete steps, stair treads, porches and patios. "Slow" work is based on a $10.25 hourly wage, "Medium" work on a $16.50 hourly wage, and "Fast" work on a $22.00 hourly wage. Other qualifications that apply to this table are on page 9.

| | Labor SF per manhour | Material coverage SF/gallon | Material cost per gallon | Labor cost per 100 SF | Labor burden 100 SF | Material cost per 100 SF | Overhead per 100 SF | Profit per 100 SF | Total price per 100 SF |
|---|---|---|---|---|---|---|---|---|---|
| **Floors, concrete, roll, interior or exterior** | | | | | | | | | |
| Masonry (concrete) paint, water base (material #31) | | | | | | | | | |
| Roll 1st coat | | | | | | | | | |
| Slow | 135 | 275 | 19.50 | 7.59 | 2.07 | 7.09 | 5.19 | 3.29 | 25.23 |
| Medium | 218 | 263 | 17.10 | 7.57 | 2.55 | 6.50 | 4.07 | 2.28 | 22.97 |
| Fast | 300 | 250 | 14.60 | 7.33 | 2.97 | 5.84 | 2.99 | 1.15 | 20.28 |
| Roll 2nd coat | | | | | | | | | |
| Slow | 195 | 350 | 19.50 | 5.26 | 1.43 | 5.57 | 3.80 | 2.41 | 18.47 |
| Medium | 268 | 325 | 17.10 | 6.16 | 2.06 | 5.26 | 3.31 | 1.85 | 18.64 |
| Fast | 340 | 300 | 14.60 | 6.47 | 2.62 | 4.87 | 2.58 | .99 | 17.53 |
| Roll 3rd or additional coats | | | | | | | | | |
| Slow | 210 | 375 | 19.50 | 4.88 | 1.33 | 5.20 | 3.54 | 2.24 | 17.19 |
| Medium | 300 | 350 | 17.10 | 5.50 | 1.84 | 4.89 | 3.00 | 1.68 | 16.91 |
| Fast | 390 | 325 | 14.60 | 5.64 | 2.28 | 4.49 | 2.30 | .88 | 15.59 |
| Masonry (concrete) paint, oil base (material #32) | | | | | | | | | |
| Roll 1st coat | | | | | | | | | |
| Slow | 135 | 370 | 24.90 | 7.59 | 2.07 | 6.73 | 5.08 | 3.22 | 24.69 |
| Medium | 218 | 345 | 21.80 | 7.57 | 2.55 | 6.32 | 4.03 | 2.25 | 22.72 |
| Fast | 300 | 320 | 18.70 | 7.33 | 2.97 | 5.84 | 2.99 | 1.15 | 20.28 |
| Roll 2nd coat | | | | | | | | | |
| Slow | 195 | 500 | 24.90 | 5.26 | 1.43 | 4.98 | 3.62 | 2.29 | 17.58 |
| Medium | 268 | 475 | 21.80 | 6.16 | 2.06 | 4.59 | 3.14 | 1.76 | 17.71 |
| Fast | 340 | 450 | 18.70 | 6.47 | 2.62 | 4.16 | 2.45 | .94 | 16.64 |
| Roll 3rd or additional coats | | | | | | | | | |
| Slow | 210 | 550 | 24.90 | 4.88 | 1.33 | 4.53 | 3.33 | 2.11 | 16.18 |
| Medium | 300 | 525 | 21.80 | 5.50 | 1.84 | 4.15 | 2.82 | 1.58 | 15.89 |
| Fast | 390 | 500 | 18.70 | 5.64 | 2.28 | 3.74 | 2.16 | .83 | 14.65 |
| Epoxy, 1 part, water base (material #28) | | | | | | | | | |
| Roll each coat | | | | | | | | | |
| Slow | 150 | 500 | 29.40 | 6.83 | 1.87 | 5.88 | 4.52 | 2.86 | 21.96 |
| Medium | 225 | 488 | 25.70 | 7.33 | 2.46 | 5.27 | 3.69 | 2.06 | 20.81 |
| Fast | 300 | 475 | 22.10 | 7.33 | 2.97 | 4.65 | 2.77 | 1.06 | 18.78 |
| Epoxy, 2 part system (material #29) | | | | | | | | | |
| Roll each coat | | | | | | | | | |
| Slow | 135 | 500 | 46.40 | 7.59 | 2.07 | 9.28 | 5.87 | 3.72 | 28.53 |
| Medium | 208 | 488 | 40.60 | 7.93 | 2.68 | 8.32 | 4.64 | 2.59 | 26.16 |
| Fast | 250 | 475 | 34.80 | 8.80 | 3.57 | 7.33 | 3.64 | 1.40 | 24.74 |

Also use these formulas for concrete steps, stair treads, porches and patios. "Slow" work is based on a $10.25 hourly wage, "Medium" work on a $16.50 hourly wage, and "Fast" work on a $22.00 hourly wage. Other qualifications that apply to this table are on page 9.

| | Labor SF per manhour | Material coverage SF/gallon | Material cost per gallon | Labor cost per 100 SF | Labor burden 100 SF | Material cost per 100 SF | Overhead per 100 SF | Profit per 100 SF | Total price per 100 SF |
|---|---|---|---|---|---|---|---|---|---|
| **Floors, concrete, spray, interior or exterior** | | | | | | | | | |
| Masonry (concrete) paint, water base (material #31) | | | | | | | | | |
| Spray 1st coat | | | | | | | | | |
| Slow | 800 | 175 | 19.50 | 1.28 | .35 | 11.14 | 3.96 | 2.51 | 19.24 |
| Medium | 900 | 163 | 17.10 | 1.83 | .62 | 10.49 | 3.17 | 1.77 | 17.88 |
| Fast | 1000 | 150 | 14.60 | 2.20 | .89 | 9.73 | 2.37 | .91 | 16.10 |
| Spray 2nd coat | | | | | | | | | |
| Slow | 900 | 275 | 19.50 | 1.14 | .31 | 7.09 | 2.65 | 1.68 | 12.87 |
| Medium | 1000 | 263 | 17.10 | 1.65 | .56 | 6.50 | 2.13 | 1.19 | 12.03 |
| Fast | 1100 | 250 | 14.60 | 2.00 | .81 | 5.84 | 1.60 | .62 | 10.87 |
| Spray 3rd or additional coats | | | | | | | | | |
| Slow | 1000 | 325 | 19.50 | 1.03 | .27 | 6.00 | 2.27 | 1.44 | 11.01 |
| Medium | 1100 | 313 | 17.10 | 1.50 | .51 | 5.46 | 1.83 | 1.02 | 10.32 |
| Fast | 1200 | 300 | 14.60 | 1.83 | .74 | 4.87 | 1.38 | .53 | 9.35 |
| Masonry (concrete) paint, oil base (material #32) | | | | | | | | | |
| Spray 1st coat | | | | | | | | | |
| Slow | 800 | 200 | 24.90 | 1.28 | .35 | 12.45 | 4.36 | 2.77 | 21.21 |
| Medium | 900 | 188 | 21.80 | 1.83 | .62 | 11.60 | 3.44 | 1.92 | 19.41 |
| Fast | 1000 | 175 | 18.70 | 2.20 | .89 | 10.69 | 2.55 | .98 | 17.31 |
| Spray 2nd coat | | | | | | | | | |
| Slow | 900 | 300 | 24.90 | 1.14 | .31 | 8.30 | 3.02 | 1.92 | 14.69 |
| Medium | 1000 | 288 | 21.80 | 1.65 | .56 | 7.57 | 2.40 | 1.34 | 13.52 |
| Fast | 1100 | 275 | 18.70 | 2.00 | .81 | 6.80 | 1.78 | .68 | 12.07 |
| Spray 3rd or additional coats | | | | | | | | | |
| Slow | 1000 | 350 | 24.90 | 1.03 | .27 | 7.11 | 2.61 | 1.65 | 12.67 |
| Medium | 1100 | 338 | 21.80 | 1.50 | .51 | 6.45 | 2.07 | 1.16 | 11.69 |
| Fast | 1200 | 325 | 18.70 | 1.83 | .74 | 5.75 | 1.54 | .59 | 10.45 |

Also use these formulas for concrete steps, stair treads, porches and patios. "Slow" work is based on a $10.25 hourly wage, "Medium" work on a $16.50 hourly wage, and "Fast" work on a $22.00 hourly wage. Other qualifications that apply to this table are on page 9.

| | Labor SF per manhour | Material coverage SF/gallon | Material cost per gallon | Labor cost per 100 SF | Labor burden 100 SF | Material cost per 100 SF | Overhead per 100 SF | Profit per 100 SF | Total price per 100 SF |
|---|---|---|---|---|---|---|---|---|---|
| **Floors, concrete, penetrating stain, interior or exterior** | | | | | | | | | |
| Penetrating oil stain (material #13) | | | | | | | | | |
| Roll 1st coat | | | | | | | | | |
| Slow | 225 | 450 | 25.00 | 4.56 | 1.23 | 5.56 | 3.52 | 2.23 | 17.10 |
| Medium | 250 | 425 | 21.80 | 6.60 | 2.22 | 5.13 | 3.42 | 1.91 | 19.28 |
| Fast | 275 | 400 | 18.70 | 8.00 | 3.25 | 4.68 | 2.95 | 1.13 | 20.01 |
| Roll 2nd coat | | | | | | | | | |
| Slow | 325 | 500 | 25.00 | 3.15 | .87 | 5.00 | 2.79 | 1.77 | 13.58 |
| Medium | 345 | 475 | 21.80 | 4.78 | 1.61 | 4.59 | 2.69 | 1.50 | 15.17 |
| Fast | 365 | 450 | 18.70 | 6.03 | 2.44 | 4.16 | 2.34 | .90 | 15.87 |
| Roll 3rd and additional coats | | | | | | | | | |
| Slow | 365 | 525 | 25.00 | 2.81 | .76 | 4.76 | 2.59 | 1.64 | 12.56 |
| Medium | 383 | 500 | 21.80 | 4.31 | 1.45 | 4.36 | 2.48 | 1.39 | 13.99 |
| Fast | 400 | 475 | 18.70 | 5.50 | 2.23 | 3.94 | 2.16 | .83 | 14.66 |

Also use these formulas for concrete steps, stair treads, porches and patios. "Slow" work is based on a $10.25 hourly wage, "Medium" work on a $16.50 hourly wage, and "Fast" work on a $22.00 hourly wage. Other qualifications that apply to this table are on page 9.

| | Labor SF per manhour | Material coverage SF/gallon | Material cost per gallon | Labor cost per 100 SF | Labor burden 100 SF | Material cost per 100 SF | Overhead per 100 SF | Profit per 100 SF | Total price per 100 SF |
|---|---|---|---|---|---|---|---|---|---|
| **Floors, wood, interior or exterior, paint grade, brush application** | | | | | | | | | |
| Undercoat, water base (material #3) | | | | | | | | | |
| Brush prime coat | | | | | | | | | |
| Slow | 275 | 450 | 18.60 | 3.73 | 1.02 | 4.13 | 2.75 | 1.74 | 13.37 |
| Medium | 300 | 425 | 16.20 | 5.50 | 1.84 | 3.81 | 2.73 | 1.53 | 15.41 |
| Fast | 325 | 400 | 13.90 | 6.77 | 2.75 | 3.48 | 2.41 | .92 | 16.33 |
| Undercoat, oil base (material #4) | | | | | | | | | |
| Brush prime coat | | | | | | | | | |
| Slow | 275 | 500 | 20.10 | 3.73 | 1.02 | 4.02 | 2.72 | 1.72 | 13.21 |
| Medium | 300 | 475 | 17.60 | 5.50 | 1.84 | 3.71 | 2.71 | 1.51 | 15.27 |
| Fast | 325 | 450 | 15.10 | 6.77 | 2.75 | 3.36 | 2.38 | .92 | 16.18 |
| Porch & deck enamel, water base (material #26) | | | | | | | | | |
| Brush 1st and additional finish coats | | | | | | | | | |
| Slow | 300 | 475 | 26.80 | 3.42 | .92 | 5.64 | 3.10 | 1.96 | 15.04 |
| Medium | 325 | 450 | 23.40 | 5.08 | 1.71 | 5.20 | 2.94 | 1.64 | 16.57 |
| Fast | 350 | 425 | 20.10 | 6.29 | 2.55 | 4.73 | 2.51 | .96 | 17.04 |
| Porch & deck enamel, oil base (material #27) | | | | | | | | | |
| Brush 1st and additional finish coats | | | | | | | | | |
| Slow | 300 | 550 | 27.10 | 3.42 | .92 | 4.93 | 2.88 | 1.82 | 13.97 |
| Medium | 325 | 525 | 23.70 | 5.08 | 1.71 | 4.51 | 2.77 | 1.55 | 15.62 |
| Fast | 350 | 500 | 20.30 | 6.29 | 2.55 | 4.06 | 2.39 | .92 | 16.21 |
| Epoxy, 1 part, water base (material #28) | | | | | | | | | |
| Brush each coat | | | | | | | | | |
| Slow | 125 | 450 | 29.40 | 8.20 | 2.23 | 6.53 | 5.26 | 3.33 | 25.55 |
| Medium | 163 | 425 | 25.70 | 10.12 | 3.40 | 6.05 | 4.80 | 2.68 | 27.05 |
| Fast | 200 | 400 | 22.10 | 11.00 | 4.46 | 5.53 | 3.88 | 1.49 | 26.36 |
| Epoxy, 2 part system (material #29) | | | | | | | | | |
| Brush each coat | | | | | | | | | |
| Slow | 100 | 425 | 46.40 | 10.25 | 2.79 | 10.92 | 7.43 | 4.71 | 36.10 |
| Medium | 138 | 400 | 40.60 | 11.96 | 4.03 | 10.15 | 6.40 | 3.58 | 36.12 |
| Fast | 175 | 375 | 34.80 | 12.57 | 5.09 | 9.28 | 4.99 | 1.92 | 33.85 |

"Slow" work is based on a $10.25 hourly wage, "Medium" work on a $16.50 hourly wage, and "Fast" work on a $22.00 hourly wage. Other qualifications that apply to this table are on page 9.

| | Labor SF per manhour | Material coverage SF/gallon | Material cost per gallon | Labor cost per 100 SF | Labor burden 100 SF | Material cost per 100 SF | Overhead per 100 SF | Profit per 100 SF | Total price per 100 SF |
|---|---|---|---|---|---|---|---|---|---|
| **Floors, wood, interior or exterior, paint grade, roll application** | | | | | | | | | |
| Undercoat, water base (material #3) | | | | | | | | | |
| Roll prime coat | | | | | | | | | |
| Slow | 400 | 425 | 18.60 | 2.56 | .70 | 4.38 | 2.37 | 1.50 | 11.51 |
| Medium | 438 | 400 | 16.20 | 3.77 | 1.26 | 4.05 | 2.23 | 1.25 | 12.56 |
| Fast | 475 | 375 | 13.90 | 4.63 | 1.89 | 3.71 | 1.89 | .73 | 12.85 |
| Undercoat, oil base (material #4) | | | | | | | | | |
| Roll prime coat | | | | | | | | | |
| Slow | 400 | 475 | 20.10 | 2.56 | .70 | 4.23 | 2.32 | 1.47 | 11.28 |
| Medium | 438 | 450 | 17.60 | 3.77 | 1.26 | 3.91 | 2.19 | 1.23 | 12.36 |
| Fast | 475 | 425 | 15.10 | 4.63 | 1.89 | 3.55 | 1.86 | .72 | 12.65 |
| Porch & deck enamel, water base (material #26) | | | | | | | | | |
| Roll 1st or additional finish coats | | | | | | | | | |
| Slow | 425 | 475 | 26.80 | 2.41 | .65 | 5.64 | 2.70 | 1.71 | 13.11 |
| Medium | 463 | 450 | 23.40 | 3.56 | 1.20 | 5.20 | 2.44 | 1.36 | 13.76 |
| Fast | 500 | 425 | 20.10 | 4.40 | 1.78 | 4.73 | 2.02 | .78 | 13.71 |
| Porch & deck enamel, oil base (material #27) | | | | | | | | | |
| Roll 1st or additional finish coats | | | | | | | | | |
| Slow | 425 | 525 | 27.10 | 2.41 | .65 | 5.16 | 2.55 | 1.62 | 12.39 |
| Medium | 463 | 500 | 23.70 | 3.56 | 1.20 | 4.74 | 2.33 | 1.30 | 13.13 |
| Fast | 500 | 475 | 20.30 | 4.40 | 1.78 | 4.27 | 1.93 | .74 | 13.12 |
| Epoxy, 1 part, water base (material #28) | | | | | | | | | |
| Brush each coat | | | | | | | | | |
| Slow | 200 | 425 | 29.40 | 5.13 | 1.39 | 6.92 | 4.17 | 2.64 | 20.25 |
| Medium | 250 | 400 | 25.70 | 6.60 | 2.22 | 6.43 | 3.74 | 2.09 | 21.08 |
| Fast | 300 | 375 | 22.10 | 7.33 | 2.97 | 5.89 | 3.00 | 1.15 | 20.34 |
| Epoxy, 2 part system (material #29) | | | | | | | | | |
| Brush each coat | | | | | | | | | |
| Slow | 175 | 400 | 46.40 | 5.86 | 1.59 | 11.60 | 5.91 | 3.75 | 28.71 |
| Medium | 225 | 375 | 40.60 | 7.33 | 2.46 | 10.83 | 5.05 | 2.82 | 28.49 |
| Fast | 275 | 350 | 34.80 | 8.00 | 3.25 | 9.94 | 3.92 | 1.51 | 26.62 |

"Slow" work is based on a $10.25 hourly wage, "Medium" work on a $16.50 hourly wage, and "Fast" work on a $22.00 hourly wage. Other qualifications that apply to this table are on page 9.

| | Labor SF per manhour | Material coverage SF/gallon | Material cost per gallon | Labor cost per 100 SF | Labor burden 100 SF | Material cost per 100 SF | Overhead per 100 SF | Profit per 100 SF | Total price per 100 SF |
|---|---|---|---|---|---|---|---|---|---|
| **Floors, wood, interior or exterior, stain grade** | | | | | | | | | |
| Wiping stain, varnish, oil base (material #30a) | | | | | | | | | |
| Stain, brush 1st coat, wipe & fill | | | | | | | | | |
| Slow | 225 | 500 | 26.10 | 4.56 | 1.23 | 5.22 | 3.42 | 2.17 | 16.60 |
| Medium | 250 | 475 | 22.90 | 6.60 | 2.22 | 4.82 | 3.34 | 1.87 | 18.85 |
| Fast | 275 | 450 | 19.60 | 8.00 | 3.25 | 4.36 | 2.89 | 1.11 | 19.61 |
| Stain, brush 2nd coat, wipe & fill | | | | | | | | | |
| Slow | 400 | 525 | 26.10 | 2.56 | .70 | 4.97 | 2.55 | 1.62 | 12.40 |
| Medium | 425 | 500 | 22.90 | 3.88 | 1.30 | 4.58 | 2.39 | 1.34 | 13.49 |
| Fast | 450 | 475 | 19.60 | 4.89 | 1.97 | 4.13 | 2.04 | .78 | 13.81 |
| Stain, brush 3rd or additional coats, wipe & fill | | | | | | | | | |
| Slow | 425 | 550 | 26.10 | 2.41 | .65 | 4.75 | 2.42 | 1.54 | 11.77 |
| Medium | 450 | 525 | 22.90 | 3.67 | 1.23 | 4.36 | 2.27 | 1.27 | 12.80 |
| Fast | 475 | 500 | 19.60 | 4.63 | 1.89 | 3.92 | 1.93 | .74 | 13.11 |
| Sanding sealer, varnish (material #30b) | | | | | | | | | |
| Maple or pine, brush 1 coat | | | | | | | | | |
| Slow | 375 | 475 | 27.40 | 2.73 | .75 | 5.77 | 2.86 | 1.82 | 13.93 |
| Medium | 400 | 450 | 24.00 | 4.13 | 1.38 | 5.33 | 2.66 | 1.49 | 14.99 |
| Fast | 425 | 425 | 20.50 | 5.18 | 2.09 | 4.82 | 2.24 | .86 | 15.19 |
| Maple or pine, brush 2nd or additional coats | | | | | | | | | |
| Slow | 425 | 550 | 27.40 | 2.41 | .65 | 4.98 | 2.50 | 1.58 | 12.12 |
| Medium | 450 | 525 | 24.00 | 3.67 | 1.23 | 4.57 | 2.32 | 1.30 | 13.09 |
| Fast | 475 | 500 | 20.50 | 4.63 | 1.89 | 4.10 | 1.96 | .75 | 13.33 |
| Oak, brush 1 coat | | | | | | | | | |
| Slow | 400 | 525 | 27.40 | 2.56 | .70 | 5.22 | 2.63 | 1.67 | 12.78 |
| Medium | 425 | 500 | 24.00 | 3.88 | 1.30 | 4.80 | 2.45 | 1.37 | 13.80 |
| Fast | 450 | 475 | 20.50 | 4.89 | 1.97 | 4.32 | 2.07 | .80 | 14.05 |
| Oak, brush 2nd or additional coats | | | | | | | | | |
| Slow | 500 | 625 | 27.40 | 2.05 | .56 | 4.38 | 2.17 | 1.37 | 10.53 |
| Medium | 525 | 600 | 24.00 | 3.14 | 1.05 | 4.00 | 2.01 | 1.12 | 11.32 |
| Fast | 550 | 575 | 20.50 | 4.00 | 1.63 | 3.57 | 1.70 | .65 | 11.55 |
| Shellac, clear (material #12) | | | | | | | | | |
| Brush 1st coat | | | | | | | | | |
| Slow | 275 | 475 | 22.60 | 3.73 | 1.02 | 4.76 | 2.95 | 1.87 | 14.33 |
| Medium | 300 | 450 | 19.80 | 5.50 | 1.84 | 4.40 | 2.88 | 1.61 | 16.23 |
| Fast | 325 | 425 | 17.00 | 6.77 | 2.75 | 4.00 | 2.50 | .96 | 16.98 |
| Brush 2nd or additional coats | | | | | | | | | |
| Slow | 400 | 500 | 22.60 | 2.56 | .70 | 4.52 | 2.41 | 1.53 | 11.72 |
| Medium | 425 | 475 | 19.80 | 3.88 | 1.30 | 4.17 | 2.29 | 1.28 | 12.92 |
| Fast | 450 | 450 | 17.00 | 4.89 | 1.97 | 3.78 | 1.97 | .76 | 13.37 |

| | Labor SF per manhour | Material coverage SF/gallon | Material cost per gallon | Labor cost per 100 SF | Labor burden 100 SF | Material cost per 100 SF | Overhead per 100 SF | Profit per 100 SF | Total price per 100 SF |
|---|---|---|---|---|---|---|---|---|---|
| **Varnish, gloss or flat (material #30c)** | | | | | | | | | |
| Brush 1st coat | | | | | | | | | |
| Slow | 275 | 475 | 33.90 | 3.73 | 1.02 | 7.14 | 3.69 | 2.34 | 17.92 |
| Medium | 300 | 450 | 29.70 | 5.50 | 1.84 | 6.60 | 3.42 | 1.91 | 19.27 |
| Fast | 325 | 425 | 25.40 | 6.77 | 2.75 | 5.98 | 2.87 | 1.10 | 19.47 |
| Brush 2nd or additional coats | | | | | | | | | |
| Slow | 350 | 600 | 33.90 | 2.93 | .80 | 5.65 | 2.91 | 1.84 | 14.13 |
| Medium | 375 | 575 | 29.70 | 4.40 | 1.49 | 5.17 | 2.71 | 1.51 | 15.28 |
| Fast | 400 | 550 | 25.40 | 5.50 | 2.23 | 4.62 | 2.28 | .88 | 15.51 |
| **Penetrating stain wax & wipe (material #14)** | | | | | | | | | |
| Stain, brush 1st coat & wipe | | | | | | | | | |
| Slow | 200 | 550 | 28.60 | 5.13 | 1.39 | 5.20 | 3.64 | 2.31 | 17.67 |
| Medium | 250 | 525 | 25.00 | 6.60 | 2.22 | 4.76 | 3.33 | 1.86 | 18.77 |
| Fast | 300 | 500 | 21.40 | 7.33 | 2.97 | 4.28 | 2.70 | 1.04 | 18.32 |
| Stain, brush 2nd or additional coats & wipe | | | | | | | | | |
| Slow | 250 | 600 | 28.60 | 4.10 | 1.12 | 4.77 | 3.10 | 1.96 | 15.05 |
| Medium | 300 | 575 | 25.00 | 5.50 | 1.84 | 4.35 | 2.87 | 1.60 | 16.16 |
| Fast | 350 | 550 | 21.40 | 6.29 | 2.55 | 3.89 | 2.36 | .91 | 16.00 |
| **Wax & polish (material #15)** | | | | | | | | | |
| Hand apply 1 coat | | | | | | | | | |
| Slow | 175 | 1000 | 7.80 | 5.86 | 1.59 | .78 | 2.55 | 1.62 | 12.40 |
| Medium | 200 | 950 | 6.80 | 8.25 | 2.78 | .72 | 2.88 | 1.61 | 16.24 |
| Fast | 225 | 900 | 5.80 | 9.78 | 3.95 | .64 | 2.66 | 1.02 | 18.05 |
| Buffing with machine | | | | | | | | | |
| Slow | 400 | -- | -- | 2.56 | .70 | -- | 1.01 | .64 | 4.91 |
| Medium | 450 | -- | -- | 3.67 | 1.23 | -- | 1.20 | .67 | 6.77 |
| Fast | 500 | -- | -- | 4.40 | 1.78 | -- | 1.14 | .44 | 7.76 |

"Slow" work is based on a $10.25 hourly wage, "Medium" work on a $16.50 hourly wage, and "Fast" work on a $22.00 hourly wage. Other qualifications that apply to this table are on page 9.

| | Manhours per door | Gallons per Door | Material cost per gallon | Labor cost per door | Labor burden door | Material cost per door | Overhead per door | Profit per door | Total price per door |
|---|---|---|---|---|---|---|---|---|---|

## Garage door backs, seal coat, spray one coat

| | Manhours per door | Gallons per Door | Material cost per gallon | Labor cost per door | Labor burden door | Material cost per door | Overhead per door | Profit per door | Total price per door |
|---|---|---|---|---|---|---|---|---|---|
| Sanding sealer, lacquer (material #11b) | | | | | | | | | |
| 1 car garage, 8' x 7' | | | | | | | | | |
| Slow | 0.30 | 0.40 | 17.30 | 3.08 | .83 | 6.92 | 3.36 | 2.13 | 16.32 |
| Medium | 0.25 | 0.50 | 15.10 | 4.13 | 1.38 | 7.55 | 3.20 | 1.79 | 18.05 |
| Fast | 0.20 | 0.60 | 13.00 | 4.40 | 1.78 | 7.80 | 2.59 | .99 | 17.56 |
| 2 car garage, 16' x 7' | | | | | | | | | |
| Slow | 0.40 | 0.80 | 17.30 | 4.10 | 1.12 | 13.84 | 5.91 | 3.75 | 28.72 |
| Medium | 0.35 | 0.90 | 15.10 | 5.78 | 1.94 | 13.59 | 5.22 | 2.92 | 29.45 |
| Fast | 0.30 | 1.00 | 13.00 | 6.60 | 2.68 | 13.00 | 4.12 | 1.58 | 27.98 |
| 3 car garage, 16' x 7' + 8' x 7' | | | | | | | | | |
| Slow | 0.60 | 1.00 | 17.30 | 6.15 | 1.67 | 17.30 | 7.79 | 4.94 | 37.85 |
| Medium | 0.55 | 1.10 | 15.10 | 9.08 | 3.05 | 16.61 | 7.04 | 3.94 | 39.72 |
| Fast | 0.50 | 1.20 | 13.00 | 11.00 | 4.46 | 15.60 | 5.75 | 2.21 | 39.02 |

Use the figures for Siding when estimating the cost of painting garage door fronts. These figures assume a one-car garage door measures 7' x 8' and a two-car garage door measures 7' x 16'. A three-car garage has one single and one double door. Government funded projects (FHA, VA, HUD) usually require sealing the garage door back on new construction projects. The doors are usually sprayed along with the cabinet sealer coat (as used in this table) or stained along with the exterior trim. "Slow" work is based on a $10.25 hourly wage, "Medium" work on a $16.50 hourly wage, and "Fast" work on a $22.00 hourly wage. Other qualifications that apply to this table are on page 9.

| | Labor LF per manhour | Material coverage LF/gallon | Material cost per gallon | Labor cost per 100 LF | Labor burden 100 LF | Material cost per 100 LF | Overhead per 100 LF | Profit per 100 LF | Total price per 100 LF |
|---|---|---|---|---|---|---|---|---|---|
| **Gutters and downspouts (galvanized), brush application** | | | | | | | | | |
| Gutters | | | | | | | | | |
| Metal prime, rust inhibitor, clean metal (material #35) | | | | | | | | | |
| Brush prime coat | | | | | | | | | |
| Slow | 80 | 400 | 28.20 | 12.81 | 3.49 | 7.05 | 7.24 | 4.59 | 35.18 |
| Medium | 90 | 375 | 24.60 | 18.33 | 6.17 | 6.56 | 7.61 | 4.25 | 42.92 |
| Fast | 100 | 350 | 21.10 | 22.00 | 8.92 | 6.03 | 6.84 | 2.63 | 46.42 |
| Metal prime, rust inhibitor, rusty metal (material #36) | | | | | | | | | |
| Brush prime coat | | | | | | | | | |
| Slow | 80 | 400 | 33.30 | 12.81 | 3.49 | 8.33 | 7.64 | 4.84 | 37.11 |
| Medium | 90 | 375 | 29.20 | 18.33 | 6.17 | 7.79 | 7.91 | 4.42 | 44.62 |
| Fast | 100 | 350 | 25.00 | 22.00 | 8.92 | 7.14 | 7.04 | 2.71 | 47.81 |
| Metal finish - synthetic enamel (off white), gloss (material #37) | | | | | | | | | |
| Brush 1st finish coat | | | | | | | | | |
| Slow | 100 | 425 | 25.90 | 10.25 | 2.79 | 6.09 | 5.93 | 3.76 | 28.82 |
| Medium | 110 | 400 | 22.70 | 15.00 | 5.04 | 5.68 | 6.30 | 3.52 | 35.54 |
| Fast | 120 | 375 | 19.50 | 18.33 | 7.43 | 5.20 | 5.73 | 2.20 | 38.89 |
| Brush 2nd or additional finish coats | | | | | | | | | |
| Slow | 120 | 450 | 25.90 | 8.54 | 2.32 | 5.76 | 5.16 | 3.27 | 25.05 |
| Medium | 130 | 425 | 22.70 | 12.69 | 4.27 | 5.34 | 5.46 | 3.05 | 30.81 |
| Fast | 140 | 400 | 19.50 | 15.71 | 6.37 | 4.88 | 4.99 | 1.92 | 33.87 |
| Metal finish - synthetic enamel (colors except orange/red), gloss (material #38) | | | | | | | | | |
| Brush 1st finish coat | | | | | | | | | |
| Slow | 100 | 425 | 35.00 | 10.25 | 2.79 | 8.24 | 6.60 | 4.18 | 32.06 |
| Medium | 110 | 400 | 30.70 | 15.00 | 5.04 | 7.68 | 6.79 | 3.80 | 38.31 |
| Fast | 120 | 375 | 26.30 | 18.33 | 7.43 | 7.01 | 6.06 | 2.33 | 41.16 |
| Brush 2nd or additional finish coats | | | | | | | | | |
| Slow | 120 | 450 | 35.00 | 8.54 | 2.32 | 7.78 | 5.78 | 3.66 | 28.08 |
| Medium | 130 | 425 | 30.70 | 12.69 | 4.27 | 7.22 | 5.92 | 3.31 | 33.41 |
| Fast | 140 | 400 | 26.30 | 15.71 | 6.37 | 6.58 | 5.30 | 2.04 | 36.00 |
| Downspouts | | | | | | | | | |
| Metal prime, rust inhibitor, clean metal (material #35) | | | | | | | | | |
| Brush prime coat | | | | | | | | | |
| Slow | 30 | 250 | 28.20 | 34.17 | 9.29 | 11.28 | 16.98 | 10.76 | 82.48 |
| Medium | 35 | 225 | 24.60 | 47.14 | 15.86 | 10.93 | 18.11 | 10.12 | 102.16 |
| Fast | 40 | 200 | 21.10 | 55.00 | 22.30 | 10.55 | 16.25 | 6.25 | 110.35 |
| Metal prime, rust inhibitor, rusty metal (material #36) | | | | | | | | | |
| Brush prime coat | | | | | | | | | |
| Slow | 30 | 250 | 33.30 | 34.17 | 9.29 | 13.32 | 17.61 | 11.16 | 85.55 |
| Medium | 35 | 225 | 29.20 | 47.14 | 15.86 | 12.98 | 18.62 | 10.41 | 105.01 |
| Fast | 40 | 200 | 25.00 | 55.00 | 22.30 | 12.50 | 16.61 | 6.38 | 112.79 |

| | Labor LF per manhour | Material coverage LF/gallon | Material cost per gallon | Labor cost per 100 LF | Labor burden 100 LF | Material cost per 100 LF | Overhead per 100 LF | Profit per 100 LF | Total price per 100 LF |
|---|---|---|---|---|---|---|---|---|---|
| **Metal finish - synthetic enamel (off white), gloss (material #37)** | | | | | | | | | |
| Brush 1st finish coat | | | | | | | | | |
| Slow | 50 | 275 | 25.90 | 20.50 | 5.58 | 9.42 | 11.01 | 6.98 | 53.49 |
| Medium | 60 | 250 | 22.70 | 27.50 | 9.26 | 9.08 | 11.23 | 6.28 | 63.35 |
| Fast | 70 | 225 | 19.50 | 31.43 | 12.75 | 8.67 | 9.78 | 3.76 | 66.39 |
| Brush 2nd or additional finish coats | | | | | | | | | |
| Slow | 70 | 300 | 25.90 | 14.64 | 3.99 | 8.63 | 8.45 | 5.36 | 41.07 |
| Medium | 80 | 275 | 22.70 | 20.63 | 6.93 | 8.25 | 8.78 | 4.91 | 49.50 |
| Fast | 90 | 250 | 19.50 | 24.44 | 9.91 | 7.80 | 7.80 | 3.00 | 52.95 |
| **Metal finish - synthetic enamel (colors except orange/red), gloss (material #38)** | | | | | | | | | |
| Brush 1st finish coat | | | | | | | | | |
| Slow | 50 | 275 | 35.00 | 20.50 | 5.58 | 12.73 | 12.03 | 7.63 | 58.47 |
| Medium | 60 | 250 | 30.70 | 27.50 | 9.26 | 12.28 | 12.01 | 6.71 | 67.76 |
| Fast | 70 | 225 | 26.30 | 31.43 | 12.75 | 11.69 | 10.33 | 3.97 | 70.17 |
| Brush 2nd or additional finish coats | | | | | | | | | |
| Slow | 70 | 300 | 35.00 | 14.64 | 3.99 | 11.67 | 9.39 | 5.95 | 45.64 |
| Medium | 80 | 275 | 30.70 | 20.63 | 6.93 | 11.16 | 9.49 | 5.30 | 53.51 |
| Fast | 90 | 250 | 26.30 | 24.44 | 9.91 | 10.52 | 8.30 | 3.19 | 56.36 |

NOTE: Oil base material is recommended for any metal surface. Although water base material is often used, it may cause oxidation, corrosion and rust. These figures assume that all exposed surfaces of 5" gutters and 4" downspouts are painted. For ornamental gutters and downspouts, multiply the linear feet by 1.5 before using these figures. "Slow" work is based on a $10.25 hourly wage, "Medium" work on a $16.75 hourly wage, and "Fast" work on a $22.00 hourly wage. Other qualifications that apply to this table are on page 9.

## High Time Difficulty Factors

Painting takes longer and may require more material when heights above the floor exceed 8 feet. The additional time and material for working at these heights and using a roller pole or a wand on a spray gun, climbing up and down a ladder or scaffolding is applied by using one of the factors listed below. The wall area above 8 feet is typically referred to as the "Clip." To apply the high time difficulty factor, measure the surface above 8 feet which is to be painted and multiply that figure by the appropriate factor. This measurement can be listed on a separate line of your take-off and the appropriate price can be applied for a total.

For labor calculations only:
- Add 30% to the area for heights between 8 and 13 feet (multiply by 1.3)
- Add 60% to the area for heights from 13 to 17 feet (multiply by 1.6)
- Add 90% to the area for heights from 17 to 19 feet (multiply by 1.9)
- Add 120% to the area for heights from 19 to 21 feet (multiply by 2.2)

EXAMPLE: A 17 x 14 living room has a vaulted ceiling 13 feet high. Your take-off sheet might look like this:

Walls to 8 feet: 136 + 112 + 136 + 112 = 496 SF

Clip: [(5 x 14) / 2] x 2 + (5 x 17) = 70 + 85 = 155 SF

area of two triangles + rectangular area

155 SF x 1.3 (high time difficulty factor) = 202 SF

Then multiply each SF total by the appropriate price per square foot.

## Mail box structures, wood, apartment type

Measure the length of each board to be painted and use the manhours and material given for Trellis or Plant-on trim or Siding.

| | Labor SF per manhour | Material coverage SF/gallon | Material cost per gallon | Labor cost per 100 SF | Labor burden 100 SF | Material cost per 100 SF | Overhead per 100 SF | Profit per 100 SF | Total price per 100 SF |
|---|---|---|---|---|---|---|---|---|---|
| **Masonry, anti-graffiti stain eliminator on smooth or rough surface** | | | | | | | | | |
| Water base primer and sealer (material #39) | | | | | | | | | |
| Roll & brush each coat | | | | | | | | | |
| Slow | 350 | 400 | 30.40 | 2.93 | .80 | 7.60 | 3.51 | 2.23 | 17.07 |
| Medium | 375 | 375 | 26.60 | 4.40 | 1.49 | 7.09 | 3.18 | 1.78 | 17.94 |
| Fast | 400 | 350 | 22.80 | 5.50 | 2.23 | 6.51 | 2.63 | 1.01 | 17.88 |
| Oil base primer and sealer (material #40) | | | | | | | | | |
| Roll & brush each coat | | | | | | | | | |
| Slow | 350 | 375 | 28.20 | 2.93 | .80 | 7.52 | 3.49 | 2.21 | 16.95 |
| Medium | 375 | 350 | 24.70 | 4.40 | 1.49 | 7.06 | 3.17 | 1.77 | 17.89 |
| Fast | 400 | 325 | 21.20 | 5.50 | 2.23 | 6.52 | 2.64 | 1.01 | 17.90 |
| Polyurethane 2 part system (material #41) | | | | | | | | | |
| Roll & brush each coat | | | | | | | | | |
| Slow | 300 | 375 | 63.30 | 3.42 | .92 | 16.88 | 6.58 | 4.17 | 31.97 |
| Medium | 325 | 350 | 55.40 | 5.08 | 1.71 | 15.83 | 5.54 | 3.10 | 31.26 |
| Fast | 350 | 325 | 47.40 | 6.29 | 2.55 | 14.58 | 4.33 | 1.67 | 29.42 |

Use these figures for new brick, used brick, or Concrete Masonry Units (CMU) where the block surfaces are either smooth or rough, porous or unfilled, with joints struck to average depth. The more porous the surface, the rougher the texture, the more time and material will be required. "Slow" work is based on a $10.25 hourly wage, "Medium" work on a $16.50 hourly wage, and "Fast" work on a $22.00 hourly wage. Other qualifications that apply to this table are on page 9.

| | Labor SF per manhour | Material coverage SF/gallon | Material cost per gallon | Labor cost per 100 SF | Labor burden 100 SF | Material cost per 100 SF | Overhead per 100 SF | Profit per 100 SF | Total price per 100 SF |
|---|---|---|---|---|---|---|---|---|---|
| **Masonry, block filler** | | | | | | | | | |
| Brush 1 coat (material #33) | | | | | | | | | |
| Slow | 95 | 75 | 14.10 | 10.79 | 2.94 | 18.80 | 10.08 | 6.39 | 49.00 |
| Medium | 125 | 65 | 12.30 | 13.20 | 4.44 | 18.92 | 8.96 | 5.01 | 50.53 |
| Fast | 155 | 55 | 10.60 | 14.19 | 5.75 | 19.27 | 7.25 | 2.79 | 49.25 |
| Roll 1 coat (material #33) | | | | | | | | | |
| Slow | 190 | 70 | 14.10 | 5.39 | 1.47 | 20.14 | 8.37 | 5.31 | 40.68 |
| Medium | 215 | 60 | 12.30 | 7.67 | 2.58 | 20.50 | 7.53 | 4.21 | 42.49 |
| Fast | 240 | 50 | 10.60 | 9.17 | 3.72 | 21.20 | 6.31 | 2.42 | 42.82 |
| Spray 1 coat (material #33) | | | | | | | | | |
| Slow | 425 | 65 | 14.10 | 2.41 | .65 | 21.69 | 7.68 | 4.87 | 37.30 |
| Medium | 525 | 55 | 12.30 | 3.14 | 1.05 | 22.36 | 6.51 | 3.64 | 36.70 |
| Fast | 625 | 45 | 10.60 | 3.52 | 1.43 | 23.56 | 5.27 | 2.03 | 35.81 |

Use these figures for using block filler on rough or porous masonry with joints struck to average depth. "Slow" work is based on a $10.25 hourly wage, "Medium" work on a $16.50 hourly wage, and "Fast" work on a $22.50 hourly wage. Other qualifications that apply to this table are on page 9.

| | Labor SF per manhour | Material coverage SF/gallon | Material cost per gallon | Labor cost per 100 SF | Labor burden 100 SF | Material cost per 100 SF | Overhead per 100 SF | Profit per 100 SF | Total price per 100 SF |
|---|---|---|---|---|---|---|---|---|---|
| **Masonry, brick, new, smooth-surface, brush** | | | | | | | | | |
| Masonry paint, water base, flat or gloss (material #31) | | | | | | | | | |
| Brush 1st coat | | | | | | | | | |
| Slow | 200 | 300 | 19.50 | 5.13 | 1.39 | 6.50 | 4.04 | 2.56 | 19.62 |
| Medium | 225 | 275 | 17.10 | 7.33 | 2.46 | 6.22 | 3.92 | 2.19 | 22.12 |
| Fast | 250 | 250 | 14.60 | 8.80 | 3.57 | 5.84 | 3.37 | 1.29 | 22.87 |
| Brush 2nd or additional coats | | | | | | | | | |
| Slow | 250 | 325 | 19.50 | 4.10 | 1.12 | 6.00 | 3.48 | 2.21 | 16.91 |
| Medium | 275 | 300 | 17.10 | 6.00 | 2.03 | 5.70 | 3.36 | 1.88 | 18.97 |
| Fast | 300 | 275 | 14.60 | 7.33 | 2.97 | 5.31 | 2.89 | 1.11 | 19.61 |
| Masonry paint, oil base (material #32) | | | | | | | | | |
| Brush 1st coat | | | | | | | | | |
| Slow | 200 | 350 | 24.90 | 5.13 | 1.39 | 7.11 | 4.23 | 2.68 | 20.54 |
| Medium | 225 | 325 | 21.80 | 7.33 | 2.46 | 6.71 | 4.04 | 2.26 | 22.80 |
| Fast | 250 | 300 | 18.70 | 8.80 | 3.57 | 6.23 | 3.44 | 1.32 | 23.36 |
| Brush 2nd or additional coats | | | | | | | | | |
| Slow | 250 | 400 | 24.90 | 4.10 | 1.12 | 6.23 | 3.55 | 2.25 | 17.25 |
| Medium | 275 | 363 | 21.80 | 6.00 | 2.03 | 6.01 | 3.44 | 1.92 | 19.40 |
| Fast | 300 | 325 | 18.70 | 7.33 | 2.97 | 5.75 | 2.97 | 1.14 | 20.16 |

Use these figures for new smooth-surface brick with joints struck to average depth. "Slow" work is based on a $10.25 hourly wage, "Medium" work on a $16.50 hourly wage, and "Fast" work on a $22.00 hourly wage. Other qualifications that apply to this table are on page 9.

| | Labor SF per manhour | Material coverage SF/gallon | Material cost per gallon | Labor cost per 100 SF | Labor burden 100 SF | Material cost per 100 SF | Overhead per 100 SF | Profit per 100 SF | Total price per 100 SF |
|---|---|---|---|---|---|---|---|---|---|
| **Masonry, brick, new, smooth-surface, roll** | | | | | | | | | |
| Masonry paint, water base, flat or gloss (material #31) | | | | | | | | | |
| Roll 1st coat | | | | | | | | | |
| Slow | 325 | 250 | 19.50 | 3.15 | .87 | 7.80 | 3.66 | 2.32 | 17.80 |
| Medium | 350 | 213 | 17.10 | 4.71 | 1.60 | 8.03 | 3.51 | 1.96 | 19.81 |
| Fast | 375 | 175 | 14.60 | 5.87 | 2.39 | 8.34 | 3.07 | 1.18 | 20.85 |
| Roll 2nd or additional coats | | | | | | | | | |
| Slow | 375 | 275 | 19.50 | 2.73 | .75 | 7.09 | 3.27 | 2.07 | 15.91 |
| Medium | 400 | 250 | 17.10 | 4.13 | 1.38 | 6.84 | 3.03 | 1.69 | 17.07 |
| Fast | 425 | 225 | 14.60 | 5.18 | 2.09 | 6.49 | 2.55 | .98 | 17.29 |
| Masonry paint, oil base (material #32) | | | | | | | | | |
| Roll 1st coat | | | | | | | | | |
| Slow | 325 | 325 | 24.90 | 3.15 | .87 | 7.66 | 3.62 | 2.29 | 17.59 |
| Medium | 350 | 288 | 21.80 | 4.71 | 1.60 | 7.57 | 3.40 | 1.90 | 19.18 |
| Fast | 375 | 250 | 18.70 | 5.87 | 2.39 | 7.48 | 2.91 | 1.12 | 19.77 |
| Roll 2nd or additional coats | | | | | | | | | |
| Slow | 375 | 350 | 24.90 | 2.73 | .75 | 7.11 | 3.28 | 2.08 | 15.95 |
| Medium | 400 | 313 | 21.80 | 4.13 | 1.38 | 6.96 | 3.06 | 1.71 | 17.24 |
| Fast | 425 | 275 | 18.70 | 5.18 | 2.09 | 6.80 | 2.60 | 1.00 | 17.67 |
| Waterproofing, clear hydro sealer, oil base (material #34) | | | | | | | | | |
| Roll 1st coat | | | | | | | | | |
| Slow | 200 | 175 | 18.00 | 5.13 | 1.39 | 10.29 | 5.21 | 3.30 | 25.32 |
| Medium | 225 | 150 | 15.80 | 7.33 | 2.46 | 10.53 | 4.98 | 2.78 | 28.08 |
| Fast | 250 | 125 | 13.50 | 8.80 | 3.57 | 10.80 | 4.29 | 1.65 | 29.11 |
| Roll 2nd or additional coats | | | | | | | | | |
| Slow | 225 | 200 | 18.00 | 4.56 | 1.23 | 9.00 | 4.59 | 2.91 | 22.29 |
| Medium | 250 | 190 | 15.80 | 6.60 | 2.22 | 8.32 | 4.20 | 2.35 | 23.69 |
| Fast | 275 | 180 | 13.50 | 8.00 | 3.25 | 7.50 | 3.47 | 1.33 | 23.55 |

Use these figures for new smooth-surface brick with joints struck to average depth. "Slow" work is based on a $10.25 hourly wage, "Medium" work on a $16.50 hourly wage, and "Fast" work on a $22.00 hourly wage. Other qualifications that apply to this table are on page 9.

| | Labor SF per manhour | Material coverage SF/gallon | Material cost per gallon | Labor cost per 100 SF | Labor burden 100 SF | Material cost per 100 SF | Overhead per 100 SF | Profit per 100 SF | Total price per 100 SF |
|---|---|---|---|---|---|---|---|---|---|
| **Masonry, brick, new, smooth-surface, spray** | | | | | | | | | |
| Masonry paint, water base, flat or gloss (material #31) | | | | | | | | | |
| Spray 1st coat | | | | | | | | | |
| Slow | 650 | 250 | 19.50 | 1.58 | .43 | 7.80 | 3.04 | 1.93 | 14.78 |
| Medium | 750 | 225 | 17.10 | 2.20 | .73 | 7.60 | 2.58 | 1.44 | 14.55 |
| Fast | 850 | 200 | 14.60 | 2.59 | 1.06 | 7.30 | 2.02 | .78 | 13.75 |
| Spray 2nd or additional coats | | | | | | | | | |
| Slow | 750 | 275 | 19.50 | 1.37 | .36 | 7.09 | 2.74 | 1.74 | 13.30 |
| Medium | 825 | 238 | 17.10 | 2.00 | .67 | 7.18 | 2.41 | 1.35 | 13.61 |
| Fast | 900 | 250 | 14.60 | 2.44 | .99 | 5.84 | 1.71 | .66 | 11.64 |
| Masonry paint, oil base (material #32) | | | | | | | | | |
| Spray 1st coat | | | | | | | | | |
| Slow | 650 | 275 | 24.90 | 1.58 | .43 | 9.05 | 3.43 | 2.17 | 16.66 |
| Medium | 750 | 250 | 21.80 | 2.20 | .73 | 8.72 | 2.86 | 1.60 | 16.11 |
| Fast | 850 | 225 | 18.70 | 2.59 | 1.06 | 8.31 | 2.21 | .85 | 15.02 |
| Spray 2nd or additional coats | | | | | | | | | |
| Slow | 750 | 300 | 24.90 | 1.37 | .36 | 8.30 | 3.11 | 1.97 | 15.11 |
| Medium | 825 | 288 | 21.80 | 2.00 | .67 | 7.57 | 2.51 | 1.40 | 14.15 |
| Fast | 900 | 275 | 18.70 | 2.44 | .99 | 6.80 | 1.89 | .73 | 12.85 |
| Waterproofing, clear hydro sealer, oil base (material #34) | | | | | | | | | |
| Spray 1st coat | | | | | | | | | |
| Slow | 700 | 120 | 18.00 | 1.46 | .40 | 15.00 | 5.23 | 3.31 | 25.40 |
| Medium | 800 | 100 | 15.80 | 2.06 | .70 | 15.80 | 4.54 | 2.54 | 25.64 |
| Fast | 900 | 80 | 13.50 | 2.44 | .99 | 16.88 | 3.76 | 1.44 | 25.51 |
| Spray 2nd or additional coats | | | | | | | | | |
| Slow | 800 | 150 | 18.00 | 1.28 | .35 | 12.00 | 4.23 | 2.68 | 20.54 |
| Medium | 900 | 138 | 15.80 | 1.83 | .62 | 11.45 | 3.41 | 1.90 | 19.21 |
| Fast | 1000 | 125 | 13.50 | 2.20 | .89 | 10.80 | 2.57 | .99 | 17.45 |

Use these figures for new smooth-surface brick with joints struck to average depth. "Slow" work is based on a $10.25 hourly wage, "Medium" work on a $16.50 hourly wage, and "Fast" work on a $22.00 hourly wage. Other qualifications that apply to this table are on page 9.

| | Labor SF per manhour | Material coverage SF/gallon | Material cost per gallon | Labor cost per 100 SF | Labor burden 100 SF | Material cost per 100 SF | Overhead per 100 SF | Profit per 100 SF | Total price per 100 SF |
|---|---|---|---|---|---|---|---|---|---|

## Masonry, brick, used, rough surface, brush

| | Labor SF per manhour | Material coverage SF/gallon | Material cost per gallon | Labor cost per 100 SF | Labor burden 100 SF | Material cost per 100 SF | Overhead per 100 SF | Profit per 100 SF | Total price per 100 SF |
|---|---|---|---|---|---|---|---|---|---|
| Masonry paint, water base, flat or gloss (material #31) | | | | | | | | | |
| Brush 1st coat | | | | | | | | | |
| Slow | 150 | 300 | 19.50 | 6.83 | 1.87 | 6.50 | 4.71 | 2.99 | 22.90 |
| Medium | 175 | 275 | 17.10 | 9.43 | 3.16 | 6.22 | 4.61 | 2.58 | 26.00 |
| Fast | 200 | 250 | 14.60 | 11.00 | 4.46 | 5.84 | 3.94 | 1.51 | 26.75 |
| Brush 2nd or additional coats | | | | | | | | | |
| Slow | 200 | 375 | 19.50 | 5.13 | 1.39 | 5.20 | 3.64 | 2.31 | 17.67 |
| Medium | 225 | 350 | 17.10 | 7.33 | 2.46 | 4.89 | 3.60 | 2.01 | 20.29 |
| Fast | 250 | 325 | 14.60 | 8.80 | 3.57 | 4.49 | 3.12 | 1.20 | 21.18 |
| Masonry paint, oil base (material #32) | | | | | | | | | |
| Brush 1st coat | | | | | | | | | |
| Slow | 150 | 325 | 24.90 | 6.83 | 1.87 | 7.66 | 5.07 | 3.21 | 24.64 |
| Medium | 175 | 300 | 21.80 | 9.43 | 3.16 | 7.27 | 4.87 | 2.72 | 27.45 |
| Fast | 200 | 275 | 18.70 | 11.00 | 4.46 | 6.80 | 4.12 | 1.58 | 27.96 |
| Brush 2nd or additional coats | | | | | | | | | |
| Slow | 200 | 400 | 24.90 | 5.13 | 1.39 | 6.23 | 3.96 | 2.51 | 19.22 |
| Medium | 225 | 375 | 21.80 | 7.33 | 2.46 | 5.81 | 3.82 | 2.14 | 21.56 |
| Fast | 250 | 350 | 18.70 | 8.80 | 3.57 | 5.34 | 3.28 | 1.26 | 22.25 |

Use these figures for dry pressed used brick, clay brick tile, or adobe block with joints struck to average depth. "Slow" work is based on a $10.25 hourly wage, "Medium" work on a $16.50 hourly wage, and "Fast" work on a $22.00 hourly wage. Other qualifications that apply to this table are on page 9.

| | Labor SF per manhour | Material coverage SF/gallon | Material cost per gallon | Labor cost per 100 SF | Labor burden 100 SF | Material cost per 100 SF | Overhead per 100 SF | Profit per 100 SF | Total price per 100 SF |
|---|---|---|---|---|---|---|---|---|---|
| **Masonry, brick, used, rough surface, roll** | | | | | | | | | |
| Masonry paint, water base, flat or gloss (material #31) | | | | | | | | | |
| Roll 1st coat | | | | | | | | | |
| Slow | 300 | 250 | 19.50 | 3.42 | .92 | 7.80 | 3.77 | 2.39 | 18.30 |
| Medium | 325 | 225 | 17.10 | 5.08 | 1.71 | 7.60 | 3.53 | 1.97 | 19.89 |
| Fast | 350 | 200 | 14.60 | 6.29 | 2.55 | 7.30 | 2.99 | 1.15 | 20.28 |
| Roll 2nd or additional coats | | | | | | | | | |
| Slow | 350 | 325 | 19.50 | 2.93 | .80 | 6.00 | 3.02 | 1.91 | 14.66 |
| Medium | 375 | 300 | 17.10 | 4.40 | 1.49 | 5.70 | 2.84 | 1.59 | 16.02 |
| Fast | 400 | 275 | 14.60 | 5.50 | 2.23 | 5.31 | 2.41 | .93 | 16.38 |
| Masonry paint, oil base (material #32) | | | | | | | | | |
| Roll 1st coat | | | | | | | | | |
| Slow | 300 | 275 | 24.90 | 3.42 | .92 | 9.05 | 4.15 | 2.63 | 20.17 |
| Medium | 325 | 250 | 21.80 | 5.08 | 1.71 | 8.72 | 3.80 | 2.12 | 21.43 |
| Fast | 350 | 225 | 18.70 | 6.29 | 2.55 | 8.31 | 3.17 | 1.22 | 21.54 |
| Roll 2nd or additional coats | | | | | | | | | |
| Slow | 350 | 350 | 24.90 | 2.93 | .80 | 7.11 | 3.36 | 2.13 | 16.33 |
| Medium | 375 | 325 | 21.80 | 4.40 | 1.49 | 6.71 | 3.08 | 1.72 | 17.40 |
| Fast | 400 | 300 | 18.70 | 5.50 | 2.23 | 6.23 | 2.58 | .99 | 17.53 |
| Waterproofing, clear hydro sealer, oil base (material #34) | | | | | | | | | |
| Roll 1st coat | | | | | | | | | |
| Slow | 150 | 125 | 18.00 | 6.83 | 1.87 | 14.40 | 7.16 | 4.54 | 34.80 |
| Medium | 175 | 113 | 15.80 | 9.43 | 3.16 | 13.98 | 6.51 | 3.64 | 36.72 |
| Fast | 200 | 100 | 13.50 | 11.00 | 4.46 | 13.50 | 5.36 | 2.06 | 36.38 |
| Roll 2nd or additional coats | | | | | | | | | |
| Slow | 175 | 150 | 18.00 | 5.86 | 1.59 | 12.00 | 6.03 | 3.82 | 29.30 |
| Medium | 200 | 138 | 15.80 | 8.25 | 2.78 | 11.45 | 5.51 | 3.08 | 31.07 |
| Fast | 225 | 125 | 13.50 | 9.78 | 3.95 | 10.80 | 4.54 | 1.75 | 30.82 |

Use these figures for dry pressed used brick, clay brick tile, or adobe block with joints struck to average depth. "Slow" work is based on a $10.25 hourly wage, "Medium" work on a $16.50 hourly wage, and "Fast" work on a $22.00 hourly wage. Other qualifications that apply to this table are on page 9.

| | Labor SF per manhour | Material coverage SF/gallon | Material cost per gallon | Labor cost per 100 SF | Labor burden 100 SF | Material cost per 100 SF | Overhead per 100 SF | Profit per 100 SF | Total price per 100 SF |
|---|---|---|---|---|---|---|---|---|---|
| **Masonry, brick, used, rough surface, spray** | | | | | | | | | |
| Masonry paint, water base, flat or gloss (material #31) | | | | | | | | | |
| Spray 1st coat | | | | | | | | | |
| Slow | 600 | 200 | 19.50 | 1.71 | .47 | 9.75 | 3.70 | 2.34 | 17.97 |
| Medium | 700 | 175 | 17.10 | 2.36 | .79 | 9.77 | 3.17 | 1.77 | 17.86 |
| Fast | 800 | 150 | 14.60 | 2.75 | 1.12 | 9.73 | 2.52 | .97 | 17.09 |
| Spray 2nd or additional coats | | | | | | | | | |
| Slow | 700 | 225 | 19.50 | 1.46 | .40 | 8.67 | 3.26 | 2.07 | 15.86 |
| Medium | 800 | 213 | 17.10 | 2.06 | .70 | 8.03 | 2.64 | 1.48 | 14.91 |
| Fast | 900 | 200 | 14.60 | 2.44 | .99 | 7.30 | 1.99 | .76 | 13.48 |
| Masonry paint, oil base (material #32) | | | | | | | | | |
| Spray 1st coat | | | | | | | | | |
| Slow | 600 | 225 | 24.90 | 1.71 | .47 | 11.07 | 4.11 | 2.60 | 19.96 |
| Medium | 700 | 200 | 21.80 | 2.36 | .79 | 10.90 | 3.44 | 1.92 | 19.41 |
| Fast | 800 | 175 | 18.70 | 2.75 | 1.12 | 10.69 | 2.69 | 1.04 | 18.29 |
| Spray 2nd or additional coats | | | | | | | | | |
| Slow | 700 | 250 | 24.90 | 1.46 | .40 | 9.96 | 3.66 | 2.32 | 17.80 |
| Medium | 800 | 238 | 21.80 | 2.06 | .70 | 9.16 | 2.92 | 1.63 | 16.47 |
| Fast | 900 | 225 | 18.70 | 2.44 | .99 | 8.31 | 2.17 | .83 | 14.74 |
| Waterproofing, clear hydro sealer, oil base (material #34) | | | | | | | | | |
| Spray 1st coat | | | | | | | | | |
| Slow | 600 | 80 | 18.00 | 1.71 | .47 | 22.50 | 7.65 | 4.85 | 37.18 |
| Medium | 700 | 75 | 15.80 | 2.36 | .79 | 21.07 | 5.93 | 3.32 | 33.47 |
| Fast | 800 | 70 | 13.50 | 2.75 | 1.12 | 19.29 | 4.28 | 1.65 | 29.09 |
| Spray 2nd coat | | | | | | | | | |
| Slow | 800 | 100 | 18.00 | 1.28 | .35 | 18.00 | 6.09 | 3.86 | 29.58 |
| Medium | 900 | 90 | 15.80 | 1.83 | .62 | 17.56 | 4.90 | 2.74 | 27.65 |
| Fast | 1000 | 80 | 13.50 | 2.20 | .89 | 16.88 | 3.69 | 1.42 | 25.08 |

Use these figures for dry pressed used brick, clay brick tile, or adobe block with joints struck to average depth. "Slow" work is based on a $10.25 hourly wage, "Medium" work on a $16.50 hourly wage, and "Fast" work on a $22.00 hourly wage. Other qualifications that apply to this table are on page 9.

| | Labor SF per manhour | Material coverage SF/gallon | Material cost per gallon | Labor cost per 100 SF | Labor burden 100 SF | Material cost per 100 SF | Overhead per 100 SF | Profit per 100 SF | Total price per 100 SF |
|---|---|---|---|---|---|---|---|---|---|
| **Masonry, Concrete Masonry Units (CMU), rough, porous surface, brush** | | | | | | | | | |
| Masonry paint, water base, flat or gloss (material #31) | | | | | | | | | |
| Brush 1st coat | | | | | | | | | |
| Slow | 110 | 100 | 19.50 | 9.32 | 2.53 | 19.50 | 9.72 | 6.16 | 47.23 |
| Medium | 130 | 88 | 17.10 | 12.69 | 4.27 | 19.43 | 8.92 | 4.98 | 50.29 |
| Fast | 150 | 75 | 14.60 | 14.67 | 5.95 | 19.47 | 7.42 | 2.85 | 50.36 |
| Brush 2nd or additional coats | | | | | | | | | |
| Slow | 185 | 180 | 19.50 | 5.54 | 1.51 | 10.83 | 5.54 | 3.51 | 26.93 |
| Medium | 210 | 168 | 17.10 | 7.86 | 2.64 | 10.18 | 5.07 | 2.83 | 28.58 |
| Fast | 230 | 155 | 14.60 | 9.57 | 3.88 | 9.42 | 4.23 | 1.63 | 28.73 |
| Masonry paint, oil base (material #32) | | | | | | | | | |
| Brush 1st coat | | | | | | | | | |
| Slow | 110 | 130 | 24.90 | 9.32 | 2.53 | 19.15 | 9.61 | 6.09 | 46.70 |
| Medium | 130 | 120 | 21.80 | 12.69 | 4.27 | 18.17 | 8.61 | 4.81 | 48.55 |
| Fast | 150 | 110 | 18.70 | 14.67 | 5.95 | 17.00 | 6.96 | 2.67 | 47.25 |
| Brush 2nd or additional coats | | | | | | | | | |
| Slow | 185 | 200 | 24.90 | 5.54 | 1.51 | 12.45 | 6.05 | 3.83 | 29.38 |
| Medium | 208 | 180 | 21.80 | 7.93 | 2.68 | 12.11 | 5.56 | 3.11 | 31.39 |
| Fast | 230 | 160 | 18.70 | 9.57 | 3.88 | 11.69 | 4.65 | 1.79 | 31.58 |
| Epoxy coating, 2 part system clear (material #51) | | | | | | | | | |
| Brush 1st coat | | | | | | | | | |
| Slow | 95 | 110 | 57.00 | 10.79 | 2.94 | 51.82 | 20.32 | 12.88 | 98.75 |
| Medium | 115 | 98 | 49.90 | 14.35 | 4.83 | 50.92 | 17.17 | 9.60 | 96.87 |
| Fast | 135 | 85 | 42.80 | 16.30 | 6.61 | 50.35 | 13.55 | 5.21 | 92.02 |
| Brush 2nd or additional coats | | | | | | | | | |
| Slow | 165 | 200 | 57.00 | 6.21 | 1.69 | 28.50 | 11.28 | 7.15 | 54.83 |
| Medium | 190 | 188 | 49.90 | 8.68 | 2.92 | 26.54 | 9.34 | 5.22 | 52.70 |
| Fast | 210 | 175 | 42.80 | 10.48 | 4.24 | 24.46 | 7.25 | 2.79 | 49.22 |
| Waterproofing, clear hydro sealer, oil base (material #34) | | | | | | | | | |
| Brush 1st coat | | | | | | | | | |
| Slow | 125 | 90 | 18.00 | 8.20 | 2.23 | 20.00 | 9.43 | 5.98 | 45.84 |
| Medium | 150 | 80 | 15.80 | 11.00 | 3.71 | 19.75 | 8.44 | 4.72 | 47.62 |
| Fast | 175 | 70 | 13.50 | 12.57 | 5.09 | 19.29 | 6.84 | 2.63 | 46.42 |
| Brush 2nd or additional coats | | | | | | | | | |
| Slow | 230 | 130 | 18.00 | 4.46 | 1.21 | 13.85 | 6.05 | 3.84 | 29.41 |
| Medium | 275 | 110 | 15.80 | 6.00 | 2.03 | 14.36 | 5.48 | 3.06 | 30.93 |
| Fast | 295 | 90 | 13.50 | 7.46 | 3.02 | 15.00 | 4.72 | 1.81 | 32.01 |

Use these figures for Concrete Masonry Units (CMU) where the block surfaces are rough, porous or unfilled, with joints struck to average depth. The more porous the surface, the rougher the texture, the more time and material will be required. For heavy waterproofing applications, see Masonry under Industrial Painting Operations. "Slow" work is based on a $10.25 hourly wage, "Medium" work on a $16.50 hourly wage, and "Fast" work on a $22.00 hourly wage. Other qualifications that apply to this table are on page 9.

| | Labor SF per manhour | Material coverage SF/gallon | Material cost per gallon | Labor cost per 100 SF | Labor burden 100 SF | Material cost per 100 SF | Overhead per 100 SF | Profit per 100 SF | Total price per 100 SF |
|---|---|---|---|---|---|---|---|---|---|
| **Masonry, Concrete Masonry Units (CMU), rough, porous surface, roll** | | | | | | | | | |
| Masonry paint, water base, flat or gloss (material #31) | | | | | | | | | |
| Roll 1st coat | | | | | | | | | |
| Slow | 245 | 90 | 19.50 | 4.18 | 1.14 | 21.67 | 8.37 | 5.30 | 40.66 |
| Medium | 300 | 78 | 17.10 | 5.50 | 1.84 | 21.92 | 7.17 | 4.01 | 40.44 |
| Fast | 350 | 65 | 14.60 | 6.29 | 2.55 | 22.46 | 5.79 | 2.23 | 39.32 |
| Roll 2nd or additional coats | | | | | | | | | |
| Slow | 275 | 160 | 19.50 | 3.73 | 1.02 | 12.19 | 5.25 | 3.33 | 25.52 |
| Medium | 325 | 143 | 17.10 | 5.08 | 1.71 | 11.96 | 4.59 | 2.57 | 25.91 |
| Fast | 420 | 125 | 14.60 | 5.24 | 2.12 | 11.68 | 3.52 | 1.35 | 23.91 |
| Masonry paint, oil base (material #32) | | | | | | | | | |
| Roll 1st coat | | | | | | | | | |
| Slow | 245 | 110 | 24.90 | 4.18 | 1.14 | 22.64 | 8.67 | 5.49 | 42.12 |
| Medium | 300 | 98 | 21.80 | 5.50 | 1.84 | 22.24 | 7.25 | 4.05 | 40.88 |
| Fast | 350 | 85 | 18.70 | 6.29 | 2.55 | 22.00 | 5.71 | 2.19 | 38.74 |
| Roll 2nd or additional coats | | | | | | | | | |
| Slow | 275 | 185 | 24.90 | 3.73 | 1.02 | 13.46 | 5.65 | 3.58 | 27.44 |
| Medium | 325 | 170 | 21.80 | 5.08 | 1.71 | 12.82 | 4.80 | 2.69 | 27.10 |
| Fast | 420 | 155 | 18.70 | 5.24 | 2.12 | 12.06 | 3.59 | 1.38 | 24.39 |
| Epoxy coating, 2 part system, clear (material #51) | | | | | | | | | |
| Roll 1st coat | | | | | | | | | |
| Slow | 220 | 100 | 57.00 | 4.66 | 1.27 | 57.00 | 19.51 | 12.37 | 94.81 |
| Medium | 275 | 88 | 49.90 | 6.00 | 2.03 | 56.70 | 15.86 | 8.86 | 89.45 |
| Fast | 325 | 75 | 42.80 | 6.77 | 2.75 | 57.07 | 12.32 | 4.73 | 83.64 |
| Roll 2nd or additional coats | | | | | | | | | |
| Slow | 250 | 175 | 57.00 | 4.10 | 1.12 | 32.57 | 11.71 | 7.43 | 56.93 |
| Medium | 300 | 160 | 49.90 | 5.50 | 1.84 | 31.19 | 9.44 | 5.28 | 53.25 |
| Fast | 395 | 145 | 42.80 | 5.57 | 2.25 | 29.52 | 6.91 | 2.66 | 46.91 |
| Waterproofing, clear hydro sealer, oil base (material #34) | | | | | | | | | |
| Roll 1st coat | | | | | | | | | |
| Slow | 170 | 110 | 18.00 | 6.03 | 1.64 | 16.36 | 7.45 | 4.72 | 36.20 |
| Medium | 200 | 98 | 15.80 | 8.25 | 2.78 | 16.12 | 6.65 | 3.72 | 37.52 |
| Fast | 245 | 85 | 13.50 | 8.98 | 3.64 | 15.88 | 5.27 | 2.03 | 35.80 |
| Roll 2nd or additional coats | | | | | | | | | |
| Slow | 275 | 175 | 18.00 | 3.73 | 1.02 | 10.29 | 4.66 | 2.96 | 22.66 |
| Medium | 300 | 145 | 15.80 | 5.50 | 1.84 | 10.90 | 4.47 | 2.50 | 25.21 |
| Fast | 325 | 115 | 13.50 | 6.77 | 2.75 | 11.74 | 3.93 | 1.51 | 26.70 |

Use these figures for Concrete Masonry Units (CMU) where the block surfaces are rough, porous or unfilled, with joints struck to average depth. The more porous the surface, the rougher the texture, the more time and material will be required. For heavy waterproofing applications, see Masonry under Industrial Painting Operations. "Slow" work is based on a $10.25 hourly wage, "Medium" work on a $16.50 hourly wage, and "Fast" work on a $22.00 hourly wage. Other qualifications that apply to this table are on page 9.

| | Labor SF per manhour | Material coverage SF/gallon | Material cost per gallon | Labor cost per 100 SF | Labor burden 100 SF | Material cost per 100 SF | Overhead per 100 SF | Profit per 100 SF | Total price per 100 SF |
|---|---|---|---|---|---|---|---|---|---|
| **Masonry, Concrete Masonry Units (CMU), rough, porous surface, spray** | | | | | | | | | |
| Masonry paint, water base, flat or gloss (material #31) | | | | | | | | | |
| Spray 1st coat | | | | | | | | | |
| Slow | 600 | 100 | 19.50 | 1.71 | .47 | 19.50 | 6.72 | 4.26 | 32.66 |
| Medium | 700 | 78 | 17.10 | 2.36 | .79 | 21.92 | 6.14 | 3.43 | 34.64 |
| Fast | 800 | 55 | 14.60 | 2.75 | 1.12 | 26.55 | 5.63 | 2.16 | 38.21 |
| Spray 2nd or additional coats | | | | | | | | | |
| Slow | 700 | 155 | 19.50 | 1.46 | .40 | 12.58 | 4.48 | 2.84 | 21.76 |
| Medium | 800 | 133 | 17.10 | 2.06 | .70 | 12.86 | 3.82 | 2.14 | 21.58 |
| Fast | 900 | 110 | 14.60 | 2.44 | .99 | 13.27 | 3.09 | 1.19 | 20.98 |
| Masonry paint, oil base (material #32) | | | | | | | | | |
| Spray 1st coat | | | | | | | | | |
| Slow | 600 | 100 | 24.90 | 1.71 | .47 | 24.90 | 8.39 | 5.32 | 40.79 |
| Medium | 700 | 83 | 21.80 | 2.36 | .79 | 26.27 | 7.21 | 4.03 | 40.66 |
| Fast | 800 | 65 | 18.70 | 2.75 | 1.12 | 28.77 | 6.04 | 2.32 | 41.00 |
| Spray 2nd or additional coats | | | | | | | | | |
| Slow | 700 | 160 | 24.90 | 1.46 | .40 | 15.56 | 5.40 | 3.42 | 26.24 |
| Medium | 800 | 143 | 21.80 | 2.06 | .70 | 15.24 | 4.41 | 2.46 | 24.87 |
| Fast | 900 | 125 | 18.70 | 2.44 | .99 | 14.96 | 3.40 | 1.31 | 23.10 |
| Epoxy coating, 2 part system, clear (material #51) | | | | | | | | | |
| Spray 1st coat | | | | | | | | | |
| Slow | 500 | 85 | 57.00 | 2.05 | .56 | 67.06 | 21.60 | 13.69 | 104.96 |
| Medium | 600 | 68 | 49.90 | 2.75 | .93 | 73.38 | 18.88 | 10.55 | 106.49 |
| Fast | 700 | 50 | 42.80 | 3.14 | 1.28 | 85.60 | 16.65 | 6.40 | 113.07 |
| Spray 2nd or additional coats | | | | | | | | | |
| Slow | 600 | 145 | 57.00 | 1.71 | .47 | 39.31 | 12.86 | 8.15 | 62.50 |
| Medium | 700 | 130 | 49.90 | 2.36 | .79 | 38.38 | 10.17 | 5.69 | 57.39 |
| Fast | 800 | 115 | 42.80 | 2.75 | 1.12 | 37.22 | 7.60 | 2.92 | 51.61 |
| Waterproofing, clear hydro sealer, oil base (material #34) | | | | | | | | | |
| Spray 1st coat | | | | | | | | | |
| Slow | 500 | 60 | 18.00 | 2.05 | .56 | 30.00 | 10.11 | 6.41 | 49.13 |
| Medium | 700 | 50 | 15.80 | 2.36 | .79 | 31.60 | 8.51 | 4.76 | 48.02 |
| Fast | 900 | 40 | 13.50 | 2.44 | .99 | 33.75 | 6.88 | 2.64 | 46.70 |
| Spray 2nd or additional coats | | | | | | | | | |
| Slow | 600 | 90 | 18.00 | 1.71 | .47 | 20.00 | 6.88 | 4.36 | 33.42 |
| Medium | 800 | 75 | 15.80 | 2.06 | .70 | 21.07 | 5.84 | 3.26 | 32.93 |
| Fast | 1000 | 60 | 13.50 | 2.20 | .89 | 22.50 | 4.73 | 1.82 | 32.14 |

Use these figures for Concrete Masonry Units (CMU) where the block surfaces are rough, porous or unfilled, with joints struck to average depth. The more porous the surface, the rougher the texture, the more time and material will be required. For heavy waterproofing applications, see Masonry under Industrial Painting Operations. "Slow" work is based on a $10.25 hourly wage, "Medium" work on a $16.50 hourly wage, and "Fast" work on a $22.00 hourly wage. Other qualifications that apply to this table are on page 9.

| | Labor SF per manhour | Material coverage SF/gallon | Material cost per gallon | Labor cost per 100 SF | Labor burden 100 SF | Material cost per 100 SF | Overhead per 100 SF | Profit per 100 SF | Total price per 100 SF |
|---|---|---|---|---|---|---|---|---|---|

## Masonry, Concrete Masonry Units (CMU), smooth-surface, brush

| | Labor SF per manhour | Material coverage SF/gallon | Material cost per gallon | Labor cost per 100 SF | Labor burden 100 SF | Material cost per 100 SF | Overhead per 100 SF | Profit per 100 SF | Total price per 100 SF |
|---|---|---|---|---|---|---|---|---|---|
| **Masonry paint, water base, flat or gloss (material #31)** | | | | | | | | | |
| Brush 1st coat | | | | | | | | | |
| Slow | 140 | 310 | 19.50 | 7.32 | 1.99 | 6.29 | 4.84 | 3.07 | 23.51 |
| Medium | 190 | 275 | 17.10 | 8.68 | 2.92 | 6.22 | 4.37 | 2.44 | 24.63 |
| Fast | 230 | 240 | 14.60 | 9.57 | 3.88 | 6.08 | 3.61 | 1.39 | 24.53 |
| Brush 2nd or additional coats | | | | | | | | | |
| Slow | 170 | 410 | 19.50 | 6.03 | 1.64 | 4.76 | 3.85 | 2.44 | 18.72 |
| Medium | 250 | 370 | 17.10 | 6.60 | 2.22 | 4.62 | 3.29 | 1.84 | 18.57 |
| Fast | 325 | 340 | 14.60 | 6.77 | 2.75 | 4.29 | 2.55 | .98 | 17.34 |
| **Masonry paint, oil base (material #32)** | | | | | | | | | |
| Brush 1st coat | | | | | | | | | |
| Slow | 140 | 350 | 24.90 | 7.32 | 1.99 | 7.11 | 5.09 | 3.23 | 24.74 |
| Medium | 190 | 320 | 21.80 | 8.68 | 2.92 | 6.81 | 4.51 | 2.52 | 25.44 |
| Fast | 230 | 290 | 18.70 | 9.57 | 3.88 | 6.45 | 3.68 | 1.41 | 24.99 |
| Brush 2nd or additional coats | | | | | | | | | |
| Slow | 170 | 450 | 24.90 | 6.03 | 1.64 | 5.53 | 4.09 | 2.59 | 19.88 |
| Medium | 250 | 420 | 21.80 | 6.60 | 2.22 | 5.19 | 3.43 | 1.92 | 19.36 |
| Fast | 325 | 390 | 18.70 | 6.77 | 2.75 | 4.79 | 2.65 | 1.02 | 17.98 |
| **Epoxy coating, 2 part system clear (material #51)** | | | | | | | | | |
| Brush 1st coat | | | | | | | | | |
| Slow | 120 | 325 | 57.00 | 8.54 | 2.32 | 17.54 | 8.81 | 5.58 | 42.79 |
| Medium | 160 | 295 | 49.90 | 10.31 | 3.47 | 16.92 | 7.52 | 4.20 | 42.42 |
| Fast | 200 | 265 | 42.80 | 11.00 | 4.46 | 16.15 | 5.85 | 2.25 | 39.71 |
| Brush 2nd or additional coats | | | | | | | | | |
| Slow | 150 | 425 | 57.00 | 6.83 | 1.87 | 13.41 | 6.85 | 4.34 | 33.30 |
| Medium | 225 | 395 | 49.90 | 7.33 | 2.46 | 12.63 | 5.50 | 3.07 | 30.99 |
| Fast | 300 | 365 | 42.80 | 7.33 | 2.97 | 11.73 | 4.08 | 1.57 | 27.68 |
| **Waterproofing, clear hydro sealer, oil base (material #34)** | | | | | | | | | |
| Brush 1st coat | | | | | | | | | |
| Slow | 150 | 100 | 18.00 | 6.83 | 1.87 | 18.00 | 8.27 | 5.24 | 40.21 |
| Medium | 175 | 88 | 15.80 | 9.43 | 3.16 | 17.95 | 7.48 | 4.18 | 42.20 |
| Fast | 200 | 75 | 13.50 | 11.00 | 4.46 | 18.00 | 6.19 | 2.38 | 42.03 |
| Brush 2nd or additional coats | | | | | | | | | |
| Slow | 230 | 140 | 18.00 | 4.46 | 1.21 | 12.86 | 5.75 | 3.64 | 27.92 |
| Medium | 275 | 120 | 15.80 | 6.00 | 2.03 | 13.17 | 5.19 | 2.90 | 29.29 |
| Fast | 295 | 100 | 13.50 | 7.46 | 3.02 | 13.50 | 4.44 | 1.71 | 30.13 |

Use these figures for Concrete Masonry Units (CMU) where the block surfaces are smooth with joints struck to average depth. For heavy waterproofing applications, see Masonry under Industrial Painting Operations. "Slow" work is based on a $10.25 hourly wage, "Medium" work on a $16.50 hourly wage, and "Fast" work on a $22.00 hourly wage. Other qualifications that apply to this table are on page 9.

| | Labor SF per manhour | Material coverage SF/gallon | Material cost per gallon | Labor cost per 100 SF | Labor burden 100 SF | Material cost per 100 SF | Overhead per 100 SF | Profit per 100 SF | Total price per 100 SF |
|---|---|---|---|---|---|---|---|---|---|
| **Masonry, Concrete Masonry Units (CMU), smooth-surface, roll** | | | | | | | | | |
| Masonry paint, water base, flat or gloss (material #31) | | | | | | | | | |
| Roll 1st coat | | | | | | | | | |
| Slow | 300 | 240 | 19.50 | 3.42 | .92 | 8.13 | 3.87 | 2.45 | 18.79 |
| Medium | 350 | 228 | 17.10 | 4.71 | 1.60 | 7.50 | 3.38 | 1.89 | 19.08 |
| Fast | 390 | 215 | 14.60 | 5.64 | 2.28 | 6.79 | 2.72 | 1.05 | 18.48 |
| Roll 2nd or additional coats | | | | | | | | | |
| Slow | 350 | 325 | 19.50 | 2.93 | .80 | 6.00 | 3.02 | 1.91 | 14.66 |
| Medium | 400 | 313 | 17.10 | 4.13 | 1.38 | 5.46 | 2.69 | 1.50 | 15.16 |
| Fast | 450 | 300 | 14.60 | 4.89 | 1.97 | 4.87 | 2.17 | .83 | 14.73 |
| Masonry paint, oil base (material #32) | | | | | | | | | |
| Roll 1st coat | | | | | | | | | |
| Slow | 300 | 285 | 24.90 | 3.42 | .92 | 8.74 | 4.06 | 2.57 | 19.71 |
| Medium | 350 | 273 | 21.80 | 4.71 | 1.60 | 7.99 | 3.50 | 1.96 | 19.76 |
| Fast | 390 | 260 | 18.70 | 5.64 | 2.28 | 7.19 | 2.80 | 1.08 | 18.99 |
| Roll 2nd or additional coats | | | | | | | | | |
| Slow | 350 | 375 | 24.90 | 2.93 | .80 | 6.64 | 3.21 | 2.04 | 15.62 |
| Medium | 400 | 358 | 21.80 | 4.13 | 1.38 | 6.09 | 2.84 | 1.59 | 16.03 |
| Fast | 450 | 340 | 18.70 | 4.89 | 1.97 | 5.50 | 2.29 | .88 | 15.53 |
| Epoxy coating, 2 part system, clear (material #51) | | | | | | | | | |
| Roll 1st coat | | | | | | | | | |
| Slow | 275 | 265 | 57.00 | 3.73 | 1.02 | 21.51 | 8.14 | 5.16 | 39.56 |
| Medium | 325 | 253 | 49.90 | 5.08 | 1.71 | 19.72 | 6.49 | 3.63 | 36.63 |
| Fast | 375 | 240 | 42.80 | 5.87 | 2.39 | 17.83 | 4.82 | 1.85 | 32.76 |
| Roll 2nd or additional coats | | | | | | | | | |
| Slow | 325 | 355 | 57.00 | 3.15 | .87 | 16.06 | 6.22 | 3.94 | 30.24 |
| Medium | 375 | 340 | 49.90 | 4.40 | 1.49 | 14.68 | 5.04 | 2.82 | 28.43 |
| Fast | 425 | 325 | 42.80 | 5.18 | 2.09 | 13.17 | 3.78 | 1.45 | 25.67 |
| Waterproofing, clear hydro sealer, oil base (material #34) | | | | | | | | | |
| Roll 1st coat | | | | | | | | | |
| Slow | 170 | 110 | 18.00 | 6.03 | 1.64 | 16.36 | 7.45 | 4.72 | 36.20 |
| Medium | 200 | 95 | 15.80 | 8.25 | 2.78 | 16.63 | 6.78 | 3.79 | 38.23 |
| Fast | 245 | 80 | 13.50 | 8.98 | 3.64 | 16.88 | 5.46 | 2.10 | 37.06 |
| Roll 2nd or additional coats | | | | | | | | | |
| Slow | 275 | 150 | 18.00 | 3.73 | 1.02 | 12.00 | 5.19 | 3.29 | 25.23 |
| Medium | 300 | 125 | 15.80 | 5.50 | 1.84 | 12.64 | 4.90 | 2.74 | 27.62 |
| Fast | 325 | 100 | 13.50 | 6.77 | 2.75 | 13.50 | 4.26 | 1.64 | 28.92 |

Use these figures for Concrete Masonry Units (CMU) where the block surfaces are smooth with joints struck to average depth. For heavy waterproofing applications, see Masonry under Industrial Painting Operations. "Slow" work is based on a $10.25 hourly wage, "Medium" work on a $16.50 hourly wage, and "Fast" work on a $22.00 hourly wage. Other qualifications that apply to this table are on page 9.

## Masonry: Concrete Masonry Units (CMU), smooth-surface, spray

| | Labor SF per manhour | Material coverage SF/gallon | Material cost per gallon | Labor cost per 100 SF | Labor burden 100 SF | Material cost per 100 SF | Overhead per 100 SF | Profit per 100 SF | Total price per 100 SF |
|---|---|---|---|---|---|---|---|---|---|
| **Masonry paint, water base, flat or gloss (material #31)** | | | | | | | | | |
| Spray 1st coat | | | | | | | | | |
| Slow | 725 | 240 | 19.50 | 1.41 | .39 | 8.13 | 3.08 | 1.95 | 14.96 |
| Medium | 788 | 215 | 17.10 | 2.09 | .71 | 7.95 | 2.63 | 1.47 | 14.85 |
| Fast | 850 | 190 | 14.60 | 2.59 | 1.06 | 7.68 | 2.09 | .80 | 14.22 |
| Spray 2nd or additional coats | | | | | | | | | |
| Slow | 800 | 320 | 19.50 | 1.28 | .35 | 6.09 | 2.39 | 1.52 | 11.63 |
| Medium | 950 | 295 | 17.10 | 1.74 | .58 | 5.80 | 1.99 | 1.11 | 11.22 |
| Fast | 1100 | 270 | 14.60 | 2.00 | .81 | 5.41 | 1.52 | .58 | 10.32 |
| **Masonry paint, oil base (material #32)** | | | | | | | | | |
| Spray 1st coat | | | | | | | | | |
| Slow | 725 | 280 | 24.90 | 1.41 | .39 | 8.89 | 3.31 | 2.10 | 16.10 |
| Medium | 788 | 255 | 21.80 | 2.09 | .71 | 8.55 | 2.78 | 1.55 | 15.68 |
| Fast | 850 | 230 | 18.70 | 2.59 | 1.06 | 8.13 | 2.18 | .84 | 14.80 |
| Spray 2nd or additional coats | | | | | | | | | |
| Slow | 800 | 345 | 24.90 | 1.28 | .35 | 7.22 | 2.74 | 1.74 | 13.33 |
| Medium | 950 | 328 | 21.80 | 1.74 | .58 | 6.65 | 2.20 | 1.23 | 12.40 |
| Fast | 1100 | 310 | 18.70 | 2.00 | .81 | 6.03 | 1.64 | .63 | 11.11 |
| **Epoxy coating, 2 part system clear (material #51)** | | | | | | | | | |
| Spray 1st coat | | | | | | | | | |
| Slow | 675 | 255 | 57.00 | 1.52 | .41 | 22.35 | 7.53 | 4.77 | 36.58 |
| Medium | 738 | 238 | 49.90 | 2.24 | .76 | 20.97 | 5.87 | 3.28 | 33.12 |
| Fast | 800 | 220 | 42.80 | 2.75 | 1.12 | 19.45 | 4.31 | 1.66 | 29.29 |
| Spray 2nd or additional coats | | | | | | | | | |
| Slow | 750 | 325 | 57.00 | 1.37 | .36 | 17.54 | 5.98 | 3.79 | 29.04 |
| Medium | 900 | 308 | 49.90 | 1.83 | .62 | 16.20 | 4.57 | 2.55 | 25.77 |
| Fast | 1050 | 290 | 42.80 | 2.10 | .84 | 14.76 | 3.28 | 1.26 | 22.24 |
| **Waterproofing, clear hydro sealer, oil base (material #34)** | | | | | | | | | |
| Spray 1st coat | | | | | | | | | |
| Slow | 500 | 75 | 18.00 | 2.05 | .56 | 24.00 | 8.25 | 5.23 | 40.09 |
| Medium | 750 | 63 | 15.80 | 2.20 | .73 | 25.08 | 6.86 | 3.84 | 38.71 |
| Fast | 900 | 50 | 13.50 | 2.44 | .99 | 27.00 | 5.63 | 2.16 | 38.22 |
| Spray 2nd or additional coats | | | | | | | | | |
| Slow | 600 | 100 | 18.00 | 1.71 | .47 | 18.00 | 6.26 | 3.97 | 30.41 |
| Medium | 800 | 88 | 15.80 | 2.06 | .70 | 17.95 | 5.07 | 2.83 | 28.61 |
| Fast | 1000 | 75 | 13.50 | 2.20 | .89 | 18.00 | 3.90 | 1.50 | 26.49 |

Use these figures for Concrete Masonry Units (CMU) where the block surfaces are smooth with joints struck to average depth. For heavy waterproofing applications, see Masonry under Industrial Painting Operations. "Slow" work is based on a $10.25 hourly wage, "Medium" work on a $16.50 hourly wage, and "Fast" work on a $22.00 hourly wage. Other qualifications that apply to this table are on page 9.

| | Labor SF per manhour | Material coverage SF/gallon | Material cost per gallon | Labor cost per 100 SF | Labor burden 100 SF | Material cost per 100 SF | Overhead per 100 SF | Profit per 100 SF | Total price per 100 SF |
|---|---|---|---|---|---|---|---|---|---|
| **Masonry, stone, marble or granite** | | | | | | | | | |
| Waterproof, clear hydro sealer, oil base (material #34) | | | | | | | | | |
| Spray 1st coat | | | | | | | | | |
| Slow | 600 | 220 | 18.00 | 1.71 | .47 | 8.18 | 3.21 | 2.04 | 15.61 |
| Medium | 700 | 200 | 15.80 | 2.36 | .79 | 7.90 | 2.71 | 1.51 | 15.27 |
| Fast | 800 | 180 | 13.50 | 2.75 | 1.12 | 7.50 | 2.10 | .81 | 14.28 |
| Spray 2nd coat | | | | | | | | | |
| Slow | 700 | 225 | 18.00 | 1.46 | .40 | 8.00 | 3.06 | 1.94 | 14.86 |
| Medium | 800 | 213 | 15.80 | 2.06 | .70 | 7.42 | 2.49 | 1.39 | 14.06 |
| Fast | 900 | 200 | 13.50 | 2.44 | .99 | 6.75 | 1.88 | .72 | 12.78 |

Use these figures for the cost to apply waterproof sealer on stone, marble or granite surfaces. "Slow" work is based on a $10.25 hourly wage, "Medium" work on a $16.50 hourly wage, and "Fast" work on a $22.50 hourly wage. Other qualifications that apply to this table are on page 9.

| | Labor LF per manhour | Material coverage LF/gallon | Material cost per gallon | Labor cost per 100 LF | Labor burden 100 LF | Material cost per 100 LF | Overhead per 100 LF | Profit per 100 LF | Total price per 100 LF |
|---|---|---|---|---|---|---|---|---|---|
| **Molding, interior or exterior, paint grade, smooth-surface** | | | | | | | | | |
| Undercoat, water or oil base (material #3 or #4) | | | | | | | | | |
| Brush prime coat | | | | | | | | | |
| Slow | 135 | 700 | 19.35 | 7.59 | 2.07 | 2.76 | 3.85 | 2.44 | 18.71 |
| Medium | 205 | 600 | 16.90 | 8.05 | 2.71 | 2.82 | 3.33 | 1.86 | 18.77 |
| Fast | 275 | 500 | 14.50 | 8.00 | 3.25 | 2.90 | 2.62 | 1.01 | 17.78 |
| Split coat (1/2 undercoat + 1/2 enamel), water or oil base (material #3 and #9, or material #4 and #10) | | | | | | | | | |
| Brush 1st or additional finish coats | | | | | | | | | |
| Slow | 125 | 750 | 21.85 | 8.20 | 2.23 | 2.91 | 4.14 | 2.62 | 20.10 |
| Medium | 135 | 675 | 19.10 | 12.22 | 4.12 | 2.83 | 4.69 | 2.62 | 26.48 |
| Fast | 145 | 600 | 16.38 | 15.17 | 6.16 | 2.73 | 4.45 | 1.71 | 30.22 |
| Enamel, interior, water or oil base (material #9 or #10) | | | | | | | | | |
| Brush 1st finish coat | | | | | | | | | |
| Slow | 125 | 750 | 24.35 | 8.20 | 2.23 | 3.25 | 4.24 | 2.69 | 20.61 |
| Medium | 160 | 675 | 21.30 | 10.31 | 3.47 | 3.16 | 4.15 | 2.32 | 23.41 |
| Fast | 200 | 600 | 18.25 | 11.00 | 4.46 | 3.04 | 3.42 | 1.32 | 23.24 |
| Brush 2nd or additional finish coats | | | | | | | | | |
| Slow | 115 | 750 | 24.35 | 8.91 | 2.43 | 3.25 | 4.52 | 2.87 | 21.98 |
| Medium | 150 | 675 | 21.30 | 11.00 | 3.71 | 3.16 | 4.38 | 2.45 | 24.70 |
| Fast | 175 | 600 | 18.25 | 12.57 | 5.09 | 3.04 | 3.83 | 1.47 | 26.00 |

| | Labor LF per manhour | Material coverage LF/gallon | Material cost per gallon | Labor cost per 100 LF | Labor burden 100 LF | Material cost per 100 LF | Overhead per 100 LF | Profit per 100 LF | Total price per 100 LF |
|---|---|---|---|---|---|---|---|---|---|
| **Enamel, exterior, water or oil base (material #24 or #25)** | | | | | | | | | |
| Brush 1st finish coat | | | | | | | | | |
| Slow | 125 | 750 | 26.10 | 8.20 | 2.23 | 3.48 | 4.31 | 2.73 | 20.95 |
| Medium | 160 | 675 | 22.80 | 10.31 | 3.47 | 3.38 | 4.20 | 2.35 | 23.71 |
| Fast | 200 | 600 | 19.55 | 11.00 | 4.46 | 3.26 | 3.46 | 1.33 | 23.51 |
| Brush 2nd or additional finish coats | | | | | | | | | |
| Slow | 115 | 750 | 26.10 | 8.91 | 2.43 | 3.48 | 4.59 | 2.91 | 22.32 |
| Medium | 150 | 675 | 22.80 | 11.00 | 3.71 | 3.38 | 4.43 | 2.48 | 25.00 |
| Fast | 175 | 600 | 19.55 | 12.57 | 5.09 | 3.26 | 3.87 | 1.49 | 26.28 |
| **Stipple finish** | | | | | | | | | |
| Slow | 80 | -- | -- | 12.81 | 3.49 | -- | 5.05 | 3.20 | 24.55 |
| Medium | 90 | -- | -- | 18.33 | 6.17 | -- | 6.00 | 3.36 | 33.86 |
| Fast | 100 | -- | -- | 22.00 | 8.92 | -- | 5.72 | 2.20 | 38.84 |
| **Glazing or mottling over enamel (material #16)** | | | | | | | | | |
| Glaze & wipe, brush 1 coat | | | | | | | | | |
| Slow | 55 | 950 | 29.00 | 18.64 | 5.07 | 3.05 | 8.30 | 5.26 | 40.32 |
| Medium | 65 | 900 | 25.40 | 25.38 | 8.53 | 2.82 | 9.00 | 5.03 | 50.76 |
| Fast | 75 | 850 | 21.80 | 29.33 | 11.89 | 2.56 | 8.10 | 3.11 | 54.99 |

Consider all trim to be at least 12" wide (even if it's much less than 12" wide) when calculating the area to be painted. Trim painted the same color as the wall or ceiling behind it may take no more time than painting the wall or ceiling itself. Use the slow rate when cutting-in is required to paint molding that's a different color from the surface behind the molding. "Slow" work is based on a $10.25 hourly wage, "Medium" work on a $16.75 hourly wage, and "Fast" work on a $22.00 hourly wage. Other qualifications that apply to this table are on page 9.

| | Labor LF per manhour | Material coverage LF/gallon | Material cost per gallon | Labor cost per 100 LF | Labor burden 100 LF | Material cost per 100 LF | Overhead per 100 LF | Profit per 100 LF | Total price per 100 LF |
|---|---|---|---|---|---|---|---|---|---|
| **Molding, interior, stain grade, smooth-surface** | | | | | | | | | |
| Stain, seal and 2 coat lacquer system (7 step process) | | | | | | | | | |
| STEP 1: Sand & putty; | | | | | | | | | |
| Slow | 150 | -- | -- | 6.83 | 1.87 | -- | 2.69 | 1.71 | 13.10 |
| Medium | 175 | -- | -- | 9.43 | 3.16 | -- | 3.09 | 1.73 | 17.41 |
| Fast | 200 | -- | -- | 11.00 | 4.46 | -- | 2.86 | 1.10 | 19.42 |
| STEP 2 & 3: Wiping stain (material 11a) & wipe | | | | | | | | | |
| Brush 1 coat & wipe | | | | | | | | | |
| Slow | 150 | 550 | 27.40 | 6.83 | 1.87 | 4.98 | 4.24 | 2.69 | 20.61 |
| Medium | 175 | 525 | 24.00 | 9.43 | 3.16 | 4.57 | 4.21 | 2.35 | 23.72 |
| Fast | 200 | 500 | 20.60 | 11.00 | 4.46 | 4.12 | 3.62 | 1.39 | 24.59 |
| Spray 1 coat & wipe | | | | | | | | | |
| Slow | 400 | 250 | 27.40 | 2.56 | .70 | 10.96 | 4.41 | 2.79 | 21.42 |
| Medium | 425 | 225 | 24.00 | 3.88 | 1.30 | 10.67 | 3.89 | 2.17 | 21.91 |
| Fast | 450 | 200 | 20.60 | 4.89 | 1.97 | 10.30 | 3.18 | 1.22 | 21.56 |
| STEP 4 & 5: Sanding sealer (material 11b) & light sand | | | | | | | | | |
| Brush 1 coat | | | | | | | | | |
| Slow | 260 | 575 | 17.30 | 3.94 | 1.08 | 3.01 | 2.49 | 1.58 | 12.10 |
| Medium | 280 | 563 | 15.10 | 5.89 | 1.98 | 2.68 | 2.58 | 1.44 | 14.57 |
| Fast | 300 | 550 | 13.00 | 7.33 | 2.97 | 2.36 | 2.34 | .90 | 15.90 |
| Spray 1 coat | | | | | | | | | |
| Slow | 450 | 250 | 17.30 | 2.28 | .61 | 6.92 | 3.04 | 1.93 | 14.78 |
| Medium | 475 | 225 | 15.10 | 3.47 | 1.18 | 6.71 | 2.78 | 1.55 | 15.69 |
| Fast | 500 | 200 | 13.00 | 4.40 | 1.78 | 6.50 | 2.35 | .90 | 15.93 |
| STEP 6 & 7: Lacquer (material 11c), 2 coats | | | | | | | | | |
| Brush 1st coat | | | | | | | | | |
| Slow | 200 | 300 | 18.00 | 5.13 | 1.39 | 6.00 | 3.88 | 2.46 | 18.86 |
| Medium | 275 | 288 | 15.70 | 6.00 | 2.03 | 5.45 | 3.30 | 1.84 | 18.62 |
| Fast | 350 | 275 | 13.50 | 6.29 | 2.55 | 4.91 | 2.54 | .98 | 17.27 |
| Brush 2nd coat | | | | | | | | | |
| Slow | 225 | 375 | 18.00 | 4.56 | 1.23 | 4.80 | 3.29 | 2.08 | 15.96 |
| Medium | 300 | 350 | 15.70 | 5.50 | 1.84 | 4.49 | 2.90 | 1.62 | 16.35 |
| Fast | 375 | 325 | 13.50 | 5.87 | 2.39 | 4.15 | 2.29 | .88 | 15.58 |
| Spray 1st coat | | | | | | | | | |
| Slow | 250 | 200 | 18.00 | 4.10 | 1.12 | 9.00 | 4.41 | 2.79 | 21.42 |
| Medium | 350 | 188 | 15.70 | 4.71 | 1.60 | 8.35 | 3.59 | 2.01 | 20.26 |
| Fast | 450 | 175 | 13.50 | 4.89 | 1.97 | 7.71 | 2.70 | 1.04 | 18.31 |
| Spray 2nd coat | | | | | | | | | |
| Slow | 300 | 250 | 18.00 | 3.42 | .92 | 7.20 | 3.58 | 2.27 | 17.39 |
| Medium | 388 | 225 | 15.70 | 4.25 | 1.44 | 6.98 | 3.10 | 1.73 | 17.50 |
| Fast | 475 | 200 | 13.50 | 4.63 | 1.89 | 6.75 | 2.45 | .94 | 16.66 |

| | Labor LF per manhour | Material coverage LF/gallon | Material cost per gallon | Labor cost per 100 LF | Labor burden 100 LF | Material cost per 100 LF | Overhead per 100 LF | Profit per 100 LF | Total price per 100 LF |
|---|---|---|---|---|---|---|---|---|---|
| Complete 7 step process, stain, seal & 2 coat lacquer system (material #11) | | | | | | | | | |
| Brush all coats | | | | | | | | | |
| Slow | 50 | 60 | 20.20 | 20.50 | 5.58 | 33.67 | 18.53 | 11.74 | 90.02 |
| Medium | 75 | 50 | 18.30 | 22.00 | 7.39 | 36.60 | 16.17 | 9.04 | 91.20 |
| Fast | 100 | 40 | 15.10 | 22.00 | 8.92 | 37.75 | 12.70 | 4.88 | 86.25 |
| Spray all coats | | | | | | | | | |
| Slow | 100 | 40 | 18.00 | 10.25 | 2.79 | 45.00 | 17.99 | 11.40 | 87.43 |
| Medium | 150 | 30 | 15.70 | 11.00 | 3.71 | 52.33 | 16.42 | 9.18 | 92.64 |
| Fast | 200 | 20 | 13.50 | 11.00 | 4.46 | 67.50 | 15.35 | 5.90 | 104.21 |

These figures are based on linear feet for all molding up to 14" wide. The spray figures are based on finishing large quantities of molding in a boneyard before it's installed. Use the brush figures for small quantities. Use the fast brush rate for molding that's finished before it's installed. If the molding is attached and has to be masked off, use the slow brush figures. Masking time is not included. See the Preparation Operations tables for masking rates. Trim stained the same color as the surface behind it will take no more time than staining the wall itself. "Slow" work is based on a $10.25 hourly wage, "Medium" work on a $16.75 hourly wage, and "Fast" work on a $22.00 hourly wage. Other qualifications that apply to this table are on page 9.

| | Labor LF per manhour | Material coverage LF/gallon | Material cost per gallon | Labor cost per 100 LF | Labor burden 100 LF | Material cost per 100 LF | Overhead per 100 LF | Profit per 100 LF | Total price per 100 LF |
|---|---|---|---|---|---|---|---|---|---|
| **Molding, interior or exterior, stain grade, smooth-surface** | | | | | | | | | |
| Stain, fill and shellac or varnish | | | | | | | | | |
| Wiping stain (material #30a) & fill | | | | | | | | | |
| Brush each coat | | | | | | | | | |
| Slow | 80 | 550 | 26.10 | 12.81 | 3.49 | 4.75 | 6.53 | 4.14 | 31.72 |
| Medium | 130 | 525 | 23.40 | 12.69 | 4.27 | 4.46 | 5.25 | 2.93 | 29.60 |
| Fast | 180 | 500 | 20.10 | 12.22 | 4.97 | 4.02 | 3.92 | 1.51 | 26.64 |
| Shellac, clear (material #12) | | | | | | | | | |
| Brush each coat | | | | | | | | | |
| Slow | 180 | 550 | 22.60 | 5.69 | 1.56 | 4.11 | 3.52 | 2.23 | 17.11 |
| Medium | 230 | 525 | 19.80 | 7.17 | 2.42 | 3.77 | 3.27 | 1.83 | 18.46 |
| Fast | 280 | 500 | 17.00 | 7.86 | 3.18 | 3.40 | 2.67 | 1.03 | 18.14 |
| Varnish, gloss (material #30c) | | | | | | | | | |
| Brush each coat | | | | | | | | | |
| Slow | 115 | 550 | 33.90 | 8.91 | 2.43 | 6.16 | 5.43 | 3.44 | 26.37 |
| Medium | 150 | 525 | 29.70 | 11.00 | 3.71 | 5.66 | 4.99 | 2.79 | 28.15 |
| Fast | 210 | 500 | 25.40 | 10.48 | 4.24 | 5.08 | 3.66 | 1.41 | 24.87 |
| Varnish, flat (material #30c) | | | | | | | | | |
| Brush each coat | | | | | | | | | |
| Slow | 125 | 550 | 33.90 | 8.20 | 2.23 | 6.16 | 5.14 | 3.26 | 24.99 |
| Medium | 160 | 525 | 29.70 | 10.31 | 3.47 | 5.66 | 4.76 | 2.66 | 26.86 |
| Fast | 220 | 500 | 25.40 | 10.00 | 4.07 | 5.08 | 3.54 | 1.36 | 24.05 |
| Penetrating stain wax (material #14) & wipe | | | | | | | | | |
| Stain, brush each coat & wipe | | | | | | | | | |
| Slow | 225 | 500 | 28.60 | 4.56 | 1.23 | 5.72 | 3.57 | 2.26 | 17.34 |
| Medium | 275 | 450 | 25.00 | 6.00 | 2.03 | 5.56 | 3.33 | 1.86 | 18.78 |
| Fast | 325 | 400 | 21.40 | 6.77 | 2.75 | 5.35 | 2.75 | 1.06 | 18.68 |
| Polish added coats of wax | | | | | | | | | |
| Slow | 150 | -- | -- | 6.83 | 1.87 | -- | 2.69 | 1.71 | 13.10 |
| Medium | 175 | -- | -- | 9.43 | 3.16 | -- | 3.09 | 1.73 | 17.41 |
| Fast | 200 | -- | -- | 11.00 | 4.46 | -- | 2.86 | 1.10 | 19.42 |

| | Labor LF per manhour | Material coverage LF/gallon | Material cost per gallon | Labor cost per 100 LF | Labor burden 100 LF | Material cost per 100 LF | Overhead per 100 LF | Profit per 100 LF | Total price per 100 LF |
|---|---|---|---|---|---|---|---|---|---|
| Steel wool buff by hand | | | | | | | | | |
| Slow | 100 | -- | -- | 10.25 | 2.79 | -- | 4.04 | 2.56 | 19.64 |
| Medium | 110 | -- | -- | 15.00 | 5.04 | -- | 4.91 | 2.75 | 27.70 |
| Fast | 125 | -- | -- | 17.60 | 7.14 | -- | 4.58 | 1.76 | 31.08 |
| Wax application & polish, hand apply 1 coat | | | | | | | | | |
| Slow | 100 | -- | -- | 10.25 | 2.79 | -- | 4.04 | 2.56 | 19.64 |
| Medium | 110 | -- | -- | 15.00 | 5.04 | -- | 4.91 | 2.75 | 27.70 |
| Fast | 125 | -- | -- | 17.60 | 7.14 | -- | 4.58 | 1.76 | 31.08 |

These figures are based on linear feet for all molding up to 14" wide. The spray figures are based on finishing large quantities of molding in a boneyard before it's installed. Use the brush figures for small quantities. Use the fast brush rate for molding that's finished before it's installed. If the molding is attached and has to be masked off, use the slow brush figures. Masking time is not included. See the Preparation Operations tables for masking rates. Trim stained the same color as the surface behind it will take no more time than staining the wall itself. "Slow" work is based on a $10.25 hourly wage, "Medium" work on a $16.75 hourly wage, and "Fast" work on a $22.00 hourly wage. Other qualifications that apply to this table are on page 9.

| | Boxed eaves One color | Boxed eaves Two color | Exposed rafters One color | Exposed rafters Two color |
|---|---|---|---|---|
| **Overhang difficulty factors, eaves, cornice** | | | | |
| One story, repaint | | | | |
| Standard | 1.5 | 2.0 | 2.0 | 2.5 |
| Ornamental | -- | -- | 2.5 | 3.0 |
| One story, new construction | -- | 1.5 | 1.5 | 2.0 |
| Two story, with scaffolding, repaint | | | | |
| Standard | 1.5 | 2.0 | 2.0 | 2.5 |
| Ornamental | -- | -- | 2.5 | 3.0 |
| Two story, without scaffolding, repaint | | | | |
| Standard | 3.0 | 3.5 | 3.5 | 4.0 |
| Ornamental | -- | -- | 4.0 | 4.5 |
| Two story, new construction | | | | |
| With scaffolding | -- | 1.5 | 1.5 | 2.0 |
| No scaffolding | 2.0 | 2.5 | 2.5 | 3.0 |

Before using the figures in the tables for overhangs, apply these difficulty factors to the surface area (length times width) to be painted. Multiply the factor in this table by the overall surface area of the overhang. This allows for slower work on high eaves and the extra time needed to paint rafter tails. This table adjusts for the kind of eaves, the eave height, the number of colors used, and whether scaffolding is available and erected on site. Boxed eaves have plywood covering the rafter tails.

| | Labor SF per manhour | Material coverage SF/gallon | Material cost per gallon | Labor cost per 100 SF | Labor burden 100 SF | Material cost per 100 SF | Overhead per 100 SF | Profit per 100 SF | Total price per 100 SF |
|---|---|---|---|---|---|---|---|---|---|
| **Overhang, carport, large continuous surface areas** | | | | | | | | | |
| Solid body or semi-transparent stain, water or oil base (material #18 or #19 or #20 or #21) | | | | | | | | | |
| Spray 1st coat | | | | | | | | | |
| Slow | 550 | 95 | 21.38 | 1.86 | .51 | 22.51 | 7.71 | 4.89 | 37.48 |
| Medium | 600 | 90 | 18.70 | 2.75 | .93 | 20.78 | 5.99 | 3.35 | 33.80 |
| Fast | 650 | 85 | 16.03 | 3.38 | 1.38 | 18.86 | 4.37 | 1.68 | 29.67 |
| Spray 2nd coat | | | | | | | | | |
| Slow | 600 | 175 | 21.38 | 1.71 | .47 | 12.22 | 4.46 | 2.83 | 21.69 |
| Medium | 650 | 163 | 18.70 | 2.54 | .86 | 11.47 | 3.64 | 2.04 | 20.55 |
| Fast | 700 | 150 | 16.03 | 3.14 | 1.28 | 10.69 | 2.79 | 1.07 | 18.97 |
| Spray 3rd or additional coats | | | | | | | | | |
| Slow | 650 | 225 | 21.38 | 1.58 | .43 | 9.50 | 3.57 | 2.26 | 17.34 |
| Medium | 700 | 213 | 18.70 | 2.36 | .79 | 8.78 | 2.92 | 1.63 | 16.48 |
| Fast | 750 | 200 | 16.03 | 2.93 | 1.18 | 8.02 | 2.25 | .86 | 15.24 |

"Slow" work is based on a $10.25 hourly wage, "Medium" work on a $16.50 hourly wage, and "Fast" work on a $22.00 hourly wage. Other qualifications that apply to this table are on page 9.

| | Labor SF per manhour | Material coverage SF/gallon | Material cost per gallon | Labor cost per 100 SF | Labor burden 100 SF | Material cost per 100 SF | Overhead per 100 SF | Profit per 100 SF | Total price per 100 SF |
|---|---|---|---|---|---|---|---|---|---|
| **Overhang, eaves or rake, widths up to 2.5 feet** | | | | | | | | | |
| Solid body stain, water or oil base (material #18 or #19) | | | | | | | | | |
| Roll & Brush 1st coat | | | | | | | | | |
| Slow | 95 | 200 | 21.25 | 10.79 | 2.94 | 10.63 | 7.55 | 4.79 | 36.70 |
| Medium | 125 | 185 | 18.60 | 13.20 | 4.44 | 10.05 | 6.78 | 3.79 | 38.26 |
| Fast | 150 | 170 | 15.90 | 14.67 | 5.95 | 9.35 | 5.54 | 2.13 | 37.64 |
| Roll & Brush 2nd coat | | | | | | | | | |
| Slow | 140 | 260 | 21.25 | 7.32 | 1.99 | 8.17 | 5.42 | 3.44 | 26.34 |
| Medium | 185 | 240 | 18.60 | 8.92 | 3.01 | 7.75 | 4.82 | 2.69 | 27.19 |
| Fast | 220 | 220 | 15.90 | 10.00 | 4.07 | 7.23 | 3.94 | 1.51 | 26.75 |
| Roll & Brush 3rd or additional coats | | | | | | | | | |
| Slow | 170 | 295 | 21.25 | 6.03 | 1.64 | 7.20 | 4.61 | 2.92 | 22.40 |
| Medium | 225 | 270 | 18.60 | 7.33 | 2.46 | 6.89 | 4.09 | 2.29 | 23.06 |
| Fast | 275 | 245 | 15.90 | 8.00 | 3.25 | 6.49 | 3.28 | 1.26 | 22.28 |
| Solid body stain or semi-transparent stain, water or oil base (material #18 or #19 or #20 or #21) | | | | | | | | | |
| Spray 1st coat | | | | | | | | | |
| Slow | 300 | 250 | 21.38 | 3.42 | .92 | 8.55 | 4.00 | 2.54 | 19.43 |
| Medium | 350 | 213 | 18.70 | 4.71 | 1.60 | 8.78 | 3.69 | 2.06 | 20.84 |
| Fast | 400 | 175 | 16.03 | 5.50 | 2.23 | 9.16 | 3.12 | 1.20 | 21.21 |
| Spray 2nd coat | | | | | | | | | |
| Slow | 375 | 275 | 21.38 | 2.73 | .75 | 7.77 | 3.48 | 2.21 | 16.94 |
| Medium | 450 | 250 | 18.70 | 3.67 | 1.23 | 7.48 | 3.03 | 1.70 | 17.11 |
| Fast | 540 | 225 | 16.03 | 4.07 | 1.65 | 7.12 | 2.38 | .91 | 16.13 |
| Spray 3rd or additional coats | | | | | | | | | |
| Slow | 450 | 275 | 21.38 | 2.28 | .61 | 7.77 | 3.31 | 2.10 | 16.07 |
| Medium | 525 | 250 | 18.70 | 3.14 | 1.05 | 7.48 | 2.86 | 1.60 | 16.13 |
| Fast | 600 | 225 | 16.03 | 3.67 | 1.49 | 7.12 | 2.27 | .87 | 15.42 |

Use this table after multiplying the overall area (length time width) by the difficulty factor listed under Overhang, difficulty factor. Remember to add preparation time for masking, caulking, sanding, waterblasting, etc. Slow work is based on a $10.25 hourly wage, "Medium" work on a $16.50 hourly wage, and "Fast" work on a $22.00 hourly wage. Other qualifications that apply to this table are on page 9.

| | Labor SF per manhour | Material coverage SF/gallon | Material cost per gallon | Labor cost per 100 SF | Labor burden 100 SF | Material cost per 100 SF | Overhead per 100 SF | Profit per 100 SF | Total price per 100 SF |
|---|---|---|---|---|---|---|---|---|---|
| **Overhang, eaves or rake, widths greater than 2.5 feet** | | | | | | | | | |
| Solid body stain, water or oil base (material #18 or #19) | | | | | | | | | |
| Roll & Brush 1st coat | | | | | | | | | |
| Slow | 125 | 225 | 21.25 | 8.20 | 2.23 | 9.44 | 6.16 | 3.90 | 29.93 |
| Medium | 180 | 210 | 18.60 | 9.17 | 3.09 | 8.86 | 5.17 | 2.89 | 29.18 |
| Fast | 225 | 195 | 15.90 | 9.78 | 3.95 | 8.15 | 4.05 | 1.56 | 27.49 |
| Roll & Brush 2nd coat | | | | | | | | | |
| Slow | 170 | 295 | 21.25 | 6.03 | 1.64 | 7.20 | 4.61 | 2.92 | 22.40 |
| Medium | 225 | 275 | 18.60 | 7.33 | 2.46 | 6.76 | 4.06 | 2.27 | 22.88 |
| Fast | 275 | 255 | 15.90 | 8.00 | 3.25 | 6.24 | 3.23 | 1.24 | 21.96 |
| Roll & Brush 3rd or additional coats | | | | | | | | | |
| Slow | 210 | 340 | 21.25 | 4.88 | 1.33 | 6.25 | 3.86 | 2.45 | 18.77 |
| Medium | 280 | 315 | 18.60 | 5.89 | 1.98 | 5.90 | 3.37 | 1.89 | 19.03 |
| Fast | 345 | 290 | 15.90 | 6.38 | 2.59 | 5.48 | 2.67 | 1.03 | 18.15 |
| Solid body stain or semi-transparent stain, water or oil base (material #18 or #19 or #20 or #21) | | | | | | | | | |
| Spray 1st coat | | | | | | | | | |
| Slow | 400 | 250 | 21.38 | 2.56 | .70 | 8.55 | 3.66 | 2.32 | 17.79 |
| Medium | 500 | 213 | 18.70 | 3.30 | 1.11 | 8.78 | 3.23 | 1.81 | 18.23 |
| Fast | 600 | 175 | 16.03 | 3.67 | 1.49 | 9.16 | 2.65 | 1.02 | 17.99 |
| Spray 2nd coat | | | | | | | | | |
| Slow | 450 | 275 | 21.38 | 2.28 | .61 | 7.77 | 3.31 | 2.10 | 16.07 |
| Medium | 550 | 250 | 18.70 | 3.00 | 1.01 | 7.48 | 2.82 | 1.57 | 15.88 |
| Fast | 650 | 225 | 16.03 | 3.38 | 1.38 | 7.12 | 2.20 | .84 | 14.92 |
| Spray 3rd or additional coats | | | | | | | | | |
| Slow | 550 | 275 | 21.38 | 1.86 | .51 | 7.77 | 3.14 | 1.99 | 15.27 |
| Medium | 650 | 250 | 18.70 | 2.54 | .86 | 7.48 | 2.66 | 1.49 | 15.03 |
| Fast | 750 | 225 | 16.03 | 2.93 | 1.18 | 7.12 | 2.08 | .80 | 14.11 |

Use this table after multiplying the overall area (length time width) by the difficulty factor listed under Overhang, difficulty factor. Remember to add preparation time for masking, caulking, sanding, waterblasting, etc. Slow work is based on a $10.25 hourly wage, "Medium" work on a $16.50 hourly wage, and "Fast" work on a $22.00 hourly wage. Other qualifications that apply to this table are on page 9.

| | Labor LF per manhour | Material coverage LF/gallon | Material cost per gallon | Labor cost per 100 LF | Labor burden 100 LF | Material cost per 100 LF | Overhead per 100 LF | Profit per 100 LF | Total price per 100 LF |
|---|---|---|---|---|---|---|---|---|---|

## Pass-through shelves, wood top & wrought iron support

| | | | | | | | | | |
|---|---|---|---|---|---|---|---|---|---|
| Metal primer, rust inhibitor, clean metal (material #35) | | | | | | | | | |
| Brush 1st coat | | | | | | | | | |
| Slow | 30 | 125 | 28.20 | 34.17 | 9.29 | 22.56 | 20.47 | 12.98 | 99.47 |
| Medium | 35 | 113 | 24.60 | 47.14 | 15.86 | 21.77 | 20.77 | 11.61 | 117.15 |
| Fast | 40 | 100 | 21.10 | 55.00 | 22.30 | 21.10 | 18.20 | 7.00 | 123.60 |
| Metal finish - synthetic enamel (colors except orange/red) (material #38) | | | | | | | | | |
| Brush 1st coat | | | | | | | | | |
| Slow | 30 | 125 | 35.00 | 34.17 | 9.29 | 28.00 | 22.16 | 14.05 | 107.67 |
| Medium | 35 | 113 | 30.70 | 47.14 | 15.86 | 27.17 | 22.09 | 12.35 | 124.61 |
| Fast | 40 | 100 | 26.30 | 55.00 | 22.30 | 26.30 | 19.17 | 7.37 | 130.14 |

Use these figures to estimate pass-through shelves which are about 12" wide and 3'0" long. The rule-of-thumb minimum time and material is .2 hours and $1.00 for material per shelf. A two-coat system using oil based material is recommended for any metal surface. Although water based material is often used, it may cause oxidation, corrosion and rust. One coat of oil based, solid body stain is often used on exterior metal, but it may crack, peel or chip without the proper prime coat application. "Slow" work is based on a $10.25 hourly wage, "Medium" work on a $16.50 hourly wage, and "Fast" work on a $22.00 hourly wage. Other qualifications that apply to this table are on page 9.

| | Labor LF per manhour | Material coverage LF/gallon | Material cost per gallon | Labor cost per 100 LF | Labor burden 100 LF | Material cost per 100 LF | Overhead per 100 LF | Profit per 100 LF | Total price per 100 LF |
|---|---|---|---|---|---|---|---|---|---|

## Plant-on trim, exterior, 2" x 2" to 2" x 4" wood

| | | | | | | | | | |
|---|---|---|---|---|---|---|---|---|---|
| 2" x 2" to 2" x 4" rough sawn or resawn wood | | | | | | | | | |
| Solid body or semi-transparent stain, water base (material #18 or #20) | | | | | | | | | |
| Roll & brush 1st coat | | | | | | | | | |
| Slow | 80 | 310 | 19.50 | 12.81 | 3.49 | 6.29 | 7.00 | 4.44 | 34.03 |
| Medium | 100 | 275 | 17.05 | 16.50 | 5.55 | 6.20 | 6.92 | 3.87 | 39.04 |
| Fast | 120 | 240 | 14.60 | 18.33 | 7.43 | 6.08 | 5.89 | 2.26 | 39.99 |
| Roll & brush 2nd coat | | | | | | | | | |
| Slow | 90 | 350 | 19.50 | 11.39 | 3.10 | 5.57 | 6.22 | 3.94 | 30.22 |
| Medium | 120 | 325 | 17.05 | 13.75 | 4.62 | 5.25 | 5.79 | 3.24 | 32.65 |
| Fast | 150 | 300 | 14.60 | 14.67 | 5.95 | 4.87 | 4.72 | 1.81 | 32.02 |
| Roll & brush 3rd or additional coats | | | | | | | | | |
| Slow | 100 | 375 | 19.50 | 10.25 | 2.79 | 5.20 | 5.65 | 3.58 | 27.47 |
| Medium | 140 | 350 | 17.05 | 11.79 | 3.95 | 4.87 | 5.05 | 2.82 | 28.48 |
| Fast | 180 | 325 | 14.60 | 12.22 | 4.97 | 4.49 | 4.01 | 1.54 | 27.23 |
| Solid body or semi-transparent stain, oil base (material #19 or #21) | | | | | | | | | |
| Roll & brush 1st coat | | | | | | | | | |
| Slow | 95 | 400 | 23.25 | 10.79 | 2.94 | 5.81 | 6.06 | 3.84 | 29.44 |
| Medium | 115 | 365 | 20.35 | 14.35 | 4.83 | 5.58 | 6.07 | 3.39 | 34.22 |
| Fast | 135 | 330 | 17.45 | 16.30 | 6.61 | 5.29 | 5.22 | 2.01 | 35.43 |

| | Labor LF per manhour | Material coverage LF/gallon | Material cost per gallon | Labor cost per 100 LF | Labor burden 100 LF | Material cost per 100 LF | Overhead per 100 LF | Profit per 100 LF | Total price per 100 LF |
|---|---|---|---|---|---|---|---|---|---|
| Roll & brush 2nd coat | | | | | | | | | |
| Slow | 105 | 475 | 23.25 | 9.76 | 2.65 | 4.89 | 5.37 | 3.40 | 26.07 |
| Medium | 135 | 438 | 20.35 | 12.22 | 4.12 | 4.65 | 5.14 | 2.87 | 29.00 |
| Fast | 165 | 400 | 17.45 | 13.33 | 5.41 | 4.36 | 4.27 | 1.64 | 29.01 |
| Roll & brush 3rd or additional coats | | | | | | | | | |
| Slow | 115 | 500 | 23.25 | 8.91 | 2.43 | 4.65 | 4.96 | 3.14 | 24.09 |
| Medium | 155 | 463 | 20.35 | 10.65 | 3.57 | 4.40 | 4.56 | 2.55 | 25.73 |
| Fast | 195 | 425 | 17.45 | 11.28 | 4.58 | 4.11 | 3.69 | 1.42 | 25.08 |
| Varnish, flat or gloss (material #30c) | | | | | | | | | |
| Roll & brush 1st coat | | | | | | | | | |
| Slow | 70 | 270 | 27.40 | 14.64 | 3.99 | 10.15 | 8.92 | 5.66 | 43.36 |
| Medium | 90 | 255 | 24.00 | 18.33 | 6.17 | 9.41 | 8.31 | 4.64 | 46.86 |
| Fast | 110 | 240 | 20.50 | 20.00 | 8.11 | 8.54 | 6.78 | 2.61 | 46.04 |
| Roll & brush 2nd coat | | | | | | | | | |
| Slow | 80 | 330 | 27.40 | 12.81 | 3.49 | 8.30 | 7.63 | 4.83 | 37.06 |
| Medium | 110 | 315 | 24.00 | 15.00 | 5.04 | 7.62 | 6.78 | 3.79 | 38.23 |
| Fast | 140 | 300 | 20.50 | 15.71 | 6.37 | 6.83 | 5.35 | 2.06 | 36.32 |
| Roll & brush 3rd or additional coats | | | | | | | | | |
| Slow | 90 | 350 | 27.40 | 11.39 | 3.10 | 7.83 | 6.92 | 4.39 | 33.63 |
| Medium | 130 | 335 | 24.00 | 12.69 | 4.27 | 7.16 | 5.91 | 3.30 | 33.33 |
| Fast | 170 | 320 | 20.50 | 12.94 | 5.24 | 6.41 | 4.55 | 1.75 | 30.89 |

Don't add additional time for plant-on trim if it's painted with the same coating as the adjacent siding. For heights above 8 feet, use the High Time Difficulty Factors on page 137. Use slow rates when cutting-in or masking adjacent surfaces. Add preparation time for masking adjacent surfaces or protecting windows. Use fast rates when plant-on trim is finished before it's installed or on new construction projects where a prime coat can be sprayed prior to stucco color coat application. "Slow" work is based on a $10.25 hourly wage, "Medium" work on a $16.50 hourly wage, and "Fast" work on a $22.00 hourly wage. Other qualifications that apply to this table are on page 9.

| | Labor LF per manhour | Material coverage LF/gallon | Material cost per gallon | Labor cost per 100 LF | Labor burden 100 LF | Material cost per 100 LF | Overhead per 100 LF | Profit per 100 LF | Total price per 100 LF |
|---|---|---|---|---|---|---|---|---|---|

## Plant-on trim, exterior, 2" x 6" to 2" x 8" wood

2" x 6" to 2" x 8" rough sawn or resawn wood

Solid body or semi-transparent stain, water base (material #18 or #20)

| | Labor LF per manhour | Material coverage LF/gallon | Material cost per gallon | Labor cost per 100 LF | Labor burden 100 LF | Material cost per 100 LF | Overhead per 100 LF | Profit per 100 LF | Total price per 100 LF |
|---|---|---|---|---|---|---|---|---|---|
| Roll & brush 1st coat | | | | | | | | | |
| Slow | 70 | 210 | 19.50 | 14.64 | 3.99 | 9.29 | 8.66 | 5.49 | 42.07 |
| Medium | 85 | 175 | 17.05 | 19.41 | 6.52 | 9.74 | 8.74 | 4.89 | 49.30 |
| Fast | 100 | 140 | 14.60 | 22.00 | 8.92 | 10.43 | 7.65 | 2.94 | 51.94 |
| Roll & brush 2nd coat | | | | | | | | | |
| Slow | 80 | 250 | 19.50 | 12.81 | 3.49 | 7.80 | 7.47 | 4.74 | 36.31 |
| Medium | 110 | 225 | 17.05 | 15.00 | 5.04 | 7.58 | 6.77 | 3.78 | 38.17 |
| Fast | 140 | 200 | 14.60 | 15.71 | 6.37 | 7.30 | 5.44 | 2.09 | 36.91 |
| Roll & brush 3rd or additional coats | | | | | | | | | |
| Slow | 90 | 275 | 19.50 | 11.39 | 3.10 | 7.09 | 6.69 | 4.24 | 32.51 |
| Medium | 130 | 250 | 17.05 | 12.69 | 4.27 | 6.82 | 5.83 | 3.26 | 32.87 |
| Fast | 170 | 225 | 14.60 | 12.94 | 5.24 | 6.49 | 4.57 | 1.76 | 31.00 |

Solid body or semi-transparent stain, oil base (material #19 or #21)

| | Labor LF per manhour | Material coverage LF/gallon | Material cost per gallon | Labor cost per 100 LF | Labor burden 100 LF | Material cost per 100 LF | Overhead per 100 LF | Profit per 100 LF | Total price per 100 LF |
|---|---|---|---|---|---|---|---|---|---|
| Roll & brush 1st coat | | | | | | | | | |
| Slow | 85 | 300 | 23.25 | 12.06 | 3.28 | 7.75 | 7.16 | 4.54 | 34.79 |
| Medium | 110 | 265 | 20.35 | 15.00 | 5.04 | 7.68 | 6.79 | 3.80 | 38.31 |
| Fast | 135 | 230 | 17.45 | 16.30 | 6.61 | 7.59 | 5.64 | 2.17 | 38.31 |
| Roll & brush 2nd coat | | | | | | | | | |
| Slow | 95 | 375 | 23.25 | 10.79 | 2.94 | 6.20 | 6.18 | 3.92 | 30.03 |
| Medium | 125 | 338 | 20.35 | 13.20 | 4.44 | 6.02 | 5.80 | 3.24 | 32.70 |
| Fast | 155 | 300 | 17.45 | 14.19 | 5.75 | 5.82 | 4.77 | 1.83 | 32.36 |
| Roll & brush 3rd or additional coats | | | | | | | | | |
| Slow | 105 | 400 | 23.25 | 9.76 | 2.65 | 5.81 | 5.65 | 3.58 | 27.45 |
| Medium | 145 | 363 | 20.35 | 11.38 | 3.83 | 5.61 | 5.10 | 2.85 | 28.77 |
| Fast | 185 | 325 | 17.45 | 11.89 | 4.84 | 5.37 | 4.08 | 1.57 | 27.75 |

Varnish, flat or gloss (material #30c)

| | Labor LF per manhour | Material coverage LF/gallon | Material cost per gallon | Labor cost per 100 LF | Labor burden 100 LF | Material cost per 100 LF | Overhead per 100 LF | Profit per 100 LF | Total price per 100 LF |
|---|---|---|---|---|---|---|---|---|---|
| Roll & brush 1st coat | | | | | | | | | |
| Slow | 60 | 170 | 27.40 | 17.08 | 4.66 | 16.12 | 11.73 | 7.44 | 57.03 |
| Medium | 70 | 155 | 24.00 | 23.57 | 7.94 | 15.48 | 11.51 | 6.43 | 64.93 |
| Fast | 90 | 140 | 20.50 | 24.44 | 9.91 | 14.64 | 9.06 | 3.48 | 61.53 |

| | Labor LF per manhour | Material coverage LF/gallon | Material cost per gallon | Labor cost per 100 LF | Labor burden 100 LF | Material cost per 100 LF | Overhead per 100 LF | Profit per 100 LF | Total price per 100 LF |
|---|---|---|---|---|---|---|---|---|---|
| Roll & brush 2nd coat | | | | | | | | | |
| Slow | 70 | 240 | 27.40 | 14.64 | 3.99 | 11.42 | 9.32 | 5.91 | 45.28 |
| Medium | 100 | 220 | 24.00 | 16.50 | 5.55 | 10.91 | 8.08 | 4.51 | 45.55 |
| Fast | 130 | 200 | 20.50 | 16.92 | 6.86 | 10.25 | 6.30 | 2.42 | 42.75 |
| Roll & brush 3rd or additional coats | | | | | | | | | |
| Slow | 80 | 260 | 27.40 | 12.81 | 3.49 | 10.54 | 8.32 | 5.27 | 40.43 |
| Medium | 115 | 240 | 24.00 | 14.35 | 4.83 | 10.00 | 7.15 | 4.00 | 40.33 |
| Fast | 150 | 220 | 20.50 | 14.67 | 5.95 | 9.32 | 5.54 | 2.13 | 37.61 |

Don't add additional time for plant-on trim if it's painted with the same coating as the adjacent siding. For heights above 8 feet, use the High Time Difficulty Factors on page 137. Use slow rates when cutting-in or masking adjacent surfaces. Add preparation time for masking adjacent surfaces or protecting windows. Use fast rates when plant-on trim is finished before it's installed or on new construction projects where a prime coat can be sprayed prior to stucco color coat application. "Slow" work is based on a $10.25 hourly wage, "Medium" work on a $16.50 hourly wage, and "Fast" work on a $22.00 hourly wage. Other qualifications that apply to this table are on page 9.

| | Labor LF per manhour | Material coverage LF/gallon | Material cost per gallon | Labor cost per 100 LF | Labor burden 100 LF | Material cost per 100 LF | Overhead per 100 LF | Profit per 100 LF | Total price per 100 LF |
|---|---|---|---|---|---|---|---|---|---|
| **Plant-on trim, exterior, 2" x 10" to 2" x 12" wood** | | | | | | | | | |
| 2" x 10" to 2" x 12" rough sawn or resawn wood | | | | | | | | | |
| Solid body or semi-transparent stain, water base (material #18 or #20) | | | | | | | | | |
| Roll & brush 1st coat | | | | | | | | | |
| Slow | 60 | 150 | 19.50 | 17.08 | 4.66 | 13.00 | 10.77 | 6.83 | 52.34 |
| Medium | 75 | 115 | 17.05 | 22.00 | 7.39 | 14.83 | 10.84 | 6.06 | 61.12 |
| Fast | 90 | 80 | 14.60 | 24.44 | 9.91 | 18.25 | 9.73 | 3.74 | 66.07 |
| Roll & brush 2nd coat | | | | | | | | | |
| Slow | 70 | 190 | 19.50 | 14.64 | 3.99 | 10.26 | 8.96 | 5.68 | 43.53 |
| Medium | 110 | 165 | 17.05 | 15.00 | 5.04 | 10.33 | 7.44 | 4.16 | 41.97 |
| Fast | 130 | 140 | 14.60 | 16.92 | 6.86 | 10.43 | 6.33 | 2.43 | 42.97 |
| Roll & brush 3rd or additional coats | | | | | | | | | |
| Slow | 80 | 215 | 19.50 | 12.81 | 3.49 | 9.07 | 7.86 | 4.98 | 38.21 |
| Medium | 120 | 190 | 17.05 | 13.75 | 4.62 | 8.97 | 6.70 | 3.75 | 37.79 |
| Fast | 160 | 165 | 14.60 | 13.75 | 5.58 | 8.85 | 5.21 | 2.00 | 35.39 |
| Solid body or semi-transparent stain, oil base (material #19 or #21) | | | | | | | | | |
| Roll & brush 1st coat | | | | | | | | | |
| Slow | 70 | 290 | 23.25 | 14.64 | 3.99 | 8.02 | 8.26 | 5.24 | 40.15 |
| Medium | 85 | 255 | 20.35 | 19.41 | 6.52 | 7.98 | 8.31 | 4.65 | 46.87 |
| Fast | 100 | 220 | 17.45 | 22.00 | 8.92 | 7.93 | 7.19 | 2.76 | 48.80 |
| Roll & brush 2nd coat | | | | | | | | | |
| Slow | 80 | 315 | 23.25 | 12.81 | 3.49 | 7.38 | 7.34 | 4.65 | 35.67 |
| Medium | 110 | 278 | 20.35 | 15.00 | 5.04 | 7.32 | 6.71 | 3.75 | 37.82 |
| Fast | 140 | 240 | 17.45 | 15.71 | 6.37 | 7.27 | 5.43 | 2.09 | 36.87 |
| Roll & brush 3rd or additional coats | | | | | | | | | |
| Slow | 90 | 340 | 23.25 | 11.39 | 3.10 | 6.84 | 6.61 | 4.19 | 32.13 |
| Medium | 130 | 303 | 20.35 | 12.69 | 4.27 | 6.72 | 5.80 | 3.24 | 32.72 |
| Fast | 170 | 265 | 17.45 | 12.94 | 5.24 | 6.58 | 4.58 | 1.76 | 31.10 |
| Varnish, flat or gloss (material #30c) | | | | | | | | | |
| Roll & brush 1st coat | | | | | | | | | |
| Slow | 50 | 120 | 27.40 | 20.50 | 5.58 | 22.83 | 15.17 | 9.61 | 73.69 |
| Medium | 65 | 100 | 24.00 | 25.38 | 8.53 | 24.00 | 14.19 | 7.93 | 80.03 |
| Fast | 80 | 80 | 20.50 | 27.50 | 11.15 | 25.63 | 11.89 | 4.57 | 80.74 |

| | Labor LF per manhour | Material coverage LF/gallon | Material cost per gallon | Labor cost per 100 LF | Labor burden 100 LF | Material cost per 100 LF | Overhead per 100 LF | Profit per 100 LF | Total price per 100 LF |
|---|---|---|---|---|---|---|---|---|---|
| Roll & brush 2nd coat | | | | | | | | | |
| Slow | 60 | 190 | 27.40 | 17.08 | 4.66 | 14.42 | 11.21 | 7.10 | 54.47 |
| Medium | 90 | 165 | 24.00 | 18.33 | 6.17 | 14.55 | 9.57 | 5.35 | 53.97 |
| Fast | 120 | 140 | 20.50 | 18.33 | 7.43 | 14.64 | 7.47 | 2.87 | 50.74 |
| Roll & brush 3rd or additional coats | | | | | | | | | |
| Slow | 70 | 200 | 27.40 | 14.64 | 3.99 | 13.70 | 10.02 | 6.35 | 48.70 |
| Medium | 110 | 175 | 24.00 | 15.00 | 5.04 | 13.71 | 8.27 | 4.62 | 46.64 |
| Fast | 150 | 150 | 20.50 | 14.67 | 5.95 | 13.67 | 6.34 | 2.44 | 43.07 |

Don't add additional time for plant-on trim if it's painted with the same coating as the adjacent siding. For heights above 8 feet, use the High Time Difficulty Factors on page 137. Use slow rates when cutting-in or masking adjacent surfaces. Add preparation time for masking adjacent surfaces or protecting windows. Use fast rates when plant-on trim is finished before it's installed or on new construction projects where a prime coat can be sprayed prior to stucco color coat application. "Slow" work is based on a $10.25 hourly wage, "Medium" work on a $16.50 hourly wage, and "Fast" work on a $22.00 hourly wage. Other qualifications that apply to this table are on page 9.

| | Labor SF per manhour | Material coverage SF/gallon | Material cost per gallon | Labor cost per 100 SF | Labor burden 100 SF | Material cost per 100 SF | Overhead per 100 SF | Profit per 100 SF | Total price per 100 SF |
|---|---|---|---|---|---|---|---|---|---|
| **Plaster or stucco, exterior, medium texture, brush application** | | | | | | | | | |
| Masonry paint, water base, flat or gloss (material #31) | | | | | | | | | |
| Brush prime coat | | | | | | | | | |
| Slow | 100 | 150 | 19.50 | 10.25 | 2.79 | 13.00 | 8.07 | 5.12 | 39.23 |
| Medium | 120 | 125 | 17.10 | 13.75 | 4.62 | 13.68 | 7.85 | 4.39 | 44.29 |
| Fast | 140 | 100 | 14.60 | 15.71 | 6.37 | 14.60 | 6.79 | 2.61 | 46.08 |
| Brush 2nd coat | | | | | | | | | |
| Slow | 150 | 250 | 19.50 | 6.83 | 1.87 | 7.80 | 5.11 | 3.24 | 24.85 |
| Medium | 163 | 225 | 17.10 | 10.12 | 3.40 | 7.60 | 5.18 | 2.89 | 29.19 |
| Fast | 175 | 200 | 14.60 | 12.57 | 5.09 | 7.30 | 4.62 | 1.78 | 31.36 |
| Brush 3rd or additional coats | | | | | | | | | |
| Slow | 160 | 270 | 19.50 | 6.41 | 1.74 | 7.22 | 4.77 | 3.02 | 23.16 |
| Medium | 173 | 245 | 17.10 | 9.54 | 3.20 | 6.98 | 4.83 | 2.70 | 27.25 |
| Fast | 185 | 220 | 14.60 | 11.89 | 4.84 | 6.64 | 4.32 | 1.66 | 29.35 |
| Masonry paint, oil base (material #32) | | | | | | | | | |
| Brush prime coat | | | | | | | | | |
| Slow | 80 | 265 | 24.90 | 12.81 | 3.49 | 9.40 | 7.97 | 5.05 | 38.72 |
| Medium | 100 | 250 | 21.80 | 16.50 | 5.55 | 8.72 | 7.54 | 4.21 | 42.52 |
| Fast | 120 | 235 | 18.70 | 18.33 | 7.43 | 7.96 | 6.24 | 2.40 | 42.36 |
| Brush 2nd coat | | | | | | | | | |
| Slow | 145 | 300 | 24.90 | 7.07 | 1.93 | 8.30 | 5.36 | 3.40 | 26.06 |
| Medium | 165 | 275 | 21.80 | 10.00 | 3.36 | 7.93 | 5.22 | 2.92 | 29.43 |
| Fast | 185 | 250 | 18.70 | 11.89 | 4.84 | 7.48 | 4.48 | 1.72 | 30.41 |
| Brush 3rd or additional coats | | | | | | | | | |
| Slow | 155 | 350 | 24.90 | 6.61 | 1.80 | 7.11 | 4.81 | 3.05 | 23.38 |
| Medium | 175 | 325 | 21.80 | 9.43 | 3.16 | 6.71 | 4.73 | 2.64 | 26.67 |
| Fast | 185 | 300 | 18.70 | 11.89 | 4.84 | 6.23 | 4.24 | 1.63 | 28.83 |

For heights above 8 feet, use the High Time Difficulty Factors on page 137. For oil base paint and clear hydro sealer, I recommend spraying. For painting interior plaster, see Walls, plaster. "Slow" work is based on a $10.25 hourly wage, "Medium" work on a $16.50 hourly wage, and "Fast" work on a $22.00 hourly wage. Other qualifications that apply to this table are on page 9.

| | Labor SF per manhour | Material coverage SF/gallon | Material cost per gallon | Labor cost per 100 SF | Labor burden 100 SF | Material cost per 100 SF | Overhead per 100 SF | Profit per 100 SF | Total price per 100 SF |
|---|---|---|---|---|---|---|---|---|---|
| **Plaster or stucco, exterior, medium texture, roll application** | | | | | | | | | |
| Masonry paint, water base, flat or gloss (material #31) | | | | | | | | | |
| Roll prime coat | | | | | | | | | |
| Slow | 245 | 225 | 19.50 | 4.18 | 1.14 | 8.67 | 4.34 | 2.75 | 21.08 |
| Medium | 273 | 188 | 17.10 | 6.04 | 2.03 | 9.10 | 4.21 | 2.35 | 23.73 |
| Fast | 300 | 150 | 14.60 | 7.33 | 2.97 | 9.73 | 3.71 | 1.42 | 25.16 |
| Roll 2nd coat | | | | | | | | | |
| Slow | 300 | 275 | 19.50 | 3.42 | .92 | 7.09 | 3.55 | 2.25 | 17.23 |
| Medium | 320 | 243 | 17.10 | 5.16 | 1.74 | 7.04 | 3.42 | 1.91 | 19.27 |
| Fast | 340 | 210 | 14.60 | 6.47 | 2.62 | 6.95 | 2.97 | 1.14 | 20.15 |
| Roll 3rd or additional coats | | | | | | | | | |
| Slow | 320 | 300 | 19.50 | 3.20 | .88 | 6.50 | 3.28 | 2.08 | 15.94 |
| Medium | 340 | 268 | 17.10 | 4.85 | 1.63 | 6.38 | 3.15 | 1.76 | 17.77 |
| Fast | 360 | 235 | 14.60 | 6.11 | 2.49 | 6.21 | 2.74 | 1.05 | 18.60 |
| Masonry paint, oil base (material #32) | | | | | | | | | |
| Roll prime coat | | | | | | | | | |
| Slow | 200 | 250 | 24.90 | 5.13 | 1.39 | 9.96 | 5.11 | 3.24 | 24.83 |
| Medium | 240 | 225 | 21.80 | 6.88 | 2.31 | 9.69 | 4.63 | 2.59 | 26.10 |
| Fast | 280 | 200 | 18.70 | 7.86 | 3.18 | 9.35 | 3.77 | 1.45 | 25.61 |
| Roll 2nd coat | | | | | | | | | |
| Slow | 220 | 275 | 24.90 | 4.66 | 1.27 | 9.05 | 4.64 | 2.94 | 22.56 |
| Medium | 265 | 250 | 21.80 | 6.23 | 2.08 | 8.72 | 4.18 | 2.34 | 23.55 |
| Fast | 305 | 225 | 18.70 | 7.21 | 2.93 | 8.31 | 3.41 | 1.31 | 23.17 |
| Roll 3rd or additional coats | | | | | | | | | |
| Slow | 235 | 350 | 24.90 | 4.36 | 1.20 | 7.11 | 3.92 | 2.49 | 19.08 |
| Medium | 285 | 325 | 21.80 | 5.79 | 1.95 | 6.71 | 3.54 | 1.98 | 19.97 |
| Fast | 335 | 300 | 18.70 | 6.57 | 2.68 | 6.23 | 2.86 | 1.10 | 19.44 |

For heights above 8 feet, use the High Time Difficulty Factors on page 137. For oil base paint and clear hydro sealer, I recommend spraying. For painting interior plaster, see Walls, plaster. "Slow" work is based on a $10.25 hourly wage, "Medium" work on a $16.50 hourly wage, and "Fast" work on a $22.00 hourly wage. Other qualifications that apply to this table are on page 9.

| | Labor SF per manhour | Material coverage SF/gallon | Material cost per gallon | Labor cost per 100 SF | Labor burden 100 SF | Material cost per 100 SF | Overhead per 100 SF | Profit per 100 SF | Total price per 100 SF |
|---|---|---|---|---|---|---|---|---|---|
| **Plaster or stucco, exterior, medium texture, spray application** | | | | | | | | | |
| Masonry paint, water base, flat or gloss (material #31) | | | | | | | | | |
| Spray prime coat | | | | | | | | | |
| Slow | 600 | 200 | 19.50 | 1.71 | .47 | 9.75 | 3.70 | 2.34 | 17.97 |
| Medium | 675 | 145 | 17.10 | 2.44 | .82 | 11.79 | 3.69 | 2.06 | 20.80 |
| Fast | 750 | 90 | 14.60 | 2.93 | 1.18 | 16.22 | 3.76 | 1.45 | 25.54 |
| Spray 2nd coat | | | | | | | | | |
| Slow | 700 | 225 | 19.50 | 1.46 | .40 | 8.67 | 3.26 | 2.07 | 15.86 |
| Medium | 800 | 175 | 17.10 | 2.06 | .70 | 9.77 | 3.07 | 1.71 | 17.31 |
| Fast | 900 | 125 | 14.60 | 2.44 | .99 | 11.68 | 2.80 | 1.07 | 18.98 |
| Spray 3rd or additional coats | | | | | | | | | |
| Slow | 750 | 235 | 19.50 | 1.37 | .36 | 8.30 | 3.11 | 1.97 | 15.11 |
| Medium | 850 | 185 | 17.10 | 1.94 | .66 | 9.24 | 2.90 | 1.62 | 16.36 |
| Fast | 950 | 135 | 14.60 | 2.32 | .93 | 10.81 | 2.60 | 1.00 | 17.66 |
| Masonry paint, oil base (material #32) | | | | | | | | | |
| Spray prime coat | | | | | | | | | |
| Slow | 550 | 225 | 24.90 | 1.86 | .51 | 11.07 | 4.17 | 2.64 | 20.25 |
| Medium | 600 | 158 | 21.80 | 2.75 | .93 | 13.80 | 4.28 | 2.39 | 24.15 |
| Fast | 650 | 90 | 18.70 | 3.38 | 1.38 | 20.78 | 4.72 | 1.82 | 32.08 |
| Spray 2nd coat | | | | | | | | | |
| Slow | 650 | 350 | 24.90 | 1.58 | .43 | 7.11 | 2.83 | 1.79 | 13.74 |
| Medium | 775 | 238 | 21.80 | 2.13 | .71 | 9.16 | 2.94 | 1.64 | 16.58 |
| Fast | 900 | 125 | 18.70 | 2.44 | .99 | 14.96 | 3.40 | 1.31 | 23.10 |
| Spray 3rd or additional coats | | | | | | | | | |
| Slow | 700 | 375 | 24.90 | 1.46 | .40 | 6.64 | 2.64 | 1.67 | 12.81 |
| Medium | 825 | 205 | 21.80 | 2.00 | .67 | 10.63 | 3.26 | 1.82 | 18.38 |
| Fast | 950 | 135 | 18.70 | 2.32 | .93 | 13.85 | 3.17 | 1.22 | 21.49 |

For heights above 8 feet, use the High Time Difficulty Factors on page 137. For oil base paint and clear hydro sealer, I recommend spraying. For painting interior plaster, see Walls, plaster. "Slow" work is based on a $10.25 hourly wage, "Medium" work on a $16.50 hourly wage, and "Fast" work on a $22.00 hourly wage. Other qualifications that apply to this table are on page 9.

| | Labor SF per manhour | Material coverage SF/gallon | Material cost per gallon | Labor cost per 100 SF | Labor burden 100 SF | Material cost per 100 SF | Overhead per 100 SF | Profit per 100 SF | Total price per 100 SF |
|---|---|---|---|---|---|---|---|---|---|
| **Plaster or stucco, exterior, medium texture, waterproofing** | | | | | | | | | |
| Waterproofing, clear hydro sealer (material #34) | | | | | | | | | |
| Brush 1st coat | | | | | | | | | |
| Slow | 125 | 100 | 18.00 | 8.20 | 2.23 | 18.00 | 8.81 | 5.59 | 42.83 |
| Medium | 150 | 88 | 15.80 | 11.00 | 3.71 | 17.95 | 8.00 | 4.47 | 45.13 |
| Fast | 175 | 75 | 13.50 | 12.57 | 5.09 | 18.00 | 6.60 | 2.54 | 44.80 |
| Brush 2nd or additional coats | | | | | | | | | |
| Slow | 175 | 135 | 18.00 | 5.86 | 1.59 | 13.33 | 6.44 | 4.08 | 31.30 |
| Medium | 200 | 118 | 15.80 | 8.25 | 2.78 | 13.39 | 5.98 | 3.34 | 33.74 |
| Fast | 225 | 100 | 13.50 | 9.78 | 3.95 | 13.50 | 5.04 | 1.94 | 34.21 |
| Roll 1st coat | | | | | | | | | |
| Slow | 325 | 150 | 18.00 | 3.15 | .87 | 12.00 | 4.96 | 3.15 | 24.13 |
| Medium | 363 | 138 | 15.80 | 4.55 | 1.51 | 11.45 | 4.29 | 2.40 | 24.20 |
| Fast | 400 | 125 | 13.50 | 5.50 | 2.23 | 10.80 | 3.43 | 1.32 | 23.28 |
| Roll 2nd or additional coats | | | | | | | | | |
| Slow | 400 | 200 | 18.00 | 2.56 | .70 | 9.00 | 3.80 | 2.41 | 18.47 |
| Medium | 425 | 190 | 15.80 | 3.88 | 1.30 | 8.32 | 3.31 | 1.85 | 18.66 |
| Fast | 450 | 180 | 13.50 | 4.89 | 1.97 | 7.50 | 2.66 | 1.02 | 18.04 |
| Spray 1st coat | | | | | | | | | |
| Slow | 650 | 225 | 18.00 | 1.58 | .43 | 8.00 | 3.10 | 1.97 | 15.08 |
| Medium | 725 | 163 | 15.80 | 2.28 | .76 | 9.69 | 3.12 | 1.74 | 17.59 |
| Fast | 800 | 100 | 13.50 | 2.75 | 1.12 | 13.50 | 3.21 | 1.23 | 21.81 |
| Spray 2nd or additional coats | | | | | | | | | |
| Slow | 800 | 250 | 18.00 | 1.28 | .35 | 7.20 | 2.74 | 1.74 | 13.31 |
| Medium | 900 | 200 | 15.80 | 1.83 | .62 | 7.90 | 2.54 | 1.42 | 14.31 |
| Fast | 1000 | 150 | 13.50 | 2.20 | .89 | 9.00 | 2.24 | .86 | 15.19 |

For heights above 8 feet, use the High Time Difficulty Factors on page 137. For oil base paint and clear hydro sealer, I recommend spraying. For painting interior plaster, see Walls, plaster. "Slow" work is based on a $10.25 hourly wage, "Medium" work on a $16.50 hourly wage, and "Fast" work on a $22.00 hourly wage. Other qualifications that apply to this table are on page 9.

## Plaster, interior: see Walls

| | Labor SF per manhour | Material coverage SF/gallon | Material cost per gallon | Labor cost per 100 SF | Labor burden 100 SF | Material cost per 100 SF | Overhead per 100 SF | Profit per 100 SF | Total price per 100 SF |
|---|---|---|---|---|---|---|---|---|---|
| **Plaster or stucco, interior/exterior, medium texture, anti-graffiti stain eliminator** | | | | | | | | | |
| Water base primer and sealer (material #39) | | | | | | | | | |
| Roll & brush each coat | | | | | | | | | |
| Slow | 350 | 400 | 30.40 | 2.93 | .80 | 7.60 | 3.51 | 2.23 | 17.07 |
| Medium | 375 | 375 | 26.60 | 4.40 | 1.49 | 7.09 | 3.18 | 1.78 | 17.94 |
| Fast | 400 | 350 | 22.80 | 5.50 | 2.23 | 6.51 | 2.63 | 1.01 | 17.88 |
| Oil base primer and sealer (material #40) | | | | | | | | | |
| Roll & brush each coat | | | | | | | | | |
| Slow | 350 | 375 | 28.20 | 2.93 | .80 | 7.52 | 3.49 | 2.21 | 16.95 |
| Medium | 375 | 350 | 24.70 | 4.40 | 1.49 | 7.06 | 3.17 | 1.77 | 17.89 |
| Fast | 400 | 325 | 21.20 | 5.50 | 2.23 | 6.52 | 2.64 | 1.01 | 17.90 |
| Polyurethane 2 part system (material #41) | | | | | | | | | |
| Roll & brush each coat | | | | | | | | | |
| Slow | 300 | 375 | 63.30 | 3.42 | .92 | 16.88 | 6.58 | 4.17 | 31.97 |
| Medium | 325 | 350 | 55.40 | 5.08 | 1.71 | 15.83 | 5.54 | 3.10 | 31.26 |
| Fast | 350 | 325 | 47.40 | 6.29 | 2.55 | 14.58 | 4.33 | 1.67 | 29.42 |

For heights above 8 feet, use the High Time Difficulty Factors on page 137. For oil base paint and clear hydro sealer, I recommend spraying. For painting interior plaster, see Walls, plaster. "Slow" work is based on a $10.25 hourly wage, "Medium" work on a $16.50 hourly wage, and "Fast" work on a $22.00 hourly wage. Other qualifications that apply to this table are on page 9.

| | Labor LF per manhour | Material coverage LF/gallon | Material cost per gallon | Labor cost per 100 LF | Labor burden 100 LF | Material cost per 100 LF | Overhead per 100 LF | Profit per 100 LF | Total price per 100 LF |
|---|---|---|---|---|---|---|---|---|---|
| **Pot shelves, 12" to 18" wide** | | | | | | | | | |
| Solid body or semi-transparent stain, water or oil base (material #18 or #19 or #20 or #21) | | | | | | | | | |
| Roll & brush each coat | | | | | | | | | |
| Slow | 50 | 50 | 21.38 | 20.50 | 5.58 | 42.76 | 21.34 | 13.53 | 103.71 |
| Medium | 68 | 45 | 18.70 | 24.26 | 8.18 | 41.56 | 18.13 | 10.13 | 102.26 |
| Fast | 75 | 40 | 16.03 | 29.33 | 11.89 | 40.08 | 15.04 | 5.78 | 102.12 |

These figures are based on painting all sides of exterior or interior pot shelves. "Slow" work is based on a $10.25 hourly wage, "Medium" work on a $16.50 hourly wage, and "Fast" work on a $22.00 hourly wage. Other qualifications that apply to this table are on page 9.

| | Labor LF per manhour | Material coverage LF/gallon | Material cost per gallon | Labor cost per 100 LF | Labor burden 100 LF | Material cost per 100 LF | Overhead per 100 LF | Profit per 100 LF | Total price per 100 LF |
|---|---|---|---|---|---|---|---|---|---|
| **Railing, exterior, rough sawn or resawn wood** | | | | | | | | | |
| Solid body or semi-transparent stain, water or oil base (material #18 or #19 or #20 or #21) | | | | | | | | | |
| Roll & brush each coat | | | | | | | | | |
| Slow | 16 | 20 | 21.38 | 64.06 | 17.44 | 106.90 | 58.41 | 37.02 | 283.83 |
| Medium | 18 | 18 | 18.70 | 91.67 | 30.84 | 103.89 | 55.47 | 31.01 | 312.88 |
| Fast | 20 | 15 | 16.03 | 110.00 | 44.60 | 106.87 | 48.37 | 18.59 | 328.43 |

Use these costs for finishing railing that's 36" to 42" high and with 2" x 2" verticals spaced 6" on center. These figures include painting the rail cap, baluster, newels and spindles. "Slow" work is based on a $10.25 hourly wage, "Medium" work on a $16.50 hourly wage, and "Fast" work on a $22.00 hourly wage. Other qualifications that apply to this table are on page 9.

| | Labor LF per manhour | Material coverage LF/gallon | Material cost per gallon | Labor cost per 100 LF | Labor burden 100 LF | Material cost per 100 LF | Overhead per 100 LF | Profit per 100 LF | Total price per 100 LF |
|---|---|---|---|---|---|---|---|---|---|
| **Railing, exterior, stain grade, decorative wood** | | | | | | | | | |
| STEP 1: Sand & putty | | | | | | | | | |
| Slow | 50 | -- | -- | 20.50 | 5.58 | -- | 8.09 | 5.13 | 39.30 |
| Medium | 60 | -- | -- | 27.50 | 9.26 | -- | 9.00 | 5.03 | 50.79 |
| Fast | 70 | -- | -- | 31.43 | 12.75 | -- | 8.17 | 3.14 | 55.49 |
| STEP 2 & 3: Stain (material #30a) & wipe | | | | | | | | | |
| Brush & wipe, 1 coat | | | | | | | | | |
| Slow | 25 | 60 | 26.10 | 41.00 | 11.16 | 43.50 | 29.66 | 18.80 | 144.12 |
| Medium | 30 | 55 | 22.90 | 55.00 | 18.49 | 41.64 | 28.21 | 15.77 | 159.11 |
| Fast | 35 | 50 | 19.60 | 62.86 | 25.48 | 39.20 | 23.60 | 9.07 | 160.21 |
| Spray & wipe, 1 coat | | | | | | | | | |
| Slow | 75 | 35 | 26.10 | 13.67 | 3.71 | 74.57 | 28.51 | 18.07 | 138.53 |
| Medium | 85 | 30 | 22.90 | 19.41 | 6.52 | 76.33 | 25.06 | 14.01 | 141.33 |
| Fast | 95 | 25 | 19.60 | 23.16 | 9.40 | 78.40 | 20.53 | 7.89 | 139.38 |
| STEP 4 & 5: Sanding sealer (material #30b) & light sand | | | | | | | | | |
| Brush 1 coat | | | | | | | | | |
| Slow | 45 | 65 | 27.40 | 22.78 | 6.19 | 42.15 | 22.05 | 13.98 | 107.15 |
| Medium | 50 | 60 | 24.00 | 33.00 | 11.10 | 40.00 | 20.60 | 11.52 | 116.22 |
| Fast | 55 | 55 | 20.50 | 40.00 | 16.21 | 37.27 | 17.30 | 6.65 | 117.43 |
| Spray 1 coat | | | | | | | | | |
| Slow | 125 | 35 | 27.40 | 8.20 | 2.23 | 78.29 | 27.50 | 17.43 | 133.65 |
| Medium | 138 | 30 | 24.00 | 11.96 | 4.03 | 80.00 | 23.52 | 13.15 | 132.66 |
| Fast | 150 | 25 | 20.50 | 14.67 | 5.95 | 82.00 | 18.98 | 7.30 | 128.90 |

| | Labor LF per manhour | Material coverage LF/gallon | Material cost per gallon | Labor cost per 100 LF | Labor burden 100 LF | Material cost per 100 LF | Overhead per 100 LF | Profit per 100 LF | Total price per 100 LF |
|---|---|---|---|---|---|---|---|---|---|
| STEP 6 & 7: Varnish 2 coats (material #30c) | | | | | | | | | |
| Varnish, flat or gloss (material #30c) | | | | | | | | | |
| Brush 1st coat | | | | | | | | | |
| Slow | 22 | 100 | 33.90 | 46.59 | 12.68 | 33.90 | 28.89 | 18.31 | 140.37 |
| Medium | 24 | 90 | 29.70 | 68.75 | 23.13 | 33.00 | 30.60 | 17.10 | 172.58 |
| Fast | 26 | 80 | 25.40 | 84.62 | 34.30 | 31.75 | 27.88 | 10.71 | 189.26 |
| Brush 2nd or additional coats | | | | | | | | | |
| Slow | 30 | 120 | 33.90 | 34.17 | 9.29 | 28.25 | 22.24 | 14.10 | 108.05 |
| Medium | 32 | 110 | 29.70 | 51.56 | 17.35 | 27.00 | 23.50 | 13.14 | 132.55 |
| Fast | 34 | 100 | 25.40 | 64.71 | 26.23 | 25.40 | 21.52 | 8.27 | 146.13 |
| Steel wool, hand buff | | | | | | | | | |
| Slow | 100 | -- | -- | 10.25 | 2.79 | -- | 4.04 | 2.56 | 19.64 |
| Medium | 113 | -- | -- | 14.60 | 4.91 | -- | 4.78 | 2.67 | 26.96 |
| Fast | 125 | -- | -- | 17.60 | 7.14 | -- | 4.58 | 1.76 | 31.08 |
| Wax & polish by hand | | | | | | | | | |
| Slow | 100 | -- | -- | 10.25 | 2.79 | -- | 4.04 | 2.56 | 19.64 |
| Medium | 113 | -- | -- | 14.60 | 4.91 | -- | 4.78 | 2.67 | 26.96 |
| Fast | 125 | -- | -- | 17.60 | 7.14 | -- | 4.58 | 1.76 | 31.08 |

Use these figures to estimate the cost of applying a natural finish on stain grade railing that's from 36" to 42" high and with spindles spaced at 6" on center. These figures include painting the rail cap, baluster, newels and spindles. "Slow" work is based on a $10.25 hourly wage, "Medium" work on a $16.50 hourly wage, and "Fast" work on a $22.00 hourly wage. Other qualifications that apply to this table are on page 9.

| | Labor LF per manhour | Material coverage LF/gallon | Material cost per gallon | Labor cost per 100 LF | Labor burden 100 LF | Material cost per 100 LF | Overhead per 100 LF | Profit per 100 LF | Total price per 100 LF |
|---|---|---|---|---|---|---|---|---|---|
| **Railing, interior, handrail, decorative wood** | | | | | | | | | |
| Paint grade | | | | | | | | | |
| Undercoat or enamel, water or oil base (material #3 or #4 or #9 or #10) | | | | | | | | | |
| Brush each coat | | | | | | | | | |
| Slow | 30 | 120 | 21.85 | 34.17 | 9.29 | 18.21 | 19.12 | 12.12 | 92.91 |
| Medium | 35 | 110 | 19.10 | 47.14 | 15.86 | 17.36 | 19.69 | 11.01 | 111.06 |
| Fast | 40 | 100 | 16.38 | 55.00 | 22.30 | 16.38 | 17.33 | 6.66 | 117.67 |
| Stain grade, 7 step process | | | | | | | | | |
| Stain, seal and 2 coat lacquer system (material #11) | | | | | | | | | |
| Spray all coats | | | | | | | | | |
| Slow | 55 | 100 | 20.20 | 18.64 | 5.07 | 20.20 | 13.62 | 8.63 | 66.16 |
| Medium | 60 | 88 | 18.30 | 27.50 | 9.26 | 20.80 | 14.10 | 7.88 | 79.54 |
| Fast | 65 | 75 | 15.10 | 33.85 | 13.70 | 20.13 | 12.53 | 4.81 | 85.02 |

Use these costs for finishing decorative wood, wall mounted handrail. "Slow" work is based on a $10.25 hourly wage, "Medium" work on a $16.50 hourly wage, and "Fast" work on a $22.00 hourly wage. Other qualifications that apply to this table are on page 9.

| | Labor LF per manhour | Material coverage LF/gallon | Material cost per gallon | Labor cost per 100 LF | Labor burden 100 LF | Material cost per 100 LF | Overhead per 100 LF | Profit per 100 LF | Total price per 100 LF |
|---|---|---|---|---|---|---|---|---|---|
| **Railing, interior, stain grade, decorative wood** | | | | | | | | | |
| Stain, seal and 2 coat lacquer system (7 step process) | | | | | | | | | |
| STEP 1: Sand & putty | | | | | | | | | |
| Slow | 50 | -- | -- | 20.50 | 5.58 | -- | 8.09 | 5.13 | 39.30 |
| Medium | 60 | -- | -- | 27.50 | 9.26 | -- | 9.00 | 5.03 | 50.79 |
| Fast | 70 | -- | -- | 31.43 | 12.75 | -- | 8.17 | 3.14 | 55.49 |
| STEP 2 & 3: Stain (material #11a) & wipe | | | | | | | | | |
| Brush & wipe, 1 coat | | | | | | | | | |
| Slow | 25 | 60 | 27.40 | 41.00 | 11.16 | 45.67 | 30.33 | 19.23 | 147.39 |
| Medium | 30 | 55 | 24.00 | 55.00 | 18.49 | 43.64 | 28.70 | 16.04 | 161.87 |
| Fast | 35 | 50 | 20.60 | 62.86 | 25.48 | 41.20 | 23.97 | 9.21 | 162.72 |
| Spray & wipe, 1 coat | | | | | | | | | |
| Slow | 75 | 35 | 27.40 | 13.67 | 3.71 | 78.29 | 29.66 | 18.80 | 144.13 |
| Medium | 85 | 30 | 24.00 | 19.41 | 6.52 | 80.00 | 25.96 | 14.51 | 146.40 |
| Fast | 95 | 25 | 20.60 | 23.16 | 9.40 | 82.40 | 21.27 | 8.17 | 144.40 |

| | Labor LF per manhour | Material coverage LF/gallon | Material cost per gallon | Labor cost per 100 LF | Labor burden 100 LF | Material cost per 100 LF | Overhead per 100 LF | Profit per 100 LF | Total price per 100 LF |
|---|---|---|---|---|---|---|---|---|---|
| STEP 4 & 5: Sanding sealer (material #11b) & light sand | | | | | | | | | |
| Brush 1 coat | | | | | | | | | |
| Slow | 45 | 65 | 17.30 | 22.78 | 6.19 | 26.62 | 17.24 | 10.93 | 83.76 |
| Medium | 50 | 60 | 15.10 | 33.00 | 11.10 | 25.17 | 16.97 | 9.49 | 95.73 |
| Fast | 55 | 55 | 13.00 | 40.00 | 16.21 | 23.64 | 14.77 | 5.68 | 100.30 |
| Spray 1 coat | | | | | | | | | |
| Slow | 125 | 35 | 17.30 | 8.20 | 2.23 | 49.43 | 18.56 | 11.76 | 90.18 |
| Medium | 138 | 30 | 15.10 | 11.96 | 4.03 | 50.33 | 16.25 | 9.08 | 91.65 |
| Fast | 150 | 25 | 13.00 | 14.67 | 5.95 | 52.00 | 13.43 | 5.16 | 91.21 |
| STEP 6 & 7: Lacquer (material #11c), 2 coats | | | | | | | | | |
| Brush 1st coat | | | | | | | | | |
| Slow | 40 | 65 | 18.00 | 25.63 | 6.97 | 27.69 | 18.69 | 11.85 | 90.83 |
| Medium | 50 | 60 | 15.70 | 33.00 | 11.10 | 26.17 | 17.22 | 9.62 | 97.11 |
| Fast | 60 | 55 | 13.50 | 36.67 | 14.87 | 24.55 | 14.08 | 5.41 | 95.58 |
| Brush 2nd coat | | | | | | | | | |
| Slow | 45 | 70 | 18.00 | 22.78 | 6.19 | 25.71 | 16.96 | 10.75 | 82.39 |
| Medium | 55 | 65 | 15.70 | 30.00 | 10.09 | 24.15 | 15.74 | 8.80 | 88.78 |
| Fast | 65 | 60 | 13.50 | 33.85 | 13.70 | 22.50 | 12.96 | 4.98 | 87.99 |
| Spray 1st coat | | | | | | | | | |
| Slow | 75 | 55 | 18.00 | 13.67 | 3.71 | 32.73 | 15.54 | 9.85 | 75.50 |
| Medium | 85 | 50 | 15.70 | 19.41 | 6.52 | 31.40 | 14.05 | 7.85 | 79.23 |
| Fast | 95 | 45 | 13.50 | 23.16 | 9.40 | 30.00 | 11.57 | 4.45 | 78.58 |
| Spray 2nd coat | | | | | | | | | |
| Slow | 85 | 60 | 18.00 | 12.06 | 3.28 | 30.00 | 14.06 | 8.91 | 68.31 |
| Medium | 95 | 55 | 15.70 | 17.37 | 5.85 | 28.55 | 12.68 | 7.09 | 71.54 |
| Fast | 105 | 50 | 13.50 | 20.95 | 8.49 | 27.00 | 10.44 | 4.01 | 70.89 |
| Complete stain, seal & 2 coat lacquer system (material #11) | | | | | | | | | |
| Brush all coats | | | | | | | | | |
| Slow | 8 | 30 | 20.20 | 128.13 | 34.87 | 67.33 | 71.42 | 45.27 | 347.02 |
| Medium | 10 | 25 | 18.30 | 165.00 | 55.50 | 73.20 | 71.96 | 40.22 | 405.88 |
| Fast | 12 | 20 | 15.10 | 183.33 | 74.33 | 75.50 | 61.64 | 23.69 | 418.49 |
| Spray all coats | | | | | | | | | |
| Slow | 16 | 20 | 18.00 | 64.06 | 17.44 | 90.00 | 53.17 | 33.70 | 258.37 |
| Medium | 20 | 15 | 15.70 | 82.50 | 27.75 | 104.67 | 52.66 | 29.43 | 297.01 |
| Fast | 24 | 10 | 13.50 | 91.67 | 37.17 | 135.00 | 48.81 | 18.76 | 331.41 |

Use these costs for applying stain, sanding sealer and lacquer to interior railings. Railing is based on a 36" to 42" height and with spindles spaced at 6" on center. These costs include painting the rail cap, baluster, newels and spindles. For rough sawn wood railing with 2" x 2" spindles spaced at 6" on center, see the tables for Exterior wood railing. "Slow" work is based on a $10.25 hourly wage, "Medium" work on a $16.75 hourly wage, and "Fast" work on a $22.00 hourly wage. Other qualifications that apply to this table are on page 9.

| | Labor LF per manhour | Material coverage LF/gallon | Material cost per gallon | Labor cost per 100 LF | Labor burden 100 LF | Material cost per 100 LF | Overhead per 100 LF | Profit per 100 LF | Total price per 100 LF |
|---|---|---|---|---|---|---|---|---|---|
| **Railing, interior, wood, paint grade, brush application** | | | | | | | | | |
| Undercoat, water or oil base (material #3 or #4) | | | | | | | | | |
| Brush 1 coat | | | | | | | | | |
| Slow | 11 | 50 | 19.35 | 93.18 | 25.37 | 38.70 | 48.75 | 30.90 | 236.90 |
| Medium | 14 | 45 | 16.90 | 117.86 | 39.64 | 37.56 | 47.79 | 26.72 | 269.57 |
| Fast | 17 | 40 | 14.50 | 129.41 | 52.46 | 36.25 | 40.36 | 15.51 | 273.99 |
| Split coat (1/2 undercoat + 1/2 enamel), water or oil base (material #3 and #9, or material #4 and #10) | | | | | | | | | |
| Brush 1 coat | | | | | | | | | |
| Slow | 15 | 70 | 21.85 | 68.33 | 18.61 | 31.21 | 36.63 | 23.22 | 178.00 |
| Medium | 18 | 65 | 19.10 | 91.67 | 30.84 | 29.38 | 37.22 | 20.80 | 209.91 |
| Fast | 21 | 60 | 16.38 | 104.76 | 42.48 | 27.30 | 32.29 | 12.41 | 219.24 |
| Enamel, water or oil base (material #9 or #10) | | | | | | | | | |
| Brush 1st finish coat | | | | | | | | | |
| Slow | 13 | 60 | 24.35 | 78.85 | 21.45 | 40.58 | 43.69 | 27.69 | 212.26 |
| Medium | 16 | 55 | 21.30 | 103.13 | 34.68 | 38.73 | 43.26 | 24.18 | 243.98 |
| Fast | 18 | 50 | 18.25 | 122.22 | 49.57 | 36.50 | 38.53 | 14.81 | 261.63 |
| Brush 2nd or additional finish coats | | | | | | | | | |
| Slow | 15 | 70 | 24.35 | 68.33 | 18.61 | 34.79 | 37.74 | 23.92 | 183.39 |
| Medium | 18 | 65 | 21.30 | 91.67 | 30.84 | 32.77 | 38.05 | 21.27 | 214.60 |
| Fast | 21 | 60 | 18.25 | 104.76 | 42.48 | 30.42 | 32.87 | 12.63 | 223.16 |

Use these costs for applying undercoat or enamel to interior railings. Railing is based on a 36" to 42" height and with spindles spaced at 6" on center. These costs include painting the rail cap, baluster, newels and spindles. For rough sawn wood railing with 2" x 2" spindles spaced at 6" on center, see the tables for exterior wood railing. "Slow" work is based on a $10.25 hourly wage, "Medium" work on a $16.50 hourly wage, and "Fast" work on a $22.00 hourly wage. Other qualifications that apply to this table are on page 9.

| | Labor LF per manhour | Material coverage LF/gallon | Material cost per gallon | Labor cost per 100 LF | Labor burden 100 LF | Material cost per 100 LF | Overhead per 100 LF | Profit per 100 LF | Total price per 100 LF |
|---|---|---|---|---|---|---|---|---|---|
| **Railing, interior, wood, paint grade, spray application** | | | | | | | | | |
| Undercoat, water or oil base (material #3 or #4) | | | | | | | | | |
| Spray 1 coat | | | | | | | | | |
| Slow | 50 | 40 | 19.35 | 20.50 | 5.58 | 48.38 | 23.09 | 14.63 | 112.18 |
| Medium | 60 | 35 | 16.90 | 27.50 | 9.26 | 48.29 | 20.83 | 11.65 | 117.53 |
| Fast | 70 | 30 | 14.50 | 31.43 | 12.75 | 48.33 | 17.11 | 6.58 | 116.20 |
| Split coat (1/2 undercoat + 1/2 enamel), water or oil base (material #3 and #9, or material #4 and #10) | | | | | | | | | |
| Spray 1 coat | | | | | | | | | |
| Slow | 65 | 50 | 21.85 | 15.77 | 4.29 | 43.70 | 19.77 | 12.53 | 96.06 |
| Medium | 75 | 45 | 19.10 | 22.00 | 7.39 | 42.44 | 17.60 | 9.84 | 99.27 |
| Fast | 85 | 40 | 16.38 | 25.88 | 10.48 | 40.95 | 14.30 | 5.50 | 97.11 |
| Enamel, water or oil base (material #9 or #10) | | | | | | | | | |
| Spray 1st finish coat | | | | | | | | | |
| Slow | 55 | 45 | 24.35 | 18.64 | 5.07 | 54.11 | 24.13 | 15.29 | 117.24 |
| Medium | 65 | 40 | 21.30 | 25.38 | 8.53 | 53.25 | 21.36 | 11.94 | 120.46 |
| Fast | 75 | 35 | 18.25 | 29.33 | 11.89 | 52.14 | 17.27 | 6.64 | 117.27 |
| Spray 2nd or additional finish coats | | | | | | | | | |
| Slow | 65 | 50 | 24.35 | 15.77 | 4.29 | 48.70 | 21.32 | 13.51 | 103.59 |
| Medium | 75 | 45 | 21.30 | 22.00 | 7.39 | 47.33 | 18.80 | 10.51 | 106.03 |
| Fast | 85 | 40 | 18.25 | 25.88 | 10.48 | 45.63 | 15.17 | 5.83 | 102.99 |

Use these costs for applying undercoat or enamel to interior railings. Railing is based on a 36" to 42" height and with spindles spaced at 6" on center. These costs include painting the rail cap, baluster, newels and spindles. For rough sawn wood railing with 2" x 2" spindles spaced at 6" on center, see the tables for exterior wood railing. "Slow" work is based on a $10.25 hourly wage, "Medium" work on a $16.50 hourly wage, and "Fast" work on a $22.00 hourly wage. Other qualifications that apply to this table are on page 9.

| | Labor LF per manhour | Material coverage LF/gallon | Material cost per gallon | Labor cost per 100 LF | Labor burden 100 LF | Material cost per 100 LF | Overhead per 100 LF | Profit per 100 LF | Total price per 100 LF |
|---|---|---|---|---|---|---|---|---|---|
| **Railing, wrought iron, 42" high bars with wood cap** | | | | | | | | | |
| Metal primer, rust inhibitor, clean metal (material #35) | | | | | | | | | |
| Brush 1 coat | | | | | | | | | |
| Slow | 15 | 90 | 28.20 | 68.33 | 18.61 | 31.33 | 36.67 | 23.24 | 178.18 |
| Medium | 18 | 85 | 24.60 | 91.67 | 30.84 | 28.94 | 37.11 | 20.74 | 209.30 |
| Fast | 20 | 80 | 21.10 | 110.00 | 44.60 | 26.38 | 33.48 | 12.87 | 227.33 |
| Metal primer, rust inhibitor, rusty metal (material #36) | | | | | | | | | |
| Brush 1 coat | | | | | | | | | |
| Slow | 15 | 90 | 33.30 | 68.33 | 18.61 | 37.00 | 38.42 | 24.36 | 186.72 |
| Medium | 18 | 85 | 29.20 | 91.67 | 30.84 | 34.35 | 38.43 | 21.48 | 216.77 |
| Fast | 20 | 80 | 25.00 | 110.00 | 44.60 | 31.25 | 34.38 | 13.21 | 233.44 |
| Metal finish, synthetic enamel, off white, gloss, interior or exterior - (material #37) | | | | | | | | | |
| Brush 1st finish coat | | | | | | | | | |
| Slow | 20 | 110 | 25.90 | 51.25 | 13.95 | 23.55 | 27.52 | 17.44 | 133.71 |
| Medium | 23 | 100 | 22.70 | 71.74 | 24.13 | 22.70 | 29.05 | 16.24 | 163.86 |
| Fast | 26 | 90 | 19.50 | 84.62 | 34.30 | 21.67 | 26.01 | 10.00 | 176.60 |
| Brush 2nd or additional finish coats | | | | | | | | | |
| Slow | 30 | 125 | 25.90 | 34.17 | 9.29 | 20.72 | 19.90 | 12.62 | 96.70 |
| Medium | 35 | 120 | 22.70 | 47.14 | 15.86 | 18.92 | 20.07 | 11.22 | 113.21 |
| Fast | 40 | 115 | 19.50 | 55.00 | 22.30 | 16.96 | 17.44 | 6.70 | 118.40 |
| Metal finish, synthetic enamel, colors (except orange/red), gloss, interior or exterior - (material #38) | | | | | | | | | |
| Brush 1st finish coat | | | | | | | | | |
| Slow | 20 | 110 | 35.00 | 51.25 | 13.95 | 31.82 | 30.08 | 19.07 | 146.17 |
| Medium | 23 | 100 | 30.70 | 71.74 | 24.13 | 30.70 | 31.01 | 17.33 | 174.91 |
| Fast | 26 | 90 | 26.30 | 84.62 | 34.30 | 29.22 | 27.41 | 10.53 | 186.08 |
| Brush 2nd or additional finish coats | | | | | | | | | |
| Slow | 30 | 125 | 35.00 | 34.17 | 9.29 | 28.00 | 22.16 | 14.05 | 107.67 |
| Medium | 35 | 120 | 30.70 | 47.14 | 15.86 | 25.58 | 21.70 | 12.13 | 122.41 |
| Fast | 40 | 115 | 26.30 | 55.00 | 22.30 | 22.87 | 18.53 | 7.12 | 125.82 |

Use these figures for painting prefabricated wrought iron railing which is 42" high with 1/2" square vertical bars at 6" to 9" on center with a stain grade wood cap supported by a 1/2" by 1-1/2" top rail, and 1" square support posts at 6' to 10' on center and with a 1/2" by 1-1/2" bottom rail, unless otherwise noted. The Metal finish figures include only minor touchup of pre-primed steel or wrought iron prefabricated railings. Note: A two coat system, prime and finish, using oil base material is recommended for any metal surface. Although water base material is often used, it may cause oxidation, corrosion and rust. Using one coat of oil base paint on exterior metal may result in cracking, peeling, or chipping without the proper prime coat application. If off white or other light colored finish paint is specified, make sure the prime coat is a light color also, or more than one finish coat will be necessary. "Slow" work is based on a $10.25 hourly wage,"Medium" work on a $16.50 hourly wage, and "Fast" work on a $22.00 hourly wage. Other qualifications that apply to this table are on page 9.

| | Labor LF per manhour | Material coverage LF/gallon | Material cost per gallon | Labor cost per 100 LF | Labor burden 100 LF | Material cost per 100 LF | Overhead per 100 LF | Profit per 100 LF | Total price per 100 LF |
|---|---|---|---|---|---|---|---|---|---|

## Railing, wrought iron, 42" high bars with wrought iron cap

| | Labor LF per manhour | Material coverage LF/gallon | Material cost per gallon | Labor cost per 100 LF | Labor burden 100 LF | Material cost per 100 LF | Overhead per 100 LF | Profit per 100 LF | Total price per 100 LF |
|---|---|---|---|---|---|---|---|---|---|
| **Metal primer, rust inhibitor, clean metal (material #35)** | | | | | | | | | |
| Brush 1 coat | | | | | | | | | |
| Slow | 20 | 110 | 28.20 | 51.25 | 13.95 | 25.64 | 28.17 | 17.85 | 136.86 |
| Medium | 25 | 105 | 24.60 | 66.00 | 22.20 | 23.43 | 27.35 | 15.29 | 154.27 |
| Fast | 30 | 100 | 21.10 | 73.33 | 29.73 | 21.10 | 22.97 | 8.83 | 155.96 |
| **Metal primer, rust inhibitor, rusty metal (material #36)** | | | | | | | | | |
| Brush 1 coat | | | | | | | | | |
| Slow | 20 | 110 | 33.30 | 51.25 | 13.95 | 30.27 | 29.60 | 18.76 | 143.83 |
| Medium | 25 | 105 | 29.20 | 66.00 | 22.20 | 27.81 | 28.42 | 15.89 | 160.32 |
| Fast | 30 | 100 | 25.00 | 73.33 | 29.73 | 25.00 | 23.69 | 9.11 | 160.86 |
| **Metal finish, synthetic enamel, off white, gloss, interior or exterior - (material #37)** | | | | | | | | | |
| Brush 1st finish coat | | | | | | | | | |
| Slow | 25 | 120 | 25.90 | 41.00 | 11.16 | 21.58 | 22.86 | 14.49 | 111.09 |
| Medium | 30 | 115 | 22.70 | 55.00 | 18.49 | 19.74 | 22.85 | 12.77 | 128.85 |
| Fast | 35 | 110 | 19.50 | 62.86 | 25.48 | 17.73 | 19.62 | 7.54 | 133.23 |
| Brush 2nd or additional finish coats | | | | | | | | | |
| Slow | 35 | 135 | 25.90 | 29.29 | 7.97 | 19.19 | 17.50 | 11.09 | 85.04 |
| Medium | 40 | 130 | 22.70 | 41.25 | 13.88 | 17.46 | 17.78 | 9.94 | 100.31 |
| Fast | 45 | 125 | 19.50 | 48.89 | 19.81 | 15.60 | 15.60 | 5.99 | 105.89 |
| **Metal finish, synthetic enamel, colors (except orange/red), gloss, interior or exterior - (material #38)** | | | | | | | | | |
| Brush 1st finish coat | | | | | | | | | |
| Slow | 25 | 120 | 35.00 | 41.00 | 11.16 | 29.17 | 25.22 | 15.98 | 122.53 |
| Medium | 30 | 115 | 30.70 | 55.00 | 18.49 | 26.70 | 24.55 | 13.72 | 138.46 |
| Fast | 35 | 110 | 26.30 | 62.86 | 25.48 | 23.91 | 20.77 | 7.98 | 141.00 |
| Brush 2nd or additional finish coats | | | | | | | | | |
| Slow | 35 | 135 | 35.00 | 29.29 | 7.97 | 25.93 | 19.59 | 12.42 | 95.20 |
| Medium | 40 | 130 | 30.70 | 41.25 | 13.88 | 23.62 | 19.29 | 10.78 | 108.82 |
| Fast | 45 | 125 | 26.30 | 48.89 | 19.81 | 21.04 | 16.60 | 6.38 | 112.72 |

Use these figures for painting prefabricated wrought iron railing which is 42" high with 1/2" square vertical bars at 6" to 9" on center and a 1/2" by 1-1/2" wrought iron cap with 1" square support posts at 6' to 10' on center, and a 1/2" by 1-1/2" bottom rail, unless otherwise noted. The Metal finish figures include only minor touchup of pre-primed steel or wrought iron prefabricated railings. Note: A two coat system, prime and finish, using oil base material is recommended for any metal surface. Although water base material is often used, it may cause oxidation, corrosion and rust. Using one coat of oil base paint on exterior metal may result in cracking, peeling, or chipping without the proper prime coat application. If off white or other light colored finish paint is specified, make sure the prime coat is a light color also, or more than one finish coat will be necessary. "Slow" work is based on a $10.25 hourly wage, "Medium" work on a $16.50 hourly wage, and "Fast" work on a $22.00 hourly wage. Other qualifications that apply to this table are on page 9.

| | Labor LF per manhour | Material coverage LF/gallon | Material cost per gallon | Labor cost per 100 LF | Labor burden 100 LF | Material cost per 100 LF | Overhead per 100 LF | Profit per 100 LF | Total price per 100 LF |
|---|---|---|---|---|---|---|---|---|---|

## Railing, wrought iron, 60" to 72" high bars with wrought iron cap

Metal primer, rust inhibitor, clean metal (material #35)

| Brush 1 coat | | | | | | | | | |
|---|---|---|---|---|---|---|---|---|---|
| Slow | 10 | 90 | 28.20 | 102.50 | 27.90 | 31.33 | 50.15 | 31.79 | 243.67 |
| Medium | 15 | 85 | 24.60 | 110.00 | 37.01 | 28.94 | 43.11 | 24.10 | 243.16 |
| Fast | 20 | 80 | 21.10 | 110.00 | 44.60 | 26.38 | 33.48 | 12.87 | 227.33 |

Metal primer, rust inhibitor, rusty metal (material #36)

| Brush 1 coat | | | | | | | | | |
|---|---|---|---|---|---|---|---|---|---|
| Slow | 10 | 90 | 33.30 | 102.50 | 27.90 | 37.00 | 51.90 | 32.90 | 252.20 |
| Medium | 15 | 85 | 29.20 | 110.00 | 37.01 | 34.35 | 44.44 | 24.84 | 250.64 |
| Fast | 20 | 80 | 25.00 | 110.00 | 44.60 | 31.25 | 34.38 | 13.21 | 233.44 |

Metal finish, synthetic enamel, off white, gloss, interior or exterior - (material #37)

| Brush 1st finish coat | | | | | | | | | |
|---|---|---|---|---|---|---|---|---|---|
| Slow | 15 | 120 | 25.90 | 68.33 | 18.61 | 21.58 | 33.64 | 21.33 | 163.49 |
| Medium | 20 | 115 | 22.70 | 82.50 | 27.75 | 19.74 | 31.85 | 17.80 | 179.64 |
| Fast | 25 | 110 | 19.50 | 88.00 | 35.68 | 17.73 | 26.16 | 10.05 | 177.62 |

| Brush 2nd or additional finish coats | | | | | | | | | |
|---|---|---|---|---|---|---|---|---|---|
| Slow | 25 | 135 | 25.90 | 41.00 | 11.16 | 19.19 | 22.12 | 14.02 | 107.49 |
| Medium | 30 | 130 | 22.70 | 55.00 | 18.49 | 17.46 | 22.29 | 12.46 | 125.70 |
| Fast | 35 | 125 | 19.50 | 62.86 | 25.48 | 15.60 | 19.23 | 7.39 | 130.56 |

Metal finish, synthetic enamel, colors (except orange/red), gloss, interior or exterior - (material #38)

| Brush 1st finish coat | | | | | | | | | |
|---|---|---|---|---|---|---|---|---|---|
| Slow | 15 | 120 | 35.00 | 68.33 | 18.61 | 29.17 | 36.00 | 22.82 | 174.93 |
| Medium | 20 | 115 | 30.70 | 82.50 | 27.75 | 26.70 | 33.56 | 18.76 | 189.27 |
| Fast | 25 | 110 | 26.30 | 88.00 | 35.68 | 23.91 | 27.30 | 10.49 | 185.38 |

| Brush 2nd or additional finish coats | | | | | | | | | |
|---|---|---|---|---|---|---|---|---|---|
| Slow | 25 | 135 | 35.00 | 41.00 | 11.16 | 25.93 | 24.21 | 15.35 | 117.65 |
| Medium | 30 | 130 | 30.70 | 55.00 | 18.49 | 23.62 | 23.80 | 13.30 | 134.21 |
| Fast | 35 | 125 | 26.30 | 62.86 | 25.48 | 21.04 | 20.24 | 7.78 | 137.40 |

Use these figures for painting prefabricated wrought iron railing which is 60" to 72" high with 1/2" square vertical bars at 6" to 9" on center with 1" square support posts at 6' to 10' on center and with a 1/2" by 1-1/2" bottom rail, unless otherwise noted. The Metal finish figures include only minor touchup of pre-primed steel or wrought iron prefabricated railings. Note: A two coat system, prime and finish, using oil base material is recommended for any metal surface. Although water base material is often used, it may cause oxidation, corrosion and rust. Using one coat of oil base paint on exterior metal may result in cracking, peeling, or chipping without the proper prime coat application. If off white or other light colored finish paint is specified, make sure the prime coat is a light color also, or more than one finish coat will be necessary. "Slow" work is based on a $10.25 hourly wage, "Medium" work on a $16.50 hourly wage, and "Fast" work on a $22.50 hourly wage. Other qualifications that apply to this table are on page 9.

| | SF floor per manhour | Material SF floor per can | Material cost per can | Labor per 100 SF floor | Labor burden 100 SF | Material per 100 SF floor | Overhead per 100 SF floor | Profit per 100 SF floor | Total per 100 SF floor |
|---|---|---|---|---|---|---|---|---|---|
| **Registers, HVAC, per 100 square feet of floor area** | | | | | | | | | |
| Repaint jobs, spray cans (material #17) | | | | | | | | | |
| Spray 1 coat | | | | | | | | | |
| Slow | 900 | 700 | 3.60 | 1.14 | .31 | .51 | .61 | .39 | 2.96 |
| Medium | 950 | 650 | 3.20 | 1.74 | .58 | .49 | .69 | .39 | 3.89 |
| Fast | 1000 | 600 | 2.70 | 2.20 | .89 | .45 | .65 | .25 | 4.44 |
| New construction projects, spray cans (material #17) | | | | | | | | | |
| Spray 1 coat | | | | | | | | | |
| Slow | 2500 | 800 | 3.60 | .41 | .11 | .45 | .30 | .19 | 1.46 |
| Medium | 2750 | 750 | 3.20 | .60 | .19 | .43 | .30 | .17 | 1.69 |
| Fast | 3000 | 700 | 2.70 | .73 | .29 | .39 | .26 | .10 | 1.77 |

These costs assume HVAC registers are painted with spray cans (bombs) to match the adjacent walls. Costs are based on square footage of the floor area of the building. These rates include time to remove, paint and replace the HVAC registers. Use the square feet of floor area divided by these rates to find manhours and the number of spray bombs needed to paint all the heat registers in a building. Rule of Thumb: 2 minutes per 100 square feet of floor is for new construction projects. "Slow" work is based on a $10.25 hourly wage, "Medium" work on a $16.50 hourly wage, and "Fast" work on a $22.00 hourly wage. Other qualifications that apply to this table are on page 9.

| | Manhours per 1500 SF roof | Material gallons 1500 SF | Material cost per gallon | Labor per 1500 SF roof | Labor burden 1500 SF | Material per 1500 SF roof | Overhead per 1500 SF roof | Profit per 1500 SF roof | Total per 1500 SF roof |
|---|---|---|---|---|---|---|---|---|---|
| **Roof jacks, per 1500 square feet of roof area** | | | | | | | | | |
| Metal primer, clean metal (material #35) | | | | | | | | | |
| 1 story building, brush prime coat | | | | | | | | | |
| Slow | 0.40 | 0.30 | 28.20 | 4.10 | 1.11 | 8.46 | 4.24 | 2.69 | 20.60 |
| Medium | 0.35 | 0.33 | 24.60 | 5.78 | 1.93 | 8.12 | 3.88 | 2.17 | 21.88 |
| Fast | 0.30 | 0.35 | 21.10 | 6.60 | 2.68 | 7.39 | 3.08 | 1.19 | 20.94 |
| 2 story building, brush prime coat | | | | | | | | | |
| Slow | 0.50 | 0.40 | 28.20 | 5.13 | 1.39 | 11.28 | 5.52 | 3.50 | 26.82 |
| Medium | 0.45 | 0.43 | 24.60 | 7.43 | 2.50 | 10.58 | 5.02 | 2.81 | 28.34 |
| Fast | 0.40 | 0.45 | 21.10 | 8.80 | 3.57 | 9.50 | 4.05 | 1.56 | 27.48 |
| Metal primer, rusty metal (material #36) | | | | | | | | | |
| 1 story building, brush prime coat | | | | | | | | | |
| Slow | 0.40 | 0.30 | 33.30 | 4.10 | 1.11 | 9.99 | 4.72 | 2.99 | 22.91 |
| Medium | 0.35 | 0.33 | 29.20 | 5.78 | 1.93 | 9.64 | 4.25 | 2.38 | 23.98 |
| Fast | 0.30 | 0.35 | 25.00 | 6.60 | 2.68 | 8.75 | 3.34 | 1.28 | 22.65 |
| 2 story building, brush prime coat | | | | | | | | | |
| Slow | 0.50 | 0.40 | 33.30 | 5.13 | 1.39 | 13.32 | 6.15 | 3.90 | 29.89 |
| Medium | 0.45 | 0.43 | 29.20 | 7.43 | 2.50 | 12.56 | 5.51 | 3.08 | 31.08 |
| Fast | 0.40 | 0.45 | 25.00 | 8.80 | 3.57 | 11.25 | 4.37 | 1.68 | 29.67 |
| Metal finish, synthetic enamel - off white, gloss, interior or exterior (material #37) | | | | | | | | | |
| 1 story building, brush each coat | | | | | | | | | |
| Slow | 0.30 | 0.20 | 25.90 | 3.08 | .83 | 5.18 | 2.82 | 1.79 | 13.70 |
| Medium | 0.25 | 0.25 | 22.70 | 4.13 | 1.39 | 5.68 | 2.74 | 1.53 | 15.47 |
| Fast | 0.20 | 0.30 | 19.50 | 4.40 | 1.78 | 5.85 | 2.23 | .86 | 15.12 |
| 2 story building, brush each coat | | | | | | | | | |
| Slow | 0.40 | 0.30 | 25.90 | 4.10 | 1.11 | 7.77 | 4.03 | 2.55 | 19.56 |
| Medium | 0.35 | 0.33 | 22.70 | 5.78 | 1.93 | 7.49 | 3.73 | 2.08 | 21.01 |
| Fast | 0.30 | 0.35 | 19.50 | 6.60 | 2.68 | 6.83 | 2.98 | 1.15 | 20.24 |

| | Manhours per 1500 SF roof | Material gallons 1500 SF | Material cost per gallon | Labor per 1500 SF roof | Labor burden 1500 SF | Material per 1500 SF roof | Overhead per 1500 SF roof | Profit per 1500 SF roof | Total per 1500 SF roof |
|---|---|---|---|---|---|---|---|---|---|
| Metal finish, synthetic enamel, colors (except orange/red), gloss, interior or exterior - (material #38) | | | | | | | | | |
| 1 story building, brush each coat | | | | | | | | | |
| Slow | 0.30 | 0.20 | 35.00 | 3.08 | .83 | 7.00 | 3.39 | 2.15 | 16.45 |
| Medium | 0.25 | 0.25 | 30.70 | 4.13 | 1.39 | 7.68 | 3.23 | 1.81 | 18.24 |
| Fast | 0.20 | 0.30 | 26.30 | 4.40 | 1.78 | 7.89 | 2.60 | 1.00 | 17.67 |
| 2 story building, brush each coat | | | | | | | | | |
| Slow | 0.40 | 0.30 | 35.00 | 4.10 | 1.11 | 10.50 | 4.87 | 3.09 | 23.67 |
| Medium | 0.35 | 0.33 | 30.70 | 5.78 | 1.93 | 10.13 | 4.37 | 2.44 | 24.65 |
| Fast | 0.30 | 0.35 | 26.30 | 6.60 | 2.68 | 9.21 | 3.42 | 1.31 | 23.22 |

Production rates and coverage figures are minimum values based on 1 or 2 story roof areas of up to 1500 square feet. For example, to apply metal primer on clean metal roof jacks on a 3000 SF one-story building at a medium rate, use two times the cost of $22.88 or $45.76. This figure includes ladder time. See the paragraphs below on Roof pitch difficulty factors and Roof area conversion factors to adjust for roof slope and type. Note: A two coat system, prime and finish, using oil base material is recommended for any metal surface. Although water base material is often used, it may cause oxidation, corrosion and rust. One coat of oil base solid body stain is often used on exterior metal but it may crack, peel or chip without the proper prime coat application. "Slow" work is based on a $10.25 hourly wage, "Medium" work on a $16.50 hourly wage, and "Fast" work on a $22.00 hourly wage. Other qualifications that apply to this table are on page 9.

## Roof area conversion factors

For an arched roof, multiply the building length by the building width, then multiply by 1.5.
For a gambrel roof, multiply the building length by the building width, then multiply by 1.33.

## Roof pitch difficulty factors

It's harder to paint on a sloped surface than a flat surface. The steeper the slope, the more difficult the work. Roof slope is usually measured in inches of rise per inch of horizontal run. For example, a 3 in 12 pitch means the roof rises 3 inches for each 12 inches of run, measuring horizontally. Use the difficulty factors that follow when estimating the time needed to paint on a sloping roof.

On a flat roof, or roof with a pitch of less than 3 in 12, calculate the roof area without modification.
If the pitch is 3 in 12, multiply the surface area by 1.1.
If the pitch is 4 in 12, multiply the surface area by 1.2.
If the pitch is 6 in 12, multiply the surface area by 1.3.

| | Labor SF per manhour | Material coverage SF/gallon | Material cost per gallon | Labor cost per 100 SF | Labor burden 100 SF | Material cost per 100 SF | Overhead per 100 SF | Profit per 100 SF | Total price per 100 SF |
|---|---|---|---|---|---|---|---|---|---|
| **Roofing, composition shingles, brush application** | | | | | | | | | |
| Solid body stain, water base (material #18) | | | | | | | | | |
| Brush 1st coat | | | | | | | | | |
| Slow | 45 | 220 | 19.40 | 22.78 | 6.19 | 8.82 | 11.72 | 7.43 | 56.94 |
| Medium | 60 | 200 | 17.00 | 27.50 | 9.26 | 8.50 | 11.09 | 6.20 | 62.55 |
| Fast | 80 | 180 | 14.50 | 27.50 | 11.15 | 8.06 | 8.64 | 3.32 | 58.67 |
| Brush 2nd coat | | | | | | | | | |
| Slow | 65 | 330 | 19.40 | 15.77 | 4.29 | 5.88 | 8.04 | 5.10 | 39.08 |
| Medium | 80 | 210 | 17.00 | 20.63 | 6.93 | 8.10 | 8.74 | 4.89 | 49.29 |
| Fast | 100 | 290 | 14.50 | 22.00 | 8.92 | 5.00 | 6.65 | 2.55 | 45.12 |
| Brush 3rd or additional coats | | | | | | | | | |
| Slow | 85 | 405 | 19.40 | 12.06 | 3.28 | 4.79 | 6.24 | 3.96 | 30.33 |
| Medium | 100 | 385 | 17.00 | 16.50 | 5.55 | 4.42 | 6.49 | 3.63 | 36.59 |
| Fast | 120 | 365 | 14.50 | 18.33 | 7.43 | 3.97 | 5.50 | 2.11 | 37.34 |
| Solid body stain, oil base (material #19) | | | | | | | | | |
| Brush 1st coat | | | | | | | | | |
| Slow | 45 | 270 | 23.10 | 22.78 | 6.19 | 8.56 | 11.64 | 7.38 | 56.55 |
| Medium | 60 | 250 | 20.20 | 27.50 | 9.26 | 8.08 | 10.98 | 6.14 | 61.96 |
| Fast | 80 | 230 | 17.30 | 27.50 | 11.15 | 7.52 | 8.54 | 3.28 | 57.99 |
| Brush 2nd coat | | | | | | | | | |
| Slow | 65 | 360 | 23.10 | 15.77 | 4.29 | 6.42 | 8.21 | 5.21 | 39.90 |
| Medium | 80 | 345 | 20.20 | 20.63 | 6.93 | 5.86 | 8.19 | 4.58 | 46.19 |
| Fast | 100 | 330 | 17.30 | 22.00 | 8.92 | 5.24 | 6.69 | 2.57 | 45.42 |
| Brush 3rd or additional coats | | | | | | | | | |
| Slow | 85 | 425 | 23.10 | 12.06 | 3.28 | 5.44 | 6.44 | 4.08 | 31.30 |
| Medium | 100 | 405 | 20.20 | 16.50 | 5.55 | 4.99 | 6.62 | 3.70 | 37.36 |
| Fast | 120 | 385 | 17.30 | 18.33 | 7.43 | 4.49 | 5.60 | 2.15 | 38.00 |

Use these figures for repaint jobs only. It has been established that asbestos fibers are a known carcinogen (cancer causing) and it is likely that no new construction projects will specify materials or products which contain asbestos. Roofing and siding products usually contain very little asbestos and are typically non-friable (hand pressure can not crumble, pulverize or reduce to powder when dry). There is danger when asbestos is being removed because of exposure to airborne particulate matter. Apparently, there is little danger when painting asbestos roofing or siding, but it is a good idea to have your painters wear respirators or particle masks for their safety. Coverage figures are based on shingles or shakes with average moisture content. See the paragraphs on Roof pitch difficulty factors and Roof area conversion factors to adjust for roof slope and type. "Slow" work is based on a $10.25 hourly wage, "Medium" work on a $16.50 hourly wage, and "Fast" work on a $22.00 hourly wage. Other qualifications that apply to this table are on page 9.

| | Labor SF per manhour | Material coverage SF/gallon | Material cost per gallon | Labor cost per 100 SF | Labor burden 100 SF | Material cost per 100 SF | Overhead per 100 SF | Profit per 100 SF | Total price per 100 SF |
|---|---|---|---|---|---|---|---|---|---|

## Roofing, composition shingles, roll application

| | Labor SF per manhour | Material coverage SF/gallon | Material cost per gallon | Labor cost per 100 SF | Labor burden 100 SF | Material cost per 100 SF | Overhead per 100 SF | Profit per 100 SF | Total price per 100 SF |
|---|---|---|---|---|---|---|---|---|---|
| Solid body stain, water base (material #18) | | | | | | | | | |
| Roll 1st coat | | | | | | | | | |
| Slow | 150 | 190 | 19.40 | 6.83 | 1.87 | 10.21 | 5.86 | 3.71 | 28.48 |
| Medium | 170 | 180 | 17.00 | 9.71 | 3.26 | 9.44 | 5.49 | 3.07 | 30.97 |
| Fast | 200 | 170 | 14.50 | 11.00 | 4.46 | 8.53 | 4.44 | 1.71 | 30.14 |
| Roll 2nd coat | | | | | | | | | |
| Slow | 250 | 300 | 19.40 | 4.10 | 1.12 | 6.47 | 3.62 | 2.30 | 17.61 |
| Medium | 305 | 290 | 17.00 | 5.41 | 1.82 | 5.86 | 3.21 | 1.79 | 18.09 |
| Fast | 360 | 280 | 14.50 | 6.11 | 2.49 | 5.18 | 2.55 | .98 | 17.31 |
| Roll 3rd or additional coats | | | | | | | | | |
| Slow | 360 | 385 | 19.40 | 2.85 | .78 | 5.04 | 2.69 | 1.70 | 13.06 |
| Medium | 385 | 375 | 17.00 | 4.29 | 1.44 | 4.53 | 2.51 | 1.40 | 14.17 |
| Fast | 420 | 365 | 14.50 | 5.24 | 2.12 | 3.97 | 2.10 | .81 | 14.24 |
| Solid body stain, oil base (material #19) | | | | | | | | | |
| Roll 1st coat | | | | | | | | | |
| Slow | 150 | 220 | 23.10 | 6.83 | 1.87 | 10.50 | 5.95 | 3.77 | 28.92 |
| Medium | 170 | 205 | 20.20 | 9.71 | 3.26 | 9.85 | 5.59 | 3.13 | 31.54 |
| Fast | 200 | 190 | 17.30 | 11.00 | 4.46 | 9.11 | 4.55 | 1.75 | 30.87 |
| Roll 2nd coat | | | | | | | | | |
| Slow | 250 | 330 | 23.10 | 4.10 | 1.12 | 7.00 | 3.79 | 2.40 | 18.41 |
| Medium | 305 | 320 | 20.20 | 5.41 | 1.82 | 6.31 | 3.32 | 1.85 | 18.71 |
| Fast | 360 | 310 | 17.30 | 6.11 | 2.49 | 5.58 | 2.62 | 1.01 | 17.81 |
| Roll 3rd or additional coats | | | | | | | | | |
| Slow | 360 | 405 | 23.10 | 2.85 | .78 | 5.70 | 2.89 | 1.83 | 14.05 |
| Medium | 385 | 395 | 20.20 | 4.29 | 1.44 | 5.11 | 2.66 | 1.49 | 14.99 |
| Fast | 420 | 385 | 17.30 | 5.24 | 2.12 | 4.49 | 2.19 | .84 | 14.88 |

| | Labor SF per manhour | Material coverage SF/gallon | Material cost per gallon | Labor cost per 100 SF | Labor burden 100 SF | Material cost per 100 SF | Overhead per 100 SF | Profit per 100 SF | Total price per 100 SF |
|---|---|---|---|---|---|---|---|---|---|
| Waterproofing, clear hydro sealer (material #34) | | | | | | | | | |
| Roll 1st coat | | | | | | | | | |
| Slow | 100 | 300 | 18.00 | 10.25 | 2.79 | 6.00 | 5.90 | 3.74 | 28.68 |
| Medium | 200 | 275 | 15.80 | 8.25 | 2.78 | 5.75 | 4.11 | 2.30 | 23.19 |
| Fast | 300 | 250 | 13.50 | 7.33 | 2.97 | 5.40 | 2.90 | 1.12 | 19.72 |
| Roll 2nd or additional coats | | | | | | | | | |
| Slow | 150 | 350 | 18.00 | 6.83 | 1.87 | 5.14 | 4.29 | 2.72 | 20.85 |
| Medium | 250 | 325 | 15.80 | 6.60 | 2.22 | 4.86 | 3.35 | 1.87 | 18.90 |
| Fast | 350 | 300 | 13.50 | 6.29 | 2.55 | 4.50 | 2.47 | .95 | 16.76 |

Use these figures for repaint jobs only. It has been established that asbestos fibers are a known carcinogen (cancer causing) and it is likely that no new construction projects will specify materials or products which contain asbestos. Roofing and siding products usually contain very little asbestos and are typically non-friable (hand pressure can not crumble, pulverize or reduce to powder when dry). There is danger when asbestos is being removed because of exposure to airborne particulate matter. Apparently, there is little danger when painting asbestos roofing or siding, but it is a good idea to have your painters wear respirators or particle masks for their safety. Coverage figures are based on shingles or shakes with average moisture content. See the paragraphs on Roof pitch difficulty factors and Roof area conversion factors to adjust for roof slope and type. "Slow" work is based on a $10.25 hourly wage, "Medium" work on a $16.50 hourly wage, and "Fast" work on a $22.00 hourly wage. Other qualifications that apply to this table are on page 9.

| | Labor SF per manhour | Material coverage SF/gallon | Material cost per gallon | Labor cost per 100 SF | Labor burden 100 SF | Material cost per 100 SF | Overhead per 100 SF | Profit per 100 SF | Total price per 100 SF |
|---|---|---|---|---|---|---|---|---|---|

## Roofing, composition shingles, spray application

| | Labor SF per manhour | Material coverage SF/gallon | Material cost per gallon | Labor cost per 100 SF | Labor burden 100 SF | Material cost per 100 SF | Overhead per 100 SF | Profit per 100 SF | Total price per 100 SF |
|---|---|---|---|---|---|---|---|---|---|
| Solid body stain, water base (material #18) | | | | | | | | | |
| Spray 1st coat | | | | | | | | | |
| Slow | 325 | 200 | 19.40 | 3.15 | .87 | 9.70 | 4.25 | 2.69 | 20.66 |
| Medium | 350 | 180 | 17.00 | 4.71 | 1.60 | 9.44 | 3.85 | 2.15 | 21.75 |
| Fast | 375 | 160 | 14.50 | 5.87 | 2.39 | 9.06 | 3.20 | 1.23 | 21.75 |
| Spray 2nd coat | | | | | | | | | |
| Slow | 425 | 290 | 19.40 | 2.41 | .65 | 6.69 | 3.03 | 1.92 | 14.70 |
| Medium | 450 | 280 | 17.00 | 3.67 | 1.23 | 6.07 | 2.69 | 1.50 | 15.16 |
| Fast | 475 | 270 | 14.50 | 4.63 | 1.89 | 5.37 | 2.20 | .84 | 14.93 |
| Spray 3rd or additional coats | | | | | | | | | |
| Slow | 500 | 365 | 19.40 | 2.05 | .56 | 5.32 | 2.46 | 1.56 | 11.95 |
| Medium | 538 | 355 | 17.00 | 3.07 | 1.03 | 4.79 | 2.18 | 1.22 | 12.29 |
| Fast | 575 | 345 | 14.50 | 3.83 | 1.55 | 4.20 | 1.77 | .68 | 12.03 |
| Solid body stain, oil base (material #19) | | | | | | | | | |
| Spray 1st coat | | | | | | | | | |
| Slow | 325 | 230 | 23.10 | 3.15 | .87 | 10.04 | 4.36 | 2.76 | 21.18 |
| Medium | 350 | 205 | 20.20 | 4.71 | 1.60 | 9.85 | 3.95 | 2.21 | 22.32 |
| Fast | 375 | 180 | 17.30 | 5.87 | 2.39 | 9.61 | 3.30 | 1.27 | 22.44 |
| Spray 2nd coat | | | | | | | | | |
| Slow | 425 | 310 | 23.10 | 2.41 | .65 | 7.45 | 3.26 | 2.07 | 15.84 |
| Medium | 450 | 300 | 20.20 | 3.67 | 1.23 | 6.73 | 2.85 | 1.59 | 16.07 |
| Fast | 475 | 290 | 17.30 | 4.63 | 1.89 | 5.97 | 2.31 | .89 | 15.69 |
| Spray 3rd or additional coats | | | | | | | | | |
| Slow | 500 | 380 | 23.10 | 2.05 | .56 | 6.08 | 2.69 | 1.71 | 13.09 |
| Medium | 538 | 370 | 20.20 | 3.07 | 1.03 | 5.46 | 2.34 | 1.31 | 13.21 |
| Fast | 575 | 360 | 17.30 | 3.83 | 1.55 | 4.81 | 1.89 | .72 | 12.80 |

| | Labor SF per manhour | Material coverage SF/gallon | Material cost per gallon | Labor cost per 100 SF | Labor burden 100 SF | Material cost per 100 SF | Overhead per 100 SF | Profit per 100 SF | Total price per 100 SF |
|---|---|---|---|---|---|---|---|---|---|
| Waterproofing, clear hydro sealer (material #34) | | | | | | | | | |
| Spray 1st coat | | | | | | | | | |
| Slow | 550 | 100 | 18.00 | 1.86 | .51 | 18.00 | 6.31 | 4.00 | 30.68 |
| Medium | 600 | 88 | 15.80 | 2.75 | .93 | 17.95 | 5.30 | 2.96 | 29.89 |
| Fast | 650 | 75 | 13.50 | 3.38 | 1.38 | 18.00 | 4.21 | 1.62 | 28.59 |
| Spray 2nd or additional coats | | | | | | | | | |
| Slow | 600 | 150 | 18.00 | 1.71 | .47 | 12.00 | 4.40 | 2.79 | 21.37 |
| Medium | 650 | 125 | 15.80 | 2.54 | .86 | 12.64 | 3.93 | 2.20 | 22.17 |
| Fast | 700 | 100 | 13.50 | 3.14 | 1.28 | 13.50 | 3.31 | 1.27 | 22.50 |

Use these figures for repaint jobs only. It has been established that asbestos fibers are a known carcinogen (cancer causing) and it is likely that no new construction projects will specify materials or products which contain asbestos. Roofing and siding products usually contain very little asbestos and are typically non-friable (hand pressure can not crumble, pulverize or reduce to powder when dry). There is danger when asbestos is being removed because of exposure to airborne particulate matter. Apparently, there is little danger when painting asbestos roofing or siding, but it is a good idea to have your painters wear respirators or particle masks for their safety. Coverage figures are based on shingles or shakes with average moisture content. See the paragraphs on Roof pitch difficulty factors and Roof area conversion factors to adjust for roof slope and type. "Slow" work is based on a $10.25 hourly wage, "Medium" work on a $16.50 hourly wage, and "Fast" work on a $22.00 hourly wage. Other qualifications that apply to this table are on page 9.

| | Labor SF per manhour | Material coverage SF/gallon | Material cost per gallon | Labor cost per 100 SF | Labor burden 100 SF | Material cost per 100 SF | Overhead per 100 SF | Profit per 100 SF | Total price per 100 SF |
|---|---|---|---|---|---|---|---|---|---|
| **Roofing, wood shingles or shakes, brush application** | | | | | | | | | |
| Solid body stain, water base (material #18) | | | | | | | | | |
| Brush 1st coat | | | | | | | | | |
| Slow | 100 | 240 | 19.40 | 10.25 | 2.79 | 8.08 | 6.55 | 4.15 | 31.82 |
| Medium | 155 | 228 | 17.00 | 10.65 | 3.57 | 7.46 | 5.31 | 2.97 | 29.96 |
| Fast | 210 | 215 | 14.50 | 10.48 | 4.24 | 6.74 | 3.97 | 1.53 | 26.96 |
| Brush 2nd or additional coats | | | | | | | | | |
| Slow | 150 | 290 | 19.40 | 6.83 | 1.87 | 6.69 | 4.77 | 3.02 | 23.18 |
| Medium | 195 | 278 | 17.00 | 8.46 | 2.85 | 6.12 | 4.27 | 2.39 | 24.09 |
| Fast | 240 | 265 | 14.50 | 9.17 | 3.72 | 5.47 | 3.40 | 1.31 | 23.07 |
| Solid body stain, oil base (material #19) | | | | | | | | | |
| Brush 1st coat | | | | | | | | | |
| Slow | 100 | 160 | 23.10 | 10.25 | 2.79 | 14.44 | 8.52 | 5.40 | 41.40 |
| Medium | 155 | 150 | 20.20 | 10.65 | 3.57 | 13.47 | 6.79 | 3.79 | 38.27 |
| Fast | 210 | 140 | 17.30 | 10.48 | 4.24 | 12.36 | 5.01 | 1.93 | 34.02 |
| Brush 2nd or additional coats | | | | | | | | | |
| Slow | 150 | 260 | 23.10 | 6.83 | 1.87 | 8.88 | 5.45 | 3.45 | 26.48 |
| Medium | 195 | 250 | 20.20 | 8.46 | 2.85 | 8.08 | 4.75 | 2.66 | 26.80 |
| Fast | 240 | 240 | 17.30 | 9.17 | 3.72 | 7.21 | 3.72 | 1.43 | 25.25 |
| Semi-transparent stain, water base (material #20) | | | | | | | | | |
| Brush 1st coat | | | | | | | | | |
| Slow | 120 | 260 | 19.60 | 8.54 | 2.32 | 7.54 | 5.71 | 3.62 | 27.73 |
| Medium | 175 | 248 | 17.10 | 9.43 | 3.16 | 6.90 | 4.78 | 2.67 | 26.94 |
| Fast | 230 | 235 | 14.70 | 9.57 | 3.88 | 6.26 | 3.65 | 1.40 | 24.76 |
| Brush 2nd or additional coats | | | | | | | | | |
| Slow | 160 | 300 | 19.60 | 6.41 | 1.74 | 6.53 | 4.55 | 2.89 | 22.12 |
| Medium | 205 | 288 | 17.10 | 8.05 | 2.71 | 5.94 | 4.09 | 2.29 | 23.08 |
| Fast | 250 | 275 | 14.70 | 8.80 | 3.57 | 5.35 | 3.28 | 1.26 | 22.26 |
| Semi-transparent stain, oil base (material #21) | | | | | | | | | |
| Brush 1st coat | | | | | | | | | |
| Slow | 120 | 180 | 23.40 | 8.54 | 2.32 | 13.00 | 7.40 | 4.69 | 35.95 |
| Medium | 175 | 170 | 20.50 | 9.43 | 3.16 | 12.06 | 6.04 | 3.38 | 34.07 |
| Fast | 230 | 160 | 17.60 | 9.57 | 3.88 | 11.00 | 4.52 | 1.74 | 30.71 |
| Brush 2nd or additional coats | | | | | | | | | |
| Slow | 160 | 280 | 23.40 | 6.41 | 1.74 | 8.36 | 5.12 | 3.25 | 24.88 |
| Medium | 205 | 270 | 20.50 | 8.05 | 2.71 | 7.59 | 4.50 | 2.51 | 25.36 |
| Fast | 250 | 260 | 17.60 | 8.80 | 3.57 | 6.77 | 3.54 | 1.36 | 24.04 |

| | Labor SF per manhour | Material coverage SF/gallon | Material cost per gallon | Labor cost per 100 SF | Labor burden 100 SF | Material cost per 100 SF | Overhead per 100 SF | Profit per 100 SF | Total price per 100 SF |
|---|---|---|---|---|---|---|---|---|---|
| Penetrating oil stain (material #13) | | | | | | | | | |
| Brush 1st coat | | | | | | | | | |
| Slow | 100 | 160 | 25.00 | 10.25 | 2.79 | 15.63 | 8.89 | 5.63 | 43.19 |
| Medium | 155 | 150 | 21.80 | 10.65 | 3.57 | 14.53 | 7.05 | 3.94 | 39.74 |
| Fast | 210 | 140 | 18.70 | 10.48 | 4.24 | 13.36 | 5.20 | 2.00 | 35.28 |
| Brush 2nd or additional coats | | | | | | | | | |
| Slow | 150 | 205 | 25.00 | 6.83 | 1.87 | 12.20 | 6.48 | 4.11 | 31.49 |
| Medium | 195 | 195 | 21.80 | 8.46 | 2.85 | 11.18 | 5.51 | 3.08 | 31.08 |
| Fast | 240 | 185 | 18.70 | 9.17 | 3.72 | 10.11 | 4.26 | 1.64 | 28.90 |

Coverage figures are based on shingles or shakes with average moisture content. See the paragraphs on Roof pitch difficulty factors and Roof area conversion factors to adjust for roof slope and type. "Slow" work is based on a $10.25 hourly wage, "Medium" work on a $16.50 hourly wage, and "Fast" work on a $22.50 hourly wage. Other qualifications that apply to this table are on page 9.

| | Labor SF per manhour | Material coverage SF/gallon | Material cost per gallon | Labor cost per 100 SF | Labor burden 100 SF | Material cost per 100 SF | Overhead per 100 SF | Profit per 100 SF | Total price per 100 SF |
|---|---|---|---|---|---|---|---|---|---|
| **Roofing, wood shingles or shakes, roll application** | | | | | | | | | |
| Solid body stain, water base (material #18) | | | | | | | | | |
| Roll 1st coat | | | | | | | | | |
| Slow | 210 | 225 | 19.40 | 4.88 | 1.33 | 8.62 | 4.60 | 2.91 | 22.34 |
| Medium | 258 | 213 | 17.00 | 6.40 | 2.16 | 7.98 | 4.05 | 2.26 | 22.85 |
| Fast | 305 | 200 | 14.50 | 7.21 | 2.93 | 7.25 | 3.22 | 1.24 | 21.85 |
| Roll 2nd or additional coats | | | | | | | | | |
| Slow | 250 | 275 | 19.40 | 4.10 | 1.12 | 7.05 | 3.80 | 2.41 | 18.48 |
| Medium | 300 | 263 | 17.00 | 5.50 | 1.84 | 6.46 | 3.38 | 1.89 | 19.07 |
| Fast | 350 | 250 | 14.50 | 6.29 | 2.55 | 5.80 | 2.71 | 1.04 | 18.39 |
| Solid body stain, oil base (material #19) | | | | | | | | | |
| Roll 1st coat | | | | | | | | | |
| Slow | 210 | 150 | 23.10 | 4.88 | 1.33 | 15.40 | 6.70 | 4.25 | 32.56 |
| Medium | 255 | 140 | 20.20 | 6.47 | 2.17 | 14.43 | 5.65 | 3.16 | 31.88 |
| Fast | 305 | 130 | 17.30 | 7.21 | 2.93 | 13.31 | 4.34 | 1.67 | 29.46 |
| Roll 2nd or additional coats | | | | | | | | | |
| Slow | 250 | 245 | 23.10 | 4.10 | 1.12 | 9.43 | 4.54 | 2.88 | 22.07 |
| Medium | 300 | 235 | 20.20 | 5.50 | 1.84 | 8.60 | 3.91 | 2.18 | 22.03 |
| Fast | 350 | 225 | 17.30 | 6.29 | 2.55 | 7.69 | 3.06 | 1.18 | 20.77 |
| Semi-transparent stain, water base (material #20) | | | | | | | | | |
| Roll 1st coat | | | | | | | | | |
| Slow | 240 | 250 | 19.60 | 4.27 | 1.17 | 7.84 | 4.11 | 2.61 | 20.00 |
| Medium | 288 | 238 | 17.10 | 5.73 | 1.92 | 7.18 | 3.64 | 2.03 | 20.50 |
| Fast | 335 | 225 | 14.70 | 6.57 | 2.68 | 6.53 | 2.92 | 1.12 | 19.82 |
| Roll 2nd or additional coats | | | | | | | | | |
| Slow | 270 | 290 | 19.60 | 3.80 | 1.02 | 6.76 | 3.60 | 2.28 | 17.46 |
| Medium | 320 | 278 | 17.10 | 5.16 | 1.74 | 6.15 | 3.20 | 1.79 | 18.04 |
| Fast | 370 | 255 | 14.70 | 5.95 | 2.40 | 5.76 | 2.61 | 1.00 | 17.72 |
| Semi-transparent stain, oil base (material #21) | | | | | | | | | |
| Roll 1st coat | | | | | | | | | |
| Slow | 240 | 175 | 23.40 | 4.27 | 1.17 | 13.37 | 5.83 | 3.69 | 28.33 |
| Medium | 288 | 165 | 20.50 | 5.73 | 1.92 | 12.42 | 4.92 | 2.75 | 27.74 |
| Fast | 335 | 155 | 17.60 | 6.57 | 2.68 | 11.35 | 3.81 | 1.46 | 25.87 |
| Roll 2nd or additional coats | | | | | | | | | |
| Slow | 270 | 260 | 23.40 | 3.80 | 1.02 | 9.00 | 4.29 | 2.72 | 20.83 |
| Medium | 320 | 250 | 20.50 | 5.16 | 1.74 | 8.20 | 3.70 | 2.07 | 20.87 |
| Fast | 370 | 240 | 17.60 | 5.95 | 2.40 | 7.33 | 2.90 | 1.12 | 19.70 |

| | Labor SF per manhour | Material coverage SF/gallon | Material cost per gallon | Labor cost per 100 SF | Labor burden 100 SF | Material cost per 100 SF | Overhead per 100 SF | Profit per 100 SF | Total price per 100 SF |
|---|---|---|---|---|---|---|---|---|---|
| Penetrating oil stain (material #13) | | | | | | | | | |
| Roll 1st coat | | | | | | | | | |
| Slow | 210 | 200 | 25.00 | 4.88 | 1.33 | 12.50 | 5.80 | 3.68 | 28.19 |
| Medium | 258 | 190 | 21.80 | 6.40 | 2.16 | 11.47 | 4.90 | 2.74 | 27.67 |
| Fast | 305 | 180 | 18.70 | 7.21 | 2.93 | 10.39 | 3.80 | 1.46 | 25.79 |
| Roll 2nd or additional coats | | | | | | | | | |
| Slow | 250 | 295 | 25.00 | 4.10 | 1.12 | 8.47 | 4.24 | 2.69 | 20.62 |
| Medium | 300 | 285 | 21.80 | 5.50 | 1.84 | 7.65 | 3.68 | 2.05 | 20.72 |
| Fast | 350 | 275 | 18.70 | 6.29 | 2.55 | 6.80 | 2.89 | 1.11 | 19.64 |

Coverage figures are based on shingles or shakes with average moisture content. See the paragraphs on Roof pitch difficulty factors and Roof area conversion factors to adjust for roof slope and type. "Slow" work is based on a $10.25 hourly wage, "Medium" work on a $16.50 hourly wage, and "Fast" work on a $22.50 hourly wage. Other qualifications that apply to this table are on page 9.

| | Labor SF per manhour | Material coverage SF/gallon | Material cost per gallon | Labor cost per 100 SF | Labor burden 100 SF | Material cost per 100 SF | Overhead per 100 SF | Profit per 100 SF | Total price per 100 SF |
|---|---|---|---|---|---|---|---|---|---|

## Roofing, wood shingles or shakes, spray application

| | Labor SF per manhour | Material coverage SF/gallon | Material cost per gallon | Labor cost per 100 SF | Labor burden 100 SF | Material cost per 100 SF | Overhead per 100 SF | Profit per 100 SF | Total price per 100 SF |
|---|---|---|---|---|---|---|---|---|---|
| **Solid body stain, water base (material #18)** | | | | | | | | | |
| Spray 1st coat | | | | | | | | | |
| Slow | 600 | 230 | 19.40 | 1.71 | .47 | 8.43 | 3.29 | 2.09 | 15.99 |
| Medium | 700 | 220 | 17.00 | 2.36 | .79 | 7.73 | 2.67 | 1.49 | 15.04 |
| Fast | 800 | 200 | 14.50 | 2.75 | 1.12 | 7.25 | 2.06 | .79 | 13.97 |
| Spray 2nd or additional coats | | | | | | | | | |
| Slow | 700 | 250 | 19.40 | 1.46 | .40 | 7.76 | 2.98 | 1.89 | 14.49 |
| Medium | 800 | 235 | 17.00 | 2.06 | .70 | 7.23 | 2.45 | 1.37 | 13.81 |
| Fast | 900 | 220 | 14.50 | 2.44 | .99 | 6.59 | 1.85 | .71 | 12.58 |
| **Solid body stain, oil base (material #19)** | | | | | | | | | |
| Spray 1st coat | | | | | | | | | |
| Slow | 600 | 170 | 23.10 | 1.71 | .47 | 13.59 | 4.89 | 3.10 | 23.76 |
| Medium | 700 | 150 | 20.20 | 2.36 | .79 | 13.47 | 4.07 | 2.28 | 22.97 |
| Fast | 800 | 130 | 17.30 | 2.75 | 1.12 | 13.31 | 3.18 | 1.22 | 21.58 |
| Spray 2nd or additional coats | | | | | | | | | |
| Slow | 700 | 230 | 23.10 | 1.46 | .40 | 10.04 | 3.69 | 2.34 | 17.93 |
| Medium | 800 | 215 | 20.20 | 2.06 | .70 | 9.40 | 2.98 | 1.66 | 16.80 |
| Fast | 900 | 200 | 17.30 | 2.44 | .99 | 8.65 | 2.23 | .86 | 15.17 |
| **Semi-transparent stain, water base (material #20)** | | | | | | | | | |
| Spray 1st coat | | | | | | | | | |
| Slow | 650 | 265 | 19.60 | 1.58 | .43 | 7.40 | 2.92 | 1.85 | 14.18 |
| Medium | 750 | 253 | 17.10 | 2.20 | .73 | 6.76 | 2.38 | 1.33 | 13.40 |
| Fast | 850 | 240 | 14.70 | 2.59 | 1.06 | 6.13 | 1.81 | .69 | 12.28 |
| Spray 2nd or additional coats | | | | | | | | | |
| Slow | 750 | 305 | 19.60 | 1.37 | .36 | 6.43 | 2.53 | 1.61 | 12.30 |
| Medium | 850 | 293 | 17.10 | 1.94 | .66 | 5.84 | 2.07 | 1.16 | 11.67 |
| Fast | 950 | 270 | 14.70 | 2.32 | .93 | 5.44 | 1.61 | .62 | 10.92 |
| **Semi-transparent stain, oil base (material #21)** | | | | | | | | | |
| Spray 1st coat | | | | | | | | | |
| Slow | 650 | 190 | 23.40 | 1.58 | .43 | 12.32 | 4.44 | 2.82 | 21.59 |
| Medium | 750 | 180 | 20.50 | 2.20 | .73 | 11.39 | 3.51 | 1.96 | 19.79 |
| Fast | 850 | 170 | 17.60 | 2.59 | 1.06 | 10.35 | 2.59 | .99 | 17.58 |
| Spray 2nd or additional coats | | | | | | | | | |
| Slow | 750 | 275 | 23.40 | 1.37 | .36 | 8.51 | 3.18 | 2.01 | 15.43 |
| Medium | 850 | 265 | 20.50 | 1.94 | .66 | 7.74 | 2.53 | 1.41 | 14.28 |
| Fast | 950 | 255 | 17.60 | 2.32 | .93 | 6.90 | 1.88 | .72 | 12.75 |

| | Labor SF per manhour | Material coverage SF/gallon | Material cost per gallon | Labor cost per 100 SF | Labor burden 100 SF | Material cost per 100 SF | Overhead per 100 SF | Profit per 100 SF | Total price per 100 SF |
|---|---|---|---|---|---|---|---|---|---|
| Penetrating oil stain (material #13) | | | | | | | | | |
| Spray 1st coat | | | | | | | | | |
| Slow | 600 | 200 | 25.00 | 1.71 | .47 | 12.50 | 4.55 | 2.88 | 22.11 |
| Medium | 700 | 170 | 21.80 | 2.36 | .79 | 12.82 | 3.91 | 2.19 | 22.07 |
| Fast | 800 | 140 | 18.70 | 2.75 | 1.12 | 13.36 | 3.19 | 1.23 | 21.65 |
| Spray 2nd or additional coats | | | | | | | | | |
| Slow | 700 | 300 | 25.00 | 1.46 | .40 | 8.33 | 3.16 | 2.00 | 15.35 |
| Medium | 800 | 260 | 21.80 | 2.06 | .70 | 8.38 | 2.73 | 1.52 | 15.39 |
| Fast | 900 | 240 | 18.70 | 2.44 | .99 | 7.79 | 2.08 | .80 | 14.10 |
| Waterproofing, clear hydro sealer (material #34) | | | | | | | | | |
| Spray 1st coat | | | | | | | | | |
| Slow | 450 | 75 | 18.00 | 2.28 | .61 | 24.00 | 8.34 | 5.29 | 40.52 |
| Medium | 475 | 63 | 15.80 | 3.47 | 1.18 | 25.08 | 7.28 | 4.07 | 41.08 |
| Fast | 500 | 50 | 13.50 | 4.40 | 1.78 | 27.00 | 6.14 | 2.36 | 41.68 |
| Spray 2nd or additional coats | | | | | | | | | |
| Slow | 500 | 150 | 18.00 | 2.05 | .56 | 12.00 | 4.53 | 2.87 | 22.01 |
| Medium | 525 | 138 | 15.80 | 3.14 | 1.05 | 11.45 | 3.83 | 2.14 | 21.61 |
| Fast | 550 | 125 | 13.50 | 4.00 | 1.63 | 10.80 | 3.04 | 1.17 | 20.64 |

Coverage figures are based on shingles or shakes with average moisture content. See the paragraphs on Roof pitch difficulty factors and Roof area conversion factors to adjust for roof slope and type. "Slow" work is based on a $10.25 hourly wage, "Medium" work on a $16.50 hourly wage, and "Fast" work on a $22.50 hourly wage. Other qualifications that apply to this table are on page 9.

| | Labor LF per manhour | Material coverage LF/gallon | Material cost per gallon | Labor cost per 100 LF | Labor burden 100 LF | Material cost per 100 LF | Overhead per 100 LF | Profit per 100 LF | Total price per 100 LF |
|---|---|---|---|---|---|---|---|---|---|
| **Sheet metal cap or flashing** | | | | | | | | | |
| 3" to 8" wide: | | | | | | | | | |
| Metal primer, clean metal (material #35) | | | | | | | | | |
| Brush prime coat | | | | | | | | | |
| Slow | 215 | 525 | 28.20 | 4.77 | 1.29 | 5.37 | 3.55 | 2.25 | 17.23 |
| Medium | 228 | 500 | 24.60 | 7.24 | 2.44 | 4.92 | 3.58 | 2.00 | 20.18 |
| Fast | 240 | 475 | 21.10 | 9.17 | 3.72 | 4.44 | 3.21 | 1.23 | 21.77 |
| Metal primer, rusty metal (material #36) | | | | | | | | | |
| Brush prime coat | | | | | | | | | |
| Slow | 215 | 525 | 33.30 | 4.77 | 1.29 | 6.34 | 3.85 | 2.44 | 18.69 |
| Medium | 228 | 500 | 29.20 | 7.24 | 2.44 | 5.84 | 3.80 | 2.13 | 21.45 |
| Fast | 240 | 475 | 25.00 | 9.17 | 3.72 | 5.26 | 3.36 | 1.29 | 22.80 |
| Metal finish, synthetic enamel - off white, gloss, interior or exterior (material #37) | | | | | | | | | |
| Brush 1st or additional finish coats | | | | | | | | | |
| Slow | 275 | 550 | 25.90 | 3.73 | 1.02 | 4.71 | 2.93 | 1.86 | 14.25 |
| Medium | 288 | 525 | 22.70 | 5.73 | 1.92 | 4.32 | 2.94 | 1.64 | 16.55 |
| Fast | 300 | 500 | 19.50 | 7.33 | 2.97 | 3.90 | 2.63 | 1.01 | 17.84 |
| Metal finish, synthetic enamel, colors (except orange/red), gloss, interior or exterior (material #38) | | | | | | | | | |
| Brush 1st or additional finish coats | | | | | | | | | |
| Slow | 275 | 550 | 35.00 | 3.73 | 1.02 | 6.36 | 3.44 | 2.18 | 16.73 |
| Medium | 288 | 525 | 30.70 | 5.73 | 1.92 | 5.85 | 3.31 | 1.85 | 18.66 |
| Fast | 300 | 500 | 26.30 | 7.33 | 2.97 | 5.26 | 2.88 | 1.11 | 19.55 |
| 8" to 12" wide: | | | | | | | | | |
| Metal primer, clean metal (material #35) | | | | | | | | | |
| Brush prime coat | | | | | | | | | |
| Slow | 200 | 500 | 28.20 | 5.13 | 1.39 | 5.64 | 3.77 | 2.39 | 18.32 |
| Medium | 215 | 475 | 24.60 | 7.67 | 2.58 | 5.18 | 3.78 | 2.11 | 21.32 |
| Fast | 230 | 450 | 21.10 | 9.57 | 3.88 | 4.69 | 3.36 | 1.29 | 22.79 |
| Metal primer, rusty metal (material #36) | | | | | | | | | |
| Brush prime coat | | | | | | | | | |
| Slow | 200 | 500 | 33.30 | 5.13 | 1.39 | 6.66 | 4.09 | 2.59 | 19.86 |
| Medium | 215 | 475 | 29.20 | 7.67 | 2.58 | 6.15 | 4.02 | 2.25 | 22.67 |
| Fast | 230 | 450 | 25.00 | 9.57 | 3.88 | 5.56 | 3.52 | 1.35 | 23.88 |

| | Labor LF per manhour | Material coverage LF/gallon | Material cost per gallon | Labor cost per 100 LF | Labor burden 100 LF | Material cost per 100 LF | Overhead per 100 LF | Profit per 100 LF | Total price per 100 LF |
|---|---|---|---|---|---|---|---|---|---|
| **Metal finish, synthetic enamel - off white, gloss, interior or exterior (material #37)** | | | | | | | | | |
| Brush 1st or additional finish coats | | | | | | | | | |
| Slow | 250 | 525 | 25.90 | 4.10 | 1.12 | 4.93 | 3.15 | 2.00 | 15.30 |
| Medium | 265 | 500 | 22.70 | 6.23 | 2.08 | 4.54 | 3.15 | 1.76 | 17.76 |
| Fast | 280 | 475 | 19.50 | 7.86 | 3.18 | 4.11 | 2.80 | 1.08 | 19.03 |
| **Metal finish, synthetic enamel, colors (except orange/red), gloss, interior or exterior (material #38)** | | | | | | | | | |
| Brush 1st or additional finish coats | | | | | | | | | |
| Slow | 250 | 525 | 35.00 | 4.10 | 1.12 | 6.67 | 3.69 | 2.34 | 17.92 |
| Medium | 265 | 500 | 30.70 | 6.23 | 2.08 | 6.14 | 3.55 | 1.98 | 19.98 |
| Fast | 280 | 475 | 26.30 | 7.86 | 3.18 | 5.54 | 3.07 | 1.18 | 20.83 |

A two coat system, prime and finish, using oil base material is recommended for any metal surface. Although water base material is often used, it may cause oxidation, corrosion and rust. One coat of oil base solid body stain is often used on exterior metal but it may crack, peel or chip without the proper prime coat application. "Slow" work is based on a $10.25 hourly wage, "Medium" work on a $16.50 hourly wage, and "Fast" work on a $22.00 hourly wage. Other qualifications that apply to this table are on page 9.

| | Labor LF per manhour | Material coverage LF/gallon | Material cost per gallon | Labor cost per 100 LF | Labor burden 100 LF | Material cost per 100 LF | Overhead per 100 LF | Profit per 100 LF | Total price per 100 LF |
|---|---|---|---|---|---|---|---|---|---|
| **Sheet metal, diverters or gravel stop** | | | | | | | | | |
| Up to 3" wide: | | | | | | | | | |
| Metal primer, clean metal (material #35) | | | | | | | | | |
| Brush prime coat | | | | | | | | | |
| Slow | 230 | 600 | 28.20 | 4.46 | 1.21 | 4.70 | 3.22 | 2.04 | 15.63 |
| Medium | 240 | 550 | 24.60 | 6.88 | 2.31 | 4.47 | 3.35 | 1.87 | 18.88 |
| Fast | 250 | 500 | 21.10 | 8.80 | 3.57 | 4.22 | 3.07 | 1.18 | 20.84 |
| Metal primer, rusty metal (material #36) | | | | | | | | | |
| Brush prime coat | | | | | | | | | |
| Slow | 230 | 600 | 33.30 | 4.46 | 1.21 | 5.55 | 3.48 | 2.21 | 16.91 |
| Medium | 240 | 550 | 29.20 | 6.88 | 2.31 | 5.31 | 3.55 | 1.99 | 20.04 |
| Fast | 250 | 500 | 25.00 | 8.80 | 3.57 | 5.00 | 3.21 | 1.23 | 21.81 |
| Metal finish, synthetic enamel - off white, gloss, interior or exterior (material #37) | | | | | | | | | |
| Brush 1st or additional finish coats | | | | | | | | | |
| Slow | 300 | 650 | 25.90 | 3.42 | .92 | 3.98 | 2.58 | 1.64 | 12.54 |
| Medium | 313 | 600 | 22.70 | 5.27 | 1.76 | 3.78 | 2.65 | 1.48 | 14.94 |
| Fast | 325 | 550 | 19.50 | 6.77 | 2.75 | 3.55 | 2.42 | .93 | 16.42 |
| Metal finish, synthetic enamel, colors (except orange/red), gloss, interior or exterior (material #38) | | | | | | | | | |
| Brush 1st or additional finish coats | | | | | | | | | |
| Slow | 300 | 650 | 35.00 | 3.42 | .92 | 5.38 | 3.02 | 1.91 | 14.65 |
| Medium | 313 | 600 | 30.70 | 5.27 | 1.76 | 5.12 | 2.98 | 1.67 | 16.80 |
| Fast | 325 | 550 | 26.30 | 6.77 | 2.75 | 4.78 | 2.65 | 1.02 | 17.97 |

A two coat system, prime and finish, using oil base material is recommended for any metal surface. Although water base material is often used, it may cause oxidation, corrosion and rust. One coat of oil base solid body stain is often used on exterior metal but it may crack, peel or chip without the proper prime coat application. "Slow" work is based on a $10.25 hourly wage, "Medium" work on a $16.50 hourly wage, and "Fast" work on a $22.00 hourly wage. Other qualifications that apply to this table are on page 9.

| | Manhours per home | Material gallons per home | Material cost per gallon | Labor cost per home | Labor burden per home | Material cost per home | Overhead per home | Profit per home | Total price per home |
|---|---|---|---|---|---|---|---|---|---|
| **Sheet metal, vents & flashing** | | | | | | | | | |
| Metal primer, clean metal (material #35) | | | | | | | | | |
| 1 story home, brush prime coat | | | | | | | | | |
| Slow | 0.7 | 0.40 | 28.20 | 7.18 | 1.95 | 11.28 | 6.33 | 4.01 | 30.75 |
| Medium | 0.6 | 0.45 | 24.60 | 9.90 | 3.33 | 11.07 | 5.95 | 3.33 | 33.58 |
| Fast | 0.5 | 0.50 | 21.10 | 11.00 | 4.46 | 10.55 | 4.81 | 1.85 | 32.67 |
| 2 story home, brush prime coat | | | | | | | | | |
| Slow | 0.9 | 0.50 | 28.20 | 9.23 | 2.51 | 14.10 | 8.01 | 5.08 | 38.93 |
| Medium | 0.8 | 0.55 | 24.60 | 13.20 | 4.44 | 13.53 | 7.64 | 4.27 | 43.08 |
| Fast | 0.7 | 0.60 | 21.10 | 15.40 | 6.24 | 12.66 | 6.35 | 2.44 | 43.09 |
| Attached units (per unit), brush 1st or additional finish coats | | | | | | | | | |
| Slow | 0.6 | 0.40 | 28.20 | 6.15 | 1.67 | 11.28 | 5.92 | 3.75 | 28.77 |
| Medium | 0.5 | 0.45 | 24.60 | 8.25 | 2.78 | 11.07 | 5.41 | 3.03 | 30.54 |
| Fast | 0.4 | 0.50 | 21.10 | 8.80 | 3.57 | 10.55 | 4.24 | 1.63 | 28.79 |
| Metal primer, rusty metal (material #36) | | | | | | | | | |
| 1 story home, brush prime coat | | | | | | | | | |
| Slow | 0.7 | 0.40 | 33.30 | 7.18 | 1.95 | 13.32 | 6.96 | 4.41 | 33.82 |
| Medium | 0.6 | 0.45 | 29.20 | 9.90 | 3.33 | 13.14 | 6.46 | 3.61 | 36.44 |
| Fast | 0.5 | 0.50 | 25.00 | 11.00 | 4.46 | 12.50 | 5.17 | 1.99 | 35.12 |
| 2 story home, brush prime coat | | | | | | | | | |
| Slow | 0.9 | 0.50 | 33.30 | 9.23 | 2.51 | 16.65 | 8.80 | 5.58 | 42.77 |
| Medium | 0.8 | 0.55 | 29.20 | 13.20 | 4.44 | 16.06 | 8.26 | 4.62 | 46.58 |
| Fast | 0.7 | 0.60 | 25.00 | 15.40 | 6.24 | 15.00 | 6.78 | 2.61 | 46.03 |
| Attached units (per unit), brush 1st or additional finish coats | | | | | | | | | |
| Slow | 0.6 | 0.40 | 33.30 | 6.15 | 1.67 | 13.32 | 6.56 | 4.16 | 31.86 |
| Medium | 0.5 | 0.45 | 29.20 | 8.25 | 2.78 | 13.14 | 5.92 | 3.31 | 33.40 |
| Fast | 0.4 | 0.50 | 25.00 | 8.80 | 3.57 | 12.50 | 4.60 | 1.77 | 31.24 |
| Metal finish, synthetic enamel - off white, gloss, interior or exterior (material #37) | | | | | | | | | |
| 1 story home, brush 1st or additional finish coats | | | | | | | | | |
| Slow | 0.7 | 0.40 | 25.90 | 7.18 | 1.95 | 10.36 | 6.05 | 3.83 | 29.37 |
| Medium | 0.6 | 0.45 | 22.70 | 9.90 | 3.33 | 10.22 | 5.75 | 3.21 | 32.41 |
| Fast | 0.5 | 0.50 | 19.50 | 11.00 | 4.46 | 9.75 | 4.66 | 1.79 | 31.66 |
| 2 story home, brush 1st or additional finish coats | | | | | | | | | |
| Slow | 0.9 | 0.50 | 25.90 | 9.23 | 2.51 | 12.95 | 7.66 | 4.85 | 37.20 |
| Medium | 0.8 | 0.55 | 22.70 | 13.20 | 4.44 | 12.49 | 7.38 | 4.13 | 41.64 |
| Fast | 0.7 | 0.60 | 19.50 | 15.40 | 6.24 | 11.70 | 6.17 | 2.37 | 41.88 |

| | Manhours per home | Material gallons per home | Material cost per gallon | Labor cost per home | Labor burden per home | Material cost per home | Overhead per home | Profit per home | Total price per home |
|---|---|---|---|---|---|---|---|---|---|
| Attached units (per unit), brush 1st or additional finish coats | | | | | | | | | |
| Slow | 0.6 | 0.40 | 25.90 | 6.15 | 1.67 | 10.36 | 5.64 | 3.57 | 27.39 |
| Medium | 0.5 | 0.45 | 22.70 | 8.25 | 2.78 | 10.22 | 5.21 | 2.91 | 29.37 |
| Fast | 0.4 | 0.50 | 19.50 | 8.80 | 3.57 | 9.75 | 4.09 | 1.57 | 27.78 |
| Metal finish, synthetic enamel - colors (except orange/red), gloss, interior or exterior (material #38) | | | | | | | | | |
| 1 story home, brush 1st or additional finish coats | | | | | | | | | |
| Slow | 0.7 | 0.40 | 35.00 | 7.18 | 1.95 | 14.00 | 7.17 | 4.55 | 34.85 |
| Medium | 0.6 | 0.45 | 30.70 | 9.90 | 3.33 | 13.82 | 6.63 | 3.70 | 37.38 |
| Fast | 0.5 | 0.50 | 26.30 | 11.00 | 4.46 | 13.15 | 5.29 | 2.03 | 35.93 |
| 2 story home, brush 1st or additional finish coats | | | | | | | | | |
| Slow | 0.9 | 0.50 | 35.00 | 9.23 | 2.51 | 17.50 | 9.07 | 5.75 | 44.06 |
| Medium | 0.8 | 0.55 | 30.70 | 13.20 | 4.44 | 16.89 | 8.46 | 4.73 | 47.72 |
| Fast | 0.7 | 0.60 | 26.30 | 15.40 | 6.24 | 15.78 | 6.92 | 2.66 | 47.00 |
| Attached units (per unit), brush 1st or additional finish coats | | | | | | | | | |
| Slow | 0.6 | 0.40 | 35.00 | 6.15 | 1.67 | 14.00 | 6.77 | 4.29 | 32.88 |
| Medium | 0.5 | 0.45 | 30.70 | 8.25 | 2.78 | 13.82 | 6.09 | 3.40 | 34.34 |
| Fast | 0.4 | 0.50 | 26.30 | 8.80 | 3.57 | 13.15 | 4.72 | 1.81 | 32.05 |

Use this table for oil base paint only. This table shows the time needed to paint sheet metal vents on an average home or attached dwelling unit. Use it to estimate the costs for residential units without having to take off each vent or piece of flashing. Use the "attached units" section based on 900 square feet of roof area to estimate commercial buildings. Note: A two coat system, prime and finish, using oil base material is recommended for any metal surface. Although water base material is often used, it may cause oxidation, corrosion and rust. One coat of oil base solid body stain is often used on exterior metal but it may crack, peel or chip without the proper prime coat application. "Slow" work is based on a $10.25 hourly wage, "Medium" work on a $16.50 hourly wage, and "Fast" work on a $22.00 hourly wage. Other qualifications that apply to this table are on page 9.

| | Shutters per manhour | Shutters per gallon | Material cost per gallon | Labor cost per shutter | Labor burden per shutter | Material cost per shutter | Overhead per shutter | Profit per shutter | Total price per shutter |
|---|---|---|---|---|---|---|---|---|---|
| **Shutters or blinds, 2' x 4' average size** | | | | | | | | | |
| Brush each coat | | | | | | | | | |
| Undercoat, water or oil base (material #3 or #4) | | | | | | | | | |
| Slow | 2.5 | 12 | 19.35 | 4.10 | 1.12 | 1.61 | 2.12 | 1.34 | 10.29 |
| Medium | 3.0 | 11 | 16.90 | 5.50 | 1.84 | 1.54 | 2.18 | 1.22 | 12.28 |
| Fast | 3.5 | 10 | 14.50 | 6.29 | 2.55 | 1.45 | 1.90 | .73 | 12.92 |
| Split coat (1/2 undercoat + 1/2 enamel), water or oil base (material #3 and #24, or material #4 and #25) | | | | | | | | | |
| Slow | 3.0 | 15 | 22.73 | 3.42 | .92 | 1.52 | 1.82 | 1.15 | 8.83 |
| Medium | 3.5 | 14 | 19.85 | 4.71 | 1.60 | 1.42 | 1.89 | 1.06 | 10.68 |
| Fast | 4.0 | 13 | 17.03 | 5.50 | 2.23 | 1.31 | 1.67 | .64 | 11.35 |
| Exterior enamel, water or oil base (material #24 or #25) | | | | | | | | | |
| Slow | 2.0 | 15 | 26.10 | 5.13 | 1.39 | 1.74 | 2.56 | 1.62 | 12.44 |
| Medium | 2.5 | 14 | 22.80 | 6.60 | 2.22 | 1.63 | 2.56 | 1.43 | 14.44 |
| Fast | 3.0 | 13 | 19.55 | 7.33 | 2.97 | 1.50 | 2.18 | .84 | 14.82 |
| Spray each coat | | | | | | | | | |
| Undercoat, water or oil base (material #3 or #4) | | | | | | | | | |
| Slow | 8 | 10 | 19.35 | 1.28 | .35 | 1.94 | 1.11 | .70 | 5.38 |
| Medium | 9 | 9 | 16.90 | 1.83 | .62 | 1.88 | 1.06 | .59 | 5.98 |
| Fast | 10 | 8 | 14.50 | 2.20 | .89 | 1.81 | .91 | .35 | 6.16 |
| Split coat (1/2 undercoat + 1/2 enamel), water or oil base (material #3 and #24, or material #4 and #25) | | | | | | | | | |
| Slow | 8 | 12 | 22.73 | 1.28 | .35 | 1.89 | 1.09 | .69 | 5.30 |
| Medium | 10 | 11 | 19.85 | 1.65 | .56 | 1.80 | .98 | .55 | 5.54 |
| Fast | 12 | 10 | 17.03 | 1.83 | .74 | 1.70 | .79 | .30 | 5.36 |
| Exterior enamel, water or oil base (material #24 or #25) | | | | | | | | | |
| Slow | 7 | 12 | 26.10 | 1.46 | .40 | 2.18 | 1.25 | .79 | 6.08 |
| Medium | 8 | 11 | 22.80 | 2.06 | .70 | 2.07 | 1.18 | .66 | 6.67 |
| Fast | 9 | 10 | 19.55 | 2.44 | .99 | 1.96 | 1.00 | .38 | 6.77 |

Use these figures to estimate the costs to paint all six sides of solid face, paint grade, interior or exterior shutters or blinds. Costs are based on the number of single-panel 2' x 4' false (solid) shutters or blinds that can be painted in one hour. For real louvered shutters, multiply the quantity of shutters by a difficulty factor of 1.5 and use this table. "Slow" work is based on a $10.25 hourly wage, "Medium" work on a $16.50 hourly wage, and "Fast" work on a $22.00 hourly wage. Other qualifications that apply to this table are on page 9.

| | Labor SF per manhour | Material coverage SF/gallon | Material cost per gallon | Labor cost per 100 SF | Labor burden 100 SF | Material cost per 100 SF | Overhead per 100 SF | Profit per 100 SF | Total price per 100 SF |
|---|---|---|---|---|---|---|---|---|---|
| **Siding, aluminum** | | | | | | | | | |
| Metal primer, clean metal (material #35) | | | | | | | | | |
| Brush prime coat | | | | | | | | | |
| Slow | 215 | 440 | 28.20 | 4.77 | 1.29 | 6.41 | 3.87 | 2.45 | 18.79 |
| Medium | 235 | 420 | 24.60 | 7.02 | 2.37 | 5.86 | 3.73 | 2.09 | 21.07 |
| Fast | 255 | 400 | 21.10 | 8.63 | 3.49 | 5.28 | 3.22 | 1.24 | 21.86 |
| Metal primer, rusty metal (material #36) | | | | | | | | | |
| Brush prime coat | | | | | | | | | |
| Slow | 215 | 440 | 33.30 | 4.77 | 1.29 | 7.57 | 4.23 | 2.68 | 20.54 |
| Medium | 235 | 420 | 29.20 | 7.02 | 2.37 | 6.95 | 4.00 | 2.24 | 22.58 |
| Fast | 255 | 400 | 25.00 | 8.63 | 3.49 | 6.25 | 3.40 | 1.31 | 23.08 |
| Metal finish, synthetic enamel - off white, gloss, interior or exterior (material #37) | | | | | | | | | |
| Brush 1st or additional finish coats | | | | | | | | | |
| Slow | 265 | 480 | 25.90 | 3.87 | 1.05 | 5.40 | 3.20 | 2.03 | 15.55 |
| Medium | 285 | 465 | 22.70 | 5.79 | 1.95 | 4.88 | 3.09 | 1.73 | 17.44 |
| Fast | 305 | 450 | 19.50 | 7.21 | 2.93 | 4.33 | 2.68 | 1.03 | 18.18 |
| Metal finish, synthetic enamel - colors (except orange/red), gloss, interior or exterior (material #38) | | | | | | | | | |
| Brush 1st or additional finish coats | | | | | | | | | |
| Slow | 265 | 480 | 35.00 | 3.87 | 1.05 | 7.29 | 3.79 | 2.40 | 18.40 |
| Medium | 285 | 465 | 30.70 | 5.79 | 1.95 | 6.60 | 3.51 | 1.96 | 19.81 |
| Fast | 305 | 450 | 26.30 | 7.21 | 2.93 | 5.84 | 2.95 | 1.14 | 20.07 |

Don't deduct for openings under 100 square feet. For heights above 8 feet, use the High Time Difficulty Factors on page 137. Note: A two coat system prime and finish, using oil base material is recommended for any metal surface. Although water base material is often used, it may cause oxidation, corrosion and rust. One coat of oil base solid body stain is often used on exterior metal but it may crack, peel or chip without the proper prime coat application. "Slow" work is based on a $10.25 hourly wage, "Medium" work on a $16.75 hourly wage, and "Fast" work on a $22.00 hourly wage. Other qualifications that apply to this table are on page 9.

| | Labor SF per manhour | Material coverage SF/gallon | Material cost per gallon | Labor cost per 100 SF | Labor burden 100 SF | Material cost per 100 SF | Overhead per 100 SF | Profit per 100 SF | Total price per 100 SF |
|---|---|---|---|---|---|---|---|---|---|
| **Siding, composition shingle, brush application** | | | | | | | | | |
| Solid body stain, water base (material #18) | | | | | | | | | |
| Brush 1st coat | | | | | | | | | |
| Slow | 65 | 240 | 19.40 | 15.77 | 4.29 | 8.08 | 8.73 | 5.53 | 42.40 |
| Medium | 85 | 220 | 17.00 | 19.41 | 6.52 | 7.73 | 8.25 | 4.61 | 46.52 |
| Fast | 105 | 200 | 14.50 | 20.95 | 8.49 | 7.25 | 6.79 | 2.61 | 46.09 |
| Brush 2nd coat | | | | | | | | | |
| Slow | 90 | 350 | 19.40 | 11.39 | 3.10 | 5.54 | 6.21 | 3.94 | 30.18 |
| Medium | 110 | 335 | 17.00 | 15.00 | 5.04 | 5.07 | 6.15 | 3.44 | 34.70 |
| Fast | 130 | 320 | 14.50 | 16.92 | 6.86 | 4.53 | 5.24 | 2.01 | 35.56 |
| Brush 3rd or additional coats | | | | | | | | | |
| Slow | 110 | 425 | 19.40 | 9.32 | 2.53 | 4.56 | 5.09 | 3.23 | 24.73 |
| Medium | 130 | 405 | 17.00 | 12.69 | 4.27 | 4.20 | 5.18 | 2.90 | 29.24 |
| Fast | 150 | 385 | 14.50 | 14.67 | 5.95 | 3.77 | 4.51 | 1.73 | 30.63 |
| Solid body stain, oil base (material #19) | | | | | | | | | |
| Brush 1st coat | | | | | | | | | |
| Slow | 65 | 290 | 23.10 | 15.77 | 4.29 | 7.97 | 8.69 | 5.51 | 42.23 |
| Medium | 85 | 270 | 20.20 | 19.41 | 6.52 | 7.48 | 8.19 | 4.58 | 46.18 |
| Fast | 105 | 250 | 17.30 | 20.95 | 8.49 | 6.92 | 6.73 | 2.59 | 45.68 |
| Brush 2nd coat | | | | | | | | | |
| Slow | 90 | 390 | 23.10 | 11.39 | 3.10 | 5.92 | 6.33 | 4.01 | 30.75 |
| Medium | 110 | 375 | 20.20 | 15.00 | 5.04 | 5.39 | 6.23 | 3.48 | 35.14 |
| Fast | 130 | 360 | 17.30 | 16.92 | 6.86 | 4.81 | 5.29 | 2.03 | 35.91 |
| Brush 3rd or additional coats | | | | | | | | | |
| Slow | 110 | 445 | 23.10 | 9.32 | 2.53 | 5.19 | 5.29 | 3.35 | 25.68 |
| Medium | 130 | 425 | 20.20 | 12.69 | 4.27 | 4.75 | 5.32 | 2.97 | 30.00 |
| Fast | 150 | 405 | 17.30 | 14.67 | 5.95 | 4.27 | 4.60 | 1.77 | 31.26 |

Use these figures for repaint jobs only. Don't deduct for openings under 100 square feet. For heights above 8 feet, use the High Time Difficulty Factors on page 137. It has been established that asbestos fibers are known carcinogens (cancer causing) and it is likely that materials or products which contain asbestos will not be specified on any new construction projects in the future. Siding and roofing products usually contain very little asbestos and are typically non-friable (hand pressure can not crumble, pulverize or reduce to powder when dry). There is danger when asbestos is being removed because of exposure to airborne particulate matter. Apparently, there is little danger when painting asbestos siding or roofing, but it is a good idea to have your painters wear respirators or particle masks for their safety. "Slow" work is based on a $10.25 hourly wage, "Medium" work a $16.50 hourly wage, and "Fast" work on a $22.00 hourly wage. Other qualifications that apply to this table are on page 9.

| | Labor SF per manhour | Material coverage SF/gallon | Material cost per gallon | Labor cost per 100 SF | Labor burden 100 SF | Material cost per 100 SF | Overhead per 100 SF | Profit per 100 SF | Total price per 100 SF |
|---|---|---|---|---|---|---|---|---|---|
| **Siding, composition shingle, roll application** | | | | | | | | | |
| Solid body stain, water base (material #18) | | | | | | | | | |
| Roll 1st coat | | | | | | | | | |
| Slow | 140 | 230 | 19.40 | 7.32 | 1.99 | 8.43 | 5.50 | 3.49 | 26.73 |
| Medium | 160 | 215 | 17.00 | 10.31 | 3.47 | 7.91 | 5.31 | 2.97 | 29.97 |
| Fast | 180 | 200 | 14.50 | 12.22 | 4.97 | 7.25 | 4.52 | 1.74 | 30.70 |
| Roll 2nd coat | | | | | | | | | |
| Slow | 190 | 330 | 19.40 | 5.39 | 1.47 | 5.88 | 3.95 | 2.50 | 19.19 |
| Medium | 210 | 315 | 17.00 | 7.86 | 2.64 | 5.40 | 3.90 | 2.18 | 21.98 |
| Fast | 230 | 300 | 14.50 | 9.57 | 3.88 | 4.83 | 3.38 | 1.30 | 22.96 |
| Roll 3rd or additional coats | | | | | | | | | |
| Slow | 250 | 395 | 19.40 | 4.10 | 1.12 | 4.91 | 3.14 | 1.99 | 15.26 |
| Medium | 280 | 385 | 17.00 | 5.89 | 1.98 | 4.42 | 3.01 | 1.68 | 16.98 |
| Fast | 300 | 375 | 14.50 | 7.33 | 2.97 | 3.87 | 2.62 | 1.01 | 17.80 |
| Solid body stain, oil base (material #19) | | | | | | | | | |
| Roll 1st coat | | | | | | | | | |
| Slow | 140 | 260 | 23.10 | 7.32 | 1.99 | 8.88 | 5.64 | 3.57 | 27.40 |
| Medium | 160 | 240 | 20.20 | 10.31 | 3.47 | 8.42 | 5.44 | 3.04 | 30.68 |
| Fast | 180 | 220 | 17.30 | 12.22 | 4.97 | 7.86 | 4.63 | 1.78 | 31.46 |
| Roll 2nd coat | | | | | | | | | |
| Slow | 190 | 360 | 23.10 | 5.39 | 1.47 | 6.42 | 4.12 | 2.61 | 20.01 |
| Medium | 210 | 345 | 20.20 | 7.86 | 2.64 | 5.86 | 4.01 | 2.24 | 22.61 |
| Fast | 230 | 330 | 17.30 | 9.57 | 3.88 | 5.24 | 3.46 | 1.33 | 23.48 |
| Roll 3rd or additional coats | | | | | | | | | |
| Slow | 250 | 420 | 23.10 | 4.10 | 1.12 | 5.50 | 3.32 | 2.11 | 16.15 |
| Medium | 280 | 408 | 20.20 | 5.89 | 1.98 | 4.95 | 3.14 | 1.76 | 17.72 |
| Fast | 300 | 395 | 17.30 | 7.33 | 2.97 | 4.38 | 2.72 | 1.04 | 18.44 |

| | Labor SF per manhour | Material coverage SF/gallon | Material cost per gallon | Labor cost per 100 SF | Labor burden 100 SF | Material cost per 100 SF | Overhead per 100 SF | Profit per 100 SF | Total price per 100 SF |
|---|---|---|---|---|---|---|---|---|---|
| Waterproofing, clear hydro seal, oil base (material #34) | | | | | | | | | |
| Roll 1st coat | | | | | | | | | |
| Slow | 75 | 275 | 18.00 | 13.67 | 3.71 | 6.55 | 7.42 | 4.71 | 36.06 |
| Medium | 150 | 250 | 15.80 | 11.00 | 3.71 | 6.32 | 5.15 | 2.88 | 29.06 |
| Fast | 225 | 225 | 13.50 | 9.78 | 3.95 | 6.00 | 3.65 | 1.40 | 24.78 |
| Roll 2nd or additional coats | | | | | | | | | |
| Slow | 125 | 325 | 18.00 | 8.20 | 2.23 | 5.54 | 4.95 | 3.14 | 24.06 |
| Medium | 200 | 300 | 15.80 | 8.25 | 2.78 | 5.27 | 3.99 | 2.23 | 22.52 |
| Fast | 275 | 275 | 13.50 | 8.00 | 3.25 | 4.91 | 2.99 | 1.15 | 20.30 |

Use these figures for repaint jobs only. Don't deduct for openings under 100 square feet. For heights above 8 feet, use the High Time Difficulty Factors on page 137. It has been established that asbestos fibers are known carcinogens (cancer causing) and it is likely that materials or products which contain asbestos will not be specified on any new construction projects in the future. Siding and roofing products usually contain very little asbestos and are typically non-friable (hand pressure can not crumble, pulverize or reduce to powder when dry). There is danger when asbestos is being removed because of exposure to airborne particulate matter. Apparently, there is little danger when painting asbestos siding or roofing, but it is a good idea to have your painters wear respirators or particle masks for their safety. "Slow" work is based on a $10.25 hourly wage, "Medium" work a $16.50 hourly wage, and "Fast" work on a $22.00 hourly wage. Other qualifications that apply to this table are on page 9.

| | Labor SF per manhour | Material coverage SF/gallon | Material cost per gallon | Labor cost per 100 SF | Labor burden 100 SF | Material cost per 100 SF | Overhead per 100 SF | Profit per 100 SF | Total price per 100 SF |
|---|---|---|---|---|---|---|---|---|---|

## Siding, composition shingle, spray application

| | Labor SF per manhour | Material coverage SF/gallon | Material cost per gallon | Labor cost per 100 SF | Labor burden 100 SF | Material cost per 100 SF | Overhead per 100 SF | Profit per 100 SF | Total price per 100 SF |
|---|---|---|---|---|---|---|---|---|---|
| Solid body stain, water base (material #18) | | | | | | | | | |
| Spray 1st coat | | | | | | | | | |
| Slow | 325 | 160 | 19.40 | 3.15 | .87 | 12.13 | 5.00 | 3.17 | 24.32 |
| Medium | 350 | 150 | 17.00 | 4.71 | 1.60 | 11.33 | 4.32 | 2.41 | 24.37 |
| Fast | 375 | 140 | 14.50 | 5.87 | 2.39 | 10.36 | 3.44 | 1.32 | 23.38 |
| Spray 2nd coat | | | | | | | | | |
| Slow | 350 | 240 | 19.40 | 2.93 | .80 | 8.08 | 3.66 | 2.32 | 17.79 |
| Medium | 375 | 228 | 17.00 | 4.40 | 1.49 | 7.46 | 3.27 | 1.83 | 18.45 |
| Fast | 400 | 215 | 14.50 | 5.50 | 2.23 | 6.74 | 2.68 | 1.03 | 18.18 |
| Spray 3rd or additional coats | | | | | | | | | |
| Slow | 425 | 300 | 19.40 | 2.41 | .65 | 6.47 | 2.96 | 1.88 | 14.37 |
| Medium | 450 | 288 | 17.00 | 3.67 | 1.23 | 5.90 | 2.65 | 1.48 | 14.93 |
| Fast | 475 | 275 | 14.50 | 4.63 | 1.89 | 5.27 | 2.18 | .84 | 14.81 |
| Solid body stain, oil base (material #19) | | | | | | | | | |
| Spray 1st coat | | | | | | | | | |
| Slow | 325 | 190 | 23.10 | 3.15 | .87 | 12.16 | 5.01 | 3.18 | 24.37 |
| Medium | 350 | 180 | 20.20 | 4.71 | 1.60 | 11.22 | 4.29 | 2.40 | 24.22 |
| Fast | 375 | 170 | 17.30 | 5.87 | 2.39 | 10.18 | 3.41 | 1.31 | 23.16 |
| Spray 2nd coat | | | | | | | | | |
| Slow | 350 | 270 | 23.10 | 2.93 | .80 | 8.56 | 3.81 | 2.42 | 18.52 |
| Medium | 375 | 258 | 20.20 | 4.40 | 1.49 | 7.83 | 3.36 | 1.88 | 18.96 |
| Fast | 400 | 245 | 17.30 | 5.50 | 2.23 | 7.06 | 2.74 | 1.05 | 18.58 |
| Spray 3rd or additional coats | | | | | | | | | |
| Slow | 425 | 320 | 23.10 | 2.41 | .65 | 7.22 | 3.19 | 2.02 | 15.49 |
| Medium | 450 | 308 | 20.20 | 3.67 | 1.23 | 6.56 | 2.81 | 1.57 | 15.84 |
| Fast | 475 | 295 | 17.30 | 4.63 | 1.89 | 5.86 | 2.29 | .88 | 15.55 |

| | Labor SF per manhour | Material coverage SF/gallon | Material cost per gallon | Labor cost per 100 SF | Labor burden 100 SF | Material cost per 100 SF | Overhead per 100 SF | Profit per 100 SF | Total price per 100 SF |
|---|---|---|---|---|---|---|---|---|---|
| Waterproofing, clear hydro seal, oil base (material #34) | | | | | | | | | |
| Spray 1st coat | | | | | | | | | |
| Slow | 475 | 100 | 18.00 | 2.16 | .59 | 18.00 | 6.43 | 4.08 | 31.26 |
| Medium | 525 | 88 | 15.80 | 3.14 | 1.05 | 17.95 | 5.43 | 3.03 | 30.60 |
| Fast | 575 | 75 | 13.50 | 3.83 | 1.55 | 18.00 | 4.33 | 1.66 | 29.37 |
| Spray 2nd or additional coats | | | | | | | | | |
| Slow | 500 | 150 | 18.00 | 2.05 | .56 | 12.00 | 4.53 | 2.87 | 22.01 |
| Medium | 550 | 125 | 15.80 | 3.00 | 1.01 | 12.64 | 4.08 | 2.28 | 23.01 |
| Fast | 600 | 100 | 13.50 | 3.67 | 1.49 | 13.50 | 3.45 | 1.33 | 23.44 |

Use these figures for repaint jobs only. Don't deduct for openings under 100 square feet. For heights above 8 feet, use the High Time Difficulty Factors on page 137. It has been established that asbestos fibers are known carcinogens (cancer causing) and it is likely that materials or products which contain asbestos will not be specified on any new construction projects in the future. Siding and roofing products usually contain very little asbestos and are typically non-friable (hand pressure can not crumble, pulverize or reduce to powder when dry). There is danger when asbestos is being removed because of exposure to airborne particulate matter. Apparently, there is little danger when painting asbestos siding or roofing, but it is a good idea to have your painters wear respirators or particle masks for their safety. "Slow" work is based on a $10.25 hourly wage, "Medium" work a $16.50 hourly wage, and "Fast" work on a $22.00 hourly wage. Other qualifications that apply to this table are on page 9.

| | Labor SF per manhour | Material coverage SF/gallon | Material cost per gallon | Labor cost per 100 SF | Labor burden 100 SF | Material cost per 100 SF | Overhead per 100 SF | Profit per 100 SF | Total price per 100 SF |
|---|---|---|---|---|---|---|---|---|---|

## Siding, rough sawn or resawn wood, brush

Solid body or semi-transparent stain, water base (material #18 or #20)

| | | | | | | | | | |
|---|---|---|---|---|---|---|---|---|---|
| Brush 1st coat | | | | | | | | | |
| Slow | 100 | 250 | 19.50 | 10.25 | 2.79 | 7.80 | 6.46 | 4.10 | 31.40 |
| Medium | 135 | 238 | 17.05 | 12.22 | 4.12 | 7.16 | 5.76 | 3.22 | 32.48 |
| Fast | 170 | 225 | 14.60 | 12.94 | 5.24 | 6.49 | 4.57 | 1.76 | 31.00 |
| Brush 2nd coat | | | | | | | | | |
| Slow | 135 | 300 | 19.50 | 7.59 | 2.07 | 6.50 | 5.01 | 3.18 | 24.35 |
| Medium | 168 | 288 | 17.05 | 9.82 | 3.30 | 5.92 | 4.66 | 2.61 | 26.31 |
| Fast | 200 | 275 | 14.60 | 11.00 | 4.46 | 5.31 | 3.84 | 1.48 | 26.09 |
| Brush 3rd or additional coats | | | | | | | | | |
| Slow | 150 | 335 | 19.50 | 6.83 | 1.87 | 5.82 | 4.50 | 2.85 | 21.87 |
| Medium | 183 | 323 | 17.05 | 9.02 | 3.02 | 5.28 | 4.25 | 2.37 | 23.94 |
| Fast | 215 | 310 | 14.60 | 10.23 | 4.15 | 4.71 | 3.53 | 1.36 | 23.98 |

Solid body or semi-transparent stain, oil base (material #19 or #21)

| | | | | | | | | | |
|---|---|---|---|---|---|---|---|---|---|
| Brush 1st coat | | | | | | | | | |
| Slow | 100 | 275 | 23.25 | 10.25 | 2.79 | 8.45 | 6.66 | 4.22 | 32.37 |
| Medium | 135 | 250 | 20.35 | 12.22 | 4.12 | 8.14 | 6.00 | 3.35 | 33.83 |
| Fast | 170 | 225 | 17.45 | 12.94 | 5.24 | 7.76 | 4.80 | 1.85 | 32.59 |
| Brush 2nd coat | | | | | | | | | |
| Slow | 135 | 350 | 23.25 | 7.59 | 2.07 | 6.64 | 5.05 | 3.20 | 24.55 |
| Medium | 168 | 325 | 20.35 | 9.82 | 3.30 | 6.26 | 4.75 | 2.65 | 26.78 |
| Fast | 200 | 300 | 17.45 | 11.00 | 4.46 | 5.82 | 3.94 | 1.51 | 26.73 |
| Brush 3rd or additional coats | | | | | | | | | |
| Slow | 150 | 400 | 23.25 | 6.83 | 1.87 | 5.81 | 4.50 | 2.85 | 21.86 |
| Medium | 183 | 375 | 20.35 | 9.02 | 3.02 | 5.43 | 4.29 | 2.40 | 24.16 |
| Fast | 215 | 350 | 17.45 | 10.23 | 4.15 | 4.99 | 3.58 | 1.38 | 24.33 |

Penetrating oil stain (material #13)

| | | | | | | | | | |
|---|---|---|---|---|---|---|---|---|---|
| Brush 1st coat | | | | | | | | | |
| Slow | 100 | 230 | 25.00 | 10.25 | 2.79 | 10.87 | 7.41 | 4.70 | 36.02 |
| Medium | 135 | 210 | 21.80 | 12.22 | 4.12 | 10.38 | 6.54 | 3.66 | 36.92 |
| Fast | 170 | 190 | 18.70 | 12.94 | 5.24 | 9.84 | 5.19 | 1.99 | 35.20 |

| | Labor SF per manhour | Material coverage SF/gallon | Material cost per gallon | Labor cost per 100 SF | Labor burden 100 SF | Material cost per 100 SF | Overhead per 100 SF | Profit per 100 SF | Total price per 100 SF |
|---|---|---|---|---|---|---|---|---|---|
| Brush 2nd coat | | | | | | | | | |
| Slow | 135 | 275 | 25.00 | 7.59 | 2.07 | 9.09 | 5.81 | 3.68 | 28.24 |
| Medium | 168 | 255 | 21.80 | 9.82 | 3.30 | 8.55 | 5.31 | 2.97 | 29.95 |
| Fast | 200 | 235 | 18.70 | 11.00 | 4.46 | 7.96 | 4.33 | 1.67 | 29.42 |
| Brush 3rd or additional coats | | | | | | | | | |
| Slow | 150 | 350 | 25.00 | 6.83 | 1.87 | 7.14 | 4.91 | 3.11 | 23.86 |
| Medium | 183 | 330 | 21.80 | 9.02 | 3.02 | 6.61 | 4.57 | 2.56 | 25.78 |
| Fast | 215 | 310 | 18.70 | 10.23 | 4.15 | 6.03 | 3.78 | 1.45 | 25.64 |

Use this table to estimate the cost of painting shingle, shake, resawn or rough sawn wood or plywood siding with average moisture content. Don't deduct for openings under 100 square feet. For heights above 8 feet, use the High Time Difficulty Factors on page 137. For wood or composition drop siding with exposed bevel edges, multiply the surface area by 1.12 to allow for the extra time and material needed to paint the underside of each board. "Slow" work is based on a $10.25 hourly wage, "Medium" work on a $16.75 hourly wage, and "Fast" work on a $22.00 hourly wage. Other qualifications that apply to this table are on page 9.

| | Labor SF per manhour | Material coverage SF/gallon | Material cost per gallon | Labor cost per 100 SF | Labor burden 100 SF | Material cost per 100 SF | Overhead per 100 SF | Profit per 100 SF | Total price per 100 SF |
|---|---|---|---|---|---|---|---|---|---|
| **Siding, rough sawn or resawn wood, roll** | | | | | | | | | |
| Solid body or semi-transparent stain, water base (material #18 or #20) | | | | | | | | | |
| Roll 1st coat | | | | | | | | | |
| Slow | 150 | 235 | 19.50 | 6.83 | 1.87 | 8.30 | 5.27 | 3.34 | 25.61 |
| Medium | 225 | 213 | 17.05 | 7.33 | 2.46 | 8.00 | 4.36 | 2.44 | 24.59 |
| Fast | 275 | 210 | 14.60 | 8.00 | 3.25 | 6.95 | 3.37 | 1.29 | 22.86 |
| Roll 2nd coat | | | | | | | | | |
| Slow | 200 | 285 | 19.50 | 5.13 | 1.39 | 6.84 | 4.14 | 2.63 | 20.13 |
| Medium | 275 | 273 | 17.05 | 6.00 | 2.03 | 6.25 | 3.50 | 1.95 | 19.73 |
| Fast | 350 | 260 | 14.60 | 6.29 | 2.55 | 5.62 | 2.68 | 1.03 | 18.17 |
| Roll 3rd or additional coats | | | | | | | | | |
| Slow | 260 | 335 | 19.50 | 3.94 | 1.08 | 5.82 | 3.36 | 2.13 | 16.33 |
| Medium | 335 | 323 | 17.05 | 4.93 | 1.66 | 5.28 | 2.91 | 1.63 | 16.41 |
| Fast | 410 | 310 | 14.60 | 5.37 | 2.17 | 4.71 | 2.27 | .87 | 15.39 |
| Solid body or semi-transparent stain, oil base (material #19 or #21) | | | | | | | | | |
| Roll 1st coat | | | | | | | | | |
| Slow | 150 | 250 | 23.25 | 6.83 | 1.87 | 9.30 | 5.58 | 3.54 | 27.12 |
| Medium | 225 | 225 | 20.35 | 7.33 | 2.46 | 9.04 | 4.62 | 2.58 | 26.03 |
| Fast | 275 | 200 | 17.45 | 8.00 | 3.25 | 8.73 | 3.69 | 1.42 | 25.09 |
| Roll 2nd coat | | | | | | | | | |
| Slow | 200 | 330 | 23.25 | 5.13 | 1.39 | 7.05 | 4.21 | 2.67 | 20.45 |
| Medium | 275 | 305 | 20.35 | 6.00 | 2.03 | 6.67 | 3.60 | 2.01 | 20.31 |
| Fast | 350 | 280 | 17.45 | 6.29 | 2.55 | 6.23 | 2.79 | 1.07 | 18.93 |
| Roll 3rd or additional coats | | | | | | | | | |
| Slow | 260 | 415 | 23.25 | 3.94 | 1.08 | 5.60 | 3.29 | 2.09 | 16.00 |
| Medium | 335 | 390 | 20.35 | 4.93 | 1.66 | 5.22 | 2.89 | 1.62 | 16.32 |
| Fast | 410 | 365 | 17.45 | 5.37 | 2.17 | 4.78 | 2.28 | .88 | 15.48 |
| Penetrating oil stain (material #13) | | | | | | | | | |
| Roll 1st coat | | | | | | | | | |
| Slow | 150 | 150 | 25.00 | 6.83 | 1.87 | 16.67 | 7.86 | 4.98 | 38.21 |
| Medium | 225 | 125 | 21.80 | 7.33 | 2.46 | 17.44 | 6.67 | 3.73 | 37.63 |
| Fast | 275 | 100 | 18.70 | 8.00 | 3.25 | 18.70 | 5.54 | 2.13 | 37.62 |

| | Labor SF per manhour | Material coverage SF/gallon | Material cost per gallon | Labor cost per 100 SF | Labor burden 100 SF | Material cost per 100 SF | Overhead per 100 SF | Profit per 100 SF | Total price per 100 SF |
|---|---|---|---|---|---|---|---|---|---|
| Roll 2nd coat | | | | | | | | | |
| Slow | 200 | 280 | 25.00 | 5.13 | 1.39 | 8.93 | 4.79 | 3.04 | 23.28 |
| Medium | 275 | 240 | 21.80 | 6.00 | 2.03 | 9.08 | 4.19 | 2.34 | 23.64 |
| Fast | 350 | 200 | 18.70 | 6.29 | 2.55 | 9.35 | 3.37 | 1.29 | 22.85 |
| Roll 3rd or additional coats | | | | | | | | | |
| Slow | 260 | 365 | 25.00 | 3.94 | 1.08 | 6.85 | 3.68 | 2.33 | 17.88 |
| Medium | 335 | 330 | 21.80 | 4.93 | 1.66 | 6.61 | 3.23 | 1.81 | 18.24 |
| Fast | 410 | 295 | 18.70 | 5.37 | 2.17 | 6.34 | 2.57 | .99 | 17.44 |

Use this table to estimate the cost of painting shingle, shake, resawn or rough sawn wood or plywood siding with average moisture content. Don't deduct for openings under 100 square feet. For heights above 8 feet, use the High Time Difficulty Factors on page 137. For wood or composition drop siding with exposed bevel edges, multiply the surface area by 1.12 to allow for the extra time and material needed to paint the underside of each board. "Slow" work is based on a $10.25 hourly wage, "Medium" work on a $16.75 hourly wage, and "Fast" work on a $22.00 hourly wage. Other qualifications that apply to this table are on page 9.

| | Labor SF per manhour | Material coverage SF/gallon | Material cost per gallon | Labor cost per 100 SF | Labor burden 100 SF | Material cost per 100 SF | Overhead per 100 SF | Profit per 100 SF | Total price per 100 SF |
|---|---|---|---|---|---|---|---|---|---|

## Siding, rough sawn or resawn wood, spray

| | Labor SF per manhour | Material coverage SF/gallon | Material cost per gallon | Labor cost per 100 SF | Labor burden 100 SF | Material cost per 100 SF | Overhead per 100 SF | Profit per 100 SF | Total price per 100 SF |
|---|---|---|---|---|---|---|---|---|---|
| Solid body or semi-transparent stain, water base (material #18 or #20) | | | | | | | | | |
| Spray 1st coat | | | | | | | | | |
| Slow | 400 | 280 | 19.50 | 2.56 | .70 | 6.96 | 3.17 | 2.01 | 15.40 |
| Medium | 500 | 260 | 17.05 | 3.30 | 1.11 | 6.56 | 2.69 | 1.50 | 15.16 |
| Fast | 600 | 240 | 14.60 | 3.67 | 1.49 | 6.08 | 2.08 | .80 | 14.12 |
| Spray 2nd coat | | | | | | | | | |
| Slow | 450 | 330 | 19.50 | 2.28 | .61 | 5.91 | 2.73 | 1.73 | 13.26 |
| Medium | 550 | 310 | 17.05 | 3.00 | 1.01 | 5.50 | 2.33 | 1.30 | 13.14 |
| Fast | 650 | 290 | 14.60 | 3.38 | 1.38 | 5.03 | 1.81 | .70 | 12.30 |
| Spray 3rd or additional coats | | | | | | | | | |
| Slow | 550 | 380 | 19.50 | 1.86 | .51 | 5.13 | 2.33 | 1.47 | 11.30 |
| Medium | 650 | 360 | 17.05 | 2.54 | .86 | 4.74 | 1.99 | 1.11 | 11.24 |
| Fast | 750 | 340 | 14.60 | 2.93 | 1.18 | 4.29 | 1.56 | .60 | 10.56 |
| Solid body or semi-transparent stain, oil base (material #19 or #21) | | | | | | | | | |
| Spray 1st coat | | | | | | | | | |
| Slow | 400 | 170 | 23.25 | 2.56 | .70 | 13.68 | 5.25 | 3.33 | 25.52 |
| Medium | 500 | 150 | 20.35 | 3.30 | 1.11 | 13.57 | 4.41 | 2.46 | 24.85 |
| Fast | 600 | 130 | 17.45 | 3.67 | 1.49 | 13.42 | 3.44 | 1.32 | 23.34 |
| Spray 2nd coat | | | | | | | | | |
| Slow | 450 | 255 | 23.25 | 2.28 | .61 | 9.12 | 3.73 | 2.36 | 18.10 |
| Medium | 550 | 235 | 20.35 | 3.00 | 1.01 | 8.66 | 3.10 | 1.73 | 17.50 |
| Fast | 650 | 215 | 17.45 | 3.38 | 1.38 | 8.12 | 2.38 | .92 | 16.18 |
| Spray 3rd or additional coats | | | | | | | | | |
| Slow | 550 | 305 | 23.25 | 1.86 | .51 | 7.62 | 3.10 | 1.96 | 15.05 |
| Medium | 650 | 285 | 20.35 | 2.54 | .86 | 7.14 | 2.58 | 1.44 | 14.56 |
| Fast | 750 | 245 | 17.45 | 2.93 | 1.18 | 7.12 | 2.08 | .80 | 14.11 |
| Penetrating oil stain (material #13) | | | | | | | | | |
| Spray 1st coat | | | | | | | | | |
| Slow | 400 | 200 | 25.00 | 2.56 | .70 | 12.50 | 4.89 | 3.10 | 23.75 |
| Medium | 500 | 180 | 21.80 | 3.30 | 1.11 | 12.11 | 4.05 | 2.26 | 22.83 |
| Fast | 600 | 160 | 18.70 | 3.67 | 1.49 | 11.69 | 3.12 | 1.20 | 21.17 |
| Spray 2nd coat | | | | | | | | | |
| Slow | 450 | 290 | 25.00 | 2.28 | .61 | 8.62 | 3.57 | 2.26 | 17.34 |
| Medium | 550 | 245 | 21.80 | 3.00 | 1.01 | 8.90 | 3.16 | 1.77 | 17.84 |
| Fast | 650 | 200 | 18.70 | 3.38 | 1.38 | 9.35 | 2.61 | 1.00 | 17.72 |
| Spray 3rd or additional coats | | | | | | | | | |
| Slow | 550 | 340 | 25.00 | 1.86 | .51 | 7.35 | 3.01 | 1.91 | 14.64 |
| Medium | 650 | 310 | 21.80 | 2.54 | .86 | 7.03 | 2.55 | 1.43 | 14.41 |
| Fast | 750 | 280 | 18.70 | 2.93 | 1.18 | 6.68 | 2.00 | .77 | 13.56 |

| | Labor SF per manhour | Material coverage SF/gallon | Material cost per gallon | Labor cost per 100 SF | Labor burden 100 SF | Material cost per 100 SF | Overhead per 100 SF | Profit per 100 SF | Total price per 100 SF |
|---|---|---|---|---|---|---|---|---|---|
| Waterproofing, clear hydro seal, oil base (material #34) | | | | | | | | | |
| Spray 1st coat | | | | | | | | | |
| Slow | 400 | 150 | 18.00 | 2.56 | .70 | 12.00 | 4.73 | 2.20 | 22.19 |
| Medium | 500 | 113 | 15.80 | 3.30 | 1.11 | 13.98 | 4.51 | 2.52 | 25.42 |
| Fast | 600 | 75 | 13.50 | 3.67 | 1.49 | 18.00 | 4.28 | 1.65 | 29.09 |
| Spray 2nd coat | | | | | | | | | |
| Slow | 650 | 175 | 18.00 | 1.58 | .43 | 10.29 | 3.81 | 2.42 | 18.53 |
| Medium | 675 | 150 | 15.80 | 2.44 | .82 | 10.53 | 3.38 | 1.89 | 19.06 |
| Fast | 700 | 125 | 13.50 | 3.14 | 1.28 | 10.80 | 2.81 | 1.08 | 19.11 |
| Spray 3rd or additional coats | | | | | | | | | |
| Slow | 700 | 200 | 18.00 | 1.46 | .40 | 9.00 | 3.37 | 2.13 | 16.36 |
| Medium | 725 | 175 | 15.80 | 2.28 | .76 | 9.03 | 2.96 | 1.65 | 16.68 |
| Fast | 750 | 150 | 13.50 | 2.93 | 1.18 | 9.00 | 2.43 | .93 | 16.47 |

Use this table to estimate the cost of painting shingle, shake, resawn or rough sawn wood or plywood siding with average moisture content. Don't deduct for openings under 100 square feet. For heights above 8 feet, use the High Time Difficulty Factors on page 137. For wood or composition drop siding with exposed bevel edges, multiply the surface area by 1.12 to allow for the extra time and material needed to paint the underside of each board. "Slow" work is based on a $10.25 hourly wage, "Medium" work on a $16.75 hourly wage, and "Fast" work on a $22.00 hourly wage. Other qualifications that apply to this table are on page 9.

| | Labor SF per manhour | Material coverage SF/gallon | Material cost per gallon | Labor cost per 100 SF | Labor burden 100 SF | Material cost per 100 SF | Overhead per 100 SF | Profit per 100 SF | Total price per 100 SF |
|---|---|---|---|---|---|---|---|---|---|
| **Siding, smooth wood, brush** | | | | | | | | | |
| Solid body or semi-transparent stain, water base (material #18 or #20) | | | | | | | | | |
| Brush 1st coat | | | | | | | | | |
| Slow | 100 | 275 | 19.50 | 10.25 | 2.79 | 7.09 | 6.24 | 3.96 | 30.33 |
| Medium | 125 | 363 | 17.05 | 13.20 | 4.44 | 4.70 | 5.47 | 3.06 | 30.87 |
| Fast | 150 | 250 | 14.60 | 14.67 | 5.95 | 5.84 | 4.90 | 1.88 | 33.24 |
| Brush 2nd coat | | | | | | | | | |
| Slow | 135 | 350 | 19.50 | 7.59 | 2.07 | 5.57 | 4.72 | 2.99 | 22.94 |
| Medium | 168 | 325 | 17.05 | 9.82 | 3.30 | 5.25 | 4.50 | 2.52 | 25.39 |
| Fast | 200 | 300 | 14.60 | 11.00 | 4.46 | 4.87 | 3.76 | 1.45 | 25.54 |
| Brush 3rd or additional coats | | | | | | | | | |
| Slow | 150 | 425 | 19.50 | 6.83 | 1.87 | 4.59 | 4.12 | 2.61 | 20.02 |
| Medium | 188 | 400 | 17.05 | 8.78 | 2.95 | 4.26 | 3.92 | 2.19 | 22.10 |
| Fast | 215 | 375 | 14.60 | 10.23 | 4.15 | 3.89 | 3.38 | 1.30 | 22.95 |
| Solid body or semi-transparent stain, oil base (material #19 or #21) | | | | | | | | | |
| Brush 1st coat | | | | | | | | | |
| Slow | 100 | 400 | 23.25 | 10.25 | 2.79 | 5.81 | 5.84 | 3.70 | 28.39 |
| Medium | 125 | 363 | 20.35 | 13.20 | 4.44 | 5.61 | 5.70 | 3.18 | 32.13 |
| Fast | 150 | 325 | 17.45 | 14.67 | 5.95 | 5.37 | 4.81 | 1.85 | 32.65 |
| Brush 2nd coat | | | | | | | | | |
| Slow | 135 | 450 | 23.25 | 7.59 | 2.07 | 5.17 | 4.60 | 2.91 | 22.34 |
| Medium | 168 | 408 | 20.35 | 9.82 | 3.30 | 4.99 | 4.44 | 2.48 | 25.03 |
| Fast | 200 | 375 | 17.45 | 11.00 | 4.46 | 4.65 | 3.72 | 1.43 | 25.26 |
| Brush 3rd or additional coats | | | | | | | | | |
| Slow | 150 | 525 | 23.25 | 6.83 | 1.87 | 4.43 | 4.07 | 2.58 | 19.78 |
| Medium | 188 | 438 | 20.35 | 8.78 | 2.95 | 4.65 | 4.01 | 2.24 | 22.63 |
| Fast | 215 | 450 | 17.45 | 10.23 | 4.15 | 3.88 | 3.38 | 1.30 | 22.94 |
| Penetrating oil stain (material #13) | | | | | | | | | |
| Brush 1st coat | | | | | | | | | |
| Slow | 100 | 315 | 25.00 | 10.25 | 2.79 | 7.94 | 6.50 | 4.12 | 31.60 |
| Medium | 125 | 303 | 21.80 | 13.20 | 4.44 | 7.19 | 6.08 | 3.40 | 34.31 |
| Fast | 150 | 290 | 18.70 | 14.67 | 5.95 | 6.45 | 5.01 | 1.92 | 34.00 |

| | Labor SF per manhour | Material coverage SF/gallon | Material cost per gallon | Labor cost per 100 SF | Labor burden 100 SF | Material cost per 100 SF | Overhead per 100 SF | Profit per 100 SF | Total price per 100 SF |
|---|---|---|---|---|---|---|---|---|---|
| Brush 2nd coat | | | | | | | | | |
| Slow | 135 | 355 | 25.00 | 7.59 | 2.07 | 7.04 | 5.18 | 3.28 | 25.16 |
| Medium | 168 | 343 | 21.80 | 9.82 | 3.30 | 6.36 | 4.77 | 2.67 | 26.92 |
| Fast | 200 | 330 | 18.70 | 11.00 | 4.46 | 5.67 | 3.91 | 1.50 | 26.54 |
| Brush 3rd or additional coats | | | | | | | | | |
| Slow | 150 | 430 | 25.00 | 6.83 | 1.87 | 5.81 | 4.50 | 2.85 | 21.86 |
| Medium | 188 | 418 | 21.80 | 8.78 | 2.95 | 5.22 | 4.15 | 2.32 | 23.42 |
| Fast | 215 | 405 | 18.70 | 10.23 | 4.15 | 4.62 | 3.52 | 1.35 | 23.87 |

Use this table for butt or tongue and groove siding, joint lap, drop, beveled or board and batten siding in redwood, plywood, fir, hemlock or pine. Don't deduct for openings under 100 square feet. For heights above 8 feet, use the High Time Difficulty Factors on page 137. For wood or composition drop siding with exposed bevel edges, multiply the surface area by 1.12 to allow for the extra time and material needed to paint the underside of each board. "Slow" work is based on a $10.25 hourly wage, "Medium" work on a $16.75 hourly wage, and "Fast" work on a $22.00 hourly wage. Other qualifications that apply to this table are on page 9.

| | Labor SF per manhour | Material coverage SF/gallon | Material cost per gallon | Labor cost per 100 SF | Labor burden 100 SF | Material cost per 100 SF | Overhead per 100 SF | Profit per 100 SF | Total price per 100 SF |
|---|---|---|---|---|---|---|---|---|---|
| **Siding, smooth wood, roll** | | | | | | | | | |
| Solid body or semi-transparent stain, water base (material #18 or #20) | | | | | | | | | |
| Roll 1st coat | | | | | | | | | |
| Slow | 175 | 275 | 19.50 | 5.86 | 1.59 | 7.09 | 4.51 | 2.86 | 21.91 |
| Medium | 225 | 268 | 17.05 | 7.33 | 2.46 | 6.36 | 3.96 | 2.21 | 22.32 |
| Fast | 275 | 250 | 14.60 | 8.00 | 3.25 | 5.84 | 3.16 | 1.21 | 21.46 |
| Roll 2nd coat | | | | | | | | | |
| Slow | 225 | 350 | 19.50 | 4.56 | 1.23 | 5.57 | 3.52 | 2.23 | 17.11 |
| Medium | 275 | 325 | 17.05 | 6.00 | 2.03 | 5.25 | 3.25 | 1.82 | 18.35 |
| Fast | 325 | 300 | 14.60 | 6.77 | 2.75 | 4.87 | 2.66 | 1.02 | 18.07 |
| Roll 3rd or additional coats | | | | | | | | | |
| Slow | 260 | 400 | 19.50 | 3.94 | 1.08 | 4.88 | 3.07 | 1.94 | 14.91 |
| Medium | 335 | 363 | 17.05 | 4.93 | 1.66 | 4.70 | 2.77 | 1.55 | 15.61 |
| Fast | 410 | 375 | 14.60 | 5.37 | 2.17 | 3.89 | 2.12 | .81 | 14.36 |
| Solid body or semi-transparent stain, oil base (material #19 or #21) | | | | | | | | | |
| Roll 1st coat | | | | | | | | | |
| Slow | 175 | 350 | 23.25 | 5.86 | 1.59 | 6.64 | 4.37 | 2.77 | 21.23 |
| Medium | 225 | 325 | 20.35 | 7.33 | 2.46 | 6.26 | 3.93 | 2.20 | 22.18 |
| Fast | 275 | 300 | 17.45 | 8.00 | 3.25 | 5.82 | 3.16 | 1.21 | 21.44 |
| Roll 2nd coat | | | | | | | | | |
| Slow | 225 | 400 | 23.25 | 4.56 | 1.23 | 5.81 | 3.60 | 2.28 | 17.48 |
| Medium | 275 | 375 | 20.35 | 6.00 | 2.03 | 5.43 | 3.30 | 1.84 | 18.60 |
| Fast | 325 | 350 | 17.45 | 6.77 | 2.75 | 4.99 | 2.68 | 1.03 | 18.22 |
| Roll 3rd or additional coats | | | | | | | | | |
| Slow | 260 | 425 | 23.25 | 3.94 | 1.08 | 5.47 | 3.25 | 2.06 | 15.80 |
| Medium | 335 | 413 | 20.35 | 4.93 | 1.66 | 4.93 | 2.82 | 1.58 | 15.92 |
| Fast | 410 | 400 | 17.45 | 5.37 | 2.17 | 4.36 | 2.20 | .85 | 14.95 |
| Penetrating oil stain (material #13) | | | | | | | | | |
| Roll 1st coat | | | | | | | | | |
| Slow | 175 | 200 | 25.00 | 5.86 | 1.59 | 12.50 | 6.19 | 3.92 | 30.06 |
| Medium | 225 | 150 | 21.80 | 7.33 | 2.46 | 14.53 | 5.96 | 3.33 | 33.61 |
| Fast | 275 | 100 | 18.70 | 8.00 | 3.25 | 18.70 | 5.54 | 2.13 | 37.62 |

| | Labor SF per manhour | Material coverage SF/gallon | Material cost per gallon | Labor cost per 100 SF | Labor burden 100 SF | Material cost per 100 SF | Overhead per 100 SF | Profit per 100 SF | Total price per 100 SF |
|---|---|---|---|---|---|---|---|---|---|
| Roll 2nd coat | | | | | | | | | |
| Slow | 225 | 250 | 25.00 | 4.56 | 1.23 | 10.00 | 4.90 | 3.11 | 23.80 |
| Medium | 275 | 200 | 21.80 | 6.00 | 2.03 | 10.90 | 4.64 | 2.59 | 26.16 |
| Fast | 325 | 150 | 18.70 | 6.77 | 2.75 | 12.47 | 4.07 | 1.56 | 27.62 |
| Roll 3rd or additional coats | | | | | | | | | |
| Slow | 260 | 300 | 25.00 | 3.94 | 1.08 | 8.33 | 4.14 | 2.62 | 20.11 |
| Medium | 335 | 250 | 21.80 | 4.93 | 1.66 | 8.72 | 3.75 | 2.10 | 21.16 |
| Fast | 410 | 200 | 18.70 | 5.37 | 2.17 | 9.35 | 3.13 | 1.20 | 21.22 |

Use this table for butt or tongue and groove siding, joint lap, drop, beveled or board and batten siding in redwood, plywood, fir, hemlock or pine. Don't deduct for openings under 100 square feet. For heights above 8 feet, use the High Time Difficulty Factors on page 137. For wood or composition drop siding with exposed bevel edges, multiply the surface area by 1.12 to allow for the extra time and material needed to paint the underside of each board. "Slow" work is based on a $10.25 hourly wage, "Medium" work on a $16.75 hourly wage, and "Fast" work on a $22.00 hourly wage. Other qualifications that apply to this table are on page 9.

| | Labor SF per manhour | Material coverage SF/gallon | Material cost per gallon | Labor cost per 100 SF | Labor burden 100 SF | Material cost per 100 SF | Overhead per 100 SF | Profit per 100 SF | Total price per 100 SF |
|---|---|---|---|---|---|---|---|---|---|

## Siding, smooth wood, spray

| | Labor SF per manhour | Material coverage SF/gallon | Material cost per gallon | Labor cost per 100 SF | Labor burden 100 SF | Material cost per 100 SF | Overhead per 100 SF | Profit per 100 SF | Total price per 100 SF |
|---|---|---|---|---|---|---|---|---|---|
| Solid body or semi-transparent stain, water base (material #18 or #20) | | | | | | | | | |
| Spray 1st coat | | | | | | | | | |
| Slow | 400 | 150 | 19.50 | 2.56 | .70 | 13.00 | 5.04 | 3.20 | 24.50 |
| Medium | 500 | 125 | 17.05 | 3.30 | 1.11 | 13.64 | 4.42 | 2.47 | 24.94 |
| Fast | 600 | 100 | 14.60 | 3.67 | 1.49 | 14.60 | 3.66 | 1.41 | 24.83 |
| Spray 2nd coat | | | | | | | | | |
| Slow | 550 | 250 | 19.50 | 1.86 | .51 | 7.80 | 3.15 | 2.00 | 15.32 |
| Medium | 725 | 225 | 17.05 | 2.28 | .76 | 7.58 | 2.60 | 1.46 | 14.68 |
| Fast | 900 | 200 | 14.60 | 2.44 | .99 | 7.30 | 1.99 | .76 | 13.48 |
| Spray 3rd or additional coats | | | | | | | | | |
| Slow | 650 | 350 | 19.50 | 1.58 | .43 | 5.57 | 2.35 | 1.49 | 11.42 |
| Medium | 825 | 325 | 17.05 | 2.00 | .67 | 5.25 | 1.94 | 1.08 | 10.94 |
| Fast | 1000 | 300 | 14.60 | 2.20 | .89 | 4.87 | 1.47 | .57 | 10.00 |
| Solid body or semi-transparent stain, oil base (material #19 or #21) | | | | | | | | | |
| Spray 1st coat | | | | | | | | | |
| Slow | 400 | 170 | 23.25 | 2.56 | .70 | 13.68 | 5.25 | 3.33 | 25.52 |
| Medium | 500 | 150 | 20.35 | 3.30 | 1.11 | 13.57 | 4.41 | 2.46 | 24.85 |
| Fast | 600 | 130 | 17.45 | 3.67 | 1.49 | 13.42 | 3.44 | 1.32 | 23.34 |
| Spray 2nd coat | | | | | | | | | |
| Slow | 550 | 300 | 23.25 | 1.86 | .51 | 7.75 | 3.14 | 1.99 | 15.25 |
| Medium | 725 | 273 | 20.35 | 2.28 | .76 | 7.45 | 2.57 | 1.44 | 14.50 |
| Fast | 900 | 245 | 17.45 | 2.44 | .99 | 7.12 | 1.95 | .75 | 13.25 |
| Spray 3rd or additional coats | | | | | | | | | |
| Slow | 650 | 400 | 23.25 | 1.58 | .43 | 5.81 | 2.42 | 1.54 | 11.78 |
| Medium | 825 | 373 | 20.35 | 2.00 | .67 | 5.46 | 1.99 | 1.11 | 11.23 |
| Fast | 1000 | 345 | 17.45 | 2.20 | .89 | 5.06 | 1.51 | .58 | 10.24 |
| Penetrating oil stain (material #13) | | | | | | | | | |
| Spray 1st coat | | | | | | | | | |
| Slow | 400 | 150 | 25.00 | 2.56 | .70 | 16.67 | 6.18 | 3.92 | 30.03 |
| Medium | 500 | 113 | 21.80 | 3.30 | 1.11 | 19.29 | 5.81 | 3.25 | 32.76 |
| Fast | 600 | 75 | 18.70 | 3.67 | 1.49 | 24.93 | 5.57 | 2.14 | 37.80 |
| Spray 2nd coat | | | | | | | | | |
| Slow | 550 | 225 | 25.00 | 1.86 | .51 | 11.11 | 4.18 | 2.65 | 20.31 |
| Medium | 725 | 188 | 21.80 | 2.28 | .76 | 11.60 | 3.59 | 2.01 | 20.24 |
| Fast | 900 | 150 | 18.70 | 2.44 | .99 | 12.47 | 2.94 | 1.13 | 19.97 |
| Spray 3rd or additional coats | | | | | | | | | |
| Slow | 650 | 250 | 25.00 | 1.58 | .43 | 10.00 | 3.72 | 2.36 | 18.09 |
| Medium | 825 | 225 | 21.80 | 2.00 | .67 | 9.69 | 3.03 | 1.69 | 17.08 |
| Fast | 1000 | 200 | 18.70 | 2.20 | .89 | 9.35 | 2.30 | .88 | 15.62 |

| | Labor SF per manhour | Material coverage SF/gallon | Material cost per gallon | Labor cost per 100 SF | Labor burden 100 SF | Material cost per 100 SF | Overhead per 100 SF | Profit per 100 SF | Total price per 100 SF |
|---|---|---|---|---|---|---|---|---|---|
| Waterproofing, clear hydro seal, oil base (material #34) | | | | | | | | | |
| Spray 1st coat | | | | | | | | | |
| Slow | 550 | 250 | 18.00 | 1.86 | .51 | 7.20 | 2.97 | 1.38 | 13.92 |
| Medium | 675 | 200 | 15.80 | 2.44 | .82 | 7.90 | 2.73 | 1.53 | 15.42 |
| Fast | 800 | 150 | 13.50 | 2.75 | 1.12 | 9.00 | 2.38 | .92 | 16.17 |
| Spray 2nd coat | | | | | | | | | |
| Slow | 625 | 300 | 18.00 | 1.64 | .45 | 6.00 | 2.51 | 1.59 | 12.19 |
| Medium | 750 | 250 | 15.80 | 2.20 | .73 | 6.32 | 2.27 | 1.27 | 12.79 |
| Fast | 875 | 200 | 13.50 | 2.51 | 1.01 | 6.75 | 1.90 | .73 | 12.90 |
| Spray 3rd or additional coats | | | | | | | | | |
| Slow | 675 | 325 | 18.00 | 1.52 | .41 | 5.54 | 2.32 | 1.47 | 11.26 |
| Medium | 800 | 275 | 15.80 | 2.06 | .70 | 5.75 | 2.08 | 1.16 | 11.75 |
| Fast | 925 | 225 | 13.50 | 2.38 | .96 | 6.00 | 1.73 | .66 | 11.73 |

Use this table for butt or tongue and groove siding, joint lap, drop, beveled or board and batten siding in redwood, plywood, fir, hemlock or pine. Don't deduct for openings under 100 square feet. For heights above 8 feet, use the High Time Difficulty Factors on page 137. For wood or composition drop siding with exposed bevel edges, multiply the surface area by 1.12 to allow for the extra time and material needed to paint the underside of each board. "Slow" work is based on a $10.25 hourly wage, "Medium" work on a $16.75 hourly wage, and "Fast" work on a $22.00 hourly wage. Other qualifications that apply to this table are on page 9.

## Stair steps, interior or exterior, wood

To estimate the cost to paint or stain stairs, find the surface area. Then use the tables for wood siding. To find the surface area of each tread and riser, multiply the length by the width. To find the tread length, add the run, the rise, and the tread nosing. For example, a tread with a 12" run, an 8" rise, and 1" nosing, has a 23" surface area (measured one side). For estimating purposes, figure any length from 14" to 26" as 2 feet. Use the actual width of the tread if the stringers are calculated separately. If the tread in the example is 3 feet wide and you use 2 feet for the length, the surface area is 6 feet. If there are 15 treads, the total top surface area is 90 square feet.

If you're calculating the area to paint the stair treads and stringers in one operation, add 2 feet to the actual tread width to include the stringers. That would make the effective width of the tread in the example 5 feet. Then multiply 5 feet by 2 feet to find the area of each tread. For 15 treads, the total surface area is 150 square feet.

| | Labor LF per manhour | Material coverage LF/gallon | Material cost per gallon | Labor cost per 100 LF | Labor burden 100 LF | Material cost per 100 LF | Overhead per 100 LF | Profit per 100 LF | Total price per 100 LF |
|---|---|---|---|---|---|---|---|---|---|

## Stair stringers, exterior, metal, shapes up to 14" wide

| | Labor LF per manhour | Material coverage LF/gallon | Material cost per gallon | Labor cost per 100 LF | Labor burden 100 LF | Material cost per 100 LF | Overhead per 100 LF | Profit per 100 LF | Total price per 100 LF |
|---|---|---|---|---|---|---|---|---|---|
| Metal primer - rust inhibitor, clean metal (material #35) | | | | | | | | | |
| Roll & brush prime coat | | | | | | | | | |
| Slow | 50 | 120 | 28.20 | 20.50 | 5.58 | 23.50 | 15.37 | 9.74 | 74.69 |
| Medium | 55 | 115 | 24.60 | 30.00 | 10.09 | 21.39 | 15.07 | 8.42 | 84.97 |
| Fast | 60 | 110 | 21.10 | 36.67 | 14.87 | 19.18 | 13.08 | 5.03 | 88.83 |
| Metal primer - rust inhibitor, rusty metal (material #36) | | | | | | | | | |
| Roll & brush prime coat | | | | | | | | | |
| Slow | 50 | 120 | 33.30 | 20.50 | 5.58 | 27.75 | 16.69 | 10.58 | 81.10 |
| Medium | 55 | 115 | 29.20 | 30.00 | 10.09 | 25.39 | 16.05 | 8.97 | 90.50 |
| Fast | 60 | 110 | 25.00 | 36.67 | 14.87 | 22.73 | 13.74 | 5.28 | 93.29 |
| Metal finish - synthetic enamel, off white (material #37) | | | | | | | | | |
| Roll & brush 1st or additional finish coats | | | | | | | | | |
| Slow | 50 | 135 | 25.90 | 20.50 | 5.58 | 19.19 | 14.04 | 8.90 | 68.21 |
| Medium | 55 | 130 | 22.70 | 30.00 | 10.09 | 17.46 | 14.10 | 7.88 | 79.53 |
| Fast | 60 | 125 | 19.50 | 36.67 | 14.87 | 15.60 | 12.42 | 4.77 | 84.33 |
| Metal finish - synthetic enamel, colors - except orange/red (material #38) | | | | | | | | | |
| Roll & brush 1st or additional finish coats | | | | | | | | | |
| Slow | 50 | 135 | 35.00 | 20.50 | 5.58 | 25.93 | 16.13 | 10.22 | 78.36 |
| Medium | 55 | 130 | 30.70 | 30.00 | 10.09 | 23.62 | 15.61 | 8.73 | 88.05 |
| Fast | 60 | 125 | 26.30 | 36.67 | 14.87 | 21.04 | 13.43 | 5.16 | 91.17 |

Use these figures to paint installed stair stringers. Measurements are based on linear feet of each stringer. Note: A two coat system, prime and finish, using oil base material is recommended for any metal surface. Although water base material is often used, it may cause oxidation, corrosion and rust. Using one coat of oil base paint on exterior metal may result in cracking, peeling or chipping without the proper prime coat application. Pre-primed steel or wrought iron generally requires only one coat to cover. The metal finish figures include minor touchup to the prime coat. If off white or other light colored finish paint is specified, make sure the prime coat is a light color also, or more than one finish coat will be necessary. "Slow" work is based on a $10.25 hourly wage, "Medium" work on a $16.50 hourly wage, and "Fast" work on a $22.00 hourly wage. Other qualifications that apply to this table are on page 9.

## Stair stringers, exterior, rough sawn wood up to 4" x 12"

| | Labor LF per manhour | Material coverage LF/gallon | Material cost per gallon | Labor cost per 100 LF | Labor burden 100 LF | Material cost per 100 LF | Overhead per 100 LF | Profit per 100 LF | Total price per 100 LF |
|---|---|---|---|---|---|---|---|---|---|
| Solid body or semi-transparent stain, water or oil base (material #18, #19, #20 or #21) | | | | | | | | | |
| Roll & brush each coat | | | | | | | | | |
| Slow | 40 | 70 | 21.38 | 25.63 | 6.97 | 30.54 | 19.58 | 12.41 | 95.13 |
| Medium | 45 | 65 | 18.70 | 36.67 | 12.33 | 28.77 | 19.06 | 10.65 | 107.48 |
| Fast | 50 | 60 | 16.03 | 44.00 | 17.84 | 26.72 | 16.38 | 6.30 | 111.24 |

Measurements are based on the linear feet of each stringer. "Slow" work is based on a $10.25 hourly wage, "Medium" work on a $16.50 hourly wage, and "Fast" work on a $22.00 hourly wage. Other qualifications that apply to this table are on page 9.

| | Labor LF per manhour | Material coverage LF/gallon | Material cost per gallon | Labor cost per 100 LF | Labor burden 100 LF | Material cost per 100 LF | Overhead per 100 LF | Profit per 100 LF | Total price per 100 LF |
|---|---|---|---|---|---|---|---|---|---|
| **Stair stringers, interior, metal, shapes up to 14" wide** | | | | | | | | | |
| Metal primer - rust inhibitor, clean metal (material #35) | | | | | | | | | |
| Roll & brush prime coat | | | | | | | | | |
| Slow | 45 | 130 | 28.20 | 22.78 | 6.19 | 21.69 | 15.71 | 9.96 | 76.33 |
| Medium | 50 | 125 | 24.60 | 33.00 | 11.10 | 19.68 | 15.63 | 8.74 | 88.15 |
| Fast | 55 | 120 | 21.10 | 40.00 | 16.21 | 17.58 | 13.65 | 5.25 | 92.69 |
| Metal primer - rust inhibitor, rusty metal (material #36) | | | | | | | | | |
| Roll & brush prime coat | | | | | | | | | |
| Slow | 45 | 130 | 33.30 | 22.78 | 6.19 | 25.62 | 16.93 | 10.73 | 82.25 |
| Medium | 50 | 125 | 29.20 | 33.00 | 11.10 | 23.36 | 16.53 | 9.24 | 93.23 |
| Fast | 55 | 120 | 25.00 | 40.00 | 16.21 | 20.83 | 14.25 | 5.48 | 96.77 |
| Metal finish - synthetic enamel, off white (material #37) | | | | | | | | | |
| Roll & brush 1st or additional finish coats | | | | | | | | | |
| Slow | 45 | 145 | 25.90 | 22.78 | 6.19 | 17.86 | 14.52 | 9.21 | 70.56 |
| Medium | 50 | 140 | 22.70 | 33.00 | 11.10 | 16.21 | 14.78 | 8.26 | 83.35 |
| Fast | 55 | 135 | 19.50 | 40.00 | 16.21 | 14.44 | 13.07 | 5.02 | 88.74 |
| Metal finish - synthetic enamel, colors - except orange/red (material #38) | | | | | | | | | |
| Roll & brush 1st or additional finish coats | | | | | | | | | |
| Slow | 45 | 145 | 35.00 | 22.78 | 6.19 | 24.14 | 16.47 | 10.44 | 80.02 |
| Medium | 50 | 140 | 30.70 | 33.00 | 11.10 | 21.93 | 16.18 | 9.04 | 91.25 |
| Fast | 55 | 135 | 26.30 | 40.00 | 16.21 | 19.48 | 14.00 | 5.38 | 95.07 |

Use these figures to paint installed stair stringers. Measurements are based on linear feet of each stringer. Note: A two coat system, prime and finish, using oil base material is recommended for any metal surface. Although water base material is often used, it may cause oxidation, corrosion and rust. Using one coat of oil base paint on exterior metal may result in cracking, peeling or chipping without the proper prime coat application. Pre-primed steel or wrought iron generally requires only one coat to cover. The metal finish figures include minor touchup to the prime coat. If off white or other light colored finish paint is specified, make sure the prime coat is a light color also, or more than one finish coat will be necessary. "Slow" work is based on a $10.25 hourly wage, "Medium" work on a $16.50 hourly wage, and "Fast" work on a $22.00 hourly wage. Other qualifications that apply to this table are on page 9.

| | Labor LF per manhour | Material coverage LF/gallon | Material cost per gallon | Labor cost per 100 LF | Labor burden 100 LF | Material cost per 100 LF | Overhead per 100 LF | Profit per 100 LF | Total price per 100 LF |
|---|---|---|---|---|---|---|---|---|---|
| **Stair stringers, interior, rough sawn wood up to 4" x 12"** | | | | | | | | | |
| Solid body or semi-transparent stain, water or oil base (material #18, #19, #20 or #21) | | | | | | | | | |
| Roll & brush each coat | | | | | | | | | |
| Slow | 35 | 60 | 21.38 | 29.29 | 7.97 | 35.63 | 22.60 | 14.33 | 109.82 |
| Medium | 40 | 53 | 18.70 | 41.25 | 13.88 | 35.28 | 22.15 | 12.38 | 124.94 |
| Fast | 45 | 45 | 16.03 | 48.89 | 19.81 | 35.62 | 19.30 | 7.42 | 131.04 |

Measurements are based on the linear feet of each stringer. "Slow" work is based on a $10.25 hourly wage, "Medium" work on a $16.50 hourly wage, and "Fast" work on a $22.00 hourly wage. Other qualifications that apply to this table are on page 9.

## Stucco: see Plaster and stucco

| | Percentage of total costs |
|---|---|
| **Touchup, brush as required** | |
| Slow | 10% |
| Medium | 9% |
| Fast | 8% |

Touchup will be required on all jobs. The skill of your paint crews and the type of job will determine the time needed. The time and materials shown here are average allowances. Figure the touchup time and material needed and then add the appropriate percentage to your estimate. These figures are based on a percentage of the total labor and material needed to touch up all interior and exterior work. If you're painting only the interior or the exterior, divide these figures by 2. "Slow" work is based on a $10.25 hourly wage, "Medium" work on a $16.50 hourly wage, and "Fast" work on a $22.00 hourly wage. Other qualifications that apply to this table are on page 9.

| | Labor LF per manhour | Material coverage LF/gallon | Material cost per gallon | Labor cost per 100 LF | Labor burden 100 LF | Material cost per 100 LF | Overhead per 100 LF | Profit per 100 LF | Total price per 100 LF |
|---|---|---|---|---|---|---|---|---|---|

## Trellis or lattice, roll and brush

2" x 2" to 2" x 6", roll & brush all sides, each coat

Solid body or semi-transparent stain, water or oil base (material #18, #19, #20 or #21)

| | Labor LF per manhour | Material coverage LF/gallon | Material cost per gallon | Labor cost per 100 LF | Labor burden 100 LF | Material cost per 100 LF | Overhead per 100 LF | Profit per 100 LF | Total price per 100 LF |
|---|---|---|---|---|---|---|---|---|---|
| Slow | 120 | 130 | 21.38 | 8.54 | 2.32 | 16.45 | 8.47 | 5.37 | 41.15 |
| Medium | 125 | 120 | 18.70 | 13.20 | 4.44 | 15.58 | 8.14 | 4.55 | 45.91 |
| Fast | 130 | 110 | 16.03 | 16.92 | 6.86 | 14.57 | 7.09 | 2.73 | 48.17 |

2" x 8" to 4" x 12", roll & brush all sides, each coat

Solid body or semi-transparent stain, water or oil base (material #18, #19, #20 or #21)

| | Labor LF per manhour | Material coverage LF/gallon | Material cost per gallon | Labor cost per 100 LF | Labor burden 100 LF | Material cost per 100 LF | Overhead per 100 LF | Profit per 100 LF | Total price per 100 LF |
|---|---|---|---|---|---|---|---|---|---|
| Slow | 100 | 100 | 21.38 | 10.25 | 2.79 | 21.38 | 10.67 | 6.76 | 51.85 |
| Medium | 110 | 90 | 18.70 | 15.00 | 5.04 | 20.78 | 10.00 | 5.59 | 56.41 |
| Fast | 120 | 80 | 16.03 | 18.33 | 7.43 | 20.04 | 8.47 | 3.26 | 57.53 |

Measurements are based on accumulated total linear feet of each trellis or lattice member. These figures are based on roll and brush staining of all four sides and the ends of each member per coat. "Slow" work is based on a $10.25 hourly wage, "Medium" work on a $16.50 hourly wage, and "Fast" work on a $22.50 hourly wage. Other qualifications that apply to this table are on page 9.

## Trellis or lattice, spray

2" x 2" at 3" on center with 2" x 8" supports, spray all sides, each coat

Solid body or semi-transparent stain, water or oil base (material #18, #19, #20 or #21)

| | SF surface area per manhour | SF surface area per gallon | Material cost per gallon | Labor cost per 100 SF | Labor burden 100 SF | Material cost per 100 SF | Overhead per 100 SF | Profit per 100 SF | Total price per 100 SF |
|---|---|---|---|---|---|---|---|---|---|
| Slow | 50 | 60 | 21.38 | 20.50 | 5.58 | 35.63 | 19.13 | 12.13 | 92.97 |
| Medium | 55 | 55 | 18.70 | 30.00 | 10.09 | 34.00 | 18.15 | 10.15 | 102.39 |
| Fast | 60 | 50 | 16.03 | 36.67 | 14.87 | 32.06 | 15.47 | 5.94 | 105.01 |

Measurements are based on the square feet of the surface area footprint of the trellis or lattice structure. (The footprint is the surface area seen from the plan or overhead view.) These figures are based on staining all four sides of each member per coat. "Slow" work is based on a $10.25 hourly wage, "Medium" work on a $16.50 hourly wage, and "Fast" work on a $22.00 hourly wage. Other qualifications that apply to this table are on page 9.

| | Labor LF per manhour | Material coverage LF/gallon | Material cost per gallon | Labor cost per 100 LF | Labor burden 100 LF | Material cost per 100 LF | Overhead per 100 LF | Profit per 100 LF | Total price per 100 LF |
|---|---|---|---|---|---|---|---|---|---|
| **Valances for light fixtures** | | | | | | | | | |
| Solid body or semi-transparent stain, water or oil base (material #18, #19, #20 or #21) | | | | | | | | | |
| Brush each coat | | | | | | | | | |
| Slow | 30 | 100 | 21.38 | 34.17 | 9.29 | 21.38 | 20.11 | 12.75 | 97.70 |
| Medium | 35 | 95 | 18.70 | 47.14 | 15.86 | 19.68 | 20.26 | 11.32 | 114.26 |
| Fast | 40 | 90 | 16.03 | 55.00 | 22.30 | 17.81 | 17.60 | 6.76 | 119.47 |

Rough sawn or resawn wood valances are commonly found in baths and kitchens surrounding light fixtures or supporting plastic cracked ice diffusers. Measurements are based on the linear feet of the valance. "Slow" work is based on a $10.25 hourly wage, "Medium" work on a $16.50 hourly wage, and "Fast" work on a $22.50 hourly wage. Other qualifications that apply to this table are on page 9.

| | Labor SF per manhour | Material coverage SF/gallon | Material cost per gallon | Labor cost per 100 SF | Labor burden 100 SF | Material cost per 100 SF | Overhead per 100 SF | Profit per 100 SF | Total price per 100 SF |
|---|---|---|---|---|---|---|---|---|---|

## Walls, gypsum drywall, anti-graffiti stain eliminator, per 100 square feet of wall

| | Labor SF per manhour | Material coverage SF/gallon | Material cost per gallon | Labor cost per 100 SF | Labor burden 100 SF | Material cost per 100 SF | Overhead per 100 SF | Profit per 100 SF | Total price per 100 SF |
|---|---|---|---|---|---|---|---|---|---|
| Water base primer and sealer (material #39) | | | | | | | | | |
| Roll & brush each coat | | | | | | | | | |
| Slow | 375 | 450 | 30.40 | 2.73 | .75 | 6.76 | 3.17 | 2.01 | 15.42 |
| Medium | 400 | 425 | 26.60 | 4.13 | 1.38 | 6.26 | 2.89 | 1.61 | 16.27 |
| Fast | 425 | 400 | 22.80 | 5.18 | 2.09 | 5.70 | 2.40 | .92 | 16.29 |
| Oil base primer and sealer (material #40) | | | | | | | | | |
| Roll & brush each coat | | | | | | | | | |
| Slow | 375 | 400 | 28.20 | 2.73 | .75 | 7.05 | 3.26 | 2.07 | 15.86 |
| Medium | 400 | 388 | 24.70 | 4.13 | 1.38 | 6.37 | 2.91 | 1.63 | 16.42 |
| Fast | 425 | 375 | 21.20 | 5.18 | 2.09 | 5.65 | 2.39 | .92 | 16.23 |
| Polyurethane 2 part system (material #41) | | | | | | | | | |
| Roll & brush each coat | | | | | | | | | |
| Slow | 325 | 400 | 63.30 | 3.15 | .87 | 15.83 | 6.15 | 3.90 | 29.90 |
| Medium | 350 | 375 | 55.40 | 4.71 | 1.60 | 14.77 | 5.16 | 2.88 | 29.12 |
| Fast | 375 | 350 | 47.40 | 5.87 | 2.39 | 13.54 | 4.03 | 1.55 | 27.38 |

Measurements are based on the square feet of wall coated. These figures assume paint products are being applied over a smooth finish. For heights above 8 feet, use the High Time Difficulty Factors on page 137. "Slow" work is based on a $10.25 hourly wage, "Medium" work on a $16.50 hourly wage, and "Fast" work on a $22.00 hourly wage. Other qualifications that apply to this table are on page 9.

| | Labor SF per manhour | Material coverage SF/gallon | Material cost per gallon | Labor cost per 100 SF | Labor burden 100 SF | Material cost per 100 SF | Overhead per 100 SF | Profit per 100 SF | Total price per 100 SF |
|---|---|---|---|---|---|---|---|---|---|
| **Walls, gypsum drywall, orange peel or knock-down, brush, per 100 SF of wall** | | | | | | | | | |
| Flat latex, water base (material #5) | | | | | | | | | |
| Brush 1st coat | | | | | | | | | |
| Slow | 150 | 300 | 16.20 | 6.83 | 1.87 | 5.40 | 4.37 | 2.77 | 21.24 |
| Medium | 175 | 288 | 14.20 | 9.43 | 3.16 | 4.93 | 4.29 | 2.40 | 24.21 |
| Fast | 200 | 275 | 12.10 | 11.00 | 4.46 | 4.40 | 3.67 | 1.41 | 24.94 |
| Brush 2nd coat | | | | | | | | | |
| Slow | 175 | 350 | 16.20 | 5.86 | 1.59 | 4.63 | 3.75 | 2.38 | 18.21 |
| Medium | 200 | 338 | 14.20 | 8.25 | 2.78 | 4.20 | 3.73 | 2.09 | 21.05 |
| Fast | 225 | 325 | 12.10 | 9.78 | 3.95 | 3.72 | 3.23 | 1.24 | 21.92 |
| Brush 3rd or additional coats | | | | | | | | | |
| Slow | 200 | 400 | 16.20 | 5.13 | 1.39 | 4.05 | 3.28 | 2.08 | 15.93 |
| Medium | 225 | 375 | 14.20 | 7.33 | 2.46 | 3.79 | 3.33 | 1.86 | 18.77 |
| Fast | 250 | 350 | 12.10 | 8.80 | 3.57 | 3.46 | 2.93 | 1.13 | 19.89 |
| Sealer (drywall), water base (material #1) | | | | | | | | | |
| Brush prime coat | | | | | | | | | |
| Slow | 125 | 300 | 15.50 | 8.20 | 2.23 | 5.17 | 4.84 | 3.07 | 23.51 |
| Medium | 163 | 288 | 13.50 | 10.12 | 3.40 | 4.69 | 4.46 | 2.49 | 25.16 |
| Fast | 200 | 275 | 11.60 | 11.00 | 4.46 | 4.22 | 3.64 | 1.40 | 24.72 |
| Sealer (drywall), oil base (material #2) | | | | | | | | | |
| Brush prime coat | | | | | | | | | |
| Slow | 125 | 250 | 20.70 | 8.20 | 2.23 | 8.28 | 5.80 | 3.68 | 28.19 |
| Medium | 163 | 238 | 18.10 | 10.12 | 3.40 | 7.61 | 5.18 | 2.90 | 29.21 |
| Fast | 200 | 225 | 15.50 | 11.00 | 4.46 | 6.89 | 4.13 | 1.59 | 28.07 |
| Enamel, water base (material #9) | | | | | | | | | |
| Brush 1st finish coat | | | | | | | | | |
| Slow | 100 | 300 | 22.80 | 10.25 | 2.79 | 7.60 | 6.40 | 4.06 | 31.10 |
| Medium | 150 | 288 | 19.90 | 11.00 | 3.71 | 6.91 | 5.29 | 2.96 | 29.87 |
| Fast | 200 | 275 | 17.10 | 11.00 | 4.46 | 6.22 | 4.01 | 1.54 | 27.23 |
| Brush 2nd or additional finish coats | | | | | | | | | |
| Slow | 125 | 350 | 22.80 | 8.20 | 2.23 | 6.51 | 5.25 | 3.33 | 25.52 |
| Medium | 163 | 325 | 19.90 | 10.12 | 3.40 | 6.12 | 4.81 | 2.69 | 27.14 |
| Fast | 200 | 300 | 17.10 | 11.00 | 4.46 | 5.70 | 3.91 | 1.50 | 26.57 |
| Enamel, oil base (material #10) | | | | | | | | | |
| Brush 1st finish coat | | | | | | | | | |
| Slow | 100 | 325 | 25.90 | 10.25 | 2.79 | 7.97 | 6.51 | 4.13 | 31.65 |
| Medium | 150 | 300 | 22.70 | 11.00 | 3.71 | 7.57 | 5.46 | 3.05 | 30.79 |
| Fast | 200 | 275 | 19.40 | 11.00 | 4.46 | 7.05 | 4.16 | 1.60 | 28.27 |

| | Labor SF per manhour | Material coverage SF/gallon | Material cost per gallon | Labor cost per 100 SF | Labor burden 100 SF | Material cost per 100 SF | Overhead per 100 SF | Profit per 100 SF | Total price per 100 SF |
|---|---|---|---|---|---|---|---|---|---|
| Brush 2nd or additional finish coats | | | | | | | | | |
| Slow | 125 | 350 | 22.80 | 8.20 | 2.23 | 6.51 | 5.25 | 3.33 | 25.52 |
| Medium | 163 | 325 | 19.90 | 10.12 | 3.40 | 6.12 | 4.81 | 2.69 | 27.14 |
| Fast | 200 | 300 | 17.10 | 11.00 | 4.46 | 5.70 | 3.91 | 1.50 | 26.57 |
| Epoxy coating, 2 part system - white (material #52) | | | | | | | | | |
| Brush 1st coat | | | | | | | | | |
| Slow | 175 | 350 | 57.40 | 5.86 | 1.59 | 16.40 | 7.40 | 4.69 | 35.94 |
| Medium | 200 | 325 | 50.20 | 8.25 | 2.78 | 15.45 | 6.49 | 3.63 | 36.60 |
| Fast | 225 | 300 | 43.10 | 9.78 | 3.95 | 14.37 | 5.20 | 2.00 | 35.30 |
| Brush 2nd or additional coats | | | | | | | | | |
| Slow | 200 | 375 | 57.40 | 5.13 | 1.39 | 15.31 | 6.77 | 4.29 | 32.89 |
| Medium | 225 | 350 | 50.20 | 7.33 | 2.46 | 14.34 | 5.91 | 3.31 | 33.35 |
| Fast | 250 | 325 | 43.10 | 8.80 | 3.57 | 13.26 | 4.74 | 1.82 | 32.19 |

Measurements are based on the square feet of wall coated. These figures assume paint products are being applied over a smooth finish. For heights above 8 feet, use the High Time Difficulty Factors on page 137. "Slow" work is based on a $10.25 hourly wage, "Medium" work on a $16.50 hourly wage, and "Fast" work on a $22.00 hourly wage. Other qualifications that apply to this table are on page 9.

## Walls, gypsum drywall, orange peel or knock down, roll, per 100 SF of wall

| | Labor SF per manhour | Material coverage SF/gallon | Material cost per gallon | Labor cost per 100 SF | Labor burden 100 SF | Material cost per 100 SF | Overhead per 100 SF | Profit per 100 SF | Total price per 100 SF |
|---|---|---|---|---|---|---|---|---|---|
| **Flat latex, water base (material #5)** | | | | | | | | | |
| Roll 1st coat | | | | | | | | | |
| Slow | 400 | 300 | 16.20 | 2.56 | .70 | 5.40 | 2.68 | 1.70 | 13.04 |
| Medium | 538 | 275 | 14.20 | 3.07 | 1.03 | 5.16 | 2.27 | 1.27 | 12.80 |
| Fast | 675 | 250 | 12.10 | 3.26 | 1.32 | 4.84 | 1.74 | .67 | 11.83 |
| Roll 2nd coat | | | | | | | | | |
| Slow | 500 | 325 | 16.20 | 2.05 | .56 | 4.98 | 2.35 | 1.49 | 11.43 |
| Medium | 600 | 313 | 14.20 | 2.75 | .93 | 4.54 | 2.01 | 1.13 | 11.36 |
| Fast | 700 | 300 | 12.10 | 3.14 | 1.28 | 4.03 | 1.56 | .60 | 10.61 |
| Roll 3rd or additional coats | | | | | | | | | |
| Slow | 550 | 350 | 16.20 | 1.86 | .51 | 4.63 | 2.17 | 1.38 | 10.55 |
| Medium | 650 | 338 | 14.20 | 2.54 | .86 | 4.20 | 1.86 | 1.04 | 10.50 |
| Fast | 750 | 325 | 12.10 | 2.93 | 1.18 | 3.72 | 1.45 | .56 | 9.84 |
| **Sealer (drywall), water base (material #1)** | | | | | | | | | |
| Roll prime coat | | | | | | | | | |
| Slow | 325 | 275 | 15.50 | 3.15 | .87 | 5.64 | 2.99 | 1.90 | 14.55 |
| Medium | 500 | 263 | 13.50 | 3.30 | 1.11 | 5.13 | 2.34 | 1.31 | 13.19 |
| Fast | 675 | 250 | 11.60 | 3.26 | 1.32 | 4.64 | 1.71 | .66 | 11.59 |
| **Sealer (drywall), oil base (material #2)** | | | | | | | | | |
| Roll prime coat | | | | | | | | | |
| Slow | 325 | 275 | 20.70 | 3.15 | .87 | 7.53 | 3.58 | 2.27 | 17.40 |
| Medium | 500 | 263 | 18.10 | 3.30 | 1.11 | 6.88 | 2.77 | 1.55 | 15.61 |
| Fast | 675 | 250 | 15.50 | 3.26 | 1.32 | 6.20 | 1.99 | .77 | 13.54 |
| **Enamel, water base (material #9)** | | | | | | | | | |
| Roll 1st finish coat | | | | | | | | | |
| Slow | 300 | 285 | 22.80 | 3.42 | .92 | 8.00 | 3.83 | 2.43 | 18.60 |
| Medium | 450 | 263 | 19.90 | 3.67 | 1.23 | 7.57 | 3.06 | 1.71 | 17.24 |
| Fast | 600 | 240 | 17.10 | 3.67 | 1.49 | 7.13 | 2.27 | .87 | 15.43 |
| Roll 2nd finish coat | | | | | | | | | |
| Slow | 325 | 300 | 22.80 | 3.15 | .87 | 7.60 | 3.60 | 2.28 | 17.50 |
| Medium | 475 | 288 | 19.90 | 3.47 | 1.18 | 6.91 | 2.83 | 1.58 | 15.97 |
| Fast | 625 | 275 | 17.10 | 3.52 | 1.43 | 6.22 | 2.07 | .79 | 14.03 |
| **Enamel, oil base (material #10)** | | | | | | | | | |
| Roll 1st finish coat | | | | | | | | | |
| Slow | 300 | 250 | 25.90 | 3.42 | .92 | 10.36 | 4.56 | 2.89 | 22.15 |
| Medium | 450 | 238 | 22.70 | 3.67 | 1.23 | 9.54 | 3.54 | 1.98 | 19.96 |
| Fast | 600 | 225 | 19.40 | 3.67 | 1.49 | 8.62 | 2.55 | .98 | 17.31 |

| | Labor SF per manhour | Material coverage SF/gallon | Material cost per gallon | Labor cost per 100 SF | Labor burden 100 SF | Material cost per 100 SF | Overhead per 100 SF | Profit per 100 SF | Total price per 100 SF |
|---|---|---|---|---|---|---|---|---|---|
| Roll 2nd finish coat | | | | | | | | | |
| Slow | 325 | 275 | 25.90 | 3.15 | .87 | 9.42 | 4.16 | 2.64 | 20.24 |
| Medium | 475 | 263 | 22.70 | 3.47 | 1.18 | 8.63 | 3.25 | 1.82 | 18.35 |
| Fast | 625 | 250 | 19.40 | 3.52 | 1.43 | 7.76 | 2.35 | .90 | 15.96 |
| Epoxy coating, 2 part system - white (material #52) | | | | | | | | | |
| Roll 1st coat | | | | | | | | | |
| Slow | 500 | 300 | 57.40 | 2.05 | .56 | 19.13 | 6.74 | 4.27 | 32.75 |
| Medium | 600 | 288 | 50.20 | 2.75 | .93 | 17.43 | 5.17 | 2.89 | 29.17 |
| Fast | 700 | 275 | 43.10 | 3.14 | 1.28 | 15.67 | 3.71 | 1.43 | 25.23 |
| Roll 2nd or additional coats | | | | | | | | | |
| Slow | 550 | 325 | 57.40 | 1.86 | .51 | 17.66 | 6.21 | 3.94 | 30.18 |
| Medium | 650 | 313 | 50.20 | 2.54 | .86 | 16.04 | 4.76 | 2.66 | 26.86 |
| Fast | 750 | 300 | 43.10 | 2.93 | 1.18 | 14.37 | 3.42 | 1.31 | 23.21 |

Measurements are based on the square feet of wall coated. These figures assume paint products are being applied over a smooth finish. For heights above 8 feet, use the High Time Difficulty Factors on page 137. "Slow" work is based on a $10.25 hourly wage, "Medium" work on a $16.50 hourly wage, and "Fast" work on a $22.00 hourly wage. Other qualifications that apply to this table are on page 9.

| | SF of floor area per manhour | SF of floor area per gallon | Material cost per gallon | Labor cost per 100 SF | Labor burden 100 SF | Material cost per 100 SF | Overhead per 100 SF | Profit per 100 SF | Total price per 100 SF |
|---|---|---|---|---|---|---|---|---|---|
| **Walls, gypsum drywall, orange peel or knock-down, roll, per 100 SF of floor area** | | | | | | | | | |
| Flat latex, water base (material #5) | | | | | | | | | |
| Roll 1st coat on walls only | | | | | | | | | |
| Slow | 250 | 175 | 16.20 | 4.10 | 1.12 | 9.26 | 4.49 | 2.85 | 21.82 |
| Medium | 325 | 158 | 14.20 | 5.08 | 1.71 | 8.99 | 3.87 | 2.16 | 21.81 |
| Fast | 400 | 140 | 12.10 | 5.50 | 2.23 | 8.64 | 3.03 | 1.16 | 20.56 |
| Roll 2nd coat on walls only | | | | | | | | | |
| Slow | 300 | 200 | 16.20 | 3.42 | .92 | 8.10 | 3.86 | 2.45 | 18.75 |
| Medium | 400 | 188 | 14.20 | 4.13 | 1.38 | 7.55 | 3.20 | 1.79 | 18.05 |
| Fast | 500 | 175 | 12.10 | 4.40 | 1.78 | 6.91 | 2.42 | .93 | 16.44 |
| Sealer (drywall), water base (material #1) on walls and ceilings | | | | | | | | | |
| Roll prime coat | | | | | | | | | |
| Slow | 100 | 100 | 15.50 | 10.25 | 2.79 | 15.50 | 8.85 | 5.61 | 43.00 |
| Medium | 170 | 88 | 13.50 | 9.71 | 3.26 | 15.34 | 6.94 | 3.88 | 39.13 |
| Fast | 240 | 75 | 11.60 | 9.17 | 3.72 | 15.47 | 5.25 | 2.02 | 35.63 |
| Sealer (drywall), oil base (material #2) on walls and ceilings | | | | | | | | | |
| Roll prime coat | | | | | | | | | |
| Slow | 100 | 100 | 20.70 | 10.25 | 2.79 | 20.70 | 10.46 | 6.63 | 50.83 |
| Medium | 170 | 88 | 18.10 | 9.71 | 3.26 | 20.57 | 8.22 | 4.59 | 46.35 |
| Fast | 240 | 75 | 15.50 | 9.17 | 3.72 | 20.67 | 6.21 | 2.39 | 42.16 |
| Enamel, water base (material #9) on walls and ceilings | | | | | | | | | |
| Roll 1st finish coat | | | | | | | | | |
| Slow | 70 | 100 | 22.80 | 14.64 | 3.99 | 22.80 | 12.84 | 8.14 | 62.41 |
| Medium | 100 | 90 | 19.90 | 16.50 | 5.55 | 22.11 | 10.82 | 6.05 | 61.03 |
| Fast | 135 | 80 | 17.10 | 16.30 | 6.61 | 21.38 | 8.19 | 3.15 | 55.63 |
| Roll 2nd finish coat | | | | | | | | | |
| Slow | 125 | 150 | 22.80 | 8.20 | 2.23 | 15.20 | 7.95 | 5.04 | 38.62 |
| Medium | 175 | 125 | 19.90 | 9.43 | 3.16 | 15.92 | 6.99 | 3.91 | 39.41 |
| Fast | 225 | 100 | 17.10 | 9.78 | 3.95 | 17.10 | 5.71 | 2.19 | 38.73 |

| | SF of floor area per manhour | SF of floor area per gallon | Material cost per gallon | Labor cost per 100 SF | Labor burden 100 SF | Material cost per 100 SF | Overhead per 100 SF | Profit per 100 SF | Total price per 100 SF |
|---|---|---|---|---|---|---|---|---|---|
| Enamel, oil base (material #10) on walls and ceilings | | | | | | | | | |
| Roll 1st finish coat | | | | | | | | | |
| Slow | 70 | 100 | 25.90 | 14.64 | 3.99 | 25.90 | 13.80 | 8.75 | 67.08 |
| Medium | 100 | 90 | 22.70 | 16.50 | 5.55 | 25.22 | 11.58 | 6.47 | 65.32 |
| Fast | 135 | 80 | 19.40 | 16.30 | 6.61 | 24.25 | 8.72 | 3.35 | 59.23 |
| Roll 2nd finish coat | | | | | | | | | |
| Slow | 125 | 150 | 25.90 | 8.20 | 2.23 | 17.27 | 8.59 | 5.44 | 41.73 |
| Medium | 175 | 125 | 22.70 | 9.43 | 3.16 | 18.16 | 7.54 | 4.21 | 42.50 |
| Fast | 225 | 100 | 19.40 | 9.78 | 3.95 | 19.40 | 6.13 | 2.36 | 41.62 |

Measurements for these costs are based on square feet of floor area. The flat wall figures are for painting walls only but the Sealer and Enamel figures are for painting walls and ceilings in wet areas, i.e. kitchens, baths, utility areas, etc. The floor area measurements are from outside wall to outside wall or from the edge of the concrete slab or edge of an interior wall. This method of figuring the costs to paint the walls and ceilings is not as accurate as measuring the actual surface area of the wall or ceiling area directly, but it is much less time consuming. For heights above 8 feet, use the High Time Difficulty Factors on page 137. "Slow" work is based on a $10.25 hourly wage, "Medium" work on a $16.50 hourly wage, and "Fast" work on a $22.00 hourly wage. Other qualifications that apply to this table are on page 9.

| | Labor SF per manhour | Material coverage SF/gallon | Material cost per gallon | Labor cost per 100 SF | Labor burden 100 SF | Material cost per 100 SF | Overhead per 100 SF | Profit per 100 SF | Total price per 100 SF |
|---|---|---|---|---|---|---|---|---|---|
| **Walls, gypsum drywall, orange peel or knock-down, spray, per 100 SF of wall** | | | | | | | | | |
| Flat latex, water base (material #5) | | | | | | | | | |
| Spray 1st coat | | | | | | | | | |
| Slow | 700 | 250 | 16.20 | 1.46 | .40 | 6.48 | 2.59 | 1.64 | 12.57 |
| Medium | 800 | 225 | 14.20 | 2.06 | .70 | 6.31 | 2.22 | 1.24 | 12.53 |
| Fast | 900 | 200 | 12.10 | 2.44 | .99 | 6.05 | 1.75 | .67 | 11.90 |
| Spray 2nd coat | | | | | | | | | |
| Slow | 800 | 300 | 16.20 | 1.28 | .35 | 5.40 | 2.18 | 1.38 | 10.59 |
| Medium | 900 | 275 | 14.20 | 1.83 | .62 | 5.16 | 1.86 | 1.04 | 10.51 |
| Fast | 1000 | 250 | 12.10 | 2.20 | .89 | 4.84 | 1.47 | .56 | 9.96 |
| Spray 3rd or additional coats | | | | | | | | | |
| Slow | 850 | 325 | 16.20 | 1.21 | .33 | 4.98 | 2.02 | 1.28 | 9.82 |
| Medium | 950 | 300 | 14.20 | 1.74 | .58 | 4.73 | 1.73 | .97 | 9.75 |
| Fast | 1050 | 275 | 12.10 | 2.10 | .84 | 4.40 | 1.36 | .52 | 9.22 |
| Sealer (drywall), water base (material #1) | | | | | | | | | |
| Spray prime coat | | | | | | | | | |
| Slow | 575 | 250 | 15.50 | 1.78 | .49 | 6.20 | 2.63 | 1.67 | 12.77 |
| Medium | 738 | 225 | 13.50 | 2.24 | .76 | 6.00 | 2.20 | 1.23 | 12.43 |
| Fast | 900 | 200 | 11.60 | 2.44 | .99 | 5.80 | 1.71 | .66 | 11.60 |
| Sealer (drywall), oil base (material #2) | | | | | | | | | |
| Spray prime coat | | | | | | | | | |
| Slow | 575 | 250 | 20.70 | 1.78 | .49 | 8.28 | 3.27 | 2.07 | 15.89 |
| Medium | 738 | 225 | 18.10 | 2.24 | .76 | 8.04 | 2.70 | 1.51 | 15.25 |
| Fast | 900 | 200 | 15.50 | 2.44 | .99 | 7.75 | 2.07 | .80 | 14.05 |
| Enamel, water base (material #9) | | | | | | | | | |
| Spray 1st finish coat | | | | | | | | | |
| Slow | 500 | 250 | 22.80 | 2.05 | .56 | 9.12 | 3.64 | 2.31 | 17.68 |
| Medium | 675 | 238 | 19.90 | 2.44 | .82 | 8.36 | 2.85 | 1.59 | 16.06 |
| Fast | 850 | 225 | 17.10 | 2.59 | 1.06 | 7.60 | 2.08 | .80 | 14.13 |
| Spray 2nd finish coat | | | | | | | | | |
| Slow | 525 | 275 | 22.80 | 1.95 | .53 | 8.29 | 3.34 | 2.12 | 16.23 |
| Medium | 700 | 263 | 19.90 | 2.36 | .79 | 7.57 | 2.63 | 1.47 | 14.82 |
| Fast | 875 | 250 | 17.10 | 2.51 | 1.01 | 6.84 | 1.92 | .74 | 13.02 |
| Spray 3rd or additional finish coats | | | | | | | | | |
| Slow | 575 | 425 | 22.80 | 1.78 | .49 | 5.36 | 2.37 | 1.50 | 11.50 |
| Medium | 775 | 388 | 19.90 | 2.13 | .71 | 5.13 | 1.96 | 1.09 | 11.02 |
| Fast | 925 | 350 | 17.10 | 2.38 | .96 | 4.89 | 1.52 | .59 | 10.34 |

| | Labor SF per manhour | Material coverage SF/gallon | Material cost per gallon | Labor cost per 100 SF | Labor burden 100 SF | Material cost per 100 SF | Overhead per 100 SF | Profit per 100 SF | Total price per 100 SF |
|---|---|---|---|---|---|---|---|---|---|
| Enamel, oil base (material #10) | | | | | | | | | |
| Spray 1st finish coat | | | | | | | | | |
| Slow | 500 | 250 | 25.90 | 2.05 | .56 | 10.36 | 4.02 | 2.55 | 19.54 |
| Medium | 675 | 225 | 22.70 | 2.44 | .82 | 10.09 | 3.27 | 1.83 | 18.45 |
| Fast | 850 | 200 | 19.40 | 2.59 | 1.06 | 9.70 | 2.47 | .95 | 16.77 |
| Spray 2nd finish coat | | | | | | | | | |
| Slow | 525 | 275 | 25.90 | 1.95 | .53 | 9.42 | 3.69 | 2.34 | 17.93 |
| Medium | 700 | 250 | 22.70 | 2.36 | .79 | 9.08 | 3.00 | 1.68 | 16.91 |
| Fast | 875 | 225 | 19.40 | 2.51 | 1.01 | 8.62 | 2.25 | .86 | 15.25 |
| Spray 3rd or additional finish coat | | | | | | | | | |
| Slow | 575 | 425 | 25.90 | 1.78 | .49 | 6.09 | 2.59 | 1.64 | 12.59 |
| Medium | 775 | 388 | 22.70 | 2.13 | .71 | 5.85 | 2.13 | 1.19 | 12.01 |
| Fast | 925 | 350 | 19.40 | 2.38 | .96 | 5.54 | 1.64 | .63 | 11.15 |

Measurements are based on the square feet of wall coated. These figures assume paint products are being applied over a smooth finish. For heights above 8 feet, use the High Time Difficulty Factors on page 137. "Slow" work is based on a $10.25 hourly wage, "Medium" work on a $16.50 hourly wage, and "Fast" work on a $22.00 hourly wage. Other qualifications that apply to this table are on page 9.

| | Labor SF per manhour | Material coverage SF/gallon | Material cost per gallon | Labor cost per 100 SF | Labor burden 100 SF | Material cost per 100 SF | Overhead per 100 SF | Profit per 100 SF | Total price per 100 SF |
|---|---|---|---|---|---|---|---|---|---|
| **Walls, gypsum drywall, skip trowel or sand finish, brush, per 100 square feet of wall** | | | | | | | | | |
| Flat latex, water base (material #5) | | | | | | | | | |
| Brush 1st coat | | | | | | | | | |
| Slow | 175 | 325 | 16.20 | 5.86 | 1.59 | 4.98 | 3.86 | 2.45 | 18.74 |
| Medium | 200 | 313 | 14.20 | 8.25 | 2.78 | 4.54 | 3.81 | 2.13 | 21.51 |
| Fast | 225 | 300 | 12.10 | 9.78 | 3.95 | 4.03 | 3.29 | 1.26 | 22.31 |
| Brush 2nd coat | | | | | | | | | |
| Slow | 200 | 400 | 16.20 | 5.13 | 1.39 | 4.05 | 3.28 | 2.08 | 15.93 |
| Medium | 225 | 375 | 14.20 | 7.33 | 2.46 | 3.79 | 3.33 | 1.86 | 18.77 |
| Fast | 250 | 350 | 12.10 | 8.80 | 3.57 | 3.46 | 2.93 | 1.13 | 19.89 |
| Brush 3rd or additional coats | | | | | | | | | |
| Slow | 225 | 425 | 16.20 | 4.56 | 1.23 | 3.81 | 2.98 | 1.89 | 14.47 |
| Medium | 250 | 400 | 14.20 | 6.60 | 2.22 | 3.55 | 3.03 | 1.69 | 17.09 |
| Fast | 275 | 375 | 12.10 | 8.00 | 3.25 | 3.23 | 2.68 | 1.03 | 18.19 |
| Sealer (drywall), water base (material #1) | | | | | | | | | |
| Brush prime coat | | | | | | | | | |
| Slow | 140 | 325 | 15.50 | 7.32 | 1.99 | 4.77 | 4.36 | 2.77 | 21.21 |
| Medium | 183 | 313 | 13.50 | 9.02 | 3.02 | 4.31 | 4.01 | 2.24 | 22.60 |
| Fast | 225 | 300 | 11.60 | 9.78 | 3.95 | 3.87 | 3.26 | 1.25 | 22.11 |
| Sealer (drywall), oil base (material #2) | | | | | | | | | |
| Brush prime coat | | | | | | | | | |
| Slow | 140 | 350 | 20.70 | 7.32 | 1.99 | 5.91 | 4.72 | 2.99 | 22.93 |
| Medium | 183 | 338 | 18.10 | 9.02 | 3.02 | 5.36 | 4.27 | 2.39 | 24.06 |
| Fast | 225 | 325 | 15.50 | 9.78 | 3.95 | 4.77 | 3.43 | 1.32 | 23.25 |
| Enamel, water base (material #9) | | | | | | | | | |
| Brush 1st finish coat | | | | | | | | | |
| Slow | 125 | 350 | 22.80 | 8.20 | 2.23 | 6.51 | 5.25 | 3.33 | 25.52 |
| Medium | 175 | 325 | 19.90 | 9.43 | 3.16 | 6.12 | 4.59 | 2.56 | 25.86 |
| Fast | 225 | 300 | 17.10 | 9.78 | 3.95 | 5.70 | 3.60 | 1.38 | 24.41 |
| Brush 2nd or additional finish coats | | | | | | | | | |
| Slow | 140 | 350 | 22.80 | 7.32 | 1.99 | 6.51 | 4.90 | 3.11 | 23.83 |
| Medium | 185 | 325 | 19.90 | 8.92 | 3.01 | 6.12 | 4.42 | 2.47 | 24.94 |
| Fast | 235 | 300 | 17.10 | 9.36 | 3.81 | 5.70 | 3.49 | 1.34 | 23.70 |
| Enamel, oil base (material #10) | | | | | | | | | |
| Brush 1st finish coat | | | | | | | | | |
| Slow | 125 | 350 | 25.90 | 8.20 | 2.23 | 7.40 | 5.53 | 3.50 | 26.86 |
| Medium | 175 | 325 | 22.70 | 9.43 | 3.16 | 6.98 | 4.80 | 2.68 | 27.05 |
| Fast | 225 | 300 | 19.40 | 9.78 | 3.95 | 6.47 | 3.74 | 1.44 | 25.38 |

| | Labor SF per manhour | Material coverage SF/gallon | Material cost per gallon | Labor cost per 100 SF | Labor burden 100 SF | Material cost per 100 SF | Overhead per 100 SF | Profit per 100 SF | Total price per 100 SF |
|---|---|---|---|---|---|---|---|---|---|
| Brush 2nd or additional finish coats | | | | | | | | | |
| Slow | 140 | 350 | 25.90 | 7.32 | 1.99 | 7.40 | 5.18 | 3.28 | 25.17 |
| Medium | 185 | 338 | 22.70 | 8.92 | 3.01 | 6.72 | 4.57 | 2.55 | 25.77 |
| Fast | 235 | 325 | 19.40 | 9.36 | 3.81 | 5.97 | 3.54 | 1.36 | 24.04 |
| Epoxy coating, 2 part system - white (material #52) | | | | | | | | | |
| Brush 1st coat | | | | | | | | | |
| Slow | 200 | 375 | 57.40 | 5.13 | 1.39 | 15.31 | 6.77 | 4.29 | 32.89 |
| Medium | 225 | 350 | 50.20 | 7.33 | 2.46 | 14.34 | 5.91 | 3.31 | 33.35 |
| Fast | 250 | 325 | 43.10 | 8.80 | 3.57 | 13.26 | 4.74 | 1.82 | 32.19 |
| Brush 2nd or additional coats | | | | | | | | | |
| Slow | 225 | 400 | 57.40 | 4.56 | 1.23 | 14.35 | 6.25 | 3.96 | 30.35 |
| Medium | 250 | 375 | 50.20 | 6.60 | 2.22 | 13.39 | 5.44 | 3.04 | 30.69 |
| Fast | 275 | 350 | 43.10 | 8.00 | 3.25 | 12.31 | 4.36 | 1.67 | 29.59 |

Measurements are based on the square feet of wall coated. These figures assume paint products are being applied over a smooth finish. For heights above 8 feet, use the High Time Difficulty Factors on page 137. "Slow" work is based on a $10.25 hourly wage, "Medium" work on a $16.50 hourly wage, and "Fast" work on a $22.00 hourly wage. Other qualifications that apply to this table are on page 9.

| | Labor SF per manhour | Material coverage SF/gallon | Material cost per gallon | Labor cost per 100 SF | Labor burden 100 SF | Material cost per 100 SF | Overhead per 100 SF | Profit per 100 SF | Total price per 100 SF |
|---|---|---|---|---|---|---|---|---|---|

## Walls, gypsum drywall, skip trowel or sand finish, roll, per 100 square feet of wall

| | Labor SF per manhour | Material coverage SF/gallon | Material cost per gallon | Labor cost per 100 SF | Labor burden 100 SF | Material cost per 100 SF | Overhead per 100 SF | Profit per 100 SF | Total price per 100 SF |
|---|---|---|---|---|---|---|---|---|---|
| Flat latex, water base (material #5) | | | | | | | | | |
| Roll 1st coat | | | | | | | | | |
| Slow | 275 | 325 | 16.20 | 3.73 | 1.02 | 4.98 | 3.02 | 1.91 | 14.66 |
| Medium | 488 | 300 | 14.20 | 3.38 | 1.14 | 4.73 | 2.27 | 1.27 | 12.79 |
| Fast | 700 | 275 | 12.10 | 3.14 | 1.28 | 4.40 | 1.63 | .63 | 11.08 |
| Roll 2nd coat | | | | | | | | | |
| Slow | 350 | 350 | 16.20 | 2.93 | .80 | 4.63 | 2.59 | 1.64 | 12.59 |
| Medium | 538 | 338 | 14.20 | 3.07 | 1.03 | 4.20 | 2.03 | 1.14 | 11.47 |
| Fast | 725 | 325 | 12.10 | 3.03 | 1.24 | 3.72 | 1.48 | .57 | 10.04 |
| Roll 3rd or additional coats | | | | | | | | | |
| Slow | 425 | 350 | 16.20 | 2.41 | .65 | 4.63 | 2.39 | 1.51 | 11.59 |
| Medium | 600 | 338 | 14.20 | 2.75 | .93 | 4.20 | 1.93 | 1.08 | 10.89 |
| Fast | 775 | 325 | 12.10 | 2.84 | 1.15 | 3.72 | 1.43 | .55 | 9.69 |
| Sealer (drywall), water base (material #1) | | | | | | | | | |
| Roll prime coat | | | | | | | | | |
| Slow | 225 | 325 | 15.50 | 4.56 | 1.23 | 4.77 | 3.28 | 2.08 | 15.92 |
| Medium | 463 | 300 | 13.50 | 3.56 | 1.20 | 4.50 | 2.27 | 1.27 | 12.80 |
| Fast | 700 | 275 | 11.60 | 3.14 | 1.28 | 4.22 | 1.60 | .61 | 10.85 |
| Sealer (drywall), oil base (material #2) | | | | | | | | | |
| Roll prime coat | | | | | | | | | |
| Slow | 225 | 325 | 20.70 | 4.56 | 1.23 | 6.37 | 3.77 | 2.39 | 18.32 |
| Medium | 463 | 313 | 18.10 | 3.56 | 1.20 | 5.78 | 2.58 | 1.44 | 14.56 |
| Fast | 700 | 300 | 15.50 | 3.14 | 1.28 | 5.17 | 1.77 | .68 | 12.04 |
| Enamel, water base (material #9) | | | | | | | | | |
| Roll 1st finish coat | | | | | | | | | |
| Slow | 225 | 300 | 22.80 | 4.56 | 1.23 | 7.60 | 4.15 | 2.63 | 20.17 |
| Medium | 400 | 288 | 19.90 | 4.13 | 1.38 | 6.91 | 3.05 | 1.70 | 17.17 |
| Fast | 600 | 275 | 17.10 | 3.67 | 1.49 | 6.22 | 2.11 | .81 | 14.30 |
| Roll 2nd or additional finish coats | | | | | | | | | |
| Slow | 275 | 300 | 22.80 | 3.73 | 1.02 | 7.60 | 3.83 | 2.43 | 18.61 |
| Medium | 450 | 288 | 19.90 | 3.67 | 1.23 | 6.91 | 2.89 | 1.62 | 16.32 |
| Fast | 650 | 275 | 17.10 | 3.38 | 1.38 | 6.22 | 2.03 | .78 | 13.79 |
| Enamel, oil base (material #10) | | | | | | | | | |
| Roll 1st finish coat | | | | | | | | | |
| Slow | 225 | 275 | 25.90 | 4.56 | 1.23 | 9.42 | 4.72 | 2.99 | 22.92 |
| Medium | 400 | 263 | 22.70 | 4.13 | 1.38 | 8.63 | 3.47 | 1.94 | 19.55 |
| Fast | 600 | 250 | 19.40 | 3.67 | 1.49 | 7.76 | 2.39 | .92 | 16.23 |

| | Labor SF per manhour | Material coverage SF/gallon | Material cost per gallon | Labor cost per 100 SF | Labor burden 100 SF | Material cost per 100 SF | Overhead per 100 SF | Profit per 100 SF | Total price per 100 SF |
|---|---|---|---|---|---|---|---|---|---|
| Roll 2nd or additional finish coats | | | | | | | | | |
| Slow | 275 | 300 | 25.90 | 3.73 | 1.02 | 8.63 | 4.15 | 2.63 | 20.16 |
| Medium | 450 | 288 | 22.70 | 3.67 | 1.23 | 7.88 | 3.13 | 1.75 | 17.66 |
| Fast | 650 | 275 | 19.40 | 3.38 | 1.38 | 7.05 | 2.18 | .84 | 14.83 |
| Epoxy coating, 2 part system - white (material #52) | | | | | | | | | |
| Roll 1st coat | | | | | | | | | |
| Slow | 350 | 350 | 57.40 | 2.93 | .80 | 16.40 | 6.24 | 3.96 | 30.33 |
| Medium | 550 | 325 | 50.20 | 3.00 | 1.01 | 15.45 | 4.77 | 2.67 | 26.90 |
| Fast | 725 | 300 | 43.10 | 3.03 | 1.24 | 14.37 | 3.45 | 1.32 | 23.41 |
| Roll 2nd or additional coats | | | | | | | | | |
| Slow | 425 | 375 | 57.40 | 2.41 | .65 | 15.31 | 5.70 | 3.61 | 27.68 |
| Medium | 600 | 350 | 50.20 | 2.75 | .93 | 14.34 | 4.41 | 2.47 | 24.90 |
| Fast | 775 | 325 | 43.10 | 2.84 | 1.15 | 13.26 | 3.19 | 1.23 | 21.67 |

Measurements are based on the square feet of wall coated. These figures assume paint products are being applied over a smooth finish. For heights above 8 feet, use the High Time Difficulty Factors on page 137. "Slow" work is based on a $10.25 hourly wage, "Medium" work on a $16.50 hourly wage, and "Fast" work on a $22.00 hourly wage. Other qualifications that apply to this table are on page 9.

| | Labor SF per manhour | Material coverage SF/gallon | Material cost per gallon | Labor cost per 100 SF | Labor burden 100 SF | Material cost per 100 SF | Overhead per 100 SF | Profit per 100 SF | Total price per 100 SF |
|---|---|---|---|---|---|---|---|---|---|
| **Walls, gypsum drywall, skip trowel or sand finish, spray, per 100 square feet of wall** | | | | | | | | | |
| Flat latex, water base (material #5) | | | | | | | | | |
| Spray 1st coat | | | | | | | | | |
| Slow | 700 | 275 | 16.20 | 1.46 | .40 | 5.89 | 2.40 | 1.52 | 11.67 |
| Medium | 800 | 250 | 14.20 | 2.06 | .70 | 5.68 | 2.07 | 1.16 | 11.67 |
| Fast | 900 | 225 | 12.10 | 2.44 | .99 | 5.38 | 1.63 | .63 | 11.07 |
| Spray 2nd coat | | | | | | | | | |
| Slow | 800 | 325 | 16.20 | 1.28 | .35 | 4.98 | 2.05 | 1.30 | 9.96 |
| Medium | 900 | 300 | 14.20 | 1.83 | .62 | 4.73 | 1.76 | .98 | 9.92 |
| Fast | 1000 | 275 | 12.10 | 2.20 | .89 | 4.40 | 1.39 | .53 | 9.41 |
| Spray 3rd or additional coats | | | | | | | | | |
| Slow | 850 | 325 | 16.20 | 1.21 | .33 | 4.98 | 2.02 | 1.28 | 9.82 |
| Medium | 950 | 313 | 14.20 | 1.74 | .58 | 4.54 | 1.68 | .94 | 9.48 |
| Fast | 1050 | 300 | 12.10 | 2.10 | .84 | 4.03 | 1.29 | .50 | 8.76 |
| Sealer (drywall), water base (material #1) | | | | | | | | | |
| Spray prime coat | | | | | | | | | |
| Slow | 575 | 275 | 15.50 | 1.78 | .49 | 5.64 | 2.45 | 1.55 | 11.91 |
| Medium | 738 | 250 | 13.50 | 2.24 | .76 | 5.40 | 2.06 | 1.15 | 11.61 |
| Fast | 900 | 225 | 11.60 | 2.44 | .99 | 5.16 | 1.59 | .61 | 10.79 |
| Sealer (drywall), oil base (material #2) | | | | | | | | | |
| Spray prime coat | | | | | | | | | |
| Slow | 575 | 275 | 20.70 | 1.78 | .49 | 7.53 | 3.04 | 1.93 | 14.77 |
| Medium | 738 | 250 | 18.10 | 2.24 | .76 | 7.24 | 2.51 | 1.40 | 14.15 |
| Fast | 900 | 225 | 15.50 | 2.44 | .99 | 6.89 | 1.91 | .73 | 12.96 |
| Enamel, water base (material #9) | | | | | | | | | |
| Spray 1st finish coat | | | | | | | | | |
| Slow | 500 | 275 | 22.80 | 2.05 | .56 | 8.29 | 3.38 | 2.14 | 16.42 |
| Medium | 675 | 250 | 19.90 | 2.44 | .82 | 7.96 | 2.75 | 1.54 | 15.51 |
| Fast | 850 | 225 | 17.10 | 2.59 | 1.06 | 7.60 | 2.08 | .80 | 14.13 |
| Spray 2nd or additional finish coats | | | | | | | | | |
| Slow | 525 | 275 | 22.80 | 1.95 | .53 | 8.29 | 3.34 | 2.12 | 16.23 |
| Medium | 700 | 263 | 19.90 | 2.36 | .79 | 7.57 | 2.63 | 1.47 | 14.82 |
| Fast | 900 | 250 | 17.10 | 2.44 | .99 | 6.84 | 1.90 | .73 | 12.90 |

| | Labor SF per manhour | Material coverage SF/gallon | Material cost per gallon | Labor cost per 100 SF | Labor burden 100 SF | Material cost per 100 SF | Overhead per 100 SF | Profit per 100 SF | Total price per 100 SF |
|---|---|---|---|---|---|---|---|---|---|
| Enamel, oil base (material #10) | | | | | | | | | |
| Spray 1st finish coat | | | | | | | | | |
| Slow | 500 | 250 | 25.90 | 2.05 | .56 | 10.36 | 4.02 | 2.55 | 19.54 |
| Medium | 675 | 238 | 22.70 | 2.44 | .82 | 9.54 | 3.14 | 1.75 | 17.69 |
| Fast | 850 | 225 | 19.40 | 2.59 | 1.06 | 8.62 | 2.27 | .87 | 15.41 |
| Spray 2nd or additional finish coats | | | | | | | | | |
| Slow | 525 | 275 | 25.90 | 1.95 | .53 | 9.42 | 3.69 | 2.34 | 17.93 |
| Medium | 700 | 263 | 22.70 | 2.36 | .79 | 8.63 | 2.89 | 1.61 | 16.28 |
| Fast | 900 | 250 | 19.40 | 2.44 | .99 | 7.76 | 2.07 | .80 | 14.06 |

Measurements are based on the square feet of wall coated. These figures assume paint products are being applied over a smooth finish. For heights above 8 feet, use the High Time Difficulty Factors on page 137. "Slow" work is based on a $10.25 hourly wage, "Medium" work on a $16.50 hourly wage, and "Fast" work on a $22.00 hourly wage. Other qualifications that apply to this table are on page 9.

| | Labor SF per manhour | Material coverage SF/gallon | Material cost per gallon | Labor cost per 100 SF | Labor burden 100 SF | Material cost per 100 SF | Overhead per 100 SF | Profit per 100 SF | Total price per 100 SF |
|---|---|---|---|---|---|---|---|---|---|

## Walls, gypsum drywall, smooth finish, brush, per 100 square feet of wall

| | Labor SF per manhour | Material coverage SF/gallon | Material cost per gallon | Labor cost per 100 SF | Labor burden 100 SF | Material cost per 100 SF | Overhead per 100 SF | Profit per 100 SF | Total price per 100 SF |
|---|---|---|---|---|---|---|---|---|---|
| Flat latex, water base (material #5) | | | | | | | | | |
| Brush 1st coat | | | | | | | | | |
| Slow | 175 | 325 | 16.20 | 5.86 | 1.59 | 4.98 | 3.86 | 2.45 | 18.74 |
| Medium | 200 | 313 | 14.20 | 8.25 | 2.78 | 4.54 | 3.81 | 2.13 | 21.51 |
| Fast | 225 | 300 | 12.10 | 9.78 | 3.95 | 4.03 | 3.29 | 1.26 | 22.31 |
| Brush 2nd coat | | | | | | | | | |
| Slow | 225 | 400 | 16.20 | 4.56 | 1.23 | 4.05 | 3.05 | 1.94 | 14.83 |
| Medium | 250 | 375 | 14.20 | 6.60 | 2.22 | 3.79 | 3.09 | 1.73 | 17.43 |
| Fast | 275 | 350 | 12.10 | 8.00 | 3.25 | 3.46 | 2.72 | 1.05 | 18.48 |
| Brush 3rd or additional coats | | | | | | | | | |
| Slow | 250 | 425 | 16.20 | 4.10 | 1.12 | 3.81 | 2.80 | 1.77 | 13.60 |
| Medium | 275 | 400 | 14.20 | 6.00 | 2.03 | 3.55 | 2.83 | 1.58 | 15.99 |
| Fast | 300 | 375 | 12.10 | 7.33 | 2.97 | 3.23 | 2.50 | .96 | 16.99 |
| Sealer (drywall), water base (material #1) | | | | | | | | | |
| Brush prime coat | | | | | | | | | |
| Slow | 150 | 325 | 15.50 | 6.83 | 1.87 | 4.77 | 4.17 | 2.64 | 20.28 |
| Medium | 188 | 313 | 13.50 | 8.78 | 2.95 | 4.31 | 3.93 | 2.20 | 22.17 |
| Fast | 225 | 300 | 11.60 | 9.78 | 3.95 | 3.87 | 3.26 | 1.25 | 22.11 |
| Sealer (drywall), oil base (material #2) | | | | | | | | | |
| Brush prime coat | | | | | | | | | |
| Slow | 150 | 350 | 20.70 | 6.83 | 1.87 | 5.91 | 4.53 | 2.87 | 22.01 |
| Medium | 188 | 338 | 18.10 | 8.78 | 2.95 | 5.36 | 4.19 | 2.34 | 23.62 |
| Fast | 225 | 325 | 15.50 | 9.78 | 3.95 | 4.77 | 3.43 | 1.32 | 23.25 |
| Enamel, water base (material #9) | | | | | | | | | |
| Brush 1st finish coat | | | | | | | | | |
| Slow | 140 | 350 | 22.80 | 7.32 | 1.99 | 6.51 | 4.90 | 3.11 | 23.83 |
| Medium | 185 | 325 | 19.90 | 8.92 | 3.01 | 6.12 | 4.42 | 2.47 | 24.94 |
| Fast | 235 | 300 | 17.10 | 9.36 | 3.81 | 5.70 | 3.49 | 1.34 | 23.70 |
| Brush 2nd or additional finish coats | | | | | | | | | |
| Slow | 150 | 350 | 22.80 | 6.83 | 1.87 | 6.51 | 4.71 | 2.99 | 22.91 |
| Medium | 200 | 333 | 19.90 | 8.25 | 2.78 | 5.98 | 4.17 | 2.33 | 23.51 |
| Fast | 250 | 315 | 17.10 | 8.80 | 3.57 | 5.43 | 3.29 | 1.27 | 22.36 |
| Enamel, oil base (material #10) | | | | | | | | | |
| Brush 1st finish coat | | | | | | | | | |
| Slow | 140 | 350 | 25.90 | 7.32 | 1.99 | 7.40 | 5.18 | 3.28 | 25.17 |
| Medium | 185 | 338 | 22.70 | 8.92 | 3.01 | 6.72 | 4.57 | 2.55 | 25.77 |
| Fast | 235 | 325 | 19.40 | 9.36 | 3.81 | 5.97 | 3.54 | 1.36 | 24.04 |

| | Labor SF per manhour | Material coverage SF/gallon | Material cost per gallon | Labor cost per 100 SF | Labor burden 100 SF | Material cost per 100 SF | Overhead per 100 SF | Profit per 100 SF | Total price per 100 SF |
|---|---|---|---|---|---|---|---|---|---|
| Brush 2nd or additional finish coats | | | | | | | | | |
| Slow | 150 | 360 | 25.90 | 6.83 | 1.87 | 7.19 | 4.92 | 3.12 | 23.93 |
| Medium | 200 | 348 | 22.70 | 8.25 | 2.78 | 6.52 | 4.30 | 2.40 | 24.25 |
| Fast | 250 | 335 | 19.40 | 8.80 | 3.57 | 5.79 | 3.36 | 1.29 | 22.81 |
| Stipple finish | | | | | | | | | |
| Slow | 225 | -- | -- | 4.56 | 1.23 | -- | 1.80 | 1.14 | 8.73 |
| Medium | 250 | -- | -- | 6.60 | 2.22 | -- | 2.16 | 1.21 | 12.19 |
| Fast | 275 | -- | -- | 8.00 | 3.25 | -- | 2.08 | .80 | 14.13 |
| Epoxy coating, 2 part system - white (material #52) | | | | | | | | | |
| Brush 1st coat | | | | | | | | | |
| Slow | 225 | 425 | 57.40 | 4.56 | 1.23 | 13.51 | 5.99 | 3.80 | 29.09 |
| Medium | 250 | 400 | 50.20 | 6.60 | 2.22 | 12.55 | 5.24 | 2.93 | 29.54 |
| Fast | 275 | 375 | 43.10 | 8.00 | 3.25 | 11.49 | 4.21 | 1.62 | 28.57 |
| Brush 2nd or additional coats | | | | | | | | | |
| Slow | 250 | 450 | 57.40 | 4.10 | 1.12 | 12.76 | 5.57 | 3.53 | 27.08 |
| Medium | 275 | 425 | 50.20 | 6.00 | 2.03 | 11.81 | 4.86 | 2.72 | 27.42 |
| Fast | 300 | 400 | 43.10 | 7.33 | 2.97 | 10.78 | 3.90 | 1.50 | 26.48 |

Measurements are based on the square feet of wall coated. These figures assume paint products are being applied over a smooth finish. For heights above 8 feet, use the High Time Difficulty Factors on page 137. "Slow" work is based on a $10.25 hourly wage, "Medium" work on a $16.50 hourly wage, and "Fast" work on a $22.00 hourly wage. Other qualifications that apply to this table are on page 9.

| | Labor SF per manhour | Material coverage SF/gallon | Material cost per gallon | Labor cost per 100 SF | Labor burden 100 SF | Material cost per 100 SF | Overhead per 100 SF | Profit per 100 SF | Total price per 100 SF |
|---|---|---|---|---|---|---|---|---|---|
| **Walls, gypsum drywall, smooth finish, roll, per 100 square feet of wall** | | | | | | | | | |
| Flat latex, water base (material #5) | | | | | | | | | |
| Roll 1st coat | | | | | | | | | |
| Slow | 300 | 325 | 16.20 | 3.42 | .92 | 4.98 | 2.89 | 1.83 | 14.04 |
| Medium | 513 | 313 | 14.20 | 3.22 | 1.08 | 4.54 | 2.17 | 1.21 | 12.22 |
| Fast | 725 | 300 | 12.10 | 3.03 | 1.24 | 4.03 | 1.53 | .59 | 10.42 |
| Roll 2nd coat | | | | | | | | | |
| Slow | 375 | 375 | 16.20 | 2.73 | .75 | 4.32 | 2.41 | 1.53 | 11.74 |
| Medium | 563 | 363 | 14.20 | 2.93 | .99 | 3.91 | 1.92 | 1.07 | 10.82 |
| Fast | 750 | 350 | 12.10 | 2.93 | 1.18 | 3.46 | 1.40 | .54 | 9.51 |
| Roll 3rd or additional coats | | | | | | | | | |
| Slow | 450 | 400 | 16.20 | 2.28 | .61 | 4.05 | 2.15 | 1.37 | 10.46 |
| Medium | 625 | 388 | 14.20 | 2.64 | .89 | 3.66 | 1.76 | .98 | 9.93 |
| Fast | 800 | 375 | 12.10 | 2.75 | 1.12 | 3.23 | 1.31 | .50 | 8.91 |
| Sealer (drywall), water base (material #1) | | | | | | | | | |
| Roll prime coat | | | | | | | | | |
| Slow | 245 | 350 | 15.50 | 4.18 | 1.14 | 4.43 | 3.02 | 1.92 | 14.69 |
| Medium | 485 | 325 | 13.50 | 3.40 | 1.14 | 4.15 | 2.13 | 1.19 | 12.01 |
| Fast | 725 | 300 | 11.60 | 3.03 | 1.24 | 3.87 | 1.50 | .58 | 10.22 |
| Sealer (drywall), oil base (material #2) | | | | | | | | | |
| Roll prime coat | | | | | | | | | |
| Slow | 245 | 300 | 20.70 | 4.18 | 1.14 | 6.90 | 3.79 | 2.40 | 18.41 |
| Medium | 485 | 288 | 18.10 | 3.40 | 1.14 | 6.28 | 2.65 | 1.48 | 14.95 |
| Fast | 725 | 275 | 15.50 | 3.03 | 1.24 | 5.64 | 1.83 | .70 | 12.44 |
| Enamel, water base (material #9) | | | | | | | | | |
| Roll 1st finish coat | | | | | | | | | |
| Slow | 235 | 325 | 22.80 | 4.36 | 1.20 | 7.02 | 3.90 | 2.47 | 18.95 |
| Medium | 438 | 313 | 19.90 | 3.77 | 1.26 | 6.36 | 2.79 | 1.56 | 15.74 |
| Fast | 640 | 300 | 17.10 | 3.44 | 1.38 | 5.70 | 1.95 | .75 | 13.22 |
| Roll 2nd or additional finish coats | | | | | | | | | |
| Slow | 280 | 350 | 22.80 | 3.66 | 1.00 | 6.51 | 3.46 | 2.19 | 16.82 |
| Medium | 465 | 338 | 19.90 | 3.55 | 1.19 | 5.89 | 2.60 | 1.46 | 14.69 |
| Fast | 680 | 325 | 17.10 | 3.24 | 1.31 | 5.26 | 1.81 | .70 | 12.32 |
| Enamel, oil base (material #10) | | | | | | | | | |
| Roll 1st finish coat | | | | | | | | | |
| Slow | 235 | 300 | 25.90 | 4.36 | 1.20 | 8.63 | 4.40 | 2.79 | 21.38 |
| Medium | 438 | 288 | 22.70 | 3.77 | 1.26 | 7.88 | 3.17 | 1.77 | 17.85 |
| Fast | 640 | 275 | 19.40 | 3.44 | 1.38 | 7.05 | 2.20 | .84 | 14.91 |

| | Labor SF per manhour | Material coverage SF/gallon | Material cost per gallon | Labor cost per 100 SF | Labor burden 100 SF | Material cost per 100 SF | Overhead per 100 SF | Profit per 100 SF | Total price per 100 SF |
|---|---|---|---|---|---|---|---|---|---|
| Roll 2nd or additional finish coats | | | | | | | | | |
| Slow | 280 | 325 | 25.90 | 3.66 | 1.00 | 7.97 | 3.92 | 2.48 | 19.03 |
| Medium | 465 | 313 | 22.70 | 3.55 | 1.19 | 7.25 | 2.94 | 1.64 | 16.57 |
| Fast | 680 | 300 | 19.40 | 3.24 | 1.31 | 6.47 | 2.04 | .78 | 13.84 |
| Epoxy coating, 2 part system - white (material #52) | | | | | | | | | |
| Roll 1st coat | | | | | | | | | |
| Slow | 375 | 400 | 57.40 | 2.73 | .75 | 14.35 | 5.52 | 3.50 | 26.85 |
| Medium | 550 | 375 | 50.20 | 3.00 | 1.01 | 13.39 | 4.26 | 2.38 | 24.04 |
| Fast | 750 | 350 | 43.10 | 2.93 | 1.18 | 12.31 | 3.04 | 1.17 | 20.63 |
| Roll 2nd or additional coats | | | | | | | | | |
| Slow | 450 | 425 | 57.40 | 2.28 | .61 | 13.51 | 5.09 | 3.23 | 24.72 |
| Medium | 625 | 400 | 50.20 | 2.64 | .89 | 12.55 | 3.94 | 2.20 | 22.22 |
| Fast | 800 | 375 | 43.10 | 2.75 | 1.12 | 11.49 | 2.84 | 1.09 | 19.29 |

Measurements are based on the square feet of wall coated. These figures assume paint products are being applied over a smooth finish. For heights above 8 feet, use the High Time Difficulty Factors on page 137. "Slow" work is based on a $10.25 hourly wage, "Medium" work on a $16.50 hourly wage, and "Fast" work on a $22.00 hourly wage. Other qualifications that apply to this table are on page 9.

| | Labor SF per manhour | Material coverage SF/gallon | Material cost per gallon | Labor cost per 100 SF | Labor burden 100 SF | Material cost per 100 SF | Overhead per 100 SF | Profit per 100 SF | Total price per 100 SF |
|---|---|---|---|---|---|---|---|---|---|

## Walls, gypsum drywall, smooth finish, spray, per 100 square feet of wall

| | Labor SF per manhour | Material coverage SF/gallon | Material cost per gallon | Labor cost per 100 SF | Labor burden 100 SF | Material cost per 100 SF | Overhead per 100 SF | Profit per 100 SF | Total price per 100 SF |
|---|---|---|---|---|---|---|---|---|---|
| **Flat latex, water base (material #5)** | | | | | | | | | |
| Spray 1st coat | | | | | | | | | |
| Slow | 750 | 300 | 16.20 | 1.37 | .36 | 5.40 | 2.21 | 1.40 | 10.74 |
| Medium | 850 | 275 | 14.20 | 1.94 | .66 | 5.16 | 1.90 | 1.06 | 10.72 |
| Fast | 950 | 250 | 12.10 | 2.32 | .93 | 4.84 | 1.50 | .58 | 10.17 |
| Spray 2nd coat | | | | | | | | | |
| Slow | 850 | 350 | 16.20 | 1.21 | .33 | 4.63 | 1.91 | 1.21 | 9.29 |
| Medium | 950 | 325 | 14.20 | 1.74 | .58 | 4.37 | 1.64 | .92 | 9.25 |
| Fast | 1050 | 300 | 12.10 | 2.10 | .84 | 4.03 | 1.29 | .50 | 8.76 |
| Spray 3rd or additional coats | | | | | | | | | |
| Slow | 950 | 350 | 16.20 | 1.08 | .29 | 4.63 | 1.86 | 1.18 | 9.04 |
| Medium | 1050 | 338 | 14.20 | 1.57 | .52 | 4.20 | 1.54 | .86 | 8.69 |
| Fast | 1150 | 325 | 12.10 | 1.91 | .78 | 3.72 | 1.18 | .45 | 8.04 |
| **Sealer (drywall), water base (material #1)** | | | | | | | | | |
| Spray prime coat | | | | | | | | | |
| Slow | 600 | 300 | 15.50 | 1.71 | .47 | 5.17 | 2.28 | 1.44 | 11.07 |
| Medium | 775 | 275 | 13.50 | 2.13 | .71 | 4.91 | 1.90 | 1.06 | 10.71 |
| Fast | 950 | 250 | 12.10 | 2.32 | .93 | 4.84 | 1.50 | .58 | 10.17 |
| **Sealer (drywall), oil base (material #2)** | | | | | | | | | |
| Spray prime coat | | | | | | | | | |
| Slow | 600 | 275 | 20.70 | 1.71 | .47 | 7.53 | 3.01 | 1.91 | 14.63 |
| Medium | 775 | 250 | 18.10 | 2.13 | .71 | 7.24 | 2.47 | 1.38 | 13.93 |
| Fast | 950 | 225 | 15.50 | 2.32 | .93 | 6.89 | 1.88 | .72 | 12.74 |
| **Enamel, water base (material #9)** | | | | | | | | | |
| Spray 1st finish coat | | | | | | | | | |
| Slow | 525 | 300 | 22.80 | 1.95 | .53 | 7.60 | 3.12 | 1.98 | 15.18 |
| Medium | 713 | 275 | 19.90 | 2.31 | .78 | 7.24 | 2.53 | 1.41 | 14.27 |
| Fast | 900 | 250 | 17.10 | 2.44 | .99 | 6.84 | 1.90 | .73 | 12.90 |
| Spray 2nd or additional finish coats | | | | | | | | | |
| Slow | 600 | 300 | 22.80 | 1.71 | .47 | 7.60 | 3.03 | 1.92 | 14.73 |
| Medium | 788 | 288 | 19.90 | 2.09 | .71 | 6.91 | 2.38 | 1.33 | 13.42 |
| Fast | 975 | 275 | 17.10 | 2.26 | .92 | 6.22 | 1.74 | .67 | 11.81 |

| | Labor SF per manhour | Material coverage SF/gallon | Material cost per gallon | Labor cost per 100 SF | Labor burden 100 SF | Material cost per 100 SF | Overhead per 100 SF | Profit per 100 SF | Total price per 100 SF |
|---|---|---|---|---|---|---|---|---|---|
| Enamel, oil base (material #10) | | | | | | | | | |
| Spray 1st finish coat | | | | | | | | | |
| Slow | 525 | 275 | 25.90 | 1.95 | .53 | 9.42 | 3.69 | 2.34 | 17.93 |
| Medium | 713 | 263 | 22.70 | 2.31 | .78 | 8.63 | 2.87 | 1.60 | 16.19 |
| Fast | 900 | 250 | 19.40 | 2.44 | .99 | 7.76 | 2.07 | .80 | 14.06 |
| Spray 2nd or additional finish coats | | | | | | | | | |
| Slow | 600 | 300 | 25.90 | 1.71 | .47 | 8.63 | 3.35 | 2.12 | 16.28 |
| Medium | 788 | 288 | 22.70 | 2.09 | .71 | 7.88 | 2.61 | 1.46 | 14.75 |
| Fast | 975 | 275 | 19.40 | 2.26 | .92 | 7.05 | 1.89 | .73 | 12.85 |

Measurements are based on the square feet of wall coated. These figures assume paint products are being applied over a smooth finish. For heights above 8 feet, use the High Time Difficulty Factors on page 137. "Slow" work is based on a $10.25 hourly wage, "Medium" work on a $16.50 hourly wage, and "Fast" work on a $22.00 hourly wage. Other qualifications that apply to this table are on page 9.

## Walls, plaster, exterior: see Plaster and stucco

| | Labor SF per manhour | Material coverage SF/gallon | Material cost per gallon | Labor cost per 100 SF | Labor burden 100 SF | Material cost per 100 SF | Overhead per 100 SF | Profit per 100 SF | Total price per 100 SF |
|---|---|---|---|---|---|---|---|---|---|
| **Walls, plaster, interior, medium texture, per 100 square feet of wall** | | | | | | | | | |
| Anti-graffiti stain eliminator | | | | | | | | | |
| Water base primer and sealer (material #39) | | | | | | | | | |
| Roll & brush each coat | | | | | | | | | |
| Slow | 375 | 425 | 30.40 | 2.73 | .75 | 7.15 | 3.29 | 2.09 | 16.01 |
| Medium | 400 | 400 | 26.60 | 4.13 | 1.38 | 6.65 | 2.98 | 1.67 | 16.81 |
| Fast | 425 | 375 | 22.80 | 5.18 | 2.09 | 6.08 | 2.47 | .95 | 16.77 |
| Oil base primer and sealer (material #40) | | | | | | | | | |
| Roll & brush each coat | | | | | | | | | |
| Slow | 375 | 400 | 28.20 | 2.73 | .75 | 7.05 | 3.26 | 2.07 | 15.86 |
| Medium | 400 | 375 | 24.70 | 4.13 | 1.38 | 6.59 | 2.97 | 1.66 | 16.73 |
| Fast | 425 | 350 | 21.20 | 5.18 | 2.09 | 6.06 | 2.47 | .95 | 16.75 |
| Polyurethane 2 part system (material #41) | | | | | | | | | |
| Roll & brush each coat | | | | | | | | | |
| Slow | 325 | 375 | 63.30 | 3.15 | .87 | 16.88 | 6.48 | 4.11 | 31.49 |
| Medium | 350 | 350 | 55.40 | 4.71 | 1.60 | 15.83 | 5.42 | 3.03 | 30.59 |
| Fast | 375 | 325 | 47.40 | 5.87 | 2.39 | 14.58 | 4.22 | 1.62 | 28.68 |

Measurements are based on the square feet of wall coated. These figures assume paint products are being applied over a smooth finish. For heights above 8 feet, use the High Time Difficulty Factors on page 137. "Slow" work is based on a $10.25 hourly wage, "Medium" work on a $16.50 hourly wage, and "Fast" work on a $22.00 hourly wage. Other qualifications that apply to this table are on page 9.

| | Labor SF per manhour | Material coverage SF/gallon | Material cost per gallon | Labor cost per 100 SF | Labor burden 100 SF | Material cost per 100 SF | Overhead per 100 SF | Profit per 100 SF | Total price per 100 SF |
|---|---|---|---|---|---|---|---|---|---|

## Walls, plaster, interior, medium texture, brush, per 100 square feet of wall

| | Labor SF per manhour | Material coverage SF/gallon | Material cost per gallon | Labor cost per 100 SF | Labor burden 100 SF | Material cost per 100 SF | Overhead per 100 SF | Profit per 100 SF | Total price per 100 SF |
|---|---|---|---|---|---|---|---|---|---|
| Flat latex, water base (material #5) | | | | | | | | | |
| Brush 1st coat | | | | | | | | | |
| Slow | 125 | 300 | 16.20 | 8.20 | 2.23 | 5.40 | 4.91 | 3.11 | 23.85 |
| Medium | 150 | 288 | 14.20 | 11.00 | 3.71 | 4.93 | 4.81 | 2.69 | 27.14 |
| Fast | 175 | 275 | 12.10 | 12.57 | 5.09 | 4.40 | 4.08 | 1.57 | 27.71 |
| Brush 2nd coat | | | | | | | | | |
| Slow | 150 | 325 | 16.20 | 6.83 | 1.87 | 4.98 | 4.24 | 2.69 | 20.61 |
| Medium | 168 | 313 | 14.20 | 9.82 | 3.30 | 4.54 | 4.33 | 2.42 | 24.41 |
| Fast | 185 | 300 | 12.10 | 11.89 | 4.84 | 4.03 | 3.84 | 1.47 | 26.07 |
| Brush 3rd or additional coats | | | | | | | | | |
| Slow | 160 | 350 | 16.20 | 6.41 | 1.74 | 4.63 | 3.96 | 2.51 | 19.25 |
| Medium | 185 | 338 | 14.20 | 8.92 | 3.01 | 4.20 | 3.95 | 2.21 | 22.29 |
| Fast | 210 | 325 | 12.10 | 10.48 | 4.24 | 3.72 | 3.41 | 1.31 | 23.16 |
| Enamel, water base (material #9) | | | | | | | | | |
| Brush 1st finish coat | | | | | | | | | |
| Slow | 100 | 300 | 22.80 | 10.25 | 2.79 | 7.60 | 6.40 | 4.06 | 31.10 |
| Medium | 125 | 288 | 19.90 | 13.20 | 4.44 | 6.91 | 6.01 | 3.36 | 33.92 |
| Fast | 150 | 275 | 17.10 | 14.67 | 5.95 | 6.22 | 4.97 | 1.91 | 33.72 |
| Brush 2nd finish coat | | | | | | | | | |
| Slow | 125 | 325 | 22.80 | 8.20 | 2.23 | 7.02 | 5.41 | 3.43 | 26.29 |
| Medium | 143 | 313 | 19.90 | 11.54 | 3.87 | 6.36 | 5.34 | 2.98 | 30.09 |
| Fast | 160 | 300 | 17.10 | 13.75 | 5.58 | 5.70 | 4.63 | 1.78 | 31.44 |
| Brush 3rd or additional finish coats | | | | | | | | | |
| Slow | 135 | 350 | 22.80 | 7.59 | 2.07 | 6.51 | 5.01 | 3.18 | 24.36 |
| Medium | 160 | 338 | 19.90 | 10.31 | 3.47 | 5.89 | 4.82 | 2.69 | 27.18 |
| Fast | 185 | 325 | 17.10 | 11.89 | 4.84 | 5.26 | 4.06 | 1.56 | 27.61 |
| Enamel, oil base (material #10) | | | | | | | | | |
| Brush 1st finish coat | | | | | | | | | |
| Slow | 100 | 325 | 25.90 | 10.25 | 2.79 | 7.97 | 6.51 | 4.13 | 31.65 |
| Medium | 125 | 313 | 22.70 | 13.20 | 4.44 | 7.25 | 6.10 | 3.41 | 34.40 |
| Fast | 150 | 300 | 19.40 | 14.67 | 5.95 | 6.47 | 5.01 | 1.93 | 34.03 |
| Brush 2nd finish coat | | | | | | | | | |
| Slow | 125 | 400 | 25.90 | 8.20 | 2.23 | 6.48 | 5.24 | 3.32 | 25.47 |
| Medium | 143 | 375 | 22.70 | 11.54 | 3.87 | 6.05 | 5.26 | 2.94 | 29.66 |
| Fast | 160 | 350 | 19.40 | 13.75 | 5.58 | 5.54 | 4.60 | 1.77 | 31.24 |

| | Labor SF per manhour | Material coverage SF/gallon | Material cost per gallon | Labor cost per 100 SF | Labor burden 100 SF | Material cost per 100 SF | Overhead per 100 SF | Profit per 100 SF | Total price per 100 SF |
|---|---|---|---|---|---|---|---|---|---|
| Brush 3rd or additional finish coats | | | | | | | | | |
| Slow | 135 | 425 | 25.90 | 7.59 | 2.07 | 6.09 | 4.88 | 3.09 | 23.72 |
| Medium | 160 | 400 | 22.70 | 10.31 | 3.47 | 5.68 | 4.77 | 2.67 | 26.90 |
| Fast | 185 | 375 | 19.40 | 11.89 | 4.84 | 5.17 | 4.05 | 1.56 | 27.51 |
| Stipple finish | | | | | | | | | |
| Slow | 125 | -- | -- | 8.20 | 2.23 | -- | 3.23 | 2.05 | 15.71 |
| Medium | 143 | -- | -- | 11.54 | 3.87 | -- | 3.78 | 2.11 | 21.30 |
| Fast | 160 | -- | -- | 13.75 | 5.58 | -- | 3.58 | 1.37 | 24.28 |
| Epoxy coating, 2 part system - white (material #52) | | | | | | | | | |
| Brush 1st coat | | | | | | | | | |
| Slow | 150 | 400 | 57.40 | 6.83 | 1.87 | 14.35 | 7.14 | 4.53 | 34.72 |
| Medium | 165 | 388 | 50.20 | 10.00 | 3.36 | 12.94 | 6.45 | 3.60 | 36.35 |
| Fast | 185 | 375 | 43.10 | 11.89 | 4.84 | 11.49 | 5.22 | 2.01 | 35.45 |
| Brush 2nd or additional coats | | | | | | | | | |
| Slow | 160 | 425 | 57.40 | 6.41 | 1.74 | 13.51 | 6.72 | 4.26 | 32.64 |
| Medium | 185 | 413 | 50.20 | 8.92 | 3.01 | 12.15 | 5.90 | 3.30 | 33.28 |
| Fast | 210 | 400 | 43.10 | 10.48 | 4.24 | 10.78 | 4.72 | 1.81 | 32.03 |
| Glazing & mottling over enamel (material #16) | | | | | | | | | |
| Brush each coat | | | | | | | | | |
| Slow | 50 | 900 | 29.00 | 20.50 | 5.58 | 3.22 | 9.09 | 5.76 | 44.15 |
| Medium | 65 | 800 | 25.40 | 25.38 | 8.53 | 3.18 | 9.09 | 5.08 | 51.26 |
| Fast | 80 | 700 | 21.80 | 27.50 | 11.15 | 3.11 | 7.73 | 2.97 | 52.46 |
| Stipple | | | | | | | | | |
| Slow | 100 | -- | -- | 10.25 | 2.79 | -- | 4.04 | 2.56 | 19.64 |
| Medium | 113 | -- | -- | 14.60 | 4.91 | -- | 4.78 | 2.67 | 26.96 |
| Fast | 125 | -- | -- | 17.60 | 7.14 | -- | 4.58 | 1.76 | 31.08 |

Measurements are based on the square feet of wall coated. These figures assume paint products are being applied over a smooth finish. For heights above 8 feet, use the High Time Difficulty Factors on page 137. "Slow" work is based on a $10.25 hourly wage, "Medium" work on a $16.50 hourly wage, and "Fast" work on a $22.00 hourly wage. Other qualifications that apply to this table are on page 9.

| | Labor SF per manhour | Material coverage SF/gallon | Material cost per gallon | Labor cost per 100 SF | Labor burden 100 SF | Material cost per 100 SF | Overhead per 100 SF | Profit per 100 SF | Total price per 100 SF |
|---|---|---|---|---|---|---|---|---|---|
| **Walls, plaster, interior, medium texture, roll, per 100 square feet of wall** | | | | | | | | | |
| Flat latex, water base (material #5) | | | | | | | | | |
| Roll 1st coat | | | | | | | | | |
| Slow | 225 | 250 | 16.20 | 4.56 | 1.23 | 6.48 | 3.81 | 2.41 | 18.49 |
| Medium | 438 | 238 | 14.20 | 3.77 | 1.26 | 5.97 | 2.70 | 1.51 | 15.21 |
| Fast | 650 | 225 | 12.10 | 3.38 | 1.38 | 5.38 | 1.87 | .72 | 12.73 |
| Roll 2nd coat | | | | | | | | | |
| Slow | 250 | 300 | 16.20 | 4.10 | 1.12 | 5.40 | 3.29 | 2.09 | 16.00 |
| Medium | 463 | 288 | 14.20 | 3.56 | 1.20 | 4.93 | 2.37 | 1.33 | 13.39 |
| Fast | 675 | 275 | 12.10 | 3.26 | 1.32 | 4.40 | 1.66 | .64 | 11.28 |
| Roll 3rd or additional coats | | | | | | | | | |
| Slow | 275 | 325 | 16.20 | 3.73 | 1.02 | 4.98 | 3.02 | 1.91 | 14.66 |
| Medium | 500 | 313 | 14.20 | 3.30 | 1.11 | 4.54 | 2.19 | 1.23 | 12.37 |
| Fast | 725 | 300 | 12.10 | 3.03 | 1.24 | 4.03 | 1.53 | .59 | 10.42 |
| Enamel, water base (material #9) | | | | | | | | | |
| Roll 1st finish coat | | | | | | | | | |
| Slow | 200 | 250 | 22.80 | 5.13 | 1.39 | 9.12 | 4.85 | 3.08 | 23.57 |
| Medium | 413 | 238 | 19.90 | 4.00 | 1.34 | 8.36 | 3.36 | 1.88 | 18.94 |
| Fast | 625 | 225 | 17.10 | 3.52 | 1.43 | 7.60 | 2.32 | .89 | 15.76 |
| Roll 2nd finish coat | | | | | | | | | |
| Slow | 225 | 300 | 22.80 | 4.56 | 1.23 | 7.60 | 4.15 | 2.63 | 20.17 |
| Medium | 438 | 288 | 19.90 | 3.77 | 1.26 | 6.91 | 2.93 | 1.64 | 16.51 |
| Fast | 650 | 275 | 17.10 | 3.38 | 1.38 | 6.22 | 2.03 | .78 | 13.79 |
| Roll 3rd or additional finish coats | | | | | | | | | |
| Slow | 250 | 325 | 22.80 | 4.10 | 1.12 | 7.02 | 3.79 | 2.40 | 18.43 |
| Medium | 475 | 313 | 19.90 | 3.47 | 1.18 | 6.36 | 2.70 | 1.51 | 15.22 |
| Fast | 700 | 300 | 17.10 | 3.14 | 1.28 | 5.70 | 1.87 | .72 | 12.71 |
| Enamel, oil base (material #10) | | | | | | | | | |
| Roll 1st finish coat | | | | | | | | | |
| Slow | 200 | 275 | 25.90 | 5.13 | 1.39 | 9.42 | 4.94 | 3.13 | 24.01 |
| Medium | 413 | 263 | 22.70 | 4.00 | 1.34 | 8.63 | 3.43 | 1.92 | 19.32 |
| Fast | 625 | 250 | 19.40 | 3.52 | 1.43 | 7.76 | 2.35 | .90 | 15.96 |
| Roll 2nd finish coat | | | | | | | | | |
| Slow | 225 | 350 | 25.90 | 4.56 | 1.23 | 7.40 | 4.09 | 2.59 | 19.87 |
| Medium | 438 | 325 | 22.70 | 3.77 | 1.26 | 6.98 | 2.94 | 1.65 | 16.60 |
| Fast | 650 | 300 | 19.40 | 3.38 | 1.38 | 6.47 | 2.08 | .80 | 14.11 |

| | Labor SF per manhour | Material coverage SF/gallon | Material cost per gallon | Labor cost per 100 SF | Labor burden 100 SF | Material cost per 100 SF | Overhead per 100 SF | Profit per 100 SF | Total price per 100 SF |
|---|---|---|---|---|---|---|---|---|---|
| Roll 3rd or additional finish coats | | | | | | | | | |
| Slow | 250 | 375 | 25.90 | 4.10 | 1.12 | 6.91 | 3.76 | 2.38 | 18.27 |
| Medium | 475 | 350 | 22.70 | 3.47 | 1.18 | 6.49 | 2.73 | 1.52 | 15.39 |
| Fast | 700 | 325 | 19.40 | 3.14 | 1.28 | 5.97 | 1.92 | .74 | 13.05 |
| Epoxy coating, 2 part system - white (material #52) | | | | | | | | | |
| Roll 1st coat | | | | | | | | | |
| Slow | 250 | 350 | 57.40 | 4.10 | 1.12 | 16.40 | 6.70 | 4.25 | 32.57 |
| Medium | 463 | 335 | 50.20 | 3.56 | 1.20 | 14.99 | 4.84 | 2.70 | 27.29 |
| Fast | 675 | 320 | 43.10 | 3.26 | 1.32 | 13.47 | 3.34 | 1.28 | 22.67 |
| Roll 2nd or additional coats | | | | | | | | | |
| Slow | 275 | 400 | 57.40 | 3.73 | 1.02 | 14.35 | 5.92 | 3.75 | 28.77 |
| Medium | 500 | 375 | 50.20 | 3.30 | 1.11 | 13.39 | 4.36 | 2.44 | 24.60 |
| Fast | 725 | 350 | 43.10 | 3.03 | 1.24 | 12.31 | 3.07 | 1.18 | 20.83 |

Measurements are based on the square feet of wall coated. These figures assume paint products are being applied over a smooth finish. For heights above 8 feet, use the High Time Difficulty Factors on page 137. "Slow" work is based on a $10.25 hourly wage, "Medium" work on a $16.50 hourly wage, and "Fast" work on a $22.00 hourly wage. Other qualifications that apply to this table are on page 9.

| | Labor SF per manhour | Material coverage SF/gallon | Material cost per gallon | Labor cost per 100 SF | Labor burden 100 SF | Material cost per 100 SF | Overhead per 100 SF | Profit per 100 SF | Total price per 100 SF |
|---|---|---|---|---|---|---|---|---|---|
| **Walls, plaster, interior, medium texture, spray, per 100 square feet of wall** | | | | | | | | | |
| Flat latex, water base (material #5) | | | | | | | | | |
| Spray 1st coat | | | | | | | | | |
| Slow | 475 | 350 | 16.20 | 2.16 | .59 | 4.63 | 2.29 | 1.45 | 11.12 |
| Medium | 600 | 313 | 14.20 | 2.75 | .93 | 4.54 | 2.01 | 1.13 | 11.36 |
| Fast | 725 | 275 | 12.10 | 3.03 | 1.24 | 4.40 | 1.60 | .62 | 10.89 |
| Spray 2nd coat | | | | | | | | | |
| Slow | 525 | 400 | 16.20 | 1.95 | .53 | 4.05 | 2.02 | 1.28 | 9.83 |
| Medium | 675 | 350 | 14.20 | 2.44 | .82 | 4.06 | 1.79 | 1.00 | 10.11 |
| Fast | 825 | 300 | 12.10 | 2.67 | 1.07 | 4.03 | 1.44 | .55 | 9.76 |
| Spray 3rd or additional coats | | | | | | | | | |
| Slow | 575 | 450 | 16.20 | 1.78 | .49 | 3.60 | 1.82 | 1.15 | 8.84 |
| Medium | 750 | 388 | 14.20 | 2.20 | .73 | 3.66 | 1.62 | .90 | 9.11 |
| Fast | 925 | 325 | 12.10 | 2.38 | .96 | 3.72 | 1.31 | .50 | 8.87 |
| Enamel, water base (material #9) | | | | | | | | | |
| Spray 1st finish coat | | | | | | | | | |
| Slow | 450 | 350 | 22.80 | 2.28 | .61 | 6.51 | 2.92 | 1.85 | 14.17 |
| Medium | 575 | 313 | 19.90 | 2.87 | .97 | 6.36 | 2.50 | 1.40 | 14.10 |
| Fast | 700 | 275 | 17.10 | 3.14 | 1.28 | 6.22 | 1.97 | .76 | 13.37 |
| Spray 2nd finish coat | | | | | | | | | |
| Slow | 500 | 400 | 22.80 | 2.05 | .56 | 5.70 | 2.58 | 1.63 | 12.52 |
| Medium | 650 | 350 | 19.90 | 2.54 | .86 | 5.69 | 2.22 | 1.24 | 12.55 |
| Fast | 800 | 300 | 17.10 | 2.75 | 1.12 | 5.70 | 1.77 | .68 | 12.02 |
| Spray 3rd or additional finish coats | | | | | | | | | |
| Slow | 550 | 450 | 22.80 | 1.86 | .51 | 5.07 | 2.31 | 1.46 | 11.21 |
| Medium | 750 | 388 | 19.90 | 2.20 | .73 | 5.13 | 1.98 | 1.11 | 11.15 |
| Fast | 900 | 325 | 17.10 | 2.44 | .99 | 5.26 | 1.61 | .62 | 10.92 |
| Enamel, oil base (material #10) | | | | | | | | | |
| Spray 1st finish coat | | | | | | | | | |
| Slow | 450 | 400 | 25.90 | 2.28 | .61 | 6.48 | 2.91 | 1.84 | 14.12 |
| Medium | 575 | 363 | 22.70 | 2.87 | .97 | 6.25 | 2.47 | 1.38 | 13.94 |
| Fast | 700 | 325 | 19.40 | 3.14 | 1.28 | 5.97 | 1.92 | .74 | 13.05 |
| Spray 2nd finish coat | | | | | | | | | |
| Slow | 500 | 425 | 25.90 | 2.05 | .56 | 6.09 | 2.70 | 1.71 | 13.11 |
| Medium | 650 | 388 | 22.70 | 2.54 | .86 | 5.85 | 2.26 | 1.27 | 12.78 |
| Fast | 800 | 350 | 19.40 | 2.75 | 1.12 | 5.54 | 1.74 | .67 | 11.82 |

| | Labor SF per manhour | Material coverage SF/gallon | Material cost per gallon | Labor cost per 100 SF | Labor burden 100 SF | Material cost per 100 SF | Overhead per 100 SF | Profit per 100 SF | Total price per 100 SF |
|---|---|---|---|---|---|---|---|---|---|
| Spray 3rd or additional finish coats | | | | | | | | | |
| Slow | 550 | 450 | 25.90 | 1.86 | .51 | 5.76 | 2.52 | 1.60 | 12.25 |
| Medium | 725 | 400 | 22.70 | 2.28 | .76 | 5.68 | 2.14 | 1.20 | 12.06 |
| Fast | 900 | 375 | 19.40 | 2.44 | .99 | 5.17 | 1.59 | .61 | 10.80 |
| Epoxy coating, 2 part system - white (material #52) | | | | | | | | | |
| Spray 1st coat | | | | | | | | | |
| Slow | 525 | 325 | 57.40 | 1.95 | .53 | 17.66 | 6.24 | 3.96 | 30.34 |
| Medium | 675 | 300 | 50.20 | 2.44 | .82 | 16.73 | 4.90 | 2.74 | 27.63 |
| Fast | 825 | 275 | 43.10 | 2.67 | 1.07 | 15.67 | 3.59 | 1.38 | 24.38 |
| Spray 2nd or additional coats | | | | | | | | | |
| Slow | 575 | 350 | 57.40 | 1.78 | .49 | 16.40 | 5.79 | 3.67 | 28.13 |
| Medium | 725 | 325 | 50.20 | 2.28 | .76 | 15.45 | 4.53 | 2.53 | 25.55 |
| Fast | 875 | 300 | 43.10 | 2.51 | 1.01 | 14.37 | 3.31 | 1.27 | 22.47 |

Measurements are based on the square feet of wall coated. These figures assume paint products are being applied over a smooth finish. For heights above 8 feet, use the High Time Difficulty Factors on page 137. "Slow" work is based on a $10.25 hourly wage, "Medium" work on a $16.50 hourly wage, and "Fast" work on a $22.00 hourly wage. Other qualifications that apply to this table are on page 9.

| | Labor SF per manhour | Material coverage SF/gallon | Material cost per gallon | Labor cost per 100 SF | Labor burden 100 SF | Material cost per 100 SF | Overhead per 100 SF | Profit per 100 SF | Total price per 100 SF |
|---|---|---|---|---|---|---|---|---|---|
| **Walls, plaster, interior, rough texture, per 100 square feet of wall** | | | | | | | | | |
| Anti-graffiti stain eliminator | | | | | | | | | |
| Water base primer and sealer (material #39) | | | | | | | | | |
| Roll & brush each coat | | | | | | | | | |
| Slow | 350 | 400 | 30.40 | 2.93 | .80 | 7.60 | 3.51 | 2.23 | 17.07 |
| Medium | 375 | 375 | 26.60 | 4.40 | 1.49 | 7.09 | 3.18 | 1.78 | 17.94 |
| Fast | 400 | 350 | 22.80 | 5.50 | 2.23 | 6.51 | 2.63 | 1.01 | 17.88 |
| Oil base primer and sealer (material #40) | | | | | | | | | |
| Roll & brush each coat | | | | | | | | | |
| Slow | 350 | 375 | 28.20 | 2.93 | .80 | 7.52 | 3.49 | 2.21 | 16.95 |
| Medium | 375 | 350 | 24.70 | 4.40 | 1.49 | 7.06 | 3.17 | 1.77 | 17.89 |
| Fast | 400 | 325 | 21.20 | 5.50 | 2.23 | 6.52 | 2.64 | 1.01 | 17.90 |
| Polyurethane 2 part system (material #41) | | | | | | | | | |
| Roll & brush each coat | | | | | | | | | |
| Slow | 300 | 350 | 63.30 | 3.42 | .92 | 18.09 | 6.96 | 4.41 | 33.80 |
| Medium | 325 | 325 | 55.40 | 5.08 | 1.71 | 17.05 | 5.84 | 3.26 | 32.94 |
| Fast | 350 | 300 | 47.40 | 6.29 | 2.55 | 15.80 | 4.56 | 1.75 | 30.95 |

Measurements are based on the square feet of wall coated. These figures assume paint products are being applied over a rough finish, sand finish, or orange peel texture finish. For heights above 8 feet, use the High Time Difficulty Factors on page 137. "Slow" work is based on a $10.25 hourly wage, "Medium" work on a $16.50 hourly wage, and "Fast" work on a $22.00 hourly wage. Other qualifications that apply to this table are on page 9.

| | Labor SF per manhour | Material coverage SF/gallon | Material cost per gallon | Labor cost per 100 SF | Labor burden 100 SF | Material cost per 100 SF | Overhead per 100 SF | Profit per 100 SF | Total price per 100 SF |
|---|---|---|---|---|---|---|---|---|---|

## Walls, plaster, interior, rough texture, brush, per 100 square feet of wall

| | Labor SF per manhour | Material coverage SF/gallon | Material cost per gallon | Labor cost per 100 SF | Labor burden 100 SF | Material cost per 100 SF | Overhead per 100 SF | Profit per 100 SF | Total price per 100 SF |
|---|---|---|---|---|---|---|---|---|---|
| Flat latex, water base (material #5) | | | | | | | | | |
| Brush 1st coat | | | | | | | | | |
| Slow | 115 | 300 | 16.20 | 8.91 | 2.43 | 5.40 | 5.19 | 3.29 | 25.22 |
| Medium | 140 | 275 | 14.20 | 11.79 | 3.95 | 5.16 | 5.13 | 2.87 | 28.90 |
| Fast | 165 | 250 | 12.10 | 13.33 | 5.41 | 4.84 | 4.36 | 1.68 | 29.62 |
| Brush 2nd coat | | | | | | | | | |
| Slow | 125 | 325 | 16.20 | 8.20 | 2.23 | 4.98 | 4.78 | 3.03 | 23.22 |
| Medium | 153 | 300 | 14.20 | 10.78 | 3.64 | 4.73 | 4.69 | 2.62 | 26.46 |
| Fast | 180 | 275 | 12.10 | 12.22 | 4.97 | 4.40 | 3.99 | 1.53 | 27.11 |
| Brush 3rd or additional coats | | | | | | | | | |
| Slow | 135 | 350 | 16.20 | 7.59 | 2.07 | 4.63 | 4.43 | 2.81 | 21.53 |
| Medium | 168 | 325 | 14.20 | 9.82 | 3.30 | 4.37 | 4.29 | 2.40 | 24.18 |
| Fast | 200 | 300 | 12.10 | 11.00 | 4.46 | 4.03 | 3.61 | 1.39 | 24.49 |
| Enamel, water base (material #9) | | | | | | | | | |
| Brush 1st finish coat | | | | | | | | | |
| Slow | 100 | 300 | 22.80 | 10.25 | 2.79 | 7.60 | 6.40 | 4.06 | 31.10 |
| Medium | 125 | 275 | 19.90 | 13.20 | 4.44 | 7.24 | 6.10 | 3.41 | 34.39 |
| Fast | 150 | 250 | 17.10 | 14.67 | 5.95 | 6.84 | 5.08 | 1.95 | 34.49 |
| Brush 2nd finish coat | | | | | | | | | |
| Slow | 115 | 325 | 22.80 | 8.91 | 2.43 | 7.02 | 5.69 | 3.61 | 27.66 |
| Medium | 140 | 300 | 19.90 | 11.79 | 3.95 | 6.63 | 5.49 | 3.07 | 30.93 |
| Fast | 165 | 275 | 17.10 | 13.33 | 5.41 | 6.22 | 4.62 | 1.77 | 31.35 |
| Brush 3rd or additional finish coats | | | | | | | | | |
| Slow | 125 | 350 | 22.80 | 8.20 | 2.23 | 6.51 | 5.25 | 3.33 | 25.52 |
| Medium | 160 | 325 | 19.90 | 10.31 | 3.47 | 6.12 | 4.88 | 2.73 | 27.51 |
| Fast | 185 | 300 | 17.10 | 11.89 | 4.84 | 5.70 | 4.15 | 1.59 | 28.17 |
| Enamel, oil base (material #10) | | | | | | | | | |
| Brush 1st finish coat | | | | | | | | | |
| Slow | 100 | 300 | 25.90 | 10.25 | 2.79 | 8.63 | 6.72 | 4.26 | 32.65 |
| Medium | 125 | 288 | 22.70 | 13.20 | 4.44 | 7.88 | 6.25 | 3.49 | 35.26 |
| Fast | 150 | 275 | 19.40 | 14.67 | 5.95 | 7.05 | 5.12 | 1.97 | 34.76 |
| Brush 2nd finish coat | | | | | | | | | |
| Slow | 115 | 375 | 25.90 | 8.91 | 2.43 | 6.91 | 5.66 | 3.59 | 27.50 |
| Medium | 140 | 350 | 22.70 | 11.79 | 3.95 | 6.49 | 5.45 | 3.05 | 30.73 |
| Fast | 165 | 325 | 19.40 | 13.33 | 5.41 | 5.97 | 4.57 | 1.76 | 31.04 |

| | Labor SF per manhour | Material coverage SF/gallon | Material cost per gallon | Labor cost per 100 SF | Labor burden 100 SF | Material cost per 100 SF | Overhead per 100 SF | Profit per 100 SF | Total price per 100 SF |
|---|---|---|---|---|---|---|---|---|---|
| Brush 3rd or additional finish coats | | | | | | | | | |
| Slow | 125 | 400 | 25.90 | 8.20 | 2.23 | 6.48 | 5.24 | 3.32 | 25.47 |
| Medium | 160 | 375 | 22.70 | 10.31 | 3.47 | 6.05 | 4.86 | 2.72 | 27.41 |
| Fast | 185 | 350 | 19.40 | 11.89 | 4.84 | 5.54 | 4.12 | 1.58 | 27.97 |
| Epoxy coating, 2 part system - white (material #52) | | | | | | | | | |
| Brush 1st coat | | | | | | | | | |
| Slow | 125 | 375 | 57.40 | 8.20 | 2.23 | 15.31 | 7.98 | 5.06 | 38.78 |
| Medium | 160 | 363 | 50.20 | 10.31 | 3.47 | 13.83 | 6.76 | 3.78 | 38.15 |
| Fast | 180 | 350 | 43.10 | 12.22 | 4.97 | 12.31 | 5.46 | 2.10 | 37.06 |
| Brush 2nd or additional coats | | | | | | | | | |
| Slow | 135 | 400 | 57.40 | 7.59 | 2.07 | 14.35 | 7.44 | 4.72 | 36.17 |
| Medium | 175 | 388 | 50.20 | 9.43 | 3.16 | 12.94 | 6.26 | 3.50 | 35.29 |
| Fast | 200 | 375 | 43.10 | 11.00 | 4.46 | 11.49 | 4.99 | 1.92 | 33.86 |
| Glazing & mottling over enamel (material #16) | | | | | | | | | |
| Brush each coat | | | | | | | | | |
| Slow | 40 | 875 | 29.00 | 25.63 | 6.97 | 3.31 | 11.14 | 7.06 | 54.11 |
| Medium | 50 | 838 | 25.40 | 33.00 | 11.10 | 3.03 | 11.55 | 6.45 | 65.13 |
| Fast | 60 | 800 | 21.80 | 36.67 | 14.87 | 2.73 | 10.04 | 3.86 | 68.17 |
| Stipple | | | | | | | | | |
| Slow | 90 | -- | -- | 11.39 | 3.10 | -- | 4.49 | 2.85 | 21.83 |
| Medium | 103 | -- | -- | 16.02 | 5.39 | -- | 5.25 | 2.93 | 29.59 |
| Fast | 115 | -- | -- | 19.13 | 7.77 | -- | 4.97 | 1.91 | 33.78 |

Measurements are based on the square feet of wall coated. These figures assume paint products are being applied over a rough finish. For heights above 8 feet, use the High Time Difficulty Factors on page 137. "Slow" work is based on a $10.25 hourly wage, "Medium" work on a $16.50 hourly wage, and "Fast" work on a $22.00 hourly wage. Other qualifications that apply to this table are on page 9.

| | Labor SF per manhour | Material coverage SF/gallon | Material cost per gallon | Labor cost per 100 SF | Labor burden 100 SF | Material cost per 100 SF | Overhead per 100 SF | Profit per 100 SF | Total price per 100 SF |
|---|---|---|---|---|---|---|---|---|---|
| **Walls, plaster, interior, rough texture, roll, per 100 square feet of wall** | | | | | | | | | |
| Flat latex, water base (material #5) | | | | | | | | | |
| Roll 1st coat | | | | | | | | | |
| Slow | 200 | 250 | 16.20 | 5.13 | 1.39 | 6.48 | 4.03 | 2.56 | 19.59 |
| Medium | 413 | 238 | 14.20 | 4.00 | 1.34 | 5.97 | 2.77 | 1.55 | 15.63 |
| Fast | 625 | 225 | 12.10 | 3.52 | 1.43 | 5.38 | 1.91 | .73 | 12.97 |
| Roll 2nd coat | | | | | | | | | |
| Slow | 225 | 300 | 16.20 | 4.56 | 1.23 | 5.40 | 3.47 | 2.20 | 16.86 |
| Medium | 438 | 288 | 14.20 | 3.77 | 1.26 | 4.93 | 3.09 | 1.96 | 15.01 |
| Fast | 650 | 275 | 12.10 | 3.38 | 1.38 | 4.40 | 1.69 | .65 | 11.50 |
| Roll 3rd or additional coats | | | | | | | | | |
| Slow | 250 | 325 | 16.20 | 4.10 | 1.12 | 4.98 | 3.16 | 2.00 | 15.36 |
| Medium | 463 | 313 | 14.20 | 3.56 | 1.20 | 4.54 | 2.28 | 1.27 | 12.85 |
| Fast | 675 | 300 | 12.10 | 3.26 | 1.32 | 4.03 | 1.59 | .61 | 10.81 |
| Enamel, water base (material #9) | | | | | | | | | |
| Roll 1st finish coat | | | | | | | | | |
| Slow | 175 | 250 | 22.80 | 5.86 | 1.59 | 9.12 | 5.14 | 3.26 | 24.97 |
| Medium | 388 | 238 | 19.90 | 4.25 | 1.44 | 8.36 | 3.44 | 1.92 | 19.41 |
| Fast | 600 | 225 | 17.10 | 3.67 | 1.49 | 7.60 | 2.36 | .91 | 16.03 |
| Roll 2nd finish coat | | | | | | | | | |
| Slow | 200 | 325 | 22.80 | 5.13 | 1.39 | 7.02 | 4.20 | 2.66 | 20.40 |
| Medium | 413 | 313 | 19.90 | 4.00 | 1.34 | 6.36 | 2.87 | 1.60 | 16.17 |
| Fast | 625 | 300 | 17.10 | 3.52 | 1.43 | 5.70 | 1.97 | .76 | 13.38 |
| Roll 3rd or additional finish coats | | | | | | | | | |
| Slow | 225 | 325 | 22.80 | 4.56 | 1.23 | 7.02 | 3.97 | 2.52 | 19.30 |
| Medium | 438 | 313 | 19.90 | 3.77 | 1.26 | 6.36 | 2.79 | 1.56 | 15.74 |
| Fast | 650 | 300 | 17.10 | 3.38 | 1.38 | 5.70 | 1.93 | .74 | 13.13 |
| Enamel, oil base (material #10) | | | | | | | | | |
| Roll 1st finish coat | | | | | | | | | |
| Slow | 175 | 300 | 25.90 | 5.86 | 1.59 | 8.63 | 4.99 | 3.16 | 24.23 |
| Medium | 388 | 288 | 22.70 | 4.25 | 1.44 | 7.88 | 3.32 | 1.86 | 18.75 |
| Fast | 600 | 275 | 19.40 | 3.67 | 1.49 | 7.05 | 2.26 | .87 | 15.34 |
| Roll 2nd finish coat | | | | | | | | | |
| Slow | 200 | 350 | 25.90 | 5.13 | 1.39 | 7.40 | 4.32 | 2.74 | 20.98 |
| Medium | 413 | 325 | 22.70 | 4.00 | 1.34 | 6.98 | 3.02 | 1.69 | 17.03 |
| Fast | 625 | 300 | 19.40 | 3.52 | 1.43 | 6.47 | 2.11 | .81 | 14.34 |

| | Labor SF per manhour | Material coverage SF/gallon | Material cost per gallon | Labor cost per 100 SF | Labor burden 100 SF | Material cost per 100 SF | Overhead per 100 SF | Profit per 100 SF | Total price per 100 SF |
|---|---|---|---|---|---|---|---|---|---|
| Roll 3rd or additional finish coats | | | | | | | | | |
| Slow | 225 | 375 | 25.90 | 4.56 | 1.23 | 6.91 | 3.94 | 2.50 | 19.14 |
| Medium | 438 | 350 | 22.70 | 3.77 | 1.26 | 6.49 | 2.82 | 1.58 | 15.92 |
| Fast | 650 | 325 | 19.40 | 3.38 | 1.38 | 5.97 | 1.98 | .76 | 13.47 |
| Epoxy coating, 2 part system - white (material #52) | | | | | | | | | |
| Roll 1st coat | | | | | | | | | |
| Slow | 225 | 350 | 57.40 | 4.56 | 1.23 | 16.40 | 6.88 | 4.36 | 33.43 |
| Medium | 438 | 325 | 50.20 | 3.77 | 1.26 | 15.45 | 5.02 | 2.81 | 28.31 |
| Fast | 650 | 300 | 43.10 | 3.38 | 1.38 | 14.37 | 3.54 | 1.36 | 24.03 |
| Roll 2nd or additional coats | | | | | | | | | |
| Slow | 250 | 400 | 57.40 | 4.10 | 1.12 | 14.35 | 6.07 | 3.85 | 29.49 |
| Medium | 463 | 375 | 50.20 | 3.56 | 1.20 | 13.39 | 4.45 | 2.49 | 25.09 |
| Fast | 675 | 350 | 43.10 | 3.26 | 1.32 | 12.31 | 3.12 | 1.20 | 21.21 |

Measurements are based on the square feet of wall coated. These figures assume paint products are being applied over a rough finish. For heights above 8 feet, use the High Time Difficulty Factors on page 137. "Slow" work is based on a $10.25 hourly wage, "Medium" work on a $16.50 hourly wage, and "Fast" work on a $22.00 hourly wage. Other qualifications that apply to this table are on page 9.

| | Labor SF per manhour | Material coverage SF/gallon | Material cost per gallon | Labor cost per 100 SF | Labor burden 100 SF | Material cost per 100 SF | Overhead per 100 SF | Profit per 100 SF | Total price per 100 SF |
|---|---|---|---|---|---|---|---|---|---|

## Walls, plaster, interior, rough texture, spray, per 100 square feet of wall

| | Labor SF per manhour | Material coverage SF/gallon | Material cost per gallon | Labor cost per 100 SF | Labor burden 100 SF | Material cost per 100 SF | Overhead per 100 SF | Profit per 100 SF | Total price per 100 SF |
|---|---|---|---|---|---|---|---|---|---|
| Flat latex, water base (material #5) | | | | | | | | | |
| Spray 1st coat | | | | | | | | | |
| Slow | 500 | 325 | 16.20 | 2.05 | .56 | 4.98 | 2.35 | 1.49 | 11.43 |
| Medium | 600 | 288 | 14.20 | 2.75 | .93 | 4.93 | 2.11 | 1.18 | 11.90 |
| Fast | 700 | 250 | 12.10 | 3.14 | 1.28 | 4.84 | 1.71 | .66 | 11.63 |
| Spray 2nd coat | | | | | | | | | |
| Slow | 600 | 400 | 16.20 | 1.71 | .47 | 4.05 | 1.93 | 1.22 | 9.38 |
| Medium | 700 | 350 | 14.20 | 2.36 | .79 | 4.06 | 1.77 | .99 | 9.97 |
| Fast | 800 | 300 | 12.10 | 2.75 | 1.12 | 4.03 | 1.46 | .56 | 9.92 |
| Spray 3rd or additional coats | | | | | | | | | |
| Slow | 700 | 425 | 16.20 | 1.46 | .40 | 3.81 | 1.76 | 1.11 | 8.54 |
| Medium | 800 | 375 | 14.20 | 2.06 | .70 | 3.79 | 1.60 | .90 | 9.05 |
| Fast | 900 | 325 | 12.10 | 2.44 | .99 | 3.72 | 1.32 | .51 | 8.98 |
| Enamel, water base (material #9) | | | | | | | | | |
| Spray 1st finish coat | | | | | | | | | |
| Slow | 450 | 325 | 22.80 | 2.28 | .61 | 7.02 | 3.08 | 1.95 | 14.94 |
| Medium | 550 | 288 | 19.90 | 3.00 | 1.01 | 6.91 | 2.68 | 1.50 | 15.10 |
| Fast | 650 | 250 | 17.10 | 3.38 | 1.38 | 6.84 | 2.14 | .82 | 14.56 |
| Spray 2nd finish coat | | | | | | | | | |
| Slow | 550 | 400 | 22.80 | 1.86 | .51 | 5.70 | 2.50 | 1.59 | 12.16 |
| Medium | 650 | 350 | 19.90 | 2.54 | .86 | 5.69 | 2.22 | 1.24 | 12.55 |
| Fast | 750 | 300 | 17.10 | 2.93 | 1.18 | 5.70 | 1.82 | .70 | 12.33 |
| Spray 3rd or additional finish coats | | | | | | | | | |
| Slow | 650 | 425 | 22.80 | 1.58 | .43 | 5.36 | 2.28 | 1.45 | 11.10 |
| Medium | 750 | 375 | 19.90 | 2.20 | .73 | 5.31 | 2.02 | 1.13 | 11.39 |
| Fast | 850 | 325 | 17.10 | 2.59 | 1.06 | 5.26 | 1.65 | .63 | 11.19 |
| Enamel, oil base (material #10) | | | | | | | | | |
| Spray 1st finish coat | | | | | | | | | |
| Slow | 450 | 325 | 25.90 | 2.28 | .61 | 7.97 | 3.37 | 2.14 | 16.37 |
| Medium | 550 | 300 | 22.70 | 3.00 | 1.01 | 7.57 | 2.84 | 1.59 | 16.01 |
| Fast | 650 | 275 | 19.40 | 3.38 | 1.38 | 7.05 | 2.18 | .84 | 14.83 |
| Spray 2nd finish coat | | | | | | | | | |
| Slow | 550 | 400 | 25.90 | 1.86 | .51 | 6.48 | 2.74 | 1.74 | 13.33 |
| Medium | 650 | 362 | 22.70 | 2.54 | .86 | 6.27 | 2.37 | 1.32 | 13.36 |
| Fast | 750 | 325 | 19.40 | 2.93 | 1.18 | 5.97 | 1.87 | .72 | 12.67 |

| | Labor SF per manhour | Material coverage SF/gallon | Material cost per gallon | Labor cost per 100 SF | Labor burden 100 SF | Material cost per 100 SF | Overhead per 100 SF | Profit per 100 SF | Total price per 100 SF |
|---|---|---|---|---|---|---|---|---|---|
| Spray 3rd or additional finish coats | | | | | | | | | |
| Slow | 650 | 425 | 25.90 | 1.58 | .43 | 6.09 | 2.51 | 1.59 | 12.20 |
| Medium | 750 | 388 | 22.70 | 2.20 | .73 | 5.85 | 2.15 | 1.20 | 12.13 |
| Fast | 850 | 350 | 19.40 | 2.59 | 1.06 | 5.54 | 1.70 | .65 | 11.54 |
| Epoxy coating, 2 part system - white (material #52) | | | | | | | | | |
| Spray 1st coat | | | | | | | | | |
| Slow | 525 | 325 | 57.40 | 1.95 | .53 | 17.66 | 6.24 | 3.96 | 30.34 |
| Medium | 663 | 313 | 50.20 | 2.49 | .84 | 16.04 | 4.75 | 2.65 | 26.77 |
| Fast | 800 | 300 | 43.10 | 2.75 | 1.12 | 14.37 | 3.37 | 1.30 | 22.91 |
| Spray 2nd or additional coats | | | | | | | | | |
| Slow | 575 | 375 | 57.40 | 1.78 | .49 | 15.31 | 5.45 | 3.45 | 26.48 |
| Medium | 713 | 363 | 50.20 | 2.31 | .78 | 13.83 | 4.15 | 2.32 | 23.39 |
| Fast | 850 | 350 | 43.10 | 2.59 | 1.06 | 12.31 | 2.95 | 1.13 | 20.04 |

Measurements are based on the square feet of wall coated. These figures assume paint products are being applied over a rough finish. For heights above 8 feet, use the High Time Difficulty Factors on page 137. "Slow" work is based on a $10.25 hourly wage, "Medium" work on a $16.50 hourly wage, and "Fast" work on a $22.00 hourly wage. Other qualifications that apply to this table are on page 9.

| | Labor SF per manhour | Material coverage SF/gallon | Material cost per gallon | Labor cost per 100 SF | Labor burden 100 SF | Material cost per 100 SF | Overhead per 100 SF | Profit per 100 SF | Total price per 100 SF |
|---|---|---|---|---|---|---|---|---|---|

## Walls, plaster, interior, smooth finish, per 100 square feet of wall

| | Labor SF per manhour | Material coverage SF/gallon | Material cost per gallon | Labor cost per 100 SF | Labor burden 100 SF | Material cost per 100 SF | Overhead per 100 SF | Profit per 100 SF | Total price per 100 SF |
|---|---|---|---|---|---|---|---|---|---|
| Anti-graffiti stain eliminator | | | | | | | | | |
| Water base primer and sealer (material #39) | | | | | | | | | |
| Roll & brush each coat | | | | | | | | | |
| Slow | 375 | 425 | 30.40 | 2.73 | .75 | 7.15 | 3.29 | 2.09 | 16.01 |
| Medium | 400 | 400 | 26.60 | 4.13 | 1.38 | 6.65 | 2.98 | 1.67 | 16.81 |
| Fast | 425 | 375 | 22.80 | 5.18 | 2.09 | 6.08 | 2.47 | .95 | 16.77 |
| Oil base primer and sealer (material #40) | | | | | | | | | |
| Roll & brush each coat | | | | | | | | | |
| Slow | 375 | 400 | 28.20 | 2.73 | .75 | 7.05 | 3.26 | 2.07 | 15.86 |
| Medium | 400 | 375 | 24.70 | 4.13 | 1.38 | 6.59 | 2.97 | 1.66 | 16.73 |
| Fast | 425 | 350 | 21.20 | 5.18 | 2.09 | 6.06 | 2.47 | .95 | 16.75 |
| Polyurethane 2 part system (material #41) | | | | | | | | | |
| Roll & brush each coat | | | | | | | | | |
| Slow | 325 | 375 | 63.30 | 3.15 | .87 | 16.88 | 6.48 | 4.11 | 31.49 |
| Medium | 350 | 350 | 55.40 | 4.71 | 1.60 | 15.83 | 5.42 | 3.03 | 30.59 |
| Fast | 375 | 325 | 47.40 | 5.87 | 2.39 | 14.58 | 4.22 | 1.62 | 28.68 |

Measurements are based on the square feet of wall coated. These figures assume paint products are being applied over a smooth finish. For heights above 8 feet, use the High Time Difficulty Factors on page 137. "Slow" work is based on a $10.25 hourly wage, "Medium" work on a $16.50 hourly wage, and "Fast" work on a $22.00 hourly wage. Other qualifications that apply to this table are on page 9.

| | Labor SF per manhour | Material coverage SF/gallon | Material cost per gallon | Labor cost per 100 SF | Labor burden 100 SF | Material cost per 100 SF | Overhead per 100 SF | Profit per 100 SF | Total price per 100 SF |
|---|---|---|---|---|---|---|---|---|---|
| **Walls, plaster, interior, smooth finish, brush, per 100 square feet of wall** | | | | | | | | | |
| Flat latex, water base (material #5) | | | | | | | | | |
| Brush 1st coat | | | | | | | | | |
| Slow | 150 | 350 | 16.20 | 6.83 | 1.87 | 4.63 | 4.13 | 2.62 | 20.08 |
| Medium | 175 | 325 | 14.20 | 9.43 | 3.16 | 4.37 | 4.16 | 2.32 | 23.44 |
| Fast | 200 | 300 | 12.10 | 11.00 | 4.46 | 4.03 | 3.61 | 1.39 | 24.49 |
| Brush 2nd coat | | | | | | | | | |
| Slow | 175 | 375 | 16.20 | 5.86 | 1.59 | 4.32 | 3.65 | 2.31 | 17.73 |
| Medium | 200 | 350 | 14.20 | 8.25 | 2.78 | 4.06 | 3.70 | 2.07 | 20.86 |
| Fast | 225 | 325 | 12.10 | 9.78 | 3.95 | 3.72 | 3.23 | 1.24 | 21.92 |
| Brush 3rd or additional coats | | | | | | | | | |
| Slow | 200 | 400 | 16.20 | 5.13 | 1.39 | 4.05 | 3.28 | 2.08 | 15.93 |
| Medium | 225 | 375 | 14.20 | 7.33 | 2.46 | 3.79 | 3.33 | 1.86 | 18.77 |
| Fast | 250 | 350 | 12.10 | 8.80 | 3.57 | 3.46 | 2.93 | 1.13 | 19.89 |
| Enamel, water base (material #9) | | | | | | | | | |
| Brush 1st finish coat | | | | | | | | | |
| Slow | 125 | 350 | 22.80 | 8.20 | 2.23 | 6.51 | 5.25 | 3.33 | 25.52 |
| Medium | 163 | 325 | 19.90 | 10.12 | 3.40 | 6.12 | 4.81 | 2.69 | 27.14 |
| Fast | 200 | 300 | 17.10 | 11.00 | 4.46 | 5.70 | 3.91 | 1.50 | 26.57 |
| Brush 2nd finish coat | | | | | | | | | |
| Slow | 150 | 375 | 22.80 | 6.83 | 1.87 | 6.08 | 4.58 | 2.90 | 22.26 |
| Medium | 175 | 350 | 19.90 | 9.43 | 3.16 | 5.69 | 4.48 | 2.50 | 25.26 |
| Fast | 200 | 325 | 17.10 | 11.00 | 4.46 | 5.26 | 3.83 | 1.47 | 26.02 |
| Brush 3rd or additional finish coats | | | | | | | | | |
| Slow | 175 | 400 | 22.80 | 5.86 | 1.59 | 5.70 | 4.08 | 2.59 | 19.82 |
| Medium | 200 | 375 | 19.90 | 8.25 | 2.78 | 5.31 | 4.00 | 2.24 | 22.58 |
| Fast | 225 | 350 | 17.10 | 9.78 | 3.95 | 4.89 | 3.45 | 1.33 | 23.40 |
| Enamel, oil base (material #10) | | | | | | | | | |
| Brush 1st finish coat | | | | | | | | | |
| Slow | 125 | 400 | 25.90 | 8.20 | 2.23 | 6.48 | 5.24 | 3.32 | 25.47 |
| Medium | 163 | 375 | 22.70 | 10.12 | 3.40 | 6.05 | 4.80 | 2.68 | 27.05 |
| Fast | 200 | 350 | 19.40 | 11.00 | 4.46 | 5.54 | 3.89 | 1.49 | 26.38 |
| Brush 2nd finish coat | | | | | | | | | |
| Slow | 150 | 425 | 22.80 | 6.83 | 1.87 | 5.36 | 4.36 | 2.76 | 21.18 |
| Medium | 175 | 400 | 19.90 | 9.43 | 3.16 | 4.98 | 4.31 | 2.41 | 24.29 |
| Fast | 200 | 375 | 17.10 | 11.00 | 4.46 | 4.56 | 3.70 | 1.42 | 25.14 |

| | Labor SF per manhour | Material coverage SF/gallon | Material cost per gallon | Labor cost per 100 SF | Labor burden 100 SF | Material cost per 100 SF | Overhead per 100 SF | Profit per 100 SF | Total price per 100 SF |
|---|---|---|---|---|---|---|---|---|---|
| Brush 3rd or additional finish coats | | | | | | | | | |
| Slow | 175 | 450 | 25.90 | 5.86 | 1.59 | 5.76 | 4.10 | 2.60 | 19.91 |
| Medium | 200 | 425 | 22.70 | 8.25 | 2.78 | 5.34 | 4.01 | 2.24 | 22.62 |
| Fast | 225 | 400 | 19.40 | 9.78 | 3.95 | 4.85 | 3.44 | 1.32 | 23.34 |
| Stipple finish | | | | | | | | | |
| Slow | 130 | -- | -- | 7.88 | 2.15 | -- | 3.11 | 1.97 | 15.11 |
| Medium | 150 | -- | -- | 11.00 | 3.71 | -- | 3.60 | 2.01 | 20.32 |
| Fast | 170 | -- | -- | 12.94 | 5.24 | -- | 3.37 | 1.29 | 22.84 |
| Epoxy coating, 2 part system - white (material #52) | | | | | | | | | |
| Brush 1st coat | | | | | | | | | |
| Slow | 175 | 400 | 57.40 | 5.86 | 1.59 | 14.35 | 6.76 | 4.29 | 32.85 |
| Medium | 200 | 388 | 50.20 | 8.25 | 2.78 | 12.94 | 5.87 | 3.28 | 33.12 |
| Fast | 225 | 375 | 43.10 | 9.78 | 3.95 | 11.49 | 4.67 | 1.79 | 31.68 |
| Brush 2nd or additional coats | | | | | | | | | |
| Slow | 200 | 425 | 57.40 | 5.13 | 1.39 | 13.51 | 6.21 | 3.94 | 30.18 |
| Medium | 225 | 413 | 50.20 | 7.33 | 2.46 | 12.15 | 5.38 | 3.01 | 30.33 |
| Fast | 250 | 400 | 43.10 | 8.80 | 3.57 | 10.78 | 4.28 | 1.65 | 29.08 |
| Glazing & mottling over enamel (material #16) | | | | | | | | | |
| Brush each coat | | | | | | | | | |
| Slow | 75 | 900 | 29.00 | 13.67 | 3.71 | 3.22 | 6.39 | 4.05 | 31.04 |
| Medium | 98 | 850 | 25.40 | 16.84 | 5.65 | 2.99 | 6.25 | 3.49 | 35.22 |
| Fast | 120 | 800 | 21.80 | 18.33 | 7.43 | 2.73 | 5.27 | 2.03 | 35.79 |
| Stipple finish | | | | | | | | | |
| Slow | 100 | -- | -- | 10.25 | 2.79 | -- | 4.04 | 2.56 | 19.64 |
| Medium | 123 | -- | -- | 13.41 | 4.52 | -- | 4.39 | 2.45 | 24.77 |
| Fast | 135 | -- | -- | 16.30 | 6.61 | -- | 4.24 | 1.63 | 28.78 |

Measurements are based on the square feet of wall coated. These figures assume paint products are being applied over a smooth finish. For heights above 8 feet, use the High Time Difficulty Factors on page 137. "Slow" work is based on a $10.25 hourly wage, "Medium" work on a $16.50 hourly wage, and "Fast" work on a $22.00 hourly wage. Other qualifications that apply to this table are on page 9.

| | Labor SF per manhour | Material coverage SF/gallon | Material cost per gallon | Labor cost per 100 SF | Labor burden 100 SF | Material cost per 100 SF | Overhead per 100 SF | Profit per 100 SF | Total price per 100 SF |
|---|---|---|---|---|---|---|---|---|---|
| **Walls, plaster, interior, smooth finish, roll, per 100 square feet of wall** | | | | | | | | | |
| Flat latex, water base (material #5) | | | | | | | | | |
| Roll 1st coat | | | | | | | | | |
| Slow | 260 | 350 | 16.20 | 3.94 | 1.08 | 4.63 | 2.99 | 1.89 | 14.53 |
| Medium | 430 | 325 | 14.20 | 3.84 | 1.30 | 4.37 | 2.33 | 1.30 | 13.14 |
| Fast | 640 | 300 | 12.10 | 3.44 | 1.38 | 4.03 | 1.64 | .63 | 11.12 |
| Roll 2nd coat | | | | | | | | | |
| Slow | 300 | 375 | 16.20 | 3.42 | .92 | 4.32 | 2.69 | 1.70 | 13.05 |
| Medium | 488 | 350 | 14.20 | 3.38 | 1.14 | 4.06 | 2.10 | 1.17 | 11.85 |
| Fast | 675 | 325 | 12.10 | 3.26 | 1.32 | 3.72 | 1.54 | .59 | 10.43 |
| Roll 3rd or additional coats | | | | | | | | | |
| Slow | 325 | 400 | 16.20 | 3.15 | .87 | 4.05 | 2.50 | 1.58 | 12.15 |
| Medium | 513 | 375 | 14.20 | 3.22 | 1.08 | 3.79 | 1.98 | 1.11 | 11.18 |
| Fast | 700 | 350 | 12.10 | 3.14 | 1.28 | 3.46 | 1.46 | .56 | 9.90 |
| Enamel, water base (material #9) | | | | | | | | | |
| Roll 1st finish coat | | | | | | | | | |
| Slow | 235 | 350 | 22.80 | 4.36 | 1.20 | 6.51 | 3.74 | 2.37 | 18.18 |
| Medium | 423 | 325 | 19.90 | 3.90 | 1.30 | 6.12 | 2.78 | 1.55 | 15.65 |
| Fast | 615 | 300 | 17.10 | 3.58 | 1.46 | 5.70 | 1.99 | .76 | 13.49 |
| Roll 2nd finish coat | | | | | | | | | |
| Slow | 275 | 375 | 22.80 | 3.73 | 1.02 | 6.08 | 3.36 | 2.13 | 16.32 |
| Medium | 453 | 350 | 19.90 | 3.64 | 1.23 | 5.69 | 2.58 | 1.44 | 14.58 |
| Fast | 630 | 325 | 17.10 | 3.49 | 1.43 | 5.26 | 1.88 | .72 | 12.78 |
| Roll 3rd or additional finish coats | | | | | | | | | |
| Slow | 300 | 400 | 22.80 | 3.42 | .92 | 5.70 | 3.12 | 1.98 | 15.14 |
| Medium | 475 | 375 | 19.90 | 3.47 | 1.18 | 5.31 | 2.44 | 1.36 | 13.76 |
| Fast | 650 | 350 | 17.10 | 3.38 | 1.38 | 4.89 | 1.78 | .69 | 12.12 |
| Enamel, oil base (material #10) | | | | | | | | | |
| Roll 1st finish coat | | | | | | | | | |
| Slow | 235 | 375 | 25.90 | 4.36 | 1.20 | 6.91 | 3.86 | 2.45 | 18.78 |
| Medium | 423 | 350 | 22.70 | 3.90 | 1.30 | 6.49 | 2.87 | 1.60 | 16.16 |
| Fast | 615 | 325 | 19.40 | 3.58 | 1.46 | 5.97 | 2.04 | .78 | 13.83 |
| Roll 2nd finish coat | | | | | | | | | |
| Slow | 275 | 400 | 25.90 | 3.73 | 1.02 | 6.48 | 3.48 | 2.21 | 16.92 |
| Medium | 453 | 375 | 22.70 | 3.64 | 1.23 | 6.05 | 2.67 | 1.49 | 15.08 |
| Fast | 630 | 350 | 19.40 | 3.49 | 1.43 | 5.54 | 1.93 | .74 | 13.13 |
| Roll 3rd or additional finish coats | | | | | | | | | |
| Slow | 300 | 425 | 25.90 | 3.42 | .92 | 6.09 | 3.24 | 2.05 | 15.72 |
| Medium | 475 | 400 | 22.70 | 3.47 | 1.18 | 5.68 | 2.53 | 1.41 | 14.27 |
| Fast | 650 | 375 | 19.40 | 3.38 | 1.38 | 5.17 | 1.84 | .71 | 12.48 |

| | Labor SF per manhour | Material coverage SF/gallon | Material cost per gallon | Labor cost per 100 SF | Labor burden 100 SF | Material cost per 100 SF | Overhead per 100 SF | Profit per 100 SF | Total price per 100 SF |
|---|---|---|---|---|---|---|---|---|---|
| Epoxy coating, 2 part system - white (material #52) | | | | | | | | | |
| Roll 1st coat | | | | | | | | | |
| Slow | 300 | 375 | 57.40 | 3.42 | .92 | 15.31 | 6.09 | 3.86 | 29.60 |
| Medium | 488 | 350 | 50.20 | 3.38 | 1.14 | 14.34 | 4.62 | 2.58 | 26.06 |
| Fast | 675 | 325 | 43.10 | 3.26 | 1.32 | 13.26 | 3.30 | 1.27 | 22.41 |
| Roll 2nd or additional coats | | | | | | | | | |
| Slow | 325 | 425 | 57.40 | 3.15 | .87 | 13.51 | 5.43 | 3.44 | 26.40 |
| Medium | 513 | 400 | 50.20 | 3.22 | 1.08 | 12.55 | 4.13 | 2.31 | 23.29 |
| Fast | 700 | 375 | 43.10 | 3.14 | 1.28 | 11.49 | 2.94 | 1.13 | 19.98 |

Measurements are based on the square feet of wall coated. These figures assume paint products are being applied over a smooth finish. For heights above 8 feet, use the High Time Difficulty Factors on page 137. "Slow" work is based on a $10.25 hourly wage, "Medium" work on a $16.50 hourly wage, and "Fast" work on a $22.00 hourly wage. Other qualifications that apply to this table are on page 9.

| | Labor SF per manhour | Material coverage SF/gallon | Material cost per gallon | Labor cost per 100 SF | Labor burden 100 SF | Material cost per 100 SF | Overhead per 100 SF | Profit per 100 SF | Total price per 100 SF |
|---|---|---|---|---|---|---|---|---|---|
| **Walls, plaster, interior, smooth finish, spray, per 100 square feet of wall** | | | | | | | | | |
| Flat latex, water base (material #5) | | | | | | | | | |
| Spray 1st coat | | | | | | | | | |
| Slow | 500 | 375 | 16.20 | 2.05 | .56 | 4.32 | 2.15 | 1.36 | 10.44 |
| Medium | 625 | 338 | 14.20 | 2.64 | .89 | 4.20 | 1.89 | 1.06 | 10.68 |
| Fast | 750 | 300 | 12.10 | 2.93 | 1.18 | 4.03 | 1.51 | .58 | 10.23 |
| Spray 2nd coat | | | | | | | | | |
| Slow | 550 | 400 | 16.20 | 1.86 | .51 | 4.05 | 1.99 | 1.26 | 9.67 |
| Medium | 700 | 363 | 14.20 | 2.36 | .79 | 3.91 | 1.73 | .97 | 9.76 |
| Fast | 850 | 325 | 12.10 | 2.59 | 1.06 | 3.72 | 1.36 | .52 | 9.25 |
| Spray 3rd or additional coats | | | | | | | | | |
| Slow | 600 | 425 | 16.20 | 1.71 | .47 | 3.81 | 1.86 | 1.18 | 9.03 |
| Medium | 775 | 388 | 14.20 | 2.13 | .71 | 3.66 | 1.59 | .89 | 8.98 |
| Fast | 950 | 350 | 12.10 | 2.32 | .93 | 3.46 | 1.24 | .48 | 8.43 |
| Enamel, water base (material #9) | | | | | | | | | |
| Spray 1st finish coat | | | | | | | | | |
| Slow | 475 | 375 | 22.80 | 2.16 | .59 | 6.08 | 2.74 | 1.74 | 13.31 |
| Medium | 600 | 338 | 19.90 | 2.75 | .93 | 5.89 | 2.34 | 1.31 | 13.22 |
| Fast | 725 | 300 | 17.10 | 3.03 | 1.24 | 5.70 | 1.84 | .71 | 12.52 |
| Spray 2nd finish coat | | | | | | | | | |
| Slow | 525 | 400 | 22.80 | 1.95 | .53 | 5.70 | 2.54 | 1.61 | 12.33 |
| Medium | 675 | 388 | 19.90 | 2.44 | .82 | 5.13 | 2.06 | 1.15 | 11.60 |
| Fast | 825 | 325 | 17.10 | 2.67 | 1.07 | 5.26 | 1.67 | .64 | 11.31 |
| Spray 3rd or additional finish coats | | | | | | | | | |
| Slow | 575 | 425 | 22.80 | 1.78 | .49 | 5.36 | 2.37 | 1.50 | 11.50 |
| Medium | 775 | 388 | 19.90 | 2.13 | .71 | 5.13 | 1.96 | 1.09 | 11.02 |
| Fast | 925 | 350 | 17.10 | 2.38 | .96 | 4.89 | 1.52 | .59 | 10.34 |
| Enamel, oil base (material #10) | | | | | | | | | |
| Spray 1st finish coat | | | | | | | | | |
| Slow | 475 | 425 | 25.90 | 2.16 | .59 | 6.09 | 2.74 | 1.74 | 13.32 |
| Medium | 575 | 388 | 22.70 | 2.87 | .97 | 5.85 | 2.37 | 1.33 | 13.39 |
| Fast | 725 | 350 | 19.40 | 3.03 | 1.24 | 5.54 | 1.81 | .70 | 12.32 |
| Spray 2nd finish coat | | | | | | | | | |
| Slow | 525 | 450 | 25.90 | 1.95 | .53 | 5.76 | 2.55 | 1.62 | 12.41 |
| Medium | 675 | 413 | 22.70 | 2.44 | .82 | 5.50 | 2.15 | 1.20 | 12.11 |
| Fast | 825 | 375 | 19.40 | 2.67 | 1.07 | 5.17 | 1.65 | .63 | 11.19 |

| | Labor SF per manhour | Material coverage SF/gallon | Material cost per gallon | Labor cost per 100 SF | Labor burden 100 SF | Material cost per 100 SF | Overhead per 100 SF | Profit per 100 SF | Total price per 100 SF |
|---|---|---|---|---|---|---|---|---|---|
| Spray 3rd or additional finish coats | | | | | | | | | |
| Slow | 575 | 475 | 25.90 | 1.78 | .49 | 5.45 | 2.39 | 1.52 | 11.63 |
| Medium | 750 | 438 | 22.70 | 2.20 | .73 | 5.18 | 1.99 | 1.11 | 11.21 |
| Fast | 925 | 400 | 19.40 | 2.38 | .96 | 4.85 | 1.52 | .58 | 10.29 |
| Epoxy coating, 2 part system - white (material #52) | | | | | | | | | |
| Spray 1st coat | | | | | | | | | |
| Slow | 550 | 325 | 57.40 | 1.86 | .51 | 17.66 | 6.21 | 3.94 | 30.18 |
| Medium | 700 | 300 | 50.20 | 2.36 | .79 | 16.73 | 4.87 | 2.72 | 27.47 |
| Fast | 850 | 275 | 43.10 | 2.59 | 1.06 | 15.67 | 3.57 | 1.37 | 24.26 |
| Spray 2nd or additional coats | | | | | | | | | |
| Slow | 600 | 375 | 57.40 | 1.71 | .47 | 15.31 | 5.42 | 3.44 | 26.35 |
| Medium | 750 | 350 | 50.20 | 2.20 | .73 | 14.34 | 4.23 | 2.37 | 23.87 |
| Fast | 900 | 325 | 43.10 | 2.44 | .99 | 13.26 | 3.09 | 1.19 | 20.97 |

Measurements are based on the square feet of wall coated. These figures assume paint products are being applied over a smooth finish. For heights above 8 feet, use the High Time Difficulty Factors on page 137. "Slow" work is based on a $10.25 hourly wage, "Medium" work on a $16.50 hourly wage, and "Fast" work on a $22.00 hourly wage. Other qualifications that apply to this table are on page 9.

| | Labor SF per manhour | Material coverage SF/gallon | Material cost per gallon | Labor cost per 100 SF | Labor burden 100 SF | Material cost per 100 SF | Overhead per 100 SF | Profit per 100 SF | Total price per 100 SF |
|---|---|---|---|---|---|---|---|---|---|
| **Walls, wood paneled, interior, paint grade, brush, per 100 square feet of wall** | | | | | | | | | |
| Undercoat, water base (material #3) | | | | | | | | | |
| Brush, 1 coat | | | | | | | | | |
| Slow | 65 | 300 | 18.60 | 15.77 | 4.29 | 6.20 | 8.14 | 5.16 | 39.56 |
| Medium | 75 | 288 | 16.20 | 22.00 | 7.39 | 5.63 | 8.58 | 4.80 | 48.40 |
| Fast | 85 | 275 | 13.90 | 25.88 | 10.48 | 5.05 | 7.66 | 2.94 | 52.01 |
| Undercoat, oil base (material #4) | | | | | | | | | |
| Brush, 1 coat | | | | | | | | | |
| Slow | 65 | 375 | 20.10 | 15.77 | 4.29 | 5.36 | 7.88 | 5.00 | 38.30 |
| Medium | 75 | 363 | 17.60 | 22.00 | 7.39 | 4.85 | 8.39 | 4.69 | 47.32 |
| Fast | 85 | 350 | 15.10 | 25.88 | 10.48 | 4.31 | 7.53 | 2.89 | 51.09 |
| Split coat (1/2 undercoat + 1/2 enamel), water base (material #3 & #9) | | | | | | | | | |
| Brush, 1 coat | | | | | | | | | |
| Slow | 55 | 300 | 20.70 | 18.64 | 5.07 | 6.90 | 9.49 | 6.02 | 46.12 |
| Medium | 65 | 288 | 18.05 | 25.38 | 8.53 | 6.27 | 9.85 | 5.50 | 55.53 |
| Fast | 75 | 275 | 15.50 | 29.33 | 11.89 | 5.64 | 8.67 | 3.33 | 58.86 |
| Split coat (1/2 undercoat + 1/2 enamel), oil base (material #4 & #10) | | | | | | | | | |
| Brush, 1 coat | | | | | | | | | |
| Slow | 55 | 375 | 23.00 | 18.64 | 5.07 | 6.13 | 9.25 | 5.87 | 44.96 |
| Medium | 65 | 363 | 20.15 | 25.38 | 8.53 | 5.55 | 9.67 | 5.41 | 54.54 |
| Fast | 75 | 350 | 17.25 | 29.33 | 11.89 | 4.93 | 8.54 | 3.28 | 57.97 |
| Enamel, water base (material #9) | | | | | | | | | |
| Brush 1st finish coat | | | | | | | | | |
| Slow | 80 | 350 | 22.80 | 12.81 | 3.49 | 6.51 | 7.07 | 4.48 | 34.36 |
| Medium | 95 | 338 | 19.90 | 17.37 | 5.85 | 5.89 | 7.13 | 3.99 | 40.23 |
| Fast | 110 | 325 | 17.10 | 20.00 | 8.11 | 5.26 | 6.17 | 2.37 | 41.91 |
| Brush 2nd or additional finish coats | | | | | | | | | |
| Slow | 100 | 375 | 22.80 | 10.25 | 2.79 | 6.08 | 5.93 | 3.76 | 28.81 |
| Medium | 110 | 363 | 19.90 | 15.00 | 5.04 | 5.48 | 6.25 | 3.50 | 35.27 |
| Fast | 120 | 350 | 17.10 | 18.33 | 7.43 | 4.89 | 5.67 | 2.18 | 38.50 |

| | Labor SF per manhour | Material coverage SF/gallon | Material cost per gallon | Labor cost per 100 SF | Labor burden 100 SF | Material cost per 100 SF | Overhead per 100 SF | Profit per 100 SF | Total price per 100 SF |
|---|---|---|---|---|---|---|---|---|---|
| Enamel, oil base (material #10) | | | | | | | | | |
| Brush 1st finish coat | | | | | | | | | |
| Slow | 80 | 400 | 25.90 | 12.81 | 3.49 | 6.48 | 7.06 | 4.48 | 34.32 |
| Medium | 95 | 388 | 22.70 | 17.37 | 5.85 | 5.85 | 7.12 | 3.98 | 40.17 |
| Fast | 110 | 375 | 19.40 | 20.00 | 8.11 | 5.17 | 6.16 | 2.37 | 41.81 |
| Brush 2nd or additional finish coats | | | | | | | | | |
| Slow | 100 | 425 | 25.90 | 10.25 | 2.79 | 6.09 | 5.93 | 3.76 | 28.82 |
| Medium | 110 | 413 | 22.70 | 15.00 | 5.04 | 5.50 | 6.26 | 3.50 | 35.30 |
| Fast | 120 | 400 | 19.40 | 18.33 | 7.43 | 4.85 | 5.66 | 2.18 | 38.45 |

These costs are based on painting interior tongue and groove, wood veneer or plain wainscot paneling. For heights above 8 feet, use the High Time Difficulty Factors on page 137. "Slow" work is based on a $10.25 hourly wage, "Medium" work on a $16.50 hourly wage, and "Fast" work on a $22.00 hourly wage. Other qualifications that apply to this table are on page 9.

| | Labor SF per manhour | Material coverage SF/gallon | Material cost per gallon | Labor cost per 100 SF | Labor burden 100 SF | Material cost per 100 SF | Overhead per 100 SF | Profit per 100 SF | Total price per 100 SF |
|---|---|---|---|---|---|---|---|---|---|
| **Walls, wood paneled, interior, paint grade, roll, per 100 square feet of wall** | | | | | | | | | |
| Undercoat, water base (material #3) | | | | | | | | | |
| Roll, 1 coat | | | | | | | | | |
| Slow | 200 | 275 | 18.60 | 5.13 | 1.39 | 6.76 | 4.12 | 2.61 | 20.01 |
| Medium | 300 | 263 | 16.20 | 5.50 | 1.84 | 6.16 | 3.31 | 1.85 | 18.66 |
| Fast | 400 | 250 | 13.90 | 5.50 | 2.23 | 5.56 | 2.46 | .95 | 16.70 |
| Undercoat, oil base (material #4) | | | | | | | | | |
| Roll, 1 coat | | | | | | | | | |
| Slow | 200 | 350 | 20.10 | 5.13 | 1.39 | 5.74 | 3.80 | 2.41 | 18.47 |
| Medium | 300 | 325 | 17.60 | 5.50 | 1.84 | 5.42 | 3.13 | 1.75 | 17.64 |
| Fast | 400 | 300 | 15.10 | 5.50 | 2.23 | 5.03 | 2.36 | .91 | 16.03 |
| Split coat (1/2 undercoat + 1/2 enamel), water base (material #3 & #9) | | | | | | | | | |
| Roll, 1 coat | | | | | | | | | |
| Slow | 175 | 275 | 20.70 | 5.86 | 1.59 | 7.53 | 4.65 | 2.95 | 22.58 |
| Medium | 275 | 263 | 18.05 | 6.00 | 2.03 | 6.86 | 3.65 | 2.04 | 20.58 |
| Fast | 375 | 250 | 15.50 | 5.87 | 2.39 | 6.20 | 2.67 | 1.03 | 18.16 |
| Split coat (1/2 undercoat + 1/2 enamel), oil base (material #4 & #10) | | | | | | | | | |
| Roll, 1 coat | | | | | | | | | |
| Slow | 175 | 350 | 23.00 | 5.86 | 1.59 | 6.57 | 4.35 | 2.76 | 21.13 |
| Medium | 275 | 325 | 20.15 | 6.00 | 2.03 | 6.20 | 3.48 | 1.95 | 19.66 |
| Fast | 375 | 300 | 17.25 | 5.87 | 2.39 | 5.75 | 2.59 | 1.00 | 17.60 |
| Enamel, water base (material #9) | | | | | | | | | |
| Roll 1st finish coat | | | | | | | | | |
| Slow | 250 | 325 | 22.80 | 4.10 | 1.12 | 7.02 | 3.79 | 2.40 | 18.43 |
| Medium | 375 | 313 | 19.90 | 4.40 | 1.49 | 6.36 | 3.00 | 1.68 | 16.93 |
| Fast | 500 | 300 | 17.10 | 4.40 | 1.78 | 5.70 | 2.20 | .84 | 14.92 |
| Roll 2nd or additional finish coats | | | | | | | | | |
| Slow | 300 | 350 | 22.80 | 3.42 | .92 | 6.51 | 3.37 | 2.13 | 16.35 |
| Medium | 425 | 338 | 19.90 | 3.88 | 1.30 | 5.89 | 2.71 | 1.52 | 15.30 |
| Fast | 550 | 325 | 17.10 | 4.00 | 1.63 | 5.26 | 2.01 | .77 | 13.67 |

| | Labor SF per manhour | Material coverage SF/gallon | Material cost per gallon | Labor cost per 100 SF | Labor burden 100 SF | Material cost per 100 SF | Overhead per 100 SF | Profit per 100 SF | Total price per 100 SF |
|---|---|---|---|---|---|---|---|---|---|
| Enamel, oil base (material #10) | | | | | | | | | |
| Roll 1st finish coat | | | | | | | | | |
| Slow | 250 | 375 | 25.90 | 4.10 | 1.12 | 6.91 | 3.76 | 2.38 | 18.27 |
| Medium | 375 | 363 | 22.70 | 4.40 | 1.49 | 6.25 | 2.97 | 1.66 | 16.77 |
| Fast | 500 | 350 | 19.40 | 4.40 | 1.78 | 5.54 | 2.17 | .83 | 14.72 |
| Roll 2nd or additional finish coats | | | | | | | | | |
| Slow | 300 | 400 | 25.90 | 3.42 | .92 | 6.48 | 3.36 | 2.13 | 16.31 |
| Medium | 425 | 388 | 22.70 | 3.88 | 1.30 | 5.85 | 2.70 | 1.51 | 15.24 |
| Fast | 550 | 375 | 19.40 | 4.00 | 1.63 | 5.17 | 2.00 | .77 | 13.57 |

These costs are based on painting interior tongue and groove, wood veneer or plain wainscot paneling. For heights above 8 feet, use the High Time Difficulty Factors on page 137. "Slow" work is based on a $10.25 hourly wage, "Medium" work on a $16.50 hourly wage, and "Fast" work on a $22.00 hourly wage. Other qualifications that apply to this table are on page 9.

| | Labor SF per manhour | Material coverage SF/gallon | Material cost per gallon | Labor cost per 100 SF | Labor burden 100 SF | Material cost per 100 SF | Overhead per 100 SF | Profit per 100 SF | Total price per 100 SF |
|---|---|---|---|---|---|---|---|---|---|
| **Walls, wood paneled, interior, paint grade, spray, per 100 square feet of wall** | | | | | | | | | |
| Undercoat, water base (material #3) | | | | | | | | | |
| Spray, 1 coat | | | | | | | | | |
| Slow | 350 | 175 | 18.60 | 2.93 | .80 | 10.63 | 4.45 | 2.82 | 21.63 |
| Medium | 425 | 150 | 16.20 | 3.88 | 1.30 | 10.80 | 3.92 | 2.19 | 22.09 |
| Fast | 500 | 125 | 13.90 | 4.40 | 1.78 | 11.12 | 3.20 | 1.23 | 21.73 |
| Undercoat, oil base (material #4) | | | | | | | | | |
| Spray, 1 coat | | | | | | | | | |
| Slow | 350 | 200 | 20.10 | 2.93 | .80 | 10.05 | 4.27 | 2.71 | 20.76 |
| Medium | 425 | 188 | 17.60 | 3.88 | 1.30 | 9.36 | 3.56 | 1.99 | 20.09 |
| Fast | 500 | 175 | 15.10 | 4.40 | 1.78 | 8.63 | 2.74 | 1.05 | 18.60 |
| Split coat (1/2 undercoat + 1/2 enamel), water base (material #3 & #9) | | | | | | | | | |
| Spray, 1 coat | | | | | | | | | |
| Slow | 325 | 175 | 20.70 | 3.15 | .87 | 11.83 | 4.91 | 3.11 | 23.87 |
| Medium | 400 | 150 | 18.05 | 4.13 | 1.38 | 12.03 | 4.30 | 2.40 | 24.24 |
| Fast | 475 | 125 | 15.50 | 4.63 | 1.89 | 12.40 | 3.50 | 1.34 | 23.76 |
| Split coat (1/2 undercoat + 1/2 enamel), oil base (material #4 & #10) | | | | | | | | | |
| Spray, 1 coat | | | | | | | | | |
| Slow | 325 | 200 | 23.00 | 3.15 | .87 | 11.50 | 4.81 | 3.05 | 23.38 |
| Medium | 400 | 188 | 20.15 | 4.13 | 1.38 | 10.72 | 3.98 | 2.22 | 22.43 |
| Fast | 475 | 175 | 17.25 | 4.63 | 1.89 | 9.86 | 3.03 | 1.16 | 20.57 |
| Enamel, water base (material #9) | | | | | | | | | |
| Spray 1st finish coat | | | | | | | | | |
| Slow | 500 | 250 | 22.80 | 2.05 | .56 | 9.12 | 3.64 | 2.31 | 17.68 |
| Medium | 550 | 225 | 19.90 | 3.00 | 1.01 | 8.84 | 3.15 | 1.76 | 17.76 |
| Fast | 600 | 200 | 17.10 | 3.67 | 1.49 | 8.55 | 2.54 | .98 | 17.23 |
| Spray 2nd or additional finish coats | | | | | | | | | |
| Slow | 600 | 350 | 22.80 | 1.71 | .47 | 6.51 | 2.69 | 1.71 | 13.09 |
| Medium | 650 | 325 | 19.90 | 2.54 | .86 | 6.12 | 2.33 | 1.30 | 13.15 |
| Fast | 700 | 300 | 17.10 | 3.14 | 1.28 | 5.70 | 1.87 | .72 | 12.71 |

| | Labor SF per manhour | Material coverage SF/gallon | Material cost per gallon | Labor cost per 100 SF | Labor burden 100 SF | Material cost per 100 SF | Overhead per 100 SF | Profit per 100 SF | Total price per 100 SF |
|---|---|---|---|---|---|---|---|---|---|
| Enamel, oil base (material #10) | | | | | | | | | |
| Spray 1st finish coat | | | | | | | | | |
| Slow | 500 | 300 | 25.90 | 2.05 | .56 | 8.63 | 3.48 | 2.21 | 16.93 |
| Medium | 550 | 275 | 22.70 | 3.00 | 1.01 | 8.25 | 3.00 | 1.68 | 16.94 |
| Fast | 600 | 250 | 19.40 | 3.67 | 1.49 | 7.76 | 2.39 | .92 | 16.23 |
| Spray 2nd or additional finish coats | | | | | | | | | |
| Slow | 600 | 400 | 25.90 | 1.71 | .47 | 6.48 | 2.68 | 1.70 | 13.04 |
| Medium | 650 | 375 | 22.70 | 2.54 | .86 | 6.05 | 2.31 | 1.29 | 13.05 |
| Fast | 700 | 350 | 19.40 | 3.14 | 1.28 | 5.54 | 1.84 | .71 | 12.51 |

These costs are based on painting interior tongue and groove, wood veneer or plain wainscot paneling. For heights above 8 feet, use the High Time Difficulty Factors on page 137. "Slow" work is based on a $10.25 hourly wage, "Medium" work on a $16.50 hourly wage, and "Fast" work on a $22.00 hourly wage. Other qualifications that apply to this table are on page 9.

| | Labor SF per manhour | Material coverage SF/gallon | Material cost per gallon | Labor cost per 100 SF | Labor burden 100 SF | Material cost per 100 SF | Overhead per 100 SF | Profit per 100 SF | Total price per 100 SF |
|---|---|---|---|---|---|---|---|---|---|

## Walls, wood paneled, interior, stain grade, per 100 square feet of wall

| | | | | | | | | | |
|---|---|---|---|---|---|---|---|---|---|
| Solid body or semi-transparent stain, water or oil base (material #18 or #19 or #20 or #21) | | | | | | | | | |
| Roll & brush each coat | | | | | | | | | |
| Slow | 225 | 250 | 21.38 | 4.56 | 1.23 | 8.55 | 4.45 | 2.82 | 21.61 |
| Medium | 263 | 238 | 18.70 | 6.27 | 2.11 | 7.86 | 3.98 | 2.22 | 22.44 |
| Fast | 300 | 225 | 16.03 | 7.33 | 2.97 | 7.12 | 3.22 | 1.24 | 21.88 |
| Solid body or semi-transparent stain, water or oil base (material #18 or #19 or #20 or #21) | | | | | | | | | |
| Spray each coat | | | | | | | | | |
| Slow | 350 | 300 | 21.38 | 2.93 | .80 | 7.13 | 3.37 | 2.13 | 16.36 |
| Medium | 400 | 250 | 18.70 | 4.13 | 1.38 | 7.48 | 3.19 | 1.78 | 17.96 |
| Fast | 450 | 200 | 16.03 | 4.89 | 1.97 | 8.02 | 2.75 | 1.06 | 18.69 |

Use these figures for quantities greater than 100 square feet. For quantities less than 100 square feet, use Fireplace siding. These costs are based on painting interior tongue and groove, wood veneer or plain wainscot paneling. Do not deduct for openings less than 100 square feet. For heights above 8 feet, use the High Time Difficulty Factors on page 137. "Slow" work is based on a $10.25 hourly wage, "Medium" work on a $16.50 hourly wage, and "Fast" work on a $22.00 hourly wage. Other qualifications that apply to this table are on page 9.

## Walls, wood paneled, interior, stain grade, per 100 square feet of wall

| | Labor SF per manhour | Material coverage SF/gallon | Material cost per gallon | Labor cost per 100 SF | Labor burden 100 SF | Material cost per 100 SF | Overhead per 100 SF | Profit per 100 SF | Total price per 100 SF |
|---|---|---|---|---|---|---|---|---|---|
| Stain, seal and 2 coat lacquer system (7 step process) | | | | | | | | | |
| STEP 1: Sand & putty | | | | | | | | | |
| Slow | 175 | -- | -- | 5.86 | 1.59 | -- | 2.31 | 1.47 | 11.23 |
| Medium | 200 | -- | -- | 8.25 | 2.78 | -- | 2.70 | 1.51 | 15.24 |
| Fast | 225 | -- | -- | 9.78 | 3.95 | -- | 2.54 | .98 | 17.25 |
| STEP 2 & 3: Wiping stain, oil base (material #11a) & wipe | | | | | | | | | |
| Brush 1 coat & wipe | | | | | | | | | |
| Slow | 100 | 400 | 27.40 | 10.25 | 2.79 | 6.85 | 6.17 | 3.91 | 29.97 |
| Medium | 125 | 375 | 24.00 | 13.20 | 4.44 | 6.40 | 5.89 | 3.29 | 33.22 |
| Fast | 150 | 350 | 20.60 | 14.67 | 5.95 | 5.89 | 4.90 | 1.88 | 33.29 |
| Spray 1 coat & wipe | | | | | | | | | |
| Slow | 350 | 175 | 27.40 | 2.93 | .80 | 15.66 | 6.01 | 3.81 | 29.21 |
| Medium | 425 | 150 | 24.00 | 3.88 | 1.30 | 16.00 | 5.19 | 2.90 | 29.27 |
| Fast | 500 | 125 | 20.60 | 4.40 | 1.78 | 16.48 | 4.19 | 1.61 | 28.46 |
| STEP 4 & 5: Sanding sealer (material #11b) & sand lightly | | | | | | | | | |
| Brush 1 coat & wipe | | | | | | | | | |
| Slow | 200 | 450 | 17.30 | 5.13 | 1.39 | 3.84 | 3.21 | 2.04 | 15.61 |
| Medium | 220 | 425 | 15.10 | 7.50 | 2.53 | 3.55 | 3.32 | 1.86 | 18.76 |
| Fast | 240 | 400 | 13.00 | 9.17 | 3.72 | 3.25 | 2.99 | 1.15 | 20.28 |
| Spray 1 coat & wipe | | | | | | | | | |
| Slow | 400 | 175 | 17.30 | 2.56 | .70 | 9.89 | 4.08 | 2.58 | 19.81 |
| Medium | 500 | 150 | 15.10 | 3.30 | 1.11 | 10.07 | 3.55 | 1.98 | 20.01 |
| Fast | 600 | 125 | 13.00 | 3.67 | 1.49 | 10.40 | 2.88 | 1.11 | 19.55 |
| STEP 6 & 7: Lacquer (material #11c) | | | | | | | | | |
| Brush 1st coat | | | | | | | | | |
| Slow | 175 | 375 | 18.00 | 5.86 | 1.59 | 4.80 | 3.80 | 2.41 | 18.46 |
| Medium | 225 | 350 | 15.70 | 7.33 | 2.46 | 4.49 | 3.50 | 1.96 | 19.74 |
| Fast | 300 | 325 | 13.50 | 7.33 | 2.97 | 4.15 | 2.67 | 1.03 | 18.15 |
| Brush 2nd coat | | | | | | | | | |
| Slow | 225 | 400 | 18.00 | 4.56 | 1.23 | 4.50 | 3.19 | 2.02 | 15.50 |
| Medium | 288 | 388 | 15.70 | 5.73 | 1.92 | 4.05 | 2.87 | 1.60 | 16.17 |
| Fast | 350 | 375 | 13.50 | 6.29 | 2.55 | 3.60 | 2.30 | .88 | 15.62 |
| Spray 1st coat | | | | | | | | | |
| Slow | 450 | 150 | 18.00 | 2.28 | .61 | 12.00 | 4.62 | 2.93 | 22.44 |
| Medium | 550 | 125 | 15.70 | 3.00 | 1.01 | 12.56 | 4.06 | 2.27 | 22.90 |
| Fast | 650 | 100 | 13.50 | 3.38 | 1.38 | 13.50 | 3.38 | 1.30 | 22.94 |

| | Labor SF per manhour | Material coverage SF/gallon | Material cost per gallon | Labor cost per 100 SF | Labor burden 100 SF | Material cost per 100 SF | Overhead per 100 SF | Profit per 100 SF | Total price per 100 SF |
|---|---|---|---|---|---|---|---|---|---|
| Spray 2nd coat | | | | | | | | | |
| Slow | 450 | 225 | 18.00 | 2.28 | .61 | 8.00 | 3.38 | 2.14 | 16.41 |
| Medium | 550 | 200 | 15.70 | 3.00 | 1.01 | 7.85 | 2.91 | 1.62 | 16.39 |
| Fast | 650 | 175 | 13.50 | 3.38 | 1.38 | 7.71 | 2.31 | .89 | 15.67 |
| Complete 7 step stain, seal & 2 coat lacquer system (material #11) | | | | | | | | | |
| Brush all coats | | | | | | | | | |
| Slow | 40 | 150 | 20.20 | 25.63 | 6.97 | 13.47 | 14.28 | 9.05 | 69.40 |
| Medium | 45 | 138 | 18.30 | 36.67 | 12.33 | 13.26 | 15.26 | 8.53 | 86.05 |
| Fast | 50 | 125 | 15.10 | 44.00 | 17.84 | 12.08 | 13.68 | 5.26 | 92.86 |
| Spray all coats | | | | | | | | | |
| Slow | 70 | 50 | 20.20 | 14.64 | 3.99 | 40.40 | 18.30 | 11.60 | 88.93 |
| Medium | 80 | 40 | 18.30 | 20.63 | 6.93 | 45.75 | 17.96 | 10.04 | 101.31 |
| Fast | 90 | 30 | 15.10 | 24.44 | 9.91 | 50.33 | 15.67 | 6.02 | 106.37 |
| Penetrating stain wax (material #14) | | | | | | | | | |
| Brush each coat | | | | | | | | | |
| Slow | 300 | 500 | 28.60 | 3.42 | .92 | 5.72 | 3.12 | 1.98 | 15.16 |
| Medium | 350 | 475 | 25.00 | 4.71 | 1.60 | 5.26 | 2.83 | 1.58 | 15.98 |
| Fast | 400 | 450 | 21.40 | 5.50 | 2.23 | 4.76 | 2.31 | .89 | 15.69 |

These costs are based on painting interior tongue and groove, wood veneer or plain wainscot paneling. Do not deduct for openings less than 100 square feet. For heights above 8 feet, use the High Time Difficulty Factors on page 137. "Slow" work is based on a $10.25 hourly wage, "Medium" work on a $16.75 hourly wage, and "Fast" work on a $22.00 hourly wage. Other qualifications that apply to this table are on page 9.

| | Frames per manhour | Frames per gallon | Material cost per gallon | Labor cost per frame | Labor burden frame | Material cost per frame | Overhead per frame | Profit per frame | Total price per frame |
|---|---|---|---|---|---|---|---|---|---|
| **Window screen frames, paint grade, per frame (15 square feet)** | | | | | | | | | |
| Undercoat, water or oil base (material #3 or #4) | | | | | | | | | |
| Brush 1 coat | | | | | | | | | |
| Slow | 5 | 50 | 19.35 | 2.05 | .56 | .39 | .93 | .59 | 4.52 |
| Medium | 6 | 45 | 16.90 | 2.75 | .93 | .38 | .99 | .56 | 5.61 |
| Fast | 7 | 40 | 14.50 | 3.14 | 1.28 | .36 | .88 | .34 | 6.00 |
| Split coat (1/2 undercoat + 1/2 enamel), water or oil base (material #3 or #4 or #9 or #10) | | | | | | | | | |
| Brush 1 coat | | | | | | | | | |
| Slow | 8 | 60 | 21.85 | 1.28 | .35 | .36 | .62 | .39 | 3.00 |
| Medium | 9 | 58 | 19.10 | 1.83 | .62 | .33 | .68 | .38 | 3.84 |
| Fast | 10 | 55 | 16.38 | 2.20 | .89 | .30 | .63 | .24 | 4.26 |
| Enamel, water or oil base (material #9 or #10) | | | | | | | | | |
| Brush each finish coat | | | | | | | | | |
| Slow | 6 | 55 | 24.35 | 1.71 | .47 | .44 | .81 | .51 | 3.94 |
| Medium | 7 | 53 | 21.30 | 2.36 | .79 | .40 | .87 | .49 | 4.91 |
| Fast | 8 | 50 | 18.25 | 2.75 | 1.12 | .37 | .78 | .30 | 5.32 |

These figures will apply when painting all sides of wood window screens up to 15 square feet (length times width). Add: Preparation time for protecting adjacent surfaces with masking tape and paper as required. "Slow" work is based on a $10.25 hourly wage, "Medium" work on a $16.50 hourly wage, and "Fast" work on a $22.00 hourly wage. Other qualifications that apply to this table are on page 9.

| | Labor SF per manhour | Material coverage SF/gallon | Material cost per gallon | Labor cost per 100 SF | Labor burden 100 SF | Material cost per 100 SF | Overhead per 100 SF | Profit per 100 SF | Total price per 100 SF |
|---|---|---|---|---|---|---|---|---|---|
| **Window seats, wood, paint grade, per 100 square feet coated** | | | | | | | | | |
| Undercoat, water or oil base (material #3 or #4) | | | | | | | | | |
| Brush 1 coat | | | | | | | | | |
| Slow | 20 | 45 | 19.35 | 51.25 | 13.95 | 43.00 | 33.55 | 21.27 | 163.02 |
| Medium | 25 | 43 | 16.90 | 66.00 | 22.20 | 39.30 | 31.24 | 17.46 | 176.20 |
| Fast | 30 | 40 | 14.50 | 73.33 | 29.73 | 36.25 | 25.77 | 9.91 | 174.99 |
| Split coat (1/2 undercoat + 1/2 enamel), water or oil base (material #3 or #4 or #9 or #10) | | | | | | | | | |
| Brush 1 coat | | | | | | | | | |
| Slow | 30 | 60 | 21.85 | 34.17 | 9.29 | 36.42 | 24.77 | 15.70 | 120.35 |
| Medium | 35 | 58 | 19.10 | 47.14 | 15.86 | 32.93 | 23.50 | 13.14 | 132.57 |
| Fast | 40 | 55 | 16.38 | 55.00 | 22.30 | 29.78 | 19.81 | 7.61 | 134.50 |
| Enamel, water or oil base (material #9 or #10) | | | | | | | | | |
| Brush each finish coat | | | | | | | | | |
| Slow | 25 | 55 | 24.35 | 41.00 | 11.16 | 44.27 | 29.90 | 18.95 | 145.28 |
| Medium | 30 | 53 | 21.30 | 55.00 | 18.49 | 40.19 | 27.86 | 15.57 | 157.11 |
| Fast | 35 | 50 | 18.25 | 62.86 | 25.48 | 36.50 | 23.10 | 8.88 | 156.82 |

Measurements are based on square feet of surface area of each window seat. "Slow" work is based on a $10.25 hourly wage, "Medium" work on a $16.50 hourly wage, and "Fast" work on a $22.00 hourly wage. Other qualifications that apply to this table are on page 9.

| | Labor LF per manhour | Material coverage LF/gallon | Material cost per gallon | Labor cost per 100 LF | Labor burden 100 LF | Material cost per 100 LF | Overhead per 100 LF | Profit per 100 LF | Total price per 100 LF |
|---|---|---|---|---|---|---|---|---|---|
| **Window sills, wood, paint grade, per 100 linear feet coated** | | | | | | | | | |
| Undercoat, water or oil base (material #3 or #4) | | | | | | | | | |
| Brush 1 coat | | | | | | | | | |
| Slow | 40 | 140 | 19.35 | 25.63 | 6.97 | 13.82 | 14.39 | 9.12 | 69.93 |
| Medium | 50 | 130 | 16.90 | 33.00 | 11.10 | 13.00 | 13.99 | 7.82 | 78.91 |
| Fast | 60 | 120 | 14.50 | 36.67 | 14.87 | 12.08 | 11.77 | 4.52 | 79.91 |
| Split coat (1/2 undercoat + 1/2 enamel), water or oil base (material #3 or #4 or #9 or #10) | | | | | | | | | |
| Brush 1 coat | | | | | | | | | |
| Slow | 60 | 180 | 21.85 | 17.08 | 4.66 | 12.14 | 10.50 | 6.66 | 51.04 |
| Medium | 70 | 170 | 19.10 | 23.57 | 7.94 | 11.24 | 10.47 | 5.85 | 59.07 |
| Fast | 80 | 160 | 16.38 | 27.50 | 11.15 | 10.24 | 9.04 | 3.48 | 61.41 |
| Enamel, water or oil base (material #9 or #10) | | | | | | | | | |
| Brush each finish coat | | | | | | | | | |
| Slow | 50 | 175 | 24.35 | 20.50 | 5.58 | 13.91 | 12.40 | 7.86 | 60.25 |
| Medium | 60 | 163 | 21.30 | 27.50 | 9.26 | 13.07 | 12.21 | 6.82 | 68.86 |
| Fast | 70 | 150 | 18.25 | 31.43 | 12.75 | 12.17 | 10.42 | 4.01 | 70.78 |

Measurements are based on linear feet of window sill. Add: Preparation time for protecting adjacent surfaces with masking tape and paper as required. "Slow" work is based on a $10.25 hourly wage, "Medium" work on a $16.50 hourly wage, and "Fast" work on a $22.00 hourly wage. Other qualifications that apply to this table are on page 9.

| | Frames per manhour | Frames per gallon | Material cost per gallon | Labor cost per frame | Labor burden frame | Material cost per frame | Overhead per frame | Profit per frame | Total price per frame |
|---|---|---|---|---|---|---|---|---|---|
| **Window storm sash, paint grade, per 15 square feet painted** | | | | | | | | | |
| Undercoat, water or oil base (material #3 or #4) | | | | | | | | | |
| Brush 1 coat | | | | | | | | | |
| Slow | 3 | 25 | 19.35 | 3.42 | .92 | .77 | 1.59 | 1.01 | 7.71 |
| Medium | 4 | 24 | 16.90 | 4.13 | 1.38 | .70 | 1.52 | .85 | 8.58 |
| Fast | 5 | 22 | 14.50 | 4.40 | 1.78 | .66 | 1.27 | .49 | 8.60 |
| Split coat (1/2 undercoat + 1/2 enamel), water or oil base (material #3 or #4 or #9 or #10) | | | | | | | | | |
| Brush 1 coat | | | | | | | | | |
| Slow | 5 | 35 | 21.85 | 2.05 | .56 | .62 | 1.00 | .63 | 4.86 |
| Medium | 6 | 33 | 19.10 | 2.75 | .93 | .58 | 1.04 | .58 | 5.88 |
| Fast | 7 | 30 | 16.38 | 3.14 | 1.28 | .55 | .92 | .35 | 6.24 |
| Enamel, water or oil base (material #9 or #10) | | | | | | | | | |
| Brush each finish coat | | | | | | | | | |
| Slow | 4 | 30 | 24.35 | 2.56 | .70 | .81 | 1.26 | .80 | 6.13 |
| Medium | 5 | 28 | 21.30 | 3.30 | 1.11 | .76 | 1.27 | .71 | 7.15 |
| Fast | 6 | 25 | 18.25 | 3.67 | 1.49 | .73 | 1.09 | .42 | 7.40 |

These figures will apply when painting all sides of two-lite wood storm sash measuring up to 15 square feet overall (length times width). Add: Preparation time for protecting adjacent surfaces with window protective coating or masking tape and paper as required. "Slow" work is based on a $10.25 hourly wage, "Medium" work on a $16.50 hourly wage, and "Fast" work on a $22.00 hourly wage. Other qualifications that apply to this table are on page 9.

## Windows, exterior

The estimates that follow include the time and material needed to paint the sash (mullions or muntins), trim, frames, jambs, sill and apron on the outside only. Manhour and material rates are the same for painting either the inside or outside of windows. These estimates are per window painted, and are based on a window size of 15 square feet overall (length times width).

| | Labor SF per manhour | Material coverage SF/gallon | Material cost per gallon | Labor cost per 100 SF | Labor burden 100 SF | Material cost per 100 SF | Overhead per 100 SF | Profit per 100 SF | Total price per 100 SF |
|---|---|---|---|---|---|---|---|---|---|
| **Windows, wood, exterior, paint grade, square foot basis, brush** | | | | | | | | | |
| Undercoat, water or oil base (material #3 or #4) | | | | | | | | | |
| Brush 1 coat | | | | | | | | | |
| Slow | 150 | 460 | 19.35 | 6.83 | 1.87 | 4.21 | 4.00 | 2.54 | 19.45 |
| Medium | 165 | 450 | 16.90 | 10.00 | 3.36 | 3.76 | 4.20 | 2.35 | 23.67 |
| Fast | 180 | 440 | 14.50 | 12.22 | 4.97 | 3.30 | 3.79 | 1.46 | 25.74 |
| Split coat (1/2 undercoat + 1/2 enamel), water or oil base (material #3 or #4 or #9 or #10) | | | | | | | | | |
| Brush 1 coat | | | | | | | | | |
| Slow | 120 | 520 | 21.85 | 8.54 | 2.32 | 4.20 | 4.67 | 2.96 | 22.69 |
| Medium | 135 | 500 | 19.10 | 12.22 | 4.12 | 3.82 | 4.94 | 2.76 | 27.86 |
| Fast | 150 | 480 | 16.38 | 14.67 | 5.95 | 3.41 | 4.45 | 1.71 | 30.19 |
| Enamel, water or oil base (material #9 or #10) | | | | | | | | | |
| Brush each finish coat | | | | | | | | | |
| Slow | 100 | 500 | 24.35 | 10.25 | 2.79 | 4.87 | 5.55 | 3.52 | 26.98 |
| Medium | 113 | 480 | 21.30 | 14.60 | 4.91 | 4.44 | 5.87 | 3.28 | 33.10 |
| Fast | 125 | 460 | 18.25 | 17.60 | 7.14 | 3.97 | 5.31 | 2.04 | 36.06 |

Use this table to estimate the costs for paint grade windows larger than 15 square feet (width times height). When calculating the window area, add 1 foot in each direction to both dimensions before multiplying width by the height. For example, a 4' x 4' window would have a calculated surface area of 36 square feet: Width of 4' + 1' + 1' = 6' times height of 4' + 1' + 1' = 6'. The total is 6' x 6', or 36 square feet per side. Add an extra 2 square feet to the calculated area for each window lite. This allows for the extra time needed to finish muntins, mullions and sash. For windows that are stained, sealed and varnished, use the previous tables based on window lites. Add: Preparation time to protect adjacent surfaces with window protective coating (wax) or masking tape and paper as required. For heights above 8 feet, use the High Time Difficulty Factors on page 137. "Slow" work is based on a $10.25 hourly wage, "Medium" work on a $16.50 hourly wage, and "Fast" work on a $22.00 hourly wage. Other qualifications that apply to this table are on page 9.

| | Manhours per window | Windows per gallon | Material cost per gallon | Labor cost per window | Labor burden window | Material cost per window | Overhead per window | Profit per window | Total price per window |
|---|---|---|---|---|---|---|---|---|---|
| **Windows, wood, exterior, 1, 2 and 3 lite, brush** | | | | | | | | | |
| Undercoat, water or oil base (material #3 or #4) | | | | | | | | | |
| Brush 1 coat | | | | | | | | | |
| Slow | .25 | 15.0 | 19.35 | 2.56 | .70 | 1.29 | 1.41 | .89 | 6.85 |
| Medium | .20 | 14.5 | 16.90 | 3.30 | 1.11 | 1.17 | 1.37 | .76 | 7.71 |
| Fast | .15 | 14.0 | 14.50 | 3.30 | 1.34 | 1.04 | 1.05 | .40 | 7.13 |
| Split coat (1/2 undercoat + 1/2 enamel), water or oil base (material #3 or #4 or #9 or #10) | | | | | | | | | |
| Brush 1 coat | | | | | | | | | |
| Slow | .30 | 17.0 | 21.85 | 3.08 | .83 | 1.29 | 1.62 | 1.02 | 7.84 |
| Medium | .25 | 16.5 | 19.10 | 4.13 | 1.38 | 1.16 | 1.64 | .92 | 9.23 |
| Fast | .20 | 16.0 | 16.38 | 4.40 | 1.78 | 1.02 | 1.33 | .51 | 9.04 |
| Enamel, water or oil base (material #9 or #10) | | | | | | | | | |
| Brush each finish coat | | | | | | | | | |
| Slow | .35 | 16.0 | 24.35 | 3.59 | .97 | 1.52 | 1.89 | 1.20 | 9.17 |
| Medium | .30 | 15.5 | 21.30 | 4.95 | 1.67 | 1.37 | 1.96 | 1.09 | 11.04 |
| Fast | .25 | 15.0 | 18.25 | 5.50 | 2.23 | 1.22 | 1.66 | .64 | 11.25 |
| Stain, seal & 1 coat varnish system (material #30) | | | | | | | | | |
| Brush each finish coat | | | | | | | | | |
| Slow | .70 | 11.0 | 29.10 | 7.18 | 1.95 | 2.65 | 3.65 | 2.32 | 17.75 |
| Medium | .65 | 10.0 | 25.50 | 10.73 | 3.60 | 2.55 | 4.14 | 2.31 | 23.33 |
| Fast | .60 | 9.0 | 21.90 | 13.20 | 5.35 | 2.43 | 3.88 | 1.49 | 26.35 |
| Varnish (material #30c) | | | | | | | | | |
| Brush additional coats of varnish | | | | | | | | | |
| Slow | .25 | 22.0 | 33.90 | 2.56 | .70 | 1.54 | 1.49 | .94 | 7.23 |
| Medium | .23 | 20.0 | 29.70 | 3.80 | 1.27 | 1.49 | 1.61 | .90 | 9.07 |
| Fast | .20 | 18.0 | 25.40 | 4.40 | 1.78 | 1.41 | 1.40 | .54 | 9.53 |
| Preparation for varnish application (material - minimal) | | | | | | | | | |
| Steel wool buff | | | | | | | | | |
| Slow | .25 | -- | -- | 2.56 | .70 | -- | 1.01 | .64 | 4.91 |
| Medium | .20 | -- | -- | 3.30 | 1.11 | -- | 1.08 | .60 | 6.09 |
| Fast | .15 | -- | -- | 3.30 | 1.34 | -- | .86 | .33 | 5.83 |
| Wax application (material - minimal) | | | | | | | | | |
| Slow | .25 | -- | -- | 2.56 | .70 | -- | 1.01 | .64 | 4.91 |
| Medium | .20 | -- | -- | 3.30 | 1.11 | -- | 1.08 | .60 | 6.09 |
| Fast | .15 | -- | -- | 3.30 | 1.34 | -- | .86 | .33 | 5.83 |

The stain, seal and 1 coat varnish system includes 1 coat of stain, sanding sealer, light sanding, putty and 1 coat of varnish. Minimum preparation for varnish usually includes a steel wool buff and wax operation with minimum material usage. Add: Preparation time for protecting adjacent surfaces with window protective coating (wax) or masking tape and paper as required. For heights above 8 feet, use the High Time Difficulty Factors on page 137. "Slow" work is based on a $10.25 hourly wage, "Medium" work on a $16.75 hourly wage, and "Fast" work on a $22.00 hourly wage. Other qualifications that apply to this table are on page 9.

| | Manhours per window | Windows per gallon | Material cost per gallon | Labor cost per window | Labor burden window | Material cost per window | Overhead per window | Profit per window | Total price per window |
|---|---|---|---|---|---|---|---|---|---|
| **Windows, wood, exterior, 4 to 6 lite, brush** | | | | | | | | | |
| Undercoat, water or oil base (material #3 or #4) | | | | | | | | | |
| Brush 1 coat | | | | | | | | | |
| Slow | .35 | 14.0 | 19.35 | 3.59 | .97 | 1.38 | 1.84 | 1.17 | 8.95 |
| Medium | .30 | 13.5 | 16.90 | 4.95 | 1.67 | 1.25 | 1.93 | 1.08 | 10.88 |
| Fast | .25 | 13.0 | 14.50 | 5.50 | 2.23 | 1.12 | 1.64 | .63 | 11.12 |
| Split coat (1/2 undercoat + 1/2 enamel), water or oil base (material #3 or #4 or #9 or #10) | | | | | | | | | |
| Brush 1 coat | | | | | | | | | |
| Slow | .40 | 16.0 | 21.85 | 4.10 | 1.12 | 1.37 | 2.04 | 1.29 | 9.92 |
| Medium | .35 | 15.5 | 19.10 | 5.78 | 1.94 | 1.23 | 2.19 | 1.23 | 12.37 |
| Fast | .30 | 15.0 | 16.38 | 6.60 | 2.68 | 1.09 | 1.92 | .74 | 13.03 |
| Enamel, water or oil base (material #9 or #10) | | | | | | | | | |
| Brush each finish coat | | | | | | | | | |
| Slow | .50 | 15.0 | 24.35 | 5.13 | 1.39 | 1.62 | 2.53 | 1.60 | 12.27 |
| Medium | .45 | 14.5 | 21.30 | 7.43 | 2.49 | 1.47 | 2.79 | 1.56 | 15.74 |
| Fast | .40 | 14.0 | 18.25 | 8.80 | 3.57 | 1.30 | 2.53 | .97 | 17.17 |
| Stain, seal & 1 coat varnish system (material #30) | | | | | | | | | |
| Brush each finish coat | | | | | | | | | |
| Slow | .83 | 10.0 | 29.10 | 8.51 | 2.31 | 2.91 | 4.26 | 2.70 | 20.69 |
| Medium | .73 | 9.0 | 25.50 | 12.05 | 4.05 | 2.83 | 4.64 | 2.59 | 26.16 |
| Fast | .63 | 8.0 | 21.90 | 13.86 | 5.62 | 2.74 | 4.11 | 1.58 | 27.91 |
| Varnish (material #30c) | | | | | | | | | |
| Brush additional coats of varnish | | | | | | | | | |
| Slow | .33 | 21.0 | 33.90 | 3.38 | .92 | 1.61 | 1.83 | 1.16 | 8.90 |
| Medium | .27 | 19.0 | 29.70 | 4.46 | 1.49 | 1.56 | 1.84 | 1.03 | 10.38 |
| Fast | .22 | 17.0 | 25.40 | 4.84 | 1.96 | 1.49 | 1.53 | .59 | 10.41 |
| Preparation for varnish application (material - minimal) | | | | | | | | | |
| Steel wool buff | | | | | | | | | |
| Slow | .30 | -- | -- | 3.08 | .83 | -- | 1.22 | .77 | 5.90 |
| Medium | .25 | -- | -- | 4.13 | 1.38 | -- | 1.35 | .76 | 7.62 |
| Fast | .20 | -- | -- | 4.40 | 1.78 | -- | 1.14 | .44 | 7.76 |
| Wax application (material - minimal) | | | | | | | | | |
| Slow | .30 | -- | -- | 3.08 | .83 | -- | 1.22 | .77 | 5.90 |
| Medium | .25 | -- | -- | 4.13 | 1.38 | -- | 1.35 | .76 | 7.62 |
| Fast | .20 | -- | -- | 4.40 | 1.78 | -- | 1.14 | .44 | 7.76 |

The stain, seal and 1 coat varnish system includes 1 coat of stain, sanding sealer, light sanding, putty and 1 coat of varnish. Minimum preparation for varnish usually includes a steel wool buff and wax operation with minimum material usage. Add: Preparation time for protecting adjacent surfaces with window protective coating (wax) or masking tape and paper as required. For heights above 8 feet, use the High Time Difficulty Factors on page 137. "Slow" work is based on a $10.25 hourly wage, "Medium" work on a $16.75 hourly wage, and "Fast" work on a $22.00 hourly wage. Other qualifications that apply to this table are on page 9.

| | Manhours per window | Windows per gallon | Material cost per gallon | Labor cost per window | Labor burden window | Material cost per window | Overhead per window | Profit per window | Total price per window |
|---|---|---|---|---|---|---|---|---|---|
| **Windows, wood, exterior, 7 to 8 lite, brush** | | | | | | | | | |
| Undercoat, water or oil base (material #3 or #4) | | | | | | | | | |
| Brush 1 coat | | | | | | | | | |
| Slow | .45 | 13 | 19.35 | 4.61 | 1.26 | 1.49 | 2.28 | 1.45 | 11.09 |
| Medium | .40 | 13 | 16.90 | 6.60 | 2.22 | 1.30 | 2.48 | 1.39 | 13.99 |
| Fast | .35 | 12 | 14.50 | 7.70 | 3.12 | 1.21 | 2.23 | .86 | 15.12 |
| Split coat (1/2 undercoat + 1/2 enamel), water or oil base (material #3 or #4 or #9 or #10) | | | | | | | | | |
| Brush 1 coat | | | | | | | | | |
| Slow | .55 | 15 | 21.85 | 5.64 | 1.53 | 1.46 | 2.68 | 1.70 | 13.01 |
| Medium | .50 | 15 | 19.10 | 8.25 | 2.78 | 1.27 | 3.01 | 1.68 | 16.99 |
| Fast | .45 | 14 | 16.38 | 9.90 | 4.01 | 1.17 | 2.79 | 1.07 | 18.94 |
| Enamel, water or oil base (material #9 or #10) | | | | | | | | | |
| Brush each finish coat | | | | | | | | | |
| Slow | .65 | 14 | 24.35 | 6.66 | 1.82 | 1.74 | 3.17 | 2.01 | 15.40 |
| Medium | .60 | 14 | 21.30 | 9.90 | 3.33 | 1.52 | 3.61 | 2.02 | 20.38 |
| Fast | .55 | 13 | 18.25 | 12.10 | 4.91 | 1.40 | 3.41 | 1.31 | 23.13 |
| Stain, seal & 1 coat varnish system (material #30) | | | | | | | | | |
| Brush each finish coat | | | | | | | | | |
| Slow | .90 | 9 | 29.10 | 9.23 | 2.51 | 3.23 | 4.64 | 2.94 | 22.55 |
| Medium | .80 | 8 | 25.50 | 13.20 | 4.44 | 3.19 | 5.10 | 2.85 | 28.78 |
| Fast | .70 | 7 | 21.90 | 15.40 | 6.24 | 3.13 | 4.58 | 1.76 | 31.11 |
| Varnish (material #30c) | | | | | | | | | |
| Brush additional coats of varnish | | | | | | | | | |
| Slow | .38 | 20 | 33.90 | 3.90 | 1.06 | 1.70 | 2.06 | 1.31 | 10.03 |
| Medium | .33 | 18 | 29.70 | 5.45 | 1.83 | 1.65 | 2.19 | 1.22 | 12.34 |
| Fast | .27 | 16 | 25.40 | 5.94 | 2.41 | 1.59 | 1.84 | .71 | 12.49 |
| Preparation for varnish application (material - minimal) | | | | | | | | | |
| Steel wool buff | | | | | | | | | |
| Slow | .35 | -- | -- | 3.59 | .97 | -- | 1.42 | .90 | 6.88 |
| Medium | .30 | -- | -- | 4.95 | 1.67 | -- | 1.62 | .91 | 9.15 |
| Fast | .25 | -- | -- | 5.50 | 2.23 | -- | 1.43 | .55 | 9.71 |
| Wax application (material - minimal) | | | | | | | | | |
| Slow | .35 | -- | -- | 3.59 | .97 | -- | 1.42 | .90 | 6.88 |
| Medium | .30 | -- | -- | 4.95 | 1.67 | -- | 1.62 | .91 | 9.15 |
| Fast | .25 | -- | -- | 5.50 | 2.23 | -- | 1.43 | .55 | 9.71 |

The stain, seal and 1 coat varnish system includes 1 coat of stain, sanding sealer, light sanding, putty and 1 coat of varnish. Minimum preparation for varnish usually includes a steel wool buff and wax operation with minimum material usage. Add: Preparation time for protecting adjacent surfaces with window protective coating (wax) or masking tape and paper as required. For heights above 8 feet, use the High Time Difficulty Factors on page 137. "Slow" work is based on a $10.25 hourly wage, "Medium" work on a $16.75 hourly wage, and "Fast" work on a $22.00 hourly wage. Other qualifications that apply to this table are on page 9.

| | Manhours per window | Windows per gallon | Material cost per gallon | Labor cost per window | Labor burden window | Material cost per window | Overhead per window | Profit per window | Total price per window |
|---|---|---|---|---|---|---|---|---|---|

## Windows, wood, exterior, 9 to 11 lite, brush

| | Manhours per window | Windows per gallon | Material cost per gallon | Labor cost per window | Labor burden window | Material cost per window | Overhead per window | Profit per window | Total price per window |
|---|---|---|---|---|---|---|---|---|---|
| Undercoat, water or oil base (material #3 or #4) | | | | | | | | | |
| Brush 1 coat | | | | | | | | | |
| Slow | .55 | 12 | 19.35 | 5.64 | 1.53 | 1.61 | 2.72 | 1.73 | 13.23 |
| Medium | .50 | 12 | 16.90 | 8.25 | 2.78 | 1.41 | 3.05 | 1.70 | 17.19 |
| Fast | .45 | 11 | 14.50 | 9.90 | 4.01 | 1.32 | 2.82 | 1.08 | 19.13 |
| Split coat (1/2 undercoat + 1/2 enamel), water or oil base (material #3 or #4 or #9 or #10) | | | | | | | | | |
| Brush 1 coat | | | | | | | | | |
| Slow | .67 | 14 | 21.85 | 6.87 | 1.87 | 1.56 | 3.19 | 2.02 | 15.51 |
| Medium | .62 | 14 | 19.10 | 10.23 | 3.44 | 1.36 | 3.68 | 2.06 | 20.77 |
| Fast | .57 | 13 | 16.38 | 12.54 | 5.08 | 1.26 | 3.49 | 1.34 | 23.71 |
| Enamel, water or oil base (material #9 or #10) | | | | | | | | | |
| Brush each finish coat | | | | | | | | | |
| Slow | .80 | 13 | 24.35 | 8.20 | 2.23 | 1.87 | 3.81 | 2.42 | 18.53 |
| Medium | .73 | 13 | 21.30 | 12.05 | 4.05 | 1.64 | 4.35 | 2.43 | 24.52 |
| Fast | .65 | 12 | 18.25 | 14.30 | 5.80 | 1.52 | 4.00 | 1.54 | 27.16 |
| Stain, seal & 1 coat varnish system (material #30) | | | | | | | | | |
| Brush each finish coat | | | | | | | | | |
| Slow | 1.00 | 7 | 29.10 | 10.25 | 2.79 | 4.16 | 5.33 | 3.38 | 25.91 |
| Medium | .90 | 6 | 25.50 | 14.85 | 5.00 | 4.25 | 5.90 | 3.30 | 33.30 |
| Fast | .80 | 5 | 21.90 | 17.60 | 7.14 | 4.38 | 5.39 | 2.07 | 36.58 |
| Varnish (material #30c) | | | | | | | | | |
| Brush additional coats of varnish | | | | | | | | | |
| Slow | .45 | 19 | 33.90 | 4.61 | 1.26 | 1.78 | 2.37 | 1.50 | 11.52 |
| Medium | .40 | 17 | 29.70 | 6.60 | 2.22 | 1.75 | 2.59 | 1.45 | 14.61 |
| Fast | .35 | 15 | 25.40 | 7.70 | 3.12 | 1.69 | 2.31 | .89 | 15.71 |
| Preparation for varnish application (material - minimal) | | | | | | | | | |
| Steel wool buff | | | | | | | | | |
| Slow | .45 | -- | -- | 4.61 | 1.26 | -- | 1.82 | 1.15 | 8.84 |
| Medium | .40 | -- | -- | 6.60 | 2.22 | -- | 2.16 | 1.21 | 12.19 |
| Fast | .35 | -- | -- | 7.70 | 3.12 | -- | 2.00 | .77 | 13.59 |
| Wax application (material - minimal) | | | | | | | | | |
| Slow | .40 | -- | -- | 4.10 | 1.12 | -- | 1.62 | 1.03 | 7.87 |
| Medium | .35 | -- | -- | 5.78 | 1.94 | -- | 1.89 | 1.06 | 10.67 |
| Fast | .30 | -- | -- | 6.60 | 2.68 | -- | 1.72 | .66 | 11.66 |

The stain, seal and 1 coat varnish system includes 1 coat of stain, sanding sealer, light sanding, putty and 1 coat of varnish. Minimum preparation for varnish usually includes a steel wool buff and wax operation with minimum material usage. Add: Preparation time for protecting adjacent surfaces with window protective coating (wax) or masking tape and paper as required. For heights above 8 feet, use the High Time Difficulty Factors on page 137. "Slow" work is based on a \$10.25 hourly wage, "Medium" work on a \$16.75 hourly wage, and "Fast" work on a \$22.00 hourly wage. Other qualifications that apply to this table are on page 9.

| | Manhours per window | Windows per gallon | Material cost per gallon | Labor cost per window | Labor burden window | Material cost per window | Overhead per window | Profit per window | Total price per window |
|---|---|---|---|---|---|---|---|---|---|
| **Windows, wood, exterior, 12 lite, brush** | | | | | | | | | |
| Undercoat, water or oil base (material #3 or #4) | | | | | | | | | |
| Brush 1 coat | | | | | | | | | |
| Slow | .67 | 11 | 19.35 | 6.87 | 1.87 | 1.76 | 3.26 | 2.06 | 15.82 |
| Medium | .62 | 11 | 16.90 | 10.23 | 3.44 | 1.54 | 3.73 | 2.08 | 21.02 |
| Fast | .57 | 10 | 14.50 | 12.54 | 5.08 | 1.45 | 3.53 | 1.36 | 23.96 |
| Split coat (1/2 undercoat + 1/2 enamel), water or oil base (material #3 or #4 or #9 or #10) | | | | | | | | | |
| Brush 1 coat | | | | | | | | | |
| Slow | .72 | 13 | 21.85 | 7.38 | 2.01 | 1.68 | 3.43 | 2.18 | 16.68 |
| Medium | .67 | 13 | 19.10 | 11.06 | 3.71 | 1.47 | 3.98 | 2.23 | 22.45 |
| Fast | .62 | 12 | 16.38 | 13.64 | 5.53 | 1.37 | 3.80 | 1.46 | 25.80 |
| Enamel, water or oil base (material #9 or #10) | | | | | | | | | |
| Brush each finish coat | | | | | | | | | |
| Slow | .85 | 12 | 24.35 | 8.71 | 2.37 | 2.03 | 4.06 | 2.58 | 19.75 |
| Medium | .80 | 12 | 21.30 | 13.20 | 4.44 | 1.78 | 4.76 | 2.66 | 26.84 |
| Fast | .75 | 11 | 18.25 | 16.50 | 6.69 | 1.66 | 4.60 | 1.77 | 31.22 |
| Stain, seal & 1 coat varnish system (material #30) | | | | | | | | | |
| Brush each finish coat | | | | | | | | | |
| Slow | 1.10 | 6 | 29.10 | 11.28 | 3.06 | 4.85 | 5.95 | 3.77 | 28.91 |
| Medium | 1.00 | 5 | 25.50 | 16.50 | 5.55 | 5.10 | 6.65 | 3.72 | 37.52 |
| Fast | .90 | 4 | 21.90 | 19.80 | 8.03 | 5.48 | 6.16 | 2.37 | 41.84 |
| Varnish (material #30c) | | | | | | | | | |
| Brush additional coats of varnish | | | | | | | | | |
| Slow | .45 | 18 | 33.90 | 4.61 | 1.26 | 1.88 | 2.40 | 1.52 | 11.67 |
| Medium | .40 | 16 | 29.70 | 6.60 | 2.22 | 1.86 | 2.62 | 1.46 | 14.76 |
| Fast | .35 | 14 | 25.40 | 7.70 | 3.12 | 1.81 | 2.34 | .90 | 15.87 |
| Preparation for varnish application (material - minimal) | | | | | | | | | |
| Steel wool buff | | | | | | | | | |
| Slow | .45 | -- | -- | 4.61 | 1.26 | -- | 1.82 | 1.15 | 8.84 |
| Medium | .40 | -- | -- | 6.60 | 2.22 | -- | 2.16 | 1.21 | 12.19 |
| Fast | .35 | -- | -- | 7.70 | 3.12 | -- | 2.00 | .77 | 13.59 |
| Wax application (material - minimal) | | | | | | | | | |
| Slow | .45 | -- | -- | 4.61 | 1.26 | -- | 1.82 | 1.15 | 8.84 |
| Medium | .40 | -- | -- | 6.60 | 2.22 | -- | 2.16 | 1.21 | 12.19 |
| Fast | .35 | -- | -- | 7.70 | 3.12 | -- | 2.00 | .77 | 13.59 |

The stain, seal and 1 coat varnish system includes 1 coat of stain, sanding sealer, light sanding, putty and 1 coat of varnish. Minimum preparation for varnish usually includes a steel wool buff and wax operation with minimum material usage. Add: Preparation time for protecting adjacent surfaces with window protective coating (wax) or masking tape and paper as required. For heights above 8 feet, use the High Time Difficulty Factors on page 137. "Slow" work is based on a $10.25 hourly wage, "Medium" work on a $16.75 hourly wage, and "Fast" work on a $22.00 hourly wage. Other qualifications that apply to this table are on page 9.

## Windows, interior

The figures that follow include costs for painting the sash (mullions or muntins), trim, frames, jambs, sill and apron on the inside only. Manhour and material rates are the same for painting either the inside or outside of windows. These estimates are per window painted, and are based on a window size of 15 square feet overall (length times width).

| | Manhour per window | Windows per gallon | Material cost per gallon | Labor cost per window | Labor burden window | Material cost per window | Overhead per window | Profit per window | Total price per window |
|---|---|---|---|---|---|---|---|---|---|
| **Windows, wood, interior, 1, 2 or 3 lite, brush** | | | | | | | | | |
| Undercoat, water or oil base (material #3 or #4) | | | | | | | | | |
| Brush 1 coat | | | | | | | | | |
| Slow | .25 | 15 | 19.35 | 2.56 | .70 | 1.29 | 1.41 | .89 | 6.85 |
| Medium | .20 | 15 | 16.90 | 3.30 | 1.11 | 1.13 | 1.36 | .76 | 7.66 |
| Fast | .15 | 14 | 14.50 | 3.30 | 1.34 | 1.04 | 1.05 | .40 | 7.13 |
| Split coat (1/2 undercoat + 1/2 enamel), water or oil base (material #3 or #4 or #9 or #10) | | | | | | | | | |
| Brush 1 coat | | | | | | | | | |
| Slow | .30 | 17 | 21.85 | 3.08 | .83 | 1.29 | 1.62 | 1.02 | 7.84 |
| Medium | .25 | 17 | 19.10 | 4.13 | 1.38 | 1.12 | 1.63 | .91 | 9.17 |
| Fast | .20 | 16 | 16.38 | 4.40 | 1.78 | 1.02 | 1.33 | .51 | 9.04 |
| Enamel, water or oil base (material #9 or #10) | | | | | | | | | |
| Brush each finish coat | | | | | | | | | |
| Slow | .35 | 16 | 24.35 | 3.59 | .97 | 1.52 | 1.89 | 1.20 | 9.17 |
| Medium | .30 | 16 | 21.30 | 4.95 | 1.67 | 1.33 | 1.95 | 1.09 | 10.99 |
| Fast | .25 | 15 | 18.25 | 5.50 | 2.23 | 1.22 | 1.66 | .64 | 11.25 |
| Stain, seal & 2 coat lacquer system, (material #11) | | | | | | | | | |
| Brush each coat | | | | | | | | | |
| Slow | .67 | 11 | 20.20 | 6.87 | 1.87 | 1.84 | 3.28 | 2.08 | 15.94 |
| Medium | .62 | 10 | 18.30 | 10.23 | 3.44 | 1.83 | 3.80 | 2.12 | 21.42 |
| Fast | .57 | 9 | 15.10 | 12.54 | 5.08 | 1.68 | 3.57 | 1.37 | 24.24 |

The stain, seal and 2 coat lacquer system includes 1 coat of stain, sanding sealer, light sanding, putty and 2 coats of lacquer. Add: Preparation time for protecting adjacent surfaces with window protective coating (wax) or masking tape and paper as required. "Slow" work is based on a $10.25 hourly wage, "Medium" work on a $16.50 hourly wage, and "Fast" work on a $22.00 hourly wage. Other qualifications that apply to this table are on page 9.

| | Manhours per window | Windows per gallon | Material cost per gallon | Labor cost per window | Labor burden window | Material cost per window | Overhead per window | Profit per window | Total price per window |
|---|---|---|---|---|---|---|---|---|---|

## Windows, wood, interior, 4 to 6 lite, brush

| | Manhours per window | Windows per gallon | Material cost per gallon | Labor cost per window | Labor burden window | Material cost per window | Overhead per window | Profit per window | Total price per window |
|---|---|---|---|---|---|---|---|---|---|
| Undercoat, water or oil base (material #3 or #4) | | | | | | | | | |
| Brush 1 coat | | | | | | | | | |
| Slow | .35 | 14 | 19.35 | 3.59 | .97 | 1.38 | 1.84 | 1.17 | 8.95 |
| Medium | .30 | 14 | 16.90 | 4.95 | 1.67 | 1.21 | 1.92 | 1.07 | 10.82 |
| Fast | .25 | 13 | 14.50 | 5.50 | 2.23 | 1.12 | 1.64 | .63 | 11.12 |
| Split coat (1/2 undercoat + 1/2 enamel), water or oil base (material #3 or #4 or #9 or #10) | | | | | | | | | |
| Brush 1 coat | | | | | | | | | |
| Slow | .40 | 16 | 21.85 | 4.10 | 1.12 | 1.37 | 2.04 | 1.29 | 9.92 |
| Medium | .35 | 16 | 19.10 | 5.78 | 1.94 | 1.19 | 2.18 | 1.22 | 12.31 |
| Fast | .30 | 15 | 16.38 | 6.60 | 2.68 | 1.09 | 1.92 | .74 | 13.03 |
| Enamel, water or oil base (material #9 or #10) | | | | | | | | | |
| Brush each finish coat | | | | | | | | | |
| Slow | .50 | 15 | 24.35 | 5.13 | 1.39 | 1.62 | 2.53 | 1.60 | 12.27 |
| Medium | .45 | 15 | 21.30 | 7.43 | 2.49 | 1.42 | 2.78 | 1.55 | 15.67 |
| Fast | .40 | 14 | 18.25 | 8.80 | 3.57 | 1.30 | 2.53 | .97 | 17.17 |
| Stain, seal & 2 coat lacquer system, (material #11) | | | | | | | | | |
| Brush each coat | | | | | | | | | |
| Slow | .80 | 10 | 20.20 | 8.20 | 2.23 | 2.02 | 3.86 | 2.45 | 18.76 |
| Medium | .73 | 9 | 18.30 | 12.05 | 4.05 | 2.03 | 4.44 | 2.48 | 25.05 |
| Fast | .65 | 8 | 15.10 | 14.30 | 5.80 | 1.89 | 4.07 | 1.56 | 27.62 |

The stain, seal and 2 coat lacquer system includes 1 coat of stain, sanding sealer, light sanding, putty and 2 coats of lacquer. Add: Preparation time for protecting adjacent surfaces with window protective coating (wax) or masking tape and paper as required. "Slow" work is based on a $10.25 hourly wage, "Medium" work on a $16.50 hourly wage, and "Fast" work on a $22.00 hourly wage. Other qualifications that apply to this table are on page 9.

| | Manhours per window | Windows per gallon | Material cost per gallon | Labor cost per window | Labor burden window | Material cost per window | Overhead per window | Profit per window | Total price per window |
|---|---|---|---|---|---|---|---|---|---|
| **Windows, wood, interior, 7 to 8 lite, brush** | | | | | | | | | |
| Undercoat, water or oil base (material #3 or #4) | | | | | | | | | |
| Brush 1 coat | | | | | | | | | |
| Slow | .45 | 13 | 19.35 | 4.61 | 1.26 | 1.49 | 2.28 | 1.45 | 11.09 |
| Medium | .40 | 13 | 16.90 | 6.60 | 2.22 | 1.30 | 2.48 | 1.39 | 13.99 |
| Fast | .35 | 12 | 14.50 | 7.70 | 3.12 | 1.21 | 2.23 | .86 | 15.12 |
| Split coat (1/2 undercoat + 1/2 enamel), water or oil base (material #3 or #4 or #9 or #10) | | | | | | | | | |
| Brush 1 coat | | | | | | | | | |
| Slow | .55 | 15 | 21.85 | 5.64 | 1.53 | 1.46 | 2.68 | 1.70 | 13.01 |
| Medium | .50 | 15 | 19.10 | 8.25 | 2.78 | 1.27 | 3.01 | 1.68 | 16.99 |
| Fast | .45 | 14 | 16.38 | 9.90 | 4.01 | 1.17 | 2.79 | 1.07 | 18.94 |
| Enamel, water or oil base (material #9 or #10) | | | | | | | | | |
| Brush each finish coat | | | | | | | | | |
| Slow | .67 | 14 | 24.35 | 6.87 | 1.87 | 1.74 | 3.25 | 2.06 | 15.79 |
| Medium | .62 | 14 | 21.30 | 10.23 | 3.44 | 1.52 | 3.72 | 2.08 | 20.99 |
| Fast | .57 | 13 | 18.25 | 12.54 | 5.08 | 1.40 | 3.52 | 1.35 | 23.89 |
| Stain, seal & 2 coat lacquer system, (material #11) | | | | | | | | | |
| Brush each coat | | | | | | | | | |
| Slow | .90 | 9 | 20.20 | 9.23 | 2.51 | 2.24 | 4.34 | 2.75 | 21.07 |
| Medium | .80 | 8 | 18.30 | 13.20 | 4.44 | 2.29 | 4.88 | 2.73 | 27.54 |
| Fast | .70 | 7 | 15.10 | 15.40 | 6.24 | 2.16 | 4.40 | 1.69 | 29.89 |

The stain, seal and 2 coat lacquer system includes 1 coat of stain, sanding sealer, light sanding, putty and 2 coats of lacquer. Add: Preparation time for protecting adjacent surfaces with window protective coating (wax) or masking tape and paper as required. "Slow" work is based on a $10.25 hourly wage, "Medium" work on a $16.50 hourly wage, and "Fast" work on a $22.00 hourly wage. Other qualifications that apply to this table are on page 9.

| | Manhours per window | Windows per gallon | Material cost per gallon | Labor cost per window | Labor burden window | Material cost per window | Overhead per window | Profit per window | Total price per window |
|---|---|---|---|---|---|---|---|---|---|

## Windows, wood, interior, 9 to 11 lite, brush

| | Manhours per window | Windows per gallon | Material cost per gallon | Labor cost per window | Labor burden window | Material cost per window | Overhead per window | Profit per window | Total price per window |
|---|---|---|---|---|---|---|---|---|---|
| Undercoat, water or oil base (material #3 or #4) | | | | | | | | | |
| Brush 1 coat | | | | | | | | | |
| Slow | .55 | 12 | 19.35 | 5.64 | 1.53 | 1.61 | 2.72 | 1.73 | 13.23 |
| Medium | .50 | 12 | 16.90 | 8.25 | 2.78 | 1.41 | 3.05 | 1.70 | 17.19 |
| Fast | .45 | 11 | 14.50 | 9.90 | 4.01 | 1.32 | 2.82 | 1.08 | 19.13 |
| Split coat (1/2 undercoat + 1/2 enamel), water or oil base (material #3 or #4 or #9 or #10) | | | | | | | | | |
| Brush 1 coat | | | | | | | | | |
| Slow | .65 | 14 | 21.85 | 6.66 | 1.82 | 1.56 | 3.11 | 1.97 | 15.12 |
| Medium | .60 | 14 | 19.10 | 9.90 | 3.33 | 1.36 | 3.57 | 2.00 | 20.16 |
| Fast | .55 | 13 | 16.38 | 12.10 | 4.91 | 1.26 | 3.38 | 1.30 | 22.95 |
| Enamel, water or oil base (material #9 or #10) | | | | | | | | | |
| Brush each finish coat | | | | | | | | | |
| Slow | .75 | 13 | 24.35 | 7.69 | 2.09 | 1.87 | 3.61 | 2.29 | 17.55 |
| Medium | .70 | 13 | 21.30 | 11.55 | 3.89 | 1.64 | 4.18 | 2.34 | 23.60 |
| Fast | .65 | 12 | 18.25 | 14.30 | 5.80 | 1.52 | 4.00 | 1.54 | 27.16 |
| Stain, seal & 2 coat lacquer system, (material #11) | | | | | | | | | |
| Brush each coat | | | | | | | | | |
| Slow | 1.00 | 8 | 20.20 | 10.25 | 2.79 | 2.53 | 4.83 | 3.06 | 23.46 |
| Medium | .90 | 7 | 18.30 | 14.85 | 5.00 | 2.61 | 5.50 | 3.08 | 31.04 |
| Fast | .80 | 6 | 15.10 | 17.60 | 7.14 | 2.52 | 5.04 | 1.94 | 34.24 |

The stain, seal and 2 coat lacquer system includes 1 coat of stain, sanding sealer, light sanding, putty and 2 coats of lacquer. Add: Preparation time for protecting adjacent surfaces with window protective coating (wax) or masking tape and paper as required. "Slow" work is based on a $10.25 hourly wage, "Medium" work on a $16.50 hourly wage, and "Fast" work on a $22.00 hourly wage. Other qualifications that apply to this table are on page 9.

| | Manhours per window | Windows per gallon | Material cost per gallon | Labor cost per window | Labor burden window | Material cost per window | Overhead per window | Profit per window | Total price per window |
|---|---|---|---|---|---|---|---|---|---|
| **Windows, wood, interior, 12 lite, brush** | | | | | | | | | |
| Undercoat, water or oil base (material #3 or #4) | | | | | | | | | |
| Brush 1 coat | | | | | | | | | |
| Slow | .65 | 11 | 19.35 | 6.66 | 1.82 | 1.76 | 3.17 | 2.01 | 15.42 |
| Medium | .60 | 11 | 16.90 | 9.90 | 3.33 | 1.54 | 3.62 | 2.02 | 20.41 |
| Fast | .55 | 10 | 14.50 | 12.10 | 4.91 | 1.45 | 3.42 | 1.31 | 23.19 |
| Split coat (1/2 undercoat + 1/2 enamel), water or oil base (material #3 or #4 or #9 or #10) | | | | | | | | | |
| Brush 1 coat | | | | | | | | | |
| Slow | .70 | 13 | 21.85 | 7.18 | 1.95 | 1.68 | 3.35 | 2.13 | 16.29 |
| Medium | .65 | 13 | 19.10 | 10.73 | 3.60 | 1.47 | 3.87 | 2.16 | 21.83 |
| Fast | .60 | 12 | 16.38 | 13.20 | 5.35 | 1.37 | 3.69 | 1.42 | 25.03 |
| Enamel, water or oil base (material #9 or #10) | | | | | | | | | |
| Brush each finish coat | | | | | | | | | |
| Slow | .80 | 12 | 24.35 | 8.20 | 2.23 | 2.03 | 3.86 | 2.45 | 18.77 |
| Medium | .75 | 12 | 21.30 | 12.38 | 4.16 | 1.78 | 4.49 | 2.51 | 25.32 |
| Fast | .70 | 11 | 18.25 | 15.40 | 6.24 | 1.66 | 4.31 | 1.66 | 29.27 |
| Stain, seal & 2 coat lacquer system, (material #11) | | | | | | | | | |
| Brush each coat | | | | | | | | | |
| Slow | 1.10 | 7 | 20.20 | 11.28 | 3.06 | 2.89 | 5.34 | 3.39 | 25.96 |
| Medium | 1.00 | 6 | 18.30 | 16.50 | 5.55 | 3.05 | 6.15 | 3.44 | 34.69 |
| Fast | .90 | 5 | 15.10 | 19.80 | 8.03 | 3.02 | 5.71 | 2.19 | 38.75 |

The stain, seal and 2 coat lacquer system includes 1 coat of stain, sanding sealer, light sanding, putty and 2 coats of lacquer. Add: Preparation time for protecting adjacent surfaces with window protective coating (wax) or masking tape and paper as required. "Slow" work is based on a $10.25 hourly wage, "Medium" work on a $16.50 hourly wage, and "Fast" work on a $22.00 hourly wage. Other qualifications that apply to this table are on page 9.

| | Labor SF per manhour | Material coverage SF/gallon | Material cost per gallon | Labor cost per 100 SF | Labor burden 100 SF | Material cost per 100 SF | Overhead per 100 SF | Profit per 100 SF | Total price per 100 SF |
|---|---|---|---|---|---|---|---|---|---|

## Windows, wood, interior, paint grade, square foot basis, brush

Undercoat, water or oil base (material #3 or #4)

| Brush, 1 coat | | | | | | | | | |
|---|---|---|---|---|---|---|---|---|---|
| Slow | 140 | 460 | 19.35 | 7.32 | 1.99 | 4.21 | 4.19 | 2.66 | 20.37 |
| Medium | 165 | 450 | 16.90 | 10.00 | 3.36 | 3.76 | 4.20 | 2.35 | 23.67 |
| Fast | 190 | 440 | 14.50 | 11.58 | 4.68 | 3.30 | 3.62 | 1.39 | 24.57 |

Split coat (1/2 undercoat + 1/2 enamel), water base (material #3 or #4 or #9 or#10)

| Brush 1 coat | | | | | | | | | |
|---|---|---|---|---|---|---|---|---|---|
| Slow | 120 | 520 | 21.85 | 8.54 | 2.32 | 4.20 | 4.67 | 2.96 | 22.69 |
| Medium | 135 | 500 | 19.10 | 12.22 | 4.12 | 3.82 | 4.94 | 2.76 | 27.86 |
| Fast | 150 | 480 | 16.38 | 14.67 | 5.95 | 3.41 | 4.45 | 1.71 | 30.19 |

Enamel, water or oil base (material #9 or #10)

| Brush each finish coat | | | | | | | | | |
|---|---|---|---|---|---|---|---|---|---|
| Slow | 95 | 500 | 24.35 | 10.79 | 2.94 | 4.87 | 5.77 | 3.66 | 28.03 |
| Medium | 110 | 480 | 21.30 | 15.00 | 5.04 | 4.44 | 6.00 | 3.35 | 33.83 |
| Fast | 125 | 460 | 18.25 | 17.60 | 7.14 | 3.97 | 5.31 | 2.04 | 36.06 |

Use this table to estimate the costs for paint grade windows larger than 15 square feet (width times height). When calculating the window area, add 1 foot in each direction to both dimensions before multiplying width by the height. For example, a 4' x 4' window would have a calculated surface area of 36 square feet: Width of 4' + 1' + 1' = 6' times height of 4' + 1' + 1' = 6'. The total is 6' x 6', or 36 square feet per side. Add an extra 2 square feet to the calculated area for each window lite. This allows for the extra time needed to finish muntins, mullions and sash. For windows that are stained, sealed and varnished, use the previous tables based on window lites. Add: Preparation time to protect adjacent surfaces with window protective coating (wax) or masking tape and paper as required. "Slow" work is based on a $10.25 hourly wage, "Medium" work on a $16.50 hourly wage, and "Fast" work on a $22.00 hourly wage. Other qualifications that apply to this table are on page 9.

## Window conversion factors

The table below gives you a faster way to figure the costs for painting windows by using the square foot basis tables above. First, calculate the square feet of window area. For example, assume you're undercoating a 3 by 5 foot window with 4 lites. Multiply length times width to find square feet. Then multiply the area by the manhour conversion factor below for a 4 lite window. Then divide the calculated square feet by the medium manhour rate from the previous table. Example: Area of a 3.0 x 5.0 window is 15 square feet, times the conversion factor for 4 lites (3.0) equals 45 square feet. Divide that by the medium rate from the above table for undercoating, (165) for a total time of .27 hours.

To find the material needed to paint the same window, multiply the 15 square feet by the factor in the right column below for a 4 lite window (2.2), for a total of 33 square feet. Now divide that figure by the light coverage rate, 460 square feet per gallon. It will take .072 gallons to undercoat that window. Divide 1 gallon by .072 to find that you can undercoat 13.9 windows with 1 gallon of undercoat material.

| Lites | Manhours per SF conversion factor | Material per SF conversion factor |
|---|---|---|
| 1, 2 or 3 lites | L x W x 2.0 | L x W x 2.0 |
| 4 to 6 lites | L x W x 3.0 | L x W x 2.2 |
| 7 to 8 lites | L x W x 4.0 | L x W x 2.4 |
| 9 to 11 lites | L x W x 5.0 | L x W x 2.6 |
| 12 lites | L x W x 6.0 | L x W x 2.9 |

| | Labor SF per manhour | Material coverage SF/gallon | Material cost per gallon | Labor cost per 100 SF | Labor burden 100 SF | Material cost per 100 SF | Overhead per 100 SF | Profit per 100 SF | Total price per 100 SF |
|---|---|---|---|---|---|---|---|---|---|
| **Wine racks, paint grade, spray, square feet of face** | | | | | | | | | |
| Undercoat, water base (material #3) | | | | | | | | | |
| Spray 1 coat | | | | | | | | | |
| Slow | 50 | 100 | 18.60 | 20.50 | 5.58 | 18.60 | 13.85 | 8.78 | 67.31 |
| Medium | 65 | 88 | 16.20 | 25.38 | 8.53 | 18.41 | 12.82 | 7.17 | 72.31 |
| Fast | 80 | 75 | 13.90 | 27.50 | 11.15 | 18.53 | 10.58 | 4.07 | 71.83 |
| Undercoat, oil base (material #4) | | | | | | | | | |
| Spray 1 coat | | | | | | | | | |
| Slow | 50 | 100 | 20.10 | 20.50 | 5.58 | 20.10 | 14.32 | 9.08 | 69.58 |
| Medium | 65 | 88 | 17.60 | 25.38 | 8.53 | 20.00 | 13.21 | 7.38 | 74.50 |
| Fast | 80 | 75 | 15.10 | 27.50 | 11.15 | 20.13 | 10.87 | 4.18 | 73.83 |
| Split coat (1/2 undercoat + 1/2 enamel), water base (material #3 or #9) | | | | | | | | | |
| Spray 1 coat | | | | | | | | | |
| Slow | 75 | 150 | 20.70 | 13.67 | 3.71 | 13.80 | 9.67 | 6.13 | 46.98 |
| Medium | 90 | 138 | 18.05 | 18.33 | 6.17 | 13.08 | 9.21 | 5.15 | 51.94 |
| Fast | 125 | 125 | 15.50 | 17.60 | 7.14 | 12.40 | 6.87 | 2.64 | 46.65 |
| Split coat (1/2 undercoat + 1/2 enamel), oil base (material #4 or #10) | | | | | | | | | |
| Spray 1 coat | | | | | | | | | |
| Slow | 75 | 150 | 23.00 | 13.67 | 3.71 | 15.33 | 10.15 | 6.43 | 49.29 |
| Medium | 90 | 138 | 20.15 | 18.33 | 6.17 | 14.60 | 9.58 | 5.35 | 54.03 |
| Fast | 125 | 125 | 17.25 | 17.60 | 7.14 | 13.80 | 7.13 | 2.74 | 48.41 |
| Enamel, water base (material #9) | | | | | | | | | |
| Spray 1st finish coat | | | | | | | | | |
| Slow | 65 | 125 | 22.80 | 15.77 | 4.29 | 18.24 | 11.88 | 7.53 | 57.71 |
| Medium | 88 | 113 | 19.90 | 18.75 | 6.30 | 17.61 | 10.45 | 5.84 | 58.95 |
| Fast | 100 | 100 | 17.10 | 22.00 | 8.92 | 17.10 | 8.88 | 3.41 | 60.31 |
| Spray additional finish coats | | | | | | | | | |
| Slow | 75 | 150 | 22.80 | 13.67 | 3.71 | 15.20 | 10.11 | 6.41 | 49.10 |
| Medium | 100 | 138 | 19.90 | 16.50 | 5.55 | 14.42 | 8.94 | 5.00 | 50.41 |
| Fast | 125 | 125 | 17.10 | 17.60 | 7.14 | 13.68 | 7.11 | 2.73 | 48.26 |
| Enamel, oil base (material #10) | | | | | | | | | |
| Spray 1st finish coat | | | | | | | | | |
| Slow | 65 | 125 | 25.90 | 15.77 | 4.29 | 20.72 | 12.64 | 8.01 | 61.43 |
| Medium | 88 | 113 | 22.70 | 18.75 | 6.30 | 20.09 | 11.06 | 6.18 | 62.38 |
| Fast | 100 | 100 | 19.40 | 22.00 | 8.92 | 19.40 | 9.31 | 3.58 | 63.21 |
| Spray additional finish coats | | | | | | | | | |
| Slow | 75 | 150 | 25.90 | 13.67 | 3.71 | 17.27 | 10.75 | 6.81 | 52.21 |
| Medium | 100 | 138 | 22.70 | 16.50 | 5.55 | 16.45 | 9.43 | 5.27 | 53.20 |
| Fast | 125 | 125 | 19.40 | 17.60 | 7.14 | 15.52 | 7.45 | 2.86 | 50.57 |

These figures include coating all interior and exterior surfaces and are based on overall dimensions (length times width) of the wine rack face. For heights above 8 feet, use the High Time Difficulty Factors on page 137. "Slow" work is based on a $10.25 hourly wage, "Medium" work on a $16.50 hourly wage, and "Fast" work on a $22.00 hourly wage. Other qualifications that apply to this table are on page 9.

| | Labor SF per manhour | Material coverage SF/gallon | Material cost per gallon | Labor cost per 100 SF | Labor burden 100 SF | Material cost per 100 SF | Overhead per 100 SF | Profit per 100 SF | Total price per 100 SF |
|---|---|---|---|---|---|---|---|---|---|
| **Wine racks, stain grade, per square feet of face** | | | | | | | | | |
| Stain, seal and 2 coat lacquer system (7 step process) | | | | | | | | | |
| STEP 1: Sand & putty | | | | | | | | | |
| Slow | 50 | -- | -- | 20.50 | 5.58 | -- | 8.09 | 5.13 | 39.30 |
| Medium | 75 | -- | -- | 22.00 | 7.39 | -- | 7.20 | 4.03 | 40.62 |
| Fast | 100 | -- | -- | 22.00 | 8.92 | -- | 5.72 | 2.20 | 38.84 |
| STEP 2 & 3: Wiping stain, oil base (material #11a) & wipe | | | | | | | | | |
| Spray 1 coat & wipe | | | | | | | | | |
| Slow | 100 | 150 | 27.40 | 10.25 | 2.79 | 18.27 | 9.71 | 6.15 | 47.17 |
| Medium | 175 | 113 | 24.00 | 9.43 | 3.16 | 21.24 | 8.29 | 4.63 | 46.75 |
| Fast | 225 | 75 | 20.60 | 9.78 | 3.95 | 27.47 | 7.63 | 2.93 | 51.76 |
| STEP 4: Sanding sealer (material #11b) | | | | | | | | | |
| Spray 1 coat & sand | | | | | | | | | |
| Slow | 130 | 150 | 17.30 | 7.88 | 2.15 | 11.53 | 6.68 | 4.24 | 32.48 |
| Medium | 208 | 113 | 15.10 | 7.93 | 2.68 | 13.36 | 5.87 | 3.28 | 33.12 |
| Fast | 275 | 75 | 13.00 | 8.00 | 3.25 | 17.33 | 5.29 | 2.03 | 35.90 |
| STEP 5: Sand lightly | | | | | | | | | |
| Slow | 75 | -- | -- | 13.67 | 3.71 | -- | 5.39 | 3.42 | 26.19 |
| Medium | 100 | -- | -- | 16.50 | 5.55 | -- | 5.40 | 3.02 | 30.47 |
| Fast | 125 | -- | -- | 17.60 | 7.14 | -- | 4.58 | 1.76 | 31.08 |
| STEP 6 & 7: Lacquer (material #11c) | | | | | | | | | |
| Spray 1st coat | | | | | | | | | |
| Slow | 100 | 100 | 18.00 | 10.25 | 2.79 | 18.00 | 9.62 | 6.10 | 46.76 |
| Medium | 200 | 75 | 15.70 | 8.25 | 2.78 | 20.93 | 7.83 | 4.38 | 44.17 |
| Fast | 300 | 50 | 13.50 | 7.33 | 2.97 | 27.00 | 6.90 | 2.65 | 46.85 |
| Spray 2nd coat | | | | | | | | | |
| Slow | 175 | 100 | 18.00 | 5.86 | 1.59 | 18.00 | 7.89 | 5.00 | 38.34 |
| Medium | 313 | 75 | 15.70 | 5.27 | 1.76 | 20.93 | 6.85 | 3.83 | 38.64 |
| Fast | 450 | 50 | 13.50 | 4.89 | 1.97 | 27.00 | 6.27 | 2.41 | 42.54 |
| Complete 7 step stain, seal & 2 coat lacquer system (material #11) | | | | | | | | | |
| Spray all coats | | | | | | | | | |
| Slow | 20 | 25 | 20.20 | 51.25 | 13.95 | 80.80 | 45.27 | 28.69 | 219.96 |
| Medium | 28 | 20 | 18.30 | 58.93 | 19.81 | 91.50 | 41.71 | 23.32 | 235.27 |
| Fast | 35 | 15 | 15.10 | 62.86 | 25.48 | 100.67 | 34.97 | 13.44 | 237.42 |

These figures include coating all interior and exterior surfaces and are based on overall dimensions (length times width) of the wine rack face. For heights above 8 feet, use the High Time Difficulty Factors on page 137. "Slow" work is based on a $10.25 hourly wage, "Medium" work on a $16.50 hourly wage, and "Fast" work on a $22.00 hourly wage. Other qualifications that apply to this table are on page 9.

# Part II

# Preparation Costs

| | Labor LF per manhour | Material coverage LF/gallon | Material cost per gallon | Labor cost per 100 LF | Labor burden 100 LF | Material cost per 100 LF | Overhead per 100 LF | Profit per 100 LF | Total price per 100 LF |
|---|---|---|---|---|---|---|---|---|---|
| **Acid wash gutters & downspouts** | | | | | | | | | |
| Acid wash, muriatic acid (material #49) | | | | | | | | | |
| Brush or mitt 1 coat | | | | | | | | | |
| Slow | 80 | 450 | 5.20 | 12.81 | 3.49 | 1.16 | 5.41 | 3.43 | 26.30 |
| Medium | 95 | 425 | 4.50 | 17.37 | 5.85 | 1.06 | 5.95 | 3.33 | 33.56 |
| Fast | 110 | 400 | 3.90 | 20.00 | 8.11 | .98 | 5.38 | 2.07 | 36.54 |

"Slow" work is based on a $10.25 hourly wage, "Medium" work on a $16.50 hourly wage, and "Fast" work on a $22.00 hourly wage. Other qualifications that apply to this table are on page 9.

| | Labor SF per manhour | Material coverage SF/gallon | Material cost per gallon | Labor cost per 100 SF | Labor burden 100 SF | Material cost per 100 SF | Overhead per 100 SF | Profit per 100 SF | Total price per 100 SF |
|---|---|---|---|---|---|---|---|---|---|
| **Airblast, compressed air** | | | | | | | | | |
| Average production | | | | | | | | | |
| Slow | 150 | -- | -- | 6.83 | 1.87 | -- | 2.69 | 1.71 | 13.10 |
| Medium | 175 | -- | -- | 9.43 | 3.16 | -- | 3.09 | 1.73 | 17.41 |
| Fast | 200 | -- | -- | 11.00 | 4.46 | -- | 2.86 | 1.10 | 19.42 |

All widths less than 12", consider as 1 square foot per linear foot. Add equipment rental costs with Overhead and Profit. "Slow" work is based on a $10.25 hourly wage, "Medium" work on a $16.50 hourly wage, and "Fast" work on a $22.00 hourly wage. Other qualifications that apply to this table are on page 9.

| | Labor SF per manhour | Material coverage SF/gallon | Material cost per gallon | Labor cost per 100 SF | Labor burden 100 SF | Material cost per 100 SF | Overhead per 100 SF | Profit per 100 SF | Total price per 100 SF |
|---|---|---|---|---|---|---|---|---|---|
| **Burn off paint** | | | | | | | | | |
| Exterior: | | | | | | | | | |
| Exterior trim | | | | | | | | | |
| Slow | 15 | -- | -- | 68.33 | 18.61 | -- | 26.95 | 17.09 | 130.98 |
| Medium | 20 | -- | -- | 82.50 | 27.75 | -- | 27.01 | 15.10 | 152.36 |
| Fast | 25 | -- | -- | 88.00 | 35.68 | -- | 22.88 | 8.79 | 155.35 |
| Plain surfaces | | | | | | | | | |
| Slow | 30 | -- | -- | 34.17 | 9.29 | -- | 13.48 | 8.54 | 65.48 |
| Medium | 40 | -- | -- | 41.25 | 13.88 | -- | 13.51 | 7.55 | 76.19 |
| Fast | 50 | -- | -- | 44.00 | 17.84 | -- | 11.44 | 4.40 | 77.68 |
| Beveled wood siding | | | | | | | | | |
| Slow | 20 | -- | -- | 51.25 | 13.95 | -- | 20.22 | 12.82 | 98.24 |
| Medium | 30 | -- | -- | 55.00 | 18.49 | -- | 18.01 | 10.07 | 101.57 |
| Fast | 40 | -- | -- | 55.00 | 22.30 | -- | 14.30 | 5.50 | 97.10 |
| Interior: | | | | | | | | | |
| Interior trim | | | | | | | | | |
| Slow | 10 | -- | -- | 102.50 | 27.90 | -- | 40.43 | 25.63 | 196.46 |
| Medium | 15 | -- | -- | 110.00 | 37.01 | -- | 36.02 | 20.13 | 203.16 |
| Fast | 20 | -- | -- | 110.00 | 44.60 | -- | 28.60 | 10.99 | 194.19 |
| Plain surfaces | | | | | | | | | |
| Slow | 15 | -- | -- | 68.33 | 18.61 | -- | 26.95 | 17.09 | 130.98 |
| Medium | 25 | -- | -- | 66.00 | 22.20 | -- | 21.61 | 12.08 | 121.89 |
| Fast | 35 | -- | -- | 62.86 | 25.48 | -- | 16.34 | 6.28 | 110.96 |

All widths less than 12", consider as 1 square foot per linear foot. Note: Because surfaces and the material being removed vary widely, it's best to quote prices for burning-off existing finishes on a Time and Material or Cost Plus Fee basis at a preset hourly rate. "Slow" work is based on a $10.25 hourly wage, "Medium" work on a $16.50 hourly wage, and "Fast" work on a $22.00 hourly wage. Other qualifications that apply to this table are on page 9.

| | Labor LF per manhour | Material LF/fluid oz ounce | Material cost per ounce | Labor cost per 100 LF | Labor burden 100 LF | Material cost per 100 LF | Overhead per 100 LF | Profit per 100 LF | Total price per 100 LF |
|---|---|---|---|---|---|---|---|---|---|
| **Caulk** | | | | | | | | | |
| 1/8" gap (material #42) | | | | | | | | | |
| Slow | 60 | 14 | .17 | 17.08 | 4.66 | 1.21 | 7.11 | 4.51 | 34.57 |
| Medium | 65 | 13 | .20 | 25.38 | 8.53 | 1.54 | 8.69 | 4.86 | 49.00 |
| Fast | 70 | 12 | .10 | 31.43 | 12.75 | .83 | 8.33 | 3.20 | 56.54 |
| 1/4" gap (material #42) | | | | | | | | | |
| Slow | 50 | 3.5 | .17 | 20.50 | 5.58 | 4.86 | 9.59 | 6.08 | 46.61 |
| Medium | 55 | 3.3 | .20 | 30.00 | 10.09 | 6.06 | 11.31 | 6.32 | 63.78 |
| Fast | 60 | 3.0 | .10 | 36.67 | 14.87 | 3.33 | 10.15 | 3.90 | 68.92 |
| 3/8" gap (material #42) | | | | | | | | | |
| Slow | 40 | 1.5 | .17 | 25.63 | 6.97 | 11.33 | 13.62 | 8.63 | 66.18 |
| Medium | 45 | 1.4 | .20 | 36.67 | 12.33 | 14.29 | 15.51 | 8.67 | 87.47 |
| Fast | 50 | 1.3 | .10 | 44.00 | 17.84 | 7.69 | 12.86 | 4.94 | 87.33 |
| 1/2" gap (material #42) | | | | | | | | | |
| Slow | 33 | 1.0 | .17 | 31.06 | 8.45 | 17.00 | 17.52 | 11.11 | 85.14 |
| Medium | 38 | 0.9 | .20 | 43.42 | 14.62 | 22.22 | 19.66 | 10.99 | 110.91 |
| Fast | 43 | 0.8 | .10 | 51.16 | 20.76 | 12.50 | 15.62 | 6.00 | 106.04 |

Caulking that's part of normal surface preparation is included in the painting cost tables. When extra caulking is required, use this cost guide. It's based on oil or latex base, silicone or urethane caulk in 10 ounce tubes. "Slow" work is based on a $10.25 hourly wage, "Medium" work on a $16.50 hourly wage, and "Fast" work on a $22.00 hourly wage. Other qualifications that apply to this table are on page 9.

| | Labor SF per manhour | Material coverage SF/gallon | Material cost per gallon | Labor cost per 100 SF | Labor burden 100 SF | Material cost per 100 SF | Overhead per 100 SF | Profit per 100 SF | Total price per 100 SF |
|---|---|---|---|---|---|---|---|---|---|
| **Cut cracks** | | | | | | | | | |
| Varnish or hard oil and repair cracks | | | | | | | | | |
| Slow | 120 | -- | -- | 8.54 | 2.32 | -- | 3.37 | 2.14 | 16.37 |
| Medium | 130 | -- | -- | 12.69 | 4.27 | -- | 4.16 | 2.32 | 23.44 |
| Fast | 140 | -- | -- | 15.71 | 6.37 | -- | 4.08 | 1.57 | 27.73 |
| Gloss painted walls and fix cracks | | | | | | | | | |
| Slow | 125 | -- | -- | 8.20 | 2.23 | -- | 3.23 | 2.05 | 15.71 |
| Medium | 135 | -- | -- | 12.22 | 4.12 | -- | 4.00 | 2.24 | 22.58 |
| Fast | 145 | -- | -- | 15.17 | 6.16 | -- | 3.94 | 1.52 | 26.79 |

All widths less than 12", consider as 1 square foot per linear foot. "Slow" work is based on a $10.25 hourly wage, "Medium" work on a $16.50 hourly wage, and "Fast" work on a $22.00 hourly wage. Other qualifications that apply to this table are on page 9.

| | Labor SF per manhour | Material coverage SF/gallon | Material cost per gallon | Labor cost per 100 SF | Labor burden 100 SF | Material cost per 100 SF | Overhead per 100 SF | Profit per 100 SF | Total price per 100 SF |
|---|---|---|---|---|---|---|---|---|---|
| **Fill wood floors** | | | | | | | | | |
| Fill and wipe wood floors (material #48) | | | | | | | | | |
| Slow | 45 | 155 | 29.50 | 22.78 | 6.19 | 19.03 | 14.89 | 9.44 | 72.33 |
| Medium | 60 | 145 | 25.80 | 27.50 | 9.26 | 17.79 | 13.36 | 7.47 | 75.38 |
| Fast | 75 | 135 | 22.10 | 29.33 | 11.89 | 16.37 | 10.65 | 4.09 | 72.33 |

All widths less than 12", consider as 1 square foot per linear foot. "Slow" work is based on a $10.25 hourly wage, "Medium" work on a $16.50 hourly wage, and "Fast" work on a $22.00 hourly wage. Other qualifications that apply to this table are on page 9.

| | Labor SF per manhour | Material coverage SF/pound | Material cost per pound | Labor cost per 100 SF | Labor burden 100 SF | Material cost per 100 SF | Overhead per 100 SF | Profit per 100 SF | Total price per 100 SF |
|---|---|---|---|---|---|---|---|---|---|
| **Putty application** | | | | | | | | | |
| Good condition, 1 coat (material #45) | | | | | | | | | |
| Slow | 60 | 150 | 6.80 | 17.08 | 4.66 | 4.53 | 8.14 | 5.16 | 39.57 |
| Medium | 90 | 135 | 6.00 | 18.33 | 6.17 | 4.44 | 7.09 | 3.96 | 39.99 |
| Fast | 120 | 120 | 5.10 | 18.33 | 7.43 | 4.25 | 5.55 | 2.13 | 37.69 |
| Average condition, 1 coat (material #45) | | | | | | | | | |
| Slow | 35 | 90 | 6.80 | 29.29 | 7.97 | 7.56 | 13.90 | 8.81 | 67.53 |
| Medium | 65 | 75 | 6.00 | 25.38 | 8.53 | 8.00 | 10.27 | 5.74 | 57.92 |
| Fast | 95 | 60 | 5.10 | 23.16 | 9.40 | 8.50 | 7.59 | 2.92 | 51.57 |
| Poor conditions, 1 coat (material #45) | | | | | | | | | |
| Slow | 15 | 40 | 6.80 | 68.33 | 18.61 | 17.00 | 32.22 | 20.43 | 156.59 |
| Medium | 30 | 30 | 6.00 | 55.00 | 18.49 | 20.00 | 22.91 | 12.81 | 129.21 |
| Fast | 45 | 20 | 5.10 | 48.89 | 19.81 | 25.50 | 17.43 | 6.70 | 118.33 |

These figures apply to either spackle or Swedish putty. All widths less than 12", consider as 1 square foot per linear foot. For flat trim or sash: Estimate 1 linear foot of trim as 1 square foot of surface. "Slow" work is based on a $10.25 hourly wage, "Medium" work on a $16.50 hourly wage, and "Fast" work on a $22.50 hourly wage. Other qualifications that apply to this table are on page 9.

| | Labor SF per manhour | Material coverage SF/gallon | Material cost per gallon | Labor cost per 100 SF | Labor burden 100 SF | Material cost per 100 SF | Overhead per 100 SF | Profit per 100 SF | Total price per 100 SF |
|---|---|---|---|---|---|---|---|---|---|
| **Sand, medium (before first coat)** | | | | | | | | | |
| Interior flatwall areas | | | | | | | | | |
| Slow | 275 | -- | -- | 3.73 | 1.02 | -- | 1.16 | .89 | 6.80 |
| Medium | 300 | -- | -- | 5.50 | 1.84 | -- | 1.80 | 1.01 | 10.15 |
| Fast | 325 | -- | -- | 6.77 | 2.75 | -- | 1.76 | .68 | 11.96 |
| Interior enamel areas | | | | | | | | | |
| Slow | 250 | -- | -- | 4.10 | 1.12 | -- | 1.62 | 1.03 | 7.87 |
| Medium | 275 | -- | -- | 6.00 | 2.03 | -- | 1.96 | 1.10 | 11.09 |
| Fast | 300 | -- | -- | 7.33 | 2.97 | -- | 1.91 | .73 | 12.94 |

All widths less than 12", consider as 1 square foot per linear foot. "Slow" work is based on a $10.25 hourly wage, "Medium" work on a $16.50 hourly wage, and "Fast" work on a $22.00 hourly wage. Other qualifications that apply to this table are on page 9.

| | Labor SF per manhour | Material coverage SF/gallon | Material cost per gallon | Labor cost per 100 SF | Labor burden 100 SF | Material cost per 100 SF | Overhead per 100 SF | Profit per 100 SF | Total price per 100 SF |
|---|---|---|---|---|---|---|---|---|---|
| **Sand & putty (before second coat)** | | | | | | | | | |
| Interior flatwall areas | | | | | | | | | |
| Slow | 190 | -- | -- | 5.39 | 1.47 | -- | 2.13 | 1.35 | 10.34 |
| Medium | 200 | -- | -- | 8.25 | 2.78 | -- | 2.70 | 1.51 | 15.24 |
| Fast | 210 | -- | -- | 10.48 | 4.24 | -- | 2.73 | 1.05 | 18.50 |
| Interior enamel areas | | | | | | | | | |
| Slow | 110 | -- | -- | 9.32 | 2.53 | -- | 3.68 | 2.33 | 17.86 |
| Medium | 125 | -- | -- | 13.20 | 4.44 | -- | 4.32 | 2.42 | 24.38 |
| Fast | 140 | -- | -- | 15.71 | 6.37 | -- | 4.08 | 1.57 | 27.73 |
| Exterior siding & trim-plain | | | | | | | | | |
| Slow | 180 | -- | -- | 5.69 | 1.56 | -- | 2.24 | 1.42 | 10.91 |
| Medium | 200 | -- | -- | 8.25 | 2.78 | -- | 2.70 | 1.51 | 15.24 |
| Fast | 220 | -- | -- | 10.00 | 4.07 | -- | 2.60 | 1.00 | 17.67 |
| Exterior trim only | | | | | | | | | |
| Slow | 100 | -- | -- | 10.25 | 2.79 | -- | 4.04 | 2.56 | 19.64 |
| Medium | 110 | -- | -- | 15.00 | 5.04 | -- | 4.91 | 2.75 | 27.70 |
| Fast | 120 | -- | -- | 18.33 | 7.43 | -- | 4.77 | 1.83 | 32.36 |
| Bookshelves | | | | | | | | | |
| Slow | 100 | -- | -- | 10.25 | 2.79 | -- | 4.04 | 2.56 | 19.64 |
| Medium | 125 | -- | -- | 13.20 | 4.44 | -- | 4.32 | 2.42 | 24.38 |
| Fast | 150 | -- | -- | 14.67 | 5.95 | -- | 3.81 | 1.47 | 25.90 |
| Cabinets | | | | | | | | | |
| Slow | 125 | -- | -- | 8.20 | 2.23 | -- | 3.23 | 2.05 | 15.71 |
| Medium | 150 | -- | -- | 11.00 | 3.71 | -- | 3.60 | 2.01 | 20.32 |
| Fast | 175 | -- | -- | 12.57 | 5.09 | -- | 3.27 | 1.26 | 22.19 |

Consider all trim which is less than 12" wide to be 12" wide. High grade work - Use the manhours equal to 1 coat of paint. Medium grade work - Use half (50%) of the manhours for 1 coat of paint. All widths less than 12", consider as 1 square foot per linear foot. "Slow" work is based on a $10.25 hourly wage, "Medium" work on a $16.50 hourly wage, and "Fast" work on a $22.00 hourly wage. Other qualifications that apply to this table are on page 9.

| | Labor SF per manhour | Material coverage SF/gallon | Material cost per gallon | Labor cost per 100 SF | Labor burden 100 SF | Material cost per 100 SF | Overhead per 100 SF | Profit per 100 SF | Total price per 100 SF |
|---|---|---|---|---|---|---|---|---|---|
| **Sand, light (before third coat)** | | | | | | | | | |
| Interior flatwall areas | | | | | | | | | |
| Slow | 335 | -- | -- | 3.06 | .84 | -- | 1.21 | .77 | 5.88 |
| Medium | 345 | -- | -- | 4.78 | 1.61 | -- | 1.57 | .88 | 8.84 |
| Fast | 355 | -- | -- | 6.20 | 2.52 | -- | 1.61 | .62 | 10.95 |
| Interior enamel areas | | | | | | | | | |
| Slow | 130 | -- | -- | 7.88 | 2.15 | -- | 3.11 | 1.97 | 15.11 |
| Medium | 140 | -- | -- | 11.79 | 3.95 | -- | 3.86 | 2.16 | 21.76 |
| Fast | 150 | -- | -- | 14.67 | 5.95 | -- | 3.81 | 1.47 | 25.90 |
| Exterior siding & trim-plain | | | | | | | | | |
| Slow | 250 | -- | -- | 4.10 | 1.12 | -- | 1.62 | 1.03 | 7.87 |
| Medium | 275 | -- | -- | 6.00 | 2.03 | -- | 1.96 | 1.10 | 11.09 |
| Fast | 300 | -- | -- | 7.33 | 2.97 | -- | 1.91 | .73 | 12.94 |
| Exterior trim only | | | | | | | | | |
| Slow | 150 | -- | -- | 6.83 | 1.87 | -- | 2.69 | 1.71 | 13.10 |
| Medium | 175 | -- | -- | 9.43 | 3.16 | -- | 3.09 | 1.73 | 17.41 |
| Fast | 200 | -- | -- | 11.00 | 4.46 | -- | 2.86 | 1.10 | 19.42 |
| Bookshelves | | | | | | | | | |
| Slow | 175 | -- | -- | 5.86 | 1.59 | -- | 2.31 | 1.47 | 11.23 |
| Medium | 225 | -- | -- | 7.33 | 2.46 | -- | 2.40 | 1.34 | 13.53 |
| Fast | 275 | -- | -- | 8.00 | 3.25 | -- | 2.08 | .80 | 14.13 |
| Cabinets | | | | | | | | | |
| Slow | 200 | -- | -- | 5.13 | 1.39 | -- | 2.02 | 1.28 | 9.82 |
| Medium | 250 | -- | -- | 6.60 | 2.22 | -- | 2.16 | 1.21 | 12.19 |
| Fast | 300 | -- | -- | 7.33 | 2.97 | -- | 1.91 | .73 | 12.94 |

For heights above 8 feet, use the High Time Difficulty Factors on page 137. "Slow" work is based on a $10.25 hourly wage, "Medium" work on a $16.50 hourly wage, and "Fast" work on a $22.00 hourly wage. Other qualifications that apply to this table are on page 9.

| | Labor SF per manhour | Material coverage SF/gallon | Material cost per gallon | Labor cost per 100 SF | Labor burden 100 SF | Material cost per 100 SF | Overhead per 100 SF | Profit per 100 SF | Total price per 100 SF |
|---|---|---|---|---|---|---|---|---|---|
| **Sand, extra fine, flat surfaces, varnish** | | | | | | | | | |
| Sand or steel wool | | | | | | | | | |
| Slow | 50 | -- | -- | 20.50 | 5.58 | -- | 8.09 | 5.13 | 39.30 |
| Medium | 88 | -- | -- | 18.75 | 6.30 | -- | 6.14 | 3.43 | 34.62 |
| Fast | 125 | -- | -- | 17.60 | 7.14 | -- | 4.58 | 1.76 | 31.08 |

All widths less than 12", consider as 1 square foot per linear foot. "Slow" work is based on a $10.25 hourly wage, "Medium" work on a $16.50 hourly wage, and "Fast" work on a $22.00 hourly wage. Other qualifications that apply to this table are on page 9.

## Sandblast, general

Sandblasting production rates may vary widely. Use the following figures as a reference for estimating and to establish performance data for your company. The abrasive material used in sandblasting is usually white silica sand although slags have recently gained in popularity. Material consumption varies with several factors:

1) Type of finish required
2) Condition of the surface
3) Quality of abrasive material (sharpness, cleanliness and hardness)
4) Nozzle size
5) Equipment arrangement and placement
6) Operator skill

All material consumption values are based on three uses of a 25 to 35 mesh white silica sand abrasive at a cost of $40 to $60 per ton. (Check the current price in your area.) Note: See the Structural steel conversion table at Figure 23 on pages 387 through 395 for converting linear feet of structural steel to square feet.

| | Labor SF per manhour | Material coverage pounds/SF | Material cost per pound | Labor cost per 100 SF | Labor burden 100 SF | Material cost per 100 SF | Overhead per 100 SF | Profit per 100 SF | Total price per 100 SF |
|---|---|---|---|---|---|---|---|---|---|
| **Sandblast, brush-off blast** | | | | | | | | | |
| Surface condition basis - large projects & surface areas (material #46) | | | | | | | | | |
| Remove cement base paint | | | | | | | | | |
| Slow | 150 | 2.0 | .03 | 6.83 | 1.87 | 6.00 | 4.55 | 2.89 | 22.14 |
| Medium | 175 | 2.5 | .03 | 9.43 | 3.16 | 7.50 | 4.92 | 2.75 | 27.76 |
| Fast | 200 | 3.0 | .02 | 11.00 | 4.46 | 6.00 | 3.97 | 1.53 | 26.96 |
| Remove oil or latex base paint | | | | | | | | | |
| Slow | 100 | 3.0 | .03 | 10.25 | 2.79 | 9.00 | 6.83 | 4.33 | 33.20 |
| Medium | 125 | 3.5 | .03 | 13.20 | 4.44 | 10.50 | 6.89 | 3.85 | 38.88 |
| Fast | 150 | 4.0 | .02 | 14.67 | 5.95 | 8.00 | 5.29 | 2.03 | 35.94 |
| Surface area basis - large projects & surface areas (material #46) | | | | | | | | | |
| Pipe up to 12" O/D | | | | | | | | | |
| Slow | 125 | 4.0 | .03 | 8.20 | 2.23 | 12.00 | 6.95 | 4.41 | 33.79 |
| Medium | 150 | 4.5 | .03 | 11.00 | 3.71 | 13.50 | 6.91 | 3.86 | 38.98 |
| Fast | 175 | 5.0 | .02 | 12.57 | 5.09 | 10.00 | 5.12 | 1.97 | 34.75 |
| Structural steel | | | | | | | | | |
| Sizes up to 2 SF/LF | | | | | | | | | |
| Slow | 150 | 4.0 | .03 | 6.83 | 1.87 | 12.00 | 6.41 | 4.07 | 31.18 |
| Medium | 175 | 4.5 | .03 | 9.43 | 3.16 | 13.50 | 6.39 | 3.57 | 36.05 |
| Fast | 200 | 5.0 | .02 | 11.00 | 4.46 | 10.00 | 4.71 | 1.81 | 31.98 |
| Sizes from 2 to 5 SF/LF | | | | | | | | | |
| Slow | 200 | 3.0 | .03 | 5.13 | 1.39 | 9.00 | 4.81 | 3.05 | 23.38 |
| Medium | 225 | 3.5 | .03 | 7.33 | 2.46 | 10.50 | 4.97 | 2.78 | 28.04 |
| Fast | 250 | 4.0 | .02 | 8.80 | 3.57 | 8.00 | 3.77 | 1.45 | 25.59 |
| Sizes over 5 SF/LF | | | | | | | | | |
| Slow | 250 | 2.0 | .03 | 4.10 | 1.12 | 6.00 | 3.48 | 2.21 | 16.91 |
| Medium | 275 | 3.0 | .03 | 6.00 | 2.03 | 9.00 | 4.17 | 2.33 | 23.53 |
| Fast | 300 | 4.0 | .02 | 7.33 | 2.97 | 8.00 | 3.39 | 1.30 | 22.99 |
| Tanks and vessels | | | | | | | | | |
| Sizes up to 12'0" O/D | | | | | | | | | |
| Slow | 200 | 3.0 | .03 | 5.13 | 1.39 | 9.00 | 4.81 | 3.05 | 23.38 |
| Medium | 225 | 3.5 | .03 | 7.33 | 2.46 | 10.50 | 4.97 | 2.78 | 28.04 |
| Fast | 250 | 4.0 | .02 | 8.80 | 3.57 | 8.00 | 3.77 | 1.45 | 25.59 |
| Sizes over 12'0" O/D | | | | | | | | | |
| Slow | 250 | 2.0 | .03 | 4.10 | 1.12 | 6.00 | 3.48 | 2.21 | 16.91 |
| Medium | 275 | 3.0 | .03 | 6.00 | 2.03 | 9.00 | 4.17 | 2.33 | 23.53 |
| Fast | 300 | 4.0 | .02 | 7.33 | 2.97 | 8.00 | 3.39 | 1.30 | 22.99 |

For heights above 8 feet, use the High Time Difficulty Factors on page 137. "Slow" work is based on a $10.25 hourly wage, "Medium" work on a $16.50 hourly wage, and "Fast" work on a $22.00 hourly wage. Other qualifications that apply to this table are on page 9.

| | Labor SF per manhour | Material coverage pounds/SF | Material cost per pound | Labor cost per 100 SF | Labor burden 100 SF | Material cost per 100 SF | Overhead per 100 SF | Profit per 100 SF | Total price per 100 SF |
|---|---|---|---|---|---|---|---|---|---|

## Sandblast, commercial blast (67% white)

| | Labor SF per manhour | Material coverage pounds/SF | Material cost per pound | Labor cost per 100 SF | Labor burden 100 SF | Material cost per 100 SF | Overhead per 100 SF | Profit per 100 SF | Total price per 100 SF |
|---|---|---|---|---|---|---|---|---|---|
| **Surface condition basis - large projects & surface areas (material #46)** | | | | | | | | | |
| Loose mill scale & fine powder rust | | | | | | | | | |
| Slow | 150 | 4.0 | .03 | 6.83 | 1.87 | 12.00 | 6.41 | 4.07 | 31.18 |
| Medium | 175 | 4.5 | .03 | 9.43 | 3.16 | 13.50 | 6.39 | 3.57 | 36.05 |
| Fast | 200 | 5.0 | .02 | 11.00 | 4.46 | 10.00 | 4.71 | 1.81 | 31.98 |
| Tight mill scale & little or no rust | | | | | | | | | |
| Slow | 125 | 5.0 | .03 | 8.20 | 2.23 | 15.00 | 7.88 | 5.00 | 38.31 |
| Medium | 150 | 5.5 | .03 | 11.00 | 3.71 | 16.50 | 7.64 | 4.27 | 43.12 |
| Fast | 175 | 6.0 | .02 | 12.57 | 5.09 | 12.00 | 5.49 | 2.11 | 37.26 |
| Hard scale, blistered, rusty surface | | | | | | | | | |
| Slow | 75 | 6.0 | .03 | 13.67 | 3.71 | 18.00 | 10.97 | 6.96 | 53.31 |
| Medium | 100 | 7.0 | .03 | 16.50 | 5.55 | 21.00 | 10.55 | 5.90 | 59.50 |
| Fast | 125 | 8.0 | .02 | 17.60 | 7.14 | 16.00 | 7.54 | 2.90 | 51.18 |
| Rust nodules and pitted surface | | | | | | | | | |
| Slow | 50 | 8.0 | .03 | 20.50 | 5.58 | 24.00 | 15.53 | 9.84 | 75.45 |
| Medium | 60 | 9.5 | .03 | 27.50 | 9.26 | 28.50 | 15.99 | 8.94 | 90.19 |
| Fast | 70 | 11.0 | .02 | 31.43 | 12.75 | 22.00 | 12.24 | 4.70 | 83.12 |
| **Surface area basis - large projects & surface areas (material #46)** | | | | | | | | | |
| Pipe up to 12" O/D | | | | | | | | | |
| Slow | 45 | 5.0 | .03 | 22.78 | 6.19 | 15.00 | 13.64 | 8.64 | 66.25 |
| Medium | 60 | 6.0 | .03 | 27.50 | 9.26 | 18.00 | 13.41 | 7.50 | 75.67 |
| Fast | 75 | 7.0 | .02 | 29.33 | 11.89 | 14.00 | 10.22 | 3.93 | 69.37 |
| Structural steel | | | | | | | | | |
| Sizes up to 2 SF/LF | | | | | | | | | |
| Slow | 70 | 5.0 | .03 | 14.64 | 3.99 | 15.00 | 10.43 | 6.61 | 50.67 |
| Medium | 85 | 6.0 | .03 | 19.41 | 6.52 | 18.00 | 10.77 | 6.02 | 60.72 |
| Fast | 100 | 7.0 | .02 | 22.00 | 8.92 | 14.00 | 8.31 | 3.19 | 56.42 |
| Sizes from 2 to 5 SF/LF | | | | | | | | | |
| Slow | 80 | 5.0 | .03 | 12.81 | 3.49 | 15.00 | 9.70 | 6.15 | 47.15 |
| Medium | 95 | 5.5 | .03 | 17.37 | 5.85 | 16.50 | 9.73 | 5.44 | 54.89 |
| Fast | 110 | 6.0 | .02 | 20.00 | 8.11 | 12.00 | 7.42 | 2.85 | 50.38 |
| Sizes over 5 SF/LF | | | | | | | | | |
| Slow | 85 | 5.0 | .03 | 12.06 | 3.28 | 15.00 | 9.41 | 5.96 | 45.71 |
| Medium | 100 | 5.5 | .03 | 16.50 | 5.55 | 16.50 | 9.44 | 5.28 | 53.27 |
| Fast | 115 | 6.0 | .02 | 19.13 | 7.77 | 12.00 | 7.19 | 2.76 | 48.85 |

| | Labor SF per manhour | Material coverage pounds/SF | Material cost per pound | Labor cost per 100 SF | Labor burden 100 SF | Material cost per 100 SF | Overhead per 100 SF | Profit per 100 SF | Total price per 100 SF |
|---|---|---|---|---|---|---|---|---|---|
| Tanks and vessels | | | | | | | | | |
| Sizes up to 12'0" O/D | | | | | | | | | |
| Slow | 80 | 6.0 | .03 | 12.81 | 3.49 | 18.00 | 10.63 | 6.74 | 51.67 |
| Medium | 95 | 6.5 | .03 | 17.37 | 5.85 | 19.50 | 10.47 | 5.85 | 59.04 |
| Fast | 110 | 7.0 | .02 | 20.00 | 8.11 | 14.00 | 7.79 | 2.99 | 52.89 |
| Sizes over 12'0" O/D | | | | | | | | | |
| Slow | 75 | 6.0 | .03 | 13.67 | 3.71 | 18.00 | 10.97 | 6.96 | 53.31 |
| Medium | 100 | 6.3 | .03 | 16.50 | 5.55 | 18.90 | 10.03 | 5.61 | 56.59 |
| Fast | 125 | 6.5 | .02 | 17.60 | 7.14 | 13.00 | 6.98 | 2.68 | 47.40 |

For heights above 8 feet, use the High Time Difficulty Factors on page 137. "Slow" work is based on a $10.25 hourly wage, "Medium" work on a $16.50 hourly wage, and "Fast" work on a $22.00 hourly wage. Other qualifications that apply to this table are on page 9.

| | Labor SF per manhour | Material coverage pounds/SF | Material cost per pound | Labor cost per 100 SF | Labor burden 100 SF | Material cost per 100 SF | Overhead per 100 SF | Profit per 100 SF | Total price per 100 SF |
|---|---|---|---|---|---|---|---|---|---|
| **Sandblast, near white blast (95% white)** | | | | | | | | | |
| Surface condition basis - large projects & surface areas (material #46) | | | | | | | | | |
| Loose mill scale & fine powder rust | | | | | | | | | |
| Slow | 125 | 5.0 | .03 | 8.20 | 2.23 | 15.00 | 7.88 | 5.00 | 38.31 |
| Medium | 150 | 6.0 | .03 | 11.00 | 3.71 | 18.00 | 8.01 | 4.48 | 45.20 |
| Fast | 175 | 7.0 | .02 | 12.57 | 5.09 | 14.00 | 5.86 | 2.25 | 39.77 |
| Tight mill scale & little or no rust | | | | | | | | | |
| Slow | 75 | 7.0 | .03 | 13.67 | 3.71 | 21.00 | 11.90 | 7.55 | 57.83 |
| Medium | 100 | 8.0 | .03 | 16.50 | 5.55 | 24.00 | 11.28 | 6.31 | 63.64 |
| Fast | 125 | 9.0 | .02 | 17.60 | 7.14 | 18.00 | 7.91 | 3.04 | 53.69 |
| Hard scale, blistered, rusty surface | | | | | | | | | |
| Slow | 50 | 9.0 | .03 | 20.50 | 5.58 | 27.00 | 16.46 | 10.43 | 79.97 |
| Medium | 75 | 11.0 | .03 | 22.00 | 7.39 | 33.00 | 15.29 | 8.55 | 86.23 |
| Fast | 100 | 13.0 | .02 | 22.00 | 8.92 | 26.00 | 10.53 | 4.05 | 71.50 |
| Rust nodules and pitted surface | | | | | | | | | |
| Slow | 35 | 12.0 | .03 | 29.29 | 7.97 | 36.00 | 22.71 | 14.40 | 110.37 |
| Medium | 50 | 14.5 | .03 | 33.00 | 11.10 | 43.50 | 21.46 | 12.00 | 121.06 |
| Fast | 65 | 17.0 | .02 | 33.85 | 13.70 | 34.00 | 15.09 | 5.80 | 102.44 |
| Surface area basis - large projects & surface areas (material #46) | | | | | | | | | |
| Pipe up to 12" O/D | | | | | | | | | |
| Slow | 30 | 8.0 | .03 | 34.17 | 9.29 | 24.00 | 20.92 | 13.26 | 101.64 |
| Medium | 45 | 9.0 | .03 | 36.67 | 12.33 | 27.00 | 18.62 | 10.41 | 105.03 |
| Fast | 60 | 10.0 | .02 | 36.67 | 14.87 | 20.00 | 13.23 | 5.09 | 89.86 |
| Structural steel | | | | | | | | | |
| Sizes up to 2 SF/LF | | | | | | | | | |
| Slow | 40 | 7.0 | .03 | 25.63 | 6.97 | 21.00 | 16.62 | 10.53 | 80.75 |
| Medium | 55 | 8.5 | .03 | 30.00 | 10.09 | 25.50 | 16.07 | 8.98 | 90.64 |
| Fast | 70 | 10.0 | .02 | 31.43 | 12.75 | 20.00 | 11.87 | 4.56 | 80.61 |
| Sizes from 2 to 5 SF/LF | | | | | | | | | |
| Slow | 45 | 7.0 | .03 | 22.78 | 6.19 | 21.00 | 15.50 | 9.82 | 75.29 |
| Medium | 60 | 8.0 | .03 | 27.50 | 9.26 | 24.00 | 14.88 | 8.32 | 83.96 |
| Fast | 75 | 9.0 | .02 | 29.33 | 11.89 | 18.00 | 10.96 | 4.21 | 74.39 |
| Sizes over 5 SF/LF | | | | | | | | | |
| Slow | 55 | 8.0 | .03 | 18.64 | 5.07 | 24.00 | 14.79 | 9.38 | 71.88 |
| Medium | 70 | 8.5 | .03 | 23.57 | 7.94 | 25.50 | 13.97 | 7.81 | 78.79 |
| Fast | 85 | 9.0 | .02 | 25.88 | 10.48 | 18.00 | 10.06 | 3.87 | 68.29 |

| | Labor SF per manhour | Material coverage pounds/SF | Material cost per pound | Labor cost per 100 SF | Labor burden 100 SF | Material cost per 100 SF | Overhead per 100 SF | Profit per 100 SF | Total price per 100 SF |
|---|---|---|---|---|---|---|---|---|---|
| Tanks and vessels | | | | | | | | | |
| Sizes up to 12'0" O/D | | | | | | | | | |
| Slow | 65 | 6.0 | .03 | 15.77 | 4.29 | 18.00 | 11.80 | 7.48 | 57.34 |
| Medium | 80 | 7.0 | .03 | 20.63 | 6.93 | 21.00 | 11.90 | 6.65 | 67.11 |
| Fast | 95 | 8.0 | .02 | 23.16 | 9.40 | 16.00 | 8.98 | 3.45 | 60.99 |
| Sizes over 12'0" O/D | | | | | | | | | |
| Slow | 70 | 6.0 | .03 | 14.64 | 3.99 | 18.00 | 11.36 | 7.20 | 55.19 |
| Medium | 85 | 7.0 | .03 | 19.41 | 6.52 | 21.00 | 11.50 | 6.43 | 64.86 |
| Fast | 100 | 8.0 | .02 | 22.00 | 8.92 | 16.00 | 8.68 | 3.34 | 58.94 |

For heights above 8 feet, use the High Time Difficulty Factors on page 137. "Slow" work is based on a $10.25 hourly wage, "Medium" work on a $16.50 hourly wage, and "Fast" work on a $22.00 hourly wage. Other qualifications that apply to this table are on page 9.

| | Labor SF per manhour | Material coverage pounds/SF | Material cost per pound | Labor cost per 100 SF | Labor burden 100 SF | Material cost per 100 SF | Overhead per 100 SF | Profit per 100 SF | Total price per 100 SF |
|---|---|---|---|---|---|---|---|---|---|

## Sandblast, white blast (100% uniform white stage)

| | Labor SF per manhour | Material coverage pounds/SF | Material cost per pound | Labor cost per 100 SF | Labor burden 100 SF | Material cost per 100 SF | Overhead per 100 SF | Profit per 100 SF | Total price per 100 SF |
|---|---|---|---|---|---|---|---|---|---|
| Surface condition basis - large projects & surface areas (material #46) | | | | | | | | | |
| Loose mill scale & fine powder rust | | | | | | | | | |
| Slow | 50 | 7.0 | .03 | 20.50 | 5.58 | 21.00 | 14.60 | 9.25 | 70.93 |
| Medium | 75 | 8.5 | .03 | 22.00 | 7.39 | 25.50 | 13.45 | 7.52 | 75.86 |
| Fast | 100 | 10.0 | .02 | 22.00 | 8.92 | 20.00 | 9.42 | 3.62 | 63.96 |
| Tight mill scale & little or no rust | | | | | | | | | |
| Slow | 40 | 8.0 | .03 | 25.63 | 6.97 | 24.00 | 17.55 | 11.12 | 85.27 |
| Medium | 60 | 9.5 | .03 | 27.50 | 9.26 | 28.50 | 15.99 | 8.94 | 90.19 |
| Fast | 80 | 11.0 | .02 | 27.50 | 11.15 | 22.00 | 11.22 | 4.31 | 76.18 |
| Hard scale, blistered, rusty surface | | | | | | | | | |
| Slow | 30 | 10.0 | .03 | 34.17 | 9.29 | 30.00 | 22.78 | 14.44 | 110.68 |
| Medium | 45 | 12.5 | .03 | 36.67 | 12.33 | 37.50 | 21.19 | 11.85 | 119.54 |
| Fast | 60 | 15.0 | .02 | 36.67 | 14.87 | 30.00 | 15.08 | 5.80 | 102.42 |
| Rust nodules and pitted surface | | | | | | | | | |
| Slow | 25 | 15.0 | .03 | 41.00 | 11.16 | 45.00 | 30.12 | 19.09 | 146.37 |
| Medium | 35 | 17.5 | .03 | 47.14 | 15.86 | 52.50 | 28.30 | 15.82 | 159.62 |
| Fast | 45 | 20.0 | .02 | 48.89 | 19.81 | 40.00 | 20.11 | 7.73 | 136.54 |
| Surface area basis - large projects & surface areas (material #46) | | | | | | | | | |
| Pipe up to 12" O/D | | | | | | | | | |
| Slow | 30 | 10.0 | .03 | 34.17 | 9.29 | 30.00 | 22.78 | 14.44 | 110.68 |
| Medium | 40 | 11.5 | .03 | 41.25 | 13.88 | 34.50 | 21.96 | 12.27 | 123.86 |
| Fast | 50 | 13.0 | .02 | 44.00 | 17.84 | 26.00 | 16.25 | 6.25 | 110.34 |
| Structural steel | | | | | | | | | |
| Sizes up to 2 SF/LF | | | | | | | | | |
| Slow | 40 | 9.0 | .03 | 25.63 | 6.97 | 27.00 | 18.48 | 11.71 | 89.79 |
| Medium | 50 | 10.5 | .03 | 33.00 | 11.10 | 31.50 | 18.52 | 10.35 | 104.47 |
| Fast | 60 | 12.0 | .02 | 36.67 | 14.87 | 24.00 | 13.97 | 5.37 | 94.88 |
| Sizes from 2 to 5 SF/LF | | | | | | | | | |
| Slow | 45 | 8.0 | .03 | 22.78 | 6.19 | 24.00 | 16.43 | 10.41 | 79.81 |
| Medium | 55 | 9.5 | .03 | 30.00 | 10.09 | 28.50 | 16.81 | 9.40 | 94.80 |
| Fast | 65 | 11.0 | .02 | 33.85 | 13.70 | 22.00 | 12.87 | 4.95 | 87.37 |
| Sizes over 5 SF/LF | | | | | | | | | |
| Slow | 50 | 8.0 | .03 | 20.50 | 5.58 | 24.00 | 15.53 | 9.84 | 75.45 |
| Medium | 60 | 9.5 | .03 | 27.50 | 9.26 | 28.50 | 15.99 | 8.94 | 90.19 |
| Fast | 70 | 11.0 | .02 | 31.43 | 12.75 | 22.00 | 12.24 | 4.70 | 83.12 |

| | Labor SF per manhour | Material coverage pounds/SF | Material cost per pound | Labor cost per 100 SF | Labor burden 100 SF | Material cost per 100 SF | Overhead per 100 SF | Profit per 100 SF | Total price per 100 SF |
|---|---|---|---|---|---|---|---|---|---|
| Tanks and vessels | | | | | | | | | |
| Sizes up to 12'0" O/D | | | | | | | | | |
| Slow | 60 | 8.0 | .03 | 17.08 | 4.66 | 24.00 | 14.18 | 8.99 | 68.91 |
| Medium | 70 | 9.5 | .03 | 23.57 | 7.94 | 28.50 | 14.70 | 8.22 | 82.93 |
| Fast | 80 | 11.0 | .02 | 27.50 | 11.15 | 22.00 | 11.22 | 4.31 | 76.18 |
| Sizes over 12'0" O/D | | | | | | | | | |
| Slow | 70 | 8.0 | .03 | 14.64 | 3.99 | 24.00 | 13.22 | 8.38 | 64.23 |
| Medium | 80 | 9.5 | .03 | 20.63 | 6.93 | 28.50 | 13.74 | 7.68 | 77.48 |
| Fast | 90 | 11.0 | .02 | 24.44 | 9.91 | 22.00 | 10.42 | 4.01 | 70.78 |

For heights above 8 feet, use the High Time Difficulty Factors on page 137. "Slow" work is based on a $10.25 hourly wage, "Medium" work on a $16.50 hourly wage, and "Fast" work on a $22.00 hourly wage. Other qualifications that apply to this table are on page 9.

| | Labor SF per manhour | Material coverage SF/gallon | Material cost per gallon | Labor cost per 100 SF | Labor burden 100 SF | Material cost per 100 SF | Overhead per 100 SF | Profit per 100 SF | Total price per 100 SF |
|---|---|---|---|---|---|---|---|---|---|
| **Strip, remove, or bleach** | | | | | | | | | |
| Remove wallcover by hand | | | | | | | | | |
| Slow | 50 | -- | -- | 20.50 | 5.58 | -- | 8.09 | 5.13 | 39.30 |
| Medium | 70 | -- | -- | 23.57 | 7.94 | -- | 7.72 | 4.31 | 43.54 |
| Fast | 90 | -- | -- | 24.44 | 9.91 | -- | 6.35 | 2.44 | 43.14 |
| Stripping flat, vertical, varnished surfaces | | | | | | | | | |
| Light duty liquid remover (material #43) | | | | | | | | | |
| Slow | 25 | 175 | 16.40 | 41.00 | 11.16 | 9.37 | 15.08 | 11.49 | 88.10 |
| Medium | 35 | 158 | 14.40 | 47.14 | 15.86 | 9.11 | 17.67 | 9.88 | 99.66 |
| Fast | 45 | 140 | 12.30 | 48.89 | 19.81 | 8.79 | 14.34 | 5.51 | 97.34 |
| Heavy duty liquid remover (material #44) | | | | | | | | | |
| Slow | 20 | 150 | 22.30 | 51.25 | 13.95 | 14.87 | 19.62 | 14.96 | 114.65 |
| Medium | 30 | 138 | 19.60 | 55.00 | 18.49 | 14.20 | 21.49 | 12.01 | 121.19 |
| Fast | 40 | 125 | 16.80 | 55.00 | 22.30 | 13.44 | 16.79 | 6.45 | 113.98 |
| Stripping flat, horizontal floor surfaces | | | | | | | | | |
| Paint removal with light duty liquid remover (material #43) | | | | | | | | | |
| Slow | 20 | 180 | 16.40 | 51.25 | 13.95 | 9.11 | 18.21 | 13.88 | 106.40 |
| Medium | 30 | 175 | 14.40 | 55.00 | 18.49 | 8.23 | 20.03 | 11.19 | 112.94 |
| Fast | 40 | 170 | 12.30 | 55.00 | 22.30 | 7.24 | 15.64 | 6.01 | 106.19 |
| Paint removal with heavy duty liquid remover (material #44) | | | | | | | | | |
| Slow | 15 | 170 | 22.30 | 68.33 | 18.61 | 13.12 | 24.52 | 18.69 | 143.27 |
| Medium | 25 | 160 | 19.60 | 66.00 | 22.20 | 12.25 | 24.61 | 13.76 | 138.82 |
| Fast | 35 | 150 | 16.80 | 62.86 | 25.48 | 11.20 | 18.42 | 7.08 | 125.04 |
| Varnish removal with light duty liquid remover (material #43) | | | | | | | | | |
| Slow | 30 | 185 | 16.40 | 34.17 | 9.29 | 8.86 | 12.82 | 9.77 | 74.91 |
| Medium | 40 | 180 | 14.40 | 41.25 | 13.88 | 8.00 | 15.47 | 8.65 | 87.25 |
| Fast | 50 | 175 | 12.30 | 44.00 | 17.84 | 7.03 | 12.74 | 4.90 | 86.51 |
| Varnish removal with heavy duty liquid remover (material #44) | | | | | | | | | |
| Slow | 25 | 180 | 22.30 | 41.00 | 11.16 | 12.39 | 15.82 | 12.06 | 92.43 |
| Medium | 35 | 170 | 19.60 | 47.14 | 15.86 | 11.53 | 18.26 | 10.21 | 103.00 |
| Fast | 45 | 160 | 16.80 | 48.89 | 19.81 | 10.50 | 14.65 | 5.63 | 99.48 |

All widths less than 12", consider as 1 square foot per linear foot. For heights above 8 feet, use the High Time Difficulty Factors on page 137. "Slow" work is based on a $10.25 hourly wage, "Medium" work on a $16.50 hourly wage, and "Fast" work on a $22.00 hourly wage. Other qualifications that apply to this table are on page 9.

## Unstick windows

On repaint jobs, test all windows during your estimating walk-through and allow approximately 15 minutes (.25 hours) for each stuck window. But don't price yourself out of the job with this extra time.

| | Labor SF per manhour | Material coverage SF/gallon | Material cost per gallon | Labor cost per 100 SF | Labor burden 100 SF | Material cost per 100 SF | Overhead per 100 SF | Profit per 100 SF | Total price per 100 SF |
|---|---|---|---|---|---|---|---|---|---|
| **Wash** | | | | | | | | | |
| Interior flatwall (smooth surfaces) | | | | | | | | | |
| Wash only | | | | | | | | | |
| Slow | 175 | -- | -- | 5.86 | 1.59 | -- | 2.31 | 1.47 | 11.23 |
| Medium | 200 | -- | -- | 8.25 | 2.78 | -- | 2.70 | 1.51 | 15.24 |
| Fast | 225 | -- | -- | 9.78 | 3.95 | -- | 2.54 | .98 | 17.25 |
| Wash & touchup | | | | | | | | | |
| Slow | 135 | -- | -- | 7.59 | 2.07 | -- | 2.99 | 1.90 | 14.55 |
| Medium | 160 | -- | -- | 10.31 | 3.47 | -- | 3.38 | 1.89 | 19.05 |
| Fast | 185 | -- | -- | 11.89 | 4.84 | -- | 3.09 | 1.19 | 21.01 |
| Interior flatwall (rough surfaces) | | | | | | | | | |
| Wash only | | | | | | | | | |
| Slow | 125 | -- | -- | 8.20 | 2.23 | -- | 3.23 | 2.05 | 15.71 |
| Medium | 150 | -- | -- | 11.00 | 3.71 | -- | 3.60 | 2.01 | 20.32 |
| Fast | 175 | -- | -- | 12.57 | 5.09 | -- | 3.27 | 1.26 | 22.19 |
| Wash & touchup | | | | | | | | | |
| Slow | 55 | -- | -- | 18.64 | 5.07 | -- | 7.35 | 4.66 | 35.72 |
| Medium | 95 | -- | -- | 17.37 | 5.85 | -- | 5.69 | 3.18 | 32.09 |
| Fast | 135 | -- | -- | 16.30 | 6.61 | -- | 4.24 | 1.63 | 28.78 |
| Interior enamel (wall surfaces) | | | | | | | | | |
| Wash only | | | | | | | | | |
| Slow | 190 | -- | -- | 5.39 | 1.47 | -- | 2.13 | 1.35 | 10.34 |
| Medium | 215 | -- | -- | 7.67 | 2.58 | -- | 2.51 | 1.40 | 14.16 |
| Fast | 240 | -- | -- | 9.17 | 3.72 | -- | 2.38 | .92 | 16.19 |
| Wash & touchup | | | | | | | | | |
| Slow | 85 | -- | -- | 12.06 | 3.28 | -- | 4.76 | 3.02 | 23.12 |
| Medium | 113 | -- | -- | 14.60 | 4.91 | -- | 4.78 | 2.67 | 26.96 |
| Fast | 140 | -- | -- | 15.71 | 6.37 | -- | 4.08 | 1.57 | 27.73 |
| Interior enamel trim | | | | | | | | | |
| Wash only | | | | | | | | | |
| Slow | 100 | -- | -- | 10.25 | 2.79 | -- | 4.04 | 2.56 | 19.64 |
| Medium | 150 | -- | -- | 11.00 | 3.71 | -- | 3.60 | 2.01 | 20.32 |
| Fast | 200 | -- | -- | 11.00 | 4.46 | -- | 2.86 | 1.10 | 19.42 |
| Wash & touchup | | | | | | | | | |
| Slow | 90 | -- | -- | 11.39 | 3.10 | -- | 4.49 | 2.85 | 21.83 |
| Medium | 120 | -- | -- | 13.75 | 4.62 | -- | 4.50 | 2.52 | 25.39 |
| Fast | 150 | -- | -- | 14.67 | 5.95 | -- | 3.81 | 1.47 | 25.90 |

| | Labor SF per manhour | Material coverage SF/gallon | Material cost per gallon | Labor cost per 100 SF | Labor burden 100 SF | Material cost per 100 SF | Overhead per 100 SF | Profit per 100 SF | Total price per 100 SF |
|---|---|---|---|---|---|---|---|---|---|
| Interior varnish trim | | | | | | | | | |
| Wash only | | | | | | | | | |
| Slow | 150 | -- | -- | 6.83 | 1.87 | -- | 2.69 | 1.71 | 13.10 |
| Medium | 195 | -- | -- | 8.46 | 2.85 | -- | 2.77 | 1.55 | 15.63 |
| Fast | 240 | -- | -- | 9.17 | 3.72 | -- | 2.38 | .92 | 16.19 |
| Wash & touchup | | | | | | | | | |
| Slow | 120 | -- | -- | 8.54 | 2.32 | -- | 3.37 | 2.14 | 16.37 |
| Medium | 140 | -- | -- | 11.79 | 3.95 | -- | 3.86 | 2.16 | 21.76 |
| Fast | 160 | -- | -- | 13.75 | 5.58 | -- | 3.58 | 1.37 | 24.28 |
| Interior varnish floors | | | | | | | | | |
| Wash only | | | | | | | | | |
| Slow | 160 | -- | -- | 6.41 | 1.74 | -- | 2.53 | 1.60 | 12.28 |
| Medium | 210 | -- | -- | 7.86 | 2.64 | -- | 2.57 | 1.44 | 14.51 |
| Fast | 260 | -- | -- | 8.46 | 3.44 | -- | 2.20 | .85 | 14.95 |
| Wash & touchup | | | | | | | | | |
| Slow | 130 | -- | -- | 7.88 | 2.15 | -- | 3.11 | 1.97 | 15.11 |
| Medium | 153 | -- | -- | 10.78 | 3.64 | -- | 3.53 | 1.97 | 19.92 |
| Fast | 175 | -- | -- | 12.57 | 5.09 | -- | 3.27 | 1.26 | 22.19 |
| Interior plaster (smooth) | | | | | | | | | |
| Wash only | | | | | | | | | |
| Slow | 150 | -- | -- | 6.83 | 1.87 | -- | 2.69 | 1.71 | 13.10 |
| Medium | 175 | -- | -- | 9.43 | 3.16 | -- | 3.09 | 1.73 | 17.41 |
| Fast | 200 | -- | -- | 11.00 | 4.46 | -- | 2.86 | 1.10 | 19.42 |
| Wash & touchup | | | | | | | | | |
| Slow | 125 | -- | -- | 8.20 | 2.23 | -- | 3.23 | 2.05 | 15.71 |
| Medium | 140 | -- | -- | 11.79 | 3.95 | -- | 3.86 | 2.16 | 21.76 |
| Fast | 155 | -- | -- | 14.19 | 5.75 | -- | 3.69 | 1.42 | 25.05 |
| Interior plaster (sand finish) | | | | | | | | | |
| Wash only | | | | | | | | | |
| Slow | 110 | -- | -- | 9.32 | 2.53 | -- | 3.68 | 2.33 | 17.86 |
| Medium | 135 | -- | -- | 12.22 | 4.12 | -- | 4.00 | 2.24 | 22.58 |
| Fast | 160 | -- | -- | 13.75 | 5.58 | -- | 3.58 | 1.37 | 24.28 |
| Wash & touchup | | | | | | | | | |
| Slow | 85 | -- | -- | 12.06 | 3.28 | -- | 4.76 | 3.02 | 23.12 |
| Medium | 110 | -- | -- | 15.00 | 5.04 | -- | 4.91 | 2.75 | 27.70 |
| Fast | 140 | -- | -- | 15.71 | 6.37 | -- | 4.08 | 1.57 | 27.73 |

Consider all trim which is less than 12" wide to be 12" wide and figured as 1 square foot per linear foot. For heights above 8 feet, use the High Time Difficulty Factors on page 137. Note: Because the type of surface and type of material being removed will alter rates, it is best to wash surfaces with calcium deposits or other unusual surface markings or debris on a Time and Material or a Cost Plus Fee basis at a pre-set hourly rate. "Slow" work is based on a $10.25 hourly wage, "Medium" work on a $16.50 hourly wage, and "Fast" work on a $22.00 hourly wage. Other qualifications that apply to this table are on page 9.

| | Labor SF per manhour | Material coverage SF/gallon | Material cost per gallon | Labor cost per 100 SF | Labor burden 100 SF | Material cost per 100 SF | Overhead per 100 SF | Profit per 100 SF | Total price per 100 SF |
|---|---|---|---|---|---|---|---|---|---|
| **Waterblast** | | | | | | | | | |
| Waterblast | | | | | | | | | |
| Slow | 450 | -- | -- | 2.28 | .61 | -- | .90 | .57 | 4.36 |
| Medium | 500 | -- | -- | 3.30 | 1.11 | -- | 1.08 | .60 | 6.09 |
| Fast | 550 | -- | -- | 4.00 | 1.63 | -- | 1.04 | .40 | 7.07 |

Use an hourly or daily rate depending on the quantity of work involved. For heights above 8 feet, use the High Time Difficulty Factors on page 137. Waterblast to remove deteriorated, cracked, flaking paint from accessible concrete, block, brick, wood, plaster or stucco surfaces. Rates assume a 1.4" diameter nozzle with 100 lbs. pressure. "Slow" work is based on a $10.25 hourly wage, "Medium" work on a $16.75 hourly wage, and "Fast" work on a $22.00 hourly wage. Other qualifications that apply to this table are on page 9.

| | Labor SF per manhour | Material coverage SF/gallon | Material cost per gallon | Labor cost per 100 SF | Labor burden 100 SF | Material cost per 100 SF | Overhead per 100 SF | Profit per 100 SF | Total price per 100 SF |
|---|---|---|---|---|---|---|---|---|---|
| **Window protective coating (paraffin wax)** | | | | | | | | | |
| Window protective coating - wax (material #47) | | | | | | | | | |
| Hand application | | | | | | | | | |
| Slow | 90 | 350 | 12.20 | 11.39 | 3.10 | 3.49 | 5.57 | 3.53 | 27.08 |
| Medium | 100 | 325 | 10.70 | 16.50 | 5.55 | 3.29 | 6.21 | 3.47 | 35.02 |
| Fast | 110 | 300 | 9.10 | 20.00 | 8.11 | 3.03 | 5.76 | 2.21 | 39.11 |
| Spray application | | | | | | | | | |
| Slow | 350 | 250 | 12.20 | 2.93 | .80 | 4.88 | 2.67 | 1.69 | 12.97 |
| Medium | 375 | 225 | 10.70 | 4.40 | 1.49 | 4.76 | 2.61 | 1.46 | 14.72 |
| Fast | 400 | 200 | 9.10 | 5.50 | 2.23 | 4.55 | 2.27 | .87 | 15.42 |

For heights above 8 feet, use the High Time Difficulty Factors on page 137. "Slow" work is based on a $10.25 hourly wage, "Medium" work on a $16.50 hourly wage, and "Fast" work on a $22.00 hourly wage. Other qualifications that apply to this table are on page 9.

| | Labor SF per manhour | Material coverage SF/gallon | Material cost per gallon | Labor cost per 100 SF | Labor burden 100 SF | Material cost per 100 SF | Overhead per 100 SF | Profit per 100 SF | Total price per 100 SF |
|---|---|---|---|---|---|---|---|---|---|
| **Wire brush** | | | | | | | | | |
| Surface area basis - large projects & surface areas | | | | | | | | | |
| Pipe up to 12" O/D | | | | | | | | | |
| Slow | 50 | -- | -- | 20.50 | 5.58 | -- | 8.09 | 5.13 | 39.30 |
| Medium | 75 | -- | -- | 22.00 | 7.39 | -- | 7.20 | 4.03 | 40.62 |
| Fast | 100 | -- | -- | 22.00 | 8.92 | -- | 5.72 | 2.20 | 38.84 |
| Structural steel | | | | | | | | | |
| Sizes up to 2 SF/LF | | | | | | | | | |
| Slow | 90 | -- | -- | 11.39 | 3.10 | -- | 4.49 | 2.85 | 21.83 |
| Medium | 110 | -- | -- | 15.00 | 5.04 | -- | 4.91 | 2.75 | 27.70 |
| Fast | 125 | -- | -- | 17.60 | 7.14 | -- | 4.58 | 1.76 | 31.08 |
| Sizes from 2 to 5 SF/LF | | | | | | | | | |
| Slow | 100 | -- | -- | 10.25 | 2.79 | -- | 4.04 | 2.56 | 19.64 |
| Medium | 120 | -- | -- | 13.75 | 4.62 | -- | 4.50 | 2.52 | 25.39 |
| Fast | 140 | -- | -- | 15.71 | 6.37 | -- | 4.08 | 1.57 | 27.73 |
| Sizes over 5 SF/LF | | | | | | | | | |
| Slow | 110 | -- | -- | 9.32 | 2.53 | -- | 3.68 | 2.33 | 17.86 |
| Medium | 130 | -- | -- | 12.69 | 4.27 | -- | 4.16 | 2.32 | 23.44 |
| Fast | 150 | -- | -- | 14.67 | 5.95 | -- | 3.81 | 1.47 | 25.90 |
| Tanks and vessels | | | | | | | | | |
| Sizes up to 12'0" O/D | | | | | | | | | |
| Slow | 110 | -- | -- | 9.32 | 2.53 | -- | 3.68 | 2.33 | 17.86 |
| Medium | 130 | -- | -- | 12.69 | 4.27 | -- | 4.16 | 2.32 | 23.44 |
| Fast | 150 | -- | -- | 14.67 | 5.95 | -- | 3.81 | 1.47 | 25.90 |
| Sizes over 12'0" O/D | | | | | | | | | |
| Slow | 120 | -- | -- | 8.54 | 2.32 | -- | 3.37 | 2.14 | 16.37 |
| Medium | 140 | -- | -- | 11.79 | 3.95 | -- | 3.86 | 2.16 | 21.76 |
| Fast | 160 | -- | -- | 13.75 | 5.58 | -- | 3.58 | 1.37 | 24.28 |

For heights above 8 feet, use the High Time Difficulty Factors on page 137. "Slow" work is based on a $10.25 hourly wage, "Medium" work on a $16.50 hourly wage, and "Fast" work on a $22.00 hourly wage. Other qualifications that apply to this table are on page 9.

# Part III

- INDUSTRIAL
- INSTITUTIONAL
- HEAVY COMMERCIAL

# Painting Costs

| | Labor SF per manhour | Material coverage SF/gallon | Material cost per gallon | Labor cost per 100 SF | Labor burden 100 SF | Material cost per 100 SF | Overhead per 100 SF | Profit per 100 SF | Total price per 100 SF |
|---|---|---|---|---|---|---|---|---|---|
| **Conduit, electric, brush application** | | | | | | | | | |
| Acid wash coat, muriatic acid (material #49) | | | | | | | | | |
| Brush each coat | | | | | | | | | |
| Slow | 60 | 700 | 5.20 | 17.08 | 4.66 | .74 | 6.97 | 4.42 | 33.87 |
| Medium | 80 | 650 | 4.50 | 20.63 | 6.93 | .69 | 6.92 | 3.87 | 39.04 |
| Fast | 100 | 600 | 3.90 | 22.00 | 8.92 | .65 | 5.84 | 2.24 | 39.65 |
| Metal primer, rust inhibitor - clean metal (material #35) | | | | | | | | | |
| Brush prime coat | | | | | | | | | |
| Slow | 60 | 450 | 28.20 | 17.08 | 4.66 | 6.27 | 8.68 | 5.50 | 42.19 |
| Medium | 80 | 425 | 24.60 | 20.63 | 6.93 | 5.79 | 8.17 | 4.57 | 46.09 |
| Fast | 100 | 400 | 21.10 | 22.00 | 8.92 | 5.28 | 6.70 | 2.57 | 45.47 |
| Metal primer, rust inhibitor - rusty metal (material #36) | | | | | | | | | |
| Brush prime coat | | | | | | | | | |
| Slow | 60 | 400 | 33.30 | 17.08 | 4.66 | 8.33 | 9.32 | 5.91 | 45.30 |
| Medium | 80 | 375 | 29.20 | 20.63 | 6.93 | 7.79 | 8.66 | 4.84 | 48.85 |
| Fast | 100 | 350 | 25.00 | 22.00 | 8.92 | 7.14 | 7.04 | 2.71 | 47.81 |
| Industrial enamel, oil base, high gloss - light colors (material #56) | | | | | | | | | |
| Brush 1st or additional finish coats | | | | | | | | | |
| Slow | 100 | 450 | 28.90 | 10.25 | 2.79 | 6.42 | 6.03 | 3.82 | 29.31 |
| Medium | 125 | 425 | 25.20 | 13.20 | 4.44 | 5.93 | 5.77 | 3.23 | 32.57 |
| Fast | 150 | 400 | 21.60 | 14.67 | 5.95 | 5.40 | 4.81 | 1.85 | 32.68 |
| Industrial enamel, oil base, high gloss - dark (OSHA) colors (material #57) | | | | | | | | | |
| Brush 1st or additional finish coats | | | | | | | | | |
| Slow | 100 | 500 | 34.90 | 10.25 | 2.79 | 6.98 | 6.21 | 3.93 | 30.16 |
| Medium | 125 | 475 | 30.50 | 13.20 | 4.44 | 6.42 | 5.89 | 3.29 | 33.24 |
| Fast | 150 | 450 | 26.10 | 14.67 | 5.95 | 5.80 | 4.89 | 1.88 | 33.19 |
| Epoxy coating, 2 part system - clear (material #51) | | | | | | | | | |
| Brush 1st coat | | | | | | | | | |
| Slow | 60 | 425 | 57.00 | 17.08 | 4.66 | 13.41 | 10.89 | 6.90 | 52.94 |
| Medium | 80 | 400 | 49.90 | 20.63 | 6.93 | 12.48 | 9.81 | 5.48 | 55.33 |
| Fast | 100 | 375 | 42.80 | 22.00 | 8.92 | 11.41 | 7.83 | 3.01 | 53.17 |
| Brush 2nd or additional finish coats | | | | | | | | | |
| Slow | 100 | 475 | 57.00 | 10.25 | 2.79 | 12.00 | 7.76 | 4.92 | 37.72 |
| Medium | 125 | 450 | 49.90 | 13.20 | 4.44 | 11.09 | 7.04 | 3.93 | 39.70 |
| Fast | 150 | 425 | 42.80 | 14.67 | 5.95 | 10.07 | 5.68 | 2.18 | 38.55 |

| | Labor SF per manhour | Material coverage SF/gallon | Material cost per gallon | Labor cost per 100 SF | Labor burden 100 SF | Material cost per 100 SF | Overhead per 100 SF | Profit per 100 SF | Total price per 100 SF |
|---|---|---|---|---|---|---|---|---|---|
| Epoxy coating, 2 part system - white (material #52) | | | | | | | | | |
| Brush 1st coat | | | | | | | | | |
| Slow | 60 | 425 | 57.40 | 17.08 | 4.66 | 13.51 | 10.92 | 6.92 | 53.09 |
| Medium | 80 | 400 | 50.20 | 20.63 | 6.93 | 12.55 | 9.83 | 5.49 | 55.43 |
| Fast | 100 | 375 | 43.10 | 22.00 | 8.92 | 11.49 | 7.85 | 3.02 | 53.28 |
| Brush 2nd or additional finish coats | | | | | | | | | |
| Slow | 100 | 475 | 57.40 | 10.25 | 2.79 | 12.08 | 7.79 | 4.94 | 37.85 |
| Medium | 125 | 450 | 50.20 | 13.20 | 4.44 | 11.16 | 7.06 | 3.94 | 39.80 |
| Fast | 150 | 425 | 43.10 | 14.67 | 5.95 | 10.14 | 5.69 | 2.19 | 38.64 |

This table is based on square feet of conduit surface area. See Figure 21 on page 321 to convert from linear feet of various conduit or pipe sizes to square feet of surface area. For heights above 8 feet, use the High Time Difficulty Factors on page 137. Note: A two coat system, prime and finish, using oil base material is recommended for any metal surface. Although water base material is often used, it may cause oxidation, corrosion and rust. One coat of oil base solid body stain is often used on exterior metal but it may crack, peel or chip without the proper prime coat application. "Slow" work is based on a $10.25 hourly wage, "Medium" work on a $16.50 hourly wage, and "Fast" work on a $22.00 hourly wage. Other qualifications that apply to this table are on page 9.

| | Labor SF per manhour | Material coverage SF/gallon | Material cost per gallon | Labor cost per 100 SF | Labor burden 100 SF | Material cost per 100 SF | Overhead per 100 SF | Profit per 100 SF | Total price per 100 SF |
|---|---|---|---|---|---|---|---|---|---|
| **Conduit, electric, roll application** | | | | | | | | | |
| Acid wash coat, muriatic acid (material #49) | | | | | | | | | |
| Roll each coat | | | | | | | | | |
| Slow | 175 | 700 | 5.20 | 5.86 | 1.59 | .74 | 2.54 | 1.61 | 12.34 |
| Medium | 200 | 650 | 4.50 | 8.25 | 2.78 | .69 | 2.87 | 1.60 | 16.19 |
| Fast | 225 | 600 | 3.90 | 9.78 | 3.95 | .65 | 2.66 | 1.02 | 18.06 |
| Metal primer, rust inhibitor - clean metal (material #35) | | | | | | | | | |
| Roll prime coat | | | | | | | | | |
| Slow | 175 | 425 | 28.20 | 5.86 | 1.59 | 6.64 | 4.37 | 2.77 | 21.23 |
| Medium | 200 | 400 | 24.60 | 8.25 | 2.78 | 6.15 | 4.21 | 2.35 | 23.74 |
| Fast | 225 | 375 | 21.10 | 9.78 | 3.95 | 5.63 | 3.59 | 1.38 | 24.33 |
| Metal primer, rust inhibitor - rusty metal (material #36) | | | | | | | | | |
| Roll prime coat | | | | | | | | | |
| Slow | 175 | 375 | 33.30 | 5.86 | 1.59 | 8.88 | 5.07 | 3.21 | 24.61 |
| Medium | 200 | 350 | 29.20 | 8.25 | 2.78 | 8.34 | 4.75 | 2.65 | 26.77 |
| Fast | 225 | 325 | 25.00 | 9.78 | 3.95 | 7.69 | 3.97 | 1.52 | 26.91 |
| Industrial enamel, oil base, high gloss - light colors (material #56) | | | | | | | | | |
| Roll 1st or additional finish coats | | | | | | | | | |
| Slow | 225 | 425 | 28.90 | 4.56 | 1.23 | 6.80 | 3.91 | 2.48 | 18.98 |
| Medium | 250 | 400 | 25.20 | 6.60 | 2.22 | 6.30 | 3.70 | 2.07 | 20.89 |
| Fast | 275 | 375 | 21.60 | 8.00 | 3.25 | 5.76 | 3.15 | 1.21 | 21.37 |
| Industrial enamel, oil base, high gloss - dark (OSHA) colors (material #57) | | | | | | | | | |
| Roll 1st or additional finish coats | | | | | | | | | |
| Slow | 225 | 475 | 34.90 | 4.56 | 1.23 | 7.35 | 4.08 | 2.58 | 19.80 |
| Medium | 250 | 450 | 30.50 | 6.60 | 2.22 | 6.78 | 3.82 | 2.14 | 21.56 |
| Fast | 275 | 425 | 26.10 | 8.00 | 3.25 | 6.14 | 3.22 | 1.24 | 21.85 |
| Epoxy coating, 2 part system - clear (material #51) | | | | | | | | | |
| Roll 1st coat | | | | | | | | | |
| Slow | 175 | 400 | 57.00 | 5.86 | 1.59 | 14.25 | 6.73 | 4.27 | 32.70 |
| Medium | 200 | 375 | 49.90 | 8.25 | 2.78 | 13.31 | 5.96 | 3.33 | 33.63 |
| Fast | 225 | 350 | 42.80 | 9.78 | 3.95 | 12.23 | 4.81 | 1.85 | 32.62 |
| Roll 2nd or additional finish coats | | | | | | | | | |
| Slow | 225 | 450 | 57.00 | 4.56 | 1.23 | 12.67 | 5.73 | 3.63 | 27.82 |
| Medium | 250 | 425 | 49.90 | 6.60 | 2.22 | 11.74 | 5.04 | 2.82 | 28.42 |
| Fast | 275 | 400 | 42.80 | 8.00 | 3.25 | 10.70 | 4.06 | 1.56 | 27.57 |

| | Labor SF per manhour | Material coverage SF/gallon | Material cost per gallon | Labor cost per 100 SF | Labor burden 100 SF | Material cost per 100 SF | Overhead per 100 SF | Profit per 100 SF | Total price per 100 SF |
|---|---|---|---|---|---|---|---|---|---|
| Epoxy coating, 2 part system - white (material #52) | | | | | | | | | |
| Roll 1st coat | | | | | | | | | |
| Slow | 175 | 400 | 57.40 | 5.86 | 1.59 | 14.35 | 6.76 | 4.29 | 32.85 |
| Medium | 200 | 375 | 50.20 | 8.25 | 2.78 | 13.39 | 5.98 | 3.34 | 33.74 |
| Fast | 225 | 350 | 43.10 | 9.78 | 3.95 | 12.31 | 4.82 | 1.85 | 32.71 |
| Roll 2nd or additional finish coats | | | | | | | | | |
| Slow | 225 | 450 | 57.40 | 4.56 | 1.23 | 12.76 | 5.75 | 3.65 | 27.95 |
| Medium | 250 | 425 | 50.20 | 6.60 | 2.22 | 11.81 | 5.05 | 2.82 | 28.50 |
| Fast | 275 | 400 | 43.10 | 8.00 | 3.25 | 10.78 | 4.07 | 1.57 | 27.67 |

This table is based on square feet of conduit surface area. See Figure 21 on page 321 to convert from linear feet of various conduit or pipe sizes to square feet of surface area. For heights above 8 feet, use the High Time Difficulty Factors on page 137. Note: A two coat system, prime and finish, using oil base material is recommended for any metal surface. Although water base material is often used, it may cause oxidation, corrosion and rust. One coat of oil base solid body stain is often used on exterior metal but it may crack, peel or chip without the proper prime coat application. "Slow" work is based on a $10.25 hourly wage, "Medium" work on a $16.50 hourly wage, and "Fast" work on a $22.00 hourly wage. Other qualifications that apply to this table are on page 9.

| | Labor SF per manhour | Material coverage SF/gallon | Material cost per gallon | Labor cost per 100 SF | Labor burden 100 SF | Material cost per 100 SF | Overhead per 100 SF | Profit per 100 SF | Total price per 100 SF |
|---|---|---|---|---|---|---|---|---|---|
| **Conduit, electric, spray application** | | | | | | | | | |
| Acid wash coat, muriatic acid (material #49) | | | | | | | | | |
| Spray each coat | | | | | | | | | |
| Slow | 350 | 350 | 5.20 | 2.93 | .80 | 1.49 | 1.62 | 1.03 | 7.87 |
| Medium | 400 | 325 | 4.50 | 4.13 | 1.38 | 1.38 | 1.69 | .94 | 9.52 |
| Fast | 450 | 300 | 3.90 | 4.89 | 1.97 | 1.30 | 1.51 | .58 | 10.25 |
| Metal primer, rust inhibitor - clean metal (material #35) | | | | | | | | | |
| Spray prime coat | | | | | | | | | |
| Slow | 350 | 275 | 28.20 | 2.93 | .80 | 10.25 | 4.33 | 2.75 | 21.06 |
| Medium | 400 | 263 | 24.60 | 4.13 | 1.38 | 9.35 | 3.64 | 2.04 | 20.54 |
| Fast | 450 | 250 | 21.10 | 4.89 | 1.97 | 8.44 | 2.83 | 1.09 | 19.22 |
| Metal primer, rust inhibitor - rusty metal (material #36) | | | | | | | | | |
| Spray prime coat | | | | | | | | | |
| Slow | 350 | 225 | 33.30 | 2.93 | .80 | 14.80 | 5.74 | 3.64 | 27.91 |
| Medium | 400 | 213 | 29.20 | 4.13 | 1.38 | 13.71 | 4.71 | 2.63 | 26.56 |
| Fast | 450 | 200 | 25.00 | 4.89 | 1.97 | 12.50 | 3.58 | 1.38 | 24.32 |
| Industrial enamel, oil base, high gloss - light colors (material #56) | | | | | | | | | |
| Spray 1st or additional finish coats | | | | | | | | | |
| Slow | 450 | 275 | 28.90 | 2.28 | .61 | 10.51 | 4.16 | 2.64 | 20.20 |
| Medium | 500 | 263 | 25.20 | 3.30 | 1.11 | 9.58 | 3.43 | 1.92 | 19.34 |
| Fast | 550 | 250 | 21.60 | 4.00 | 1.63 | 8.64 | 2.64 | 1.01 | 17.92 |
| Industrial enamel, oil base, high gloss - dark (OSHA) colors (material #57) | | | | | | | | | |
| Spray 1st or additional finish coats | | | | | | | | | |
| Slow | 450 | 300 | 34.90 | 2.28 | .61 | 11.63 | 4.50 | 2.85 | 21.87 |
| Medium | 500 | 288 | 30.50 | 3.30 | 1.11 | 10.59 | 3.68 | 2.05 | 20.73 |
| Fast | 550 | 275 | 26.10 | 4.00 | 1.63 | 9.49 | 2.80 | 1.07 | 18.99 |
| Epoxy coating, 2 part system - clear (material #51) | | | | | | | | | |
| Spray 1st coat | | | | | | | | | |
| Slow | 350 | 250 | 57.00 | 2.93 | .80 | 22.80 | 8.22 | 5.21 | 39.96 |
| Medium | 400 | 238 | 49.90 | 4.13 | 1.38 | 20.97 | 6.49 | 3.63 | 36.60 |
| Fast | 450 | 225 | 42.80 | 4.89 | 1.97 | 19.02 | 4.79 | 1.84 | 32.51 |
| Spray 2nd or additional finish coats | | | | | | | | | |
| Slow | 450 | 285 | 57.00 | 2.28 | .61 | 20.00 | 7.10 | 4.50 | 34.49 |
| Medium | 500 | 273 | 49.90 | 3.30 | 1.11 | 18.28 | 5.56 | 3.11 | 31.36 |
| Fast | 550 | 260 | 42.80 | 4.00 | 1.63 | 16.46 | 4.08 | 1.57 | 27.74 |

| | Labor SF per manhour | Material coverage SF/gallon | Material cost per gallon | Labor cost per 100 SF | Labor burden 100 SF | Material cost per 100 SF | Overhead per 100 SF | Profit per 100 SF | Total price per 100 SF |
|---|---|---|---|---|---|---|---|---|---|
| Epoxy coating, 2 part system - white (material #52) | | | | | | | | | |
| Spray 1st coat | | | | | | | | | |
| Slow | 350 | 250 | 57.40 | 2.93 | .80 | 22.96 | 8.27 | 5.24 | 40.20 |
| Medium | 400 | 238 | 50.20 | 4.13 | 1.38 | 21.09 | 6.52 | 3.64 | 36.76 |
| Fast | 450 | 225 | 43.10 | 4.89 | 1.97 | 19.16 | 4.82 | 1.85 | 32.69 |
| Spray 2nd or additional finish coats | | | | | | | | | |
| Slow | 450 | 285 | 57.40 | 2.28 | .61 | 20.14 | 7.14 | 4.53 | 34.70 |
| Medium | 500 | 273 | 50.20 | 3.30 | 1.11 | 18.39 | 5.59 | 3.12 | 31.51 |
| Fast | 550 | 260 | 43.10 | 4.00 | 1.63 | 16.58 | 4.11 | 1.58 | 27.90 |

This table is based on square feet of conduit surface area. See Figure 21 on page 321 to convert from linear feet of various conduit or pipe sizes to square feet of surface area. For heights above 8 feet, use the High Time Difficulty Factors on page 137. Note: A two coat system, prime and finish, using oil base material is recommended for any metal surface. Although water base material is often used, it may cause oxidation, corrosion and rust. One coat of oil base solid body stain is often used on exterior metal but it may crack, peel or chip without the proper prime coat application. "Slow" work is based on a $10.25 hourly wage, "Medium" work on a $16.50 hourly wage, and "Fast" work on a $22.00 hourly wage. Other qualifications that apply to this table are on page 9.

| | Labor SF per manhour | Material coverage SF/gallon | Material cost per gallon | Labor cost per 100 SF | Labor burden 100 SF | Material cost per 100 SF | Overhead per 100 SF | Profit per 100 SF | Total price per 100 SF |
|---|---|---|---|---|---|---|---|---|---|
| **Conduit, electric, mitt or glove application** | | | | | | | | | |
| Acid wash coat, muriatic acid (material #49) | | | | | | | | | |
| Mitt or glove each coat | | | | | | | | | |
| Slow | 175 | 700 | 5.20 | 5.86 | 1.59 | .74 | 2.54 | 1.61 | 12.34 |
| Medium | 200 | 650 | 4.50 | 8.25 | 2.78 | .69 | 2.87 | 1.60 | 16.19 |
| Fast | 225 | 600 | 3.90 | 9.78 | 3.95 | .65 | 2.66 | 1.02 | 18.06 |
| Metal primer, rust inhibitor - clean metal (material #35) | | | | | | | | | |
| Mitt or glove prime coat | | | | | | | | | |
| Slow | 175 | 450 | 28.20 | 5.86 | 1.59 | 6.27 | 4.26 | 2.70 | 20.68 |
| Medium | 200 | 438 | 24.60 | 8.25 | 2.78 | 5.62 | 4.08 | 2.28 | 23.01 |
| Fast | 225 | 425 | 21.10 | 9.78 | 3.95 | 4.96 | 3.46 | 1.33 | 23.48 |
| Metal primer, rust inhibitor - rusty metal (material #36) | | | | | | | | | |
| Mitt or glove prime coat | | | | | | | | | |
| Slow | 175 | 400 | 33.30 | 5.86 | 1.59 | 8.33 | 4.89 | 3.10 | 23.77 |
| Medium | 200 | 388 | 29.20 | 8.25 | 2.78 | 7.53 | 4.55 | 2.54 | 25.65 |
| Fast | 225 | 375 | 25.00 | 9.78 | 3.95 | 6.67 | 3.78 | 1.45 | 25.63 |
| Industrial enamel, oil base, high gloss - light colors (material #56) | | | | | | | | | |
| Mitt or glove 1st or additional finish coats | | | | | | | | | |
| Slow | 225 | 450 | 28.90 | 4.56 | 1.23 | 6.42 | 3.79 | 2.40 | 18.40 |
| Medium | 250 | 438 | 25.20 | 6.60 | 2.22 | 5.75 | 3.57 | 2.00 | 20.14 |
| Fast | 275 | 425 | 21.60 | 8.00 | 3.25 | 5.08 | 3.02 | 1.16 | 20.51 |
| Industrial enamel, oil base, high gloss - dark (OSHA) colors (material #57) | | | | | | | | | |
| Mitt or glove 1st or additional finish coats | | | | | | | | | |
| Slow | 225 | 500 | 34.90 | 4.56 | 1.23 | 6.98 | 3.96 | 2.51 | 19.24 |
| Medium | 250 | 488 | 30.50 | 6.60 | 2.22 | 6.25 | 3.69 | 2.06 | 20.82 |
| Fast | 275 | 475 | 26.10 | 8.00 | 3.25 | 5.49 | 3.10 | 1.19 | 21.03 |
| Epoxy coating, 2 part system - clear (material #51) | | | | | | | | | |
| Mitt or glove 1st coat | | | | | | | | | |
| Slow | 175 | 425 | 57.00 | 5.86 | 1.59 | 13.41 | 6.47 | 4.10 | 31.43 |
| Medium | 200 | 413 | 49.90 | 8.25 | 2.78 | 12.08 | 5.66 | 3.16 | 31.93 |
| Fast | 225 | 400 | 42.80 | 9.78 | 3.95 | 10.70 | 4.52 | 1.74 | 30.69 |
| Mitt or glove 2nd or additional finish coats | | | | | | | | | |
| Slow | 225 | 475 | 57.00 | 4.56 | 1.23 | 12.00 | 5.52 | 3.50 | 26.81 |
| Medium | 250 | 463 | 49.90 | 6.60 | 2.22 | 10.78 | 4.80 | 2.68 | 27.08 |
| Fast | 275 | 450 | 42.80 | 8.00 | 3.25 | 9.51 | 3.84 | 1.48 | 26.08 |

| | Labor SF per manhour | Material coverage SF/gallon | Material cost per gallon | Labor cost per 100 SF | Labor burden 100 SF | Material cost per 100 SF | Overhead per 100 SF | Profit per 100 SF | Total price per 100 SF |
|---|---|---|---|---|---|---|---|---|---|
| Epoxy coating, 2 part system - white (material #52) | | | | | | | | | |
| Mitt or glove 1st coat | | | | | | | | | |
| Slow | 175 | 425 | 57.40 | 5.86 | 1.59 | 13.51 | 6.50 | 4.12 | 31.58 |
| Medium | 200 | 413 | 50.20 | 8.25 | 2.78 | 12.15 | 5.68 | 3.17 | 32.03 |
| Fast | 225 | 400 | 43.10 | 9.78 | 3.95 | 10.78 | 4.54 | 1.74 | 30.79 |
| Mitt or glove 2nd or additional finish coats | | | | | | | | | |
| Slow | 225 | 475 | 57.40 | 4.56 | 1.23 | 12.08 | 5.54 | 3.51 | 26.92 |
| Medium | 250 | 463 | 50.20 | 6.60 | 2.22 | 10.84 | 4.82 | 2.69 | 27.17 |
| Fast | 275 | 450 | 43.10 | 8.00 | 3.25 | 9.58 | 3.85 | 1.48 | 26.16 |

This table is based on square feet of conduit surface area. See Figure 21 below to convert from linear feet of various conduit or pipe sizes to square feet of surface area. For heights above 8 feet, use the High Time Difficulty Factors on page 137. Note: A two coat system, prime and finish, using oil base material is recommended for any metal surface. Although water base material is often used, it may cause oxidation, corrosion and rust. One coat of oil base solid body stain is often used on exterior metal but it may crack, peel or chip without the proper prime coat application. "Slow" work is based on a $10.25 hourly wage, "Medium" work on a $16.50 hourly wage, and "Fast" work on a $22.00 hourly wage. Other qualifications that apply to this table are on page 9.

| Pipe O/D (inches) | Conversion Factor (SF per measured LF) |
|---|---|
| 1" to 3" | 1 SF for each 1 LF |
| 4" to 7" | 2 SF for each 1 LF |
| 8" to 11" | 3 SF for each 1 LF |
| 12" to 15" | 4 SF for each 1 LF |
| 16" to 19" | 5 SF for each 1 LF |
| 20" to 22" | 6 SF for each 1 LF |
| 23" to 26" | 7 SF for each 1 LF |
| 27" to 30" | 8 SF for each 1 LF |

**Figure 21**
Conduit/pipe area conversions

| | Labor SF per manhour | Material coverage SF/gallon | Material cost per gallon | Labor cost per 100 SF | Labor burden 100 SF | Material cost per 100 SF | Overhead per 100 SF | Profit per 100 SF | Total price per 100 SF |
|---|---|---|---|---|---|---|---|---|---|
| **Decking and siding, metal, corrugated metal** | | | | | | | | | |
| Acid wash coat, muriatic acid (material #49) | | | | | | | | | |
| Spray each coat | | | | | | | | | |
| Slow | 700 | 500 | 5.20 | 1.46 | .40 | 1.04 | .90 | .57 | 4.37 |
| Medium | 750 | 450 | 4.50 | 2.20 | .73 | 1.00 | .97 | .54 | 5.44 |
| Fast | 800 | 400 | 3.90 | 2.75 | 1.12 | .98 | .90 | .35 | 6.10 |
| Metal primer, rust inhibitor - clean metal (material #35) | | | | | | | | | |
| Spray prime coat | | | | | | | | | |
| Slow | 700 | 325 | 28.20 | 1.46 | .40 | 8.68 | 3.27 | 2.07 | 15.88 |
| Medium | 750 | 300 | 24.60 | 2.20 | .73 | 8.20 | 2.73 | 1.53 | 15.39 |
| Fast | 800 | 275 | 21.10 | 2.75 | 1.12 | 7.67 | 2.13 | .82 | 14.49 |
| Metal primer, rust inhibitor - rusty metal (material #36) | | | | | | | | | |
| Spray prime coat | | | | | | | | | |
| Slow | 700 | 275 | 33.30 | 1.46 | .40 | 12.11 | 4.33 | 2.75 | 21.05 |
| Medium | 750 | 250 | 29.20 | 2.20 | .73 | 11.68 | 3.58 | 2.00 | 20.19 |
| Fast | 800 | 225 | 25.00 | 2.75 | 1.12 | 11.11 | 2.77 | 1.07 | 18.82 |
| Industrial enamel, oil base, high gloss - light colors (material #56) | | | | | | | | | |
| Spray 1st or additional finish coats | | | | | | | | | |
| Slow | 850 | 325 | 28.90 | 1.21 | .33 | 8.89 | 3.23 | 2.05 | 15.71 |
| Medium | 900 | 300 | 25.20 | 1.83 | .62 | 8.40 | 2.66 | 1.49 | 15.00 |
| Fast | 950 | 275 | 21.60 | 2.32 | .93 | 7.85 | 2.06 | .79 | 13.95 |
| Industrial enamel, oil base, high gloss - dark (OSHA) colors (material #57) | | | | | | | | | |
| Spray 1st or additional finish coats | | | | | | | | | |
| Slow | 850 | 425 | 34.90 | 1.21 | .33 | 8.21 | 3.02 | 1.92 | 14.69 |
| Medium | 900 | 400 | 30.50 | 1.83 | .62 | 7.63 | 2.47 | 1.38 | 13.93 |
| Fast | 950 | 375 | 26.10 | 2.32 | .93 | 6.96 | 1.89 | .73 | 12.83 |
| Epoxy coating, 2 part system - clear (material #51) | | | | | | | | | |
| Spray 1st coat | | | | | | | | | |
| Slow | 700 | 300 | 57.00 | 1.46 | .40 | 19.00 | 6.47 | 4.10 | 31.43 |
| Medium | 750 | 275 | 49.90 | 2.20 | .73 | 18.15 | 5.17 | 2.89 | 29.14 |
| Fast | 800 | 250 | 42.80 | 2.75 | 1.12 | 17.12 | 3.88 | 1.49 | 26.36 |
| Spray 2nd or additional finish coats | | | | | | | | | |
| Slow | 850 | 400 | 57.00 | 1.21 | .33 | 14.25 | 4.89 | 3.10 | 23.78 |
| Medium | 900 | 375 | 49.90 | 1.83 | .62 | 13.31 | 3.86 | 2.16 | 21.78 |
| Fast | 950 | 350 | 42.80 | 2.32 | .93 | 12.23 | 2.87 | 1.10 | 19.45 |

| | Labor SF per manhour | Material coverage SF/gallon | Material cost per gallon | Labor cost per 100 SF | Labor burden 100 SF | Material cost per 100 SF | Overhead per 100 SF | Profit per 100 SF | Total price per 100 SF |
|---|---|---|---|---|---|---|---|---|---|
| Epoxy coating, 2 part system - white (material #52) | | | | | | | | | |
| Spray 1st coat | | | | | | | | | |
| Slow | 700 | 300 | 57.40 | 1.46 | .40 | 19.13 | 6.51 | 4.13 | 31.63 |
| Medium | 750 | 275 | 50.20 | 2.20 | .73 | 18.25 | 5.19 | 2.90 | 29.27 |
| Fast | 800 | 250 | 43.10 | 2.75 | 1.12 | 17.24 | 3.91 | 1.50 | 26.52 |
| Spray 2nd or additional finish coats | | | | | | | | | |
| Slow | 850 | 400 | 57.40 | 1.21 | .33 | 14.35 | 4.93 | 3.12 | 23.94 |
| Medium | 900 | 375 | 50.20 | 1.83 | .62 | 13.39 | 3.88 | 2.17 | 21.89 |
| Fast | 950 | 350 | 43.10 | 2.32 | .93 | 12.31 | 2.88 | 1.11 | 19.55 |

The figures in the table above are based on overall dimensions (length times width). But all decking and siding has corrugations, peaks and valleys that increase the surface that has to be painted. For example, corrugated siding with 2-1/2" center to center corrugations has a surface area 10% greater than the width times length dimension. For corrugated siding with 1-1/4" center to center corrugations, increase the surface area by 15%. For square corner decking, Figure 22 below shows how much area must be added to allow for peaks and valleys. For heights above 8 feet, use the High Time Difficulty Factors on page 137. Note: A two coat system, prime and finish, using oil base material is recommended for any metal surface. Although water base material is often used, it may cause oxidation, corrosion and rust. One coat of oil base solid body stain is often used on exterior metal but it may crack, peel or chip without the proper prime coat application. "Slow" work is based on a $10.25 hourly wage, "Medium" work on a $16.50 hourly wage, and "Fast" work on a $22.00 hourly wage. Other qualifications that apply to this table are on page 9.

| C | 2C | D | PW | VW | Factor |
|---|---|---|---|---|---|
| -- | 12" | 4-1/2" | 3" | 2" | 2.50 |
| -- | 12" | 1-1/2" | 3-1/8" | 2" | 1.50 |
| -- | 12" | 1-1/2" | 5-1/16" | 1" | 1.45 |
| 12" | -- | 3" | 9-5/8" | 1" | 1.50 |
| 12" | -- | 4-1/2" | 9-5/8" | 1" | 1.75 |
| 24" | -- | 4-1/2" | 12" | 12" | 1.60 |
| 24" | -- | 6" | 12" | 12" | 1.75 |
| 24" | -- | 8" | 12" | 12" | 1.95 |

For square corner decking, calculate the overall (length times width) deck area. Then measure the peaks and valleys on the deck. Select the row in the table above that most nearly matches the deck you're painting. Multiply the overall deck area by the number listed in the column headed "factor" to find the actual area you're painting. Use this actual area when calculating labor and material requirements. In the table above, figures in the column headed C show the distance between the center of the peaks. Column 2C shows the distance between every second center of peak (2 centers). Column D shows the depth of corrugation. Column PW shows the peak width. Column VW shows the valley width. If the deck you're painting doesn't match any deck listed in this table, use the factor for the most similar deck in the table.

**Figure 22**
Square corner decking factors

| | Labor SF per manhour | Material coverage SF/gallon | Material cost per gallon | Labor cost per 100 SF | Labor burden 100 SF | Material cost per 100 SF | Overhead per 100 SF | Profit per 100 SF | Total price per 100 SF |
|---|---|---|---|---|---|---|---|---|---|
| **Decking and siding, metal, flat pan metal** | | | | | | | | | |
| Acid wash coat, muriatic acid (material #49) | | | | | | | | | |
| Spray each coat | | | | | | | | | |
| Slow | 800 | 600 | 5.20 | 1.28 | .35 | .87 | .78 | .49 | 3.77 |
| Medium | 850 | 550 | 4.50 | 1.94 | .66 | .82 | .84 | .47 | 4.73 |
| Fast | 900 | 500 | 3.90 | 2.44 | .99 | .78 | .78 | .30 | 5.29 |
| Metal primer, rust inhibitor - clean metal (material #35) | | | | | | | | | |
| Spray prime coat | | | | | | | | | |
| Slow | 800 | 375 | 28.20 | 1.28 | .35 | 7.52 | 2.84 | 1.80 | 13.79 |
| Medium | 850 | 350 | 24.60 | 1.94 | .66 | 7.03 | 2.36 | 1.32 | 13.31 |
| Fast | 900 | 325 | 21.10 | 2.44 | .99 | 6.49 | 1.84 | .71 | 12.47 |
| Metal primer, rust inhibitor - rusty metal (material #36) | | | | | | | | | |
| Spray prime coat | | | | | | | | | |
| Slow | 800 | 300 | 33.30 | 1.28 | .35 | 11.10 | 3.95 | 2.50 | 19.18 |
| Medium | 850 | 275 | 29.20 | 1.94 | .66 | 10.62 | 3.24 | 1.81 | 18.27 |
| Fast | 900 | 250 | 25.00 | 2.44 | .99 | 10.00 | 2.48 | .95 | 16.86 |
| Industrial enamel, oil base, high gloss - light colors (material #56) | | | | | | | | | |
| Spray 1st or additional finish coats | | | | | | | | | |
| Slow | 1000 | 375 | 28.90 | 1.03 | .27 | 7.71 | 2.80 | 1.77 | 13.58 |
| Medium | 1050 | 350 | 25.20 | 1.57 | .52 | 7.20 | 2.28 | 1.27 | 12.84 |
| Fast | 1100 | 325 | 21.60 | 2.00 | .81 | 6.65 | 1.75 | .67 | 11.88 |
| Industrial enamel, oil base, high gloss - dark (OSHA) colors (material #57) | | | | | | | | | |
| Spray 1st or additional finish coats | | | | | | | | | |
| Slow | 1000 | 425 | 34.90 | 1.03 | .27 | 8.21 | 2.95 | 1.87 | 14.33 |
| Medium | 1050 | 400 | 30.50 | 1.57 | .52 | 7.63 | 2.38 | 1.33 | 13.43 |
| Fast | 1100 | 375 | 26.10 | 2.00 | .81 | 6.96 | 1.81 | .69 | 12.27 |
| Epoxy coating, 2 part system - clear (material #51) | | | | | | | | | |
| Spray 1st coat | | | | | | | | | |
| Slow | 800 | 350 | 57.00 | 1.28 | .35 | 16.29 | 5.56 | 3.52 | 27.00 |
| Medium | 850 | 325 | 49.90 | 1.94 | .66 | 15.35 | 4.40 | 2.46 | 24.81 |
| Fast | 900 | 300 | 42.80 | 2.44 | .99 | 14.27 | 3.27 | 1.26 | 22.23 |
| Spray 2nd or additional finish coats | | | | | | | | | |
| Slow | 1000 | 400 | 57.00 | 1.03 | .27 | 14.25 | 4.82 | 3.06 | 23.43 |
| Medium | 1050 | 375 | 49.90 | 1.57 | .52 | 13.31 | 3.78 | 2.11 | 21.29 |
| Fast | 1100 | 350 | 42.80 | 2.00 | .81 | 12.23 | 2.78 | 1.07 | 18.89 |

| | Labor SF per manhour | Material coverage SF/gallon | Material cost per gallon | Labor cost per 100 SF | Labor burden 100 SF | Material cost per 100 SF | Overhead per 100 SF | Profit per 100 SF | Total price per 100 SF |
|---|---|---|---|---|---|---|---|---|---|
| Epoxy coating, 2 part system - white (material #52) | | | | | | | | | |
| Spray 1st coat | | | | | | | | | |
| Slow | 800 | 350 | 57.40 | 1.28 | .35 | 16.40 | 5.59 | 3.54 | 27.16 |
| Medium | 850 | 325 | 50.20 | 1.94 | .66 | 15.45 | 4.42 | 2.47 | 24.94 |
| Fast | 900 | 300 | 43.10 | 2.44 | .99 | 14.37 | 3.29 | 1.27 | 22.36 |
| Spray 2nd or additional coats | | | | | | | | | |
| Slow | 1000 | 400 | 57.40 | 1.03 | .27 | 14.35 | 4.85 | 3.08 | 23.58 |
| Medium | 1050 | 375 | 50.20 | 1.57 | .52 | 13.39 | 3.80 | 2.12 | 21.40 |
| Fast | 1100 | 350 | 43.10 | 2.00 | .81 | 12.31 | 2.80 | 1.08 | 19.00 |

The figures in table above are based on overall dimensions (length times width). But all decking and siding has corrugations, peaks and valleys that increase the surface that has to be painted. For example, corrugated siding with 2-1/2" center to center corrugations has a surface area 10% greater than the width times length dimension. For corrugated siding with 1-1/4" center to center corrugations, increase the surface area by 15%. For square corner decking, Figure 22 on page 323 shows how much area must be added to allow for peaks and valleys. For heights above 8 feet, use the High Time Difficulty Factors on page 137. Note: A two coat system, prime and finish, using oil base material is recommended for any metal surface. Although water base material is often used, it may cause oxidation, corrosion and rust. One coat of oil base solid body stain is often used on exterior metal but it may crack, peel or chip without the proper prime coat application. "Slow" work is based on a $10.25 hourly wage, "Medium" work on a $16.50 hourly wage, and "Fast" work on a $22.00 hourly wage. Other qualifications that apply to this table are on page 9.

| | Labor SF per manhour | Material coverage SF/gallon | Material cost per gallon | Labor cost per 100 SF | Labor burden 100 SF | Material cost per 100 SF | Overhead per 100 SF | Profit per 100 SF | Total price per 100 SF |
|---|---|---|---|---|---|---|---|---|---|
| **Doors, hollow metal, brush application, square foot basis** | | | | | | | | | |
| Metal primer, rust inhibitor - clean metal (material #35) | | | | | | | | | |
| Roll and brush prime coat | | | | | | | | | |
| Slow | 160 | 450 | 28.20 | 6.41 | 1.74 | 6.27 | 4.47 | 2.84 | 21.73 |
| Medium | 180 | 438 | 24.60 | 9.17 | 3.09 | 5.62 | 4.38 | 2.45 | 24.71 |
| Fast | 200 | 425 | 21.10 | 11.00 | 4.46 | 4.96 | 3.78 | 1.45 | 25.65 |
| Metal primer, rust inhibitor - rusty metal (material #36) | | | | | | | | | |
| Roll and brush prime coat | | | | | | | | | |
| Slow | 160 | 425 | 33.30 | 6.41 | 1.74 | 7.84 | 4.96 | 3.14 | 24.09 |
| Medium | 180 | 413 | 29.20 | 9.17 | 3.09 | 7.07 | 4.74 | 2.65 | 26.72 |
| Fast | 200 | 400 | 25.00 | 11.00 | 4.46 | 6.25 | 4.02 | 1.54 | 27.27 |
| Metal finish - synthetic enamel, gloss, interior or exterior - off white (material #37) | | | | | | | | | |
| Roll and brush 1st or additional finish coats | | | | | | | | | |
| Slow | 175 | 450 | 25.90 | 5.86 | 1.59 | 5.76 | 4.10 | 2.60 | 19.91 |
| Medium | 195 | 438 | 22.70 | 8.46 | 2.85 | 5.18 | 4.04 | 2.26 | 22.79 |
| Fast | 215 | 425 | 19.50 | 10.23 | 4.15 | 4.59 | 3.51 | 1.35 | 23.83 |
| Metal finish - synthetic enamel, gloss, interior or exterior - colors, except orange & red (material #38) | | | | | | | | | |
| Roll and brush 1st or additional finish coats | | | | | | | | | |
| Slow | 175 | 475 | 35.00 | 5.86 | 1.59 | 7.37 | 4.60 | 2.91 | 22.33 |
| Medium | 195 | 463 | 30.70 | 8.46 | 2.85 | 6.63 | 4.40 | 2.46 | 24.80 |
| Fast | 215 | 450 | 26.30 | 10.23 | 4.15 | 5.84 | 3.74 | 1.44 | 25.40 |

To calculate the cost per door, figure the square feet for each side based on a 3 x 7 door. Add 3 feet to the width and 1 foot to the top of the door to allow for the time and material necessary to finish the door edges, frames and jambs. The result is a 6 x 8 door or 48 square feet per side, times 2 is 96 square feet per door. The cost per door is simply the 96 square feet times the cost per square foot indicated in the table above. Typically, hollow metal doors are pre-primed by the manufacturer and shipped ready to install and paint. These doors usually require only one coat of finish paint to cover. The metal finish figures include minor touchup of the prime coat. If off white or other light colored finish paint is specified, make sure the prime coat is a light color also, or more than one finish coat will be necessary. Note: A two coat system, prime and finish, using oil base material is recommended for any metal surface. Although water base material is often used, it may cause oxidation, corrosion and rust. One coat of oil base solid body stain is often used on exterior metal but it may crack, peel or chip without the proper prime coat application. "Slow" work is based on a $10.25 hourly wage, "Medium" work on a $16.50 hourly wage, and "Fast" work on a $22.00 hourly wage. Other qualifications that apply to this table are on page 9.

| | Labor SF per manhour | Material coverage SF/gallon | Material cost per gallon | Labor cost per 100 SF | Labor burden 100 SF | Material cost per 100 SF | Overhead per 100 SF | Profit per 100 SF | Total price per 100 SF |
|---|---|---|---|---|---|---|---|---|---|
| **Ductwork, bare duct, brush application** | | | | | | | | | |
| Acid wash coat, muriatic acid (material #49) | | | | | | | | | |
| Brush each coat | | | | | | | | | |
| Slow | 80 | 750 | 5.20 | 12.81 | 3.49 | .69 | 5.27 | 3.34 | 25.60 |
| Medium | 100 | 700 | 4.50 | 16.50 | 5.55 | .64 | 5.56 | 3.11 | 31.36 |
| Fast | 120 | 650 | 3.90 | 18.33 | 7.43 | .60 | 4.88 | 1.87 | 33.11 |
| Metal primer, rust inhibitor - clean metal (material #35) | | | | | | | | | |
| Brush prime coat | | | | | | | | | |
| Slow | 80 | 400 | 28.20 | 12.81 | 3.49 | 7.05 | 7.24 | 4.59 | 35.18 |
| Medium | 100 | 375 | 24.60 | 16.50 | 5.55 | 6.56 | 7.01 | 3.92 | 39.54 |
| Fast | 120 | 350 | 21.10 | 18.33 | 7.43 | 6.03 | 5.88 | 2.26 | 39.93 |
| Metal primer, rust inhibitor - rusty metal (material #36) | | | | | | | | | |
| Brush prime coat | | | | | | | | | |
| Slow | 80 | 350 | 33.30 | 12.81 | 3.49 | 9.51 | 8.00 | 5.07 | 38.88 |
| Medium | 100 | 325 | 29.20 | 16.50 | 5.55 | 8.98 | 7.60 | 4.25 | 42.88 |
| Fast | 120 | 300 | 25.00 | 18.33 | 7.43 | 8.33 | 6.31 | 2.42 | 42.82 |
| Industrial enamel, oil base, high gloss - light colors (material #56) | | | | | | | | | |
| Brush 1st or additional finish coats | | | | | | | | | |
| Slow | 90 | 400 | 28.90 | 11.39 | 3.10 | 7.23 | 6.73 | 4.27 | 32.72 |
| Medium | 115 | 375 | 25.20 | 14.35 | 4.83 | 6.72 | 6.35 | 3.55 | 35.80 |
| Fast | 140 | 350 | 21.60 | 15.71 | 6.37 | 6.17 | 5.23 | 2.01 | 35.49 |
| Industrial enamel, oil base, high gloss - dark (OSHA) colors (material #57) | | | | | | | | | |
| Brush 1st or additional finish coats | | | | | | | | | |
| Slow | 90 | 450 | 34.90 | 11.39 | 3.10 | 7.76 | 6.90 | 4.37 | 33.52 |
| Medium | 115 | 425 | 30.50 | 14.35 | 4.83 | 7.18 | 6.46 | 3.61 | 36.43 |
| Fast | 140 | 400 | 26.10 | 15.71 | 6.37 | 6.53 | 5.29 | 2.03 | 35.93 |
| Epoxy coating, 2 part system - clear (material #51) | | | | | | | | | |
| Brush 1st coat | | | | | | | | | |
| Slow | 80 | 350 | 57.00 | 12.81 | 3.49 | 16.29 | 10.10 | 6.40 | 49.09 |
| Medium | 100 | 325 | 49.90 | 16.50 | 5.55 | 15.35 | 9.16 | 5.12 | 51.68 |
| Fast | 120 | 300 | 42.80 | 18.33 | 7.43 | 14.27 | 7.41 | 2.85 | 50.29 |
| Brush 2nd or additional finish coats | | | | | | | | | |
| Slow | 90 | 425 | 57.00 | 11.39 | 3.10 | 13.41 | 8.65 | 5.48 | 42.03 |
| Medium | 115 | 400 | 49.90 | 14.35 | 4.83 | 12.48 | 7.76 | 4.34 | 43.76 |
| Fast | 140 | 375 | 42.80 | 15.71 | 6.37 | 11.41 | 6.20 | 2.38 | 42.07 |

| | Labor SF per manhour | Material coverage SF/gallon | Material cost per gallon | Labor cost per 100 SF | Labor burden 100 SF | Material cost per 100 SF | Overhead per 100 SF | Profit per 100 SF | Total price per 100 SF |
|---|---|---|---|---|---|---|---|---|---|
| Epoxy coating, 2 part system - white (material #52) | | | | | | | | | |
| Brush 1st coat | | | | | | | | | |
| Slow | 80 | 350 | 57.40 | 12.81 | 3.49 | 16.40 | 10.14 | 6.43 | 49.27 |
| Medium | 100 | 325 | 50.20 | 16.50 | 5.55 | 15.45 | 9.19 | 5.14 | 51.83 |
| Fast | 120 | 300 | 43.10 | 18.33 | 7.43 | 14.37 | 7.42 | 2.85 | 50.40 |
| Brush 2nd or additional finish coats | | | | | | | | | |
| Slow | 90 | 425 | 57.40 | 11.39 | 3.10 | 13.51 | 8.68 | 5.50 | 42.18 |
| Medium | 115 | 400 | 50.20 | 14.35 | 4.83 | 12.55 | 7.77 | 4.35 | 43.85 |
| Fast | 140 | 375 | 43.10 | 15.71 | 6.37 | 11.49 | 6.21 | 2.39 | 42.17 |

For heights over 8 feet, use the High Time Difficulty Factors on page 137. Note: A two coat system, prime and finish, using oil base material is recommended for any metal surface. Although water base material is often used, it may cause oxidation, corrosion and rust. One coat of oil base solid body stain is often used on exterior metal but it may crack, peel or chip without the proper prime coat application. "Slow" work is based on a $10.25 hourly wage, "Medium" work on a $16.50 hourly wage, and "Fast" work on a $22.50 hourly wage. Other qualifications that apply to this table are on page 9.

| | Labor SF per manhour | Material coverage SF/gallon | Material cost per gallon | Labor cost per 100 SF | Labor burden 100 SF | Material cost per 100 SF | Overhead per 100 SF | Profit per 100 SF | Total price per 100 SF |
|---|---|---|---|---|---|---|---|---|---|
| **Ductwork, bare duct, roll application** | | | | | | | | | |
| Acid wash coat, muriatic acid (material #49) | | | | | | | | | |
| Roll each coat | | | | | | | | | |
| Slow | 225 | 700 | 5.20 | 4.56 | 1.23 | .74 | 2.03 | 1.29 | 9.85 |
| Medium | 250 | 650 | 4.50 | 6.60 | 2.22 | .69 | 2.33 | 1.30 | 13.14 |
| Fast | 275 | 600 | 3.90 | 8.00 | 3.25 | .65 | 2.20 | .85 | 14.95 |
| Metal primer, rust inhibitor - clean metal (material #35) | | | | | | | | | |
| Roll prime coat | | | | | | | | | |
| Slow | 225 | 375 | 28.20 | 4.56 | 1.23 | 7.52 | 4.13 | 2.62 | 20.06 |
| Medium | 250 | 350 | 24.60 | 6.60 | 2.22 | 7.03 | 3.88 | 2.17 | 21.90 |
| Fast | 275 | 325 | 21.10 | 8.00 | 3.25 | 6.49 | 3.28 | 1.26 | 22.28 |
| Metal primer, rust inhibitor - rusty metal (material #36) | | | | | | | | | |
| Roll prime coat | | | | | | | | | |
| Slow | 225 | 325 | 33.30 | 4.56 | 1.23 | 10.25 | 4.98 | 3.15 | 24.17 |
| Medium | 250 | 300 | 29.20 | 6.60 | 2.22 | 9.73 | 4.54 | 2.54 | 25.63 |
| Fast | 275 | 275 | 25.00 | 8.00 | 3.25 | 9.09 | 3.76 | 1.45 | 25.55 |
| Industrial enamel, oil base, high gloss - light colors (material #56) | | | | | | | | | |
| Roll 1st or additional finish coats | | | | | | | | | |
| Slow | 275 | 425 | 28.90 | 3.73 | 1.02 | 6.80 | 3.58 | 2.27 | 17.40 |
| Medium | 300 | 400 | 25.20 | 5.50 | 1.84 | 6.30 | 3.34 | 1.87 | 18.85 |
| Fast | 325 | 375 | 21.60 | 6.77 | 2.75 | 5.76 | 2.83 | 1.09 | 19.20 |
| Industrial enamel, oil base, high gloss - dark (OSHA) colors (material #57) | | | | | | | | | |
| Roll 1st or additional finish coats | | | | | | | | | |
| Slow | 275 | 475 | 34.90 | 3.73 | 1.02 | 7.35 | 3.75 | 2.38 | 18.23 |
| Medium | 300 | 450 | 30.50 | 5.50 | 1.84 | 6.78 | 3.46 | 1.93 | 19.51 |
| Fast | 325 | 425 | 26.10 | 6.77 | 2.75 | 6.14 | 2.90 | 1.11 | 19.67 |
| Epoxy coating, 2 part system - clear (material #51) | | | | | | | | | |
| Roll 1st coat | | | | | | | | | |
| Slow | 225 | 350 | 57.00 | 4.56 | 1.23 | 16.29 | 6.85 | 4.34 | 33.27 |
| Medium | 250 | 325 | 49.90 | 6.60 | 2.22 | 15.35 | 5.92 | 3.31 | 33.40 |
| Fast | 275 | 300 | 42.80 | 8.00 | 3.25 | 14.27 | 4.72 | 1.81 | 32.05 |
| Roll 2nd or additional finish coats | | | | | | | | | |
| Slow | 275 | 450 | 57.00 | 3.73 | 1.02 | 12.67 | 5.40 | 3.42 | 26.24 |
| Medium | 300 | 425 | 49.90 | 5.50 | 1.84 | 11.74 | 4.68 | 2.61 | 26.37 |
| Fast | 325 | 400 | 42.80 | 6.77 | 2.75 | 10.70 | 3.74 | 1.44 | 25.40 |

| | Labor SF per manhour | Material coverage SF/gallon | Material cost per gallon | Labor cost per 100 SF | Labor burden 100 SF | Material cost per 100 SF | Overhead per 100 SF | Profit per 100 SF | Total price per 100 SF |
|---|---|---|---|---|---|---|---|---|---|
| Epoxy coating, 2 part system - white (material #52) | | | | | | | | | |
| Roll 1st coat | | | | | | | | | |
| Slow | 225 | 350 | 57.40 | 4.56 | 1.23 | 16.40 | 6.88 | 4.36 | 33.43 |
| Medium | 250 | 325 | 50.20 | 6.60 | 2.22 | 15.45 | 5.95 | 3.32 | 33.54 |
| Fast | 275 | 300 | 43.10 | 8.00 | 3.25 | 14.37 | 4.74 | 1.82 | 32.18 |
| Roll 2nd or additional finish coats | | | | | | | | | |
| Slow | 275 | 450 | 57.40 | 3.73 | 1.02 | 12.76 | 5.43 | 3.44 | 26.38 |
| Medium | 300 | 425 | 50.20 | 5.50 | 1.84 | 11.81 | 4.69 | 2.62 | 26.46 |
| Fast | 325 | 400 | 43.10 | 6.77 | 2.75 | 10.78 | 3.76 | 1.44 | 25.50 |

For heights over 8 feet, use the High Time Difficulty Factors on page 137. Note: A two coat system, prime and finish, using oil base material is recommended for any metal surface. Although water base material is often used, it may cause oxidation, corrosion and rust. One coat of oil base solid body stain is often used on exterior metal but it may crack, peel or chip without the proper prime coat application. "Slow" work is based on a $10.25 hourly wage, "Medium" work on a $16.50 hourly wage, and "Fast" work on a $22.50 hourly wage. Other qualifications that apply to this table are on page 9.

| | Labor SF per manhour | Material coverage SF/gallon | Material cost per gallon | Labor cost per 100 SF | Labor burden 100 SF | Material cost per 100 SF | Overhead per 100 SF | Profit per 100 SF | Total price per 100 SF |
|---|---|---|---|---|---|---|---|---|---|
| **Ductwork, bare duct, spray application** | | | | | | | | | |
| Acid wash coat, muriatic acid (material #49) | | | | | | | | | |
| Spray each coat | | | | | | | | | |
| Slow | 550 | 450 | 5.20 | 1.86 | .51 | 1.16 | 1.09 | .69 | 5.31 |
| Medium | 600 | 400 | 4.50 | 2.75 | .93 | 1.13 | 1.18 | .66 | 6.65 |
| Fast | 650 | 350 | 3.90 | 3.38 | 1.38 | 1.11 | 1.08 | .42 | 7.37 |
| Metal primer, rust inhibitor - clean metal (material #35) | | | | | | | | | |
| Spray prime coat | | | | | | | | | |
| Slow | 550 | 250 | 28.20 | 1.86 | .51 | 11.28 | 4.23 | 2.68 | 20.56 |
| Medium | 600 | 225 | 24.60 | 2.75 | .93 | 10.93 | 3.58 | 2.00 | 20.19 |
| Fast | 650 | 200 | 21.10 | 3.38 | 1.38 | 10.55 | 2.83 | 1.09 | 19.23 |
| Metal primer, rust inhibitor - rusty metal (material #36) | | | | | | | | | |
| Spray prime coat | | | | | | | | | |
| Slow | 550 | 200 | 33.30 | 1.86 | .51 | 16.65 | 5.90 | 3.74 | 28.66 |
| Medium | 600 | 188 | 29.20 | 2.75 | .93 | 15.53 | 4.71 | 2.63 | 26.55 |
| Fast | 650 | 175 | 25.00 | 3.38 | 1.38 | 14.29 | 3.52 | 1.35 | 23.92 |
| Industrial enamel, oil base, high gloss - light colors (material #56) | | | | | | | | | |
| Spray 1st or additional finish coats | | | | | | | | | |
| Slow | 700 | 225 | 28.90 | 1.46 | .40 | 12.84 | 4.56 | 2.89 | 22.15 |
| Medium | 750 | 213 | 25.20 | 2.20 | .73 | 11.83 | 3.62 | 2.02 | 20.40 |
| Fast | 800 | 200 | 21.60 | 2.75 | 1.12 | 10.80 | 2.71 | 1.04 | 18.42 |
| Industrial enamel, oil base, high gloss - dark (OSHA) colors (material #57) | | | | | | | | | |
| Spray 1st or additional finish coats | | | | | | | | | |
| Slow | 700 | 275 | 34.90 | 1.46 | .40 | 12.69 | 4.51 | 2.86 | 21.92 |
| Medium | 750 | 250 | 30.50 | 2.20 | .73 | 12.20 | 3.71 | 2.07 | 20.91 |
| Fast | 800 | 225 | 26.10 | 2.75 | 1.12 | 11.60 | 2.86 | 1.10 | 19.43 |
| Epoxy coating, 2 part system - clear (material #51) | | | | | | | | | |
| Spray 1st coat | | | | | | | | | |
| Slow | 550 | 225 | 57.00 | 1.86 | .51 | 25.33 | 8.59 | 5.44 | 41.73 |
| Medium | 600 | 213 | 49.90 | 2.75 | .93 | 23.43 | 6.64 | 3.71 | 37.46 |
| Fast | 650 | 200 | 42.80 | 3.38 | 1.38 | 21.40 | 4.84 | 1.86 | 32.86 |
| Spray 2nd or additional finish coats | | | | | | | | | |
| Slow | 700 | 250 | 57.00 | 1.46 | .40 | 22.80 | 7.64 | 4.85 | 37.15 |
| Medium | 750 | 238 | 49.90 | 2.20 | .73 | 20.97 | 5.86 | 3.27 | 33.03 |
| Fast | 800 | 225 | 42.80 | 2.75 | 1.12 | 19.02 | 4.23 | 1.63 | 28.75 |

| | Labor SF per manhour | Material coverage SF/gallon | Material cost per gallon | Labor cost per 100 SF | Labor burden 100 SF | Material cost per 100 SF | Overhead per 100 SF | Profit per 100 SF | Total price per 100 SF |
|---|---|---|---|---|---|---|---|---|---|
| Epoxy coating, 2 part system - white (material #52) | | | | | | | | | |
| Spray 1st coat | | | | | | | | | |
| Slow | 550 | 225 | 57.40 | 1.86 | .51 | 25.51 | 8.64 | 5.48 | 42.00 |
| Medium | 600 | 213 | 50.20 | 2.75 | .93 | 23.57 | 6.68 | 3.73 | 37.66 |
| Fast | 650 | 200 | 43.10 | 3.38 | 1.38 | 21.55 | 4.87 | 1.87 | 33.05 |
| Spray 2nd or additional finish coats | | | | | | | | | |
| Slow | 700 | 250 | 57.40 | 1.46 | .40 | 22.96 | 7.69 | 4.88 | 37.39 |
| Medium | 750 | 238 | 50.20 | 2.20 | .73 | 21.09 | 5.89 | 3.29 | 33.20 |
| Fast | 800 | 225 | 43.10 | 2.75 | 1.12 | 19.16 | 4.26 | 1.64 | 28.93 |

For heights over 8 feet, use the High Time Difficulty Factors on page 137. Note: A two coat system, prime and finish, using oil base material is recommended for any metal surface. Although water base material is often used, it may cause oxidation, corrosion and rust. One coat of oil base solid body stain is often used on exterior metal but it may crack, peel or chip without the proper prime coat application. "Slow" work is based on a $10.25 hourly wage, "Medium" work on a $16.50 hourly wage, and "Fast" work on a $22.50 hourly wage. Other qualifications that apply to this table are on page 9.

| | Labor SF per manhour | Material coverage SF/gallon | Material cost per gallon | Labor cost per 100 SF | Labor burden 100 SF | Material cost per 100 SF | Overhead per 100 SF | Profit per 100 SF | Total price per 100 SF |
|---|---|---|---|---|---|---|---|---|---|
| **Ductwork, bare duct, mitt or glove application** | | | | | | | | | |
| Acid wash coat, muriatic acid (material #49) | | | | | | | | | |
| Mitt or glove each coat | | | | | | | | | |
| Slow | 175 | 700 | 5.20 | 5.86 | 1.59 | .74 | 2.54 | 1.61 | 12.34 |
| Medium | 200 | 650 | 4.50 | 8.25 | 2.78 | .69 | 2.87 | 1.60 | 16.19 |
| Fast | 225 | 600 | 3.90 | 9.78 | 3.95 | .65 | 2.66 | 1.02 | 18.06 |
| Metal primer, rust inhibitor - clean metal (material #35) | | | | | | | | | |
| Mitt or glove prime coat | | | | | | | | | |
| Slow | 175 | 400 | 28.20 | 5.86 | 1.59 | 7.05 | 4.50 | 2.85 | 21.85 |
| Medium | 200 | 388 | 24.60 | 8.25 | 2.78 | 6.34 | 4.26 | 2.38 | 24.01 |
| Fast | 225 | 375 | 21.10 | 9.78 | 3.95 | 5.63 | 3.59 | 1.38 | 24.33 |
| Metal primer, rust inhibitor - rusty metal (material #36) | | | | | | | | | |
| Mitt or glove prime coat | | | | | | | | | |
| Slow | 175 | 400 | 33.30 | 5.86 | 1.59 | 8.33 | 4.89 | 3.10 | 23.77 |
| Medium | 200 | 350 | 29.20 | 8.25 | 2.78 | 8.34 | 4.75 | 2.65 | 26.77 |
| Fast | 225 | 325 | 25.00 | 9.78 | 3.95 | 7.69 | 3.97 | 1.52 | 26.91 |
| Industrial enamel, oil base, high gloss - light colors (material #56) | | | | | | | | | |
| Mitt or glove 1st or additional finish coats | | | | | | | | | |
| Slow | 225 | 425 | 28.90 | 4.56 | 1.23 | 6.80 | 3.91 | 2.48 | 18.98 |
| Medium | 250 | 400 | 25.20 | 6.60 | 2.22 | 6.30 | 3.70 | 2.07 | 20.89 |
| Fast | 275 | 375 | 21.60 | 8.00 | 3.25 | 5.76 | 3.15 | 1.21 | 21.37 |
| Industrial enamel, oil base, high gloss - dark (OSHA) colors (material #57) | | | | | | | | | |
| Mitt or glove 1st or additional finish coats | | | | | | | | | |
| Slow | 225 | 475 | 34.90 | 4.56 | 1.23 | 7.35 | 4.08 | 2.58 | 19.80 |
| Medium | 250 | 450 | 30.50 | 6.60 | 2.22 | 6.78 | 3.82 | 2.14 | 21.56 |
| Fast | 275 | 425 | 26.10 | 8.00 | 3.25 | 6.14 | 3.22 | 1.24 | 21.85 |
| Epoxy coating, 2 part system - clear (material #51) | | | | | | | | | |
| Mitt or glove 1st coat | | | | | | | | | |
| Slow | 175 | 400 | 57.00 | 5.86 | 1.59 | 14.25 | 6.73 | 4.27 | 32.70 |
| Medium | 200 | 375 | 49.90 | 8.25 | 2.78 | 13.31 | 5.96 | 3.33 | 33.63 |
| Fast | 225 | 350 | 42.80 | 9.78 | 3.95 | 12.23 | 4.81 | 1.85 | 32.62 |
| Mitt or glove 2nd or additional finish coats | | | | | | | | | |
| Slow | 225 | 425 | 57.00 | 4.56 | 1.23 | 13.41 | 5.96 | 3.78 | 28.94 |
| Medium | 250 | 400 | 49.90 | 6.60 | 2.22 | 12.48 | 5.22 | 2.92 | 29.44 |
| Fast | 275 | 375 | 42.80 | 8.00 | 3.25 | 11.41 | 4.19 | 1.61 | 28.46 |

| | Labor SF per manhour | Material coverage SF/gallon | Material cost per gallon | Labor cost per 100 SF | Labor burden 100 SF | Material cost per 100 SF | Overhead per 100 SF | Profit per 100 SF | Total price per 100 SF |
|---|---|---|---|---|---|---|---|---|---|
| **Epoxy coating, 2 part system - white (material #52)** | | | | | | | | | |
| Mitt or glove 1st coat | | | | | | | | | |
| Slow | 175 | 400 | 57.40 | 5.86 | 1.59 | 14.35 | 6.76 | 4.29 | 32.85 |
| Medium | 200 | 375 | 50.20 | 8.25 | 2.78 | 13.39 | 5.98 | 3.34 | 33.74 |
| Fast | 225 | 350 | 43.10 | 9.78 | 3.95 | 12.31 | 4.82 | 1.85 | 32.71 |
| Mitt or glove 2nd or additional finish coats | | | | | | | | | |
| Slow | 225 | 425 | 57.40 | 4.56 | 1.23 | 13.51 | 5.99 | 3.80 | 29.09 |
| Medium | 250 | 400 | 50.20 | 6.60 | 2.22 | 12.55 | 5.24 | 2.93 | 29.54 |
| Fast | 275 | 375 | 43.10 | 8.00 | 3.25 | 11.49 | 4.21 | 1.62 | 28.57 |

For heights over 8 feet, use the High Time Difficulty Factors on page 137. Note: A two coat system, prime and finish, using oil base material is recommended for any metal surface. Although water base material is often used, it may cause oxidation, corrosion and rust. One coat of oil base solid body stain is often used on exterior metal but it may crack, peel or chip without the proper prime coat application. "Slow" work is based on a $10.25 hourly wage, "Medium" work on a $16.50 hourly wage, and "Fast" work on a $22.50 hourly wage. Other qualifications that apply to this table are on page 9.

| | Labor SF per manhour | Material coverage SF/gallon | Material cost per gallon | Labor cost per 100 SF | Labor burden 100 SF | Material cost per 100 SF | Overhead per 100 SF | Profit per 100 SF | Total price per 100 SF |
|---|---|---|---|---|---|---|---|---|---|
| **Ductwork, canvas insulated, brush application** | | | | | | | | | |
| Flat latex, water base (material #5) | | | | | | | | | |
| Brush 1st coat | | | | | | | | | |
| Slow | 60 | 250 | 16.20 | 17.08 | 4.66 | 6.48 | 8.75 | 5.54 | 42.51 |
| Medium | 75 | 238 | 14.20 | 22.00 | 7.39 | 5.97 | 8.67 | 4.84 | 48.87 |
| Fast | 90 | 225 | 12.10 | 24.44 | 9.91 | 5.38 | 7.35 | 2.82 | 49.90 |
| Brush 2nd coat | | | | | | | | | |
| Slow | 85 | 275 | 16.20 | 12.06 | 3.28 | 5.89 | 6.58 | 4.17 | 31.98 |
| Medium | 105 | 238 | 14.20 | 15.71 | 5.28 | 5.97 | 6.61 | 3.69 | 37.26 |
| Fast | 125 | 250 | 12.10 | 17.60 | 7.14 | 4.84 | 5.47 | 2.10 | 37.15 |
| Brush 3rd or additional coats | | | | | | | | | |
| Slow | 100 | 325 | 16.20 | 10.25 | 2.79 | 4.98 | 5.59 | 3.54 | 27.15 |
| Medium | 125 | 313 | 14.20 | 13.20 | 4.44 | 4.54 | 5.43 | 3.04 | 30.65 |
| Fast | 150 | 300 | 12.10 | 14.67 | 5.95 | 4.03 | 4.56 | 1.75 | 30.96 |
| Sealer, off white, water base (material #1) | | | | | | | | | |
| Brush 1 coat | | | | | | | | | |
| Slow | 60 | 250 | 15.50 | 17.08 | 4.66 | 6.20 | 8.66 | 5.49 | 42.09 |
| Medium | 75 | 238 | 13.50 | 22.00 | 7.39 | 5.67 | 8.59 | 4.80 | 48.45 |
| Fast | 90 | 225 | 11.60 | 24.44 | 9.91 | 5.16 | 7.31 | 2.81 | 49.63 |
| Sealer, off white, oil base (material #2) | | | | | | | | | |
| Brush 1 coat | | | | | | | | | |
| Slow | 60 | 275 | 20.70 | 17.08 | 4.66 | 7.53 | 9.07 | 5.75 | 44.09 |
| Medium | 75 | 263 | 18.10 | 22.00 | 7.39 | 6.88 | 8.89 | 4.97 | 50.13 |
| Fast | 90 | 250 | 15.50 | 24.44 | 9.91 | 6.20 | 7.50 | 2.88 | 50.93 |
| Enamel, water base (material #9) | | | | | | | | | |
| Brush 1st finish coat | | | | | | | | | |
| Slow | 85 | 275 | 22.80 | 12.06 | 3.28 | 8.29 | 7.33 | 4.65 | 35.61 |
| Medium | 105 | 238 | 19.90 | 15.71 | 5.28 | 8.36 | 7.19 | 4.02 | 40.56 |
| Fast | 125 | 250 | 17.10 | 17.60 | 7.14 | 6.84 | 5.84 | 2.25 | 39.67 |
| Brush 2nd or additional finish coats | | | | | | | | | |
| Slow | 100 | 325 | 22.80 | 10.25 | 2.79 | 7.02 | 6.22 | 3.94 | 30.22 |
| Medium | 125 | 313 | 19.90 | 13.20 | 4.44 | 6.36 | 5.88 | 3.29 | 33.17 |
| Fast | 150 | 300 | 17.10 | 14.67 | 5.95 | 5.70 | 4.87 | 1.87 | 33.06 |
| Industrial enamel, oil base, high gloss - light colors (material #56) | | | | | | | | | |
| Brush 1st finish coat | | | | | | | | | |
| Slow | 85 | 300 | 28.90 | 12.06 | 3.28 | 9.63 | 7.74 | 4.91 | 37.62 |
| Medium | 105 | 288 | 25.20 | 15.71 | 5.28 | 8.75 | 7.29 | 4.07 | 41.10 |
| Fast | 125 | 275 | 21.60 | 17.60 | 7.14 | 7.85 | 6.03 | 2.32 | 40.94 |

| | Labor SF per manhour | Material coverage SF/gallon | Material cost per gallon | Labor cost per 100 SF | Labor burden 100 SF | Material cost per 100 SF | Overhead per 100 SF | Profit per 100 SF | Total price per 100 SF |
|---|---|---|---|---|---|---|---|---|---|
| Brush 2nd or additional finish coats | | | | | | | | | |
| Slow | 100 | 350 | 28.90 | 10.25 | 2.79 | 8.26 | 6.60 | 4.19 | 32.09 |
| Medium | 125 | 338 | 25.20 | 13.20 | 4.44 | 7.46 | 6.15 | 3.44 | 34.69 |
| Fast | 150 | 325 | 21.60 | 14.67 | 5.95 | 6.65 | 5.04 | 1.94 | 34.25 |
| Industrial enamel, oil base, high gloss - dark (OSHA) colors (material #57) | | | | | | | | | |
| Brush 1st finish coat | | | | | | | | | |
| Slow | 85 | 300 | 34.90 | 12.06 | 3.28 | 11.63 | 8.36 | 5.30 | 40.63 |
| Medium | 105 | 288 | 30.50 | 15.71 | 5.28 | 10.59 | 7.74 | 4.33 | 43.65 |
| Fast | 125 | 275 | 26.10 | 17.60 | 7.14 | 9.49 | 6.33 | 2.43 | 42.99 |
| Brush 2nd or additional finish coats | | | | | | | | | |
| Slow | 100 | 350 | 34.90 | 10.25 | 2.79 | 9.97 | 7.13 | 4.52 | 34.66 |
| Medium | 125 | 338 | 30.50 | 13.20 | 4.44 | 9.02 | 6.53 | 3.65 | 36.84 |
| Fast | 150 | 325 | 26.10 | 14.67 | 5.95 | 8.03 | 5.30 | 2.04 | 35.99 |
| Epoxy coating, 2 part system - clear (material #51) | | | | | | | | | |
| Brush 1st coat | | | | | | | | | |
| Slow | 60 | 260 | 57.00 | 17.08 | 4.66 | 21.92 | 13.53 | 8.58 | 65.77 |
| Medium | 75 | 243 | 49.90 | 22.00 | 7.39 | 20.53 | 12.23 | 6.84 | 68.99 |
| Fast | 90 | 235 | 42.80 | 24.44 | 9.91 | 18.21 | 9.72 | 3.74 | 66.02 |
| Brush 2nd coat | | | | | | | | | |
| Slow | 85 | 285 | 57.00 | 12.06 | 3.28 | 20.00 | 10.96 | 6.95 | 53.25 |
| Medium | 105 | 273 | 49.90 | 15.71 | 5.28 | 18.28 | 9.62 | 5.38 | 54.27 |
| Fast | 125 | 260 | 42.80 | 17.60 | 7.14 | 16.46 | 7.62 | 2.93 | 51.75 |
| Brush 3rd or additional coats | | | | | | | | | |
| Slow | 100 | 335 | 57.00 | 10.25 | 2.79 | 17.01 | 9.32 | 5.91 | 45.28 |
| Medium | 125 | 323 | 49.90 | 13.20 | 4.44 | 15.45 | 8.11 | 4.53 | 45.73 |
| Fast | 150 | 310 | 42.80 | 14.67 | 5.95 | 13.81 | 6.37 | 2.45 | 43.25 |

| | Labor SF per manhour | Material coverage SF/gallon | Material cost per gallon | Labor cost per 100 SF | Labor burden 100 SF | Material cost per 100 SF | Overhead per 100 SF | Profit per 100 SF | Total price per 100 SF |
|---|---|---|---|---|---|---|---|---|---|
| Epoxy coating, 2 part system - white (material #52) | | | | | | | | | |
| Brush 1st coat | | | | | | | | | |
| Slow | 60 | 260 | 57.40 | 17.08 | 4.66 | 22.08 | 13.58 | 8.61 | 66.01 |
| Medium | 75 | 243 | 50.20 | 22.00 | 7.39 | 20.66 | 12.26 | 6.86 | 69.17 |
| Fast | 90 | 235 | 43.10 | 24.44 | 9.91 | 18.34 | 9.75 | 3.75 | 66.19 |
| Brush 2nd coat | | | | | | | | | |
| Slow | 85 | 285 | 57.40 | 12.06 | 3.28 | 20.14 | 11.00 | 6.97 | 53.45 |
| Medium | 100 | 273 | 50.20 | 16.50 | 5.55 | 18.39 | 9.91 | 5.54 | 55.89 |
| Fast | 125 | 260 | 43.10 | 17.60 | 7.14 | 16.58 | 7.64 | 2.94 | 51.90 |
| Brush 3rd or additional coats | | | | | | | | | |
| Slow | 100 | 335 | 57.40 | 10.25 | 2.79 | 17.13 | 9.35 | 5.93 | 45.45 |
| Medium | 125 | 323 | 50.20 | 13.20 | 4.44 | 15.54 | 8.13 | 4.54 | 45.85 |
| Fast | 150 | 310 | 43.10 | 14.67 | 5.95 | 13.90 | 6.39 | 2.45 | 43.36 |

For heights over 8 feet, use the High Time Difficulty Factors on page 137. "Slow" work is based on a $10.25 hourly wage, "Medium" work on a $16.50 hourly wage, and "Fast" work on a $22.00 hourly wage. Other qualifications that apply to this table are on page 9.

| | Labor SF per manhour | Material coverage SF/gallon | Material cost per gallon | Labor cost per 100 SF | Labor burden 100 SF | Material cost per 100 SF | Overhead per 100 SF | Profit per 100 SF | Total price per 100 SF |
|---|---|---|---|---|---|---|---|---|---|

## Ductwork, canvas insulated, roll application

| | Labor SF per manhour | Material coverage SF/gallon | Material cost per gallon | Labor cost per 100 SF | Labor burden 100 SF | Material cost per 100 SF | Overhead per 100 SF | Profit per 100 SF | Total price per 100 SF |
|---|---|---|---|---|---|---|---|---|---|
| Flat latex, water base (material #5) | | | | | | | | | |
| Roll 1st coat | | | | | | | | | |
| Slow | 125 | 250 | 16.20 | 8.20 | 2.23 | 6.48 | 5.24 | 3.32 | 25.47 |
| Medium | 150 | 238 | 14.20 | 11.00 | 3.71 | 5.97 | 5.06 | 2.83 | 28.57 |
| Fast | 175 | 225 | 12.10 | 12.57 | 5.09 | 5.38 | 4.26 | 1.64 | 28.94 |
| Roll 2nd coat | | | | | | | | | |
| Slow | 175 | 275 | 16.20 | 5.86 | 1.59 | 5.89 | 4.14 | 2.62 | 20.10 |
| Medium | 200 | 263 | 14.20 | 8.25 | 2.78 | 5.40 | 4.03 | 2.25 | 22.71 |
| Fast | 225 | 250 | 12.10 | 9.78 | 3.95 | 4.84 | 3.44 | 1.32 | 23.33 |
| Roll 3rd or additional coats | | | | | | | | | |
| Slow | 225 | 400 | 16.20 | 4.56 | 1.23 | 4.05 | 3.05 | 1.94 | 14.83 |
| Medium | 250 | 388 | 14.20 | 6.60 | 2.22 | 3.66 | 3.06 | 1.71 | 17.25 |
| Fast | 275 | 375 | 12.10 | 8.00 | 3.25 | 3.23 | 2.68 | 1.03 | 18.19 |
| Sealer, off white, water base (material #1) | | | | | | | | | |
| Roll 1 coat | | | | | | | | | |
| Slow | 125 | 250 | 15.50 | 8.20 | 2.23 | 6.20 | 5.16 | 3.27 | 25.06 |
| Medium | 150 | 238 | 13.50 | 11.00 | 3.71 | 5.67 | 4.99 | 2.79 | 28.16 |
| Fast | 175 | 225 | 11.60 | 12.57 | 5.09 | 5.16 | 4.22 | 1.62 | 28.66 |
| Sealer, off white, oil base (material #2) | | | | | | | | | |
| Roll 1 coat | | | | | | | | | |
| Slow | 125 | 275 | 20.70 | 8.20 | 2.23 | 7.53 | 5.57 | 3.53 | 27.06 |
| Medium | 150 | 263 | 18.10 | 11.00 | 3.71 | 6.88 | 5.29 | 2.96 | 29.84 |
| Fast | 175 | 250 | 15.50 | 12.57 | 5.09 | 6.20 | 4.42 | 1.70 | 29.98 |
| Enamel, water base (material #9) | | | | | | | | | |
| Roll 1st finish coat | | | | | | | | | |
| Slow | 175 | 350 | 22.80 | 5.86 | 1.59 | 6.51 | 4.33 | 2.75 | 21.04 |
| Medium | 200 | 325 | 19.90 | 8.25 | 2.78 | 6.12 | 4.20 | 2.35 | 23.70 |
| Fast | 225 | 300 | 17.10 | 9.78 | 3.95 | 5.70 | 3.60 | 1.38 | 24.41 |
| Roll 2nd or additional finish coats | | | | | | | | | |
| Slow | 225 | 400 | 22.80 | 4.56 | 1.23 | 5.70 | 3.57 | 2.26 | 17.32 |
| Medium | 250 | 388 | 19.90 | 6.60 | 2.22 | 5.13 | 3.42 | 1.91 | 19.28 |
| Fast | 275 | 375 | 17.10 | 8.00 | 3.25 | 4.56 | 2.92 | 1.12 | 19.85 |

| | Labor SF per manhour | Material coverage SF/gallon | Material cost per gallon | Labor cost per 100 SF | Labor burden 100 SF | Material cost per 100 SF | Overhead per 100 SF | Profit per 100 SF | Total price per 100 SF |
|---|---|---|---|---|---|---|---|---|---|
| **Industrial enamel, oil base, high gloss - light colors (material #56)** | | | | | | | | | |
| Roll 1st finish coat | | | | | | | | | |
| Slow | 175 | 300 | 28.90 | 5.86 | 1.59 | 9.63 | 5.30 | 3.36 | 25.74 |
| Medium | 200 | 288 | 25.20 | 8.25 | 2.78 | 8.75 | 4.85 | 2.71 | 27.34 |
| Fast | 225 | 275 | 21.60 | 9.78 | 3.95 | 7.85 | 4.00 | 1.54 | 27.12 |
| Roll 2nd or additional finish coats | | | | | | | | | |
| Slow | 225 | 450 | 28.90 | 4.56 | 1.23 | 6.42 | 3.79 | 2.40 | 18.40 |
| Medium | 250 | 438 | 25.20 | 6.60 | 2.22 | 5.75 | 3.57 | 2.00 | 20.14 |
| Fast | 275 | 425 | 21.60 | 8.00 | 3.25 | 5.08 | 3.02 | 1.16 | 20.51 |
| **Industrial enamel, oil base, high gloss - dark (OSHA) colors (material #57)** | | | | | | | | | |
| Roll 1st finish coat | | | | | | | | | |
| Slow | 175 | 300 | 34.90 | 5.86 | 1.59 | 11.63 | 5.92 | 3.75 | 28.75 |
| Medium | 200 | 288 | 30.50 | 8.25 | 2.78 | 10.59 | 5.30 | 2.96 | 29.88 |
| Fast | 225 | 275 | 26.10 | 9.78 | 3.95 | 9.49 | 4.30 | 1.65 | 29.17 |
| Roll 2nd or additional finish coats | | | | | | | | | |
| Slow | 225 | 450 | 34.90 | 4.56 | 1.23 | 7.76 | 4.20 | 2.66 | 20.41 |
| Medium | 250 | 438 | 30.50 | 6.60 | 2.22 | 6.96 | 3.87 | 2.16 | 21.81 |
| Fast | 275 | 425 | 26.10 | 8.00 | 3.25 | 6.14 | 3.22 | 1.24 | 21.85 |
| **Epoxy coating, 2 part system - clear (material #51)** | | | | | | | | | |
| Roll 1st coat | | | | | | | | | |
| Slow | 125 | 250 | 57.00 | 8.20 | 2.23 | 22.80 | 10.30 | 6.53 | 50.06 |
| Medium | 150 | 238 | 49.90 | 11.00 | 3.71 | 20.97 | 8.74 | 4.89 | 49.31 |
| Fast | 175 | 225 | 42.80 | 12.57 | 5.09 | 19.02 | 6.79 | 2.61 | 46.08 |
| Roll 2nd coat | | | | | | | | | |
| Slow | 175 | 375 | 57.00 | 5.86 | 1.59 | 15.20 | 7.02 | 4.45 | 34.12 |
| Medium | 200 | 338 | 49.90 | 8.25 | 2.78 | 14.76 | 6.32 | 3.53 | 35.64 |
| Fast | 225 | 300 | 42.80 | 9.78 | 3.95 | 14.27 | 5.18 | 1.99 | 35.17 |
| Roll 3rd or additional coats | | | | | | | | | |
| Slow | 225 | 425 | 57.00 | 4.56 | 1.23 | 13.41 | 5.96 | 3.78 | 28.94 |
| Medium | 250 | 413 | 49.90 | 6.60 | 2.22 | 12.08 | 5.12 | 2.86 | 28.88 |
| Fast | 275 | 400 | 42.80 | 8.00 | 3.25 | 10.70 | 4.06 | 1.56 | 27.57 |

| | Labor SF per manhour | Material coverage SF/gallon | Material cost per gallon | Labor cost per 100 SF | Labor burden 100 SF | Material cost per 100 SF | Overhead per 100 SF | Profit per 100 SF | Total price per 100 SF |
|---|---|---|---|---|---|---|---|---|---|
| **Epoxy coating, 2 part system - white (material #52)** | | | | | | | | | |
| Roll 1st coat | | | | | | | | | |
| Slow | 125 | 250 | 57.40 | 8.20 | 2.23 | 22.96 | 10.35 | 6.56 | 50.30 |
| Medium | 150 | 238 | 50.20 | 11.00 | 3.71 | 21.09 | 8.77 | 4.90 | 49.47 |
| Fast | 175 | 225 | 43.10 | 12.57 | 5.09 | 19.16 | 6.81 | 2.62 | 46.25 |
| Roll 2nd coat | | | | | | | | | |
| Slow | 175 | 375 | 57.40 | 5.86 | 1.59 | 15.31 | 7.06 | 4.47 | 34.29 |
| Medium | 200 | 338 | 50.20 | 8.25 | 2.78 | 14.85 | 6.34 | 3.54 | 35.76 |
| Fast | 225 | 300 | 43.10 | 9.78 | 3.95 | 14.37 | 5.20 | 2.00 | 35.30 |
| Roll 3rd or additional coats | | | | | | | | | |
| Slow | 225 | 425 | 57.40 | 4.56 | 1.23 | 13.51 | 5.99 | 3.80 | 29.09 |
| Medium | 250 | 413 | 50.20 | 6.60 | 2.22 | 12.15 | 5.14 | 2.87 | 28.98 |
| Fast | 275 | 400 | 43.10 | 8.00 | 3.25 | 10.78 | 4.07 | 1.57 | 27.67 |

For heights over 8 feet, use the High Time Difficulty Factors on page 137. "Slow" work is based on a $10.25 hourly wage, "Medium" work on a $16.50 hourly wage, and "Fast" work on a $22.00 hourly wage. Other qualifications that apply to this table are on page 9.

| | Labor SF per manhour | Material coverage SF/gallon | Material cost per gallon | Labor cost per 100 SF | Labor burden 100 SF | Material cost per 100 SF | Overhead per 100 SF | Profit per 100 SF | Total price per 100 SF |
|---|---|---|---|---|---|---|---|---|---|
| **Ductwork, canvas insulated, spray application** | | | | | | | | | |
| Flat latex, water base (material #5) | | | | | | | | | |
| Spray prime coat | | | | | | | | | |
| Slow | 450 | 200 | 16.20 | 2.28 | .61 | 8.10 | 3.41 | 2.16 | 16.56 |
| Medium | 500 | 188 | 14.20 | 3.30 | 1.11 | 7.55 | 2.93 | 1.64 | 16.53 |
| Fast | 550 | 175 | 12.10 | 4.00 | 1.63 | 6.91 | 2.32 | .89 | 15.75 |
| Spray 1st finish coat | | | | | | | | | |
| Slow | 550 | 225 | 16.20 | 1.86 | .51 | 7.20 | 2.97 | 1.88 | 14.42 |
| Medium | 625 | 213 | 14.20 | 2.64 | .89 | 6.67 | 2.50 | 1.40 | 14.10 |
| Fast | 700 | 200 | 12.10 | 3.14 | 1.28 | 6.05 | 1.94 | .74 | 13.15 |
| Spray 2nd or additional finish coats | | | | | | | | | |
| Slow | 700 | 250 | 16.20 | 1.46 | .40 | 6.48 | 2.59 | 1.64 | 12.57 |
| Medium | 750 | 238 | 14.20 | 2.20 | .73 | 5.97 | 2.18 | 1.22 | 12.30 |
| Fast | 800 | 225 | 12.10 | 2.75 | 1.12 | 5.38 | 1.71 | .66 | 11.62 |
| Sealer, off white, water base (material #1) | | | | | | | | | |
| Spray 1 coat | | | | | | | | | |
| Slow | 450 | 200 | 15.50 | 2.28 | .61 | 7.75 | 3.30 | 2.09 | 16.03 |
| Medium | 500 | 188 | 13.50 | 3.30 | 1.11 | 7.18 | 2.84 | 1.59 | 16.02 |
| Fast | 550 | 175 | 11.60 | 4.00 | 1.63 | 6.63 | 2.27 | .87 | 15.40 |
| Sealer, off white, oil base (material #2) | | | | | | | | | |
| Spray 1 coat | | | | | | | | | |
| Slow | 450 | 225 | 20.70 | 2.28 | .61 | 9.20 | 3.75 | 2.38 | 18.22 |
| Medium | 500 | 213 | 18.10 | 3.30 | 1.11 | 8.50 | 3.16 | 1.77 | 17.84 |
| Fast | 550 | 200 | 15.50 | 4.00 | 1.63 | 7.75 | 2.47 | .95 | 16.80 |
| Enamel, water base (material #9) | | | | | | | | | |
| Spray 1st finish coat | | | | | | | | | |
| Slow | 550 | 225 | 22.80 | 1.86 | .51 | 10.13 | 3.88 | 2.46 | 18.84 |
| Medium | 625 | 213 | 19.90 | 2.64 | .89 | 9.34 | 3.15 | 1.76 | 17.78 |
| Fast | 700 | 200 | 17.10 | 3.14 | 1.28 | 8.55 | 2.40 | .92 | 16.29 |
| Spray 2nd or additional finish coats | | | | | | | | | |
| Slow | 700 | 250 | 22.80 | 1.46 | .40 | 9.12 | 3.40 | 2.16 | 16.54 |
| Medium | 750 | 238 | 19.90 | 2.20 | .73 | 8.36 | 2.77 | 1.55 | 15.61 |
| Fast | 800 | 225 | 17.10 | 2.75 | 1.12 | 7.60 | 2.12 | .82 | 14.41 |

| | Labor SF per manhour | Material coverage SF/gallon | Material cost per gallon | Labor cost per 100 SF | Labor burden 100 SF | Material cost per 100 SF | Overhead per 100 SF | Profit per 100 SF | Total price per 100 SF |
|---|---|---|---|---|---|---|---|---|---|
| Industrial enamel, oil base, high gloss - light colors (material #56) | | | | | | | | | |
| Spray 1st finish coat | | | | | | | | | |
| Slow | 700 | 250 | 28.90 | 1.46 | .40 | 11.56 | 4.16 | 2.64 | 20.22 |
| Medium | 625 | 238 | 25.20 | 2.64 | .89 | 10.59 | 3.46 | 1.93 | 19.51 |
| Fast | 550 | 225 | 21.60 | 4.00 | 1.63 | 9.60 | 2.82 | 1.08 | 19.13 |
| Spray 2nd or additional finish coats | | | | | | | | | |
| Slow | 800 | 275 | 28.90 | 1.28 | .35 | 10.51 | 3.76 | 2.39 | 18.29 |
| Medium | 750 | 263 | 25.20 | 2.20 | .73 | 9.58 | 3.07 | 1.71 | 17.29 |
| Fast | 700 | 250 | 21.60 | 3.14 | 1.28 | 8.64 | 2.41 | .93 | 16.40 |
| Industrial enamel, oil base, high gloss - dark (OSHA) colors (material #57) | | | | | | | | | |
| Spray 1st finish coat | | | | | | | | | |
| Slow | 700 | 250 | 34.90 | 1.46 | .40 | 13.96 | 4.90 | 3.11 | 23.83 |
| Medium | 625 | 238 | 30.50 | 2.64 | .89 | 12.82 | 4.01 | 2.24 | 22.60 |
| Fast | 550 | 225 | 26.10 | 4.00 | 1.63 | 11.60 | 3.19 | 1.22 | 21.64 |
| Spray 2nd or additional finish coats | | | | | | | | | |
| Slow | 800 | 275 | 34.90 | 1.28 | .35 | 12.69 | 4.44 | 2.81 | 21.57 |
| Medium | 750 | 263 | 30.50 | 2.20 | .73 | 11.60 | 3.56 | 1.99 | 20.08 |
| Fast | 700 | 250 | 26.10 | 3.14 | 1.28 | 10.44 | 2.75 | 1.06 | 18.67 |
| Epoxy coating, 2 part system - clear (material #51) | | | | | | | | | |
| Spray prime coat | | | | | | | | | |
| Slow | 450 | 200 | 57.00 | 2.28 | .61 | 28.50 | 9.73 | 6.17 | 47.29 |
| Medium | 500 | 188 | 49.90 | 3.30 | 1.11 | 26.54 | 7.58 | 4.24 | 42.77 |
| Fast | 550 | 175 | 42.80 | 4.00 | 1.63 | 24.46 | 5.56 | 2.14 | 37.79 |
| Spray 1st finish coat | | | | | | | | | |
| Slow | 550 | 235 | 57.00 | 1.86 | .51 | 24.26 | 8.26 | 5.23 | 40.12 |
| Medium | 625 | 225 | 49.90 | 2.64 | .89 | 22.18 | 6.30 | 3.52 | 35.53 |
| Fast | 700 | 215 | 42.80 | 3.14 | 1.28 | 19.91 | 4.50 | 1.73 | 30.56 |
| Spray 2nd or additional finish coats | | | | | | | | | |
| Slow | 700 | 265 | 57.00 | 1.46 | .40 | 21.51 | 7.24 | 4.59 | 35.20 |
| Medium | 750 | 253 | 49.90 | 2.20 | .73 | 19.72 | 5.55 | 3.10 | 31.30 |
| Fast | 800 | 240 | 42.80 | 2.75 | 1.12 | 17.83 | 4.01 | 1.54 | 27.25 |

| | Labor SF per manhour | Material coverage SF/gallon | Material cost per gallon | Labor cost per 100 SF | Labor burden 100 SF | Material cost per 100 SF | Overhead per 100 SF | Profit per 100 SF | Total price per 100 SF |
|---|---|---|---|---|---|---|---|---|---|
| Epoxy coating, 2 part system - white (material #52) | | | | | | | | | |
| Spray prime coat | | | | | | | | | |
| Slow | 450 | 200 | 57.40 | 2.28 | .61 | 28.70 | 9.80 | 6.21 | 47.60 |
| Medium | 500 | 188 | 50.20 | 3.30 | 1.11 | 26.70 | 7.62 | 4.26 | 42.99 |
| Fast | 550 | 175 | 43.10 | 4.00 | 1.63 | 24.63 | 5.60 | 2.15 | 38.01 |
| Spray 1st finish coat | | | | | | | | | |
| Slow | 550 | 235 | 57.40 | 1.86 | .51 | 24.43 | 8.31 | 5.27 | 40.38 |
| Medium | 625 | 225 | 50.20 | 2.64 | .89 | 22.31 | 6.33 | 3.54 | 35.71 |
| Fast | 700 | 215 | 43.10 | 3.14 | 1.28 | 20.05 | 4.53 | 1.74 | 30.74 |
| Spray 2nd or additional finish coats | | | | | | | | | |
| Slow | 700 | 265 | 57.40 | 1.46 | .40 | 21.66 | 7.29 | 4.62 | 35.43 |
| Medium | 750 | 253 | 50.20 | 2.20 | .73 | 19.84 | 5.58 | 3.12 | 31.47 |
| Fast | 800 | 240 | 43.10 | 2.75 | 1.12 | 17.96 | 4.04 | 1.55 | 27.42 |

For heights over 8 feet, use the High Time Difficulty Factors on page 137. "Slow" work is based on a $10.25 hourly wage, "Medium" work on a $16.50 hourly wage, and "Fast" work on a $22.00 hourly wage. Other qualifications that apply to this table are on page 9.

| | Manhours per flight | Flights per gallon | Material cost per gallon | Labor cost per flight | Labor burden flight | Material cost per flight | Overhead per flight | Profit per flight | Total price per flight |
|---|---|---|---|---|---|---|---|---|---|
| **Fire escapes** | | | | | | | | | |
| Solid (plain) deck | | | | | | | | | |
| Spray each coat | | | | | | | | | |
| Metal primer, rust inhibitor - clean metal (material #35) | | | | | | | | | |
| Slow | 2.0 | 1.25 | 28.20 | 20.50 | 5.58 | 22.56 | 15.08 | 9.56 | 73.28 |
| Medium | 1.5 | 1.00 | 24.60 | 24.75 | 8.33 | 24.60 | 14.13 | 7.90 | 79.71 |
| Fast | 1.0 | 0.75 | 21.10 | 22.00 | 8.92 | 28.13 | 10.92 | 4.20 | 74.17 |
| Metal primer, rust inhibitor - rusty metal (material #36) | | | | | | | | | |
| Slow | 2.0 | 1.25 | 33.30 | 20.50 | 5.58 | 26.64 | 16.35 | 10.36 | 79.43 |
| Medium | 1.5 | 1.00 | 29.20 | 24.75 | 8.33 | 29.20 | 15.26 | 8.53 | 86.07 |
| Fast | 1.0 | 0.75 | 25.00 | 22.00 | 8.92 | 33.33 | 11.89 | 4.57 | 80.71 |
| Industrial enamel, oil base, high gloss - light colors (material #56) | | | | | | | | | |
| Slow | 2.25 | 1.50 | 28.90 | 23.06 | 6.28 | 19.27 | 15.07 | 9.55 | 73.23 |
| Medium | 1.75 | 1.25 | 25.20 | 28.88 | 9.71 | 20.16 | 14.40 | 8.05 | 81.20 |
| Fast | 1.25 | 1.00 | 21.60 | 27.50 | 11.15 | 21.60 | 11.15 | 4.28 | 75.68 |
| Industrial enamel, oil base, high gloss - dark (OSHA) colors (material #57) | | | | | | | | | |
| Slow | 2.25 | 1.50 | 34.90 | 23.06 | 6.28 | 23.27 | 16.31 | 10.34 | 79.26 |
| Medium | 1.75 | 1.25 | 30.50 | 28.88 | 9.71 | 24.40 | 15.44 | 8.63 | 87.06 |
| Fast | 1.25 | 1.00 | 26.10 | 27.50 | 11.15 | 26.10 | 11.98 | 4.60 | 81.33 |
| Grating deck | | | | | | | | | |
| Spray each coat | | | | | | | | | |
| Metal primer, rust inhibitor - clean metal (material #35) | | | | | | | | | |
| Slow | 3.0 | 1.75 | 28.20 | 30.75 | 8.37 | 16.11 | 17.12 | 10.85 | 83.20 |
| Medium | 2.5 | 1.50 | 24.60 | 41.25 | 13.88 | 16.40 | 17.52 | 9.80 | 98.85 |
| Fast | 2.0 | 1.25 | 21.10 | 44.00 | 17.84 | 16.88 | 14.56 | 5.60 | 98.88 |
| Metal primer, rust inhibitor - rusty metal (material #36) | | | | | | | | | |
| Slow | 3.0 | 1.75 | 33.30 | 30.75 | 8.37 | 19.03 | 18.03 | 11.43 | 87.61 |
| Medium | 2.5 | 1.50 | 29.20 | 41.25 | 13.88 | 19.47 | 18.28 | 10.22 | 103.10 |
| Fast | 2.0 | 1.25 | 25.00 | 44.00 | 17.84 | 20.00 | 15.14 | 5.82 | 102.80 |

| | Manhours per flight | Flights per gallon | Material cost per gallon | Labor cost per flight | Labor burden flight | Material cost per flight | Overhead per flight | Profit per flight | Total price per flight |
|---|---|---|---|---|---|---|---|---|---|
| Industrial enamel, oil base, high gloss - light colors (material #56) | | | | | | | | | |
| Slow | 3.25 | 2.00 | 28.90 | 33.31 | 9.07 | 14.45 | 17.62 | 11.17 | 85.62 |
| Medium | 2.75 | 1.75 | 25.20 | 45.38 | 15.26 | 14.40 | 18.39 | 10.28 | 103.71 |
| Fast | 2.25 | 1.50 | 21.60 | 49.50 | 20.07 | 14.40 | 15.53 | 5.97 | 105.47 |
| Industrial enamel, oil base, high gloss - dark (OSHA) colors (material #57) | | | | | | | | | |
| Slow | 3.25 | 2.00 | 34.90 | 33.31 | 9.07 | 17.45 | 18.55 | 11.76 | 90.14 |
| Medium | 2.75 | 1.75 | 30.50 | 45.38 | 15.26 | 17.43 | 19.13 | 10.69 | 107.89 |
| Fast | 2.25 | 1.50 | 26.10 | 49.50 | 20.07 | 17.40 | 16.09 | 6.18 | 109.24 |

Fire escapes can also be estimated by the square foot. Calculate the actual area to be coated. For continuous solid (plain) deck, use the rates listed under Decking and siding. For continuous grating deck, use the rates listed under Grates, steel. Note: A two coat system, prime and finish, using oil base material is recommended for any metal surface. Although water base material is often used, it may cause oxidation, corrosion and rust. One coat of oil base solid body stain is often used on exterior metal but it may crack, peel or chip without the proper prime coat application. "Slow" work is based on a $10.25 hourly wage, "Medium" work on a $16.50 hourly wage, and "Fast" work on a $22.00 hourly wage. Other qualifications that apply to this table are on page 9.

## Fire sprinkler systems

Use the costs listed for 1" to 4" pipe. For painting sprinkler heads at 12 feet on center at a ceiling height of 12 feet, figure 3 minutes per head (20 per hour). Very little paint is needed. Your material estimate for the sprinkler pipe will include enough to cover the heads. Note: A two coat system, prime and finish, using oil base material is recommended for any metal surface. Although water base material is often used, it may cause oxidation, corrosion and rust. One coat of oil base solid body stain is often used on exterior metal but it may crack, peel or chip without the proper prime coat application.

| | Labor SF per manhour | Material coverage SF/gallon | Material cost per gallon | Labor cost per 100 SF | Labor burden 100 SF | Material cost per 100 SF | Overhead per 100 SF | Profit per 100 SF | Total price per 100 SF |
|---|---|---|---|---|---|---|---|---|---|
| **Grates, steel, over 1" thick** | | | | | | | | | |
| Without supports | | | | | | | | | |
| Brush each coat | | | | | | | | | |
| Metal primer, rust inhibitor - clean metal (material #35) | | | | | | | | | |
| Slow | 60 | 175 | 28.20 | 17.08 | 4.66 | .16 | 6.79 | 4.30 | 32.99 |
| Medium | 85 | 150 | 24.60 | 19.41 | 6.52 | .16 | 6.39 | 3.57 | 36.05 |
| Fast | 110 | 125 | 21.10 | 20.00 | 8.11 | .17 | 5.23 | 2.01 | 35.52 |
| Metal primer, rust inhibitor - rusty metal (material #36) | | | | | | | | | |
| Slow | 60 | 175 | 33.30 | 17.08 | 4.66 | .19 | 6.80 | 4.31 | 33.04 |
| Medium | 85 | 150 | 29.20 | 19.41 | 6.52 | .19 | 6.40 | 3.58 | 36.10 |
| Fast | 110 | 125 | 25.00 | 20.00 | 8.11 | .20 | 5.24 | 2.01 | 35.56 |
| Industrial enamel, oil base, high gloss - light colors (material #56) | | | | | | | | | |
| Slow | 75 | 225 | 28.90 | 13.67 | 3.71 | .13 | 5.43 | 3.44 | 26.38 |
| Medium | 100 | 200 | 25.20 | 16.50 | 5.55 | .13 | 5.43 | 3.04 | 30.65 |
| Fast | 125 | 175 | 21.60 | 17.60 | 7.14 | .12 | 4.60 | 1.77 | 31.23 |
| Industrial enamel, oil base, high gloss - dark (OSHA) colors (material #57) | | | | | | | | | |
| Slow | 75 | 225 | 34.90 | 13.67 | 3.71 | .16 | 5.44 | 3.45 | 26.43 |
| Medium | 100 | 200 | 30.50 | 16.50 | 5.55 | .15 | 5.44 | 3.04 | 30.68 |
| Fast | 125 | 175 | 26.10 | 17.60 | 7.14 | .15 | 4.60 | 1.77 | 31.26 |
| Spray each coat | | | | | | | | | |
| Metal primer, rust inhibitor - clean metal (material #35) | | | | | | | | | |
| Slow | 190 | 125 | 28.20 | 5.39 | 1.47 | .23 | 2.20 | 1.39 | 10.68 |
| Medium | 208 | 113 | 24.60 | 7.93 | 2.68 | .22 | 2.65 | 1.48 | 14.96 |
| Fast | 225 | 100 | 21.10 | 9.78 | 3.95 | .21 | 2.58 | .99 | 17.51 |
| Metal primer, rust inhibitor - rusty metal (material #36) | | | | | | | | | |
| Slow | 190 | 125 | 33.30 | 5.39 | 1.47 | .27 | 2.21 | 1.40 | 10.74 |
| Medium | 208 | 113 | 29.20 | 7.93 | 2.68 | .26 | 2.66 | 1.49 | 15.02 |
| Fast | 225 | 100 | 25.00 | 9.78 | 3.95 | .25 | 2.59 | 1.00 | 17.57 |
| Industrial enamel, oil base, high gloss - light colors (material #56) | | | | | | | | | |
| Slow | 215 | 160 | 28.90 | 4.77 | 1.29 | .18 | 1.94 | 1.23 | 9.41 |
| Medium | 233 | 148 | 25.20 | 7.08 | 2.38 | .17 | 2.36 | 1.32 | 13.31 |
| Fast | 250 | 135 | 21.60 | 8.80 | 3.57 | .16 | 2.32 | .89 | 15.74 |
| Industrial enamel, oil base, high gloss - dark (OSHA) colors (material #57) | | | | | | | | | |
| Slow | 215 | 160 | 34.90 | 4.77 | 1.29 | .22 | 1.95 | 1.24 | 9.47 |
| Medium | 233 | 148 | 30.50 | 7.08 | 2.38 | .21 | 2.37 | 1.32 | 13.36 |
| Fast | 250 | 135 | 26.10 | 8.80 | 3.57 | .19 | 2.32 | .89 | 15.77 |

| | Labor SF per manhour | Material coverage SF/gallon | Material cost per gallon | Labor cost per 100 SF | Labor burden 100 SF | Material cost per 100 SF | Overhead per 100 SF | Profit per 100 SF | Total price per 100 SF |
|---|---|---|---|---|---|---|---|---|---|
| Including typical supports | | | | | | | | | |
| Brush each coat | | | | | | | | | |
| Metal primer, rust inhibitor - clean metal (material #35) | | | | | | | | | |
| Slow | 40 | 125 | 28.20 | 25.63 | 6.97 | .23 | 10.18 | 6.45 | 49.46 |
| Medium | 55 | 113 | 24.60 | 30.00 | 10.09 | .22 | 9.88 | 5.52 | 55.71 |
| Fast | 70 | 100 | 21.10 | 31.43 | 12.75 | .21 | 8.21 | 3.16 | 55.76 |
| Metal primer, rust inhibitor - rusty metal (material #36) | | | | | | | | | |
| Slow | 40 | 125 | 33.30 | 25.63 | 6.97 | .27 | 10.19 | 6.46 | 49.52 |
| Medium | 55 | 113 | 29.20 | 30.00 | 10.09 | .26 | 9.89 | 5.53 | 55.77 |
| Fast | 70 | 100 | 25.00 | 31.43 | 12.75 | .25 | 8.22 | 3.16 | 55.81 |
| Industrial enamel, oil base, high gloss - light colors (material #56) | | | | | | | | | |
| Slow | 50 | 160 | 28.90 | 20.50 | 5.58 | .18 | 8.14 | 5.16 | 39.56 |
| Medium | 75 | 148 | 25.20 | 22.00 | 7.39 | .17 | 7.24 | 4.05 | 40.85 |
| Fast | 100 | 135 | 21.60 | 22.00 | 8.92 | .16 | 5.75 | 2.21 | 39.04 |
| Industrial enamel, oil base, high gloss - dark (OSHA) colors (material #57) | | | | | | | | | |
| Slow | 50 | 160 | 34.90 | 20.50 | 5.58 | .22 | 8.16 | 5.17 | 39.63 |
| Medium | 75 | 148 | 30.50 | 22.00 | 7.39 | .21 | 7.25 | 4.05 | 40.90 |
| Fast | 100 | 135 | 26.10 | 22.00 | 8.92 | .19 | 5.76 | 2.21 | 39.08 |
| Spray each coat | | | | | | | | | |
| Metal primer, rust inhibitor - clean metal (material #35) | | | | | | | | | |
| Slow | 120 | 100 | 28.20 | 8.54 | 2.32 | .28 | 3.46 | 2.19 | 16.79 |
| Medium | 135 | 88 | 24.60 | 12.22 | 4.12 | .28 | 4.07 | 2.27 | 22.96 |
| Fast | 150 | 75 | 21.10 | 14.67 | 5.95 | .28 | 3.87 | 1.49 | 26.26 |
| Metal primer, rust inhibitor - rusty metal (material #36) | | | | | | | | | |
| Slow | 120 | 100 | 33.30 | 8.54 | 2.32 | .33 | 3.47 | 2.20 | 16.86 |
| Medium | 135 | 88 | 29.20 | 12.22 | 4.12 | .33 | 4.08 | 2.28 | 23.03 |
| Fast | 150 | 75 | 25.00 | 14.67 | 5.95 | .33 | 3.88 | 1.49 | 26.32 |
| Industrial enamel, oil base, high gloss - light colors (material #56) | | | | | | | | | |
| Slow | 150 | 125 | 28.90 | 6.83 | 1.87 | .23 | 2.77 | 1.75 | 13.45 |
| Medium | 163 | 113 | 25.20 | 10.12 | 3.40 | .22 | 3.37 | 1.88 | 18.99 |
| Fast | 175 | 100 | 21.60 | 12.57 | 5.09 | .22 | 3.31 | 1.27 | 22.46 |
| Industrial enamel, oil base, high gloss - dark (OSHA) colors (material #57) | | | | | | | | | |
| Slow | 150 | 125 | 34.90 | 6.83 | 1.87 | .28 | 2.78 | 1.76 | 13.52 |
| Medium | 163 | 113 | 30.50 | 10.12 | 3.40 | .27 | 3.38 | 1.89 | 19.06 |
| Fast | 175 | 100 | 26.10 | 12.57 | 5.09 | .26 | 3.32 | 1.28 | 22.52 |

Use these figures when estimating steel grates over 1" thick. The figures will apply when both sides are painted with oil or water base paint. Square feet calculations for grates are based on overall (length times width) dimensions. For grills under 1" thick, see the following table. Note: A two coat system, prime and finish, using oil base material is recommended for any metal surface. Although water base material is often used, it may cause oxidation, corrosion and rust. Using one coat of oil base paint on exterior metal may result in cracking, peeling, or chipping without the proper prime coat application. "Slow" work is based on a $10.25 hourly wage, "Medium" work on a $16.50 hourly wage, and "Fast" work on a $22.00 hourly wage. Other qualifications that apply to this table are on page 9.

| | Labor SF per manhour | Material coverage SF/gallon | Material cost per gallon | Labor cost per 100 SF | Labor burden 100 SF | Material cost per 100 SF | Overhead per 100 SF | Profit per 100 SF | Total price per 100 SF |
|---|---|---|---|---|---|---|---|---|---|
| **Grates, steel, under 1" thick** | | | | | | | | | |
| Without supports | | | | | | | | | |
| Brush each coat | | | | | | | | | |
| Metal primer, rust inhibitor, clean metal (material #35) | | | | | | | | | |
| Slow | 175 | 200 | 28.20 | 5.86 | 1.59 | 14.10 | 6.68 | 4.24 | 32.47 |
| Medium | 200 | 175 | 24.60 | 8.25 | 2.78 | 14.06 | 6.15 | 3.44 | 34.68 |
| Fast | 225 | 150 | 21.10 | 9.78 | 3.95 | 14.07 | 5.15 | 1.98 | 34.93 |
| Metal primer, rust inhibitor, rusty metal (material #36) | | | | | | | | | |
| Slow | 175 | 200 | 33.30 | 5.86 | 1.59 | 16.65 | 7.47 | 4.74 | 36.31 |
| Medium | 200 | 175 | 29.20 | 8.25 | 2.78 | 16.69 | 6.79 | 3.80 | 38.31 |
| Fast | 225 | 150 | 25.00 | 9.78 | 3.95 | 16.67 | 5.63 | 2.16 | 38.19 |
| Industrial enamel, oil base, high gloss, light colors (material #56) | | | | | | | | | |
| Slow | 200 | 250 | 28.90 | 5.13 | 1.39 | 11.56 | 5.61 | 3.56 | 27.25 |
| Medium | 225 | 225 | 25.20 | 7.33 | 2.46 | 11.20 | 5.15 | 2.88 | 29.02 |
| Fast | 250 | 200 | 21.60 | 8.80 | 3.57 | 10.80 | 4.29 | 1.65 | 29.11 |
| Industrial enamel, oil base, high gloss, dark (OSHA) colors (material #57) | | | | | | | | | |
| Slow | 200 | 250 | 34.90 | 5.13 | 1.39 | 13.96 | 6.35 | 4.03 | 30.86 |
| Medium | 225 | 225 | 30.50 | 7.33 | 2.46 | 13.56 | 5.72 | 3.20 | 32.27 |
| Fast | 250 | 200 | 26.10 | 8.80 | 3.57 | 13.05 | 4.70 | 1.81 | 31.93 |
| Spray each coat | | | | | | | | | |
| Metal primer, rust inhibitor, clean metal (material #35) | | | | | | | | | |
| Slow | 400 | 150 | 28.20 | 2.56 | .70 | 18.80 | 6.84 | 4.34 | 33.24 |
| Medium | 450 | 138 | 24.60 | 3.67 | 1.23 | 17.83 | 5.57 | 3.11 | 31.41 |
| Fast | 500 | 125 | 21.10 | 4.40 | 1.78 | 16.88 | 4.27 | 1.64 | 28.97 |
| Metal primer, rust inhibitor, rusty metal (material #36) | | | | | | | | | |
| Slow | 400 | 150 | 33.30 | 2.56 | .70 | 22.20 | 7.89 | 5.00 | 38.35 |
| Medium | 450 | 138 | 29.20 | 3.67 | 1.23 | 21.16 | 6.38 | 3.57 | 36.01 |
| Fast | 500 | 125 | 25.00 | 4.40 | 1.78 | 20.00 | 4.84 | 1.86 | 32.88 |
| Industrial enamel, oil base, high gloss, light colors (material #56) | | | | | | | | | |
| Slow | 425 | 175 | 28.90 | 2.41 | .65 | 16.51 | 6.07 | 3.85 | 29.49 |
| Medium | 475 | 163 | 25.20 | 3.47 | 1.18 | 15.46 | 4.92 | 2.75 | 27.78 |
| Fast | 525 | 150 | 21.60 | 4.19 | 1.68 | 14.40 | 3.75 | 1.44 | 25.46 |
| Industrial enamel, oil base, high gloss, dark (OSHA) colors (material #57) | | | | | | | | | |
| Slow | 425 | 175 | 34.90 | 2.41 | .65 | 19.94 | 7.13 | 4.52 | 34.65 |
| Medium | 475 | 163 | 30.50 | 3.47 | 1.18 | 18.71 | 5.72 | 3.20 | 32.28 |
| Fast | 525 | 150 | 26.10 | 4.19 | 1.68 | 17.40 | 4.31 | 1.66 | 29.24 |

| | Labor SF per manhour | Material coverage SF/gallon | Material cost per gallon | Labor cost per 100 SF | Labor burden 100 SF | Material cost per 100 SF | Overhead per 100 SF | Profit per 100 SF | Total price per 100 SF |
|---|---|---|---|---|---|---|---|---|---|
| Including typical supports | | | | | | | | | |
| Brush each coat | | | | | | | | | |
| Metal primer, rust inhibitor, clean metal (material #35) | | | | | | | | | |
| Slow | 125 | 150 | 28.20 | 8.20 | 2.23 | 18.80 | 9.06 | 5.74 | 44.03 |
| Medium | 150 | 125 | 24.60 | 11.00 | 3.71 | 19.68 | 8.42 | 4.71 | 47.52 |
| Fast | 175 | 100 | 21.10 | 12.57 | 5.09 | 21.10 | 7.17 | 2.76 | 48.69 |
| Metal primer, rust inhibitor, rusty metal (material #36) | | | | | | | | | |
| Slow | 125 | 150 | 33.30 | 8.20 | 2.23 | 22.20 | 10.12 | 6.41 | 49.16 |
| Medium | 150 | 125 | 29.20 | 11.00 | 3.71 | 23.36 | 9.32 | 5.21 | 52.60 |
| Fast | 175 | 100 | 25.00 | 12.57 | 5.09 | 25.00 | 7.89 | 3.03 | 53.58 |
| Industrial enamel, oil base, high gloss, light colors (material #56) | | | | | | | | | |
| Slow | 150 | 200 | 28.90 | 6.83 | 1.87 | 14.45 | 7.17 | 4.55 | 34.87 |
| Medium | 175 | 175 | 25.20 | 9.43 | 3.16 | 14.40 | 6.62 | 3.70 | 37.31 |
| Fast | 200 | 150 | 21.60 | 11.00 | 4.46 | 14.40 | 5.52 | 2.12 | 37.50 |
| Industrial enamel, oil base, high gloss, dark (OSHA) colors (material #57) | | | | | | | | | |
| Slow | 150 | 200 | 34.90 | 6.83 | 1.87 | 17.45 | 8.10 | 5.14 | 39.39 |
| Medium | 175 | 175 | 30.50 | 9.43 | 3.16 | 17.43 | 7.36 | 4.11 | 41.49 |
| Fast | 200 | 150 | 26.10 | 11.00 | 4.46 | 17.40 | 6.08 | 2.34 | 41.28 |
| Spray each coat | | | | | | | | | |
| Metal primer, rust inhibitor, clean metal (material #35) | | | | | | | | | |
| Slow | 325 | 100 | 28.20 | 3.15 | .87 | 28.20 | 9.99 | 6.33 | 48.54 |
| Medium | 375 | 88 | 24.60 | 4.40 | 1.49 | 27.95 | 8.29 | 4.63 | 46.76 |
| Fast | 425 | 75 | 21.10 | 5.18 | 2.09 | 28.13 | 6.55 | 2.52 | 44.47 |
| Metal primer, rust inhibitor, rusty metal (material #36) | | | | | | | | | |
| Slow | 325 | 100 | 33.30 | 3.15 | .87 | 33.30 | 11.57 | 7.33 | 56.22 |
| Medium | 375 | 88 | 29.20 | 4.40 | 1.49 | 33.18 | 9.57 | 5.35 | 53.99 |
| Fast | 425 | 75 | 25.00 | 5.18 | 2.09 | 33.33 | 7.51 | 2.89 | 51.00 |
| Industrial enamel, oil base, high gloss, light colors (material #56) | | | | | | | | | |
| Slow | 350 | 125 | 28.90 | 2.93 | .80 | 23.12 | 8.32 | 5.28 | 40.45 |
| Medium | 400 | 113 | 25.20 | 4.13 | 1.38 | 22.30 | 6.82 | 3.81 | 38.44 |
| Fast | 450 | 100 | 21.60 | 4.89 | 1.97 | 21.60 | 5.27 | 2.02 | 35.75 |
| Industrial enamel, oil base, high gloss, dark (OSHA) colors (material #57) | | | | | | | | | |
| Slow | 350 | 125 | 34.90 | 2.93 | .80 | 27.92 | 9.81 | 6.22 | 47.68 |
| Medium | 400 | 113 | 30.50 | 4.13 | 1.38 | 26.99 | 7.96 | 4.45 | 44.91 |
| Fast | 450 | 100 | 26.10 | 4.89 | 1.97 | 26.10 | 6.10 | 2.34 | 41.40 |

Use these figures to estimate steel grates under 1" thick. The figures will apply when both sides are painted with oil or water base paint. Square foot calculations for grates are based on overall (length times width) dimensions. For grills over 1" thick, see the previous table. Note: A two coat system, prime and finish, using oil base material is recommended for any metal surface. Although water base material is often used, it may cause oxidation, corrosion and rust. Using one coat of oil base paint on exterior metal may result in cracking, peeling, or chipping without the proper prime coat application. "Slow" work is based on a $10.25 hourly wage, "Medium" work on a $16.50 hourly wage, and "Fast" work on a $22.00 hourly wage. Other qualifications that apply to this table are on page 9.

## Ladders

Measure the length of the ladder rungs and vertical members. Then multiply by a difficulty factor of 1.5 (Length x 1.5) to allow for limited access to the back of the ladder. Then use the rates in the Bare pipe tables to figure the labor and material costs.

| | Labor SF per manhour | Material coverage SF/gallon | Material cost per gallon | Labor cost per 100 SF | Labor burden 100 SF | Material cost per 100 SF | Overhead per 100 SF | Profit per 100 SF | Total price per 100 SF |
|---|---|---|---|---|---|---|---|---|---|
| **Masonry, Concrete Masonry Units (CMU), rough, porous surface** | | | | | | | | | |
| Industrial bonding & penetrating oil paint (material #55) | | | | | | | | | |
| Brush 1st coat | | | | | | | | | |
| Slow | 225 | 200 | 26.90 | 4.56 | 1.23 | 13.45 | 5.97 | 3.78 | 28.99 |
| Medium | 250 | 188 | 23.50 | 6.60 | 2.22 | 12.50 | 5.22 | 2.92 | 29.46 |
| Fast | 275 | 175 | 20.20 | 8.00 | 3.25 | 11.54 | 4.21 | 1.62 | 28.62 |
| Brush 2nd coat | | | | | | | | | |
| Slow | 230 | 275 | 26.90 | 4.46 | 1.21 | 9.78 | 4.79 | 3.04 | 23.28 |
| Medium | 260 | 250 | 23.50 | 6.35 | 2.14 | 9.40 | 4.38 | 2.45 | 24.72 |
| Fast | 290 | 225 | 20.20 | 7.59 | 3.08 | 8.98 | 3.64 | 1.40 | 24.69 |
| Industrial waterproofing (material #58) | | | | | | | | | |
| Brush 1st coat | | | | | | | | | |
| Slow | 65 | 65 | 21.60 | 15.77 | 4.29 | 33.23 | 16.52 | 10.47 | 80.28 |
| Medium | 90 | 55 | 18.90 | 18.33 | 6.17 | 34.36 | 14.42 | 8.06 | 81.34 |
| Fast | 115 | 45 | 16.20 | 19.13 | 7.77 | 36.00 | 11.63 | 4.47 | 79.00 |
| Brush 2nd or additional coats | | | | | | | | | |
| Slow | 90 | 150 | 21.60 | 11.39 | 3.10 | 14.40 | 8.96 | 5.68 | 43.53 |
| Medium | 115 | 125 | 18.90 | 14.35 | 4.83 | 15.12 | 8.40 | 4.70 | 47.40 |
| Fast | 145 | 100 | 16.20 | 15.17 | 6.16 | 16.20 | 6.94 | 2.67 | 47.14 |
| Roll 1st coat | | | | | | | | | |
| Slow | 100 | 125 | 21.60 | 10.25 | 2.79 | 17.28 | 9.40 | 5.96 | 45.68 |
| Medium | 125 | 108 | 18.90 | 13.20 | 4.44 | 17.50 | 8.61 | 4.81 | 48.56 |
| Fast | 150 | 90 | 16.20 | 14.67 | 5.95 | 18.00 | 7.14 | 2.75 | 48.51 |
| Roll 2nd or additional coats | | | | | | | | | |
| Slow | 150 | 175 | 21.60 | 6.83 | 1.87 | 12.34 | 6.52 | 4.13 | 31.69 |
| Medium | 180 | 150 | 18.90 | 9.17 | 3.09 | 12.60 | 6.09 | 3.40 | 34.35 |
| Fast | 210 | 125 | 16.20 | 10.48 | 4.24 | 12.96 | 5.12 | 1.97 | 34.77 |

Use these figures for Concrete Masonry Units (CMU) where the block surfaces are rough, porous or unfilled, with joints struck to average depth. The more porous the surface, the rougher the texture, the more time and material will be required. For heights above 8 feet, use the High Time Difficulty Factors on page 137. Also see Masonry in the "General Painting Operations" section of this book. "Slow" work is based on a $10.25 hourly wage, "Medium" work on a $16.50 hourly wage, "Fast" work on a $22.00 hourly wage. Other qualifications that apply to this table are on page 9.

| | Labor SF per manhour | Material coverage SF/gallon | Material cost per gallon | Labor cost per 100 SF | Labor burden 100 SF | Material cost per 100 SF | Overhead per 100 SF | Profit per 100 SF | Total price per 100 SF |
|---|---|---|---|---|---|---|---|---|---|

## Masonry, Concrete Masonry Units (CMU), smooth surface

| | Labor SF per manhour | Material coverage SF/gallon | Material cost per gallon | Labor cost per 100 SF | Labor burden 100 SF | Material cost per 100 SF | Overhead per 100 SF | Profit per 100 SF | Total price per 100 SF |
|---|---|---|---|---|---|---|---|---|---|
| Industrial bonding & penetrating oil paint (material #55) | | | | | | | | | |
| Brush 1st coat | | | | | | | | | |
| Slow | 325 | 240 | 26.90 | 3.15 | .87 | 11.21 | 4.72 | 2.99 | 22.94 |
| Medium | 350 | 230 | 23.50 | 4.71 | 1.60 | 10.22 | 4.04 | 2.26 | 22.83 |
| Fast | 375 | 220 | 20.20 | 5.87 | 2.39 | 9.18 | 3.22 | 1.24 | 21.90 |
| Brush 2nd coat | | | | | | | | | |
| Slow | 340 | 300 | 26.90 | 3.01 | .82 | 8.97 | 3.97 | 2.52 | 19.29 |
| Medium | 370 | 275 | 23.50 | 4.46 | 1.49 | 8.55 | 3.55 | 1.99 | 20.04 |
| Fast | 400 | 250 | 20.20 | 5.50 | 2.23 | 8.08 | 2.92 | 1.12 | 19.85 |
| Industrial waterproofing (material #58) | | | | | | | | | |
| Brush 1st coat | | | | | | | | | |
| Slow | 75 | 100 | 21.60 | 13.67 | 3.71 | 21.60 | 12.09 | 7.66 | 58.73 |
| Medium | 100 | 95 | 18.90 | 16.50 | 5.55 | 19.89 | 10.28 | 5.74 | 57.96 |
| Fast | 125 | 90 | 16.20 | 17.60 | 7.14 | 18.00 | 7.91 | 3.04 | 53.69 |
| Brush 2nd or additional coats | | | | | | | | | |
| Slow | 100 | 150 | 21.60 | 10.25 | 2.79 | 14.40 | 8.51 | 5.39 | 41.34 |
| Medium | 125 | 138 | 18.90 | 13.20 | 4.44 | 13.70 | 7.68 | 4.29 | 43.31 |
| Fast | 150 | 125 | 16.20 | 14.67 | 5.95 | 12.96 | 6.21 | 2.39 | 42.18 |
| Roll 1st coat | | | | | | | | | |
| Slow | 125 | 150 | 21.60 | 8.20 | 2.23 | 14.40 | 7.70 | 4.88 | 37.41 |
| Medium | 150 | 138 | 18.90 | 11.00 | 3.71 | 13.70 | 6.96 | 3.89 | 39.26 |
| Fast | 175 | 125 | 16.20 | 12.57 | 5.09 | 12.96 | 5.67 | 2.18 | 38.47 |
| Roll 2nd or additional coats | | | | | | | | | |
| Slow | 175 | 200 | 21.60 | 5.86 | 1.59 | 10.80 | 5.66 | 3.59 | 27.50 |
| Medium | 200 | 175 | 18.90 | 8.25 | 2.78 | 10.80 | 5.35 | 2.99 | 30.17 |
| Fast | 225 | 150 | 16.20 | 9.78 | 3.95 | 10.80 | 4.54 | 1.75 | 30.82 |

Use these figures for Concrete Masonry Units (CMU), precision block, filled block or slump stone with joints struck to average depth. The more porous the surface, the rougher the texture, the more time and material will be required. For heights above 8 feet, use the High Time Difficulty Factors on page 137. Also see Masonry in the "General Painting Operations" section of this book. "Slow" work is based on a $10.25 hourly wage, "Medium" work on a $16.50 hourly wage, "Fast" work on a $22.00 hourly wage. Other qualifications that apply to this table are on page 9.

| | Labor SF per manhour | Material coverage SF/gallon | Material cost per gallon | Labor cost per 100 SF | Labor burden 100 SF | Material cost per 100 SF | Overhead per 100 SF | Profit per 100 SF | Total price per 100 SF |
|---|---|---|---|---|---|---|---|---|---|

## Mechanical equipment

| | Labor SF per manhour | Material coverage SF/gallon | Material cost per gallon | Labor cost per 100 SF | Labor burden 100 SF | Material cost per 100 SF | Overhead per 100 SF | Profit per 100 SF | Total price per 100 SF |
|---|---|---|---|---|---|---|---|---|---|
| Brush each coat | | | | | | | | | |
| Metal primer, rust inhibitor, clean metal (material #35) | | | | | | | | | |
| Slow | 175 | 275 | 28.20 | 5.86 | 1.59 | 10.25 | 5.49 | 3.48 | 26.67 |
| Medium | 200 | 263 | 24.60 | 8.25 | 2.78 | 9.35 | 4.99 | 2.79 | 28.16 |
| Fast | 225 | 250 | 21.10 | 9.78 | 3.95 | 8.44 | 4.11 | 1.58 | 27.86 |
| Metal primer, rust inhibitor, rusty metal (material #36) | | | | | | | | | |
| Slow | 175 | 275 | 33.30 | 5.86 | 1.59 | 12.11 | 6.07 | 3.85 | 29.48 |
| Medium | 200 | 263 | 29.20 | 8.25 | 2.78 | 11.10 | 5.42 | 3.03 | 30.58 |
| Fast | 225 | 250 | 25.00 | 9.78 | 3.95 | 10.00 | 4.39 | 1.69 | 29.81 |
| Industrial enamel, oil base, high gloss, light colors (material #56) | | | | | | | | | |
| Slow | 200 | 375 | 28.90 | 5.13 | 1.39 | 7.71 | 4.41 | 2.80 | 21.44 |
| Medium | 225 | 363 | 25.20 | 7.33 | 2.46 | 6.94 | 4.10 | 2.29 | 23.12 |
| Fast | 250 | 350 | 21.60 | 8.80 | 3.57 | 6.17 | 3.43 | 1.32 | 23.29 |
| Industrial enamel, oil base, high gloss, dark (OSHA) colors (material #57) | | | | | | | | | |
| Slow | 200 | 375 | 34.90 | 5.13 | 1.39 | 9.31 | 4.91 | 3.11 | 23.85 |
| Medium | 225 | 363 | 30.50 | 7.33 | 2.46 | 8.40 | 4.46 | 2.49 | 25.14 |
| Fast | 250 | 350 | 26.10 | 8.80 | 3.57 | 7.46 | 3.67 | 1.41 | 24.91 |
| Spray each coat | | | | | | | | | |
| Metal primer, rust inhibitor, clean metal (material #35) | | | | | | | | | |
| Slow | 350 | 200 | 28.20 | 2.93 | .80 | 14.10 | 5.53 | 3.50 | 26.86 |
| Medium | 375 | 175 | 24.60 | 4.40 | 1.49 | 14.06 | 4.89 | 2.73 | 27.57 |
| Fast | 400 | 150 | 21.10 | 5.50 | 2.23 | 14.07 | 4.03 | 1.55 | 27.38 |
| Metal primer, rust inhibitor, rusty metal (material #36) | | | | | | | | | |
| Slow | 350 | 200 | 33.30 | 2.93 | .80 | 16.65 | 6.32 | 4.01 | 30.71 |
| Medium | 375 | 175 | 29.20 | 4.40 | 1.49 | 16.69 | 5.53 | 3.09 | 31.20 |
| Fast | 400 | 150 | 25.00 | 5.50 | 2.23 | 16.67 | 4.51 | 1.73 | 30.64 |
| Industrial enamel, oil base, high gloss, light colors (material #56) | | | | | | | | | |
| Slow | 375 | 275 | 28.90 | 2.73 | .75 | 10.51 | 4.33 | 2.75 | 21.07 |
| Medium | 400 | 250 | 25.20 | 4.13 | 1.38 | 10.08 | 3.82 | 2.14 | 21.55 |
| Fast | 425 | 225 | 21.60 | 5.18 | 2.09 | 9.60 | 3.12 | 1.20 | 21.19 |
| Industrial enamel, oil base, high gloss, dark (OSHA) colors (material #57) | | | | | | | | | |
| Slow | 375 | 275 | 34.90 | 2.73 | .75 | 12.69 | 5.01 | 3.18 | 24.36 |
| Medium | 400 | 250 | 30.50 | 4.13 | 1.38 | 12.20 | 4.34 | 2.43 | 24.48 |
| Fast | 425 | 225 | 26.10 | 5.18 | 2.09 | 11.60 | 3.49 | 1.34 | 23.70 |

Use these figures to estimate the cost of painting mechanical equipment (such as compressors and mixing boxes). Measurements are based on square feet of surface area covered. For heights above 8 feet, use the High Time Difficulty Factors on page 137. Note: A two coat system, prime and finish, using oil base material is recommended for any metal surface. Although water base material is often used, it may cause oxidation, corrosion and rust. Using one coat of oil base paint on exterior metal may result in cracking, peeling or chipping without the proper prime coat application. "Slow" work is based on a $10.25 hourly wage, "Medium" work on a $16.50 hourly wage, "Fast" work on a $22.00 hourly wage. Other qualifications that apply to this table are on page 9.

## Mechanical equipment, boiler room

Don't bother figuring the exact area of boiler room equipment that has to be coated. Instead, take the area as equal to 1/2 the wall height times the ceiling area in the room. Figure a painter will coat 125 square feet per hour and a gallon of paint will cover 300 square feet. This rate does not include time needed to paint walls, ceiling or floor around mechanical equipment. In any case, you'll need to rely on judgment when painting boiler room equipment.

| | Labor SF per manhour | Material coverage SF/gallon | Material cost per gallon | Labor cost per 100 SF | Labor burden 100 SF | Material cost per 100 SF | Overhead per 100 SF | Profit per 100 SF | Total price per 100 SF |
|---|---|---|---|---|---|---|---|---|---|
| **Piping, bare pipe, brush application** | | | | | | | | | |
| Metal primer, rust inhibitor, clean metal (material #35) | | | | | | | | | |
| Brush prime coat | | | | | | | | | |
| Slow | 75 | 360 | 28.20 | 13.67 | 3.71 | 7.83 | 7.82 | 4.96 | 37.99 |
| Medium | 100 | 335 | 24.60 | 16.50 | 5.55 | 7.34 | 7.20 | 4.02 | 40.61 |
| Fast | 125 | 310 | 21.10 | 17.60 | 7.14 | 6.81 | 5.84 | 2.24 | 39.63 |
| Metal primer, rust inhibitor, rusty metal (material #36) | | | | | | | | | |
| Brush prime coat | | | | | | | | | |
| Slow | 75 | 360 | 33.30 | 13.67 | 3.71 | 9.25 | 8.26 | 5.24 | 40.13 |
| Medium | 100 | 335 | 29.20 | 16.50 | 5.55 | 8.72 | 7.54 | 4.21 | 42.52 |
| Fast | 125 | 310 | 25.00 | 17.60 | 7.14 | 8.06 | 6.07 | 2.33 | 41.20 |
| Industrial enamel, oil base, high gloss, light colors (material #56) | | | | | | | | | |
| Brush 1st finish coat | | | | | | | | | |
| Slow | 90 | 450 | 28.90 | 11.39 | 3.10 | 6.42 | 6.48 | 4.11 | 31.50 |
| Medium | 115 | 425 | 25.20 | 14.35 | 4.83 | 5.93 | 6.15 | 3.44 | 34.70 |
| Fast | 140 | 400 | 21.60 | 15.71 | 6.37 | 5.40 | 5.08 | 1.95 | 34.51 |
| Brush 2nd or additional finish coats | | | | | | | | | |
| Slow | 125 | 500 | 28.90 | 8.20 | 2.23 | 5.78 | 5.03 | 3.19 | 24.43 |
| Medium | 150 | 475 | 25.20 | 11.00 | 3.71 | 5.31 | 4.90 | 2.74 | 27.66 |
| Fast | 175 | 450 | 21.60 | 12.57 | 5.09 | 4.80 | 4.16 | 1.60 | 28.22 |
| Industrial enamel, oil base, high gloss, dark (OSHA) colors (material #57) | | | | | | | | | |
| Brush 1st finish coat | | | | | | | | | |
| Slow | 90 | 450 | 34.90 | 11.39 | 3.10 | 7.76 | 6.90 | 4.37 | 33.52 |
| Medium | 115 | 425 | 30.50 | 14.35 | 4.83 | 7.18 | 6.46 | 3.61 | 36.43 |
| Fast | 140 | 400 | 26.10 | 15.71 | 6.37 | 6.53 | 5.29 | 2.03 | 35.93 |
| Brush 2nd or additional finish coats | | | | | | | | | |
| Slow | 125 | 500 | 34.90 | 8.20 | 2.23 | 6.98 | 5.40 | 3.42 | 26.23 |
| Medium | 150 | 475 | 30.50 | 11.00 | 3.71 | 6.42 | 5.17 | 2.89 | 29.19 |
| Fast | 175 | 450 | 26.10 | 12.57 | 5.09 | 5.80 | 4.34 | 1.67 | 29.47 |

| | Labor SF per manhour | Material coverage SF/gallon | Material cost per gallon | Labor cost per 100 SF | Labor burden 100 SF | Material cost per 100 SF | Overhead per 100 SF | Profit per 100 SF | Total price per 100 SF |
|---|---|---|---|---|---|---|---|---|---|
| **Epoxy coating, 2 part system, clear (material #51)** | | | | | | | | | |
| Brush 1st coat | | | | | | | | | |
| Slow | 75 | 375 | 57.00 | 13.67 | 3.71 | 15.20 | 10.11 | 6.41 | 49.10 |
| Medium | 100 | 350 | 49.90 | 16.50 | 5.55 | 14.26 | 8.90 | 4.97 | 50.18 |
| Fast | 125 | 325 | 42.80 | 17.60 | 7.14 | 13.17 | 7.01 | 2.70 | 47.62 |
| Brush 2nd coat | | | | | | | | | |
| Slow | 90 | 425 | 57.00 | 11.39 | 3.10 | 13.41 | 8.65 | 5.48 | 42.03 |
| Medium | 115 | 400 | 49.90 | 14.35 | 4.83 | 12.48 | 7.76 | 4.34 | 43.76 |
| Fast | 140 | 375 | 42.80 | 15.71 | 6.37 | 11.41 | 6.20 | 2.38 | 42.07 |
| Brush 3rd or additional coats | | | | | | | | | |
| Slow | 125 | 475 | 57.00 | 8.20 | 2.23 | 12.00 | 6.95 | 4.41 | 33.79 |
| Medium | 150 | 450 | 49.90 | 11.00 | 3.71 | 11.09 | 6.32 | 3.53 | 35.65 |
| Fast | 175 | 425 | 42.80 | 12.57 | 5.09 | 10.07 | 5.13 | 1.97 | 34.83 |
| **Epoxy coating, 2 part system, white (material #52)** | | | | | | | | | |
| Brush 1st coat | | | | | | | | | |
| Slow | 75 | 375 | 57.40 | 13.67 | 3.71 | 15.31 | 10.14 | 6.43 | 49.26 |
| Medium | 100 | 350 | 50.20 | 16.50 | 5.55 | 14.34 | 8.92 | 4.98 | 50.29 |
| Fast | 125 | 325 | 43.10 | 17.60 | 7.14 | 13.26 | 7.03 | 2.70 | 47.73 |
| Brush 2nd coat | | | | | | | | | |
| Slow | 90 | 425 | 57.40 | 11.39 | 3.10 | 13.51 | 8.68 | 5.50 | 42.18 |
| Medium | 115 | 400 | 50.20 | 14.35 | 4.83 | 12.55 | 7.77 | 4.35 | 43.85 |
| Fast | 140 | 375 | 43.10 | 15.71 | 6.37 | 11.49 | 6.21 | 2.39 | 42.17 |
| Brush 3rd or additional coats | | | | | | | | | |
| Slow | 125 | 475 | 57.40 | 8.20 | 2.23 | 12.08 | 6.98 | 4.42 | 33.91 |
| Medium | 150 | 450 | 50.20 | 11.00 | 3.71 | 11.16 | 6.34 | 3.54 | 35.75 |
| Fast | 175 | 425 | 43.10 | 12.57 | 5.09 | 10.14 | 5.14 | 1.98 | 34.92 |
| **Aluminum base paint (material #50)** | | | | | | | | | |
| Brush each coat | | | | | | | | | |
| Slow | 50 | 600 | 33.90 | 20.50 | 5.58 | 5.65 | 9.84 | 6.24 | 47.81 |
| Medium | 75 | 575 | 29.70 | 22.00 | 7.39 | 5.17 | 8.47 | 4.73 | 47.76 |
| Fast | 100 | 550 | 25.40 | 22.00 | 8.92 | 4.62 | 6.57 | 2.53 | 44.64 |

| | Labor SF per manhour | Material coverage SF/gallon | Material cost per gallon | Labor cost per 100 SF | Labor burden 100 SF | Material cost per 100 SF | Overhead per 100 SF | Profit per 100 SF | Total price per 100 SF |
|---|---|---|---|---|---|---|---|---|---|
| **Heat resistant enamel, 800 to 1200 degree range (material #53)** | | | | | | | | | |
| Brush each coat | | | | | | | | | |
| Slow | 50 | 600 | 69.20 | 20.50 | 5.58 | 11.53 | 11.66 | 7.39 | 56.66 |
| Medium | 75 | 575 | 60.60 | 22.00 | 7.39 | 10.54 | 9.79 | 5.47 | 55.19 |
| Fast | 100 | 550 | 51.90 | 22.00 | 8.92 | 9.44 | 7.47 | 2.87 | 50.70 |
| **Heat resistant enamel, 300 to 800 degree range (material #54)** | | | | | | | | | |
| Brush each coat | | | | | | | | | |
| Slow | 50 | 600 | 52.10 | 20.50 | 5.58 | 8.68 | 10.78 | 6.83 | 52.37 |
| Medium | 75 | 575 | 45.60 | 22.00 | 7.39 | 7.93 | 9.15 | 5.11 | 51.58 |
| Fast | 100 | 550 | 39.10 | 22.00 | 8.92 | 7.11 | 7.04 | 2.70 | 47.77 |

Use the pipe conversion factors in Figure 21 on page 321 to convert linear feet of pipe to square feet of surface. Vertical pipe runs require 2 to 3 times the manhours plus 10% more material. Solid color coded piping requires 15% to 25% more labor and material. For color bands on piping at 10' to 15' intervals, add the cost of an additional 1st coat. For heights above 8 feet, use the High Time Difficulty Factors on page 137. Note: A two coat system, prime and finish, using oil base material is recommended for any metal surface. Although water base material is often used, it may cause oxidation, corrosion and rust. One coat of oil base solid body stain is often used on interior or exterior metal but it may crack, peel or chip without the proper prime coat application. "Slow" work is based on a $10.25 hourly wage, "Medium" work on a $16.50 hourly wage, "Fast" work on a $22.00 hourly wage. Other qualifications that apply to this table are on page 9.

| | Labor SF per manhour | Material coverage SF/gallon | Material cost per gallon | Labor cost per 100 SF | Labor burden 100 SF | Material cost per 100 SF | Overhead per 100 SF | Profit per 100 SF | Total price per 100 SF |
|---|---|---|---|---|---|---|---|---|---|
| **Piping, bare pipe, roll application** | | | | | | | | | |
| Metal primer, rust inhibitor, clean metal (material #35) | | | | | | | | | |
| Roll prime coat | | | | | | | | | |
| Slow | 175 | 400 | 28.20 | 5.86 | 1.59 | 7.05 | 4.50 | 2.85 | 21.85 |
| Medium | 200 | 375 | 24.60 | 8.25 | 2.78 | 6.56 | 4.31 | 2.41 | 24.31 |
| Fast | 225 | 350 | 21.10 | 9.78 | 3.95 | 6.03 | 3.66 | 1.41 | 24.83 |
| Metal primer, rust inhibitor, rusty metal (material #36) | | | | | | | | | |
| Roll prime coat | | | | | | | | | |
| Slow | 175 | 400 | 33.30 | 5.86 | 1.59 | 8.33 | 4.89 | 3.10 | 23.77 |
| Medium | 200 | 375 | 29.20 | 8.25 | 2.78 | 7.79 | 4.61 | 2.58 | 26.01 |
| Fast | 225 | 350 | 25.00 | 9.78 | 3.95 | 7.14 | 3.86 | 1.49 | 26.22 |
| Industrial enamel, oil base, high gloss, light colors (material #56) | | | | | | | | | |
| Roll 1st finish coat | | | | | | | | | |
| Slow | 200 | 450 | 28.90 | 5.13 | 1.39 | 6.42 | 4.01 | 2.54 | 19.49 |
| Medium | 225 | 425 | 25.20 | 7.33 | 2.46 | 5.93 | 3.85 | 2.15 | 21.72 |
| Fast | 250 | 400 | 21.60 | 8.80 | 3.57 | 5.40 | 3.29 | 1.26 | 22.32 |
| Roll 2nd or additional finish coats | | | | | | | | | |
| Slow | 275 | 500 | 28.90 | 3.73 | 1.02 | 5.78 | 3.26 | 2.07 | 15.86 |
| Medium | 300 | 475 | 25.20 | 5.50 | 1.84 | 5.31 | 3.10 | 1.73 | 17.48 |
| Fast | 325 | 450 | 21.60 | 6.77 | 2.75 | 4.80 | 2.65 | 1.02 | 17.99 |
| Industrial enamel, oil base, high gloss, dark (OSHA) colors (material #57) | | | | | | | | | |
| Roll 1st finish coat | | | | | | | | | |
| Slow | 200 | 450 | 34.90 | 5.13 | 1.39 | 7.76 | 4.43 | 2.81 | 21.52 |
| Medium | 225 | 425 | 30.50 | 7.33 | 2.46 | 7.18 | 4.16 | 2.33 | 23.46 |
| Fast | 250 | 400 | 26.10 | 8.80 | 3.57 | 6.53 | 3.50 | 1.34 | 23.74 |
| Roll 2nd or additional finish coats | | | | | | | | | |
| Slow | 275 | 500 | 34.90 | 3.73 | 1.02 | 6.98 | 3.64 | 2.31 | 17.68 |
| Medium | 300 | 475 | 30.50 | 5.50 | 1.84 | 6.42 | 3.37 | 1.89 | 19.02 |
| Fast | 325 | 450 | 26.10 | 6.77 | 2.75 | 5.80 | 2.83 | 1.09 | 19.24 |

Use the pipe conversion factors in Figure 21 on page 321 to convert linear feet of pipe to square feet of surface. Vertical pipe runs require 2 to 3 times the manhours plus 10% more material. Solid color coded piping requires 15% to 25% more labor and material. For color bands on piping at 10' to 15' intervals, add the cost of an additional 1st coat. For heights above 8 feet, use the High Time Difficulty Factors on page 137. Note: A two coat system, prime and finish, using oil base material is recommended for any metal surface. Although water base material is often used, it may cause oxidation, corrosion and rust. One coat of oil base solid body stain is often used on interior or exterior metal but it may crack, peel or chip without the proper prime coat application. "Slow" work is based on a $10.25 hourly wage, "Medium" work on a $16.50 hourly wage, "Fast" work on a $22.00 hourly wage. Other qualifications that apply to this table are on page 9.

| | Labor SF per manhour | Material coverage SF/gallon | Material cost per gallon | Labor cost per 100 SF | Labor burden 100 SF | Material cost per 100 SF | Overhead per 100 SF | Profit per 100 SF | Total price per 100 SF |
|---|---|---|---|---|---|---|---|---|---|

## Piping, bare pipe, spray application

| | Labor SF per manhour | Material coverage SF/gallon | Material cost per gallon | Labor cost per 100 SF | Labor burden 100 SF | Material cost per 100 SF | Overhead per 100 SF | Profit per 100 SF | Total price per 100 SF |
|---|---|---|---|---|---|---|---|---|---|
| **Metal primer, rust inhibitor, clean metal (material #35)** | | | | | | | | | |
| Spray prime coat | | | | | | | | | |
| Slow | 300 | 175 | 28.20 | 3.42 | .92 | 16.11 | 6.34 | 4.02 | 30.81 |
| Medium | 350 | 163 | 24.60 | 4.71 | 1.60 | 15.09 | 5.24 | 2.93 | 29.57 |
| Fast | 400 | 150 | 21.10 | 5.50 | 2.23 | 14.07 | 4.03 | 1.55 | 27.38 |
| **Metal primer, rust inhibitor, rusty metal (material #36)** | | | | | | | | | |
| Spray prime coat | | | | | | | | | |
| Slow | 300 | 175 | 33.30 | 3.42 | .92 | 19.03 | 7.25 | 4.59 | 35.21 |
| Medium | 350 | 163 | 29.20 | 4.71 | 1.60 | 17.91 | 5.93 | 3.31 | 33.46 |
| Fast | 400 | 150 | 25.00 | 5.50 | 2.23 | 16.67 | 4.51 | 1.73 | 30.64 |
| **Industrial enamel, oil base, high gloss, light colors (material #56)** | | | | | | | | | |
| Spray 1st finish coat | | | | | | | | | |
| Slow | 375 | 210 | 28.90 | 2.73 | .75 | 13.76 | 5.34 | 3.39 | 25.97 |
| Medium | 425 | 198 | 25.20 | 3.88 | 1.30 | 12.73 | 4.39 | 2.45 | 24.75 |
| Fast | 475 | 185 | 21.60 | 4.63 | 1.89 | 11.68 | 3.37 | 1.29 | 22.86 |
| Spray 2nd or additional finish coats | | | | | | | | | |
| Slow | 425 | 300 | 28.90 | 2.41 | .65 | 9.63 | 3.94 | 2.50 | 19.13 |
| Medium | 475 | 288 | 25.20 | 3.47 | 1.18 | 8.75 | 3.28 | 1.83 | 18.51 |
| Fast | 525 | 275 | 21.60 | 4.19 | 1.68 | 7.85 | 2.54 | .98 | 17.24 |
| **Industrial enamel, oil base, high gloss, dark (OSHA) colors (material #57)** | | | | | | | | | |
| Spray 1st finish coat | | | | | | | | | |
| Slow | 375 | 210 | 34.90 | 2.73 | .75 | 16.62 | 6.23 | 3.95 | 30.28 |
| Medium | 425 | 198 | 30.50 | 3.88 | 1.30 | 15.40 | 5.04 | 2.82 | 28.44 |
| Fast | 475 | 185 | 26.10 | 4.63 | 1.89 | 14.11 | 3.81 | 1.47 | 25.91 |
| Spray 2nd or additional finish coats | | | | | | | | | |
| Slow | 425 | 300 | 34.90 | 2.41 | .65 | 11.63 | 4.56 | 2.89 | 22.14 |
| Medium | 475 | 288 | 30.50 | 3.47 | 1.18 | 10.59 | 3.73 | 2.09 | 21.06 |
| Fast | 525 | 275 | 26.10 | 4.19 | 1.68 | 9.49 | 2.85 | 1.09 | 19.30 |
| **Epoxy coating, 2 part system, clear (material #51)** | | | | | | | | | |
| Spray 1st coat | | | | | | | | | |
| Slow | 300 | 185 | 57.00 | 3.42 | .92 | 30.81 | 10.90 | 6.91 | 52.96 |
| Medium | 350 | 173 | 49.90 | 4.71 | 1.60 | 28.84 | 8.61 | 4.81 | 48.57 |
| Fast | 400 | 160 | 42.80 | 5.50 | 2.23 | 26.75 | 6.38 | 2.45 | 43.31 |
| Spray 2nd coat | | | | | | | | | |
| Slow | 375 | 200 | 57.00 | 2.73 | .75 | 28.50 | 9.91 | 6.28 | 48.17 |
| Medium | 425 | 188 | 49.90 | 3.88 | 1.30 | 26.54 | 7.77 | 4.35 | 43.84 |
| Fast | 475 | 175 | 42.80 | 4.63 | 1.89 | 24.46 | 5.73 | 2.20 | 38.91 |

| | Labor SF per manhour | Material coverage SF/gallon | Material cost per gallon | Labor cost per 100 SF | Labor burden 100 SF | Material cost per 100 SF | Overhead per 100 SF | Profit per 100 SF | Total price per 100 SF |
|---|---|---|---|---|---|---|---|---|---|
| Spray 3rd or additional coats | | | | | | | | | |
| Slow | 425 | 275 | 57.00 | 2.41 | .65 | 20.73 | 7.38 | 4.68 | 35.85 |
| Medium | 475 | 263 | 49.90 | 3.47 | 1.18 | 18.97 | 5.78 | 3.23 | 32.63 |
| Fast | 525 | 250 | 42.80 | 4.19 | 1.68 | 17.12 | 4.26 | 1.64 | 28.89 |
| Epoxy coating, 2 part system, white (material #52) | | | | | | | | | |
| Spray 1st coat | | | | | | | | | |
| Slow | 300 | 185 | 57.40 | 3.42 | .92 | 31.03 | 10.97 | 6.95 | 53.29 |
| Medium | 350 | 173 | 50.20 | 4.71 | 1.60 | 29.02 | 8.65 | 4.84 | 48.82 |
| Fast | 400 | 160 | 43.10 | 5.50 | 2.23 | 26.94 | 6.41 | 2.46 | 43.54 |
| Spray 2nd coat | | | | | | | | | |
| Slow | 375 | 200 | 57.40 | 2.73 | .75 | 28.70 | 9.97 | 6.32 | 48.47 |
| Medium | 425 | 188 | 50.20 | 3.88 | 1.30 | 26.70 | 7.81 | 4.37 | 44.06 |
| Fast | 475 | 175 | 43.10 | 4.63 | 1.89 | 24.63 | 5.76 | 2.21 | 39.12 |
| Spray 3rd or additional coats | | | | | | | | | |
| Slow | 425 | 275 | 57.40 | 2.41 | .65 | 20.87 | 7.42 | 4.70 | 36.05 |
| Medium | 475 | 263 | 50.20 | 3.47 | 1.18 | 19.09 | 5.81 | 3.25 | 32.80 |
| Fast | 525 | 250 | 43.10 | 4.19 | 1.68 | 17.24 | 4.28 | 1.64 | 29.03 |

Use the pipe conversion factors in Figure 21 on page 321 to convert linear feet of pipe to square feet of surface. Vertical pipe runs require 2 to 3 times the manhours plus 10% more material. Solid color coded piping requires 15% to 25% more labor and material. For color bands on piping at 10' to 15' intervals, add the cost of an additional 1st coat. For heights above 8 feet, use the High Time Difficulty Factors on page 137. Note: A two coat system, prime and finish, using oil base material is recommended for any metal surface. Although water base material is often used, it may cause oxidation, corrosion and rust. One coat of oil base solid body stain is often used on interior or exterior metal but it may crack, peel or chip without the proper prime coat application. "Slow" work is based on a $10.25 hourly wage, "Medium" work on a $16.50 hourly wage, "Fast" work on a $22.00 hourly wage. Other qualifications that apply to this table are on page 9.

| | Labor SF per manhour | Material coverage SF/gallon | Material cost per gallon | Labor cost per 100 SF | Labor burden 100 SF | Material cost per 100 SF | Overhead per 100 SF | Profit per 100 SF | Total price per 100 SF |
|---|---|---|---|---|---|---|---|---|---|
| **Piping, bare pipe, mitt or glove application** | | | | | | | | | |
| Metal primer, rust inhibitor, clean metal (material #35) | | | | | | | | | |
| Mitt or glove prime coat | | | | | | | | | |
| Slow | 175 | 325 | 28.20 | 5.86 | 1.59 | 8.68 | 5.00 | 3.17 | 24.30 |
| Medium | 200 | 313 | 24.60 | 8.25 | 2.78 | 7.86 | 4.63 | 2.59 | 26.11 |
| Fast | 225 | 300 | 21.10 | 9.78 | 3.95 | 7.03 | 3.84 | 1.48 | 26.08 |
| Metal primer, rust inhibitor, rusty metal (material #36) | | | | | | | | | |
| Mitt or glove prime coat | | | | | | | | | |
| Slow | 175 | 325 | 33.30 | 5.86 | 1.59 | 10.25 | 5.49 | 3.48 | 26.67 |
| Medium | 200 | 313 | 29.20 | 8.25 | 2.78 | 9.33 | 4.99 | 2.79 | 28.14 |
| Fast | 225 | 300 | 25.00 | 9.78 | 3.95 | 8.33 | 4.08 | 1.57 | 27.71 |
| Industrial enamel, oil base, high gloss, light colors (material #56) | | | | | | | | | |
| Mitt or glove 1st finish coat | | | | | | | | | |
| Slow | 200 | 375 | 28.90 | 5.13 | 1.39 | 7.71 | 4.41 | 2.80 | 21.44 |
| Medium | 225 | 363 | 25.20 | 7.33 | 2.46 | 6.94 | 4.10 | 2.29 | 23.12 |
| Fast | 250 | 350 | 21.60 | 8.80 | 3.57 | 6.17 | 3.43 | 1.32 | 23.29 |
| Mitt or glove 2nd or additional finish coats | | | | | | | | | |
| Slow | 275 | 400 | 28.90 | 3.73 | 1.02 | 7.23 | 3.71 | 2.35 | 18.04 |
| Medium | 300 | 388 | 25.20 | 5.50 | 1.84 | 6.49 | 3.39 | 1.90 | 19.12 |
| Fast | 325 | 375 | 21.60 | 6.77 | 2.75 | 5.76 | 2.83 | 1.09 | 19.20 |
| Industrial enamel, oil base, high gloss, dark (OSHA) colors (material #57) | | | | | | | | | |
| Mitt or glove 1st finish coat | | | | | | | | | |
| Slow | 200 | 375 | 34.90 | 5.13 | 1.39 | 9.31 | 4.91 | 3.11 | 23.85 |
| Medium | 225 | 363 | 30.50 | 7.33 | 2.46 | 8.40 | 4.46 | 2.49 | 25.14 |
| Fast | 250 | 350 | 26.10 | 8.80 | 3.57 | 7.46 | 3.67 | 1.41 | 24.91 |
| Mitt or glove 2nd or additional finish coats | | | | | | | | | |
| Slow | 275 | 400 | 34.90 | 3.73 | 1.02 | 8.73 | 4.18 | 2.65 | 20.31 |
| Medium | 300 | 388 | 30.50 | 5.50 | 1.84 | 7.86 | 3.73 | 2.08 | 21.01 |
| Fast | 325 | 375 | 26.10 | 6.77 | 2.75 | 6.96 | 3.05 | 1.17 | 20.70 |
| Epoxy coating, 2 part system, clear (material #51) | | | | | | | | | |
| Mitt or glove 1st coat | | | | | | | | | |
| Slow | 175 | 360 | 57.00 | 5.86 | 1.59 | 15.83 | 7.22 | 4.58 | 35.08 |
| Medium | 200 | 348 | 49.90 | 8.25 | 2.78 | 14.34 | 6.22 | 3.47 | 35.06 |
| Fast | 225 | 335 | 42.80 | 9.78 | 3.95 | 12.78 | 4.91 | 1.89 | 33.31 |
| Mitt or glove 2nd coat | | | | | | | | | |
| Slow | 200 | 375 | 57.00 | 5.13 | 1.39 | 15.20 | 6.74 | 4.27 | 32.73 |
| Medium | 225 | 363 | 49.90 | 7.33 | 2.46 | 13.75 | 5.77 | 3.23 | 32.54 |
| Fast | 250 | 350 | 42.80 | 8.80 | 3.57 | 12.23 | 4.55 | 1.75 | 30.90 |

| | Labor SF per manhour | Material coverage SF/gallon | Material cost per gallon | Labor cost per 100 SF | Labor burden 100 SF | Material cost per 100 SF | Overhead per 100 SF | Profit per 100 SF | Total price per 100 SF |
|---|---|---|---|---|---|---|---|---|---|
| Mitt or glove 3rd or additional coats | | | | | | | | | |
| Slow | 275 | 390 | 57.00 | 3.73 | 1.02 | 14.62 | 6.00 | 3.81 | 29.18 |
| Medium | 300 | 378 | 49.90 | 5.50 | 1.84 | 13.20 | 5.03 | 2.81 | 28.38 |
| Fast | 325 | 365 | 42.80 | 6.77 | 2.75 | 11.73 | 3.93 | 1.51 | 26.69 |
| Epoxy coating, 2 part system, white (material #52) | | | | | | | | | |
| Mitt or glove 1st coat | | | | | | | | | |
| Slow | 175 | 360 | 57.40 | 5.86 | 1.59 | 15.94 | 7.25 | 4.60 | 35.24 |
| Medium | 200 | 348 | 50.20 | 8.25 | 2.78 | 14.43 | 6.24 | 3.49 | 35.19 |
| Fast | 225 | 335 | 43.10 | 9.78 | 3.95 | 12.87 | 4.92 | 1.89 | 33.41 |
| Mitt or glove 2nd coat | | | | | | | | | |
| Slow | 200 | 375 | 57.40 | 5.13 | 1.39 | 15.31 | 6.77 | 4.29 | 32.89 |
| Medium | 225 | 363 | 50.20 | 7.33 | 2.46 | 13.83 | 5.79 | 3.24 | 32.65 |
| Fast | 250 | 350 | 43.10 | 8.80 | 3.57 | 12.31 | 4.57 | 1.76 | 31.01 |
| Mitt or glove 3rd or additional coats | | | | | | | | | |
| Slow | 275 | 390 | 57.40 | 3.73 | 1.02 | 14.72 | 6.04 | 3.83 | 29.34 |
| Medium | 300 | 378 | 50.20 | 5.50 | 1.84 | 13.28 | 5.05 | 2.82 | 28.49 |
| Fast | 325 | 365 | 43.10 | 6.77 | 2.75 | 11.81 | 3.95 | 1.52 | 26.80 |

Use the pipe conversion factors in Figure 21 on page 321 to convert linear feet of pipe to square feet of surface. Vertical pipe runs require 2 to 3 times the manhours plus 10% more material. Solid color coded piping requires 15% to 25% more labor and material. For color bands on piping at 10' to 15' intervals, add the cost of an additional 1st coat. For heights above 8 feet, use the High Time Difficulty Factors on page 137. Note: A two coat system, prime and finish, using oil base material is recommended for any metal surface. Although water base material is often used, it may cause oxidation, corrosion and rust. One coat of oil base solid body stain is often used on interior or exterior metal but it may crack, peel or chip without the proper prime coat application. "Slow" work is based on a $10.25 hourly wage, "Medium" work on a $16.50 hourly wage, "Fast" work on a $22.00 hourly wage. Other qualifications that apply to this table are on page 9.

| | Labor SF per manhour | Material coverage SF/gallon | Material cost per gallon | Labor cost per 100 SF | Labor burden 100 SF | Material cost per 100 SF | Overhead per 100 SF | Profit per 100 SF | Total price per 100 SF |
|---|---|---|---|---|---|---|---|---|---|
| **Piping, insulated, canvas jacket, brush** | | | | | | | | | |
| Flat latex, water base (material #5) | | | | | | | | | |
| Brush 1st coat | | | | | | | | | |
| Slow | 60 | 150 | 16.20 | 17.08 | 4.66 | 10.80 | 10.08 | 6.39 | 49.01 |
| Medium | 80 | 138 | 14.20 | 20.63 | 6.93 | 10.29 | 9.28 | 5.19 | 52.32 |
| Fast | 100 | 125 | 12.10 | 22.00 | 8.92 | 9.68 | 7.51 | 2.89 | 51.00 |
| Brush 2nd coat | | | | | | | | | |
| Slow | 75 | 300 | 16.20 | 13.67 | 3.71 | 5.40 | 7.07 | 4.48 | 34.33 |
| Medium | 100 | 288 | 14.20 | 16.50 | 5.55 | 4.93 | 6.61 | 3.69 | 37.28 |
| Fast | 125 | 275 | 12.10 | 17.60 | 7.14 | 4.40 | 5.39 | 2.07 | 36.60 |
| Brush 3rd or additional coats | | | | | | | | | |
| Slow | 100 | 400 | 16.20 | 10.25 | 2.79 | 4.05 | 5.30 | 3.36 | 25.75 |
| Medium | 138 | 375 | 14.20 | 11.96 | 4.03 | 3.79 | 4.84 | 2.71 | 27.33 |
| Fast | 175 | 350 | 12.10 | 12.57 | 5.09 | 3.46 | 3.91 | 1.50 | 26.53 |
| Sealer, off white, water base (material #1) | | | | | | | | | |
| Brush 1 coat | | | | | | | | | |
| Slow | 60 | 150 | 15.50 | 17.08 | 4.66 | 10.33 | 9.94 | 6.30 | 48.31 |
| Medium | 80 | 138 | 13.50 | 20.63 | 6.93 | 9.78 | 9.15 | 5.12 | 51.61 |
| Fast | 100 | 125 | 11.60 | 22.00 | 8.92 | 9.28 | 7.44 | 2.86 | 50.50 |
| Sealer, off white, oil base (material #2) | | | | | | | | | |
| Brush 1 coat | | | | | | | | | |
| Slow | 60 | 200 | 20.70 | 17.08 | 4.66 | 10.35 | 9.94 | 6.30 | 48.33 |
| Medium | 80 | 188 | 18.10 | 20.63 | 6.93 | 9.63 | 9.11 | 5.09 | 51.39 |
| Fast | 100 | 175 | 15.50 | 22.00 | 8.92 | 8.86 | 7.36 | 2.83 | 49.97 |
| Enamel, water base latex (material #9) | | | | | | | | | |
| Brush 1st finish coat | | | | | | | | | |
| Slow | 75 | 300 | 22.80 | 13.67 | 3.71 | 7.60 | 7.75 | 4.91 | 37.64 |
| Medium | 100 | 288 | 19.90 | 16.50 | 5.55 | 6.91 | 7.10 | 3.97 | 40.03 |
| Fast | 125 | 275 | 17.10 | 17.60 | 7.14 | 6.22 | 5.73 | 2.20 | 38.89 |
| Brush 2nd or additional finish coats | | | | | | | | | |
| Slow | 100 | 400 | 22.80 | 10.25 | 2.79 | 5.70 | 5.81 | 3.68 | 28.23 |
| Medium | 138 | 375 | 19.90 | 11.96 | 4.03 | 5.31 | 5.22 | 2.92 | 29.44 |
| Fast | 175 | 350 | 17.10 | 12.57 | 5.09 | 4.89 | 4.17 | 1.60 | 28.32 |
| Industrial enamel, oil base, high gloss, light colors (material #56) | | | | | | | | | |
| Brush 1st finish coat | | | | | | | | | |
| Slow | 75 | 335 | 28.90 | 13.67 | 3.71 | 8.63 | 8.07 | 5.12 | 39.20 |
| Medium | 100 | 323 | 25.20 | 16.50 | 5.55 | 7.80 | 7.31 | 4.09 | 41.25 |
| Fast | 125 | 310 | 21.60 | 17.60 | 7.14 | 6.97 | 5.87 | 2.25 | 39.83 |

| | Labor SF per manhour | Material coverage SF/gallon | Material cost per gallon | Labor cost per 100 SF | Labor burden 100 SF | Material cost per 100 SF | Overhead per 100 SF | Profit per 100 SF | Total price per 100 SF |
|---|---|---|---|---|---|---|---|---|---|
| Brush 2nd or additional finish coats | | | | | | | | | |
| Slow | 100 | 450 | 28.90 | 10.25 | 2.79 | 6.42 | 6.03 | 3.82 | 29.31 |
| Medium | 150 | 438 | 25.20 | 11.00 | 3.71 | 5.75 | 5.01 | 2.80 | 28.27 |
| Fast | 175 | 425 | 21.60 | 12.57 | 5.09 | 5.08 | 4.21 | 1.62 | 28.57 |
| Industrial enamel, oil base, high gloss, dark (OSHA) colors (material #57) | | | | | | | | | |
| Brush 1st finish coat | | | | | | | | | |
| Slow | 75 | 335 | 34.90 | 13.67 | 3.71 | 10.42 | 8.62 | 5.47 | 41.89 |
| Medium | 100 | 323 | 30.50 | 16.50 | 5.55 | 9.44 | 7.72 | 4.31 | 43.52 |
| Fast | 125 | 310 | 26.10 | 17.60 | 7.14 | 8.42 | 6.13 | 2.36 | 41.65 |
| Brush 2nd or additional finish coats | | | | | | | | | |
| Slow | 100 | 450 | 34.90 | 10.25 | 2.79 | 7.76 | 6.45 | 4.09 | 31.34 |
| Medium | 150 | 438 | 30.50 | 11.00 | 3.71 | 6.96 | 5.31 | 2.97 | 29.95 |
| Fast | 175 | 425 | 26.10 | 12.57 | 5.09 | 6.14 | 4.40 | 1.69 | 29.89 |
| Epoxy coating, 2 part system, clear (material #51) | | | | | | | | | |
| Brush 1st coat | | | | | | | | | |
| Slow | 75 | 325 | 57.00 | 13.67 | 3.71 | 17.54 | 10.83 | 6.87 | 52.62 |
| Medium | 100 | 313 | 49.90 | 16.50 | 5.55 | 15.94 | 9.31 | 5.20 | 52.50 |
| Fast | 125 | 300 | 42.80 | 17.60 | 7.14 | 14.27 | 7.22 | 2.77 | 49.00 |
| Brush 2nd or additional coats | | | | | | | | | |
| Slow | 100 | 450 | 57.00 | 10.25 | 2.79 | 12.67 | 7.97 | 5.05 | 38.73 |
| Medium | 150 | 438 | 49.90 | 11.00 | 3.71 | 11.39 | 6.39 | 3.57 | 36.06 |
| Fast | 175 | 425 | 42.80 | 12.57 | 5.09 | 10.07 | 5.13 | 1.97 | 34.83 |
| Epoxy coating, 2 part system, white (material #52) | | | | | | | | | |
| Brush 1st coat | | | | | | | | | |
| Slow | 75 | 325 | 57.40 | 13.67 | 3.71 | 17.66 | 10.87 | 6.89 | 52.80 |
| Medium | 100 | 313 | 50.20 | 16.50 | 5.55 | 16.04 | 9.33 | 5.22 | 52.64 |
| Fast | 125 | 300 | 43.10 | 17.60 | 7.14 | 14.37 | 7.24 | 2.78 | 49.13 |
| Brush 2nd or additional coats | | | | | | | | | |
| Slow | 100 | 450 | 57.40 | 10.25 | 2.79 | 12.76 | 8.00 | 5.07 | 38.87 |
| Medium | 150 | 438 | 50.20 | 11.00 | 3.71 | 11.46 | 6.41 | 3.58 | 36.16 |
| Fast | 175 | 425 | 43.10 | 12.57 | 5.09 | 10.14 | 5.14 | 1.98 | 34.92 |

Use the pipe conversion factors in Figure 21 on page 321 to convert linear feet of pipe to square feet of surface. Vertical pipe runs require 2 to 3 times the manhours plus 10% more material. Solid color coded piping requires 15% to 25% more labor and material. For color bands on piping at 10' to 15' intervals, add the cost of an additional 1st coat. For heights above 8 feet, use the High Time Difficulty Factors on page 137. "Slow" work is based on a $10.25 hourly wage, "Medium" work on a $16.50 hourly wage, "Fast" work on a $22.50 hourly wage. Other qualifications that apply to this table are on page 9.

| | Labor SF per manhour | Material coverage SF/gallon | Material cost per gallon | Labor cost per 100 SF | Labor burden 100 SF | Material cost per 100 SF | Overhead per 100 SF | Profit per 100 SF | Total price per 100 SF |
|---|---|---|---|---|---|---|---|---|---|
| **Piping, insulated, canvas jacket, roll** | | | | | | | | | |
| Flat latex, water base (material #5) | | | | | | | | | |
| Roll 1st coat | | | | | | | | | |
| Slow | 135 | 150 | 16.20 | 7.59 | 2.07 | 10.80 | 6.34 | 4.02 | 30.82 |
| Medium | 160 | 138 | 14.20 | 10.31 | 3.47 | 10.29 | 5.90 | 3.30 | 33.27 |
| Fast | 185 | 125 | 12.10 | 11.89 | 4.84 | 9.68 | 4.88 | 1.88 | 33.17 |
| Roll 2nd coat | | | | | | | | | |
| Slow | 175 | 300 | 16.20 | 5.86 | 1.59 | 5.40 | 3.99 | 2.53 | 19.37 |
| Medium | 200 | 288 | 14.20 | 8.25 | 2.78 | 4.93 | 3.91 | 2.19 | 22.06 |
| Fast | 225 | 275 | 12.10 | 9.78 | 3.95 | 4.40 | 3.36 | 1.29 | 22.78 |
| Roll 3rd or additional coats | | | | | | | | | |
| Slow | 275 | 400 | 16.20 | 3.73 | 1.02 | 4.05 | 2.73 | 1.73 | 13.26 |
| Medium | 300 | 388 | 14.20 | 5.50 | 1.84 | 3.66 | 2.70 | 1.51 | 15.21 |
| Fast | 325 | 375 | 12.10 | 6.77 | 2.75 | 3.23 | 2.36 | .91 | 16.02 |
| Sealer, off white, water base (material #1) | | | | | | | | | |
| Roll 1 coat | | | | | | | | | |
| Slow | 135 | 150 | 15.50 | 7.59 | 2.07 | 10.33 | 6.20 | 3.93 | 30.12 |
| Medium | 160 | 138 | 13.50 | 10.31 | 3.47 | 9.78 | 5.77 | 3.23 | 32.56 |
| Fast | 185 | 125 | 11.60 | 11.89 | 4.84 | 9.28 | 4.81 | 1.85 | 32.67 |
| Sealer, off white, oil base (material #2) | | | | | | | | | |
| Roll 1 coat | | | | | | | | | |
| Slow | 135 | 225 | 20.70 | 7.59 | 2.07 | 9.20 | 5.85 | 3.71 | 28.42 |
| Medium | 160 | 213 | 18.10 | 10.31 | 3.47 | 8.50 | 5.46 | 3.05 | 30.79 |
| Fast | 185 | 200 | 15.50 | 11.89 | 4.84 | 7.75 | 4.53 | 1.74 | 30.75 |
| Enamel, water base latex (material #9) | | | | | | | | | |
| Roll 1st finish coat | | | | | | | | | |
| Slow | 175 | 300 | 22.80 | 5.86 | 1.59 | 7.60 | 4.67 | 2.96 | 22.68 |
| Medium | 200 | 288 | 19.90 | 8.25 | 2.78 | 6.91 | 4.40 | 2.46 | 24.80 |
| Fast | 225 | 275 | 17.10 | 9.78 | 3.95 | 6.22 | 3.69 | 1.42 | 25.06 |
| Roll 2nd or additional finish coats | | | | | | | | | |
| Slow | 275 | 400 | 22.80 | 3.73 | 1.02 | 5.70 | 3.24 | 2.05 | 15.74 |
| Medium | 300 | 388 | 19.90 | 5.50 | 1.84 | 5.13 | 3.06 | 1.71 | 17.24 |
| Fast | 325 | 375 | 17.10 | 6.77 | 2.75 | 4.56 | 2.60 | 1.00 | 17.68 |
| Industrial enamel, oil base, high gloss, light colors (material #56) | | | | | | | | | |
| Roll 1st finish coat | | | | | | | | | |
| Slow | 175 | 335 | 28.90 | 5.86 | 1.59 | 8.63 | 4.99 | 3.16 | 24.23 |
| Medium | 200 | 323 | 25.20 | 8.25 | 2.78 | 7.80 | 4.61 | 2.58 | 26.02 |
| Fast | 225 | 310 | 21.60 | 9.78 | 3.95 | 6.97 | 3.83 | 1.47 | 26.00 |

| | Labor SF per manhour | Material coverage SF/gallon | Material cost per gallon | Labor cost per 100 SF | Labor burden 100 SF | Material cost per 100 SF | Overhead per 100 SF | Profit per 100 SF | Total price per 100 SF |
|---|---|---|---|---|---|---|---|---|---|
| Roll 2nd or additional finish coats | | | | | | | | | |
| Slow | 275 | 450 | 28.90 | 3.73 | 1.02 | 6.42 | 3.46 | 2.19 | 16.82 |
| Medium | 300 | 438 | 25.20 | 5.50 | 1.84 | 5.75 | 3.21 | 1.79 | 18.09 |
| Fast | 325 | 425 | 21.60 | 6.77 | 2.75 | 5.08 | 2.70 | 1.04 | 18.34 |
| Industrial enamel, oil base, high gloss, dark (OSHA) colors (material #57) | | | | | | | | | |
| Roll 1st finish coat | | | | | | | | | |
| Slow | 175 | 335 | 34.90 | 5.86 | 1.59 | 10.42 | 5.54 | 3.51 | 26.92 |
| Medium | 200 | 323 | 30.50 | 8.25 | 2.78 | 9.44 | 5.02 | 2.80 | 28.29 |
| Fast | 225 | 310 | 26.10 | 9.78 | 3.95 | 8.42 | 4.10 | 1.58 | 27.83 |
| Roll 2nd or additional finish coats | | | | | | | | | |
| Slow | 275 | 450 | 34.90 | 3.73 | 1.02 | 7.76 | 3.88 | 2.46 | 18.85 |
| Medium | 300 | 438 | 30.50 | 5.50 | 1.84 | 6.96 | 3.51 | 1.96 | 19.77 |
| Fast | 325 | 425 | 26.10 | 6.77 | 2.75 | 6.14 | 2.90 | 1.11 | 19.67 |
| Epoxy coating, 2 part system, clear (material #51) | | | | | | | | | |
| Roll 1st coat | | | | | | | | | |
| Slow | 175 | 325 | 57.00 | 5.86 | 1.59 | 17.54 | 7.75 | 4.91 | 37.65 |
| Medium | 200 | 313 | 49.90 | 8.25 | 2.78 | 15.94 | 6.61 | 3.69 | 37.27 |
| Fast | 225 | 300 | 42.80 | 9.78 | 3.95 | 14.27 | 5.18 | 1.99 | 35.17 |
| Roll 2nd or additional coats | | | | | | | | | |
| Slow | 275 | 425 | 57.00 | 3.73 | 1.02 | 13.41 | 5.63 | 3.57 | 27.36 |
| Medium | 300 | 413 | 49.90 | 5.50 | 1.84 | 12.08 | 4.76 | 2.66 | 26.84 |
| Fast | 325 | 400 | 42.80 | 6.77 | 2.75 | 10.70 | 3.74 | 1.44 | 25.40 |
| Epoxy coating, 2 part system, white (material #52) | | | | | | | | | |
| Roll 1st coat | | | | | | | | | |
| Slow | 175 | 325 | 57.40 | 5.86 | 1.59 | 17.66 | 7.79 | 4.94 | 37.84 |
| Medium | 200 | 313 | 50.20 | 8.25 | 2.78 | 16.04 | 6.63 | 3.71 | 37.41 |
| Fast | 225 | 300 | 43.10 | 9.78 | 3.95 | 14.37 | 5.20 | 2.00 | 35.30 |
| Roll 2nd or additional coats | | | | | | | | | |
| Slow | 275 | 425 | 57.40 | 3.73 | 1.02 | 13.51 | 5.66 | 3.59 | 27.51 |
| Medium | 300 | 413 | 50.20 | 5.50 | 1.84 | 12.15 | 4.78 | 2.67 | 26.94 |
| Fast | 325 | 400 | 43.10 | 6.77 | 2.75 | 10.78 | 3.76 | 1.44 | 25.50 |

Use the pipe conversion factors in Figure 21 on page 321 to convert linear feet of pipe to square feet of surface. Vertical pipe runs require 2 to 3 times the manhours plus 10% more material. Solid color coded piping requires 15% to 25% more labor and material. For color bands on piping at 10' to 15' intervals, add the cost of an additional 1st coat. For heights above 8 feet, use the High Time Difficulty Factors on page 137. "Slow" work is based on a $10.25 hourly wage, "Medium" work on a $16.50 hourly wage, "Fast" work on a $22.50 hourly wage. Other qualifications that apply to this table are on page 9.

| | Labor SF per manhour | Material coverage SF/gallon | Material cost per gallon | Labor cost per 100 SF | Labor burden 100 SF | Material cost per 100 SF | Overhead per 100 SF | Profit per 100 SF | Total price per 100 SF |
|---|---|---|---|---|---|---|---|---|---|
| **Piping, insulated, canvas jacket, spray** | | | | | | | | | |
| Flat latex, water base (material #5) | | | | | | | | | |
| Spray 1st coat | | | | | | | | | |
| Slow | 225 | 100 | 16.20 | 4.56 | 1.23 | 16.20 | 6.82 | 4.32 | 33.13 |
| Medium | 250 | 88 | 14.20 | 6.60 | 2.22 | 16.14 | 6.12 | 3.42 | 34.50 |
| Fast | 275 | 75 | 12.10 | 8.00 | 3.25 | 16.13 | 5.06 | 1.95 | 34.39 |
| Spray 2nd coat | | | | | | | | | |
| Slow | 275 | 200 | 16.20 | 3.73 | 1.02 | 8.10 | 3.98 | 2.52 | 19.35 |
| Medium | 300 | 188 | 14.20 | 5.50 | 1.84 | 7.55 | 3.65 | 2.04 | 20.58 |
| Fast | 325 | 175 | 12.10 | 6.77 | 2.75 | 6.91 | 3.04 | 1.17 | 20.64 |
| Spray 3rd or additional coats | | | | | | | | | |
| Slow | 375 | 300 | 16.20 | 2.73 | .75 | 5.40 | 2.75 | 1.74 | 13.37 |
| Medium | 400 | 288 | 14.20 | 4.13 | 1.38 | 4.93 | 2.56 | 1.43 | 14.43 |
| Fast | 425 | 275 | 12.10 | 5.18 | 2.09 | 4.40 | 2.16 | .83 | 14.66 |
| Sealer, off white, water base (material #1) | | | | | | | | | |
| Spray 1 coat | | | | | | | | | |
| Slow | 225 | 100 | 15.50 | 4.56 | 1.23 | 15.50 | 6.60 | 4.19 | 32.08 |
| Medium | 250 | 88 | 13.50 | 6.60 | 2.22 | 15.34 | 5.92 | 3.31 | 33.39 |
| Fast | 275 | 75 | 11.60 | 8.00 | 3.25 | 15.47 | 4.94 | 1.90 | 33.56 |
| Sealer, off white, oil base (material #2) | | | | | | | | | |
| Spray 1 coat | | | | | | | | | |
| Slow | 225 | 150 | 20.70 | 4.56 | 1.23 | 13.80 | 6.08 | 3.85 | 29.52 |
| Medium | 250 | 138 | 18.10 | 6.60 | 2.22 | 13.12 | 5.38 | 3.01 | 30.33 |
| Fast | 275 | 125 | 15.50 | 8.00 | 3.25 | 12.40 | 4.37 | 1.68 | 29.70 |
| Enamel, water base latex (material #9) | | | | | | | | | |
| Spray 1st finish coat | | | | | | | | | |
| Slow | 275 | 200 | 22.80 | 3.73 | 1.02 | 11.40 | 5.01 | 3.17 | 24.33 |
| Medium | 300 | 188 | 19.90 | 5.50 | 1.84 | 10.59 | 4.40 | 2.46 | 24.79 |
| Fast | 325 | 175 | 17.10 | 6.77 | 2.75 | 9.77 | 3.57 | 1.37 | 24.23 |
| Spray 2nd or additional finish coats | | | | | | | | | |
| Slow | 375 | 300 | 22.80 | 2.73 | .75 | 7.60 | 3.43 | 2.18 | 16.69 |
| Medium | 400 | 288 | 19.90 | 4.13 | 1.38 | 6.91 | 3.05 | 1.70 | 17.17 |
| Fast | 425 | 275 | 17.10 | 5.18 | 2.09 | 6.22 | 2.50 | .96 | 16.95 |
| Industrial enamel, oil base, high gloss, light colors (material #56) | | | | | | | | | |
| Spray 1st finish coat | | | | | | | | | |
| Slow | 275 | 235 | 28.90 | 3.73 | 1.02 | 12.30 | 5.29 | 3.35 | 25.69 |
| Medium | 300 | 223 | 25.20 | 5.50 | 1.84 | 11.30 | 4.57 | 2.55 | 25.76 |
| Fast | 325 | 210 | 21.60 | 6.77 | 2.75 | 10.29 | 3.66 | 1.41 | 24.88 |

| | Labor SF per manhour | Material coverage SF/gallon | Material cost per gallon | Labor cost per 100 SF | Labor burden 100 SF | Material cost per 100 SF | Overhead per 100 SF | Profit per 100 SF | Total price per 100 SF |
|---|---|---|---|---|---|---|---|---|---|
| Spray 2nd or additional finish coats | | | | | | | | | |
| Slow | 375 | 335 | 28.90 | 2.73 | .75 | 8.63 | 3.75 | 2.38 | 18.24 |
| Medium | 400 | 323 | 25.20 | 4.13 | 1.38 | 7.80 | 3.26 | 1.82 | 18.39 |
| Fast | 425 | 310 | 21.60 | 5.18 | 2.09 | 6.97 | 2.64 | 1.01 | 17.89 |
| Industrial enamel, oil base, high gloss, dark (OSHA) colors (material #57) | | | | | | | | | |
| Spray 1st finish coat | | | | | | | | | |
| Slow | 275 | 235 | 34.90 | 3.73 | 1.02 | 14.85 | 6.08 | 3.85 | 29.53 |
| Medium | 300 | 223 | 30.50 | 5.50 | 1.84 | 13.68 | 5.15 | 2.88 | 29.05 |
| Fast | 325 | 210 | 26.10 | 6.77 | 2.75 | 12.43 | 4.06 | 1.56 | 27.57 |
| Spray 2nd or additional finish coats | | | | | | | | | |
| Slow | 375 | 335 | 34.90 | 2.73 | .75 | 10.42 | 4.31 | 2.73 | 20.94 |
| Medium | 400 | 323 | 30.50 | 4.13 | 1.38 | 9.44 | 3.67 | 2.05 | 20.67 |
| Fast | 425 | 310 | 26.10 | 5.18 | 2.09 | 8.42 | 2.90 | 1.12 | 19.71 |
| Epoxy coating, 2 part system, clear (material #51) | | | | | | | | | |
| Spray 1st coat | | | | | | | | | |
| Slow | 275 | 225 | 57.00 | 3.73 | 1.02 | 25.33 | 9.32 | 5.91 | 45.31 |
| Medium | 300 | 213 | 49.90 | 5.50 | 1.84 | 23.43 | 7.54 | 4.22 | 42.53 |
| Fast | 325 | 200 | 42.80 | 6.77 | 2.75 | 21.40 | 5.72 | 2.20 | 38.84 |
| Spray 2nd or additional coats | | | | | | | | | |
| Slow | 375 | 325 | 57.00 | 2.73 | .75 | 17.54 | 6.51 | 4.13 | 31.66 |
| Medium | 400 | 313 | 49.90 | 4.13 | 1.38 | 15.94 | 5.26 | 2.94 | 29.65 |
| Fast | 425 | 300 | 42.80 | 5.18 | 2.09 | 14.27 | 3.99 | 1.53 | 27.06 |
| Epoxy coating, 2 part system, white (material #52) | | | | | | | | | |
| Spray 1st coat | | | | | | | | | |
| Slow | 275 | 225 | 57.40 | 3.73 | 1.02 | 25.51 | 9.38 | 5.95 | 45.59 |
| Medium | 300 | 213 | 50.20 | 5.50 | 1.84 | 23.57 | 7.58 | 4.24 | 42.73 |
| Fast | 325 | 200 | 43.10 | 6.77 | 2.75 | 21.55 | 5.75 | 2.21 | 39.03 |
| Spray 2nd or additional coats | | | | | | | | | |
| Slow | 375 | 325 | 57.40 | 2.73 | .75 | 17.66 | 6.55 | 4.15 | 31.84 |
| Medium | 400 | 313 | 50.20 | 4.13 | 1.38 | 16.04 | 5.28 | 2.95 | 29.78 |
| Fast | 425 | 300 | 43.10 | 5.18 | 2.09 | 14.37 | 4.01 | 1.54 | 27.19 |

Use the pipe conversion factors in Figure 21 on page 321 to convert linear feet of pipe to square feet of surface. Vertical pipe runs require 2 to 3 times the manhours plus 10% more material. Solid color coded piping requires 15% to 25% more labor and material. For color bands on piping at 10' to 15' intervals, add the cost of an additional 1st coat. For heights above 8 feet, use the High Time Difficulty Factors on page 137. "Slow" work is based on a $10.25 hourly wage, "Medium" work on a $16.50 hourly wage, "Fast" work on a $22.50 hourly wage. Other qualifications that apply to this table are on page 9.

| | Labor SF per manhour | Material coverage SF/gallon | Material cost per gallon | Labor cost per 100 SF | Labor burden 100 SF | Material cost per 100 SF | Overhead per 100 SF | Profit per 100 SF | Total price per 100 SF |
|---|---|---|---|---|---|---|---|---|---|
| **Radiators** | | | | | | | | | |
| Brush each coat | | | | | | | | | |
| Metal primer, rust inhibitor, clean metal (material #35) | | | | | | | | | |
| Slow | 50 | 100 | 28.20 | 20.50 | 5.58 | 28.20 | 16.83 | 10.67 | 81.78 |
| Medium | 70 | 95 | 24.60 | 23.57 | 7.94 | 25.89 | 14.06 | 7.86 | 79.32 |
| Fast | 90 | 90 | 21.10 | 24.44 | 9.91 | 23.44 | 10.69 | 4.11 | 72.59 |
| Metal primer, rust inhibitor, rusty metal (material #36) | | | | | | | | | |
| Slow | 50 | 100 | 33.30 | 20.50 | 5.58 | 33.30 | 18.41 | 11.67 | 89.46 |
| Medium | 70 | 95 | 29.20 | 23.57 | 7.94 | 30.74 | 15.25 | 8.52 | 86.02 |
| Fast | 90 | 90 | 25.00 | 24.44 | 9.91 | 27.78 | 11.49 | 4.42 | 78.04 |
| Industrial enamel, oil base, high gloss, light colors (material #56) | | | | | | | | | |
| Slow | 60 | 150 | 28.90 | 17.08 | 4.66 | 19.27 | 12.71 | 8.06 | 61.78 |
| Medium | 80 | 138 | 25.20 | 20.63 | 6.93 | 18.26 | 11.23 | 6.28 | 63.33 |
| Fast | 100 | 125 | 21.60 | 22.00 | 8.92 | 17.28 | 8.92 | 3.43 | 60.55 |
| Industrial enamel, oil base, high gloss, dark (OSHA) colors (material #57) | | | | | | | | | |
| Slow | 60 | 150 | 34.90 | 17.08 | 4.66 | 23.27 | 13.95 | 8.84 | 67.80 |
| Medium | 80 | 138 | 30.50 | 20.63 | 6.93 | 22.10 | 12.17 | 6.80 | 68.63 |
| Fast | 100 | 125 | 26.10 | 22.00 | 8.92 | 20.88 | 9.58 | 3.68 | 65.06 |
| Spray each coat | | | | | | | | | |
| Metal primer, rust inhibitor, clean metal (material #35) | | | | | | | | | |
| Slow | 225 | 90 | 28.20 | 4.56 | 1.23 | 31.33 | 11.51 | 7.30 | 55.93 |
| Medium | 250 | 83 | 24.60 | 6.60 | 2.22 | 29.64 | 9.42 | 5.27 | 53.15 |
| Fast | 275 | 75 | 21.10 | 8.00 | 3.25 | 28.13 | 7.28 | 2.80 | 49.46 |
| Metal primer, rust inhibitor, rusty metal (material #36) | | | | | | | | | |
| Slow | 225 | 90 | 33.30 | 4.56 | 1.23 | 37.00 | 13.27 | 8.41 | 64.47 |
| Medium | 250 | 83 | 29.20 | 6.60 | 2.22 | 35.18 | 10.78 | 6.03 | 60.81 |
| Fast | 275 | 75 | 25.00 | 8.00 | 3.25 | 33.33 | 8.25 | 3.17 | 56.00 |
| Industrial enamel, oil base, high gloss, light colors (material #56) | | | | | | | | | |
| Slow | 250 | 110 | 28.90 | 4.10 | 1.12 | 26.27 | 9.76 | 6.19 | 47.44 |
| Medium | 275 | 100 | 25.20 | 6.00 | 2.03 | 25.20 | 8.14 | 4.55 | 45.92 |
| Fast | 300 | 90 | 21.60 | 7.33 | 2.97 | 24.00 | 6.35 | 2.44 | 43.09 |
| Industrial enamel, oil base, high gloss, dark (OSHA) colors (material #57) | | | | | | | | | |
| Slow | 250 | 110 | 34.90 | 4.10 | 1.12 | 31.73 | 11.45 | 7.26 | 55.66 |
| Medium | 275 | 100 | 30.50 | 6.00 | 2.03 | 30.50 | 9.44 | 5.28 | 53.25 |
| Fast | 300 | 90 | 26.10 | 7.33 | 2.97 | 29.00 | 7.27 | 2.79 | 49.36 |

Use these figures to estimate the cost of painting both sides of 6" to 18" deep hot water or steam radiators with oil or water base paint. Measurements are per square foot of area measured, one side (length times width). For heights above 8 feet, use the High Time Difficulty Factors on page 137. Note: A two coat system, prime and finish, using oil base material is recommended for any metal surface. Although water base material is often used, it may cause oxidation, corrosion and rust. One coat of oil base solid body stain is often used on interior or exterior metal but it may crack, peel or chip without the proper prime coat application. "Slow" work is based on a $10.25 hourly wage, "Medium" work on a $16.50 hourly wage, "Fast" work on a $22.50 hourly wage. Other qualifications that apply to this table are on page 9.

## Structural steel

Fabrication and erection estimates are usually based on weight of the steel in tons. As a paint estimator you need to convert tons of steel to square feet of surface. Of course, the conversion factor depends on the size of the steel members. On larger jobs and where accuracy is essential, use the Structural Steel conversion table in Figure 23 on pages 387 through 395 at the end of this Structural steel section to make exact conversions. On smaller jobs, a rule of thumb is that there are 225 square feet of paintable surface per ton of steel.

| | Labor SF per manhour | Material coverage SF/gallon | Material cost per gallon | Labor cost per 100 SF | Labor burden 100 SF | Material cost per 100 SF | Overhead per 100 SF | Profit per 100 SF | Total price per 100 SF |
|---|---|---|---|---|---|---|---|---|---|
| **Structural steel, heavy, brush application** | | | | | | | | | |
| Metal primer, rust inhibitor - clean metal (material #35) | | | | | | | | | |
| Brush prime coat | | | | | | | | | |
| Slow | 130 | 400 | 28.20 | 7.88 | 2.15 | 7.05 | 5.29 | 3.36 | 25.73 |
| Medium | 150 | 388 | 24.60 | 11.00 | 3.71 | 6.34 | 5.15 | 2.88 | 29.08 |
| Fast | 170 | 375 | 21.10 | 12.94 | 5.24 | 5.63 | 4.41 | 1.69 | 29.91 |
| Metal primer, rust inhibitor - rusty metal (material #36) | | | | | | | | | |
| Brush prime coat | | | | | | | | | |
| Slow | 130 | 400 | 33.30 | 7.88 | 2.15 | 8.33 | 5.69 | 3.61 | 27.66 |
| Medium | 150 | 388 | 29.20 | 11.00 | 3.71 | 7.53 | 5.45 | 3.04 | 30.73 |
| Fast | 170 | 375 | 25.00 | 12.94 | 5.24 | 6.67 | 4.60 | 1.77 | 31.22 |
| Industrial enamel, oil base, high gloss - light colors (material #56) | | | | | | | | | |
| Brush 1st or additional finish coats | | | | | | | | | |
| Slow | 175 | 425 | 28.90 | 5.86 | 1.59 | 6.80 | 4.42 | 2.80 | 21.47 |
| Medium | 200 | 413 | 25.20 | 8.25 | 2.78 | 6.10 | 4.20 | 2.35 | 23.68 |
| Fast | 225 | 400 | 21.60 | 9.78 | 3.95 | 5.40 | 3.54 | 1.36 | 24.03 |
| Industrial enamel, oil base, high gloss - dark (OSHA) colors (material #57) | | | | | | | | | |
| Brush 1st or additional finish coats | | | | | | | | | |
| Slow | 175 | 475 | 34.90 | 5.86 | 1.59 | 7.35 | 4.59 | 2.91 | 22.30 |
| Medium | 200 | 463 | 30.50 | 8.25 | 2.78 | 6.59 | 4.32 | 2.41 | 24.35 |
| Fast | 225 | 450 | 26.10 | 9.78 | 3.95 | 5.80 | 3.62 | 1.39 | 24.54 |
| Epoxy coating, 2 part system, clear (material #51) | | | | | | | | | |
| Brush 1st coat | | | | | | | | | |
| Slow | 130 | 425 | 57.00 | 7.88 | 2.15 | 13.41 | 7.27 | 4.61 | 35.32 |
| Medium | 150 | 413 | 49.90 | 11.00 | 3.71 | 12.08 | 6.56 | 3.67 | 37.02 |
| Fast | 170 | 400 | 42.80 | 12.94 | 5.24 | 10.70 | 5.34 | 2.05 | 36.27 |
| Brush 2nd or additional coats | | | | | | | | | |
| Slow | 175 | 450 | 57.00 | 5.86 | 1.59 | 12.67 | 6.24 | 3.96 | 30.32 |
| Medium | 200 | 438 | 49.90 | 8.25 | 2.78 | 11.39 | 5.49 | 3.07 | 30.98 |
| Fast | 225 | 425 | 42.80 | 9.78 | 3.95 | 10.07 | 4.41 | 1.69 | 29.90 |

| | Labor SF per manhour | Material coverage SF/gallon | Material cost per gallon | Labor cost per 100 SF | Labor burden 100 SF | Material cost per 100 SF | Overhead per 100 SF | Profit per 100 SF | Total price per 100 SF |
|---|---|---|---|---|---|---|---|---|---|
| Epoxy coating, 2 part system, white (material #52) | | | | | | | | | |
| Brush 1st coat | | | | | | | | | |
| Slow | 130 | 425 | 57.40 | 7.88 | 2.15 | 13.51 | 7.30 | 4.63 | 35.47 |
| Medium | 150 | 413 | 50.20 | 11.00 | 3.71 | 12.15 | 6.58 | 3.68 | 37.12 |
| Fast | 170 | 400 | 43.10 | 12.94 | 5.24 | 10.78 | 5.36 | 2.06 | 36.38 |
| Brush 2nd or additional coats | | | | | | | | | |
| Slow | 175 | 450 | 57.40 | 5.86 | 1.59 | 12.76 | 6.27 | 3.97 | 30.45 |
| Medium | 200 | 438 | 50.20 | 8.25 | 2.78 | 11.46 | 5.51 | 3.08 | 31.08 |
| Fast | 225 | 425 | 43.10 | 9.78 | 3.95 | 10.14 | 4.42 | 1.70 | 29.99 |

For field painting at heights above 8 feet, use the High Time Difficulty Factors on page 137. Heavy structural steel has from 100 to 150 square feet of surface area per ton. Extra heavy structural steel has from 50 to 100 square feet of surface area per ton. Rule of thumb: When coatings are applied by brush, a journeyman painter will apply a first coat on 6 to 7 tons per 8 hour day. When coatings are applied by spray, figure output at 0.2 hours per ton and material use at about 0.2 gallons per ton. Use Figure 23 on pages 387 to 395 to convert structural steel linear feet or tonnage to surface area. Note: A two coat system, prime and finish, using oil base material is recommended for any metal surface. Although water base material is often used, it may cause oxidation, corrosion and rust. One coat of oil base solid body stain is often used on interior or exterior metal but it may crack, peel or chip without the proper prime coat application. "Slow" work is based on a $10.25 hourly wage, "Medium" work on a $16.50 hourly wage, and "Fast" work on a $22.50 hourly wage. Other qualifications that apply to this table are on page 9.

| | Labor SF per manhour | Material coverage SF/gallon | Material cost per gallon | Labor cost per 100 SF | Labor burden 100 SF | Material cost per 100 SF | Overhead per 100 SF | Profit per 100 SF | Total price per 100 SF |
|---|---|---|---|---|---|---|---|---|---|
| **Structural steel, heavy, roll application** | | | | | | | | | |
| Metal primer, rust inhibitor - clean metal (material #35) | | | | | | | | | |
| Roll prime coat | | | | | | | | | |
| Slow | 250 | 390 | 28.20 | 4.10 | 1.12 | 7.23 | 3.86 | 2.45 | 18.76 |
| Medium | 275 | 378 | 24.60 | 6.00 | 2.03 | 6.51 | 3.56 | 1.99 | 20.09 |
| Fast | 300 | 365 | 21.10 | 7.33 | 2.97 | 5.78 | 2.97 | 1.14 | 20.19 |
| Metal primer, rust inhibitor - rusty metal (material #36) | | | | | | | | | |
| Roll prime coat | | | | | | | | | |
| Slow | 250 | 380 | 33.30 | 4.10 | 1.12 | 8.76 | 4.33 | 2.75 | 21.06 |
| Medium | 275 | 363 | 29.20 | 6.00 | 2.03 | 8.04 | 3.93 | 2.20 | 22.20 |
| Fast | 300 | 355 | 25.00 | 7.33 | 2.97 | 7.04 | 3.21 | 1.23 | 21.78 |
| Industrial enamel, oil base, high gloss - light colors (material #56) | | | | | | | | | |
| Roll 1st or additional finish coats | | | | | | | | | |
| Slow | 275 | 425 | 28.90 | 3.73 | 1.02 | 6.80 | 3.58 | 2.27 | 17.40 |
| Medium | 300 | 413 | 25.20 | 5.50 | 1.84 | 6.10 | 3.30 | 1.84 | 18.58 |
| Fast | 325 | 400 | 21.60 | 6.77 | 2.75 | 5.40 | 2.76 | 1.06 | 18.74 |
| Industrial enamel, oil base, high gloss - dark (OSHA) colors (material #57) | | | | | | | | | |
| Roll 1st or additional finish coats | | | | | | | | | |
| Slow | 275 | 440 | 34.90 | 3.73 | 1.02 | 7.93 | 3.93 | 2.49 | 19.10 |
| Medium | 300 | 428 | 30.50 | 5.50 | 1.84 | 7.13 | 3.55 | 1.98 | 20.00 |
| Fast | 325 | 415 | 26.10 | 6.77 | 2.75 | 6.29 | 2.92 | 1.12 | 19.85 |
| Epoxy coating, 2 part system, clear (material #51) | | | | | | | | | |
| Roll 1st coat | | | | | | | | | |
| Slow | 250 | 400 | 57.00 | 4.10 | 1.12 | 14.25 | 6.04 | 3.83 | 29.34 |
| Medium | 275 | 388 | 49.90 | 6.00 | 2.03 | 12.86 | 5.12 | 2.86 | 28.87 |
| Fast | 300 | 375 | 42.80 | 7.33 | 2.97 | 11.41 | 4.02 | 1.54 | 27.27 |
| Roll 2nd or additional coats | | | | | | | | | |
| Slow | 275 | 425 | 57.00 | 3.73 | 1.02 | 13.41 | 5.63 | 3.57 | 27.36 |
| Medium | 300 | 413 | 49.90 | 5.50 | 1.84 | 12.08 | 4.76 | 2.66 | 26.84 |
| Fast | 325 | 400 | 42.80 | 6.77 | 2.75 | 10.70 | 3.74 | 1.44 | 25.40 |

| | Labor SF per manhour | Material coverage SF/gallon | Material cost per gallon | Labor cost per 100 SF | Labor burden 100 SF | Material cost per 100 SF | Overhead per 100 SF | Profit per 100 SF | Total price per 100 SF |
|---|---|---|---|---|---|---|---|---|---|
| Epoxy coating, 2 part system, white (material #52) | | | | | | | | | |
| Roll 1st coat | | | | | | | | | |
| Slow | 250 | 400 | 57.40 | 4.10 | 1.12 | 14.35 | 6.07 | 3.85 | 29.49 |
| Medium | 275 | 388 | 50.20 | 6.00 | 2.03 | 12.94 | 5.14 | 2.87 | 28.98 |
| Fast | 300 | 375 | 43.10 | 7.33 | 2.97 | 11.49 | 4.03 | 1.55 | 27.37 |
| Roll 2nd or additional coats | | | | | | | | | |
| Slow | 275 | 425 | 57.40 | 3.73 | 1.02 | 13.51 | 5.66 | 3.59 | 27.51 |
| Medium | 300 | 413 | 50.20 | 5.50 | 1.84 | 12.15 | 4.78 | 2.67 | 26.94 |
| Fast | 325 | 400 | 43.10 | 6.77 | 2.75 | 10.78 | 3.76 | 1.44 | 25.50 |

For field painting at heights above 8 feet, use the High Time Difficulty Factors on page 137. Heavy structural steel has from 100 to 150 square feet of surface area per ton. Extra heavy structural steel has from 50 to 100 square feet of surface area per ton. Rule of thumb: When coatings are applied by brush, a journeyman painter will apply a first coat on from 6 to 7 tons per 8 hour day. When coatings are applied by spray, figure output at 0.2 hours per ton and material use at about 0.2 gallons per ton. Use Figure 23 on pages 387 to 395 to convert structural steel linear feet or tonnage to surface area. Note: A two coat system, prime and finish, using oil base material is recommended for any metal surface. Although water base material is often used, it may cause oxidation, corrosion and rust. One coat of oil base solid body stain is often used on interior or exterior metal but it may crack, peel or chip without the proper prime coat application. "Slow" work is based on a $10.25 hourly wage, "Medium" work on a $16.50 hourly wage, and "Fast" work on a $22.50 hourly wage. Other qualifications that apply to this table are on page 9.

| | Labor SF per manhour | Material coverage SF/gallon | Material cost per gallon | Labor cost per 100 SF | Labor burden 100 SF | Material cost per 100 SF | Overhead per 100 SF | Profit per 100 SF | Total price per 100 SF |
|---|---|---|---|---|---|---|---|---|---|
| **Structural steel, heavy, spray application** | | | | | | | | | |
| Metal primer, rust inhibitor - clean metal (material #35) | | | | | | | | | |
| Spray prime coat | | | | | | | | | |
| Slow | 650 | 325 | 28.20 | 1.58 | .43 | 8.68 | 3.31 | 2.10 | 16.10 |
| Medium | 750 | 313 | 24.60 | 2.20 | .73 | 7.86 | 2.65 | 1.48 | 14.92 |
| Fast | 850 | 300 | 21.10 | 2.59 | 1.06 | 7.03 | 1.97 | .76 | 13.41 |
| Metal primer, rust inhibitor - rusty metal (material #36) | | | | | | | | | |
| Spray prime coat | | | | | | | | | |
| Slow | 650 | 300 | 33.30 | 1.58 | .43 | 11.10 | 4.06 | 2.58 | 19.75 |
| Medium | 750 | 288 | 29.20 | 2.20 | .73 | 10.14 | 3.20 | 1.79 | 18.06 |
| Fast | 850 | 275 | 25.00 | 2.59 | 1.06 | 9.09 | 2.36 | .91 | 16.01 |
| Industrial enamel, oil base, high gloss - light colors (material #56) | | | | | | | | | |
| Spray 1st or additional finish coats | | | | | | | | | |
| Slow | 750 | 340 | 28.90 | 1.37 | .36 | 8.50 | 3.17 | 2.01 | 15.41 |
| Medium | 850 | 325 | 25.20 | 1.94 | .66 | 7.75 | 2.53 | 1.42 | 14.30 |
| Fast | 950 | 310 | 21.60 | 2.32 | .93 | 6.97 | 1.89 | .73 | 12.84 |
| Industrial enamel, oil base, high gloss - dark (OSHA) colors (material #57) | | | | | | | | | |
| Spray 1st or additional finish coats | | | | | | | | | |
| Slow | 750 | 365 | 34.90 | 1.37 | .36 | 9.56 | 3.50 | 2.22 | 17.01 |
| Medium | 850 | 350 | 30.50 | 1.94 | .66 | 8.71 | 2.77 | 1.55 | 15.63 |
| Fast | 950 | 335 | 26.10 | 2.32 | .93 | 7.79 | 2.04 | .79 | 13.87 |
| Epoxy coating, 2 part system, clear (material #51) | | | | | | | | | |
| Spray 1st coat | | | | | | | | | |
| Slow | 650 | 325 | 57.00 | 1.58 | .43 | 17.54 | 6.06 | 3.84 | 29.45 |
| Medium | 750 | 313 | 49.90 | 2.20 | .73 | 15.94 | 4.63 | 2.59 | 26.09 |
| Fast | 850 | 300 | 42.80 | 2.59 | 1.06 | 14.27 | 3.31 | 1.27 | 22.50 |
| Spray 2nd or additional coats | | | | | | | | | |
| Slow | 750 | 350 | 57.00 | 1.37 | .36 | 16.29 | 5.59 | 3.54 | 27.15 |
| Medium | 850 | 338 | 49.90 | 1.94 | .66 | 14.76 | 4.25 | 2.38 | 23.99 |
| Fast | 950 | 325 | 42.80 | 2.32 | .93 | 13.17 | 3.04 | 1.17 | 20.63 |

| | Labor SF per manhour | Material coverage SF/gallon | Material cost per gallon | Labor cost per 100 SF | Labor burden 100 SF | Material cost per 100 SF | Overhead per 100 SF | Profit per 100 SF | Total price per 100 SF |
|---|---|---|---|---|---|---|---|---|---|
| Epoxy coating, 2 part system, white (material #52) | | | | | | | | | |
| Spray 1st coat | | | | | | | | | |
| Slow | 650 | 325 | 57.40 | 1.58 | .43 | 17.66 | 6.10 | 3.87 | 29.64 |
| Medium | 750 | 313 | 50.20 | 2.20 | .73 | 16.04 | 4.65 | 2.60 | 26.22 |
| Fast | 850 | 300 | 43.10 | 2.59 | 1.06 | 14.37 | 3.33 | 1.28 | 22.63 |
| Spray 2nd or additional coats | | | | | | | | | |
| Slow | 750 | 350 | 57.40 | 1.37 | .36 | 16.40 | 5.62 | 3.56 | 27.31 |
| Medium | 850 | 338 | 50.20 | 1.94 | .66 | 14.85 | 4.27 | 2.39 | 24.11 |
| Fast | 950 | 325 | 43.10 | 2.32 | .93 | 13.26 | 3.06 | 1.17 | 20.74 |

For field painting at heights above 8 feet, use the High Time Difficulty Factors on page 137. Heavy structural steel has from 100 to 150 square feet of surface area per ton. Extra heavy structural steel has from 50 to 100 square feet of surface area per ton. Rule of thumb: When coatings are applied by brush, a journeyman painter will apply a first coat on from 6 to 7 tons per 8 hour day. When coatings are applied by spray, figure output at 0.2 hours per ton and material use at about 0.2 gallons per ton. Use Figure 23 on pages 387 to 395 to convert structural steel linear feet or tonnage to surface area. Note: A two coat system, prime and finish, using oil base material is recommended for any metal surface. Although water base material is often used, it may cause oxidation, corrosion and rust. One coat of oil base solid body stain is often used on interior or exterior metal but it may crack, peel or chip without the proper prime coat application. "Slow" work is based on a $10.25 hourly wage, "Medium" work on a $16.50 hourly wage, and "Fast" work on a $22.50 hourly wage. Other qualifications that apply to this table are on page 9.

| | Labor SF per manhour | Material coverage SF/gallon | Material cost per gallon | Labor cost per 100 SF | Labor burden 100 SF | Material cost per 100 SF | Overhead per 100 SF | Profit per 100 SF | Total price per 100 SF |
|---|---|---|---|---|---|---|---|---|---|
| **Structural steel, light, brush application** | | | | | | | | | |
| Metal primer, rust inhibitor, clean metal (material #35) | | | | | | | | | |
| Brush prime coat | | | | | | | | | |
| Slow | 60 | 425 | 28.20 | 17.08 | 4.66 | 6.64 | 8.79 | 5.57 | 42.74 |
| Medium | 80 | 413 | 24.60 | 20.63 | 6.93 | 5.96 | 8.21 | 4.59 | 46.32 |
| Fast | 100 | 400 | 21.10 | 22.00 | 8.92 | 5.28 | 6.70 | 2.57 | 45.47 |
| Metal primer, rust inhibitor, rusty metal (material #36) | | | | | | | | | |
| Brush prime coat | | | | | | | | | |
| Slow | 60 | 400 | 33.30 | 17.08 | 4.66 | 8.33 | 9.32 | 5.91 | 45.30 |
| Medium | 80 | 388 | 29.20 | 20.63 | 6.93 | 7.53 | 8.60 | 4.81 | 48.50 |
| Fast | 100 | 375 | 25.00 | 22.00 | 8.92 | 6.67 | 6.95 | 2.67 | 47.21 |
| Industrial enamel, oil base, high gloss, light colors (material #56) | | | | | | | | | |
| Brush 1st or additional finish coats | | | | | | | | | |
| Slow | 80 | 425 | 28.90 | 12.81 | 3.49 | 6.80 | 7.16 | 4.54 | 34.80 |
| Medium | 100 | 413 | 25.20 | 16.50 | 5.55 | 6.10 | 6.90 | 3.86 | 38.91 |
| Fast | 120 | 400 | 21.60 | 18.33 | 7.43 | 5.40 | 5.76 | 2.22 | 39.14 |
| Industrial enamel, oil base, high gloss, dark (OSHA) colors (material #57) | | | | | | | | | |
| Brush 1st or additional finish coats | | | | | | | | | |
| Slow | 80 | 475 | 34.90 | 12.81 | 3.49 | 7.35 | 7.33 | 4.65 | 35.63 |
| Medium | 100 | 463 | 30.50 | 16.50 | 5.55 | 6.59 | 7.02 | 3.92 | 39.58 |
| Fast | 120 | 450 | 26.10 | 18.33 | 7.43 | 5.80 | 5.84 | 2.24 | 39.64 |
| Epoxy coating, 2 part system, clear (material #51) | | | | | | | | | |
| Brush 1st coat | | | | | | | | | |
| Slow | 60 | 425 | 57.00 | 17.08 | 4.66 | 13.41 | 10.89 | 6.90 | 52.94 |
| Medium | 80 | 413 | 49.90 | 20.63 | 6.93 | 12.08 | 9.71 | 5.43 | 54.78 |
| Fast | 100 | 400 | 42.80 | 22.00 | 8.92 | 10.70 | 7.70 | 2.96 | 52.28 |
| Brush 2nd or additional coats | | | | | | | | | |
| Slow | 80 | 450 | 57.00 | 12.81 | 3.49 | 12.67 | 8.98 | 5.69 | 43.64 |
| Medium | 100 | 438 | 49.90 | 16.50 | 5.55 | 11.39 | 8.19 | 4.58 | 46.21 |
| Fast | 120 | 425 | 42.80 | 18.33 | 7.43 | 10.07 | 6.63 | 2.55 | 45.01 |

| | Labor SF per manhour | Material coverage SF/gallon | Material cost per gallon | Labor cost per 100 SF | Labor burden 100 SF | Material cost per 100 SF | Overhead per 100 SF | Profit per 100 SF | Total price per 100 SF |
|---|---|---|---|---|---|---|---|---|---|
| Epoxy coating, 2 part system, white (material #52) | | | | | | | | | |
| Brush 1st coat | | | | | | | | | |
| Slow | 60 | 425 | 57.40 | 17.08 | 4.66 | 13.51 | 10.92 | 6.92 | 53.09 |
| Medium | 80 | 413 | 50.20 | 20.63 | 6.93 | 12.15 | 9.73 | 5.44 | 54.88 |
| Fast | 100 | 400 | 43.10 | 22.00 | 8.92 | 10.78 | 7.71 | 2.96 | 52.37 |
| Brush 2nd or additional coats | | | | | | | | | |
| Slow | 80 | 450 | 57.40 | 12.81 | 3.49 | 12.76 | 9.01 | 5.71 | 43.78 |
| Medium | 100 | 438 | 50.20 | 16.50 | 5.55 | 11.46 | 8.21 | 4.59 | 46.31 |
| Fast | 120 | 425 | 43.10 | 18.33 | 7.43 | 10.14 | 6.64 | 2.55 | 45.09 |

For field painting at heights above 8 feet, use the High Time Difficulty Factors on page 137. Light structural steel has from 300 to 500 square feet of surface per ton. As a comparison, when coatings are applied by brush, a journeyman painter will apply a first coat on from 2 to 3 tons per 8 hour day. A second and subsequent coats can be applied on 3 to 4 tons per day. Use Figure 23 on pages 387 to 395 to convert structural steel linear feet or tonnage to surface area. Note: A two coat system, prime and finish, using oil base material is recommended for any metal surface. Although water base material is often used, it may cause oxidation, corrosion and rust. One coat of oil base solid body stain is often used on interior or exterior metal but it may crack, peel or chip without the proper prime coat application. "Slow" work is based on a $10.25 hourly wage, "Medium" work on a $16.50 hourly wage, "Fast" work on a $22.00 hourly wage. Other qualifications that apply to this table are on page 9.

| | Labor SF per manhour | Material coverage SF/gallon | Material cost per gallon | Labor cost per 100 SF | Labor burden 100 SF | Material cost per 100 SF | Overhead per 100 SF | Profit per 100 SF | Total price per 100 SF |
|---|---|---|---|---|---|---|---|---|---|
| **Structural steel, light, roll and brush application** | | | | | | | | | |
| Metal primer, rust inhibitor, clean metal (material #35) | | | | | | | | | |
| Roll and brush prime coat | | | | | | | | | |
| Slow | 125 | 390 | 28.20 | 8.20 | 2.23 | 7.23 | 5.47 | 3.47 | 26.60 |
| Medium | 150 | 378 | 24.60 | 11.00 | 3.71 | 6.51 | 5.20 | 2.91 | 29.33 |
| Fast | 175 | 365 | 21.10 | 12.57 | 5.09 | 5.78 | 4.34 | 1.67 | 29.45 |
| Metal primer, rust inhibitor, rusty metal (material #36) | | | | | | | | | |
| Roll and brush prime coat | | | | | | | | | |
| Slow | 125 | 380 | 33.30 | 8.20 | 2.23 | 8.76 | 5.95 | 3.77 | 28.91 |
| Medium | 150 | 368 | 29.20 | 11.00 | 3.71 | 7.93 | 5.54 | 3.10 | 31.28 |
| Fast | 175 | 355 | 25.00 | 12.57 | 5.09 | 7.04 | 4.57 | 1.76 | 31.03 |
| Industrial enamel, oil base, high gloss, light colors (material #56) | | | | | | | | | |
| Roll and brush 1st or additional finish coats | | | | | | | | | |
| Slow | 175 | 390 | 28.90 | 5.86 | 1.59 | 7.41 | 4.61 | 2.92 | 22.39 |
| Medium | 200 | 378 | 25.20 | 8.25 | 2.78 | 6.67 | 4.34 | 2.42 | 24.46 |
| Fast | 225 | 365 | 21.60 | 9.78 | 3.95 | 5.92 | 3.64 | 1.40 | 24.69 |
| Industrial enamel, oil base, high gloss, dark (OSHA) colors (material #57) | | | | | | | | | |
| Roll and brush 1st or additional finish coats | | | | | | | | | |
| Slow | 175 | 440 | 34.90 | 5.86 | 1.59 | 7.93 | 4.77 | 3.02 | 23.17 |
| Medium | 200 | 428 | 30.50 | 8.25 | 2.78 | 7.13 | 4.45 | 2.49 | 25.10 |
| Fast | 225 | 415 | 26.10 | 9.78 | 3.95 | 6.29 | 3.71 | 1.43 | 25.16 |
| Epoxy coating, 2 part system, clear (material #51) | | | | | | | | | |
| Roll and brush 1st coat | | | | | | | | | |
| Slow | 125 | 400 | 57.00 | 8.20 | 2.23 | 14.25 | 7.65 | 4.85 | 37.18 |
| Medium | 150 | 388 | 49.90 | 11.00 | 3.71 | 12.86 | 6.75 | 3.77 | 38.09 |
| Fast | 175 | 375 | 42.80 | 12.57 | 5.09 | 11.41 | 5.38 | 2.07 | 36.52 |
| Roll and brush 2nd or additional coats | | | | | | | | | |
| Slow | 175 | 425 | 57.00 | 5.86 | 1.59 | 13.41 | 6.47 | 4.10 | 31.43 |
| Medium | 200 | 413 | 49.90 | 8.25 | 2.78 | 12.08 | 5.66 | 3.16 | 31.93 |
| Fast | 225 | 400 | 42.80 | 9.78 | 3.95 | 10.70 | 4.52 | 1.74 | 30.69 |

| | Labor SF per manhour | Material coverage SF/gallon | Material cost per gallon | Labor cost per 100 SF | Labor burden 100 SF | Material cost per 100 SF | Overhead per 100 SF | Profit per 100 SF | Total price per 100 SF |
|---|---|---|---|---|---|---|---|---|---|
| Epoxy coating, 2 part system, white (material #52) | | | | | | | | | |
| Roll and brush 1st coat | | | | | | | | | |
| Slow | 125 | 400 | 57.40 | 8.20 | 2.23 | 14.35 | 7.68 | 4.87 | 37.33 |
| Medium | 150 | 388 | 50.20 | 11.00 | 3.71 | 12.94 | 6.77 | 3.79 | 38.21 |
| Fast | 175 | 375 | 43.10 | 12.57 | 5.09 | 11.49 | 5.39 | 2.07 | 36.61 |
| Roll and brush 2nd or additional coats | | | | | | | | | |
| Slow | 175 | 425 | 57.40 | 5.86 | 1.59 | 13.51 | 6.50 | 4.12 | 31.58 |
| Medium | 200 | 413 | 50.20 | 8.25 | 2.78 | 12.15 | 5.68 | 3.17 | 32.03 |
| Fast | 225 | 400 | 43.10 | 9.78 | 3.95 | 10.78 | 4.54 | 1.74 | 30.79 |

For field painting at heights above 8 feet, use the High Time Difficulty Factors on page 137. Light structural steel has from 300 to 500 square feet of surface per ton. As a comparison, when coatings are applied by brush, a journeyman painter will apply a first coat on from 2 to 3 tons per 8 hour day. A second and subsequent coats can be applied on 3 to 4 tons per day. Use Figure 23 on pages 387 to 395 to convert structural steel linear feet or tonnage to surface area. Note: A two coat system, prime and finish, using oil base material is recommended for any metal surface. Although water base material is often used, it may cause oxidation, corrosion and rust. One coat of oil base solid body stain is often used on interior or exterior metal but it may crack, peel or chip without the proper prime coat application. "Slow" work is based on a $10.25 hourly wage, "Medium" work on a $16.50 hourly wage, "Fast" work on a $22.00 hourly wage. Other qualifications that apply to this table are on page 9.

| | Labor SF per manhour | Material coverage SF/gallon | Material cost per gallon | Labor cost per 100 SF | Labor burden 100 SF | Material cost per 100 SF | Overhead per 100 SF | Profit per 100 SF | Total price per 100 SF |
|---|---|---|---|---|---|---|---|---|---|
| **Structural steel, light, spray application** | | | | | | | | | |
| Metal primer, rust inhibitor, clean metal (material #35) | | | | | | | | | |
| Spray prime coat | | | | | | | | | |
| Slow | 400 | 325 | 28.20 | 2.56 | .70 | 8.68 | 3.70 | 2.35 | 17.99 |
| Medium | 500 | 313 | 24.60 | 3.30 | 1.11 | 7.86 | 3.01 | 1.68 | 16.96 |
| Fast | 600 | 300 | 21.10 | 3.67 | 1.49 | 7.03 | 2.26 | .87 | 15.32 |
| Metal primer, rust inhibitor, rusty metal (material #36) | | | | | | | | | |
| Spray prime coat | | | | | | | | | |
| Slow | 400 | 300 | 33.30 | 2.56 | .70 | 11.10 | 4.45 | 2.82 | 21.63 |
| Medium | 500 | 288 | 29.20 | 3.30 | 1.11 | 10.14 | 3.56 | 1.99 | 20.10 |
| Fast | 600 | 275 | 25.00 | 3.67 | 1.49 | 9.09 | 2.64 | 1.01 | 17.90 |
| Industrial enamel, oil base, high gloss, light colors (material #56) | | | | | | | | | |
| Spray 1st or additional finish coats | | | | | | | | | |
| Slow | 500 | 325 | 28.90 | 2.05 | .56 | 8.89 | 3.57 | 2.26 | 17.33 |
| Medium | 600 | 313 | 25.20 | 2.75 | .93 | 8.05 | 2.87 | 1.61 | 16.21 |
| Fast | 700 | 300 | 21.60 | 3.14 | 1.28 | 7.20 | 2.15 | .83 | 14.60 |
| Industrial enamel, oil base, high gloss, dark (OSHA) colors (material #57) | | | | | | | | | |
| Spray 1st or additional finish coats | | | | | | | | | |
| Slow | 500 | 365 | 34.90 | 2.05 | .56 | 9.56 | 3.77 | 2.39 | 18.33 |
| Medium | 600 | 350 | 30.50 | 2.75 | .93 | 8.71 | 3.04 | 1.70 | 17.13 |
| Fast | 700 | 335 | 26.10 | 3.14 | 1.28 | 7.79 | 2.26 | .87 | 15.34 |
| Epoxy coating, 2 part system, clear (material #51) | | | | | | | | | |
| Spray 1st coat | | | | | | | | | |
| Slow | 400 | 325 | 57.00 | 2.56 | .70 | 17.54 | 6.45 | 4.09 | 31.34 |
| Medium | 500 | 313 | 49.90 | 3.30 | 1.11 | 15.94 | 4.99 | 2.79 | 28.13 |
| Fast | 600 | 300 | 42.80 | 3.67 | 1.49 | 14.27 | 3.59 | 1.38 | 24.40 |
| Spray 2nd or additional coats | | | | | | | | | |
| Slow | 500 | 350 | 57.00 | 2.05 | .56 | 16.29 | 5.86 | 3.71 | 28.47 |
| Medium | 600 | 338 | 49.90 | 2.75 | .93 | 14.76 | 4.52 | 2.53 | 25.49 |
| Fast | 700 | 325 | 42.80 | 3.14 | 1.28 | 13.17 | 3.25 | 1.25 | 22.09 |

| | Labor SF per manhour | Material coverage SF/gallon | Material cost per gallon | Labor cost per 100 SF | Labor burden 100 SF | Material cost per 100 SF | Overhead per 100 SF | Profit per 100 SF | Total price per 100 SF |
|---|---|---|---|---|---|---|---|---|---|
| Epoxy coating, 2 part system, white (material #52) | | | | | | | | | |
| Spray 1st coat | | | | | | | | | |
| Slow | 400 | 325 | 57.40 | 2.56 | .70 | 17.66 | 6.49 | 4.11 | 31.52 |
| Medium | 500 | 313 | 50.20 | 3.30 | 1.11 | 16.04 | 5.01 | 2.80 | 28.26 |
| Fast | 600 | 300 | 43.10 | 3.67 | 1.49 | 14.37 | 3.61 | 1.39 | 24.53 |
| Spray 2nd or additional coats | | | | | | | | | |
| Slow | 500 | 350 | 57.40 | 2.05 | .56 | 16.40 | 5.89 | 3.74 | 28.64 |
| Medium | 600 | 338 | 50.20 | 2.75 | .93 | 14.85 | 4.54 | 2.54 | 25.61 |
| Fast | 700 | 325 | 43.10 | 3.14 | 1.28 | 13.26 | 3.27 | 1.26 | 22.21 |

For field painting at heights above 8 feet, use the High Time Difficulty Factors on page 137. Light structural steel has from 300 to 500 square feet of surface per ton. The rule of thumb for labor output and material usage for spray application on light structural steel is shown in the table on the next page. Use Figure 23 on pages 387 to 395 to convert structural steel linear feet or tonnage to surface area. Note: A two coat system, prime and finish, using oil base material is recommended for any metal surface. Although water base material is often used, it may cause oxidation, corrosion and rust. One coat of oil base solid body stain is often used on interior or exterior metal but it may crack, peel or chip without the proper prime coat application. "Slow" work is based on a $10.25 hourly wage, "Medium" work on a $16.50 hourly wage, "Fast" work on a $22.50 hourly wage. Other qualifications that apply to this table are on page 9.

| | Labor manhours per ton | Material gallons per ton | Material cost per gallon | Labor cost per ton | Labor burden per ton | Material cost per ton | Overhead per ton | Profit per ton | Total price per ton |
|---|---|---|---|---|---|---|---|---|---|
| **Structural steel, light, coating rule of thumb, spray, per ton** | | | | | | | | | |
| Metal primer, rust inhibitor, clean metal (material #35) | | | | | | | | | |
| Spray prime coat | | | | | | | | | |
| Slow | 1.8 | 1.0 | 28.20 | 18.45 | 5.02 | 28.20 | 16.02 | 10.16 | 77.85 |
| Medium | 1.6 | 1.1 | 24.60 | 26.40 | 8.88 | 27.06 | 15.27 | 8.54 | 86.15 |
| Fast | 1.4 | 1.2 | 21.10 | 30.80 | 12.49 | 25.32 | 12.69 | 4.88 | 86.18 |
| Metal primer, rust inhibitor, rusty metal (material #36) | | | | | | | | | |
| Spray prime coat | | | | | | | | | |
| Slow | 1.8 | 1.0 | 33.30 | 18.45 | 5.02 | 33.30 | 17.60 | 11.16 | 85.53 |
| Medium | 1.6 | 1.1 | 29.20 | 26.40 | 8.88 | 32.12 | 16.51 | 9.23 | 93.14 |
| Fast | 1.4 | 1.2 | 25.00 | 30.80 | 12.49 | 30.00 | 13.56 | 5.21 | 92.06 |
| Industrial enamel, oil base, high gloss, light colors (material #56) | | | | | | | | | |
| Spray 1st finish coat | | | | | | | | | |
| Slow | 1.5 | 0.9 | 28.90 | 15.38 | 4.18 | 26.01 | 14.13 | 8.96 | 68.66 |
| Medium | 1.3 | 1.0 | 25.20 | 21.45 | 7.22 | 25.20 | 13.20 | 7.38 | 74.45 |
| Fast | 1.1 | 1.1 | 21.60 | 24.20 | 9.81 | 23.76 | 10.69 | 4.11 | 72.57 |
| Spray 2nd or additional finish coats | | | | | | | | | |
| Slow | 1.1 | 0.8 | 28.90 | 11.28 | 3.06 | 23.12 | 11.62 | 7.36 | 56.44 |
| Medium | 1.0 | 0.9 | 25.20 | 16.50 | 5.55 | 22.68 | 10.96 | 6.13 | 61.82 |
| Fast | 0.9 | 1.0 | 21.60 | 19.80 | 8.03 | 21.60 | 9.14 | 3.51 | 62.08 |
| Industrial enamel, oil base, high gloss, dark (OSHA) colors (material #57) | | | | | | | | | |
| Spray 1st finish coat | | | | | | | | | |
| Slow | 1.5 | 0.9 | 34.90 | 15.38 | 4.18 | 31.41 | 15.80 | 10.02 | 76.79 |
| Medium | 1.3 | 1.0 | 30.50 | 21.45 | 7.22 | 30.50 | 14.50 | 8.10 | 81.77 |
| Fast | 1.1 | 1.1 | 26.10 | 24.20 | 9.81 | 28.71 | 11.60 | 4.46 | 78.78 |
| Spray 2nd or additional finish coats | | | | | | | | | |
| Slow | 1.1 | 0.8 | 34.90 | 11.28 | 3.06 | 27.92 | 13.10 | 8.31 | 63.67 |
| Medium | 1.0 | 0.9 | 30.50 | 16.50 | 5.55 | 27.45 | 12.13 | 6.78 | 68.41 |
| Fast | 0.9 | 1.0 | 26.10 | 19.80 | 8.03 | 26.10 | 9.98 | 3.83 | 67.74 |

For field painting at heights above 8 feet, use the High Time Difficulty Factors on page 137. Note: A two coat system, prime and finish, using oil base material is recommended for any metal surface. Although water base material is often used, it may cause oxidation, corrosion and rust. One coat of oil base solid body stain is often used on interior or exterior metal but it may crack, peel or chip without the proper prime coat application. "Slow" work is based on a $10.25 hourly wage, "Medium" work on a $16.50 hourly wage, "Fast" work on a $22.00 hourly wage. Other qualifications that apply to this table are on page 9.

| | Labor SF per manhour | Material coverage SF/gallon | Material cost per gallon | Labor cost per 100 SF | Labor burden 100 SF | Material cost per 100 SF | Overhead per 100 SF | Profit per 100 SF | Total price per 100 SF |
|---|---|---|---|---|---|---|---|---|---|
| **Structural steel, medium, brush application** | | | | | | | | | |
| Metal primer, rust inhibitor, clean metal (material #35) | | | | | | | | | |
| Brush prime coat | | | | | | | | | |
| Slow | 80 | 425 | 28.20 | 12.81 | 3.49 | 6.64 | 7.11 | 4.51 | 34.56 |
| Medium | 100 | 413 | 24.60 | 16.50 | 5.55 | 5.96 | 6.86 | 3.84 | 38.71 |
| Fast | 120 | 400 | 21.10 | 18.33 | 7.43 | 5.28 | 5.74 | 2.21 | 38.99 |
| Metal primer, rust inhibitor, rusty metal (material #36) | | | | | | | | | |
| Brush prime coat | | | | | | | | | |
| Slow | 80 | 400 | 33.30 | 12.81 | 3.49 | 8.33 | 7.64 | 4.84 | 37.11 |
| Medium | 100 | 388 | 29.20 | 16.50 | 5.55 | 7.53 | 7.25 | 4.05 | 40.88 |
| Fast | 120 | 375 | 25.00 | 18.33 | 7.43 | 6.67 | 6.00 | 2.31 | 40.74 |
| Industrial enamel, oil base, high gloss, light colors (material #56) | | | | | | | | | |
| Brush 1st or additional finish coats | | | | | | | | | |
| Slow | 100 | 425 | 28.90 | 10.25 | 2.79 | 6.80 | 6.15 | 3.90 | 29.89 |
| Medium | 125 | 413 | 25.20 | 13.20 | 4.44 | 6.10 | 5.82 | 3.25 | 32.81 |
| Fast | 150 | 400 | 21.60 | 14.67 | 5.95 | 5.40 | 4.81 | 1.85 | 32.68 |
| Industrial enamel, oil base, high gloss, dark (OSHA) colors (material #57) | | | | | | | | | |
| Brush 1st or additional finish coats | | | | | | | | | |
| Slow | 100 | 475 | 34.90 | 10.25 | 2.79 | 7.35 | 6.32 | 4.01 | 30.72 |
| Medium | 125 | 463 | 30.50 | 13.20 | 4.44 | 6.59 | 5.94 | 3.32 | 33.49 |
| Fast | 150 | 450 | 26.10 | 14.67 | 5.95 | 5.80 | 4.89 | 1.88 | 33.19 |
| Epoxy coating, 2 part system, clear (material #51) | | | | | | | | | |
| Brush 1st coat | | | | | | | | | |
| Slow | 80 | 425 | 57.00 | 12.81 | 3.49 | 13.41 | 9.21 | 5.84 | 44.76 |
| Medium | 100 | 413 | 49.90 | 16.50 | 5.55 | 12.08 | 8.36 | 4.67 | 47.16 |
| Fast | 120 | 400 | 42.80 | 18.33 | 7.43 | 10.70 | 6.75 | 2.59 | 45.80 |
| Brush 2nd or additional coats | | | | | | | | | |
| Slow | 100 | 450 | 57.00 | 10.25 | 2.79 | 12.67 | 7.97 | 5.05 | 38.73 |
| Medium | 125 | 438 | 49.90 | 13.20 | 4.44 | 11.39 | 7.11 | 3.98 | 40.12 |
| Fast | 150 | 425 | 42.80 | 14.67 | 5.95 | 10.07 | 5.68 | 2.18 | 38.55 |

| | Labor SF per manhour | Material coverage SF/gallon | Material cost per gallon | Labor cost per 100 SF | Labor burden 100 SF | Material cost per 100 SF | Overhead per 100 SF | Profit per 100 SF | Total price per 100 SF |
|---|---|---|---|---|---|---|---|---|---|
| Epoxy coating, 2 part system, white (material #52) | | | | | | | | | |
| Brush 1st coat | | | | | | | | | |
| Slow | 80 | 425 | 57.40 | 12.81 | 3.49 | 13.51 | 9.24 | 5.86 | 44.91 |
| Medium | 100 | 413 | 50.20 | 16.50 | 5.55 | 12.15 | 8.38 | 4.68 | 47.26 |
| Fast | 120 | 400 | 43.10 | 18.33 | 7.43 | 10.78 | 6.76 | 2.60 | 45.90 |
| Brush 2nd or additional coats | | | | | | | | | |
| Slow | 100 | 450 | 57.40 | 10.25 | 2.79 | 12.76 | 8.00 | 5.07 | 38.87 |
| Medium | 125 | 438 | 50.20 | 13.20 | 4.44 | 11.46 | 7.13 | 3.99 | 40.22 |
| Fast | 150 | 425 | 43.10 | 14.67 | 5.95 | 10.14 | 5.69 | 2.19 | 38.64 |

For field painting at heights above 8 feet, use the High Time Difficulty Factors on page 137. Medium structural steel has from 150 to 300 square feet of surface per ton. As a comparison, when coatings are applied by brush, a journeyman painter will apply a first coat on 4 to 5 tons per 8 hour day. A second and subsequent coat can be applied on 5 to 6 tons per day. When coatings are applied by spray, figure output at 0.6 hours per ton and material use at about 0.6 gallons per ton. Use Figure 23 on pages 387 to 395 to convert structural steel linear feet or tonnage to surface area. Note: A two coat system, prime and finish, using oil base material is recommended for any metal surface. Although water base material is often used, it may cause oxidation, corrosion and rust. One coat of oil base solid body stain is often used on interior or exterior metal but it may crack, peel or chip without the proper prime coat application. "Slow" work is based on a $10.25 hourly wage, "Medium" work on a $16.50 hourly wage, "Fast" work on a $22.00 hourly wage. Other qualifications that apply to this table are on page 9.

| | Labor SF per manhour | Material coverage SF/gallon | Material cost per gallon | Labor cost per 100 SF | Labor burden 100 SF | Material cost per 100 SF | Overhead per 100 SF | Profit per 100 SF | Total price per 100 SF |
|---|---|---|---|---|---|---|---|---|---|
| **Structural steel, medium, roll and brush application** | | | | | | | | | |
| Metal primer, rust inhibitor, clean metal (material #35) | | | | | | | | | |
| Roll and brush prime coat | | | | | | | | | |
| Slow | 200 | 390 | 28.20 | 5.13 | 1.39 | 7.23 | 4.27 | 2.70 | 20.72 |
| Medium | 225 | 378 | 24.60 | 7.33 | 2.46 | 6.51 | 4.00 | 2.23 | 22.53 |
| Fast | 250 | 365 | 21.10 | 8.80 | 3.57 | 5.78 | 3.36 | 1.29 | 22.80 |
| Metal primer, rust inhibitor, rusty metal (material #36) | | | | | | | | | |
| Roll and brush prime coat | | | | | | | | | |
| Slow | 200 | 380 | 33.30 | 5.13 | 1.39 | 8.76 | 4.74 | 3.00 | 23.02 |
| Medium | 225 | 368 | 29.20 | 7.33 | 2.46 | 7.93 | 4.34 | 2.43 | 24.49 |
| Fast | 250 | 355 | 25.00 | 8.80 | 3.57 | 7.04 | 3.59 | 1.38 | 24.38 |
| Industrial enamel, oil base, high gloss, light colors (material #56) | | | | | | | | | |
| Roll and brush 1st or additional finish coats | | | | | | | | | |
| Slow | 225 | 390 | 28.90 | 4.56 | 1.23 | 7.41 | 4.10 | 2.60 | 19.90 |
| Medium | 250 | 378 | 25.20 | 6.60 | 2.22 | 6.67 | 3.80 | 2.12 | 21.41 |
| Fast | 275 | 365 | 21.60 | 8.00 | 3.25 | 5.92 | 3.17 | 1.22 | 21.56 |
| Industrial enamel, oil base, high gloss, dark (OSHA) colors (material #57) | | | | | | | | | |
| Roll and brush 1st or additional finish coats | | | | | | | | | |
| Slow | 225 | 440 | 34.90 | 4.56 | 1.23 | 7.93 | 4.26 | 2.70 | 20.68 |
| Medium | 250 | 428 | 30.50 | 6.60 | 2.22 | 7.13 | 3.91 | 2.18 | 22.04 |
| Fast | 275 | 415 | 26.10 | 8.00 | 3.25 | 6.29 | 3.24 | 1.25 | 22.03 |
| Epoxy coating, 2 part system, clear (material #51) | | | | | | | | | |
| Roll and brush 1st coat | | | | | | | | | |
| Slow | 200 | 400 | 57.00 | 5.13 | 1.39 | 14.25 | 6.44 | 4.08 | 31.29 |
| Medium | 225 | 388 | 49.90 | 7.33 | 2.46 | 12.86 | 5.55 | 3.10 | 31.30 |
| Fast | 250 | 375 | 42.80 | 8.80 | 3.57 | 11.41 | 4.40 | 1.69 | 29.87 |
| Roll and brush 2nd or additional coats | | | | | | | | | |
| Slow | 225 | 425 | 57.00 | 4.56 | 1.23 | 13.41 | 5.96 | 3.78 | 28.94 |
| Medium | 250 | 413 | 49.90 | 6.60 | 2.22 | 12.08 | 5.12 | 2.86 | 28.88 |
| Fast | 275 | 400 | 42.80 | 8.00 | 3.25 | 10.70 | 4.06 | 1.56 | 27.57 |

| | Labor SF per manhour | Material coverage SF/gallon | Material cost per gallon | Labor cost per 100 SF | Labor burden 100 SF | Material cost per 100 SF | Overhead per 100 SF | Profit per 100 SF | Total price per 100 SF |
|---|---|---|---|---|---|---|---|---|---|
| Epoxy coating, 2 part system, white (material #52) | | | | | | | | | |
| Roll and brush 1st coat | | | | | | | | | |
| Slow | 200 | 400 | 57.40 | 5.13 | 1.39 | 14.35 | 6.47 | 4.10 | 31.44 |
| Medium | 225 | 388 | 50.20 | 7.33 | 2.46 | 12.94 | 5.57 | 3.11 | 31.41 |
| Fast | 250 | 375 | 43.10 | 8.80 | 3.57 | 11.49 | 4.41 | 1.70 | 29.97 |
| Roll and brush 2nd or additional coats | | | | | | | | | |
| Slow | 225 | 425 | 57.40 | 4.56 | 1.23 | 13.51 | 5.99 | 3.80 | 29.09 |
| Medium | 250 | 413 | 50.20 | 6.60 | 2.22 | 12.15 | 5.14 | 2.87 | 28.98 |
| Fast | 275 | 400 | 43.10 | 8.00 | 3.25 | 10.78 | 4.07 | 1.57 | 27.67 |

For field painting at heights above 8 feet, use the High Time Difficulty Factors on page 137. Medium structural steel has from 150 to 300 square feet of surface per ton. As a comparison, when coatings are applied by brush, a journeyman painter will apply a first coat on 4 to 5 tons per 8 hour day. A second and subsequent coat can be applied on 5 to 6 tons per day. When coatings are applied by spray, figure output at 0.6 hours per ton and material use at about 0.6 gallons per ton. Use Figure 23 on pages 387 to 395 to convert structural steel linear feet or tonnage to surface area. Note: A two coat system, prime and finish, using oil base material is recommended for any metal surface. Although water base material is often used, it may cause oxidation, corrosion and rust. One coat of oil base solid body stain is often used on interior or exterior metal but it may crack, peel or chip without the proper prime coat application. "Slow" work is based on a $10.25 hourly wage, "Medium" work on a $16.50 hourly wage, "Fast" work on a $22.00 hourly wage. Other qualifications that apply to this table are on page 9.

| | Labor SF per manhour | Material coverage SF/gallon | Material cost per gallon | Labor cost per 100 SF | Labor burden 100 SF | Material cost per 100 SF | Overhead per 100 SF | Profit per 100 SF | Total price per 100 SF |
|---|---|---|---|---|---|---|---|---|---|

## Structural steel, medium, spray application

| Metal primer, rust inhibitor, clean metal (material #35) | | | | | | | | | |
|---|---|---|---|---|---|---|---|---|---|
| Spray prime coat | | | | | | | | | |
| Slow | 500 | 325 | 28.20 | 2.05 | .56 | 8.68 | 3.50 | 2.22 | 17.01 |
| Medium | 600 | 313 | 24.60 | 2.75 | .93 | 7.86 | 2.83 | 1.58 | 15.95 |
| Fast | 700 | 300 | 21.10 | 3.14 | 1.28 | 7.03 | 2.12 | .81 | 14.38 |
| Metal primer, rust inhibitor, rusty metal (material #36) | | | | | | | | | |
| Spray prime coat | | | | | | | | | |
| Slow | 500 | 300 | 33.30 | 2.05 | .56 | 11.10 | 4.25 | 2.69 | 20.65 |
| Medium | 600 | 288 | 29.20 | 2.75 | .93 | 10.14 | 3.39 | 1.89 | 19.10 |
| Fast | 700 | 275 | 25.00 | 3.14 | 1.28 | 9.09 | 2.50 | .96 | 16.97 |
| Industrial enamel, oil base, high gloss, light colors (material #56) | | | | | | | | | |
| Spray 1st or additional finish coats | | | | | | | | | |
| Slow | 600 | 325 | 28.90 | 1.71 | .47 | 8.89 | 3.43 | 2.18 | 16.68 |
| Medium | 700 | 313 | 25.20 | 2.36 | .79 | 8.05 | 2.74 | 1.53 | 15.47 |
| Fast | 800 | 300 | 21.60 | 2.75 | 1.12 | 7.20 | 2.05 | .79 | 13.91 |
| Industrial enamel, oil base, high gloss, dark (OSHA) colors (material #57) | | | | | | | | | |
| Spray 1st or additional finish coats | | | | | | | | | |
| Slow | 600 | 365 | 34.90 | 1.71 | .47 | 9.56 | 3.64 | 2.31 | 17.69 |
| Medium | 700 | 350 | 30.50 | 2.36 | .79 | 8.71 | 2.91 | 1.62 | 16.39 |
| Fast | 800 | 335 | 26.10 | 2.75 | 1.12 | 7.79 | 2.16 | .83 | 14.65 |
| Epoxy coating, 2 part system, clear (material #51) | | | | | | | | | |
| Spray 1st coat | | | | | | | | | |
| Slow | 500 | 325 | 57.00 | 2.05 | .56 | 17.54 | 6.25 | 3.96 | 30.36 |
| Medium | 600 | 313 | 49.90 | 2.75 | .93 | 15.94 | 4.81 | 2.69 | 27.12 |
| Fast | 700 | 300 | 42.80 | 3.14 | 1.28 | 14.27 | 3.46 | 1.33 | 23.48 |
| Spray 2nd or additional coats | | | | | | | | | |
| Slow | 600 | 350 | 57.00 | 1.71 | .47 | 16.29 | 5.73 | 3.63 | 27.83 |
| Medium | 700 | 338 | 49.90 | 2.36 | .79 | 14.76 | 4.39 | 2.45 | 24.75 |
| Fast | 800 | 325 | 42.80 | 2.75 | 1.12 | 13.17 | 3.15 | 1.21 | 21.40 |

| | Labor SF per manhour | Material coverage SF/gallon | Material cost per gallon | Labor cost per 100 SF | Labor burden 100 SF | Material cost per 100 SF | Overhead per 100 SF | Profit per 100 SF | Total price per 100 SF |
|---|---|---|---|---|---|---|---|---|---|
| Epoxy coating, 2 part system, white (material #52) | | | | | | | | | |
| Spray 1st coat | | | | | | | | | |
| Slow | 500 | 325 | 57.40 | 2.05 | .56 | 17.66 | 6.28 | 3.98 | 30.53 |
| Medium | 600 | 313 | 50.20 | 2.75 | .93 | 16.04 | 4.83 | 2.70 | 27.25 |
| Fast | 700 | 300 | 43.10 | 3.14 | 1.28 | 14.37 | 3.47 | 1.34 | 23.60 |
| Spray 2nd or additional coats | | | | | | | | | |
| Slow | 600 | 350 | 57.40 | 1.71 | .47 | 16.40 | 5.76 | 3.65 | 27.99 |
| Medium | 700 | 338 | 50.20 | 2.36 | .79 | 14.85 | 4.41 | 2.47 | 24.88 |
| Fast | 800 | 325 | 43.10 | 2.75 | 1.12 | 13.26 | 3.17 | 1.22 | 21.52 |

For field painting at heights above 8 feet, use the High Time Difficulty Factors on page 137. Medium structural steel has from 150 to 300 square feet of surface per ton. As a comparison, when coatings are applied by brush, a journeyman painter will apply a first coat on 4 to 5 tons per 8 hour day. A second and subsequent coat can be applied on 5 to 6 tons per day. When coatings are applied by spray, figure output at 0.6 hours per ton and material use at about 0.6 gallons per ton. Use Figure 23 on pages 387 to 395 to convert structural steel linear feet or tonnage to surface area. Note: A two coat system, prime and finish, using oil base material is recommended for any metal surface. Although water base material is often used, it may cause oxidation, corrosion and rust. One coat of oil base solid body stain is often used on interior or exterior metal but it may crack, peel or chip without the proper prime coat application. "Slow" work is based on a $10.25 hourly wage, "Medium" work on a $16.50 hourly wage, "Fast" work on a $22.00 hourly wage. Other qualifications that apply to this table are on page 9.

| Section designation | | Square feet of surface area per foot of length: Minus one flange side | Square feet of surface area per foot of length: All around | Square feet of surface area per ton: Minus one flange side | Square feet of surface area per ton: All around |
|---|---|---|---|---|---|
| W 30 | x 99 | 7.56 | 8.43 | 152.7 | 170.3 |
| W 27 | x 94 | 6.99 | 7.81 | 148.7 | 166.2 |
| | x 84 | 6.99 | 7.79 | 166.4 | 185.5 |
| W 24 | x 100 | 7.00 | 8.00 | 140.0 | 160.0 |
| | x 94 | 6.29 | 7.04 | 133.8 | 149.8 |
| | x 84 | 6.27 | 7.02 | 149.3 | 167.1 |
| | x 76 | 6.23 | 6.98 | 163.9 | 183.7 |
| | x 68 | 6.21 | 6.96 | 182.6 | 204.7 |
| | x 61 | 5.71 | 6.29 | 187.2 | 206.2 |
| | x 55 | 5.67 | 6.25 | 206.2 | 227.2 |
| W 21 | x 96 | 5.77 | 6.52 | 120.2 | 135.8 |
| | x 82 | 5.73 | 6.48 | 139.8 | 158.0 |
| | x 73 | 5.60 | 6.29 | 153.4 | 172.3 |
| | x 68 | 5.58 | 6.57 | 164.1 | 184.4 |
| | x 62 | 5.56 | 6.25 | 179.4 | 201.6 |
| | x 55 | 5.52 | 6.21 | 200.7 | 225.8 |
| | x 49 | 5.10 | 5.65 | 208.2 | 230.6 |
| | x 44 | 5.08 | 5.63 | 230.9 | 255.9 |
| W 18 | x 96 | 5.96 | 6.94 | 124.2 | 144.6 |
| | x 85 | 5.28 | 6.02 | 124.2 | 141.6 |
| | x 77 | 5.22 | 5.95 | 135.6 | 154.5 |
| | x 70 | 5.19 | 5.92 | 148.3 | 169.1 |
| | x 64 | 5.17 | 5.90 | 161.6 | 184.4 |
| | x 60 | 4.92 | 5.54 | 164.0 | 184.7 |
| | x 55 | 4.90 | 5.52 | 178.1 | 200.7 |
| | x 50 | 4.88 | 5.50 | 195.2 | 220.0 |
| | x 45 | 4.85 | 5.48 | 215.6 | 243.6 |
| | x 40 | 4.48 | 4.98 | 224.0 | 249.0 |
| | x 35 | 4.46 | 4.96 | 254.9 | 283.4 |
| W 16 | x 96 | 5.60 | 6.56 | 116.7 | 136.7 |
| | x 88 | 5.57 | 6.53 | 126.6 | 148.4 |
| | x 78 | 4.88 | 5.60 | 125.1 | 143.6 |
| | x 71 | 4.82 | 5.53 | 135.8 | 155.8 |
| | x 64 | 4.79 | 5.50 | 149.7 | 171.9 |
| | x 58 | 4.77 | 5.48 | 164.5 | 189.0 |
| | x 50 | 4.51 | 5.10 | 180.4 | 204.0 |
| | x 45 | 4.45 | 5.03 | 197.8 | 223.6 |
| | x 40 | 4.42 | 5.00 | 221.0 | 250.0 |
| | x 36 | 4.40 | 4.98 | 244.4 | 276.7 |
| | x 31 | 4.02 | 4.48 | 259.4 | 289.0 |
| | x 26 | 3.98 | 4.44 | 306.2 | 341.5 |

| Section designation | | Square feet of surface area per foot of length: Minus one flange side | Square feet of surface area per foot of length: All around | Square feet of surface area per ton: Minus one flange side | Square feet of surface area per ton: All around |
|---|---|---|---|---|---|
| W 14 | x 95 | 5.99 | 7.20 | 126.1 | 151.6 |
| | x 87 | 5.96 | 7.17 | 137.0 | 164.8 |
| | x 84 | 5.36 | 6.36 | 127.6 | 151.4 |
| | x 78 | 5.33 | 6.33 | 136.7 | 162.3 |
| | x 74 | 4.92 | 5.77 | 133.0 | 155.9 |
| | x 68 | 4.83 | 5.67 | 142.1 | 166.8 |
| | x 61 | 4.81 | 5.65 | 157.7 | 185.2 |
| | x 53 | 4.33 | 5.00 | 163.4 | 188.7 |
| | x 48 | 4.29 | 4.96 | 178.8 | 206.7 |
| | x 43 | 4.27 | 4.94 | 198.6 | 229.8 |
| | x 38 | 4.04 | 4.60 | 212.6 | 242.1 |
| | x 34 | 4.02 | 4.58 | 236.5 | 269.4 |
| | x 30 | 4.00 | 4.56 | 266.7 | 304.0 |
| | x 26 | 3.56 | 3.98 | 273.8 | 306.2 |
| | x 22 | 3.54 | 3.96 | 321.8 | 360.0 |
| W 12 | x 99 | 5.19 | 6.21 | 104.8 | 125.4 |
| | x 92 | 5.14 | 6.15 | 111.7 | 133.7 |
| | x 85 | 5.12 | 6.13 | 120.5 | 144.2 |
| | x 79 | 5.09 | 6.10 | 128.9 | 154.4 |
| | x 72 | 5.04 | 6.04 | 140.0 | 167.8 |
| | x 65 | 5.02 | 6.02 | 154.5 | 185.2 |
| | x 58 | 4.54 | 5.38 | 156.6 | 185.5 |
| | x 53 | 4.50 | 5.33 | 169.8 | 201.1 |
| | x 50 | 4.07 | 4.75 | 162.8 | 190.0 |
| | x 45 | 4.00 | 4.67 | 177.8 | 207.6 |
| | x 40 | 4.00 | 4.67 | 200.0 | 233.5 |
| | x 36 | 3.70 | 4.25 | 205.6 | 236.1 |
| | x 31 | 3.65 | 4.19 | 235.5 | 270.3 |
| | x 27 | 3.63 | 4.17 | 268.9 | 308.9 |
| | x 22 | 3.04 | 3.38 | 276.4 | 307.3 |
| | x 19 | 3.02 | 3.35 | 317.9 | 352.6 |
| | x 16.5 | 3.00 | 3.33 | 363.6 | 403.6 |
| | x 14 | 2.98 | 3.31 | 425.7 | 472.9 |
| W 10 | x 100 | 4.45 | 5.31 | 89.0 | 106.2 |
| | x 89 | 4.38 | 5.23 | 98.4 | 117.5 |
| | x 77 | 4.33 | 5.19 | 112.5 | 134.8 |
| | x 72 | 4.28 | 5.13 | 118.9 | 142.5 |
| | x 66 | 4.26 | 5.10 | 129.1 | 154.5 |
| | x 60 | 4.24 | 5.08 | 141.3 | 169.3 |

**Figure 23**
Structural steel conversion table

| Section designation | | Square feet of surface area per foot of length | | Square feet of surface area per ton | |
|---|---|---|---|---|---|
| | | Minus one flange side | All around | Minus one flange side | All around |
| W 10 | x 54 | 4.19 | 5.02 | 155.2 | 185.9 |
| | x 49 | 4.17 | 5.00 | 170.2 | 204.0 |
| | x 45 | 3.69 | 4.35 | 164.0 | 193.3 |
| | x 39 | 3.67 | 4.33 | 188.2 | 222.1 |
| | x 33 | 3.63 | 4.29 | 220.0 | 260.0 |
| | x 29 | 3.15 | 3.63 | 217.2 | 250.3 |
| | x 25 | 3.13 | 3.60 | 250.4 | 288.0 |
| | x 21 | 3.08 | 3.56 | 293.3 | 339.0 |
| | x 19 | 2.71 | 3.04 | 285.3 | 320.0 |
| | x 17 | 2.69 | 3.02 | 316.5 | 355.3 |
| | x 15 | 2.67 | 3.00 | 356.0 | 400.0 |
| | x 11.5 | 2.65 | 2.98 | 460.9 | 518.3 |
| W 8 | x 67 | 3.56 | 4.25 | 106.3 | 126.9 |
| | x 58 | 3.52 | 4.21 | 121.4 | 145.2 |
| | x 48 | 3.45 | 4.13 | 143.8 | 172.1 |
| | x 40 | 3.41 | 4.08 | 170.5 | 204.0 |
| | x 35 | 3.35 | 4.02 | 191.4 | 229.7 |
| | x 31 | 3.33 | 4.00 | 214.8 | 258.1 |
| | x 28 | 2.96 | 3.50 | 211.4 | 250.0 |
| | x 24 | 2.94 | 3.48 | 245.0 | 290.0 |
| | x 20 | 2.67 | 3.10 | 267.0 | 310.0 |
| | x 17 | 2.65 | 3.08 | 311.8 | 362.4 |
| | x 15 | 2.35 | 2.69 | 313.3 | 358.7 |
| | x 13 | 2.33 | 2.67 | 358.5 | 410.8 |
| | x 10 | 2.31 | 2.65 | 462.0 | 530.0 |
| W 6 | x 25 | 2.59 | 3.10 | 207.2 | 248.0 |
| | x 20 | 2.54 | 3.04 | 254.0 | 304.0 |
| | x 15.5 | 2.50 | 3.00 | 322.6 | 387.1 |
| | x 16 | 2.04 | 2.38 | 255.0 | 297.5 |
| | x 12 | 2.00 | 2.33 | 333.3 | 388.3 |
| | x 8.5 | 1.98 | 2.31 | 465.9 | 543.5 |
| W 5 | x 18.5 | 2.10 | 2.52 | 227.0 | 272.4 |
| | x 16 | 2.08 | 2.50 | 260.0 | 312.5 |
| W 4 | x 13 | 1.69 | 2.02 | 260.0 | 310.8 |
| S-24 | x 90 | 5.78 | 6.38 | 128.4 | 141.8 |
| | x 79.9 | 5.75 | 6.33 | 143.9 | 158.4 |

| Section designation | | Square feet of surface area per foot of length | | Square feet of surface area per ton | |
|---|---|---|---|---|---|
| | | Minus one flange side | All around | Minus one flange side | All around |
| S-20 | x 95 | 5.15 | 5.75 | 108.4 | 121.1 |
| | x 85 | 5.08 | 5.67 | 119.5 | 133.4 |
| | x 75 | 4.93 | 5.46 | 131.5 | 145.6 |
| | x 65.4 | 4.90 | 5.42 | 149.8 | 165.7 |
| S-18 | x 70 | 4.56 | 5.08 | 130.3 | 145.1 |
| | x 54.7 | 4.50 | 5.00 | 164.5 | 182.8 |
| S-15 | x 50 | 3.91 | 4.38 | 156.4 | 175.2 |
| | x 42.9 | 3.88 | 4.33 | 180.9 | 201.9 |
| S-12 | x 50 | 3.38 | 3.83 | 135.2 | 153.2 |
| | x 40.8 | 3.31 | 3.75 | 162.3 | 183.8 |
| | x 35 | 3.28 | 3.71 | 187.4 | 212.0 |
| | x 31.8 | 3.25 | 3.67 | 204.4 | 230.8 |
| S-10 | x 35 | 2.92 | 3.33 | 166.9 | 190.2 |
| | x 25.4 | 2.82 | 3.21 | 222.0 | 252.8 |
| S-8 | x 23 | 2.36 | 2.71 | 205.2 | 235.7 |
| | x 18.4 | 2.33 | 2.67 | 253.3 | 290.2 |
| S-7 | x 20 | 2.14 | 2.46 | 214.0 | 246.0 |
| | x 15.3 | 2.07 | 2.38 | 270.6 | 311.1 |
| S-6 | x 17.25 | 1.91 | 2.21 | 221.4 | 256.2 |
| | x 12.5 | 1.84 | 2.13 | 294.4 | 340.8 |
| S-5 | x 14.75 | 1.65 | 1.92 | 223.7 | 260.3 |
| | x 10 | 1.58 | 1.83 | 316.0 | 366.0 |
| S-4 | x 9.5 | 1.35 | 1.58 | 284.2 | 332.6 |
| | x 7.7 | 1.32 | 1.54 | 342.9 | 400.0 |
| S-3 | x 7.5 | 1.13 | 1.33 | 301.3 | 354.7 |
| | x 5.7 | 1.09 | 1.29 | 382.5 | 452.6 |
| **Miscellaneous shape** | | | | | |
| M-5 | x 18.9 | 2.08 | 2.50 | 220.1 | 264.6 |

**Figure 23 (cont'd)**
Structural steel conversion table

| Section designation | Square feet of surface area per foot of length: Minus one flange side | Square feet of surface area per foot of length: All around | Square feet of surface area per ton: Minus one flange side | Square feet of surface area per ton: All around |
|---|---|---|---|---|
| C-15 x 50 | 3.44 | 3.75 | 137.6 | 150.0 |
| x 40 | 3.38 | 3.67 | 169.0 | 183.5 |
| x 33.9 | 3.34 | 3.63 | 197.1 | 214.2 |
| C-12 x 30 | 2.78 | 3.04 | 185.3 | 202.7 |
| x 25 | 2.75 | 3.00 | 220.0 | 240.0 |
| x 20.7 | 2.75 | 3.00 | 265.7 | 289.9 |
| C-10 x 30 | 2.42 | 2.67 | 161.3 | 178.0 |
| x 25 | 2.39 | 2.63 | 191.2 | 210.4 |
| x 20 | 2.35 | 2.58 | 235.0 | 258.0 |
| x 15.3 | 2.32 | 2.54 | 305.3 | 334.2 |
| C-9 x 20 | 2.16 | 2.38 | 216.0 | 238.0 |
| x 15 | 2.13 | 2.33 | 284.0 | 310.7 |
| x 13.4 | 2.09 | 2.29 | 311.9 | 341.8 |
| C-8 x 18.75 | 1.96 | 2.17 | 209.1 | 231.5 |
| x 13.75 | 1.93 | 2.13 | 280.7 | 309.8 |
| x 11.5 | 1.90 | 2.08 | 330.4 | 361.7 |
| C-7 x 14.75 | 1.73 | 1.92 | 234.6 | 260.3 |
| x 12.25 | 1.73 | 1.92 | 282.4 | 313.5 |
| x 9.8 | 1.70 | 1.88 | 346.9 | 383.7 |
| C-6 x 13 | 1.53 | 1.71 | 235.4 | 263.1 |
| x 10.5 | 1.50 | 1.67 | 285.7 | 318.1 |
| x 8.2 | 1.47 | 1.63 | 358.5 | 397.6 |
| C-5 x 9 | 1.30 | 1.46 | 288.9 | 324.4 |
| x 6.7 | 1.27 | 1.42 | 379.1 | 423.9 |
| C-4 x 7.25 | 1.10 | 1.25 | 295.2 | 344.8 |
| x 5.4 | 1.07 | 1.21 | 396.3 | 448.1 |
| C-3 x 6 | .91 | 1.04 | 303.3 | 346.7 |
| x 5 | .88 | 1.00 | 352.0 | 400.0 |
| x 4.1 | .84 | .96 | 409.8 | 468.3 |
| MC-18 x 58 | 4.06 | 4.42 | 140.0 | 152.4 |
| x 51.9 | 4.03 | 4.38 | 155.3 | 168.8 |
| x 45.8 | 4.00 | 4.33 | 176.7 | 189.1 |
| x 42.7 | 4.00 | 4.33 | 187.4 | 202.8 |
| MC-13 x 50 | 3.26 | 3.63 | 130.4 | 145.2 |
| x 40 | 3.20 | 3.54 | 160.0 | 177.0 |
| x 35 | 3.20 | 3.54 | 182.9 | 202.3 |
| x 31.8 | 3.17 | 3.50 | 199.4 | 220.1 |
| MC-12 x 50 | 3.03 | 3.38 | 121.2 | 135.2 |
| x 45 | 3.00 | 3.33 | 133.3 | 148.0 |
| x 40 | 2.97 | 3.29 | 148.5 | 164.5 |
| x 35 | 2.94 | 3.25 | 168.0 | 185.7 |
| x 37 | 2.91 | 3.21 | 157.3 | 173.5 |
| x 32.9 | 2.88 | 3.17 | 175.1 | 192.7 |
| x 30.9 | 2.88 | 3.17 | 186.4 | 205.2 |
| MC-10 x 41.1 | 2.76 | 3.13 | 134.3 | 152.3 |
| x 33.6 | 2.70 | 3.04 | 160.7 | 181.0 |
| x 28.5 | 2.67 | 3.00 | 187.4 | 210.5 |
| x 28.3 | 2.54 | 2.83 | 179.5 | 200.0 |
| x 25.3 | 2.54 | 2.83 | 200.8 | 223.7 |
| x 24.9 | 2.51 | 2.79 | 201.6 | 224.1 |
| x 21.9 | 2.54 | 2.83 | 232.0 | 258.4 |
| MC-9 x 25.4 | 2.38 | 2.67 | 187.4 | 210.2 |
| x 23.9 | 2.38 | 2.67 | 199.2 | 223.4 |
| MC-8 x 22.8 | 2.21 | 2.50 | 193.9 | 219.3 |
| x 21.4 | 2.21 | 2.50 | 206.5 | 233.6 |
| x 20 | 2.08 | 2.33 | 208.0 | 233.0 |
| x 18.7 | 2.08 | 2.33 | 222.5 | 249.2 |
| MC-7 x 22.7 | 2.07 | 2.38 | 182.4 | 209.7 |
| x 19.1 | 2.04 | 2.33 | 213.6 | 244.0 |
| x 17.6 | 1.92 | 2.17 | 218.2 | 246.6 |
| MC-6 x 18 | 1.88 | 2.17 | 208.9 | 241.1 |
| x 15.3 | 1.88 | 2.17 | 245.8 | 283.7 |
| x 16.3 | 1.75 | 2.00 | 214.7 | 245.4 |
| x 15.1 | 1.75 | 2.00 | 231.8 | 264.9 |
| x 12 | 1.63 | 1.83 | 271.7 | 305.0 |
| MC-3 x 9 | 1.03 | 1.21 | 228.9 | 268.9 |
| x 7.1 | 1.00 | 1.17 | 281.7 | 329.6 |

**Figure 23 (cont'd)**
Structural steel conversion table

| Section designation | | Surface area per foot of length | Surface area per ton |
|---|---|---|---|
| ST 18 | x 97 | 5.06 | 104.3 |
| | x 91 | 5.04 | 110.8 |
| | x 85 | 5.02 | 118.1 |
| | x 80 | 5.00 | 125.0 |
| | x 75 | 4.98 | 132.8 |
| | x 67.5 | 4.95 | 147.0 |
| ST 16.5 | x 76 | 4.72 | 124.2 |
| | x 70.5 | 4.70 | 133.3 |
| | x 65 | 4.38 | 144.0 |
| | x 59 | 4.65 | 157.6 |
| ST 15 | x 95 | 5.02 | 105.7 |
| | x 86 | 4.99 | 116.0 |
| | x 66 | 4.28 | 129.7 |
| | x 62 | 4.27 | 137.7 |
| | x 58 | 4.25 | 146.6 |
| | x 54 | 4.23 | 156.7 |
| | x 49.5 | 4.21 | 170.1 |
| ST 13.5 | x 88.5 | 4.63 | 104.6 |
| | x 80 | 4.59 | 114.8 |
| | x 72.5 | 4.57 | 126.1 |
| | x 57 | 3.95 | 138.6 |
| | x 51 | 3.93 | 154.1 |
| | x 47 | 3.91 | 166.4 |
| | x 42 | 3.89 | 185.2 |
| ST 12 | x 80 | 4.41 | 110.3 |
| | x 72.5 | 4.38 | 120.8 |
| | x 65 | 4.36 | 134.1 |
| | x 60 | 4.04 | 134.7 |
| | x 55 | 4.02 | 146.2 |
| | x 50 | 4.00 | 160.0 |
| | x 47 | 3.54 | 150.6 |
| | x 42 | 3.51 | 167.1 |
| | x 38 | 3.49 | 183.7 |
| | x 34 | 3.47 | 204.1 |
| | x 30.5 | 3.15 | 206.6 |
| | x 27.5 | 3.13 | 227.6 |

| Section designation | | Surface area per foot of length | Surface area per ton |
|---|---|---|---|
| ST 10.5 | x 71 | 3.97 | 111.8 |
| | x 63.5 | 3.95 | 124.4 |
| | x 56 | 3.92 | 140.0 |
| | x 48 | 3.27 | 136.3 |
| | x 41 | 3.23 | 157.6 |
| | x 36.5 | 3.15 | 172.6 |
| | x 34 | 3.14 | 184.7 |
| | x 31 | 3.12 | 201.3 |
| | x 27.5 | 3.10 | 225.5 |
| | x 24.5 | 2.82 | 230.2 |
| | x 22 | 2.81 | 255.5 |
| ST 9 | x 57 | 3.51 | 123.2 |
| | x 52.5 | 3.49 | 133.0 |
| | x 48 | 3.47 | 144.6 |
| | x 42.5 | 3.00 | 141.2 |
| | x 38.5 | 2.98 | 154.8 |
| | x 35 | 2.96 | 169.1 |
| | x 32 | 2.94 | 183.8 |
| | x 30 | 2.78 | 185.3 |
| | x 27.5 | 2.77 | 201.5 |
| | x 25 | 2.75 | 220.0 |
| | x 22.5 | 2.73 | 242.7 |
| | x 20 | 2.49 | 249.0 |
| | x 17.5 | 2.48 | 283.4 |
| ST 8 | x 48 | 3.28 | 136.7 |
| | x 44 | 3.26 | 148.2 |
| | x 39 | 2.79 | 143.1 |
| | x 35.5 | 2.77 | 156.1 |
| | x 32 | 2.76 | 172.5 |
| | x 29 | 2.73 | 188.3 |
| | x 25 | 2.53 | 202.4 |
| | x 22.5 | 2.52 | 224.0 |
| | x 20 | 2.50 | 250.0 |
| | x 18 | 2.49 | 276.7 |
| | x 15.5 | 2.24 | 289.0 |
| | x 13 | 2.22 | 341.5 |

**Figure 23 (cont'd)**
Structural steel conversion table

| Section designation | | Surface area per foot of length | Surface area per ton |
|---|---|---|---|
| ST 7 | x 88 | 3.88 | 88.2 |
| | x 83.5 | 3.86 | 92.5 |
| | x 79 | 3.84 | 97.2 |
| | x 75 | 3.83 | 102.1 |
| | x 71 | 3.81 | 107.3 |
| | x 68 | 3.69 | 108.5 |
| | x 63.5 | 3.67 | 115.6 |
| | x 59.5 | 3.65 | 122.7 |
| | x 55.5 | 3.64 | 131.2 |
| | x 51.5 | 3.62 | 140.6 |
| | x 47.5 | 3.60 | 151.6 |
| | x 43.5 | 3.58 | 164.6 |
| | x 42 | 3.19 | 151.9 |
| | x 39 | 3.17 | 162.6 |
| | x 37 | 2.86 | 154.6 |
| | x 34 | 2.85 | 167.6 |
| | x 30.5 | 2.83 | 185.6 |
| | x 26.5 | 2.51 | 189.4 |
| | x 24 | 2.49 | 207.5 |
| | x 21.5 | 2.47 | 229.8 |
| | x 19 | 2.31 | 243.2 |
| | x 17 | 2.29 | 269.4 |
| | x 15 | 2.28 | 304.0 |
| | x 13 | 2.00 | 307.7 |
| | x 11 | 1.98 | 360.0 |

| Section designation | | Surface area per foot of length | Surface area per ton |
|---|---|---|---|
| ST 6 | x 95 | 3.31 | 69.7 |
| | x 80.5 | 3.24 | 80.5 |
| | x 66.5 | 3.18 | 97.8 |
| | x 60 | 3.15 | 105.6 |
| | x 53 | 3.11 | 117.4 |
| | x 49.5 | 3.10 | 125.3 |
| | x 46 | 3.08 | 133.9 |
| | x 42.5 | 3.06 | 144.0 |
| | x 39.5 | 3.05 | 154.4 |
| | x 36 | 3.03 | 168.3 |
| | x 32.5 | 3.01 | 185.2 |
| | x 29 | 2.69 | 185.5 |
| | x 26.5 | 2.67 | 201.5 |
| | x 25 | 2.36 | 188.8 |
| | x 22.5 | 2.35 | 208.9 |

| Section designation | | Surface area per foot of length | Surface area per ton |
|---|---|---|---|
| ST 6 | x 20 | 2.33 | 233.0 |
| | x 18 | 2.11 | 234.4 |
| | x 15.5 | 2.10 | 271.0 |
| | x 13.5 | 2.08 | 308.1 |
| | x 11 | 1.70 | 309.1 |
| | x 9.5 | 1.68 | 353.7 |
| | x 8.25 | 1.67 | 404.8 |
| | x 7 | 1.65 | 471.4 |

| Section designation | | Surface area per foot of length | Surface area per ton |
|---|---|---|---|
| ST 5 | x 56 | 2.68 | 95.7 |
| | x 50 | 2.65 | 106.0 |
| | x 44.5 | 2.62 | 117.8 |
| | x 38.5 | 2.58 | 134.0 |
| | x 36 | 2.57 | 142.8 |
| | x 33 | 2.55 | 154.5 |
| | x 30 | 2.53 | 168.7 |
| | x 27 | 2.51 | 185.9 |
| | x 24.5 | 2.50 | 204.1 |
| | x 22.5 | 2.18 | 193.8 |
| | x 19.5 | 2.16 | 221.5 |
| | x 16.5 | 2.14 | 259.4 |
| | x 14.5 | 1.82 | 251.0 |
| | x 12.5 | 1.80 | 288.0 |
| | x 10.5 | 1.78 | 339.0 |
| | x 9.5 | 1.53 | 322.1 |
| | x 8.5 | 1.51 | 355.3 |
| | x 7.5 | 1.50 | 400.0 |
| | x 5.75 | 1.48 | 514.8 |

| Section designation | | Surface area per foot of length | Surface area per ton |
|---|---|---|---|
| ST 4 | x 33.5 | 2.13 | 127.7 |
| | x 29 | 2.10 | 144.8 |
| | x 24 | 2.06 | 171.7 |
| | x 20 | 2.03 | 203.0 |
| | x 17.5 | 2.01 | 229.7 |
| | x 15.5 | 2.00 | 258.1 |
| | x 14 | 1.76 | 251.4 |
| | x 12 | 1.75 | 291.7 |
| | x 10 | 1.56 | 312.0 |
| | x 8.5 | 1.54 | 362.4 |
| | x 7.5 | 1.35 | 360.0 |
| | x 6.5 | 1.33 | 409.2 |
| | x 5 | 1.31 | 524.0 |

**Figure 23 (cont'd)**
Structural steel conversion table

| Section designation | | Surface area per foot of length | Surface area per ton |
|---|---|---|---|
| ST 3 | x 12.5 | 1.55 | 248.0 |
| | x 10 | 1.52 | 304.0 |
| | x 7.75 | 1.50 | 387.1 |
| | x 8 | 1.19 | 297.5 |
| | x 6 | 1.17 | 390.0 |
| | x 4.25 | 1.14 | 536.5 |
| ST 2.5 | x 9.25 | 1.26 | 272.4 |
| | x 8 | 1.25 | 312.5 |
| ST 2 | x 6.5 | 1.02 | 313.8 |

**Tees cut from American standard shapes**

| Section designation | | Surface area per foot of length | Surface area per ton |
|---|---|---|---|
| ST 12 | x 60 | 3.34 | 111.3 |
| | x 52.95 | 3.31 | 125.0 |
| | x 50 | 3.21 | 128.4 |
| | x 45 | 3.19 | 141.8 |
| | x 39.95 | 3.17 | 158.7 |
| ST 10 | x47.5 | 2.87 | 120.8 |
| | x 42.5 | 2.84 | 133.6 |
| | x 37.5 | 2.73 | 145.6 |
| | x 32.7 | 2.70 | 165.1 |
| ST 9 | x 35 | 2.54 | 145.1 |
| | x 27.35 | 2.50 | 182.8 |
| ST 7.5 | x 25 | 2.19 | 175.2 |
| | x 21.45 | 2.17 | 202.3 |

| Section designation | | Surface area per foot of length | Surface area per ton |
|---|---|---|---|
| ST 6 | x 25 | 1.91 | 152.8 |
| | x 20.4 | 1.88 | 184.3 |
| | x 17.5 | 1.85 | 211.4 |
| | x 15.9 | 1.83 | 230.2 |
| ST 5 | x 17.5 | 1.66 | 189.7 |
| | x 12.7 | 1.61 | 253.5 |
| ST 4 | x 11.5 | 1.36 | 236.5 |
| | x 9.2 | 1.33 | 289.1 |
| ST 3.5 | x 10 | 1.23 | 246.0 |
| | x 7.65 | 1.19 | 311.0 |
| ST 3 | x 8.625 | 1.09 | 252.8 |
| | x 6.25 | 1.06 | 339.2 |
| ST 2.5 | x 7.375 | .96 | 260.3 |
| | x 5 | .92 | 368.0 |
| ST 2 | x 4.75 | .80 | 336.8 |
| | x 3.85 | .78 | 405.2 |

**Miscellaneous tee (cut from M5 x 18.9)**

| Section designation | | Surface area per foot of length | Surface area per ton |
|---|---|---|---|
| MT 2.5 | x 9.45 | .83 | 175.7 |

**Figure 23 (cont'd)**
Structural steel conversion table

| Section designation | | | Surface area per foot of length | Surface area per ton |
|---|---|---|---|---|
| L 8 | x 8 | x 1-1/8 | 2.67 | 93.8 |
| | x 8 | x 1 | 2.67 | 104.7 |
| | x 8 | x 1 | 2.67 | 118.7 |
| | x 8 | x 7/8 | 2.67 | 118.7 |
| | x 8 | x 3/4 | 2.67 | 137.3 |
| | x 8 | x 5/8 | 2.67 | 163.3 |
| | x 8 | x 9/16 | 2.67 | 180.4 |
| | x 8 | x 1/2 | 2.67 | 202.3 |
| L 6 | x 6 | x 1 | 2.00 | 107.0 |
| | x 6 | x 7/8 | 2.00 | 120.8 |
| | x 6 | x 3/4 | 2.00 | 139.4 |
| | x 6 | x 5/8 | 2.00 | 165.3 |
| | x 6 | x 9/16 | 2.00 | 182.6 |
| | x 6 | x 1/2 | 2.00 | 204.1 |
| | x 6 | x 7/16 | 2.00 | 232.6 |
| | x 6 | x 3/8 | 2.00 | 268.5 |
| | x 6 | x 5/16 | 2.00 | 322.6 |
| L5 | x 5 | x 7/8 | 1.67 | 122.8 |
| | x 5 | x 3/4 | 1.67 | 141.5 |
| | x 5 | x 5/8 | 1.67 | 167.0 |
| | x 5 | x 1/2 | 1.67 | 206.2 |
| | x 5 | x 7/16 | 1.67 | 233.6 |
| | x 5 | x 3/8 | 1.67 | 271.5 |
| | x 5 | x 5/16 | 1.67 | 324.3 |
| L 4 | x 4 | x 3/4 | 1.33 | 143.8 |
| | x 4 | x 5/8 | 1.33 | 169.4 |
| | x 4 | x 1/2 | 1.33 | 207.8 |
| | x 4 | x 7/16 | 1.33 | 235.4 |
| | x 4 | x 3/8 | 1.33 | 271.4 |
| | x 4 | x 5/16 | 1.33 | 324.4 |
| | x 4 | x 1/4 | 1.33 | 403.0 |
| L 3-1/2 | x 3/1/2 | x 1/2 | 1.17 | 210.8 |
| | x 3-1/2 | x 7/16 | 1.17 | 238.8 |
| | x 3-1/2 | x 3/8 | 1.17 | 275.3 |
| | x 3-1/2 | x 5/16 | 1.17 | 325.0 |
| | x 3-1/2 | x 1/4 | 1.17 | 403.4 |

| Section designation | | | Surface area per foot of length | Surface area per ton |
|---|---|---|---|---|
| L 3 | x 3 | x 1/2 | 1.00 | 212.8 |
| | x 3 | x 7/16 | 1.00 | 241.0 |
| | x 3 | x 3/8 | 1.00 | 277.8 |
| | x 3 | x 5/16 | 1.00 | 327.9 |
| | x 3 | x 1/4 | 1.00 | 408.2 |
| | x 3 | x 3/16 | 1.00 | 539.1 |
| L 2-1/2 | x 2-1/2 | x 1/2 | .83 | 215.6 |
| | x 2-1/2 | x 3/8 | .83 | 281.4 |
| | x 2-1/2 | x 5/16 | .83 | 332.0 |
| | x 2-1/2 | x 1/4 | .83 | 404.9 |
| | x 2-1/2 | x 3/16 | .83 | 540.7 |
| L 2 | x 2 | x 3/8 | .67 | 285.1 |
| | x 2 | x 5/16 | .67 | 341.8 |
| | x 2 | x 1/4 | .67 | 420.1 |
| | x 2 | x 3/16 | .67 | 549.2 |
| | x 2 | x 1/8 | .67 | 812.1 |
| L 1-3/4 | x 1-3/4 | x 1/4 | .58 | 418.8 |
| | x 1-3/4 | x 3/16 | .58 | 547.2 |
| | x 1-3/4 | x 1/8 | .58 | 805.6 |
| L 1-1/2 | x 1-1/2 | x 1/4 | .50 | 427.4 |
| | x 1-1/2 | x 3/16 | .50 | 555.6 |
| | x 1-1/2 | x 5/32 | .50 | 657.9 |
| | x 1-1/2 | x 1/8 | .50 | 813.0 |
| L 1-1/4 | x 1-1/4 | x 1/4 | .42 | 437.5 |
| | x 1-1/4 | x 3/16 | .42 | 567.6 |
| | x 1-1/4 | x 1/8 | .42 | 831.7 |
| L 1 | x 1 | x 1/4 | .33 | 443.0 |
| | x 1 | x 3/16 | .33 | 569.0 |
| | x 1 | x 1/8 | .33 | 825.0 |

**Figure 23 (cont'd)**
Structural steel conversion table

| Section designation | | | Surface area per foot of length | Surface area per ton |
|---|---|---|---|---|
| L 9 | x 4 | x 1 | 2.17 | 106.4 |
| | x 4 | x 7/8 | 2.17 | 120.2 |
| | x 4 | x 3/4 | 2.17 | 138.7 |
| | x 4 | x 5/8 | 2.17 | 165.0 |
| | x 4 | x 9/16 | 2.17 | 182.4 |
| | x 4 | x 1/2 | 2.17 | 203.8 |
| L 8 | x 6 | x 1 | 2.33 | 105.4 |
| | x 6 | x 7/8 | 2.33 | 119.2 |
| | x 6 | x 3/4 | 2.33 | 137.9 |
| | x 6 | x 5/8 | 2.33 | 163.5 |
| | x 6 | x 9/16 | 2.33 | 181.3 |
| | x 6 | x 1/2 | 2.33 | 202.6 |
| | x 6 | x 7/16 | 2.33 | 230.7 |
| L 8 | x 4 | x 1 | 2.00 | 107.0 |
| | x 4 | x 7/8 | 2.00 | 120.8 |
| | x 4 | x 3/4 | 2.00 | 139.4 |
| | x 4 | x 5/8 | 2.00 | 165.3 |
| | x 4 | x 9/16 | 2.00 | 182.6 |
| | x 4 | x 1/2 | 2.00 | 204.1 |
| | x 4 | x 7/16 | 2.00 | 232.6 |
| L 7 | x 4 | x 7/8 | 1.83 | 121.2 |
| | x 4 | x 3/4 | 1.83 | 139.7 |
| | x 4 | x 5/8 | 1.83 | 165.6 |
| | x 4 | x 9/16 | 1.83 | 183.0 |
| | x 4 | x 1/2 | 1.83 | 204.5 |
| | x 4 | x 7/16 | 1.83 | 231.6 |
| | x 4 | x 3/8 | 1.83 | 269.1 |
| L 6 | x 4 | x 7/8 | 1.67 | 122.8 |
| | x 4 | x 3/4 | 1.67 | 141.5 |
| | x 4 | x 5/8 | 1.67 | 167.0 |
| | x 4 | x 9/16 | 1.67 | 184.5 |
| | x 4 | x 1/2 | 1.67 | 206.2 |
| | x 4 | x 7/16 | 1.67 | 233.6 |
| | x 4 | x 3/8 | 1.67 | 271.5 |
| | x 4 | x 5/16 | 1.67 | 324.3 |
| | x 4 | x 1/4 | 1.67 | 402.4 |
| L 6 | x 3-1/2 | x 1/2 | 1.58 | 206.5 |
| | x 3-1/2 | x 3/8 | 1.58 | 270.1 |
| | x 3-1/2 | x 5/16 | 1.58 | 322.4 |
| | x 3-1/2 | x 1/4 | 1.58 | 400.0 |

| Section designation | | | Surface area per foot of length | Surface area per ton |
|---|---|---|---|---|
| L 5 | x 3-1/2 | x 3/4 | 1.42 | 143.4 |
| | x 3-1/2 | x 5/8 | 1.42 | 169.0 |
| | x 3-1/2 | x 1/2 | 1.42 | 208.8 |
| | x 3-1/2 | x 7/16 | 1.42 | 236.7 |
| | x 3-1/2 | x 3/8 | 1.42 | 273.1 |
| | x 3-1/2 | x 5/16 | 1.42 | 326.4 |
| | x 3-1/2 | x 1/4 | 1.42 | 405.7 |
| L 5 | x 3 | x 1/2 | 1.33 | 207.8 |
| | x 3 | x 7/16 | 1.33 | 235.4 |
| | x 3 | x 3/8 | 1.33 | 271.4 |
| | x 3 | x 5/16 | 1.33 | 324.4 |
| | x 3 | x 1/4 | 1.33 | 403.0 |
| L 4 | x 3-1/2 | x 5/8 | 1.25 | 170.1 |
| | x 3-1/2 | x 1/2 | 1.25 | 210.1 |
| | x 3-1/2 | x 7/16 | 1.25 | 235.8 |
| | x 3-1/2 | x 3/8 | 1.25 | 274.7 |
| | x 3-1/2 | x 5/16 | 1.25 | 324.7 |
| | x 3-1/2 | x 1/4 | 1.25 | 403.2 |
| L 4 | x 3 | x 5/8 | 1.17 | 172.1 |
| | x 3 | x 1/2 | 1.17 | 210.8 |
| | x 3 | x 7/16 | 1.17 | 238.8 |
| | x 3 | x 3/8 | 1.17 | 275.3 |
| | x 3 | x 5/16 | 1.17 | 325.0 |
| | x 3 | x 1/4 | 1.17 | 403.4 |
| L 3-1/2 | x 3 | x 1/2 | 1.08 | 211.8 |
| | x 3 | x 7/16 | 1.08 | 237.4 |
| | x 3 | x 3/8 | 1.08 | 273.4 |
| | x 3 | x 5/16 | 1.08 | 327.3 |
| | x 3 | x 1/4 | 1.08 | 400.0 |
| L 3-1/2 | x 2-1/2 | x 1/2 | 1.00 | 212.8 |
| | x 2-1/2 | x 7/16 | 1.00 | 241.0 |
| | x 2-1/2 | x 3/8 | 1.00 | 277.8 |
| | x 2-1/2 | x 5/16 | 1.00 | 327.9 |
| | x 2-1/2 | x 1/4 | 1.00 | 408.2 |
| L 3 | x 2-1/2 | x 1/2 | .92 | 216.5 |
| | x 2-1/2 | x 7/16 | .92 | 242.1 |
| | x 2-1/2 | x 3/8 | .92 | 278.8 |
| | x 2-1/2 | x 5/16 | .92 | 328.6 |
| | x 2-1/2 | x 1/4 | .92 | 408.9 |
| | x 2-1/2 | x 3/16 | .92 | 542.8 |

**Figure 23 (cont'd)**
Structural steel conversion table

| Section designation | | | Surface area per foot of length | Surface area per ton |
|---|---|---|---|---|
| L 3 | x 2 | x 1/2 | .83 | 215.6 |
| | x 2 | x 7/16 | .83 | 244.1 |
| | x 2 | x 3/8 | .83 | 281.4 |
| | x 2 | x 5/16 | .83 | 332.0 |
| | x 2 | x 1/4 | .83 | 404.9 |
| | x 2 | x 3/16 | .83 | 540.7 |
| L 2-1/2 | x 2 | x 3/8 | .75 | 283.0 |
| | x 2 | x 5/16 | .75 | 333.3 |
| | x 2 | x 1/4 | .75 | 414.4 |
| | x 2 | x 3/16 | .75 | 545.5 |
| L 2-1/2 | x 1-1/2 | x 5/16 | .67 | 341.8 |
| | x 1-1/2 | x 1/4 | .67 | 420.1 |
| | x 1-1/2 | x 3/16 | .67 | 549.2 |

Courtesy: Richardson Engineering Services, Inc.

| Section designation | | | Surface area per foot of length | Surface area per ton |
|---|---|---|---|---|
| L 2 | x 1-1/2 | x 1/4 | .58 | 418.8 |
| | x 1-1/2 | x 3/16 | .58 | 547.2 |
| | x 1-1/2 | x 1/8 | .58 | 805.6 |
| L 2 | x 1-1/4 | x 1/4 | .54 | 423.5 |
| | x 1-1/4 | x 3/16 | .54 | 551.0 |
| L 1-3/4 | x 1-1/4 | x 1/4 | .50 | 427.4 |
| | x 1-1/4 | x 3/16 | .50 | 555.6 |
| | x 1-1/4 | x 1/8 | .50 | 813.0 |

**Figure 23 (cont'd)**
Structural steel conversion table

| Diameter (in feet) | Area (SF) |
|---|---|
| 10 | 314 |
| 15 | 707 |
| 20 | 1,257 |
| 25 | 1,963 |
| 30 | 2,827 |
| 35 | 3,848 |
| 40 | 5,027 |
| 45 | 6,362 |
| 50 | 7,854 |
| 55 | 9,503 |
| 60 | 11,310 |
| 65 | 13,273 |
| 70 | 15,394 |

**Figure 24**
Surface area of spheres

| | Labor SF per manhour | Material coverage SF/gallon | Material cost per gallon | Labor cost per 100 SF | Labor burden 100 SF | Material cost per 100 SF | Overhead per 100 SF | Profit per 100 SF | Total price per 100 SF |
|---|---|---|---|---|---|---|---|---|---|
| **Tank, silo, vessel, or hopper, brush, exterior walls only** | | | | | | | | | |
| Metal primer, rust inhibitor - clean metal (material #35) | | | | | | | | | |
| Brush prime coat | | | | | | | | | |
| Slow | 150 | 425 | 28.20 | 6.83 | 1.87 | 6.64 | 4.75 | 3.01 | 23.10 |
| Medium | 175 | 400 | 24.60 | 9.43 | 3.16 | 6.15 | 4.59 | 2.57 | 25.90 |
| Fast | 200 | 375 | 21.10 | 11.00 | 4.46 | 5.63 | 3.90 | 1.50 | 26.49 |
| Metal primer, rust inhibitor - rusty metal (material #36) | | | | | | | | | |
| Brush prime coat | | | | | | | | | |
| Slow | 150 | 400 | 33.30 | 6.83 | 1.87 | 8.33 | 5.28 | 3.35 | 25.66 |
| Medium | 175 | 375 | 29.20 | 9.43 | 3.16 | 7.79 | 5.00 | 2.79 | 28.17 |
| Fast | 200 | 350 | 25.00 | 11.00 | 4.46 | 7.14 | 4.18 | 1.61 | 28.39 |
| Industrial enamel, oil base, high gloss - light colors (material #56) | | | | | | | | | |
| Brush 1st or additional finish coats | | | | | | | | | |
| Slow | 200 | 450 | 28.90 | 5.13 | 1.39 | 6.42 | 4.01 | 2.54 | 19.49 |
| Medium | 225 | 425 | 25.20 | 7.33 | 2.46 | 5.93 | 3.85 | 2.15 | 21.72 |
| Fast | 250 | 400 | 21.60 | 8.80 | 3.57 | 5.40 | 3.29 | 1.26 | 22.32 |
| Industrial enamel, oil base, high gloss - dark (OSHA) colors (material #57) | | | | | | | | | |
| Brush 1st or additional finish coats | | | | | | | | | |
| Slow | 200 | 475 | 34.90 | 5.13 | 1.39 | 7.35 | 4.30 | 2.73 | 20.90 |
| Medium | 225 | 450 | 30.50 | 7.33 | 2.46 | 6.78 | 4.06 | 2.27 | 22.90 |
| Fast | 250 | 425 | 26.10 | 8.80 | 3.57 | 6.14 | 3.42 | 1.32 | 23.25 |
| Epoxy coating, 2 part system, clear (material #51) | | | | | | | | | |
| Brush 1st coat | | | | | | | | | |
| Slow | 150 | 425 | 57.00 | 6.83 | 1.87 | 13.41 | 6.85 | 4.34 | 33.30 |
| Medium | 175 | 400 | 49.90 | 9.43 | 3.16 | 12.48 | 6.14 | 3.43 | 34.64 |
| Fast | 200 | 375 | 42.80 | 11.00 | 4.46 | 11.41 | 4.97 | 1.91 | 33.75 |
| Brush 2nd or additional coats | | | | | | | | | |
| Slow | 200 | 450 | 57.00 | 5.13 | 1.39 | 12.67 | 5.95 | 3.77 | 28.91 |
| Medium | 225 | 425 | 49.90 | 7.33 | 2.46 | 11.74 | 5.28 | 2.95 | 29.76 |
| Fast | 250 | 400 | 42.80 | 8.80 | 3.57 | 10.70 | 4.27 | 1.64 | 28.98 |
| Epoxy coating, 2 part system, white (material #52) | | | | | | | | | |
| Brush 1st coat | | | | | | | | | |
| Slow | 150 | 425 | 57.40 | 6.83 | 1.87 | 13.51 | 6.88 | 4.36 | 33.45 |
| Medium | 175 | 400 | 50.20 | 9.43 | 3.16 | 12.55 | 6.16 | 3.44 | 34.74 |
| Fast | 200 | 375 | 43.10 | 11.00 | 4.46 | 11.49 | 4.99 | 1.92 | 33.86 |
| Brush 2nd or additional coats | | | | | | | | | |
| Slow | 200 | 450 | 57.40 | 5.13 | 1.39 | 12.76 | 5.98 | 3.79 | 29.05 |
| Medium | 225 | 425 | 50.20 | 7.33 | 2.46 | 11.81 | 5.29 | 2.96 | 29.85 |
| Fast | 250 | 400 | 43.10 | 8.80 | 3.57 | 10.78 | 4.28 | 1.65 | 29.08 |

| | Labor SF per manhour | Material coverage SF/gallon | Material cost per gallon | Labor cost per 100 SF | Labor burden 100 SF | Material cost per 100 SF | Overhead per 100 SF | Profit per 100 SF | Total price per 100 SF |
|---|---|---|---|---|---|---|---|---|---|
| Vinyl coating (material #59) | | | | | | | | | |
| Brush 1st coat | | | | | | | | | |
| Slow | 125 | 250 | 59.20 | 8.20 | 2.23 | 23.68 | 10.57 | 6.70 | 51.38 |
| Medium | 150 | 238 | 51.80 | 11.00 | 3.71 | 21.76 | 8.93 | 4.99 | 50.39 |
| Fast | 175 | 225 | 44.40 | 12.57 | 5.09 | 19.73 | 6.92 | 2.66 | 46.97 |
| Brush 2nd or additional coats | | | | | | | | | |
| Slow | 165 | 150 | 59.20 | 6.21 | 1.69 | 39.47 | 14.68 | 9.31 | 71.36 |
| Medium | 190 | 125 | 51.80 | 8.68 | 2.92 | 41.44 | 12.99 | 7.26 | 73.29 |
| Fast | 215 | 100 | 44.40 | 10.23 | 4.15 | 44.40 | 10.87 | 4.18 | 73.83 |

See Figure 24 on page 395 to find the surface area of a spherical vessel. Use this table when estimating walls only. The cost tables for painting steel tank, silo, vessel or hopper roofs follow those for painting walls. For heights above 8 feet, use the High Time Difficulty Factors on page 137. Note: A two coat system, prime and finish, using oil base material is recommended for any metal surface. Although water base material is often used, it may cause oxidation, corrosion and rust. One coat of oil base solid body stain is often used on interior or exterior metal but it may crack, peel or chip without the proper prime coat application. "Slow" work is based on a $10.25 hourly wage, "Medium" work on a $16.50 hourly wage, and "Fast" work on a $22.00 hourly wage. Other qualifications that apply to this table are on page 9.

| | Labor SF per manhour | Material coverage SF/gallon | Material cost per gallon | Labor cost per 100 SF | Labor burden 100 SF | Material cost per 100 SF | Overhead per 100 SF | Profit per 100 SF | Total price per 100 SF |
|---|---|---|---|---|---|---|---|---|---|
| **Tank, silo, vessel, or hopper, roll, exterior walls only** | | | | | | | | | |
| Metal primer, rust inhibitor - clean metal (material #35) | | | | | | | | | |
| Roll prime coat | | | | | | | | | |
| Slow | 275 | 400 | 28.20 | 3.73 | 1.02 | 7.05 | 3.66 | 2.32 | 17.78 |
| Medium | 300 | 375 | 24.60 | 5.50 | 1.84 | 6.56 | 3.41 | 1.91 | 19.22 |
| Fast | 325 | 350 | 21.10 | 6.77 | 2.75 | 6.03 | 2.88 | 1.11 | 19.54 |
| Metal primer, rust inhibitor - rusty metal (material #36) | | | | | | | | | |
| Roll prime coat | | | | | | | | | |
| Slow | 275 | 380 | 33.30 | 3.73 | 1.02 | 8.76 | 4.19 | 2.66 | 20.36 |
| Medium | 300 | 355 | 29.20 | 5.50 | 1.84 | 8.23 | 3.82 | 2.13 | 21.52 |
| Fast | 325 | 330 | 25.00 | 6.77 | 2.75 | 7.58 | 3.16 | 1.22 | 21.48 |
| Industrial enamel, oil base, high gloss - light colors (material #56) | | | | | | | | | |
| Roll 1st or additional finish coats | | | | | | | | | |
| Slow | 375 | 425 | 28.90 | 2.73 | .75 | 6.80 | 3.18 | 2.02 | 15.48 |
| Medium | 400 | 400 | 25.20 | 4.13 | 1.38 | 6.30 | 2.90 | 1.62 | 16.33 |
| Fast | 425 | 375 | 21.60 | 5.18 | 2.09 | 5.76 | 2.41 | .93 | 16.37 |
| Industrial enamel, oil base, high gloss - dark (OSHA) colors (material #57) | | | | | | | | | |
| Roll 1st or additional finish coats | | | | | | | | | |
| Slow | 375 | 450 | 34.90 | 2.73 | .75 | 7.76 | 3.48 | 2.21 | 16.93 |
| Medium | 400 | 425 | 30.50 | 4.13 | 1.38 | 7.18 | 3.11 | 1.74 | 17.54 |
| Fast | 425 | 400 | 26.10 | 5.18 | 2.09 | 6.53 | 2.55 | .98 | 17.33 |
| Epoxy coating, 2 part system, clear (material #51) | | | | | | | | | |
| Roll 1st coat | | | | | | | | | |
| Slow | 275 | 425 | 57.00 | 3.73 | 1.02 | 13.41 | 5.63 | 3.57 | 27.36 |
| Medium | 300 | 400 | 49.90 | 5.50 | 1.84 | 12.48 | 4.86 | 2.72 | 27.40 |
| Fast | 325 | 375 | 42.80 | 6.77 | 2.75 | 11.41 | 3.87 | 1.49 | 26.29 |
| Roll 2nd or additional coats | | | | | | | | | |
| Slow | 375 | 450 | 57.00 | 2.73 | .75 | 12.67 | 5.00 | 3.17 | 24.32 |
| Medium | 400 | 425 | 49.90 | 4.13 | 1.38 | 11.74 | 4.23 | 2.36 | 23.84 |
| Fast | 425 | 400 | 42.80 | 5.18 | 2.09 | 10.70 | 3.33 | 1.28 | 22.58 |

| | Labor SF per manhour | Material coverage SF/gallon | Material cost per gallon | Labor cost per 100 SF | Labor burden 100 SF | Material cost per 100 SF | Overhead per 100 SF | Profit per 100 SF | Total price per 100 SF |
|---|---|---|---|---|---|---|---|---|---|
| Epoxy coating, 2 part system, white (material #52) | | | | | | | | | |
| Roll 1st coat | | | | | | | | | |
| Slow | 275 | 425 | 57.40 | 3.73 | 1.02 | 13.51 | 5.66 | 3.59 | 27.51 |
| Medium | 300 | 400 | 50.20 | 5.50 | 1.84 | 12.55 | 4.88 | 2.73 | 27.50 |
| Fast | 325 | 375 | 43.10 | 6.77 | 2.75 | 11.49 | 3.89 | 1.49 | 26.39 |
| Roll 2nd or additional coats | | | | | | | | | |
| Slow | 375 | 450 | 57.40 | 2.73 | .75 | 12.76 | 5.03 | 3.19 | 24.46 |
| Medium | 400 | 425 | 50.20 | 4.13 | 1.38 | 11.81 | 4.25 | 2.37 | 23.94 |
| Fast | 425 | 400 | 43.10 | 5.18 | 2.09 | 10.78 | 3.34 | 1.28 | 22.67 |

See Figure 24 on page 395 to find the surface area of a spherical vessel. Use this table when estimating walls only. The cost tables for painting steel tank, silo, vessel or hopper roofs follow those for painting walls. For heights above 8 feet, use the High Time Difficulty Factors on page 137. Note: A two coat system, prime and finish, using oil base material is recommended for any metal surface. Although water base material is often used, it may cause oxidation, corrosion and rust. One coat of oil base solid body stain is often used on interior or exterior metal but it may crack, peel or chip without the proper prime coat application. "Slow" work is based on a $10.25 hourly wage, "Medium" work on a $16.50 hourly wage, and "Fast" work on a $22.00 hourly wage. Other qualifications that apply to this table are on page 9.

| | Labor SF per manhour | Material coverage SF/gallon | Material cost per gallon | Labor cost per 100 SF | Labor burden 100 SF | Material cost per 100 SF | Overhead per 100 SF | Profit per 100 SF | Total price per 100 SF |
|---|---|---|---|---|---|---|---|---|---|
| **Tank, silo, vessel, or hopper, spray, exterior walls only** | | | | | | | | | |
| Metal primer, rust inhibitor - clean metal (material #35) | | | | | | | | | |
| Spray prime coat | | | | | | | | | |
| Slow | 700 | 325 | 28.20 | 1.46 | .40 | 8.68 | 3.27 | 2.07 | 15.88 |
| Medium | 750 | 300 | 24.60 | 2.20 | .73 | 8.20 | 2.73 | 1.53 | 15.39 |
| Fast | 800 | 275 | 21.10 | 2.75 | 1.12 | 7.67 | 2.13 | .82 | 14.49 |
| Metal primer, rust inhibitor - rusty metal (material #36) | | | | | | | | | |
| Spray prime coat | | | | | | | | | |
| Slow | 700 | 300 | 33.30 | 1.46 | .40 | 11.10 | 4.02 | 2.55 | 19.53 |
| Medium | 750 | 275 | 29.20 | 2.20 | .73 | 10.62 | 3.32 | 1.86 | 18.73 |
| Fast | 800 | 250 | 25.00 | 2.75 | 1.12 | 10.00 | 2.57 | .99 | 17.43 |
| Industrial enamel, oil base, high gloss - light colors (material #56) | | | | | | | | | |
| Spray 1st or additional finish coats | | | | | | | | | |
| Slow | 850 | 350 | 28.90 | 1.21 | .33 | 8.26 | 3.04 | 1.93 | 14.77 |
| Medium | 900 | 325 | 25.20 | 1.83 | .62 | 7.75 | 2.50 | 1.40 | 14.10 |
| Fast | 950 | 300 | 21.60 | 2.32 | .93 | 7.20 | 1.94 | .74 | 13.13 |
| Industrial enamel, oil base, high gloss - dark (OSHA) colors (material #57) | | | | | | | | | |
| Spray 1st or additional finish coats | | | | | | | | | |
| Slow | 850 | 375 | 34.90 | 1.21 | .33 | 9.31 | 3.36 | 2.13 | 16.34 |
| Medium | 900 | 350 | 30.50 | 1.83 | .62 | 8.71 | 2.73 | 1.53 | 15.42 |
| Fast | 950 | 325 | 26.10 | 2.32 | .93 | 8.03 | 2.09 | .80 | 14.17 |
| Epoxy coating, 2 part system, clear (material #51) | | | | | | | | | |
| Spray 1st coat | | | | | | | | | |
| Slow | 700 | 325 | 57.00 | 1.46 | .40 | 17.54 | 6.01 | 3.81 | 29.22 |
| Medium | 750 | 313 | 49.90 | 2.20 | .73 | 15.94 | 4.63 | 2.59 | 26.09 |
| Fast | 800 | 300 | 42.80 | 2.75 | 1.12 | 14.27 | 3.36 | 1.29 | 22.79 |
| Spray 2nd or additional coats | | | | | | | | | |
| Slow | 850 | 350 | 57.00 | 1.21 | .33 | 16.29 | 5.53 | 3.50 | 26.86 |
| Medium | 900 | 338 | 49.90 | 1.83 | .62 | 14.76 | 4.22 | 2.36 | 23.79 |
| Fast | 950 | 325 | 42.80 | 2.32 | .93 | 13.17 | 3.04 | 1.17 | 20.63 |
| Epoxy coating, 2 part system, white (material #52) | | | | | | | | | |
| Spray 1st coat | | | | | | | | | |
| Slow | 700 | 325 | 57.40 | 1.46 | .40 | 17.66 | 6.05 | 3.84 | 29.41 |
| Medium | 750 | 313 | 50.20 | 2.20 | .73 | 16.04 | 4.65 | 2.60 | 26.22 |
| Fast | 800 | 300 | 43.10 | 2.75 | 1.12 | 14.37 | 3.37 | 1.30 | 22.91 |
| Spray 2nd or additional coats | | | | | | | | | |
| Slow | 850 | 350 | 57.40 | 1.21 | .33 | 16.40 | 5.56 | 3.53 | 27.03 |
| Medium | 900 | 338 | 50.20 | 1.83 | .62 | 14.85 | 4.24 | 2.37 | 23.91 |
| Fast | 950 | 325 | 43.10 | 2.32 | .93 | 13.26 | 3.06 | 1.17 | 20.74 |

| | Labor SF per manhour | Material coverage SF/gallon | Material cost per gallon | Labor cost per 100 SF | Labor burden 100 SF | Material cost per 100 SF | Overhead per 100 SF | Profit per 100 SF | Total price per 100 SF |
|---|---|---|---|---|---|---|---|---|---|
| Vinyl coating (material #59) | | | | | | | | | |
| Spray 1st coat | | | | | | | | | |
| Slow | 600 | 225 | 59.20 | 1.71 | .47 | 26.31 | 8.83 | 5.60 | 42.92 |
| Medium | 625 | 213 | 51.80 | 2.64 | .89 | 24.32 | 6.82 | 3.81 | 38.48 |
| Fast | 650 | 200 | 44.40 | 3.38 | 1.38 | 22.20 | 4.99 | 1.92 | 33.87 |
| Spray 2nd or additional coats | | | | | | | | | |
| Slow | 725 | 130 | 59.20 | 1.41 | .39 | 45.54 | 14.67 | 9.30 | 71.31 |
| Medium | 750 | 105 | 51.80 | 2.20 | .73 | 49.33 | 12.81 | 7.16 | 72.23 |
| Fast | 775 | 80 | 44.40 | 2.84 | 1.15 | 55.50 | 11.01 | 4.23 | 74.73 |

See Figure 24 on page 395 to find the surface area of a spherical vessel. Use this table when estimating walls only. The cost tables for painting steel tank, silo, vessel or hopper roofs follow those for painting walls. For heights above 8 feet, use the High Time Difficulty Factors on page 137. Note: A two coat system, prime and finish, using oil base material is recommended for any metal surface. Although water base material is often used, it may cause oxidation, corrosion and rust. One coat of oil base solid body stain is often used on interior or exterior metal but it may crack, peel or chip without the proper prime coat application. "Slow" work is based on a $10.25 hourly wage, "Medium" work on a $16.50 hourly wage, and "Fast" work on a $22.00 hourly wage. Other qualifications that apply to this table are on page 9.

| | Labor SF per manhour | Material coverage SF/gallon | Material cost per gallon | Labor cost per 100 SF | Labor burden 100 SF | Material cost per 100 SF | Overhead per 100 SF | Profit per 100 SF | Total price per 100 SF |
|---|---|---|---|---|---|---|---|---|---|
| **Tank, silo, vessel, or hopper, brush, exterior roof only** | | | | | | | | | |
| Metal primer, rust inhibitor - clean metal (material #35) | | | | | | | | | |
| Brush prime coat | | | | | | | | | |
| Slow | 175 | 425 | 28.20 | 5.86 | 1.59 | 6.64 | 4.37 | 2.77 | 21.23 |
| Medium | 200 | 400 | 24.60 | 8.25 | 2.78 | 6.15 | 4.21 | 2.35 | 23.74 |
| Fast | 225 | 375 | 21.10 | 9.78 | 3.95 | 5.63 | 3.59 | 1.38 | 24.33 |
| Metal primer, rust inhibitor - rusty metal (material #36) | | | | | | | | | |
| Brush prime coat | | | | | | | | | |
| Slow | 175 | 400 | 33.30 | 5.86 | 1.59 | 8.33 | 4.89 | 3.10 | 23.77 |
| Medium | 200 | 375 | 29.20 | 8.25 | 2.78 | 7.79 | 4.61 | 2.58 | 26.01 |
| Fast | 225 | 350 | 25.00 | 9.78 | 3.95 | 7.14 | 3.86 | 1.49 | 26.22 |
| Industrial enamel, oil base, high gloss - light colors (material #56) | | | | | | | | | |
| Brush 1st or additional finish coats | | | | | | | | | |
| Slow | 225 | 450 | 28.90 | 4.56 | 1.23 | 6.42 | 3.79 | 2.40 | 18.40 |
| Medium | 250 | 425 | 25.20 | 6.60 | 2.22 | 5.93 | 3.61 | 2.02 | 20.38 |
| Fast | 275 | 400 | 21.60 | 8.00 | 3.25 | 5.40 | 3.08 | 1.18 | 20.91 |
| Industrial enamel, oil base, high gloss - dark (OSHA) colors (material #57) | | | | | | | | | |
| Brush 1st or additional finish coats | | | | | | | | | |
| Slow | 225 | 475 | 34.90 | 4.56 | 1.23 | 7.35 | 4.08 | 2.58 | 19.80 |
| Medium | 250 | 450 | 30.50 | 6.60 | 2.22 | 6.78 | 3.82 | 2.14 | 21.56 |
| Fast | 275 | 425 | 26.10 | 8.00 | 3.25 | 6.14 | 3.22 | 1.24 | 21.85 |
| Epoxy coating, 2 part system, clear (material #51) | | | | | | | | | |
| Brush 1st coat | | | | | | | | | |
| Slow | 175 | 425 | 57.00 | 5.86 | 1.59 | 13.41 | 6.47 | 4.10 | 31.43 |
| Medium | 200 | 400 | 49.90 | 8.25 | 2.78 | 12.48 | 5.76 | 3.22 | 32.49 |
| Fast | 225 | 375 | 42.80 | 9.78 | 3.95 | 11.41 | 4.65 | 1.79 | 31.58 |
| Brush 2nd or additional coats | | | | | | | | | |
| Slow | 225 | 450 | 57.00 | 4.56 | 1.23 | 12.67 | 5.73 | 3.63 | 27.82 |
| Medium | 250 | 425 | 49.90 | 6.60 | 2.22 | 11.74 | 5.04 | 2.82 | 28.42 |
| Fast | 275 | 400 | 42.80 | 8.00 | 3.25 | 10.70 | 4.06 | 1.56 | 27.57 |
| Epoxy coating, 2 part system, white (material #52) | | | | | | | | | |
| Brush 1st coat | | | | | | | | | |
| Slow | 175 | 425 | 57.40 | 5.86 | 1.59 | 13.51 | 6.50 | 4.12 | 31.58 |
| Medium | 200 | 400 | 50.20 | 8.25 | 2.78 | 12.55 | 5.78 | 3.23 | 32.59 |
| Fast | 225 | 375 | 43.10 | 9.78 | 3.95 | 11.49 | 4.67 | 1.79 | 31.68 |
| Brush 2nd or additional coats | | | | | | | | | |
| Slow | 225 | 450 | 57.40 | 4.56 | 1.23 | 12.76 | 5.75 | 3.65 | 27.95 |
| Medium | 250 | 425 | 50.20 | 6.60 | 2.22 | 11.81 | 5.05 | 2.82 | 28.50 |
| Fast | 275 | 400 | 43.10 | 8.00 | 3.25 | 10.78 | 4.07 | 1.57 | 27.67 |

| | Labor SF per manhour | Material coverage SF/gallon | Material cost per gallon | Labor cost per 100 SF | Labor burden 100 SF | Material cost per 100 SF | Overhead per 100 SF | Profit per 100 SF | Total price per 100 SF |
|---|---|---|---|---|---|---|---|---|---|
| Vinyl coating (material #59) | | | | | | | | | |
| Brush 1st coat | | | | | | | | | |
| Slow | 125 | 250 | 59.20 | 8.20 | 2.23 | 23.68 | 10.57 | 6.70 | 51.38 |
| Medium | 150 | 238 | 51.80 | 11.00 | 3.71 | 21.76 | 8.93 | 4.99 | 50.39 |
| Fast | 175 | 225 | 44.40 | 12.57 | 5.09 | 19.73 | 6.92 | 2.66 | 46.97 |
| Brush 2nd or additional coats | | | | | | | | | |
| Slow | 200 | 150 | 59.20 | 5.13 | 1.39 | 39.47 | 14.26 | 9.04 | 69.29 |
| Medium | 225 | 125 | 51.80 | 7.33 | 2.46 | 41.44 | 12.55 | 7.02 | 70.80 |
| Fast | 250 | 100 | 44.40 | 8.80 | 3.57 | 44.40 | 10.50 | 4.04 | 71.31 |

Use these figures to estimate labor and material costs for painting the exterior surface of a flat roof on a steel tank, silo, vessel or hopper. Rule of thumb: For a vaulted, peaked or sloping roof, figure the roof area as though it were flat and add 5%. Note: A two coat system, prime and finish, using oil base material is recommended for any metal surface. Although water base material is often used, it may cause oxidation, corrosion and rust. One coat of oil base solid body stain is often used on interior or exterior metal but it may crack, peel or chip without the proper prime coat application. "Slow" work is based on a $10.25 hourly wage, "Medium" work on a $16.50 hourly wage, and "Fast" work on a $22.00 hourly wage. Other qualifications that apply to this table are on page 9.

| | Labor SF per manhour | Material coverage SF/gallon | Material cost per gallon | Labor cost per 100 SF | Labor burden 100 SF | Material cost per 100 SF | Overhead per 100 SF | Profit per 100 SF | Total price per 100 SF |
|---|---|---|---|---|---|---|---|---|---|
| **Tank, silo, vessel, or hopper, roll, exterior roof only** | | | | | | | | | |
| Metal primer, rust inhibitor - clean metal (material #35) | | | | | | | | | |
| Roll prime coat | | | | | | | | | |
| Slow | 325 | 400 | 28.20 | 3.15 | .87 | 7.05 | 3.43 | 2.17 | 16.67 |
| Medium | 350 | 375 | 24.60 | 4.71 | 1.60 | 6.56 | 3.15 | 1.76 | 17.78 |
| Fast | 375 | 350 | 21.10 | 5.87 | 2.39 | 6.03 | 2.64 | 1.02 | 17.95 |
| Metal primer, rust inhibitor - rusty metal (material #36) | | | | | | | | | |
| Roll prime coat | | | | | | | | | |
| Slow | 325 | 380 | 33.30 | 3.15 | .87 | 8.76 | 3.96 | 2.51 | 19.25 |
| Medium | 350 | 355 | 29.20 | 4.71 | 1.60 | 8.23 | 3.56 | 1.99 | 20.09 |
| Fast | 375 | 330 | 25.00 | 5.87 | 2.39 | 7.58 | 2.93 | 1.13 | 19.90 |
| Industrial enamel, oil base, high gloss - light colors (material #56) | | | | | | | | | |
| Roll 1st or additional finish coats | | | | | | | | | |
| Slow | 400 | 425 | 28.90 | 2.56 | .70 | 6.80 | 3.12 | 1.98 | 15.16 |
| Medium | 425 | 400 | 25.20 | 3.88 | 1.30 | 6.30 | 2.82 | 1.57 | 15.87 |
| Fast | 450 | 375 | 21.60 | 4.89 | 1.97 | 5.76 | 2.34 | .90 | 15.86 |
| Industrial enamel, oil base, high gloss - dark (OSHA) colors (material #57) | | | | | | | | | |
| Roll 1st or additional finish coats | | | | | | | | | |
| Slow | 400 | 450 | 34.90 | 2.56 | .70 | 7.76 | 3.42 | 2.17 | 16.61 |
| Medium | 425 | 425 | 30.50 | 3.88 | 1.30 | 7.18 | 3.03 | 1.69 | 17.08 |
| Fast | 450 | 400 | 26.10 | 4.89 | 1.97 | 6.53 | 2.48 | .95 | 16.82 |
| Epoxy coating, 2 part system, clear (material #51) | | | | | | | | | |
| Roll 1st coat | | | | | | | | | |
| Slow | 325 | 425 | 57.00 | 3.15 | .87 | 13.41 | 5.40 | 3.42 | 26.25 |
| Medium | 350 | 400 | 49.90 | 4.71 | 1.60 | 12.48 | 4.60 | 2.57 | 25.96 |
| Fast | 375 | 375 | 42.80 | 5.87 | 2.39 | 11.41 | 3.64 | 1.40 | 24.71 |
| Roll 2nd or additional coats | | | | | | | | | |
| Slow | 400 | 450 | 57.00 | 2.56 | .70 | 12.67 | 4.94 | 3.13 | 24.00 |
| Medium | 425 | 425 | 49.90 | 3.88 | 1.30 | 11.74 | 4.15 | 2.32 | 23.39 |
| Fast | 450 | 400 | 42.80 | 4.89 | 1.97 | 10.70 | 3.25 | 1.25 | 22.06 |

| | Labor SF per manhour | Material coverage SF/gallon | Material cost per gallon | Labor cost per 100 SF | Labor burden 100 SF | Material cost per 100 SF | Overhead per 100 SF | Profit per 100 SF | Total price per 100 SF |
|---|---|---|---|---|---|---|---|---|---|
| Epoxy coating, 2 part system, white (material #52) | | | | | | | | | |
| Roll 1st coat | | | | | | | | | |
| Slow | 325 | 425 | 57.40 | 3.15 | .87 | 13.51 | 5.43 | 3.44 | 26.40 |
| Medium | 350 | 400 | 50.20 | 4.71 | 1.60 | 12.55 | 4.62 | 2.58 | 26.06 |
| Fast | 375 | 375 | 43.10 | 5.87 | 2.39 | 11.49 | 3.65 | 1.40 | 24.80 |
| Roll 2nd or additional coats | | | | | | | | | |
| Slow | 400 | 450 | 57.40 | 2.56 | .70 | 12.76 | 4.97 | 3.15 | 24.14 |
| Medium | 425 | 425 | 50.20 | 3.88 | 1.30 | 11.81 | 4.17 | 2.33 | 23.49 |
| Fast | 450 | 400 | 43.10 | 4.89 | 1.97 | 10.78 | 3.27 | 1.26 | 22.17 |

Use these figures to estimate labor and material costs for painting the exterior surface of a flat roof on a steel tank, silo, vessel or hopper. Rule of thumb: For a vaulted, peaked or sloping roof, figure the roof area as though it were flat and add 5%. Note: A two coat system, prime and finish, using oil base material is recommended for any metal surface. Although water base material is often used, it may cause oxidation, corrosion and rust. One coat of oil base solid body stain is often used on interior or exterior metal but it may crack, peel or chip without the proper prime coat application. "Slow" work is based on a $10.25 hourly wage, "Medium" work on a $16.50 hourly wage, and "Fast" work on a $22.00 hourly wage. Other qualifications that apply to this table are on page 9.

| | Labor SF per manhour | Material coverage SF/gallon | Material cost per gallon | Labor cost per 100 SF | Labor burden 100 SF | Material cost per 100 SF | Overhead per 100 SF | Profit per 100 SF | Total price per 100 SF |
|---|---|---|---|---|---|---|---|---|---|
| **Tank, silo, vessel, or hopper, spray, exterior roof only** | | | | | | | | | |
| Metal primer, rust inhibitor - clean metal (material #35) | | | | | | | | | |
| Spray prime coat | | | | | | | | | |
| Slow | 850 | 325 | 28.20 | 1.21 | .33 | 8.68 | 3.17 | 2.01 | 15.40 |
| Medium | 900 | 300 | 24.60 | 1.83 | .62 | 8.20 | 2.61 | 1.46 | 14.72 |
| Fast | 950 | 275 | 21.10 | 2.32 | .93 | 7.67 | 2.02 | .78 | 13.72 |
| Metal primer, rust inhibitor - rusty metal (material #36) | | | | | | | | | |
| Spray prime coat | | | | | | | | | |
| Slow | 850 | 300 | 33.30 | 1.21 | .33 | 11.10 | 3.92 | 2.48 | 19.04 |
| Medium | 900 | 275 | 29.20 | 1.83 | .62 | 10.62 | 3.20 | 1.79 | 18.06 |
| Fast | 950 | 250 | 25.00 | 2.32 | .93 | 10.00 | 2.45 | .94 | 16.64 |
| Industrial enamel, oil base, high gloss - light colors (material #56) | | | | | | | | | |
| Spray 1st or additional finish coats | | | | | | | | | |
| Slow | 950 | 300 | 26.50 | 1.08 | .29 | 8.83 | 3.16 | 2.00 | 15.36 |
| Medium | 1025 | 275 | 23.20 | 1.61 | .55 | 8.44 | 2.59 | 1.45 | 14.64 |
| Fast | 1100 | 250 | 29.90 | 2.00 | .81 | 11.96 | 2.73 | 1.05 | 18.55 |
| Industrial enamel, oil base, high gloss - dark (OSHA) colors (material #57) | | | | | | | | | |
| Spray 1st or additional finish coats | | | | | | | | | |
| Slow | 950 | 325 | 31.30 | 1.08 | .29 | 9.63 | 3.41 | 2.16 | 16.57 |
| Medium | 1025 | 300 | 27.40 | 1.61 | .55 | 9.13 | 2.76 | 1.54 | 15.59 |
| Fast | 1100 | 275 | 23.50 | 2.00 | .81 | 8.55 | 2.10 | .81 | 14.27 |
| Epoxy coating, 2 part system, clear (material #51) | | | | | | | | | |
| Spray 1st coat | | | | | | | | | |
| Slow | 850 | 325 | 57.00 | 1.21 | .33 | 17.54 | 5.91 | 3.75 | 28.74 |
| Medium | 900 | 313 | 49.90 | 1.83 | .62 | 15.94 | 4.51 | 2.52 | 25.42 |
| Fast | 950 | 300 | 42.80 | 2.32 | .93 | 14.27 | 3.24 | 1.25 | 22.01 |
| Spray 2nd or additional coats | | | | | | | | | |
| Slow | 950 | 350 | 57.00 | 1.08 | .29 | 16.29 | 5.47 | 3.47 | 26.60 |
| Medium | 1025 | 338 | 49.90 | 1.61 | .55 | 14.76 | 4.14 | 2.32 | 23.38 |
| Fast | 1100 | 325 | 42.80 | 2.00 | .81 | 13.17 | 2.96 | 1.14 | 20.08 |
| Epoxy coating, 2 part system, white (material #52) | | | | | | | | | |
| Spray 1st coat | | | | | | | | | |
| Slow | 850 | 325 | 57.40 | 1.21 | .33 | 17.66 | 5.95 | 3.77 | 28.92 |
| Medium | 900 | 313 | 50.20 | 1.83 | .62 | 16.04 | 4.53 | 2.53 | 25.55 |
| Fast | 950 | 300 | 43.10 | 2.32 | .93 | 14.37 | 3.26 | 1.25 | 22.13 |
| Spray 2nd or additional coats | | | | | | | | | |
| Slow | 950 | 350 | 57.40 | 1.08 | .29 | 16.40 | 5.51 | 3.49 | 26.77 |
| Medium | 1025 | 338 | 50.20 | 1.61 | .55 | 14.85 | 4.17 | 2.33 | 23.51 |
| Fast | 1100 | 325 | 43.10 | 2.00 | .81 | 13.26 | 2.97 | 1.14 | 20.18 |

| | Labor SF per manhour | Material coverage SF/gallon | Material cost per gallon | Labor cost per 100 SF | Labor burden 100 SF | Material cost per 100 SF | Overhead per 100 SF | Profit per 100 SF | Total price per 100 SF |
|---|---|---|---|---|---|---|---|---|---|
| Vinyl coating (material #59) | | | | | | | | | |
| Spray 1st coat | | | | | | | | | |
| Slow | 750 | 225 | 59.20 | 1.37 | .36 | 26.31 | 8.70 | 5.51 | 42.25 |
| Medium | 775 | 213 | 51.80 | 2.13 | .71 | 24.32 | 6.66 | 3.72 | 37.54 |
| Fast | 800 | 200 | 44.40 | 2.75 | 1.12 | 22.20 | 4.82 | 1.85 | 32.74 |
| Spray 2nd or additional coats | | | | | | | | | |
| Slow | 900 | 130 | 59.20 | 1.14 | .31 | 45.54 | 14.57 | 9.23 | 70.79 |
| Medium | 950 | 105 | 51.80 | 1.74 | .58 | 49.33 | 12.66 | 7.08 | 71.39 |
| Fast | 1000 | 80 | 44.40 | 2.20 | .89 | 55.50 | 10.84 | 4.17 | 73.60 |

Use these figures to estimate labor and material costs for painting the exterior surface of a flat roof on a steel tank, silo, vessel or hopper. Rule of thumb: For a vaulted, peaked or sloping roof, figure the roof area as though it were flat and add 5%. Note: A two coat system, prime and finish, using oil base material is recommended for any metal surface. Although water base material is often used, it may cause oxidation, corrosion and rust. One coat of oil base solid body stain is often used on interior or exterior metal but it may crack, peel or chip without the proper prime coat application. "Slow" work is based on a $10.25 hourly wage, "Medium" work on a $16.50 hourly wage, and "Fast" work on a $22.00 hourly wage. Other qualifications that apply to this table are on page 9.

| | Labor SF per manhour | Material coverage SF/gallon | Material cost per gallon | Labor cost per 100 SF | Labor burden 100 SF | Material cost per 100 SF | Overhead per 100 SF | Profit per 100 SF | Total price per 100 SF |
|---|---|---|---|---|---|---|---|---|---|
| **Walls, concrete tilt-up, brush application** | | | | | | | | | |
| Flat latex, water base (material #5) | | | | | | | | | |
| Brush 1st coat | | | | | | | | | |
| Slow | 150 | 275 | 16.20 | 6.83 | 1.87 | 5.89 | 4.52 | 2.87 | 21.98 |
| Medium | 188 | 238 | 14.20 | 8.78 | 2.95 | 5.97 | 4.34 | 2.42 | 24.46 |
| Fast | 225 | 200 | 12.10 | 9.78 | 3.95 | 6.05 | 3.66 | 1.41 | 24.85 |
| Brush 2nd or additional coats | | | | | | | | | |
| Slow | 200 | 360 | 16.20 | 5.13 | 1.39 | 4.50 | 3.42 | 2.17 | 16.61 |
| Medium | 225 | 243 | 14.20 | 7.33 | 2.46 | 5.84 | 3.83 | 2.14 | 21.60 |
| Fast | 250 | 225 | 12.10 | 8.80 | 3.57 | 5.38 | 3.28 | 1.26 | 22.29 |
| Enamel, water base (material #9) | | | | | | | | | |
| Brush 1st coat | | | | | | | | | |
| Slow | 150 | 275 | 22.80 | 6.83 | 1.87 | 8.29 | 5.26 | 3.34 | 25.59 |
| Medium | 188 | 238 | 19.90 | 8.78 | 2.95 | 8.36 | 4.92 | 2.75 | 27.76 |
| Fast | 225 | 200 | 17.10 | 9.78 | 3.95 | 8.55 | 4.13 | 1.59 | 28.00 |
| Brush 2nd or additional coats | | | | | | | | | |
| Slow | 200 | 360 | 22.80 | 5.13 | 1.39 | 6.33 | 3.99 | 2.53 | 19.37 |
| Medium | 225 | 243 | 19.90 | 7.33 | 2.46 | 8.19 | 4.41 | 2.46 | 24.85 |
| Fast | 250 | 225 | 17.10 | 8.80 | 3.57 | 7.60 | 3.69 | 1.42 | 25.08 |
| Enamel, oil base (material #10) | | | | | | | | | |
| Brush 1st coat | | | | | | | | | |
| Slow | 150 | 300 | 25.90 | 6.83 | 1.87 | 8.63 | 5.37 | 3.40 | 26.10 |
| Medium | 188 | 250 | 22.70 | 8.78 | 2.95 | 9.08 | 5.10 | 2.85 | 28.76 |
| Fast | 225 | 200 | 19.40 | 9.78 | 3.95 | 9.70 | 4.34 | 1.67 | 29.44 |
| Brush 2nd or additional coats | | | | | | | | | |
| Slow | 200 | 400 | 25.90 | 5.13 | 1.39 | 6.48 | 4.03 | 2.56 | 19.59 |
| Medium | 225 | 325 | 22.70 | 7.33 | 2.46 | 6.98 | 4.11 | 2.30 | 23.18 |
| Fast | 250 | 250 | 19.40 | 8.80 | 3.57 | 7.76 | 3.72 | 1.43 | 25.28 |
| Epoxy coating, 2 part system, clear (material #51) | | | | | | | | | |
| Brush 1st coat | | | | | | | | | |
| Slow | 150 | 330 | 57.00 | 6.83 | 1.87 | 17.27 | 8.05 | 5.10 | 39.12 |
| Medium | 188 | 290 | 49.90 | 8.78 | 2.95 | 17.21 | 7.09 | 3.96 | 39.99 |
| Fast | 225 | 250 | 42.80 | 9.78 | 3.95 | 17.12 | 5.71 | 2.19 | 38.75 |
| Brush 2nd or additional coats | | | | | | | | | |
| Slow | 200 | 380 | 57.00 | 5.13 | 1.39 | 15.00 | 6.67 | 4.23 | 32.42 |
| Medium | 225 | 340 | 49.90 | 7.33 | 2.46 | 14.68 | 6.00 | 3.35 | 33.82 |
| Fast | 250 | 300 | 42.80 | 8.80 | 3.57 | 14.27 | 4.93 | 1.89 | 33.46 |

| | Labor SF per manhour | Material coverage SF/gallon | Material cost per gallon | Labor cost per 100 SF | Labor burden 100 SF | Material cost per 100 SF | Overhead per 100 SF | Profit per 100 SF | Total price per 100 SF |
|---|---|---|---|---|---|---|---|---|---|
| **Epoxy coating, 2 part system, white (material #52)** | | | | | | | | | |
| Brush 1st coat | | | | | | | | | |
| Slow | 150 | 330 | 57.40 | 6.83 | 1.87 | 17.39 | 8.08 | 5.12 | 39.29 |
| Medium | 188 | 290 | 50.20 | 8.78 | 2.95 | 17.31 | 7.11 | 3.98 | 40.13 |
| Fast | 225 | 250 | 43.10 | 9.78 | 3.95 | 17.24 | 5.73 | 2.20 | 38.90 |
| Brush 2nd or additional coats | | | | | | | | | |
| Slow | 200 | 380 | 57.40 | 5.13 | 1.39 | 15.11 | 6.71 | 4.25 | 32.59 |
| Medium | 225 | 340 | 50.20 | 7.33 | 2.46 | 14.76 | 6.02 | 3.36 | 33.93 |
| Fast | 250 | 300 | 43.10 | 8.80 | 3.57 | 14.37 | 4.95 | 1.90 | 33.59 |
| **Waterproofing, clear hydro sealer (material #34)** | | | | | | | | | |
| Brush 1st coat | | | | | | | | | |
| Slow | 150 | 160 | 18.00 | 6.83 | 1.87 | 11.25 | 6.18 | 3.92 | 30.05 |
| Medium | 175 | 140 | 15.80 | 9.43 | 3.16 | 11.29 | 5.85 | 3.27 | 33.00 |
| Fast | 200 | 120 | 13.50 | 11.00 | 4.46 | 11.25 | 4.94 | 1.90 | 33.55 |
| Brush 2nd or additional coats | | | | | | | | | |
| Slow | 230 | 200 | 18.00 | 4.46 | 1.21 | 9.00 | 4.55 | 2.88 | 22.10 |
| Medium | 275 | 188 | 15.80 | 6.00 | 2.03 | 8.40 | 4.02 | 2.25 | 22.70 |
| Fast | 295 | 175 | 13.50 | 7.46 | 3.02 | 7.71 | 3.37 | 1.29 | 22.85 |
| **Industrial waterproofing (material #58)** | | | | | | | | | |
| Brush 1st coat | | | | | | | | | |
| Slow | 90 | 100 | 21.60 | 11.39 | 3.10 | 21.60 | 11.19 | 7.09 | 54.37 |
| Medium | 100 | 95 | 18.90 | 16.50 | 5.55 | 19.89 | 10.28 | 5.74 | 57.96 |
| Fast | 110 | 90 | 16.20 | 20.00 | 8.11 | 18.00 | 8.53 | 3.28 | 57.92 |
| Brush 2nd or additional coats | | | | | | | | | |
| Slow | 150 | 200 | 21.60 | 6.83 | 1.87 | 10.80 | 6.04 | 3.83 | 29.37 |
| Medium | 180 | 188 | 18.90 | 9.17 | 3.09 | 10.05 | 5.47 | 3.06 | 30.84 |
| Fast | 195 | 175 | 16.20 | 11.28 | 4.58 | 9.26 | 4.65 | 1.79 | 31.56 |

Use these figures to estimate the costs for finishing concrete walls which have a smooth surface (trowel), rough texture, or exposed aggregate finish. For wall heights above 10', increase the computed area by 50%. "Slow" work is based on a $10.25 hourly wage, "Medium" work on a $16.50 hourly wage, and "Fast" work on a $22.00 hourly wage. Other qualifications that apply to this table are on page 9.

| | Labor SF per manhour | Material coverage SF/gallon | Material cost per gallon | Labor cost per 100 SF | Labor burden 100 SF | Material cost per 100 SF | Overhead per 100 SF | Profit per 100 SF | Total price per 100 SF |
|---|---|---|---|---|---|---|---|---|---|
| **Walls, concrete tilt-up, roll application** | | | | | | | | | |
| Flat latex, water base (material #5) | | | | | | | | | |
| Roll 1st coat | | | | | | | | | |
| Slow | 275 | 275 | 16.20 | 3.73 | 1.02 | 5.89 | 3.30 | 2.09 | 16.03 |
| Medium | 300 | 250 | 14.20 | 5.50 | 1.84 | 5.68 | 3.19 | 1.78 | 17.99 |
| Fast | 325 | 225 | 12.10 | 6.77 | 2.75 | 5.38 | 2.76 | 1.06 | 18.72 |
| Roll 2nd or additional coats | | | | | | | | | |
| Slow | 300 | 375 | 16.20 | 3.42 | .92 | 4.32 | 2.69 | 1.70 | 13.05 |
| Medium | 338 | 325 | 14.20 | 4.88 | 1.65 | 4.37 | 2.67 | 1.49 | 15.06 |
| Fast | 375 | 275 | 12.10 | 5.87 | 2.39 | 4.40 | 2.34 | .90 | 15.90 |
| Enamel, water base (material #9) | | | | | | | | | |
| Roll 1st coat | | | | | | | | | |
| Slow | 275 | 275 | 22.80 | 3.73 | 1.02 | 8.29 | 4.04 | 2.56 | 19.64 |
| Medium | 300 | 250 | 19.90 | 5.50 | 1.84 | 7.96 | 3.75 | 2.10 | 21.15 |
| Fast | 325 | 225 | 17.10 | 6.77 | 2.75 | 7.60 | 3.17 | 1.22 | 21.51 |
| Roll 2nd or additional coats | | | | | | | | | |
| Slow | 300 | 375 | 22.80 | 3.42 | .92 | 6.08 | 3.23 | 2.05 | 15.70 |
| Medium | 338 | 325 | 19.90 | 4.88 | 1.65 | 6.12 | 3.10 | 1.73 | 17.48 |
| Fast | 375 | 275 | 17.10 | 5.87 | 2.39 | 6.22 | 2.68 | 1.03 | 18.19 |
| Enamel, oil base (material #10) | | | | | | | | | |
| Roll 1st coat | | | | | | | | | |
| Slow | 275 | 300 | 25.90 | 3.73 | 1.02 | 8.63 | 4.15 | 2.63 | 20.16 |
| Medium | 300 | 250 | 22.70 | 5.50 | 1.84 | 9.08 | 4.03 | 2.25 | 22.70 |
| Fast | 325 | 200 | 19.40 | 6.77 | 2.75 | 9.70 | 3.56 | 1.37 | 24.15 |
| Roll 2nd or additional coats | | | | | | | | | |
| Slow | 300 | 350 | 25.90 | 3.42 | .92 | 7.40 | 3.64 | 2.31 | 17.69 |
| Medium | 338 | 300 | 22.70 | 4.88 | 1.65 | 7.57 | 3.45 | 1.93 | 19.48 |
| Fast | 375 | 250 | 19.40 | 5.87 | 2.39 | 7.76 | 2.96 | 1.14 | 20.12 |
| Epoxy coating, 2 part system, clear (material #51) | | | | | | | | | |
| Roll 1st coat | | | | | | | | | |
| Slow | 275 | 290 | 57.00 | 3.73 | 1.02 | 19.66 | 7.57 | 4.80 | 36.78 |
| Medium | 300 | 250 | 49.90 | 5.50 | 1.84 | 19.96 | 6.69 | 3.74 | 37.73 |
| Fast | 325 | 210 | 42.80 | 6.77 | 2.75 | 20.38 | 5.53 | 2.13 | 37.56 |
| Roll 2nd or additional coats | | | | | | | | | |
| Slow | 300 | 380 | 57.00 | 3.42 | .92 | 15.00 | 6.00 | 3.80 | 29.14 |
| Medium | 338 | 340 | 49.90 | 4.88 | 1.65 | 14.68 | 5.19 | 2.90 | 29.30 |
| Fast | 375 | 300 | 42.80 | 5.87 | 2.39 | 14.27 | 4.17 | 1.60 | 28.30 |

| | Labor SF per manhour | Material coverage SF/gallon | Material cost per gallon | Labor cost per 100 SF | Labor burden 100 SF | Material cost per 100 SF | Overhead per 100 SF | Profit per 100 SF | Total price per 100 SF |
|---|---|---|---|---|---|---|---|---|---|
| **Epoxy coating, 2 part system, white (material #52)** | | | | | | | | | |
| Roll 1st coat | | | | | | | | | |
| Slow | 275 | 290 | 57.40 | 3.73 | 1.02 | 19.79 | 7.61 | 4.82 | 36.97 |
| Medium | 300 | 250 | 50.20 | 5.50 | 1.84 | 20.08 | 6.72 | 3.76 | 37.90 |
| Fast | 325 | 210 | 43.10 | 6.77 | 2.75 | 20.52 | 5.56 | 2.14 | 37.74 |
| Roll 2nd or additional coats | | | | | | | | | |
| Slow | 300 | 380 | 57.40 | 3.42 | .92 | 15.11 | 6.03 | 3.82 | 29.30 |
| Medium | 338 | 340 | 50.20 | 4.88 | 1.65 | 14.76 | 5.21 | 2.91 | 29.41 |
| Fast | 375 | 300 | 43.10 | 5.87 | 2.39 | 14.37 | 4.18 | 1.61 | 28.42 |
| **Waterproofing, clear hydro sealer (material #34)** | | | | | | | | | |
| Roll 1st coat | | | | | | | | | |
| Slow | 170 | 200 | 18.00 | 6.03 | 1.64 | 9.00 | 5.17 | 3.28 | 25.12 |
| Medium | 200 | 165 | 15.80 | 8.25 | 2.78 | 9.58 | 5.05 | 2.82 | 28.48 |
| Fast | 245 | 130 | 13.50 | 8.98 | 3.64 | 10.38 | 4.26 | 1.64 | 28.90 |
| Roll 2nd or additional coats | | | | | | | | | |
| Slow | 275 | 325 | 18.00 | 3.73 | 1.02 | 5.54 | 3.19 | 2.02 | 15.50 |
| Medium | 300 | 275 | 15.80 | 5.50 | 1.84 | 5.75 | 3.21 | 1.79 | 18.09 |
| Fast | 325 | 225 | 13.50 | 6.77 | 2.75 | 6.00 | 2.87 | 1.10 | 19.49 |
| **Industrial waterproofing (material #58)** | | | | | | | | | |
| Roll 1st coat | | | | | | | | | |
| Slow | 100 | 125 | 21.60 | 10.25 | 2.79 | 17.28 | 9.40 | 5.96 | 45.68 |
| Medium | 113 | 113 | 18.90 | 14.60 | 4.91 | 16.73 | 8.88 | 4.96 | 50.08 |
| Fast | 125 | 100 | 16.20 | 17.60 | 7.14 | 16.20 | 7.57 | 2.91 | 51.42 |
| Roll 2nd or additional coats | | | | | | | | | |
| Slow | 180 | 200 | 21.60 | 5.69 | 1.56 | 10.80 | 5.59 | 3.54 | 27.18 |
| Medium | 198 | 188 | 18.90 | 8.33 | 2.81 | 10.05 | 5.19 | 2.90 | 29.28 |
| Fast | 215 | 175 | 16.20 | 10.23 | 4.15 | 9.26 | 4.37 | 1.68 | 29.69 |

Use these figures to estimate the costs for finishing concrete walls which have a smooth surface (trowel), rough texture, or exposed aggregate finish. For wall heights above 10', increase the computed area by 50%. "Slow" work is based on a $10.25 hourly wage, "Medium" work on a $16.50 hourly wage, and "Fast" work on a $22.00 hourly wage. Other qualifications that apply to this table are on page 9.

| | Labor SF per manhour | Material coverage SF/gallon | Material cost per gallon | Labor cost per 100 SF | Labor burden 100 SF | Material cost per 100 SF | Overhead per 100 SF | Profit per 100 SF | Total price per 100 SF |
|---|---|---|---|---|---|---|---|---|---|
| **Walls, concrete tilt-up, spray application** | | | | | | | | | |
| Flat latex, water base (material #5) | | | | | | | | | |
| Spray 1st coat | | | | | | | | | |
| Slow | 500 | 225 | 16.20 | 2.05 | .56 | 7.20 | 3.04 | 1.93 | 14.78 |
| Medium | 700 | 188 | 14.20 | 2.36 | .79 | 7.55 | 2.62 | 1.47 | 14.79 |
| Fast | 900 | 150 | 12.10 | 2.44 | .99 | 8.07 | 2.13 | .82 | 14.45 |
| Spray 2nd or additional coats | | | | | | | | | |
| Slow | 600 | 275 | 16.20 | 1.71 | .47 | 5.89 | 2.50 | 1.59 | 12.16 |
| Medium | 800 | 238 | 14.20 | 2.06 | .70 | 5.97 | 2.14 | 1.19 | 12.06 |
| Fast | 1000 | 200 | 12.10 | 2.20 | .89 | 6.05 | 1.69 | .65 | 11.48 |
| Enamel, water base (material #9) | | | | | | | | | |
| Spray 1st coat | | | | | | | | | |
| Slow | 500 | 225 | 22.80 | 2.05 | .56 | 10.13 | 3.95 | 2.50 | 19.19 |
| Medium | 700 | 188 | 19.90 | 2.36 | .79 | 10.59 | 3.37 | 1.88 | 18.99 |
| Fast | 900 | 150 | 17.10 | 2.44 | .99 | 11.40 | 2.74 | 1.05 | 18.62 |
| Spray 2nd or additional coats | | | | | | | | | |
| Slow | 600 | 275 | 22.80 | 1.71 | .47 | 8.29 | 3.25 | 2.06 | 15.78 |
| Medium | 800 | 238 | 19.90 | 2.06 | .70 | 8.36 | 2.72 | 1.52 | 15.36 |
| Fast | 1000 | 200 | 17.10 | 2.20 | .89 | 8.55 | 2.15 | .83 | 14.62 |
| Enamel, oil base (material #10) | | | | | | | | | |
| Spray 1st coat | | | | | | | | | |
| Slow | 500 | 200 | 25.90 | 2.05 | .56 | 12.95 | 4.82 | 3.06 | 23.44 |
| Medium | 700 | 163 | 22.70 | 2.36 | .79 | 13.93 | 4.18 | 2.34 | 23.60 |
| Fast | 900 | 125 | 19.40 | 2.44 | .99 | 15.52 | 3.51 | 1.35 | 23.81 |
| Spray 2nd or additional coats | | | | | | | | | |
| Slow | 600 | 300 | 25.90 | 1.71 | .47 | 8.63 | 3.35 | 2.12 | 16.28 |
| Medium | 800 | 243 | 22.70 | 2.06 | .70 | 9.34 | 2.96 | 1.66 | 16.72 |
| Fast | 1000 | 175 | 19.40 | 2.20 | .89 | 11.09 | 2.62 | 1.01 | 17.81 |
| Epoxy coating, 2 part system, clear (material #51) | | | | | | | | | |
| Spray 1st coat | | | | | | | | | |
| Slow | 500 | 270 | 57.00 | 2.05 | .56 | 21.11 | 7.35 | 4.66 | 35.73 |
| Medium | 700 | 253 | 49.90 | 2.36 | .79 | 19.72 | 5.60 | 3.13 | 31.60 |
| Fast | 900 | 235 | 42.80 | 2.44 | .99 | 18.21 | 4.00 | 1.54 | 27.18 |
| Spray 2nd or additional coats | | | | | | | | | |
| Slow | 600 | 375 | 57.00 | 1.71 | .47 | 15.20 | 5.39 | 3.42 | 26.19 |
| Medium | 800 | 338 | 49.90 | 2.06 | .70 | 14.76 | 4.29 | 2.40 | 24.21 |
| Fast | 1000 | 300 | 42.80 | 2.20 | .89 | 14.27 | 3.21 | 1.23 | 21.80 |

| | Labor SF per manhour | Material coverage SF/gallon | Material cost per gallon | Labor cost per 100 SF | Labor burden 100 SF | Material cost per 100 SF | Overhead per 100 SF | Profit per 100 SF | Total price per 100 SF |
|---|---|---|---|---|---|---|---|---|---|
| **Epoxy coating, 2 part system, white (material #52)** | | | | | | | | | |
| Spray 1st coat | | | | | | | | | |
| Slow | 500 | 270 | 57.40 | 2.05 | .56 | 21.26 | 7.40 | 4.69 | 35.96 |
| Medium | 700 | 253 | 50.20 | 2.36 | .79 | 19.84 | 5.63 | 3.15 | 31.77 |
| Fast | 900 | 235 | 43.10 | 2.44 | .99 | 18.34 | 4.03 | 1.55 | 27.35 |
| Spray 2nd or additional coats | | | | | | | | | |
| Slow | 600 | 375 | 57.40 | 1.71 | .47 | 15.31 | 5.42 | 3.44 | 26.35 |
| Medium | 800 | 338 | 50.20 | 2.06 | .70 | 14.85 | 4.31 | 2.41 | 24.33 |
| Fast | 1000 | 300 | 43.10 | 2.20 | .89 | 14.37 | 3.23 | 1.24 | 21.93 |
| **Waterproofing, clear hydro sealer (material #34)** | | | | | | | | | |
| Spray 1st coat | | | | | | | | | |
| Slow | 500 | 125 | 18.00 | 2.05 | .56 | 14.40 | 5.27 | 3.34 | 25.62 |
| Medium | 700 | 113 | 15.80 | 2.36 | .79 | 13.98 | 4.20 | 2.35 | 23.68 |
| Fast | 900 | 100 | 13.50 | 2.44 | .99 | 13.50 | 3.13 | 1.20 | 21.26 |
| Spray 2nd or additional coats | | | | | | | | | |
| Slow | 600 | 200 | 18.00 | 1.71 | .47 | 9.00 | 3.47 | 2.20 | 16.85 |
| Medium | 800 | 170 | 15.80 | 2.06 | .70 | 9.29 | 2.95 | 1.65 | 16.65 |
| Fast | 1000 | 140 | 13.50 | 2.20 | .89 | 9.64 | 2.36 | .91 | 16.00 |

Use these figures to estimate the costs for finishing concrete walls which have a smooth surface (trowel), rough texture, or exposed aggregate finish. For wall heights above 10', increase the computed area by 50%. "Slow" work is based on a $10.25 hourly wage, "Medium" work on a $16.50 hourly wage, and "Fast" work on a $22.00 hourly wage. Other qualifications that apply to this table are on page 9.

| | Labor LF per manhour | Material coverage LF/gallon | Material cost per gallon | Labor cost per 100 LF | Labor burden 100 LF | Material cost per 100 LF | Overhead per 100 LF | Profit per 100 LF | Total price per 100 LF |
|---|---|---|---|---|---|---|---|---|---|
| **Windows, steel factory sash, brush application** | | | | | | | | | |
| Metal primer, rust inhibitor - clean metal (material #35) | | | | | | | | | |
| Brush prime coat | | | | | | | | | |
| Slow | 100 | 850 | 28.20 | 10.25 | 2.79 | 3.32 | 5.07 | 3.21 | 24.64 |
| Medium | 125 | 800 | 24.60 | 13.20 | 4.44 | 3.08 | 5.08 | 2.84 | 28.64 |
| Fast | 150 | 750 | 21.10 | 14.67 | 5.95 | 2.81 | 4.33 | 1.67 | 29.43 |
| Metal primer, rust inhibitor - rusty metal (material #36) | | | | | | | | | |
| Brush prime coat | | | | | | | | | |
| Slow | 100 | 800 | 33.30 | 10.25 | 2.79 | 4.16 | 5.33 | 3.38 | 25.91 |
| Medium | 125 | 750 | 29.20 | 13.20 | 4.44 | 3.89 | 5.27 | 2.95 | 29.75 |
| Fast | 150 | 700 | 25.00 | 14.67 | 5.95 | 3.57 | 4.48 | 1.72 | 30.39 |
| Metal finish - synthetic enamel, gloss, interior or exterior, off white (material #37) | | | | | | | | | |
| Brush 1st coat | | | | | | | | | |
| Slow | 125 | 900 | 25.90 | 8.20 | 2.23 | 2.88 | 4.13 | 2.62 | 20.06 |
| Medium | 150 | 850 | 22.70 | 11.00 | 3.71 | 2.67 | 4.26 | 2.38 | 24.02 |
| Fast | 175 | 800 | 19.50 | 12.57 | 5.09 | 2.44 | 3.72 | 1.43 | 25.25 |
| Brush 2nd or additional coats | | | | | | | | | |
| Slow | 150 | 1000 | 25.90 | 6.83 | 1.87 | 2.59 | 3.50 | 2.22 | 17.01 |
| Medium | 175 | 950 | 22.70 | 9.43 | 3.16 | 2.39 | 3.67 | 2.05 | 20.70 |
| Fast | 200 | 900 | 19.50 | 11.00 | 4.46 | 2.17 | 3.26 | 1.25 | 22.14 |
| Metal finish - synthetic enamel, gloss, interior or exterior, colors, except orange & red (material #38) | | | | | | | | | |
| Brush 1st coat | | | | | | | | | |
| Slow | 125 | 950 | 35.00 | 8.20 | 2.23 | 3.68 | 4.37 | 2.77 | 21.25 |
| Medium | 150 | 900 | 30.70 | 11.00 | 3.71 | 3.41 | 4.44 | 2.48 | 25.04 |
| Fast | 175 | 850 | 26.30 | 12.57 | 5.09 | 3.09 | 3.84 | 1.48 | 26.07 |
| Brush 2nd or additional coats | | | | | | | | | |
| Slow | 150 | 1050 | 35.00 | 6.83 | 1.87 | 3.33 | 3.73 | 2.36 | 18.12 |
| Medium | 175 | 1000 | 30.70 | 9.43 | 3.16 | 3.07 | 3.84 | 2.15 | 21.65 |
| Fast | 200 | 950 | 26.30 | 11.00 | 4.46 | 2.77 | 3.37 | 1.30 | 22.90 |

These figures will apply when painting steel factory sash but do not include work on glazing or frame. For heights above 8 feet, use the High Time Difficulty Factors on page 137. Note: A two coat system, prime and finish, using oil base material is recommended for any metal surface. Although water base material is often used, it may cause oxidation, corrosion and rust. One coat of oil base solid body stain is often used on interior or exterior metal but it may crack, peel or chip without the proper prime coat application. "Slow" work is based on a $10.25 hourly wage, "Medium" work on a $16.50 hourly wage, and "Fast" work on a $22.50 hourly wage. Other qualifications that apply to this table are on page 9.

# Part IV

# Wallcovering Costs

| | American rolls per gallon | Linear yards per gallon | Adhesive cost per gallon |
|---|---|---|---|
| **Adhesive coverage, rolls to yards conversion (2.5 yards per roll)** | | | |
| Ready-mix | | | |
| Light weight vinyl (material #60) | | | |
| Slow | 8.0 | 20.0 | 6.40 |
| Medium | 7.5 | 18.8 | 5.60 |
| Fast | 7.0 | 17.5 | 4.80 |
| Heavy weight vinyl (material #61) | | | |
| Slow | 6.0 | 15.0 | 6.60 |
| Medium | 5.5 | 13.8 | 5.80 |
| Fast | 5.0 | 12.5 | 5.00 |
| Cellulose (material #62) | | | |
| Slow | 8.0 | 20.0 | 6.50 |
| Medium | 7.0 | 17.5 | 5.70 |
| Fast | 6.0 | 15.0 | 4.80 |
| Vinyl to vinyl (material #63) | | | |
| Slow | 7.0 | 17.5 | 15.30 |
| Medium | 6.5 | 16.3 | 13.40 |
| Fast | 6.0 | 15.0 | 11.50 |
| Powdered cellulose (material #64) | | | |
| Slow | 12.0 | 30.0 | 2.10 |
| Medium | 11.0 | 27.5 | 1.90 |
| Fast | 10.0 | 25.0 | 1.60 |
| Powdered vinyl (material #65) | | | |
| Slow | 12.0 | 30.0 | 2.50 |
| Medium | 11.0 | 27.5 | 2.20 |
| Fast | 10.0 | 25.0 | 1.90 |
| Powdered wheat paste (material #66) | | | |
| Slow | 13.0 | 32.5 | 2.20 |
| Medium | 12.0 | 30.0 | 1.90 |
| Fast | 11.0 | 27.5 | 1.60 |

These figures are based on rolls or yards per gallon of liquid paste. Vinyl to vinyl ready-mix is usually distributed in pint containers which have been converted to gallons. Powdered adhesive coverage is based on water added to powder to prepare gallon quantities.

| | Labor SF per manhour | Material coverage SF/gallon | Material cost per gallon | Labor cost per 100 SF | Labor burden 100 SF | Material cost per 100 SF | Overhead per 100 SF | Profit per 100 SF | Total cost per 100 SF |
|---|---|---|---|---|---|---|---|---|---|
| **Adhesive coverage, per gallon of liquid paste** | | | | | | | | | |
| Ready-mix | | | | | | | | | |
| Light weight vinyl (material #60) | | | | | | | | | |
| Slow | 350 | 275 | 6.40 | 2.79 | .76 | 2.33 | 1.82 | 1.16 | 8.86 |
| Medium | 375 | 250 | 5.60 | 4.13 | 1.39 | 2.24 | 1.90 | 1.06 | 10.72 |
| Fast | 400 | 225 | 4.80 | 5.13 | 2.08 | 2.13 | 1.73 | .66 | 11.73 |
| Heavy weight vinyl (material #61) | | | | | | | | | |
| Slow | 275 | 200 | 6.60 | 3.55 | .97 | 3.30 | 2.42 | 1.54 | 11.78 |
| Medium | 300 | 175 | 5.80 | 5.17 | 1.74 | 3.31 | 2.50 | 1.40 | 14.12 |
| Fast | 325 | 150 | 5.00 | 6.31 | 2.56 | 3.33 | 2.26 | .87 | 15.33 |
| Cellulose (material #62) | | | | | | | | | |
| Slow | 325 | 250 | 6.50 | 3.00 | .82 | 2.60 | 1.99 | 1.26 | 9.67 |
| Medium | 350 | 225 | 5.70 | 4.43 | 1.49 | 2.53 | 2.07 | 1.16 | 11.68 |
| Fast | 375 | 200 | 4.80 | 5.47 | 2.22 | 2.40 | 1.87 | .72 | 12.68 |
| Vinyl to vinyl (material #63) | | | | | | | | | |
| Slow | 300 | 225 | 15.30 | 3.25 | .89 | 6.80 | 3.39 | 2.15 | 16.48 |
| Medium | 325 | 200 | 13.40 | 4.77 | 1.61 | 6.70 | 3.20 | 1.79 | 18.07 |
| Fast | 350 | 175 | 11.50 | 5.86 | 2.38 | 6.57 | 2.74 | 1.05 | 18.60 |
| Powdered cellulose (material #64) | | | | | | | | | |
| Slow | 425 | 400 | 2.10 | 2.29 | .62 | .53 | 1.07 | .68 | 5.19 |
| Medium | 450 | 367 | 1.90 | 3.44 | 1.16 | .52 | 1.25 | .70 | 7.07 |
| Fast | 475 | 333 | 1.60 | 4.32 | 1.75 | .48 | 1.21 | .47 | 8.23 |
| Powdered vinyl (material #65) | | | | | | | | | |
| Slow | 425 | 400 | 2.50 | 2.29 | .62 | .63 | 1.10 | .70 | 5.34 |
| Medium | 450 | 367 | 2.20 | 3.44 | 1.16 | .60 | 1.27 | .71 | 7.18 |
| Fast | 475 | 333 | 1.90 | 4.32 | 1.75 | .57 | 1.23 | .47 | 8.34 |
| Powdered wheat paste (material #66) | | | | | | | | | |
| Slow | 450 | 425 | 2.20 | 2.17 | .59 | .52 | 1.02 | .65 | 4.95 |
| Medium | 475 | 400 | 1.90 | 3.26 | 1.10 | .48 | 1.19 | .66 | 6.69 |
| Fast | 500 | 375 | 1.60 | 4.10 | 1.66 | .43 | 1.15 | .44 | 7.78 |

These figures are based on gallon quantities of liquid paste. Vinyl to vinyl ready-mix is usually distributed in pint containers which have been converted to gallons. Powdered adhesive coverage is based on water added to powder to prepare gallon quantities. Typically, powdered, ready-mix material is in 2 to 4 ounce packages which will adhere 6 to 12 rolls of wallcovering. See the Adhesive coverage table on the previous page for conversion to rolls and yards. "Slow" work is based on a $9.75 hourly rate, "Medium" work on a $15.50 hourly rate, and "Fast" work on a $20.50 hourly rate.

| | Labor rolls per day | Material by others | Labor cost per 10 rolls | Labor burden 10 rolls | Overhead per 10 rolls | Profit per 10 rolls | Total cost per 10 rolls | Total price per roll |
|---|---|---|---|---|---|---|---|---|

## Wallcovering application, average rates, medium rooms

| | Labor rolls per day | Material by others | Labor cost per 10 rolls | Labor burden 10 rolls | Overhead per 10 rolls | Profit per 10 rolls | Total cost per 10 rolls | Total price per roll |
|---|---|---|---|---|---|---|---|---|
| Walls | | | | | | | | |
| Slow | 17 | -- | 45.88 | 12.52 | 18.10 | 11.47 | 87.97 | 8.80 |
| Medium | 19 | -- | 65.26 | 21.99 | 21.37 | 11.94 | 120.56 | 12.05 |
| Fast | 21 | -- | 78.10 | 31.67 | 20.31 | 7.80 | 137.88 | 13.79 |
| Ceilings | | | | | | | | |
| Slow | 15 | -- | 52.00 | 14.18 | 20.51 | 13.00 | 99.69 | 9.97 |
| Medium | 17 | -- | 72.94 | 24.57 | 23.88 | 13.35 | 134.74 | 13.47 |
| Fast | 19 | -- | 86.32 | 35.00 | 22.44 | 8.63 | 152.39 | 15.24 |

The table above assumes that residential rolls are hand pasted. Add surface preparation time on page 420 as needed. For heights above 8 feet, use the High Time Difficulty Factors on page 137. "Slow" work is based on a $9.75 hourly wage, "Medium" work on a $15.50 hourly wage, and "Fast" work on a $20.50 hourly wage.

## Wallcovering application, average rates, small rooms

| | Labor rolls per day | Material by others | Labor cost per 10 rolls | Labor burden 10 rolls | Overhead per 10 rolls | Profit per 10 rolls | Total cost per 10 rolls | Total price per roll |
|---|---|---|---|---|---|---|---|---|
| Walls | | | | | | | | |
| Slow | 7 | -- | 111.43 | 30.40 | 43.95 | 27.86 | 213.64 | 21.36 |
| Medium | 9 | -- | 137.78 | 46.40 | 45.11 | 25.22 | 254.51 | 25.45 |
| Fast | 11 | -- | 149.09 | 60.45 | 38.77 | 14.90 | 263.21 | 26.32 |
| Ceilings | | | | | | | | |
| Slow | 6 | -- | 130.00 | 35.46 | 51.28 | 32.51 | 249.25 | 24.92 |
| Medium | 8 | -- | 155.00 | 52.20 | 50.75 | 28.37 | 286.32 | 28.63 |
| Fast | 10 | -- | 164.00 | 66.48 | 42.64 | 16.39 | 289.51 | 28.95 |

The table above assumes that residential rolls are hand pasted. Add surface preparation time on page 420 as needed. For heights above 8 feet, use the High Time Difficulty Factors on page 137. "Slow" work is based on a $9.75 hourly wage, "Medium" work on a $15.50 hourly wage, and "Fast" work on a $20.50 hourly wage.

## Borders 3" to 8" wide, commercial, machine pasted

| | Labor LF per manhour | Material by others | Labor cost per 100 LF | Labor burden 100 LF | Overhead per 100 LF | Profit per 100 LF | Total cost per 100 LF | Total price per LF |
|---|---|---|---|---|---|---|---|---|
| Medium size rooms (10 x 10 range) | | | | | | | | |
| Slow | 140 | -- | 6.25 | 1.70 | 2.46 | 1.56 | 11.97 | .12 |
| Medium | 158 | -- | 8.86 | 2.98 | 3.67 | 2.33 | 17.84 | .18 |
| Fast | 175 | -- | 10.57 | 4.28 | 4.61 | 2.92 | 22.38 | .22 |

The table above assumes that commercial rolls are machine pasted. Add surface preparation time on page 420 as needed. For heights above 8 feet, use the High Time Difficulty Factors on page 137. "Slow" work is based on a $8.75 hourly wage, "Medium" work on a $14.00 hourly wage, and "Fast" work on a $18.50 hourly wage.

| | Labor LF per manhour | Material by others | Labor cost per 100 LF | Labor burden 100 LF | Overhead per 100 LF | Profit per 100 LF | Total cost per 100 LF | Total price per LF |
|---|---|---|---|---|---|---|---|---|
| **Borders 3" to 8" wide, residential, hand pasted** | | | | | | | | |
| Medium size rooms (bedrooms, dining rooms) | | | | | | | | |
| Slow | 100 | -- | 9.75 | 2.66 | 3.85 | 2.44 | 18.70 | .19 |
| Medium | 113 | -- | 13.72 | 4.62 | 4.49 | 2.51 | 25.34 | .25 |
| Fast | 125 | -- | 16.40 | 6.65 | 4.26 | 1.64 | 28.95 | .29 |

The table above assumes that residential rolls are hand pasted. Add surface preparation time on page 420 as needed. For heights above 8 feet, use the High Time Difficulty Factors on page 137. "Slow" work is based on a $9.75 hourly wage, "Medium" work on a $15.50 hourly wage, and "Fast" work on a $20.50 hourly wage.

| | Labor SF per manhour | Material by others | Labor cost per 100 SF | Labor burden 100 SF | Overhead per 100 SF | Profit per 100 SF | Total cost per 100 SF | Total price per SF |
|---|---|---|---|---|---|---|---|---|
| **Flexible wood sheet and veneer** | | | | | | | | |
| Wood veneer flexwood (residential or commercial) | | | | | | | | |
| Medium rooms (bedrooms, dining rooms, offices, reception areas) | | | | | | | | |
| Slow | 14 | -- | 66.07 | 18.00 | 26.06 | 16.52 | 126.65 | 1.27 |
| Medium | 20 | -- | 73.75 | 24.80 | 24.15 | 13.50 | 136.20 | 1.36 |
| Fast | 26 | -- | 75.00 | 30.42 | 19.50 | 7.49 | 132.41 | 1.32 |
| Flexi-wall systems (residential or commercial) | | | | | | | | |
| Medium rooms (bedrooms, dining rooms, offices, reception areas) | | | | | | | | |
| Slow | 12 | -- | 77.08 | 21.00 | 30.40 | 19.27 | 147.75 | 1.48 |
| Medium | 18 | -- | 81.94 | 27.57 | 26.83 | 15.00 | 151.34 | 1.51 |
| Fast | 24 | -- | 81.25 | 32.97 | 21.13 | 8.12 | 143.47 | 1.43 |

Flexible wood sheet and veneer appears under section 09980 in the Construction Specifications Institute (CSI) indexing system. For heights above 8 feet, use the High Time Difficulty Factors on page 137. The labor rates in the table above are an average of the residential and commercial rates. Thus, "Slow" work is based on a $9.25 hourly rate, "Medium" work on a $14.75 hourly rate, and "Fast" work on a $19.50 hourly rate.

| | Labor SF per manhour | Material by others | Labor cost per 100 SF | Labor burden 100 SF | Overhead per 100 SF | Profit per 100 SF | Total cost per 100 SF | Total price per SF |
|---|---|---|---|---|---|---|---|---|
| **Surface preparation, wallcovering** | | | | | | | | |
| Rule of thumb, typical preparation | | | | | | | | |
| Slow | 100 | -- | 9.75 | 2.66 | 2.30 | .88 | 15.59 | .16 |
| Medium | 125 | -- | 12.40 | 4.18 | 3.07 | 1.18 | 20.83 | .21 |
| Fast | 150 | -- | 13.67 | 5.55 | 3.55 | 1.37 | 24.14 | .24 |
| Putty cracks, sand and wash | | | | | | | | |
| Slow | 120 | -- | 8.13 | 2.21 | 1.91 | .74 | 12.99 | .13 |
| Medium | 135 | -- | 11.48 | 3.87 | 2.84 | 1.09 | 19.28 | .19 |
| Fast | 150 | -- | 13.67 | 5.55 | 3.55 | 1.37 | 24.14 | .24 |

For additional preparation tasks see the Preparation operation tables beginning on page 293. For heights above 8 feet, use the High Time Difficulty Factors on page 137. "Slow" work is based on a $9.75 hourly rate, "Medium" work on a $15.50 hourly rate, and "Fast" work on a $20.50 hourly rate.

| | Labor yards per day | Material by others | Labor cost per 10 yards | Labor burden 10 yards | Overhead per 10 yards | Profit per 10 yards | Total cost per 10 yards | Total price per yard |
|---|---|---|---|---|---|---|---|---|
| **Vinyl wallcover, commercial, machine pasted, yards per day, 48" to 54" width** | | | | | | | | |
| Cut-up areas (stair halls, landing areas) | | | | | | | | |
| Walls | | | | | | | | |
| Slow | 40 | -- | 17.50 | 4.76 | 6.90 | 4.38 | 33.54 | 3.36 |
| Medium | 45 | -- | 24.89 | 8.38 | 8.15 | 4.56 | 45.98 | 4.60 |
| Fast | 50 | -- | 29.60 | 12.00 | 7.70 | 2.96 | 52.26 | 5.23 |
| Ceilings | | | | | | | | |
| Slow | 33 | -- | 21.21 | 5.77 | 8.37 | 5.30 | 40.65 | 4.07 |
| Medium | 38 | -- | 29.47 | 9.91 | 9.65 | 5.39 | 54.42 | 5.44 |
| Fast | 43 | -- | 34.42 | 13.94 | 8.95 | 3.44 | 60.75 | 6.08 |
| Small rooms (restrooms, utility rooms) | | | | | | | | |
| Walls | | | | | | | | |
| Slow | 42 | -- | 16.67 | 4.53 | 6.58 | 4.17 | 31.95 | 3.20 |
| Medium | 50 | -- | 22.40 | 7.54 | 7.34 | 4.10 | 41.38 | 4.14 |
| Fast | 55 | -- | 26.91 | 10.92 | 7.00 | 2.69 | 47.52 | 4.75 |
| Ceilings | | | | | | | | |
| Slow | 38 | -- | 18.42 | 5.01 | 7.27 | 4.61 | 35.31 | 3.53 |
| Medium | 46 | -- | 24.35 | 8.19 | 7.97 | 4.46 | 44.97 | 4.50 |
| Fast | 51 | -- | 29.02 | 11.77 | 7.55 | 2.90 | 51.24 | 5.12 |
| Medium rooms (bedrooms, offices) | | | | | | | | |
| Walls | | | | | | | | |
| Slow | 60 | -- | 11.67 | 3.17 | 4.60 | 2.92 | 22.36 | 2.24 |
| Medium | 75 | -- | 14.93 | 5.03 | 4.89 | 2.73 | 27.58 | 2.76 |
| Fast | 90 | -- | 16.44 | 6.67 | 4.28 | 1.64 | 29.03 | 2.90 |
| Ceilings | | | | | | | | |
| Slow | 56 | -- | 12.50 | 3.40 | 4.93 | 3.13 | 23.96 | 2.40 |
| Medium | 69 | -- | 16.23 | 5.45 | 5.31 | 2.97 | 29.96 | 3.00 |
| Fast | 81 | -- | 18.27 | 7.42 | 4.75 | 1.83 | 32.27 | 3.23 |
| Large rooms (conference rooms) | | | | | | | | |
| Walls | | | | | | | | |
| Slow | 76 | -- | 9.21 | 2.51 | 3.63 | 2.30 | 17.65 | 1.77 |
| Medium | 89 | -- | 12.58 | 4.24 | 4.12 | 2.30 | 23.24 | 2.32 |
| Fast | 102 | -- | 14.51 | 5.87 | 3.77 | 1.45 | 25.60 | 2.56 |
| Ceilings | | | | | | | | |
| Slow | 66 | -- | 10.61 | 2.88 | 4.19 | 2.65 | 20.33 | 2.03 |
| Medium | 79 | -- | 14.18 | 4.77 | 4.64 | 2.59 | 26.18 | 2.62 |
| Fast | 91 | -- | 16.26 | 6.59 | 4.23 | 1.62 | 28.70 | 2.87 |

| | Labor yards per day | Material by others | Labor cost per 10 yards | Labor burden 10 yards | Overhead per 10 yards | Profit per 10 yards | Total cost per 10 yards | Total price per yard |
|---|---|---|---|---|---|---|---|---|
| Large wall areas (corridors, long hallways) | | | | | | | | |
| Walls | | | | | | | | |
| Slow | 89 | -- | 7.87 | 2.14 | 3.10 | 1.97 | 15.08 | 1.51 |
| Medium | 102 | -- | 10.98 | 3.69 | 3.59 | 2.01 | 20.27 | 2.03 |
| Fast | 114 | -- | 12.98 | 5.27 | 3.37 | 1.30 | 22.92 | 2.29 |
| Ceilings | | | | | | | | |
| Slow | 76 | -- | 9.21 | 2.51 | 3.63 | 2.30 | 17.65 | 1.77 |
| Medium | 89 | -- | 12.58 | 4.24 | 4.12 | 2.30 | 23.24 | 2.32 |
| Fast | 102 | -- | 14.51 | 5.87 | 3.77 | 1.45 | 25.60 | 2.56 |
| Paper-backed vinyl on medium room walls | | | | | | | | |
| Bedrooms, dining rooms | | | | | | | | |
| Slow | 43 | -- | 16.28 | 4.42 | 6.42 | 4.07 | 31.19 | 3.12 |
| Medium | 53 | -- | 21.13 | 7.10 | 6.92 | 3.87 | 39.02 | 3.90 |
| Fast | 64 | -- | 23.13 | 9.37 | 6.01 | 2.31 | 40.82 | 4.08 |
| Cork wallcovering on medium room walls | | | | | | | | |
| Bedrooms, dining rooms | | | | | | | | |
| Slow | 43 | -- | 16.28 | 4.42 | 6.42 | 4.07 | 31.19 | 3.12 |
| Medium | 53 | -- | 21.13 | 7.10 | 6.92 | 3.87 | 39.02 | 3.90 |
| Fast | 64 | -- | 23.13 | 9.37 | 6.01 | 2.31 | 40.82 | 4.08 |

Vinyl coated wallcovering appears under section 09955 in the Construction Specifications Institute (CSI) indexing system. Vinyl wallcovering appears in section 09960. Cork wallcovering appears in section 09965. The table above assumes that commercial rolls are machine pasted. Add surface preparation time on page 420 as needed. For heights above 8 feet, use the High Time Difficulty Factors on page 137. "Slow" work is based on a $8.75 hourly wage, "Medium" work on a $14.00 hourly wage, and "Fast" work on a $18.50 hourly wage.

| | Labor rolls per day | Material by others | Labor cost per 10 rolls | Labor burden 10 rolls | Overhead per 10 rolls | Profit per 10 rolls | Total cost per 10 rolls | Total price per roll |
|---|---|---|---|---|---|---|---|---|
| **Vinyl wallcovering, residential, hand pasted, single rolls per day 18" to 27" wide** | | | | | | | | |
| Cut-up areas (stair halls, landing areas) | | | | | | | | |
| Walls | | | | | | | | |
| Slow | 10 | -- | 78.00 | 21.28 | 30.77 | 19.50 | 149.55 | 14.95 |
| Medium | 11 | -- | 112.73 | 37.97 | 36.91 | 20.63 | 208.24 | 20.82 |
| Fast | 12 | -- | 136.67 | 55.41 | 35.54 | 13.66 | 241.28 | 24.13 |
| Ceilings | | | | | | | | |
| Slow | 7 | -- | 111.43 | 30.40 | 43.95 | 27.86 | 213.64 | 21.36 |
| Medium | 8 | -- | 155.00 | 52.20 | 50.75 | 28.37 | 286.32 | 28.63 |
| Fast | 9 | -- | 182.22 | 73.87 | 47.38 | 18.21 | 321.68 | 32.17 |
| Small rooms (baths, utility rooms) | | | | | | | | |
| Walls | | | | | | | | |
| Slow | 11 | -- | 70.91 | 19.35 | 27.97 | 17.73 | 135.96 | 13.59 |
| Medium | 12 | -- | 103.33 | 34.81 | 33.83 | 18.91 | 190.88 | 19.08 |
| Fast | 13 | -- | 126.15 | 51.15 | 32.80 | 12.61 | 222.71 | 22.27 |
| Ceilings | | | | | | | | |
| Slow | 9 | -- | 86.67 | 23.64 | 34.19 | 21.67 | 166.17 | 16.62 |
| Medium | 10 | -- | 124.00 | 41.76 | 40.60 | 22.70 | 229.06 | 22.90 |
| Fast | 11 | -- | 149.09 | 60.45 | 38.77 | 14.90 | 263.21 | 26.32 |
| Medium rooms (bedrooms, dining rooms) | | | | | | | | |
| Walls | | | | | | | | |
| Slow | 17 | -- | 45.88 | 12.52 | 18.10 | 11.47 | 87.97 | 8.80 |
| Medium | 19 | -- | 65.26 | 21.99 | 21.37 | 11.94 | 120.56 | 12.05 |
| Fast | 21 | -- | 78.10 | 31.67 | 20.31 | 7.80 | 137.88 | 13.79 |
| Ceilings | | | | | | | | |
| Slow | 15 | -- | 52.00 | 14.18 | 20.51 | 13.00 | 99.69 | 9.97 |
| Medium | 17 | -- | 72.94 | 24.57 | 23.88 | 13.35 | 134.74 | 13.47 |
| Fast | 19 | -- | 86.32 | 35.00 | 22.44 | 8.63 | 152.39 | 15.24 |
| Large rooms (living rooms) | | | | | | | | |
| Walls | | | | | | | | |
| Slow | 20 | -- | 39.00 | 10.64 | 15.39 | 9.75 | 74.78 | 7.48 |
| Medium | 23 | -- | 53.91 | 18.15 | 17.65 | 9.87 | 99.58 | 9.96 |
| Fast | 26 | -- | 63.08 | 25.57 | 16.40 | 6.30 | 111.35 | 11.14 |
| Ceilings | | | | | | | | |
| Slow | 18 | -- | 43.33 | 11.82 | 17.09 | 10.83 | 83.07 | 8.31 |
| Medium | 20 | -- | 62.00 | 20.88 | 20.30 | 11.35 | 114.53 | 11.45 |
| Fast | 22 | -- | 74.55 | 30.20 | 19.38 | 7.45 | 131.58 | 13.16 |

| | Labor rolls per day | Material by others | Labor cost per 10 rolls | Labor burden 10 rolls | Overhead per 10 rolls | Profit per 10 rolls | Total cost per 10 rolls | Total price per roll |
|---|---|---|---|---|---|---|---|---|
| Large wall areas (corridors, long hallways) | | | | | | | | |
| Walls | | | | | | | | |
| Slow | 24 | -- | 32.50 | 8.86 | 12.82 | 8.13 | 62.31 | 6.23 |
| Medium | 27 | -- | 45.93 | 15.46 | 15.04 | 8.41 | 84.84 | 8.48 |
| Fast | 30 | -- | 54.67 | 22.17 | 14.22 | 5.46 | 96.52 | 9.65 |
| Ceilings | | | | | | | | |
| Slow | 21 | -- | 37.14 | 10.14 | 14.65 | 9.29 | 71.22 | 7.12 |
| Medium | 23 | -- | 53.91 | 18.15 | 17.65 | 9.87 | 99.58 | 9.96 |
| Fast | 25 | -- | 65.60 | 26.59 | 17.06 | 6.56 | 115.81 | 11.58 |
| Paper-backed vinyl on medium room walls | | | | | | | | |
| Bedrooms, dining rooms | | | | | | | | |
| Slow | 8 | -- | 97.50 | 26.60 | 38.46 | 24.38 | 186.94 | 18.69 |
| Medium | 10 | -- | 124.00 | 41.76 | 40.60 | 22.70 | 229.06 | 22.90 |
| Fast | 12 | -- | 136.67 | 55.41 | 35.54 | 13.66 | 241.28 | 24.13 |
| Cork wallcovering on medium room walls | | | | | | | | |
| Bedrooms, dining rooms | | | | | | | | |
| Slow | 8 | -- | 97.50 | 26.60 | 38.46 | 24.38 | 186.94 | 18.69 |
| Medium | 10 | -- | 124.00 | 41.76 | 40.60 | 22.70 | 229.06 | 22.90 |
| Fast | 12 | -- | 136.67 | 55.41 | 35.54 | 13.66 | 241.28 | 24.13 |

Vinyl coated wallcovering appears under section 09955 in the Construction Specifications Institute (CSI) indexing system. Vinyl wallcovering appears in section 09960. Cork wallcovering appears in section 09965. The table above assumes that residential rolls are hand pasted. Add surface preparation time on page 420 as needed. For heights above 8 feet, use the High Time Difficulty Factors on page 137. "Slow" work is based on a $9.75 hourly wage, "Medium" work on a $15.50 hourly wage, and "Fast" work on a $20.50 hourly wage. Add for ready-mix paste.

| | Labor yards per day | Material by others | Labor cost per 10 yards | Labor burden 10 yards | Overhead per 10 yards | Profit per 10 yards | Total cost per 10 yards | Total price per yard |
|---|---|---|---|---|---|---|---|---|
| **Wall fabric, commercial, machine pasted, yards per day, 48" to 54" width** | | | | | | | | |
| Coated fabric | | | | | | | | |
| Cut-up areas (stair halls, landing areas) | | | | | | | | |
| Walls | | | | | | | | |
| Slow | 58 | -- | 12.07 | 3.28 | 4.76 | 3.02 | 23.13 | 2.31 |
| Medium | 63 | -- | 17.78 | 5.98 | 5.82 | 3.25 | 32.83 | 3.28 |
| Fast | 68 | -- | 21.76 | 8.82 | 5.66 | 2.17 | 38.41 | 3.84 |
| Ceilings | | | | | | | | |
| Slow | 35 | -- | 20.00 | 5.44 | 7.89 | 5.00 | 38.33 | 3.83 |
| Medium | 40 | -- | 28.00 | 9.42 | 9.17 | 5.12 | 51.71 | 5.17 |
| Fast | 45 | -- | 32.89 | 13.34 | 8.55 | 3.29 | 58.07 | 5.81 |
| Small rooms (restrooms, utility rooms) | | | | | | | | |
| Walls | | | | | | | | |
| Slow | 60 | -- | 11.67 | 3.17 | 4.60 | 2.92 | 22.36 | 2.24 |
| Medium | 65 | -- | 17.23 | 5.80 | 5.64 | 3.15 | 31.82 | 3.18 |
| Fast | 70 | -- | 21.14 | 8.58 | 5.50 | 2.11 | 37.33 | 3.73 |
| Ceilings | | | | | | | | |
| Slow | 40 | -- | 17.50 | 4.76 | 6.90 | 4.38 | 33.54 | 3.36 |
| Medium | 48 | -- | 23.33 | 7.86 | 7.64 | 4.27 | 43.10 | 4.31 |
| Fast | 55 | -- | 26.91 | 10.92 | 7.00 | 2.69 | 47.52 | 4.75 |
| Medium rooms (bedrooms, offices) | | | | | | | | |
| Walls | | | | | | | | |
| Slow | 68 | -- | 10.29 | 2.80 | 4.06 | 2.57 | 19.72 | 1.97 |
| Medium | 80 | -- | 14.00 | 4.71 | 4.58 | 2.56 | 25.85 | 2.59 |
| Fast | 93 | -- | 15.91 | 6.45 | 4.14 | 1.59 | 28.09 | 2.81 |
| Ceilings | | | | | | | | |
| Slow | 60 | -- | 11.67 | 3.17 | 4.60 | 2.92 | 22.36 | 2.24 |
| Medium | 73 | -- | 15.34 | 5.17 | 5.02 | 2.81 | 28.34 | 2.83 |
| Fast | 85 | -- | 17.41 | 7.06 | 4.53 | 1.74 | 30.74 | 3.07 |
| Large rooms (conference rooms) | | | | | | | | |
| Walls | | | | | | | | |
| Slow | 85 | -- | 8.24 | 2.23 | 3.25 | 2.06 | 15.78 | 1.58 |
| Medium | 98 | -- | 11.43 | 3.84 | 3.74 | 2.09 | 21.10 | 2.11 |
| Fast | 110 | -- | 13.45 | 5.45 | 3.50 | 1.34 | 23.74 | 2.37 |
| Ceilings | | | | | | | | |
| Slow | 68 | -- | 10.29 | 2.80 | 4.06 | 2.57 | 19.72 | 1.97 |
| Medium | 80 | -- | 14.00 | 4.71 | 4.58 | 2.56 | 25.85 | 2.59 |
| Fast | 93 | -- | 15.91 | 6.45 | 4.14 | 1.59 | 28.09 | 2.81 |

| | Labor yards per day | Material by others | Labor cost per 10 yards | Labor burden 10 yards | Overhead per 10 yards | Profit per 10 yards | Total cost per 10 yards | Total price per yard |
|---|---|---|---|---|---|---|---|---|
| Large wall areas (corridors, long hallways) | | | | | | | | |
| Walls | | | | | | | | |
| Slow | 80 | -- | 8.75 | 2.38 | 3.45 | 2.19 | 16.77 | 1.68 |
| Medium | 110 | -- | 10.18 | 3.42 | 3.33 | 1.86 | 18.79 | 1.88 |
| Fast | 125 | -- | 11.84 | 4.80 | 3.08 | 1.18 | 20.90 | 2.09 |
| Ceilings | | | | | | | | |
| Slow | 78 | -- | 8.97 | 2.45 | 3.54 | 2.24 | 17.20 | 1.72 |
| Medium | 90 | -- | 12.44 | 4.19 | 4.07 | 2.28 | 22.98 | 2.30 |
| Fast | 103 | -- | 14.37 | 5.83 | 3.74 | 1.44 | 25.38 | 2.54 |

Canvas sheeting: see Wall fabric, residential, hand pasted

Grasscloth: see Wall fabric, residential, hand pasted

Burlap: see Wall fabric, residential, hand pasted

Natural fabric, silk: see Wall fabric, residential, hand pasted

Natural fabric, felt, linen, cotton: see Wall fabric, residential, hand pasted

Wall fabric appears under section 09975 in the Construction Specifications Institute (CSI) indexing system. The table above assumes that commercial rolls are machine pasted. Add surface preparation time on page 420 as needed. For heights above 8 feet, use the High Time Difficulty Factors on page 137. "Slow" work is based on a $8.75 hourly wage "Medium" work on a $14.00 hourly wage, and "Fast" work on a $18.50 hourly wage. Add for ready-mix paste.

| | Labor rolls per day | Material by others | Labor cost per 10 rolls | Labor burden 10 rolls | Overhead per 10 rolls | Profit per 10 rolls | Total cost per 10 rolls | Total price per roll |
|---|---|---|---|---|---|---|---|---|
| **Wall fabric, residential, hand pasted, single rolls per day 18" to 27" wide** | | | | | | | | |
| Coated fabrics | | | | | | | | |
| Cut-up areas (stair halls, landing areas) | | | | | | | | |
| Walls | | | | | | | | |
| Slow | 12 | -- | 65.00 | 17.74 | 25.64 | 16.25 | 124.63 | 12.46 |
| Medium | 13 | -- | 95.38 | 32.13 | 31.23 | 17.46 | 176.20 | 17.62 |
| Fast | 14 | -- | 117.14 | 47.48 | 30.46 | 11.71 | 206.79 | 20.68 |
| Ceilings | | | | | | | | |
| Slow | 8 | -- | 97.50 | 26.60 | 38.46 | 24.38 | 186.94 | 18.69 |
| Medium | 9 | -- | 137.78 | 46.40 | 45.11 | 25.22 | 254.51 | 25.45 |
| Fast | 10 | -- | 164.00 | 66.48 | 42.64 | 16.39 | 289.51 | 28.95 |
| Small rooms (baths, utility rooms) | | | | | | | | |
| Walls | | | | | | | | |
| Slow | 13 | -- | 60.00 | 16.37 | 23.67 | 15.00 | 115.04 | 11.50 |
| Medium | 14 | -- | 88.57 | 29.82 | 29.00 | 16.21 | 163.60 | 16.36 |
| Fast | 15 | -- | 109.33 | 44.31 | 28.43 | 10.93 | 193.00 | 19.30 |
| Ceilings | | | | | | | | |
| Slow | 10 | -- | 78.00 | 21.28 | 30.77 | 19.50 | 149.55 | 14.95 |
| Medium | 11 | -- | 112.73 | 37.97 | 36.91 | 20.63 | 208.24 | 20.82 |
| Fast | 12 | -- | 136.67 | 55.41 | 35.54 | 13.66 | 241.28 | 24.13 |
| Medium rooms (bedrooms, dining rooms) | | | | | | | | |
| Walls | | | | | | | | |
| Slow | 18 | -- | 43.33 | 11.82 | 17.09 | 10.83 | 83.07 | 8.31 |
| Medium | 20 | -- | 62.00 | 20.88 | 20.30 | 11.35 | 114.53 | 11.45 |
| Fast | 22 | -- | 74.55 | 30.20 | 19.38 | 7.45 | 131.58 | 13.16 |
| Ceilings | | | | | | | | |
| Slow | 16 | -- | 48.75 | 13.30 | 19.23 | 12.19 | 93.47 | 9.35 |
| Medium | 18 | -- | 68.89 | 23.19 | 22.56 | 12.61 | 127.25 | 12.72 |
| Fast | 20 | -- | 82.00 | 33.24 | 21.32 | 8.19 | 144.75 | 14.48 |
| Large rooms (living rooms) | | | | | | | | |
| Walls | | | | | | | | |
| Slow | 22 | -- | 35.45 | 9.67 | 13.98 | 8.86 | 67.96 | 6.80 |
| Medium | 25 | -- | 49.60 | 16.70 | 16.24 | 9.08 | 91.62 | 9.16 |
| Fast | 28 | -- | 58.57 | 23.74 | 15.23 | 5.85 | 103.39 | 10.34 |
| Ceilings | | | | | | | | |
| Slow | 19 | -- | 41.05 | 11.21 | 16.19 | 10.26 | 78.71 | 7.87 |
| Medium | 21 | -- | 59.05 | 19.89 | 19.34 | 10.81 | 109.09 | 10.91 |
| Fast | 23 | -- | 71.30 | 28.90 | 18.54 | 7.13 | 125.87 | 12.59 |

| | Labor rolls per day | Material by others | Labor cost per 10 rolls | Labor burden 10 rolls | Overhead per 10 rolls | Profit per 10 rolls | Total cost per 10 rolls | Total price per roll |
|---|---|---|---|---|---|---|---|---|
| Large wall areas (corridors, long hallways) | | | | | | | | |
| Walls | | | | | | | | |
| Slow | 26 | -- | 30.00 | 8.19 | 11.84 | 7.50 | 57.53 | 5.75 |
| Medium | 29 | -- | 42.76 | 14.41 | 14.00 | 7.83 | 79.00 | 7.90 |
| Fast | 32 | -- | 51.25 | 20.78 | 13.33 | 5.12 | 90.48 | 9.05 |
| Ceilings | | | | | | | | |
| Slow | 22 | -- | 35.45 | 9.67 | 13.98 | 8.86 | 67.96 | 6.80 |
| Medium | 24 | -- | 51.67 | 17.39 | 16.92 | 9.46 | 95.44 | 9.54 |
| Fast | 26 | -- | 63.08 | 25.57 | 16.40 | 6.30 | 111.35 | 11.14 |
| Canvas sheeting | | | | | | | | |
| Medium room walls (bedrooms, dining rooms) | | | | | | | | |
| Walls | | | | | | | | |
| Slow | 14 | -- | 55.71 | 15.20 | 21.98 | 13.93 | 106.82 | 10.68 |
| Medium | 16 | -- | 77.50 | 26.10 | 25.38 | 14.19 | 143.17 | 14.32 |
| Fast | 17 | -- | 96.47 | 39.11 | 25.08 | 9.64 | 170.30 | 17.03 |
| Grasscloth | | | | | | | | |
| Medium room walls (bedrooms, dining rooms) | | | | | | | | |
| Walls | | | | | | | | |
| Slow | 15 | -- | 52.00 | 14.18 | 20.51 | 13.00 | 99.69 | 9.97 |
| Medium | 20 | -- | 62.00 | 20.88 | 20.30 | 11.35 | 114.53 | 11.45 |
| Fast | 24 | -- | 68.33 | 27.69 | 17.77 | 6.83 | 120.62 | 12.06 |
| Burlap | | | | | | | | |
| Medium rooms (bedrooms, dining rooms) | | | | | | | | |
| Walls | | | | | | | | |
| Slow | 10 | -- | 78.00 | 21.28 | 30.77 | 19.50 | 149.55 | 14.95 |
| Medium | 14 | -- | 88.57 | 29.82 | 29.00 | 16.21 | 163.60 | 16.36 |
| Fast | 18 | -- | 91.11 | 36.92 | 23.69 | 9.11 | 160.83 | 16.09 |
| Natural fabric, silk | | | | | | | | |
| Medium rooms (bedrooms, dining rooms) | | | | | | | | |
| Walls | | | | | | | | |
| Slow | 8 | -- | 97.50 | 26.60 | 38.46 | 24.38 | 186.94 | 18.69 |
| Medium | 12 | -- | 103.33 | 34.81 | 33.83 | 18.91 | 190.88 | 19.08 |
| Fast | 15 | -- | 109.33 | 44.31 | 28.43 | 10.93 | 193.00 | 19.30 |
| Natural fabric, felt, linen, cotton | | | | | | | | |
| Medium rooms (bedrooms, dining rooms) | | | | | | | | |
| Walls | | | | | | | | |
| Slow | 7 | -- | 111.43 | 30.40 | 43.95 | 27.86 | 213.64 | 21.36 |
| Medium | 9 | -- | 137.78 | 46.40 | 45.11 | 25.22 | 254.51 | 25.45 |
| Fast | 11 | -- | 149.09 | 60.45 | 38.77 | 14.90 | 263.21 | 26.32 |

Wall fabric appears under section 09975 in the Construction Specifications Institute (CSI) indexing system. The table above assumes that residential rolls are hand pasted. Add surface preparation time on page 420 as needed. For heights above 8 feet, use the High Time Difficulty Factors on page 137. "Slow" work is based on a $9.75 hourly wage "Medium" work on a $15.50 hourly wage, and "Fast" work on a $20.50 hourly wage. Add for ready-mix paste.

| | Labor rolls per day | Material by others | Labor cost per 10 rolls | Labor burden 10 rolls | Overhead per 10 rolls | Profit per 10 rolls | Total cost per 10 rolls | Total price per roll |
|---|---|---|---|---|---|---|---|---|
| **Wallpaper, commercial, machine pasted, single rolls per day, 27" width** | | | | | | | | |
| Blind stock (lining) | | | | | | | | |
| Cut-up areas (stair halls, landing areas) | | | | | | | | |
| Walls | | | | | | | | |
| Slow | 19 | -- | 36.84 | 10.03 | 14.53 | 9.21 | 70.61 | 7.06 |
| Medium | 21 | -- | 53.33 | 17.96 | 17.46 | 9.76 | 98.51 | 9.85 |
| Fast | 23 | -- | 64.35 | 26.08 | 16.73 | 6.43 | 113.59 | 11.36 |
| Ceilings | | | | | | | | |
| Slow | 16 | -- | 43.75 | 11.90 | 17.26 | 10.94 | 83.85 | 8.39 |
| Medium | 18 | -- | 62.22 | 20.93 | 20.37 | 11.39 | 114.91 | 11.49 |
| Fast | 20 | -- | 74.00 | 30.00 | 19.24 | 7.40 | 130.64 | 13.07 |
| Small rooms (restrooms, utility rooms) | | | | | | | | |
| Walls | | | | | | | | |
| Slow | 21 | -- | 33.33 | 9.08 | 13.15 | 8.33 | 63.89 | 6.39 |
| Medium | 25 | -- | 44.80 | 15.07 | 14.67 | 8.20 | 82.74 | 8.28 |
| Fast | 29 | -- | 51.03 | 20.70 | 13.27 | 5.10 | 90.10 | 9.01 |
| Ceilings | | | | | | | | |
| Slow | 17 | -- | 41.18 | 11.20 | 16.24 | 10.30 | 78.92 | 7.89 |
| Medium | 20 | -- | 56.00 | 18.84 | 18.34 | 10.25 | 103.43 | 10.34 |
| Fast | 22 | -- | 67.27 | 27.27 | 17.49 | 6.72 | 118.75 | 11.88 |
| Medium rooms (bedrooms, offices) | | | | | | | | |
| Walls | | | | | | | | |
| Slow | 29 | -- | 24.14 | 6.57 | 9.52 | 6.04 | 46.27 | 4.63 |
| Medium | 35 | -- | 32.00 | 10.77 | 10.48 | 5.86 | 59.11 | 5.91 |
| Fast | 41 | -- | 36.10 | 14.63 | 9.39 | 3.61 | 63.73 | 6.37 |
| Ceilings | | | | | | | | |
| Slow | 25 | -- | 28.00 | 7.62 | 11.05 | 7.00 | 53.67 | 5.37 |
| Medium | 30 | -- | 37.33 | 12.57 | 12.22 | 6.83 | 68.95 | 6.89 |
| Fast | 34 | -- | 43.53 | 17.65 | 11.32 | 4.35 | 76.85 | 7.69 |
| Large rooms (conference rooms) | | | | | | | | |
| Walls | | | | | | | | |
| Slow | 35 | -- | 20.00 | 5.44 | 7.89 | 5.00 | 38.33 | 3.83 |
| Medium | 41 | -- | 27.32 | 9.18 | 8.94 | 5.00 | 50.44 | 5.05 |
| Fast | 47 | -- | 31.49 | 12.76 | 8.19 | 3.15 | 55.59 | 5.56 |
| Ceilings | | | | | | | | |
| Slow | 29 | -- | 24.14 | 6.57 | 9.52 | 6.04 | 46.27 | 4.63 |
| Medium | 34 | -- | 32.94 | 11.08 | 10.78 | 6.03 | 60.83 | 6.08 |
| Fast | 38 | -- | 38.95 | 15.78 | 10.13 | 3.89 | 68.75 | 6.88 |

| | Labor rolls per day | Material by others | Labor cost per 10 rolls | Labor burden 10 rolls | Overhead per 10 rolls | Profit per 10 rolls | Total cost per 10 rolls | Total price per roll |
|---|---|---|---|---|---|---|---|---|
| Large wall areas (corridors, long hallways) | | | | | | | | |
| Walls | | | | | | | | |
| Slow | 40 | -- | 17.50 | 4.76 | 6.90 | 4.38 | 33.54 | 3.36 |
| Medium | 45 | -- | 24.89 | 8.38 | 8.15 | 4.56 | 45.98 | 4.60 |
| Fast | 50 | -- | 29.60 | 12.00 | 7.70 | 2.96 | 52.26 | 5.23 |
| Ceilings | | | | | | | | |
| Slow | 33 | -- | 21.21 | 5.77 | 8.37 | 5.30 | 40.65 | 4.07 |
| Medium | 38 | -- | 29.47 | 9.91 | 9.65 | 5.39 | 54.42 | 5.44 |
| Fast | 42 | -- | 35.24 | 14.29 | 9.16 | 3.52 | 62.21 | 6.22 |
| Ordinary pre-trimmed wallpaper or butt joint work | | | | | | | | |
| Medium room walls (bedrooms, offices) | | | | | | | | |
| Walls | | | | | | | | |
| Slow | 26 | -- | 26.92 | 7.33 | 10.62 | 6.73 | 51.60 | 5.16 |
| Medium | 31 | -- | 36.13 | 12.16 | 11.83 | 6.61 | 66.73 | 6.67 |
| Fast | 36 | -- | 41.11 | 16.66 | 10.69 | 4.11 | 72.57 | 7.26 |
| Ceilings | | | | | | | | |
| Slow | 23 | -- | 30.43 | 8.28 | 12.00 | 7.61 | 58.32 | 5.83 |
| Medium | 28 | -- | 40.00 | 13.45 | 13.10 | 7.32 | 73.87 | 7.39 |
| Fast | 32 | -- | 46.25 | 18.75 | 12.03 | 4.62 | 81.65 | 8.17 |
| Hand-crafted wallpaper | | | | | | | | |
| Medium room walls (bedrooms, offices) | | | | | | | | |
| Walls | | | | | | | | |
| Slow | 20 | -- | 35.00 | 9.52 | 13.81 | 8.75 | 67.08 | 6.71 |
| Medium | 24 | -- | 46.67 | 15.69 | 15.28 | 8.54 | 86.18 | 8.62 |
| Fast | 28 | -- | 52.86 | 21.42 | 13.74 | 5.28 | 93.30 | 9.33 |
| Ceilings | | | | | | | | |
| Slow | 18 | -- | 38.89 | 10.57 | 15.34 | 9.72 | 74.52 | 7.46 |
| Medium | 21 | -- | 53.33 | 17.96 | 17.46 | 9.76 | 98.51 | 9.85 |
| Fast | 23 | -- | 64.35 | 26.08 | 16.73 | 6.43 | 113.59 | 11.36 |
| Flock wallpaper, medium rooms (bedrooms, offices) | | | | | | | | |
| Walls | | | | | | | | |
| Slow | 14 | -- | 50.00 | 13.60 | 19.73 | 12.50 | 95.83 | 9.59 |
| Medium | 17 | -- | 65.88 | 22.17 | 21.57 | 12.06 | 121.68 | 12.17 |
| Fast | 20 | -- | 74.00 | 30.00 | 19.24 | 7.40 | 130.64 | 13.07 |
| Foil wallpaper, medium rooms (bedrooms, offices) | | | | | | | | |
| Walls | | | | | | | | |
| Slow | 13 | -- | 53.85 | 14.64 | 21.24 | 13.46 | 103.19 | 10.32 |
| Medium | 16 | -- | 70.00 | 23.55 | 22.92 | 12.81 | 129.28 | 12.93 |
| Fast | 19 | -- | 77.89 | 31.60 | 20.25 | 7.78 | 137.52 | 13.75 |

| | Labor rolls per day | Material by others | Labor cost per 10 rolls | Labor burden 10 rolls | Overhead per 10 rolls | Profit per 10 rolls | Total cost per 10 rolls | Total price per roll |
|---|---|---|---|---|---|---|---|---|
| **Canvas wallpaper, medium rooms (bedrooms, offices)** | | | | | | | | |
| Walls | | | | | | | | |
| Slow | 12 | -- | 58.33 | 15.87 | 23.01 | 14.58 | 111.79 | 11.18 |
| Medium | 14 | -- | 80.00 | 26.91 | 26.20 | 14.64 | 147.75 | 14.78 |
| Fast | 16 | -- | 92.50 | 37.50 | 24.05 | 9.24 | 163.29 | 16.33 |
| **Scenic wallpaper, medium rooms (bedrooms, offices)** | | | | | | | | |
| Walls | | | | | | | | |
| Slow | 16 | -- | 43.75 | 11.90 | 17.26 | 10.94 | 83.85 | 8.39 |
| Medium | 18 | -- | 62.22 | 20.93 | 20.37 | 11.39 | 114.91 | 11.49 |
| Fast | 20 | -- | 74.00 | 30.00 | 19.24 | 7.40 | 130.64 | 13.07 |

Wallpaper appears under section 09970 in the Construction Specifications Institute (CSI) indexing system. The table above assumes that commercial rolls are machine pasted. Add surface preparation time on page 420 as needed. For heights above 8 feet, use the High Time Difficulty Factors on page 137. "Slow" work is based on a $8.75 hourly wage, "Medium" work on a $14.00 hourly wage, and "Fast" work on a $18.50 hourly wage. Add for ready-mix paste.

# National Estimator Quick Start

National Estimator is a construction cost estimating program with all the cost estimates in this book. You can page through these costs one screen at a time or use the electronic index to search by keyword for exactly the information you need. When you find what you're looking for, split the screen in two so your estimate is on the bottom half of the screen and the database is on the top half. Then copy and paste into your estimate. Type the estimated quantity. The program extends prices and totals columns automatically. Add your overhead and profit as a percentage and then print the estimate. It's as easy as that.

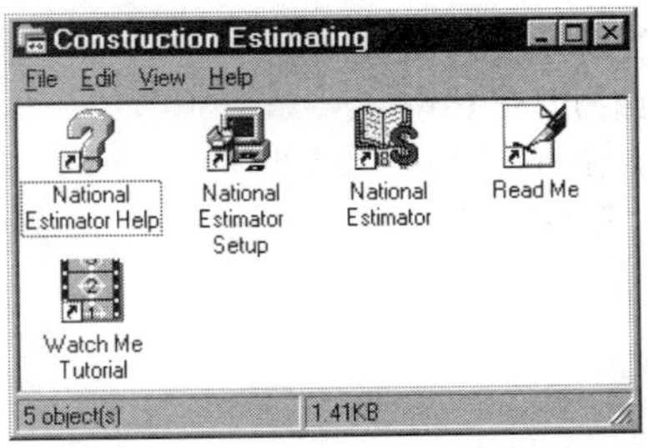


*Construction Estimating program group*

If you're a Windows user, you know most of the National Estimator program already. But even computer novices can be writing labor and material cost estimates in an hour or two. The first step is installing the National Estimator disk.

The installation program creates a Construction Estimating program group and puts several icons in that group. If you have trouble installing National Estimator, call Craftsman tech support at 760-438-7828. If your computer can't use the CD, exchange it for National Estimator on 3½" disk. Send the CD to Craftsman, Box 6500, Carlsbad, CA 92018. There's no charge for this exchange.

*Click on File. Click on Run in Windows 3.1*

## Installing National Estimator

Put the National Estimator disk in your CD drive (such as D:). Start Windows.

| In **Windows 3.1 or 3.11** go to the Program Manager. | In **Windows 95** or **NT** |
|---|---|
| 1. Click on File | 1. Click on Start |
| 2. Click on Run . . . | 2. Click on Run . . . |
| 3. Type D:\SETUP | 3. Type D:\SETUP |
| 4. Press Enter | 4. Press Enter |

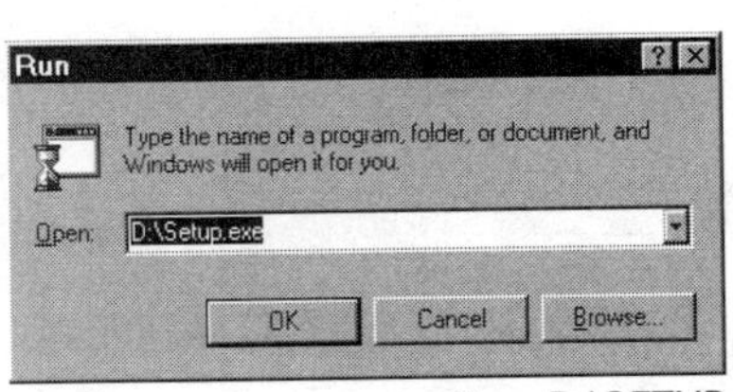


*Type D:\SETUP in Windows 95 or NT*

We recommend accepting the installation defaults:

1. National directory
2. "Typical" setup

When installation is complete, click on *Finish*.

## Uninstalling National Estimator

In Windows 3.1 or 3.11: Select the Construction Estimating program group. Then click on the Uninstall icon.

In Windows 95 or NT: Click on *Start*, *Settings*, *Control Panel*. Double click on *Add/Remove Programs*. Then click on *National Estimator* and *Add/Remove*.

## Learn the Easy Way!

*Estimating with Jay and Julie*

*Watch Me icon*

The **Watch Me** tutorial is an interactive video guide to National Estimator. Let Jay and Julie help you get started using the National Estimator program. Sit back and relax and let the video run — or click on buttons and menus to jump from topic to topic inside the video. Exit any time you want. Then go back to Watch Me later to brush up on some topic. Watch Me takes up very little space on your hard drive and is part of the "Typical" installation. Watch Me requires 8Mb of RAM, a Windows-compatible sound system and at least a 33 MHz 486 computer. Click on the Watch Me icon to begin Watch Me.

## Using National Estimator

*National Estimator icon*

National Estimator begins when you click on the National Estimator icon (in Windows 3.1 or 3.11) or click on *Start*, *Programs*, the *Construction Estimating* group and then *National Estimator* (in Windows 95 or NT).

On the title bar at the top of the screen you see the program name, National Estimator, and [Estimate1], showing that Estimate1 is on the screen. Let's take a closer look at the other information at the top of your screen.

*Your estimating form ready to begin the first estimate.*

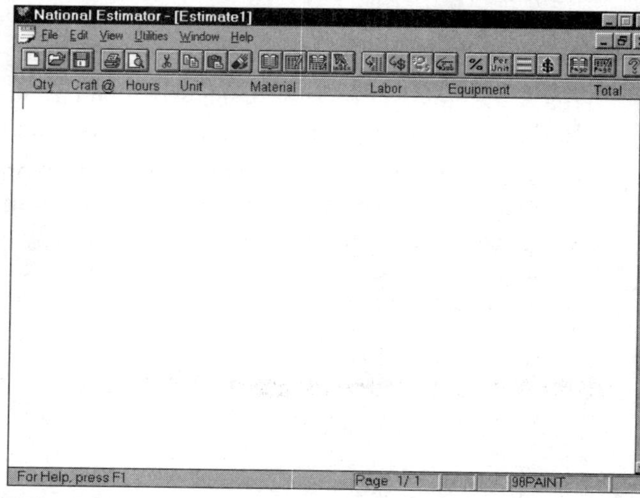


## The Menu Bar

Below the title bar you see the menu bar: Every option in National Estimator is available on the menu bar. Click with your left mouse button on any item on the menu to open a list of available commands.

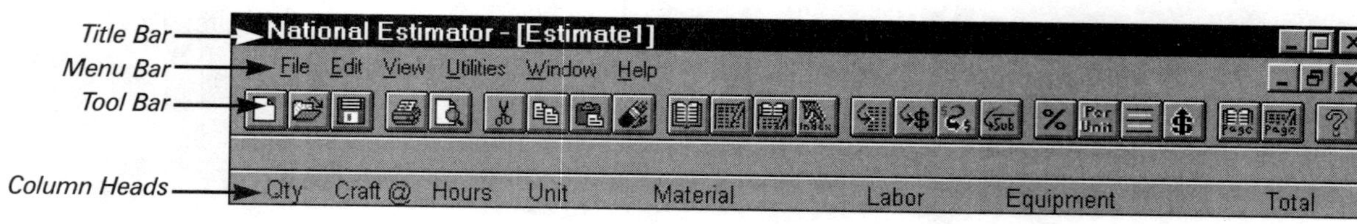


## Buttons on the Tool Bar

Below the menu bar you see 24 buttons that make up the tool bar. The options you use most in National Estimator are only a mouse click away on the tool bar.

## Column Headings

Below the tool bar you'll see column headings for your estimate form:

**Qty** for quantity

**Craft@Hours** for craft (the crew doing the work) and manhours (to complete the task)

**Unit** for unit of measure, such as per linear foot or per square foot

**Material** for material cost

**Labor** for labor cost

**Equipment** for equipment cost

**Total** for the total of all cost columns

## The Status Bar

The bottom line on your screen is the status bar. Here you'll find helpful information about the choices available. Notice "Page 1/1" near the right end of the status line. That's a clue that you're looking at page 1 of a one-page estimate.

*Check the status bar occasionally for helpful tips and explanations of what you see on screen.*

## Beginning an Estimate

Let's start by putting a heading on this estimate.

| The Blinking Cursor (insert point)

Mouse Pointer

Mouse Pointer

1. Press Enter once to space down one line.
2. Press Tab four times (or hold the space bar down) to move the Blinking Cursor (the insert point) near the middle of the line.
3. Type "Estimate One" and press Enter. That's the title of this estimate, "Estimate One."
4. Press Enter again to move the cursor down a line. That opens up a little space below the title.

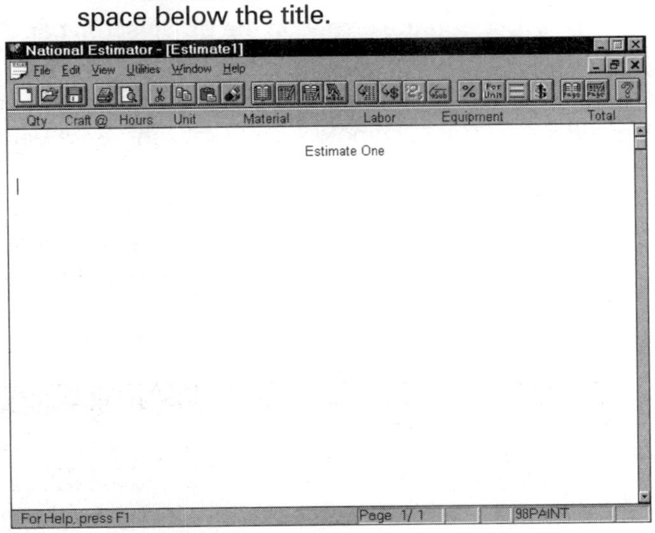


*Begin by putting a title on your estimate, such as "Estimate One."*

## The Costbook

Let's leave your estimating form for a moment to take a look at estimates stored in the costbook. To switch to the costbook, either:

- Click on the (Open Costbook) button, -or-
- Click on View on the menu bar. Then click on Costbook Window, -or-

- Tap the Alt key, tap the letter V (for View) and tap the letter C (for Costbook Window), -or-
- Press Esc. Press the ↓ arrow key to highlight Costbook Window. Then press Enter↵.

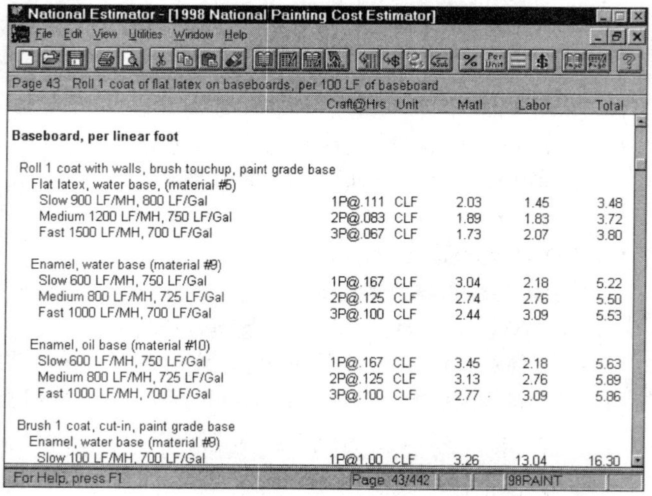


*The costbook window has the entire National Painting Cost Estimator.*

The entire 1998 National Painting Cost Estimator is available in the Costbook Window. Notice the words *Page 43 Roll 1 coat of flat latex on baseboards* at the left side of the screen just below the tool bar. That's your clue that the baseboard section of page 43 is on the screen. To turn to the next page, either:

- Press PgDn (with Num Lock off), -or-
- Click on the lower half of the scroll bar at the right edge of the screen.

To move down one line at a time, either:

- Press the ↓ arrow key (with Num Lock off), -or-
- Click on the arrow on the down scroll bar at the lower right corner of the screen.

*Type the page number you want to see.*

Press PgDn about 1,400 times and you'll page through the entire National Painting Cost Estimator. Obviously, there's a better way. To turn quickly to any page, either:

- Click on the (Turn to Costbook Page) button near the right end of the tool bar, -or-
- Click on View on the menu bar. Then click on Turn to Costbook Page, -or-
- Tap the Alt key, tap the letter V (for View) and tap the letter T (for Turn to Costbook Page), -or-
- Press Esc. Press the ↓ arrow key to highlight Turn to Costbook Page. Then press Enter↵.

Type the number of the page you want to see and press Enter. National Estimator will turn to the top of the page you requested.

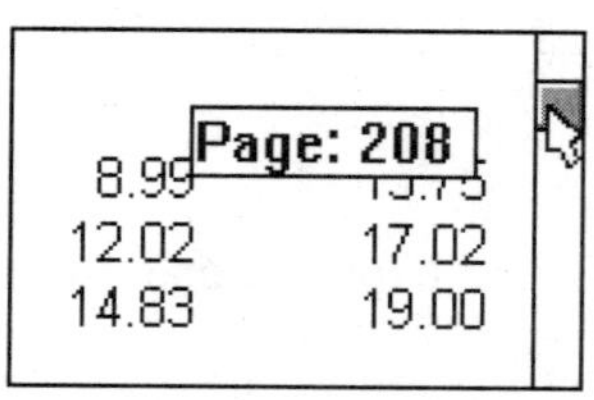


*Drag the square to see any page.*

## An Even Better Way

Find the small square in the slide bar at the right side of the Costbook Window. Click and hold on that square while rolling the mouse up or down. Keep dragging the square until you see the page you want in the Page: box. Release the mouse button to turn to the top of that page.

## A Still Better Way: Keyword Search

*Use the electronic index to find cost estimates for any item.*

To find any cost estimate in seconds, search by keyword in the index. To go to the index, either:

- Click on the (Index) button near the center of the tool bar, -or-
- Press [Esc] . Press [Enter←], -or-
- Tap the [Alt] key, tap the letter V (for View) and press [Enter←], -or-
- Click on View on the menu bar. Then press [Enter←].

Notice that the cursor is blinking in the Enter Keyword to Locate box at the right of the screen. Obviously, the index is ready to begin a search.

## Your First Estimate

*Enter keyword to locate.*

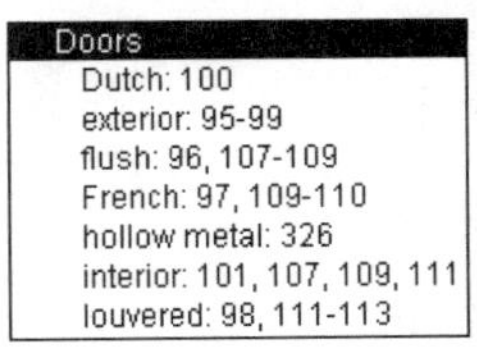


*The index jumps to Doors.*

Suppose we're estimating the cost of painting eight interior doors with brush and roller. Let's put the index to work with a search for doors. In the box under Enter Keyword to Locate, type *doors*. The index jumps to the heading *Doors*.

The sixth item under Doors is *interior: 101, 107, 109, 111*. Either:

- Click once on that line and press [Enter←], -or-
- Double click on that line, -or-
- Press [Tab] and the [↓] arrow key to move the highlight to *interior: 101, 107, 109, 111*. Then press [Enter←].

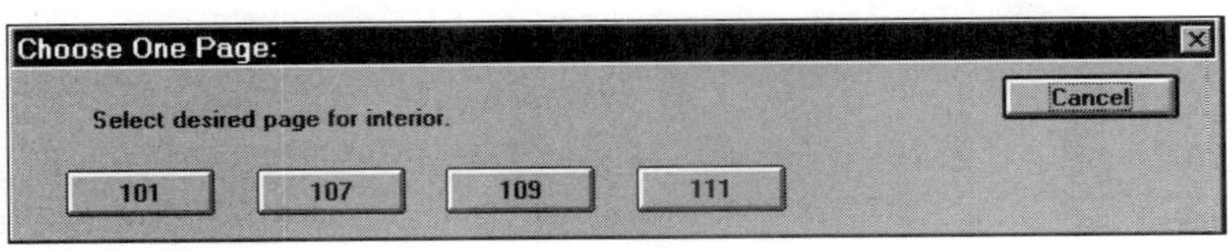


*If costs appear on several pages, click on the page you prefer.*

To select the page you want to see (page 107 in this case), either:

- Click on the number 107, -or-
- Press [Tab] to highlight 107. Then press [Enter←].

National Estimator turns to the top of page 107. See the example below. Notice that estimates at the top of this page are for painting interior flush doors.

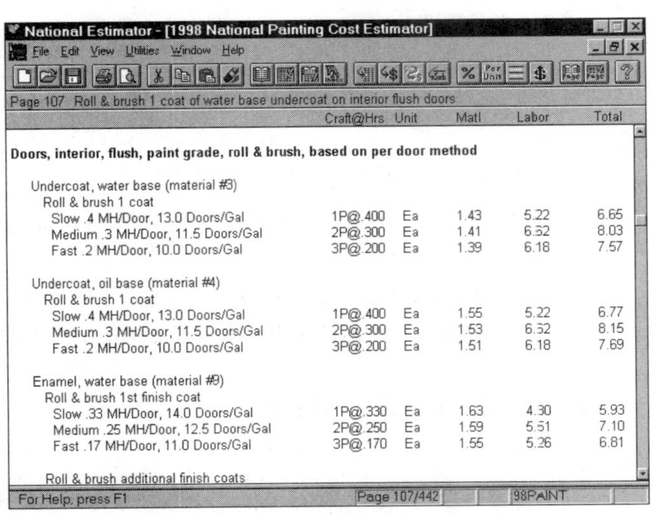

National Estimator - [1998 National Painting Cost Estimator]

File Edit View Utilities Window Help

Page 107 Roll & brush 1 coat of water base undercoat on interior flush doors

| | Craft@Hrs | Unit | Matl | Labor | Total |
|---|---|---|---|---|---|
| **Doors, interior, flush, paint grade, roll & brush, based on per door method** | | | | | |
| Undercoat, water base (material #3) | | | | | |
| Roll & brush 1 coat | | | | | |
| Slow .4 MH/Door, 13.0 Doors/Gal | 1P@.400 | Ea | 1.43 | 5.22 | 6.65 |
| Medium .3 MH/Door, 11.5 Doors/Gal | 2P@.300 | Ea | 1.41 | 6.52 | 8.03 |
| Fast .2 MH/Door, 10.0 Doors/Gal | 3P@.200 | Ea | 1.39 | 6.18 | 7.57 |
| Undercoat, oil base (material #4) | | | | | |
| Roll & brush 1 coat | | | | | |
| Slow .4 MH/Door, 13.0 Doors/Gal | 1P@.400 | Ea | 1.55 | 5.22 | 6.77 |
| Medium .3 MH/Door, 11.5 Doors/Gal | 2P@.300 | Ea | 1.53 | 6.52 | 8.15 |
| Fast .2 MH/Door, 10.0 Doors/Gal | 3P@.200 | Ea | 1.51 | 6.18 | 7.69 |
| Enamel, water base (material #9) | | | | | |
| Roll & brush 1st finish coat | | | | | |
| Slow .33 MH/Door, 14.0 Doors/Gal | 1P@.330 | Ea | 1.63 | 4.30 | 5.93 |
| Medium .25 MH/Door, 12.5 Doors/Gal | 2P@.250 | Ea | 1.59 | 5.51 | 7.10 |
| Fast .17 MH/Door, 11.0 Doors/Gal | 3P@.170 | Ea | 1.55 | 5.26 | 6.81 |
| Roll & brush additional finish coats | | | | | |

For Help, press F1 — Page 107/442 — 98PAINT

*Costs for painting interior flush doors on page 107.*

## Splitting the Screen

Most of the time you'll want to see what's in both the costbook and your estimate. To split the screen into two halves, either:

- Click on the (Split Window) button near the center of the tool bar, -or-
- Press Esc and the ↓ arrow key to move the selection bar to Split Window, -or-
- Tap the Alt key, tap the letter V (for View), tap the letter S (for Split Window), -or-
- Click on View on the menu bar. Then click on Split Window and your screen should look like the example below.

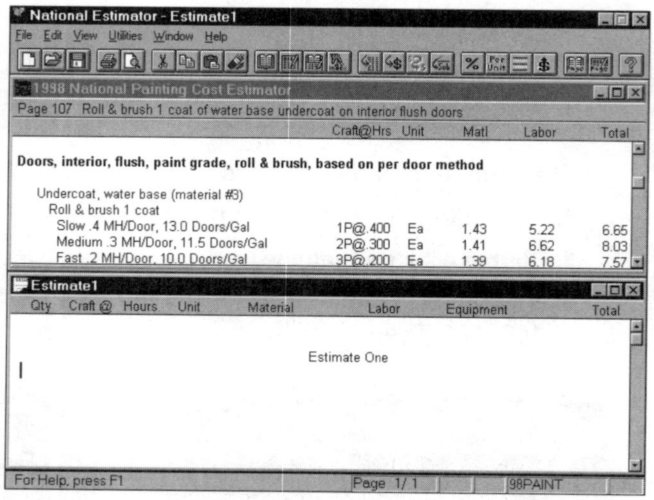


*The split window: Costbook above and estimate below.*

Notice the eight lines of the costbook are at the top of the screen and your estimate is at the bottom. You should recognize "Estimate One." It's your title for the estimate. Column headings are at the top of the costbook and across the middle of the screen (for your estimate).

## To Switch from Window to Window

- Click in the window of your choice, -or-
- Hold the Ctrl key down and press Tab.

Notice that a window title bar turns dark when that window is selected. The selected window is where keystrokes appear as you type.

## Copying Costs to Your Estimate

Next, we'll estimate the cost of painting eight doors. Click the (Split Window) button on the tool bar to be sure you're in the split window. Click anywhere in the costbook (the top half of your screen). Then press the ↓ arrow key until the cursor is on the line:

| Medium .3 MH/Door, 11.5 Doors/Gal | 2P@.300 | Ea | 1.41 | 6.62 | 8.03 |
|---|---|---|---|---|---|

To copy this line to your estimate:

1. Click on the line.
2. Click the (Copy) button.
3. Click on the (Paste) button to open the Enter Cost Information dialog box.

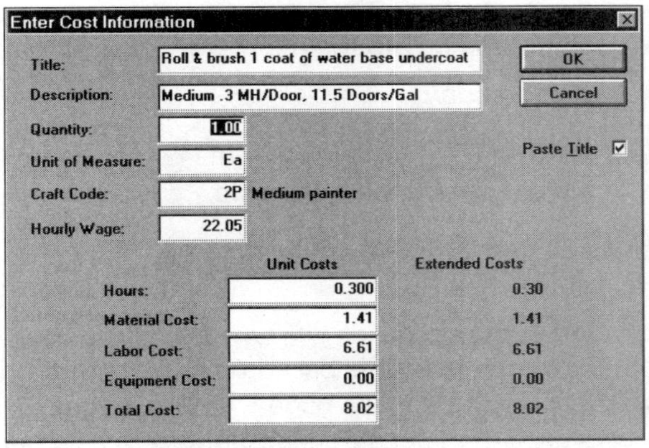


*Use the Enter Cost Information dialog box to copy or change costs.*

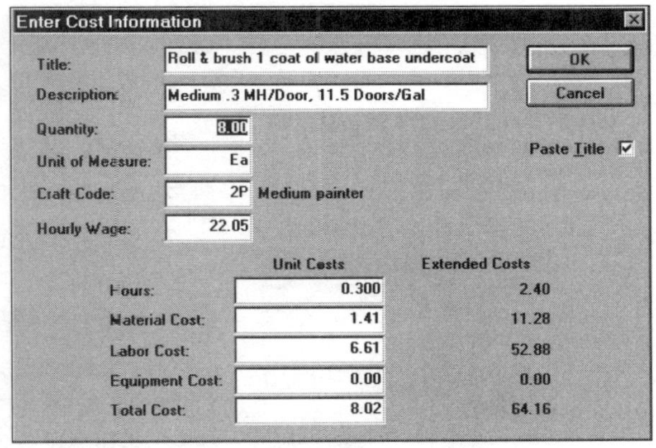


*Costs for eight doors (extended costs) are on the right.*

Notice that the blinking cursor is in the Quantity box:

1. Type a quantity of 8 because the job has eight interior doors.
2. Press Tab and check the estimate for accuracy.
3. Notice that the column headed Unit Costs shows costs per unit, per "Ea" (each) in this case.
4. The column headed Extended Costs shows costs for the entire eight doors.
5. The lines opposite Title and Description show what's getting painted and how you plan to do the work. You can change the words in either of these boxes. Just click on what you want to change and start typing or deleting.
6. You can also change any numbers in the Unit Cost column. Just click and start typing.
7. When the words and costs are exactly right, press Enter or click on OK to copy these figures to the end of your estimate.

The new line at the bottom of your estimate shows:

*Extended costs for eight doors as they appear on your estimate form.*

```
Roll & brush 1 coat of water base undercoat on interior flush doors
Medium .3 MH/Door, 11.5 Doors/Gal
  8.00    2P@2.400    Ea      11.28      52.88      0.00      64.16
```

**8.00** is the quantity of doors

**2P** is the recommended crew, a "medium" productivity painter

**@2.40** shows the manhours required for the work

**Ea** is the unit of measure, each in this case

**11.28** is the material cost (the paint)

**52.88** is labor cost for the job

**0.00** shows there is no equipment cost

**64.16** is the total of material, labor and equipment columns

## Copy Anything to Anywhere in Your Estimate

Anything in the costbook can be copied to your estimate. Just click on the line (or select the words) you want to copy and press the F8 key. It's copied to the last line of your estimating form. If your selection includes costs, you'll have a chance to enter the quantity. To copy to the *middle* of your estimate:

1. Select what you want to copy.
2. Click on the (Copy) button.
3. Click in the estimate where you want to paste.
4. Click on the (Paste) button.

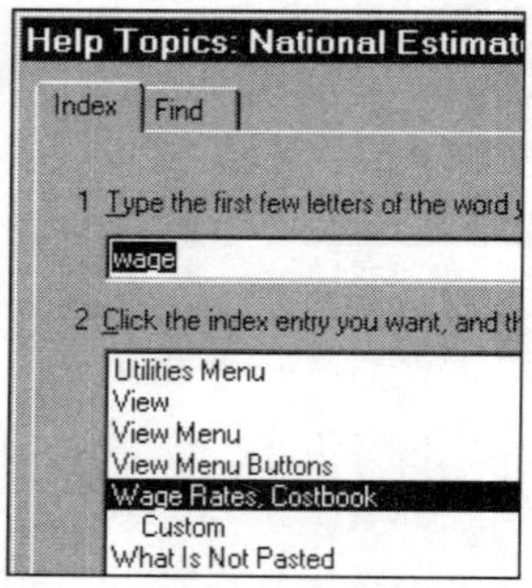


*Search for information on setting wage rates.*

## Changing Wage Rates

The labor cost in the example above is based on a medium productivity painter working at a cost of $22.05 per hour. (See page 6 in the National Painting Cost Estimator for crew rates used in the costbook.) Suppose $22.05 per hour isn't right for your estimate. What then? No problem! It's easy to use your own wage rate for any crew or even make up your own crew codes. To get more information on setting wage rates, press [F1]. At National Estimator Help Contents, click on the Search button. Type *wage* then double click on "*custom*" under Wage Rates. To return to your estimate, click on File on the National Estimator Help menu bar. Then click on Exit.

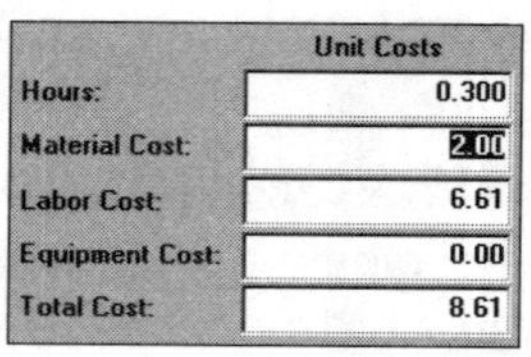


*Change the material cost to two dollars.*

## Changing Cost Estimates

With Num Lock off, use the [↑] or [↓] arrow key to move the cursor to the line you want to change (or click on that line). In this case, move to the line that begins with a quantity of 8. To open the Enter Cost Information Dialog box, either:

- Press [Enter↵], -or-
- Click on the (Change Cost) button on the tool bar.

To make a change, either

- Click on what you want to change, -or-
- Press [Tab] until the cursor advances to what you want to change.

Then type the correct figure. In this case, change the material cost to 2.00 (two dollars).

Press [Tab] and check the Extended Costs column. If it looks OK, press [Enter↵] and the change is made on your estimating form.

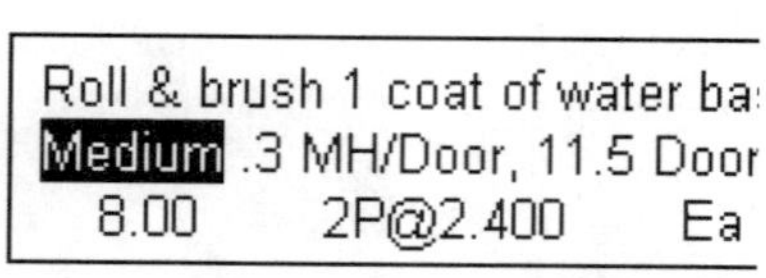


*To select, click and hold the mouse button while dragging the mouse.*

## Changing Text (Descriptions)

Click on the (Estimate Window) button on the tool bar to be sure you're in the estimate. With Num Lock off, use the [↑] or [↓] arrow key or click the mouse button to put the cursor where you want to make a change. In this case, we're going to make a change on the line that begins "Medium . . ."

To make a change, click where the change is needed. Then either:

- Press the [Del] or [←Bksp] key to erase what needs deleting, -or-
- Select what needs deleting and click on the (Cut) button on the tool bar.
- Type what needs to be added.

In this case, click just before the word "Medium." Then hold the left mouse button down and drag the mouse to the right until you've put a dark background behind the word "Medium." The dark background shows that this word is selected and ready for editing.

Press the [Del] key, or click on the (Cut) button on the tool bar, and the selection is cut from the estimate. If that's not what you wanted, click on the (Undo) button and the word "Medium" is back again.

## Adding Text (Descriptions)

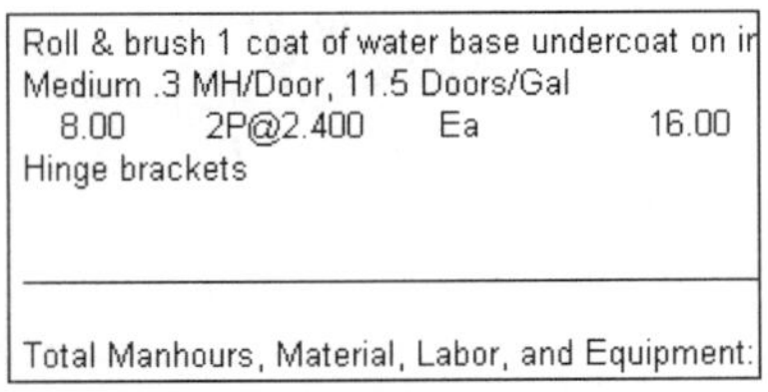


*Adding "Hinge brackets."*

Some of your estimates will require descriptions (text) and costs that can't be found in the National Painting Cost Estimator. What then? With National Estimator it's easy to add descriptions and costs of your choice anywhere in the estimate. For practice, let's add an estimate for four hinge brackets to Estimate One.

Click on the (Estimate Window) button to be sure the estimate window is maximized. We can add lines anywhere on the estimate. But in this case, let's make the addition at the end. Press the ↓ arrow key to move the cursor down until it's just above the horizontal line that separates estimate detail lines from estimate totals. To open a blank line, either:

- Press Enter, -or-
- Click on the (Insert Text) button on the tool bar, -or-
- Click on Edit on the menu bar. Then click on Insert a Text Line.

Type "Hinge brackets" and press Enter.

## Adding a Cost Estimate Line

Now let's add a cost for "Hinge brackets" to your estimate. Begin by opening the Enter Cost Information dialog box. Either:

- Click on the (Insert Cost) button on the tool bar, -or-
- Click on Edit on the menu bar. Then click on Insert a Cost Line.

1. The cursor is in the Quantity box. Type the number of units (4 in this case) and press Tab.
2. The cursor moves to the next box, Unit of Measure.
3. In the Unit of Measure box, type *Each* and press Tab.
4. Press Tab twice to leave the Craft Code blank and Hourly Wage at zero.
5. Since these brackets will be installed by the supplier, there's no material, labor or equipment cost. So press Tab four times to skip over the Hours, Material Cost, Labor Cost and Equipment Cost boxes.
6. In the Total Cost box, type 20.00. That's the cost per bracket quoted by your supplier.
7. Press Tab once more to advance to OK.
8. Press Enter and the cost of four hinge brackets is written to your estimate.

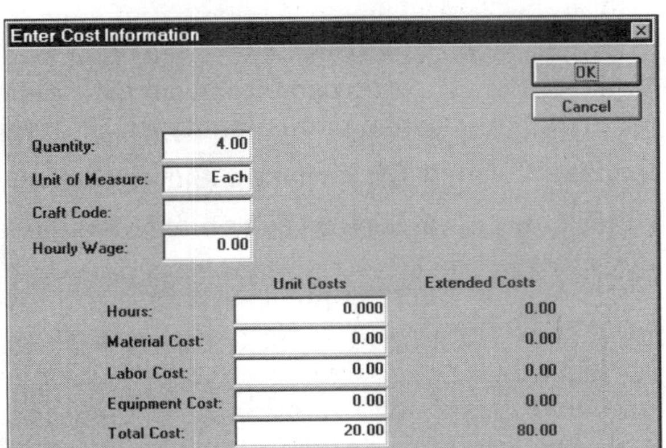


*Unit and extended costs for four hinge brackets.*

**Note:** The sum of material, labor and equipment costs appears automatically in the Total Cost box. If there's no cost entered in the Material Cost, Labor Cost or Equipment Cost boxes (such as for a subcontracted item), you can enter any figure in the Total Cost box.

## Adding Lines to the Costbook

Add lines or make changes in the costbook the same way you add lines or make changes in an estimate. The additions and changes you make become part of the user costbook. For more information on user costbooks, press F1. Click on Search. Type "user" and press [Enter ↵].

## Adding Tax

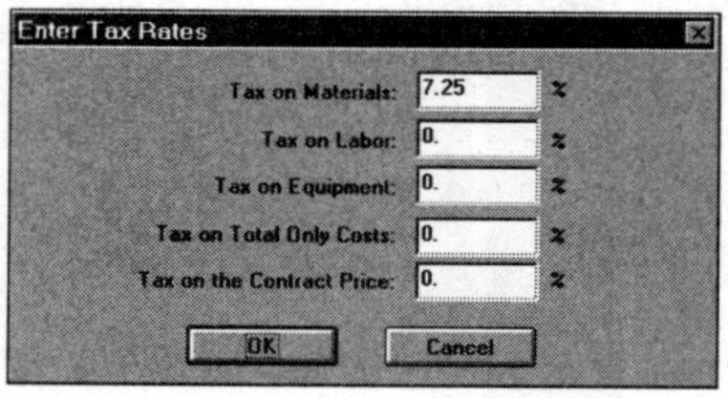


*Type the tax rate that applies.*

To include sales tax in your estimate:

1. Click on Edit.
2. Click on Tax Rates.
3. Type the tax rate in the appropriate box.
4. Press [Tab ⇆] to advance to the next box.
5. Press [Enter ↵] or click on OK when done.

In this case, the tax rate is 7.25% on materials only. Tax will appear as the last line of the estimate.

## Adding Overhead and Profit

Set markup percentages in the Add for Overhead & Profit dialog box. To open the box, either:

- Click on the [$] (Markup) button on the tool bar, -or-
- Click on Edit on the menu bar. Then click on Markup.

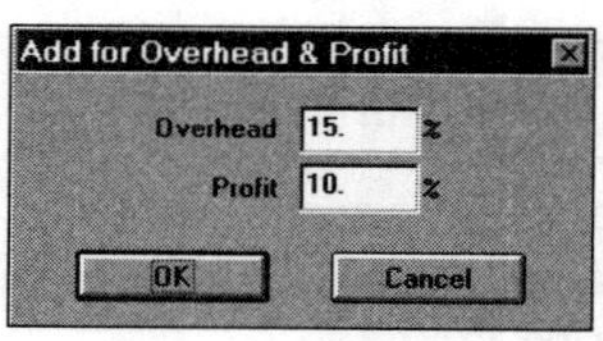


*Adding overhead & profit.*

Type the percentages you want to add for overhead. For this estimate:

1. Type 15 on the Overhead line.
2. Press [Tab ⇆] to advance to Profit.
3. Type 10 on the Profit line.
4. Press [Enter ↵].

Markup percentages can be changed at any time. Just reopen the Add for Overhead & Profit dialog box and type the correct figure.

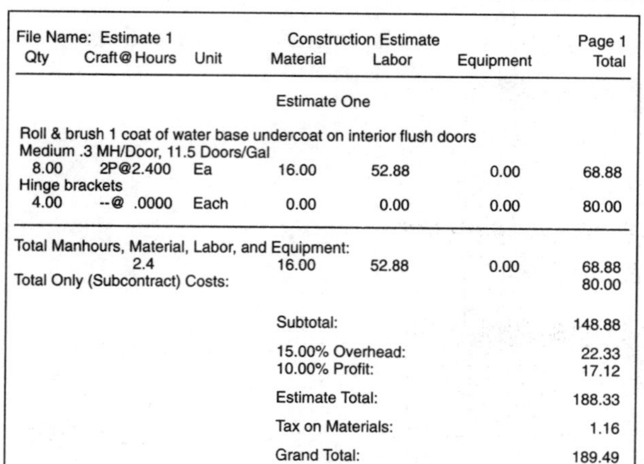

File Name: Estimate 1 — Construction Estimate — Page 1

| Qty | Craft@Hours | Unit | Material | Labor | Equipment | Total |
|---|---|---|---|---|---|---|
| | | | Estimate One | | | |
| Roll & brush 1 coat of water base undercoat on interior flush doors | | | | | | |
| Medium .3 MH/Door, 11.5 Doors/Gal | | | | | | |
| 8.00 | 2P@2.400 | Ea | 16.00 | 52.88 | 0.00 | 68.88 |
| Hinge brackets | | | | | | |
| 4.00 | --@ .0000 | Each | 0.00 | 0.00 | 0.00 | 80.00 |
| Total Manhours, Material, Labor, and Equipment: | | | | | | |
| | 2.4 | | 16.00 | 52.88 | 0.00 | 68.88 |
| Total Only (Subcontract) Costs: | | | | | | 80.00 |
| | | | Subtotal: | | | 148.88 |
| | | | 15.00% Overhead: | | | 22.33 |
| | | | 10.00% Profit: | | | 17.12 |
| | | | Estimate Total: | | | 188.33 |
| | | | Tax on Materials: | | | 1.16 |
| | | | Grand Total: | | | 189.49 |

*A preview of Estimate One.*

## Preview Your Estimate

You can display an estimate on screen just the way it will look when printed on paper. To preview your estimate, either:

- Click on the [🔍] (Print Preview) button on the tool bar, -or-
- Click on File on the menu bar. Then click on Print Preview.

In print preview:

- Click on Next Page or Prev Page to turn pages.
- Click on Two Page to see two estimate pages side by side.
- Click on Zoom In to get a closer look.
- Click on Close when you've seen enough.

*Use buttons on Print Preview to see your estimate as it will look when printed.*

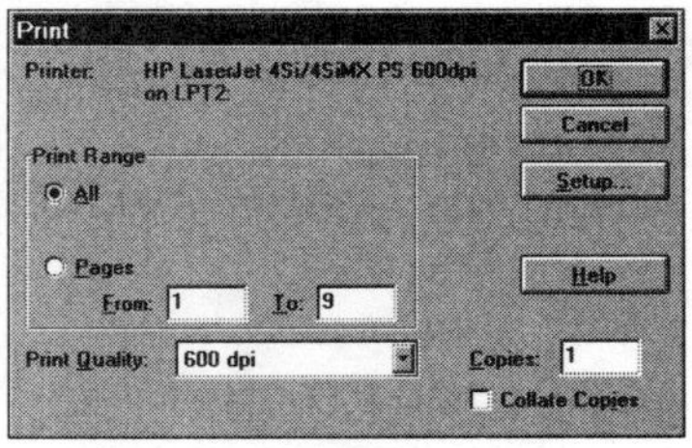


*Options available depend on the printer you're using.*

## Printing Your Estimate

When you're ready to print the estimate, either:

- Click on the (Print) button on the tool bar, -or-
- Click on File on the menu bar. Then click on Print, -or-
- Hold the Ctrl key down and type the letter P.

Press Enter↵ or click on OK to begin printing.

## Save Your Estimate to Disk

To store your estimate on the hard disk where it can be re-opened and changed at any time, either:

- Click on the (Save) button on the tool bar, -or-
- Click on File on the menu bar. Then click on Save, -or-
- Hold the Ctrl key down and type the letter S.

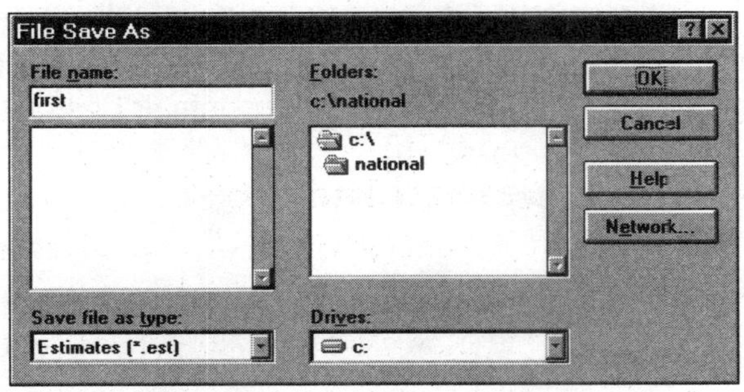


*Type the estimate name in the File Name box to assign a file name.*

The cursor is in the File Name box. Type the name you want to give this estimate, such as *First*. The name can be up to eight letters and numbers, but don't use symbols or spaces. Press Enter↵ or click on OK and the estimate is written to disk.

## Opening Other Costbooks

Many construction cost estimating databases are available for the National Estimator program. For example, a National Estimator costbook comes with each of these estimating manuals:

*National Construction Estimator*

*National Electrical Estimator*

*National Plumbing & HVAC Estimator*

*National Painting Cost Estimator*

*National Renovation & Insurance Repair Estimator*

*National Repair & Remodeling Estimator*

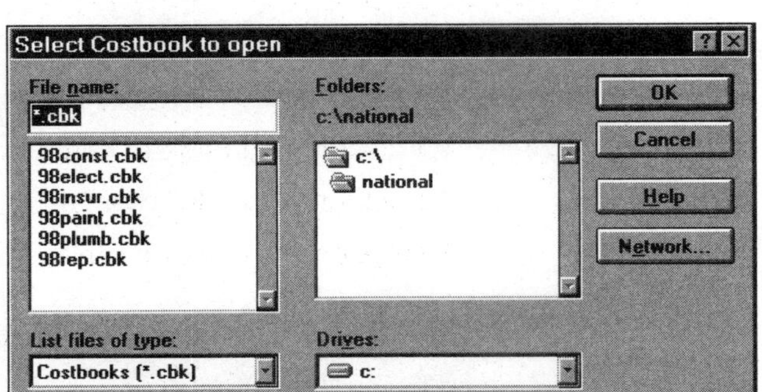


*Open the costbook of your choice.*

If you own several of these manuals and have installed the database that comes with each, you can open several costbooks at the same time. To open another costbook:

- Click on File.
- Click on Open Costbook.
- Be sure the drive and directory are correct, usually *c:\national*.
- Double click on the costbook of your choice.

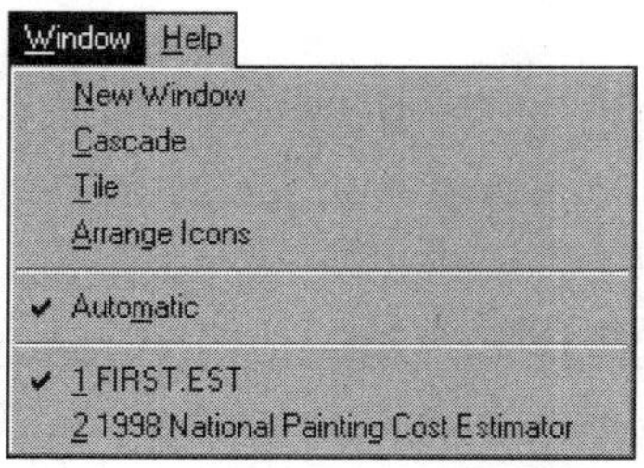


*Click to switch costbooks.*

To see a list of the costbooks open, click on Window. The name of the current estimate or costbook will be checked. Click on any other costbook name to display that costbook. Click on Window, then click on Tile to display all open costbooks and estimates.

## Select Your Default Costbook

Your default costbook opens automatically every time you begin using National Estimator. Save time by making the default costbook the one you use most.

To change your default costbook, click on Utilities on the menu bar. Then click on Options. Next, click on Select Default Costbook. Click on the costbook of your choice. Click on OK. Then click on OK again.

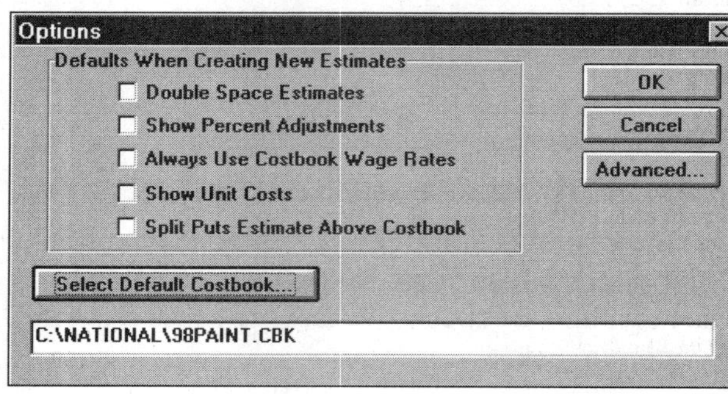


*Select the default costbook.*

## Use National Estimator Help

That completes the basics of National Estimator. You've learned enough to complete most estimates. When you need more information about the fine points, use National Estimator Help. Click on the (Help) button to see Help Contents. Then click on the menu selection of your choice. To print 28 pages of instructions for National Estimator, go to Help Contents. Click on Print All Topics (at the bottom of Help Contents). Click on File on the Help menu bar. Then click on Print Topic. Click on OK.

*Click on File and then Print Topic.*

# Index

# Practical References for Builders

### Builder's Guide to Room Additions

How to tackle problems that are unique to additions, such as requirements for basement conversions, reinforcing ceiling joists for second-story conversions, handling problems in attic conversions, what's required for footings, foundations, and slabs, how to design the best bathroom for the space, and much more. Besides actual construction, you'll even find help in designing, planning, and estimating your room addition jobs. **352 pages, 8$^{1}/_{2}$ x 11, $27.25**

### National Renovation & Insurance Repair Estimator

Current prices in dollars and cents for hard-to-find items needed on most insurance, repair, remodeling, and renovation jobs. All price items include labor, material, and equipment breakouts, plus special charts that tell you exactly how these costs are calculated. Includes a CD-ROM with an electronic version of the book with *National Estimator*, a stand-alone Windows™ estimating program, plus an interactive multimedia video that shows how to use the disk to compile construction cost estimates.
**560 pages, 8$^{1}/_{2}$ x 11, $49.50. Revised annually**

### Finish Carpenter's Manual

Everything you need to know to be a finish carpenter: assessing a job before you begin, and tricks of the trade from a master finish carpenter. Easy-to-follow instructions for installing doors and windows, ceiling treatments (including fancy beams, corbels, cornices and moldings), wall treatments (including wainscoting and sheet paneling), and the finishing touches of chair, picture, and plate rails. Specialized interior work includes cabinetry and built-ins, stair finish work, and closets. Also covers exterior trims and porches. Includes manhour tables for finish work, and hundreds of illustrations and photos. **208 pages, 8$^{1}/_{2}$ x 11, $22.50**

### Handbook of Construction Contracting

**Volume 1:** Everything you need to know to start and run your construction business; the pros and cons of each type of contracting, the records you'll need to keep, and how to read and understand house plans and specs so you find any problems before the actual work begins. All aspects of construction are covered in detail, including all-weather wood foundations, practical math for the job site, and elementary surveying.
**416 pages, 8$^{1}/_{2}$ x 11, $28.75**

**Volume 2:** Everything you need to know to keep your construction business profitable; different methods of estimating, keeping and controlling costs, estimating excavation, concrete, masonry, rough carpentry, roof covering, insulation, doors and windows, exterior finishes, specialty finishes, scheduling work flow, managing workers, advertising and sales, spec building and land development, and selecting the best legal structure for your business. **320 pages, 8$^{1}/_{2}$ x 11, $30.75**

### CD Estimator

If your computer has Windows™ and a CD-ROM drive, CD Estimator puts at your fingertips 85,000 construction costs for new construction, remodeling, renovation & insurance repair, electrical, plumbing, HVAC and painting. You'll also have the *National Estimator* program — a stand-alone estimating program for Windows™ that Remodeling magazine called a "computer wiz." Quarterly cost updates are available at no charge on the Internet. To help you create professional-looking estimates, the disk includes over 40 construction estimating and bidding forms in a format that's perfect for nearly any word processing or spreadsheet program for Windows™. And to top it off, a 70-minute interactive video teaches you how to use this CD-ROM to estimate construction costs. **CD Estimator is $68.50.**

### CD Estimator — Heavy

CD Estimator — Heavy has a complete 780-page heavy construction cost estimating volume for each of the 50 states. Select the cost database for the state where the work will be done. Includes thousands of cost estimates you won't find anywhere else, and in-depth coverage of demolition, hazardous materials remediation, tunneling, site utilities, precast concrete, structural framing, heavy timber construction, membrane waterproofing, industrial windows and doors, specialty finishes, built-in commercial and industrial equipment, and HVAC and electrical systems for commercial and industrial buildings. **CD Estimator — Heavy is $69.00**

### Video: Paint Contractor's Manual 1

How to run a paint contracting business, set up jobs for good production, and prepare surfaces for painting. Topics include planning, scheduling, using job work orders, setting production targets, advising on color selection, patching, textures, and the most common reasons for paint failure.
**40 minutes, VHS, $24.75**

### Video: Paint Contractor's Manual 2

How to select the right tools for the best results, get the most from spray paint equipment, lay down a glass-like finish in enamel, paint doors, jambs, windows, and cabinets quickly, and much more. **40 minutes, VHS, $24.75**

### Construction Forms & Contracts

125 forms you can copy and use — or load into your computer (from the FREE disk enclosed). Then you can customize the forms to fit your company, fill them out, and print. Loads into *Word for Windows, Lotus 1-2-3, WordPerfect*, or *Excel* programs. You'll find forms covering accounting, estimating, fieldwork, contracts, and general office. Each form comes with complete instructions on when to use it and how to fill it out. These forms were designed, tested and used by contractors, and will help keep your business organized, profitable and out of legal, accounting and collection troubles. Includes a 3$^{1}/_{2}$" disk for your PC. For Macintosh disks, add $15.
**432 pages, 8$^{1}/_{2}$ x 11, $39.75**

### Contractor's Guide to the Building Code Revised

This completely revised edition explains in plain English exactly what the *Uniform Building Code* requires. Based on the newly-expanded 1994 code, it explains many of the changes made. Also covers the *Uniform Mechanical Code* and the *Uniform Plumbing Code*. Shows how to design and construct residential and light commercial buildings that'll pass inspection the first time. Suggests how to work with an inspector to minimize construction costs, what common building shortcuts are likely to be cited, and where exceptions are granted. **384 pages, 8$^{1}/_{2}$ x 11, $39.00**

### Renovating & Restyling Vintage Homes

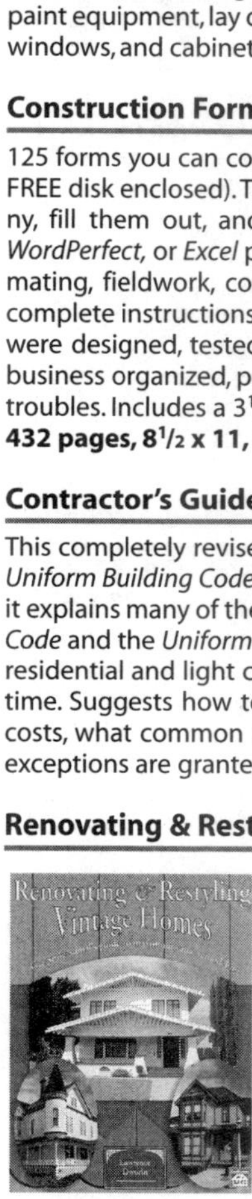


Any builder can turn a run-down old house into a showcase of perfection — if the customer has unlimited funds to spend. Unfortunately, most customers are on a tight budget. They usually want more improvements than they can afford — and they expect you to deliver. This book shows how to add economical improvements that can increase the property value by two, five or even ten times the cost of the remodel. Sound impossible? Here you'll find the secrets of a builder who has been putting these techniques to work on Victorian and Craftsman-style houses for twenty years. You'll see what to repair, what to replace and what to leave, so you can remodel or restyle older homes for the least amount of money and the greatest increase in value. **416 pages, 8$^{1}/_{2}$ x 11, $33.50**

### National Repair & Remodeling Estimator

The complete pricing guide for dwelling reconstruction costs. Reliable, specific data you can apply on every repair and remodeling job. Up-to-date material costs and labor figures based on thousands of jobs across the country. Provides recommended crew sizes; average production rates; exact material, equipment, and labor costs; a total unit cost and a total price including overhead and profit. Separate listings for high- and low-volume builders, so prices shown are specific for any size business. Estimating tips specific to repair and remodeling work to make your bids complete, realistic, and profitable. Includes a CD-ROM with an electronic version of the book with *National Estimator*, a stand-alone Windows™ estimating program, plus an interactive multi-media video that shows how to use the disk to compile construction cost estimates.
**256 pages, 8$^{1}/_{2}$ x 11, $48.50. Revised annually**